Introduction to Nuclear Engineering

Third Edition

John R. Lamarsh
Late Professor with the New York Polytechnic Institute

Anthony J. Baratta
Pennsylvania State University

Prentice Hall
Upper Saddle River, New Jersey 07458

Library of Congress Cataloging-in-Publication Data is on file.

Vice President and Editorial Director, ECS: *Marcia J. Horton*
Acquisitions Editor: *Laura Curless*
Editorial Assistant: *Erin Katchmar*
Vice President and Director of Production and Manufacturing, ESM: *David W. Riccardi*
Executive Managing Editor: *Vince O'Brien*
Managing Editor: *David A. George*
Production Editor: *Leslie Galen*
Director of Creative Services: *Paul Belfanti*
Creative Director: *Carole Anson*
Art Director: *Jayne Conte*
Art Editor: *Adam Velthaus*
Cover Designer: *Bruce Kenselaar*
Manufacturing manager: *Trudy Pisciotti*
Marketing Manager: *Holly Stark*
Marketing Assistant: *Karen Moon*
Cover image: *Courtesy of Framatome Technologies*

© 2001 by Prentice-Hall, Inc.
Upper Saddle River, New Jersey 07458

The author and publisher of this book have used their best efforts in preparing this book. These efforts include the development, research, and testing of the theories and programs to determine their effectiveness.

Printed in the United States of America

25   16

ISBN 0-201-82498-1

Prentice-Hall International (UK) Limited, *London*
Prentice-Hall of Australia Pty. Limited, *Sydney*
Prentice-Hall of Canada Inc., *Toronto*
Prentice-Hall Hispanoamericana, S.A., *Mexico*
Prentice-Hall of India Private Limited, *New Delhi*
Prentice-Hall of Japan, Inc., *Tokyo*
Pearson Education Asia Pte. Ltd., *Singapore*
Editora Prentice-Hall do Brasil, Ltda., *Rio de Janeiro*

Preface to Third Edition

This revision is derived from personal experiences in teaching introductory and advanced level nuclear engineering courses at the undergraduate level. In keeping with the original intent of John Lamarsh, every attempt is made to retain his style and approach to nuclear engineering education. Since the last edition, however, considerable changes have occurred in the industry. The changes include the development of advanced plant designs, the significant scale-back in plant construction, the extensive use of high speed computers, and the opening of the former Eastern Block countries and of the Soviet Union. From a pedagogical view, the World Wide Web allows access to many resources formerly only available in libraries. Attempts are made to include some of these resources in this edition.

In an attempt to update the text to include these technologies and to make the text useful for the study of non-western design reactors, extensive changes are made to Chapter 4, *Nuclear Reactors and Nuclear Power*. The chapter is revised to include a discussion of Soviet-design reactors and technology. The use, projection, and cost of nuclear power worldwide is updated to the latest available information.

In Chapter 11, *Reactor Licensing and Safety*, the Chernobyl accident is discussed along with the latest reactor safety study, NUREG 1150. A section is also included that describes non-power nuclear accidents such as Tokai-Mura.

The basic material in Chapters 2-7 is updated to include newer references and to reflect the author's experience in teaching nuclear engineering.

Throughout the text, the references are updated were possible to include more recent publications. In many topic areas, references to books that are dated and often out of print had to be retained, since there are no newer ones available. Since these books are usually available in college libraries, they should be available to most readers.

Chapter 9 is retained in much its same form but is updated to include a more complete discussion of the SI system of units and of changes in philosophy that have occurred in radiation protection. Since many of these changes have yet to reach general usage, however, the older discussions are still included.

As in the second edition, several errors were corrected and undoubtedly new ones introduced. Gremlins never sleep!

Preface to Second Edition

At his untimely death in July 1981, John R. Lamarsh had almost completed a revision of the first edition of Introduction to Nuclear Engineering. The major part of his effort went into considerable expansion of Chapters 4, 9, and 11 and into the addition of numerous examples and problems in many of the chapters. However, the original structure of that edition has been unchanged.

Chapter 4, Nuclear Reactors and Nuclear Power, has been completely restructured and much new material has been added. Detailed descriptions of additional types of reactors are presented. Extensive new sections include discussion of the nuclear fuel cycle, resource utilization, isotope separation, fuel reprocessing, and radioactive waste disposal.

In Chapter 9, Radiation Protection, considerable new material has been added on the biological effects of radiation, and there is a new section on the calculation of radiation effect. The section on the sources of radiation, both artificial and natural, has been expanded, and the sections on standards of radiation protection and computation of exposure have been brought up to date. A section on standards for intake of radionuclides has also been added.

In Chapter 11, Reactor Licensing, Safety, and the Environment, the sections on dispersion of effluents and radiation doses from nuclear facilities have been considerable expanded to cover new concepts and situations. Included in this chapter is a discussion of the 1979 accident at Three Mile Island. The structure of this chapter has been kept as it was in the first edition in spite of the earlier suggestion that it be broken up into two chapters treating environmental effects separately from safety and licensing.

Several errors that were still present in the last printing of the first edition have been corrected, including those in Example 6.7 and in the table of Bessel functions in Appendix V.

We are indebted to many of John Lamarsh's friends and colleagues who helped in many ways to see this revision completed. Particularly, we wish to thank Norman C. Rasmussen, Raphael Aronson, Marvin M. Miller, and Edward Melkonian for their assistance in the final stages of this revision.

Finally, we are grateful for comments and suggestions received from users of the earlier edition of this book. Although all their suggestions could not be incorporated, the book is greatly improved as a result of their comments.

November 1982   Addison-Wesley Publishing Company
                                                   Reading, Massachusetts

Preface to First Edition

This book is derived from classroom notes which were prepared for three courses offered by the Department of Nuclear Engineering at New York University and the Polytechnic Institute of New York. These are a one-year introductory course in nuclear engineering (Chapters 1-8), a one-term course in radiation protection (Chapters 9 and 10), and a one-term course in reactor licensing, safety, and the environment (Chapter 11). These courses are offered to juniors and seniors in the Department's undergraduate program and to beginning graduate students who have not had previous training in nuclear engineering.

Nuclear engineering is an extremely broad field, and it is not possible in a book of finite size and reasonable depth to cover all aspects of the profession. Needless to say, the present book is largely concerned with nuclear power plants, since most nuclear engineers are currently involved in the application of nuclear energy. Nevertheless, I have attempted in Chapter 1 to convey some feeling for the enormous breadth of the nuclear engineering profession.

In my experience, the courses in atomic and nuclear physics given by physics professors are becoming increasingly theoretical. I have found it necessary, therefore, to review these subjects at some length for nuclear engineering students. Chapters 2 and 3 are the substance of this review. Chapter 4 begins the consideration of some of the practical aspects of nuclear power, and includes a description of most of the reactors currently in production or under development.

Neutron diffusion and moderation are handled together in Chapter 5. Moderation is treated in a simple way by the group diffusion method, which avoids the usually tedious and relatively difficult calculations of slowing-down density and Fermi age theory. While such computations are essential, in my judgment, for a thorough understanding of the fundamentals of neutron moderation, they are probably not necessary in a book at this level. Chapters 6, 7, and 8 are intended to give sufficient background in reactor design methods to satisfy the needs of nuclear engineers not specifically involved in design problems and also to provide a base for more advanced courses in nuclear reactor theory and design.

Chapters 9 and 10 deal with the practical aspects of radiation protection. Both chapters rely heavily on the earlier parts of the book.

Chapter 11 was originally intended to be two chapters—one on safety and licensing, and a second on all of the environmental effects of nuclear power. However, in order to meet a publication deadline, the discussion of environmental effects had to be confined to those associated with radioactive effluents.

When the book was first conceived, I had planned to utilize only the modern metric system (the SI system). However, the U.S. Congress has been more reluctant to abandon the English system than had been expected. Especially, therefore, in connection with heat transfer calculations, I have felt compelled to introduce English units. A discussion of units and tables of conversion factors is given in Appendix I.

Most of the data required to solve the problems at the end of each chapter are given in the body of the relevant chapter or in the appendixes at the rear of the book. Data which are too voluminous to be included in the appendixes, such as atomic masses, isotopic abundances, etc., will generally be found on the Chart of Nuclides, available from the U.S. Government Printing Office. It is also helpful if the reader has access to the second and third editions of "Neutron Cross Sections," Brookhaven National Laboratory Report BNL-325. Title 10 of the Code of Federal Regulations should be obtained from the Printing Office in connection with Chapter 11.

I would like to acknowledge the assistance of several persons who read and commented upon various parts of the manuscript. I especially wish to thank R. Aronson, C. F. Bonilla, H. Chelemer, W. R. Clancey, R. J. Deland, H. Goldstein, H. C. Hopkins, F. R. Hubbard III, R. W. Kupp, G. Lear, Bruno Paelo, R. S. Thorsen, and M. E. Wrenn. I also wish to thank the personnel of the National Neutron Cross Section Center at Brookhaven National Laboratory, especially D. I. Garber, B. A.

Magurno, and S. Pearlstein, for furnishing various nuclear data and plots of neutron cross sections.

Finally, I wish to express my gratitude to my wife Barbara and daughter Michele for their priceless assistance in preparation of the manuscript.

Larchmont, New York   J. R. L.
January 1975

Acknowledgement

The author wishes to thank the following for their contribution: Nuclear Energy Institute, Westinghouse Electric Corporation, Framatome, Division of Naval Reactors, the Penn State students in Nuclear Engineering 301/302, R. Guida, G. Robinson, F. Remmick, S. Johnson, L. Curless, L. Galen, K. Almenas, R. Knief, E. Klevans, B. Lamarsh, V. Whisker, T. Beam, A. Barnes, L. Hochreiter, K. Ivanov, C. Caldwell, C. Wilson, F. Ducceschi, F. Duhamel, E. Supko, R. Scalise, F. Castrillo, R. McWaid, L. Pasquini, J. Kiell, L. Prowett, and E. Sartori.

Dedication

To all those who had faith in this project particularly my family.

Tony

Contents

Contents

APPENDIXES

1

Nuclear Engineering

Nuclear engineering is an endeavor that makes use of radiation and radioactive material for the benefit of mankind. Nuclear engineers, like their counterparts in chemical engineering, endeavor to improve the quality of life by manipulating basic building blocks of matter. Unlike chemical engineers, however, nuclear engineers work with reactions that produce millions of times more energy per reaction than any other known material. Originating from the nucleus of an atom, nuclear energy has proved to be a tremendous source of energy.

Despite its association with the atomic bombs dropped during World War II and the arms race of the cold war, nuclear energy today provides a significant amount of energy on a global scale. Many are now heralding it as a source free from the problems of fossil fuels—greenhouse gas emissions. Despite these benefits, there is still the association of nuclear power with the tremendous destructive force exhibited by the bombings in Japan.

With the end of the cold war, nuclear engineering is largely focused on the use of nuclear reactions to either generate power or on its application in the medical field. Nuclear power applications generally involve the use of the fission reactions in large central power stations and smaller mobile power plants used primarily for ship propulsion. The world demand for electricity is again increasing and with it the need for new generation facilities. For those areas of the world that have little in the way of fossil fuels or have chosen to use these for feedstock in the petro-chemical industry, nuclear power is considered the source of choice for electricity generation. In the United States alone, nuclear power generates nearly 22.8% of the electricity. In other countries, notably France, the proportion approaches 100%.

Recent concerns over the emission of nitrous oxides and carbon dioxide have increased the concern about continued use of fossil fuels as a source of energy. The Kyoto accords, developed in 1997, require a reduction in emissions below current values. These targets can be reached in the United States only by lowering the living standards or by continuing use of nuclear power for the generation of electricity. A typical 1,000-megawatt coal-burning plant may emit in 1 year as much as 100,000 tons of sulphur dioxide, 75,000 tons of nitrogen oxides, and 5,000 tons of fly ash. Nuclear power plants produce none of these air pollutants and emit only trace amounts of radioactive gasses. As a result, in 1999, the use of nuclear power to generate 20% of the electricity in the United States avoided the emission of 150 million tonnes of CO_2.

To date, the widest application of nuclear power in mobile systems has been for the propulsion of naval vessels, especially submarines and aircraft carriers. Here the tremendous advantages of nuclear power are utilized to allow extended operations without support ships. In the case of the submarine, the ability to cruise without large amounts of oxygen for combustion enables the submarine to remain at sea underwater for almost limitless time. In the case of an aircraft carrier, the large quantity of space that was taken up by fuel oil in a conventionally-powered aircraft carrier can be devoted to aviation fuel and other supplies on a nuclear-powered aircraft carrier.

In addition to naval vessels, nuclear-powered merchant ships were also developed. The U.S. ship *Savannah*, which operated briefly in the late 1960s and early 1970s, showed that nuclear power for a merchant ship while practical was not economical. Other countries including Japan, Germany, and the former Soviet Union have also used nuclear power for civilian surface ship propulsion. Of these, the German ore carrier, *Otto Hahn*, operated successfully for 10 years but was retired since it too proved uneconomical. The icebreaker *Lenin* of the former Soviet Union demonstrated another useful application of nuclear power. The trial was so successful that the Soviets built additional ships of this type.

Nuclear power has also been developed for aircraft and space applications. From 1949 to 1961, when the project was terminated, the United States spent approximately $1 billion to develop a nuclear-powered airplane. The project, the Aircraft Nuclear Propulsion Project (ANP), was begun at a time when the United States did not have aircraft that could fly roundtrip from the United States mainland to a distant adversary. Because of the enormous range that could be expected from a nuclear-powered airplane, the range problem would have been easily resolved. With the advent of long-range ballistic missiles, which could be fired from the mainland or from submarines, the need for such an aircraft disappeared and the program was terminated.

Nuclear-powered spacecraft have been developed and are in use today. Typically, a nuclear reaction is used to provide electricity for probes that are intended for use in deep space. There, photovoltaic systems cannot provide sufficient energy because of the weak solar radiation found in deep space. Typically, a radioactive source is used and the energy emitted is converted into heat and then electricity using thermocouples. Nuclear-powered rockets are under consideration as well. The long duration of a manned flight to Mars, for example, suggests that nuclear power would be useful if not essential. The desirability of a nuclear rocket for such long-distance missions stems from the fact that the total vehicular mass required for a long-distance mission is considerably less if the vehicle is powered by a nuclear rocket, rather than by a conventional chemical rocket. For instance, the estimated mass of a chemical rocket required for a manned mission from a stationary parking orbit to an orbit around Mars is approximately 4,100,000 kg. The mass of a nuclear rocket for the same mission is estimated to be only 430,000 kg. Nuclear rockets have been under active development in the United States for many years.

The application of radiation and nuclear reactions is not limited to nuclear explosives and nuclear power. Radiation and radioactive isotopes are useful in a wide range of important applications. The production of radioisotopes, whether from reactors or accelerators, is a major industry in its own right. The applications of radiation and radioisotopes range from life-saving medical procedures to material characterization to food preservation.

Radioactive tracing is one such method. In this technique, one of the atoms in a molecule is replaced by a radioactive atom of the same element. For example, a radioactive carbon atom may be substituted for a normal carbon atom at a particular location in a molecule when the molecule is synthesized. Later, after the molecule has reacted chemically, either in a laboratory experiment or a biological system, it is possible to determine the disposition of the atom in question by observing the radiation emanating from the radioactive atom. This technique has proved to be of enormous value in studies of chemical reaction processes and in research in the life sciences. A similar procedure is used in industry to measure, and sometimes to control, the flow and mixing of fluids. A small quantity of radioactive material is placed in the moving fluid and the radiation is monitored downstream. By proper calibration, it is possible to relate the downstream radiation level with the fluid's rate of flow or the extent of its dilution. In a similar way, radioactive atoms may be incorporated at the time of fabrication into various moving parts of machinery, such as pistons, tool bits, and so on. The radioactivity observed in the lubricating fluid then becomes an accurate measure of the rate of wear of the part under study.

A related technique, known as *activation analysis*, is based on the fact that every species of radioactive atom emits its own characteristic radiations. The chemical composition of a substance can therefore be determined by observing the radiation emitted when a small sample of the substance is caused to become radioactive. This may be done by exposing the sample to beams of either neutrons or charged particles. Because it is possible to determine extremely minute concentrations in this way (in some cases, one part in 10^{12}), activation analysis has proved to be a valuable tool in medicine, law enforcement, pollution control, and other fields in which trace concentrations of certain elements play an important role.

2

Atomic and Nuclear Physics

A knowledge of atomic and nuclear physics is essential to the nuclear engineer because these subjects form the scientific foundation on which the nuclear engineering profession is based. The relevant parts of atomic and nuclear physics are reviewed in this chapter and the next.

2.1 FUNDAMENTAL PARTICLES

The physical world is composed of combinations of various subatomic or fundamental particles. A number of fundamental particles have been discovered. This led to the discovery that these fundamental particles are in turn made up of quarks bound together by gluons.

In current theory, particles of interest to the nuclear engineer may be divided into leptons and hadrons. The electron, positron, and neutrino are leptons. Hadrons of interest are the proton and neutron, which belong to a subclass of hadrons called *baryons*. The leptons are subject to the weak nuclear forces, whereas hadrons and baryons in particular experience both the weak and strong nuclear forces. It is the hadrons that are composed of quarks, and it is the exchange of gluons between collections of quarks that is responsible for the strong nuclear force.

5

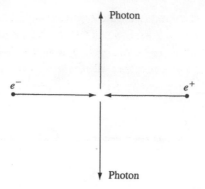

Figure 2.1   The annihilation of a negatron and positron with the release of two photons.

For the understanding of nuclear reactors and the reactions of interest to their operation, it is only important to consider a class of these particles and not explore their structure. Of these, only the following are important in nuclear engineering.[1]

Electron   The electron has a rest-mass[2] $m_e = 9.10954 \times 10^{-31}$ Kg[3] and carries a charge $e = 1.60219 \times 10^{-19}$ coulombs. There are two types of electrons: one carrying a negative charge $-e$, the other carrying a positive charge $+e$. Except for the difference in the sign of their charge, these two particles are identical. The negative electrons, or *negatrons* as they are sometimes called, are the normal electrons encountered in this world. Positive electrons, or *positrons*, are relatively rare. When, under the proper circumstances, a positron collides with a negatron, the two electrons disappear and two (and occasionally more) photons (particles of electromagnetic radiation) are emitted as shown in Fig. 2.1. This process is known as *electron annihilation*, and the photons that appear are called *annihilation radiation*.

Proton   This particle has a rest mass $m_p = 1.67265 \times 10^{-27}$ Kg and carries a positive charge equal in magnitude to the charge on the electron. Protons with negative charge have also been discovered, but these particles are of no importance in nuclear engineering.

---

[1]A discussion of quark theory may be found in several of the particle physics references at the end of this chapter.

[2]According to the theory of relativity, the mass of a particle is a function of its speed relative to the observer. In giving the masses of the fundamental particles, it is necessary to specify that the particle is at rest with respect to the observer—hence, the term *rest mass*.

[3]A discussion of units, their symbols, and abbreviations, together with tables of conversion factors, are found in Appendix I at the end of this book. Tabulations of fundamental constants and nuclear data are given in Appendix I.

Neutron   The mass of the neutron is slightly larger than the mass of the proton—namely, $m_n = 1.67495 \times 10^{-27}$ Kg, and it is electrically neutral. The neutron is not a stable particle except when it is bound into an atomic nucleus. A free neutron decays to a proton with the emission of a negative electron (β-decay; see Section 2.8) and an antineutrino, a process that takes, on the average, about 12 minutes.

Photon   It is a curious fact that all particles in nature behave sometimes like particles and sometimes like waves. Thus, certain phenomena that are normally thought of as being strictly wavelike in character also appear to have an associated corpuscular or particlelike behavior. Electromagnetic waves fall into this category. The particle associated with electromagnetic waves is called the *photon*. This is a particle with zero rest mass and zero charge, which travels in a vacuum at only one speed—namely, the speed of light, $c = 2.9979 \times 10^8$ m/sec.

Neutrino   This is another particle with zero rest mass and no electrical charge that appears in the decay of certain nuclei. There are at least six types of neutrinos, only two of which (called *electron neutrinos* and *electron antineutrinos*) are important in the atomic process and are of interest in nuclear engineering. For most purposes, it is not necessary to make a distinction between the two, and they are lumped together as neutrinos.

2.2 ATOMIC AND NUCLEAR STRUCTURE

As the reader is doubtless aware, atoms are the building blocks of gross matter as it is seen and felt. The atom, in turn, consists of a small but massive nucleus surrounded by a cloud of rapidly moving (negative) electrons. The nucleus is composed of protons and neutrons. The total number of protons in the nucleus is called the *atomic number* of the atom and is given the symbol Z. The total electrical charge of the nucleus is therefore $+Ze$. In a neutral atom, there are as many electrons as protons—namely, Z—moving about the nucleus. The electrons are responsible for the chemical behavior of atoms and identify the various chemical elements. For example, hydrogen (H) has one electron, helium (He) has two, lithium (Li) has three, and so on.

The number of neutrons in a nucleus is known as the *neutron number* and is denoted by N. The total number of *nucleons*—that is, protons and neutrons in a nucleus—is equal to $Z + N = A$, where A is called the *atomic mass number* or *nucleon number*.

The various species of atoms whose nuclei contain particular numbers of protons and neutrons are called *nuclides*. Each nuclide is denoted by the chemical symbol of the element (this specifies Z) with the atomic mass number as superscript (this determines N, since $N = A - Z$). Thus, the symbol ^{1}H refers to the nuclide of hydrogen ($Z = 1$) with a single proton as nucleus; ^{2}H is the hydrogen nuclide with a neutron as well as a proton in the nucleus (^{2}H is also called *deuterium* or *heavy hydrogen*); ^{4}He is the helium ($Z = 2$) nuclide whose nucleus consists of two protons and two neutrons; and so on. For greater clarity, Z is sometimes written as a subscript, as in $^1_1$H, $^2_1$H, $^4_2$He, and so on.

Atoms such as ^{1}H and ^{2}H, whose nuclei contain the same number of protons but different numbers of neutrons (same Z but different N—therefore different A), are known as *isotopes*. Oxygen, for instance, has three stable isotopes, ^{16}O, ^{17}O, ^{18}O ($Z = 8$; $N = 8, 9, 10$), and five known unstable (i.e., *radioactive*) isotopes, ^{13}O, ^{14}O, ^{15}O, ^{19}O, and ^{20}O ($Z = 8$; $N = 5, 6, 7, 11, 12$).

The stable isotopes (and a few of the unstable ones) are the atoms that are found in the naturally occurring elements in nature. However, they are not found in equal amounts; some isotopes of a given element are more abundant than others. For example, 99.8% of naturally occurring oxygen atoms are the isotope ^{16}O, .037% are the isotope ^{17}O, and .204% are ^{18}O. A table of some of the more important isotopes and their abundance is given in Appendix II. It should be noted that isotopic abundances are given in atom percent—that is, the percentages of the atoms of an element that are particular isotopes. Atom percent is often abbreviated as *a/o*.

Example 2.1

A glass of water is known to contain 6.6×10^{24} atoms of hydrogen. How many atoms of deuterium (^{2}H) are present?

Solution. According to Table II.2 in Appendix II, the isotopic abundance of ^{2}H is 0.015 *a/o*. The fraction of the hydrogen, which is ^{2}H, is therefore 1.5×10^{-4}. The total number of ^{2}H atoms in the glass is then $1.5 \times 10^{-4} \times 6.6 \times 10^{24} = 9.9 \times 10^{20}$. [*Ans.*]

2.3 ATOMIC AND MOLECULAR WEIGHT

The *atomic weight* of an atom is defined as the mass of the neutral atom relative to the mass of a neutral ^{12}C atom on a scale in which the atomic weight of ^{12}C is arbitrarily taken to be precisely 12. In symbols, let $m(^AZ)$ be the mass of the neutral atom denoted by ^AZ and $m(^{12}$C) be the mass of neutral ^{12}C. Then the atomic weight of ^AZ, $M(^AZ)$, is given by

$$M(^AZ) = 12 \times \frac{m(^AZ)}{m(^{12}\text{C})}. \qquad (2.1)$$

Suppose that some atom was precisely twice as heavy as ^{12}C. Then according to Eq. (2.1), this atom would have the atomic weight of $12 \times 2 = 24$.

As noted in Section 2.2, the elements found in nature often consist of a number of isotopes. The atomic weight of the element is then defined as the *average* atomic weight of the mixture. Thus, if γ_i is the isotopic abundance in atom percent of the ith isotope of atomic weight M_i, then the atomic weight of the element is

$$M = \sum_i \gamma_i M_i / 100. \tag{2.2}$$

The total mass of a molecule relative to the mass of a neutral ^{12}C atom is called the *molecular weight*. This is merely the sum of the atomic weights of the constituent atoms. For instance, oxygen gas consists of the molecule O_2, and its molecular weight is therefore $2 \times 15.99938 = 31.99876$.

Example 2.2

Using the data in the following table, compute the atomic weight of naturally occurring oxygen.

Isotope	Abundance (a/o)	Atomic weight
^{16}O	99.759	15.99492
^{17}O	0.037	16.99913
^{18}O	0.204	17.99916

Solution. From Eq. (2.2), it follows that

$$M(O) = 0.01[\gamma(^{16}O)M(^{17}O) + \gamma(^{17}O)M(^{17}O) + \gamma(^{18}O)M(^{18}O)]$$

$$= 15.99938. \ [Ans.]$$

It must be emphasized that atomic and molecular weights are unitless numbers, being ratios of the masses of atoms or molecules. By contrast, the *gram atomic weight* and *gram molecular weight* are defined as the amount of a substance having a mass, in grams, equal to the atomic or molecular weight of the substance. This amount of material is also called a *mole*. Thus, one gram atomic weight or one mole of ^{12}C is exactly 12 g of this isotope, one mole of O_2 gas is 31.99876 g, and so on.

Since atomic weight is a ratio of atomic masses and a mole is an atomic weight in grams, it follows that the number of atoms or molecules in a mole of any substance is a constant, independent of the nature of the substance. For instance, suppose that a hypothetical nuclide has an atomic weight of 24.0000. It follows that the individual atoms of this substance are exactly twice as massive as ^{12}C.

Therefore, there must be the same number of atoms in 24.0000 g of this nuclide as in 12.0000 g of ^{12}C. This state of affairs is known as *Avogadro's law*, and the number of atoms or molecules in a mole is called *Avogadro's number*. This number is denoted by N_A and is equal to $N_A = 0.6022045 \times 10^{24}$.[4]

Using Avogadro's number, it is possible to compute the mass of a single atom or molecule. For example, since one gram mole of ^{12}C has a mass of 12 g and contains N_A atoms, it follows that the mass of one atom in ^{12}C is

$$m(^{12}C) = \frac{12}{0.6022045 \times 10^{24}} = 1.99268 \times 10^{23} \text{ g.}$$

There is, however, a more natural unit in terms of which the masses of individual atoms are usually expressed. This is the *atomic mass unit*, abbreviated amu, which is defined as one twelfth the mass of the neutral ^{12}C atom, that is

$$1 \text{ amu} = \tfrac{1}{12} \times m(^{12}C).$$

Inverting this equation gives

$$m(^{12}C) = 12 \text{ amu.}$$

Introducing $m^{12}C$ from the preceding paragraph gives

$$1 \text{ amu} = \tfrac{1}{12} \times 1.99268 \times 10^{-23}\text{g} = 1/N_A \text{ g}$$
$$= 1.66057 \times 10^{-24} \text{ g.}$$

Also from Eq. (2.1),

$$m(^A Z) = \frac{m(^{12}C)}{12} \times M(^A Z),$$

so that

$$m(^A Z) = M(^A Z) \text{ amu.}$$

Thus, the mass of any atom in amu is numerically equal to the atomic weight of the atom in question.

---

[4]Ordinarily a number of this type would be written as 6.0222045×10^{23}. However, in nuclear engineering problems, for reasons given in Chap. 3 (Example 3.1), Avogadro's number should always be written as the numerical factor times 10^{24}.

2.4 ATOMIC AND NUCLEAR RADII

The size of an atom is somewhat difficult to define because the atomic electron cloud does not have a well-defined outer edge. Electrons may occasionally move far from the nucleus, while at other times they pass close to the nucleus. A reasonable measure of atomic size is given by the average distance from the nucleus that the outermost electron is to be found. Except for a few of the lightest atoms, these average radii are approximately the same for all atoms—namely, about 2×10^{-10}m. Since the number of atomic electrons increases with increasing atomic number, it is evident that the average electron density in the electron cloud also increases with atomic number.

The nucleus, like the atom, does not have a sharp outer boundary. Its surface, too, is diffuse, although somewhat less than that of an atom. Measurements in which neutrons are scattered from nuclei (see Section 3.5) show that to a first approximation the nucleus may be considered to be a sphere with a radius given by the following formula:

$$R = 1.25\,fm \times A^{1/3}, \qquad (2.3)$$

where R is in femtometers (fm) and A is the atomic mass number. One femtometer is 10^{-13} centimeters.

Since the volume of a sphere is proportional to the cube of the radius, it follows from Eq. (2.3) that the volume V of a nucleus is proportional to A. This also means that the ratio A/V—that is, the number of nucleons per unit volume—is a constant for all nuclei. This uniform density of nuclear matter suggests that nuclei are similar to liquid drops, which also have the same density whether they are large or small. This liquid-drop model of the nucleus accounts for many of the physical properties of nuclei.

2.5 MASS AND ENERGY

One of the striking results of Einstein's theory of relativity is that mass and energy are equivalent and convertible, one to the other. In particular, the complete annihilation of a particle or other body of rest mass m_0 releases an amount of energy, E_{rest}, which is given by Einstein's famous formula

$$E_{\text{rest}} = m_0 c^2, \qquad (2.4)$$

where c is the speed of light. For example, the annihilation of 1 g of matter would lead to a release of $E = 1 \times (2.9979 \times 10^{10})^2 = 8.9874 \times 10^{20}$ergs $= 8.9874 \times 10^{13}$ joules. This is a substantial amount of energy, which in more common units is equal to about 25 million kilowatt-hours.

Another unit of energy that is often used in nuclear engineering is the *electron volt*, denoted by eV. This is defined as the increase in the kinetic energy of an electron when it falls through an electrical potential of one volt. This, in turn, is equal to the charge of the electron multiplied by the potential drop—that is,

$$1 \text{ eV} = 1.60219 \times 10^{-19} \text{ coulomb} \times 1 \text{ volt}$$

$$= 1.60219 \times 10^{-19} \text{ joule.}$$

Other energy units frequently encountered are the MeV (10^6 eV) and the keV (10^3 eV).

Example 2.3

Calculate the rest-mass energy of the electron in MeV.

Solution. From Eq. (2.4), the rest-mass energy of the electron is

$$m_e c^2 = 9.1095 \times 10^{-28} \times (2.9979 \times 10^{10})^2$$

$$= 8.1871 \times 10^{-7} \text{ ergs} = 8.1871 \times 10^{-14} \text{ joule.}$$

Expressed in MeV this is

$$8.1871 \times 10^{-14} \text{ joule} \div 1.6022 \times 10^{-13} \text{ joule/MeV} = 0.5110 \text{ MeV. } [Ans.]$$

Example 2.4

Compute the energy equivalent of the atomic mass unit.

Solution. This can most easily be computed using the result of the previous example. Thus, since according to Section 2.3, 1 amu $= 1.6606 \times 10^{-24}$ g, it follows that 1 amu is equivalent to

$$\frac{1.6606 \times 10^{-24} \text{ g/amu}}{9.1095 \times 10^{-28} \text{ g/electron}} \times 0.5110 \text{ MeV/electron} = 931.5 \text{ MeV. } [Ans.]$$

When a body is in motion, its mass increases relative to an observer at rest according to the formula

$$m = \frac{m_0}{\sqrt{1 - v^2/c^2}}, \tag{2.5}$$

where m_0 is its rest mass and v is its speed. From Eq. (2.5), it is seen that m reduces to m_0 as v goes to zero. However, as v approaches c, m increases without limit. The *total energy* of a particle, that is, its rest-mass energy plus its kinetic energy, is given by

$$E_{\text{total}} = mc^2, \tag{2.6}$$

where m is as given in Eq. (2.5). Finally, the *kinetic energy* E is the difference between the total energy and the rest-mass energy. That is,

$$E = mc^2 - m_0 c^2 \tag{2.7}$$

$$= m_0 c^2 \left[\frac{1}{\sqrt{1 - v^2/c^2}} - 1 \right]. \tag{2.8}$$

The radical in the first term in Eq. (2.8) can be expanded in powers of $(v/c)^2$ using the binomial theorem. When $v \ll c$, the series may be truncated after the first term. The resulting expression for E is

$$E = \tfrac{1}{2} m_0 v^2, \tag{2.9}$$

which is the familiar formula for kinetic energy in classical mechanics. It should be noted that Eq. (2.9) may be used instead of Eq. (2.8) only when the kinetic energy computed from Eq. (2.9) is small compared with the rest-mass energy. That is, Eq. (2.9) is valid provided

$$\tfrac{1}{2} m_0 v^2 \ll m_0 c^2. \tag{2.10}$$

As a practical matter, Eq. (2.9) is usually accurate enough for most purposes provided $v \leq 0.2c$ or

$$E \leq 0.02 E_{\text{rest}}. \tag{2.11}$$

According to Example 2.3, the rest-mass energy of an electron is 0.511 MeV. From Eq. (2.11), it follows that the relativistic formula Eq. (2.8) must be used for electrons with kinetic energies greater than about 0.02×0.511 MeV= 0.010 MeV= 10 keV. Since many of the electrons encountered in nuclear engineering have kinetic energies greater than this, it is often necessary to use Eq. (2.8) for electrons.

By contrast, the rest mass of the neutron is almost 1,000 MeV and $0.02\, E_{\text{rest}} = 20$ MeV. In practice, neutrons rarely have kinetic energies in excess of 20 MeV. It is permissible, therefore, in all nuclear engineering problems to calculate the kinetic energy of neutrons from Eq. (2.9). When the neutron mass is inserted into Eq. (2.9), the following handy formula is obtained:

$$v = 1.383 \times 10^6 \sqrt{E}, \tag{2.12}$$

where v is in cm/sec and E is the kinetic energy of the neutron in eV.

It is important to recognize that Eqs. (2.8) and (2.9) are valid only for particles with nonzero rest mass; for example, they do not apply to photons. (It should be understood that photons have no rest-mass energy, and it is not proper to use the term *kinetic energy* in referring to such particles.) Photons only travel at the speed of light, and their total energy is given by quite a different formula—namely,

$$E = h\nu, \tag{2.13}$$

where h is Planck's constant and ν is the frequency of the electromagnetic wave associated with the photon. Planck's constant has units of energy × time; if E is to be expressed in eV, h is equal to 4.136×10^{-15} eV-sec.

2.6 PARTICLE WAVELENGTHS

It was pointed out in Section 2.1 that all of the particles in nature have an associated wavelength. The wavelength λ associated with a particle having momentum p is

$$\lambda = \frac{h}{p}, \tag{2.14}$$

where h is again Planck's constant. For particles of nonzero rest mass, p is given by

$$p = mv, \tag{2.15}$$

where m is the mass of the particle and v is its speed. At nonrelativistic energies, p can be written as

$$p = \sqrt{2m_0 E},$$

where E is the kinetic energy. When this expression is introduced into Eq. (2.14), the particle wavelength becomes

$$\lambda = \frac{h}{\sqrt{2m_0 E}}. \tag{2.16}$$

This formula is valid for the neutrons encountered in nuclear engineering. Introducing the value of the neutron mass gives the following expression for the neutron wavelength:

$$\lambda = \frac{2.860 \times 10^{-9}}{\sqrt{E}}, \tag{2.17}$$

where λ is in centimeters and E is the kinetic energy of the neutron in eV. For the relativistic case, it is convenient to compute p directly by solving the relativistic equations in the preceding section. This gives

$$p = \frac{1}{c}\sqrt{E_{\text{total}}^2 - E_{\text{rest}}^2}, \tag{2.18}$$

and so

$$\lambda = \frac{hc}{\sqrt{E_{\text{total}} - E_{\text{rest}}}}. \tag{2.19}$$

The momentum of a particle of zero rest mass is not given by Eq. (2.15), but rather by the expression

$$p = \frac{E}{c}, \tag{2.20}$$

in which E is the energy of the particle. When Eq. (2.20) is inserted into Eq. (2.14), the result is

$$\lambda = \frac{hc}{E}. \tag{2.21}$$

Introducing numerical values for h and c in the appropriate units gives finally

$$\lambda = \frac{1.240 \times 10^{-6}}{E}, \tag{2.22}$$

where λ is in meters and E is in eV. Equation (2.22) is valid for photons and all other particles of zero rest mass.

2.7 EXCITED STATES AND RADIATION

The Z atomic electrons that cluster about the nucleus move in more or less well-defined orbits. However, some of these electrons are more tightly bound in the atom than others. For example, only 7.38 eV is required to remove the outermost electron from a lead atom ($Z = 82$), whereas 88 keV (88,000 eV) is required to remove the innermost or *K-electron*. The process of removing an electron from an atom is called *ionization*, and the energies 7.38 eV and 88 keV are known as the *ionization energies* for the electrons in question.

In a neutral atom, it is possible for the electrons to be in a variety of different orbits or states. The state of lowest energy is the one in which an atom is normally found, and this is called the *ground state*. When the atom possesses more energy than its ground state energy, it is said to be in an *excited state* or an *energy level*. The ground state and the various excited states can conveniently be depicted by an *energy-level diagram*, like the one shown in Fig. 2.2 for hydrogen. The highest energy state corresponds to the situation in which the electron has been completely removed from the atom and the atom is ionized.

An atom cannot remain in an excited state indefinitely; it eventually decays to one or another of the states at lower energy, and in this way the atom eventually returns to the ground state. When such a transition occurs, a photon is emitted by

Figure 2.2   The energy levels of the hydrogen atom (not to scale).

the atom with an energy equal to the difference in the energies of the two states. For example, when a hydrogen atom in the first excited state at 10.19 eV (see Fig. 2.2) decays to the ground state, a photon with an energy of 10.19 eV is emitted. From Eq. (2.22), this photon has a wavelength of $\lambda = 1.240 \times 10^{-6}/10.19 = 1.217 \times 10^{-7}$ m. Radiation of this wavelength lies in the ultraviolet region of the electromagnetic spectrum.

Example 2.5

A high-energy electron strikes a lead atom and ejects one of the K-electrons from the atom. What wavelength radiation is emitted when an outer electron drops into the vacancy?

Solution. The ionization energy of the K-electron is 88 keV, and so the atom minus this electron is actually in an excited state 88 keV above the ground state. When the outer electron drops into the K position, the resulting atom still lacks an electron, but now this is an outer, weakly bound electron. In its final state, therefore, the atom is excited by only 7.38 eV, much less than its initial 88 keV. Thus, the photon in this transition is emitted with an energy of slightly less than 88 keV. The corresponding wavelength is

$$\lambda = 1.240 \times 10^{-6}/8.8 \times 10^4 = 1.409 \times 10^{-11} \text{m. } [Ans.]$$

Such a photon is in the x-ray region of the electromagnetic spectrum. This process, the ejection of an inner, tightly bound electron, followed by the transition of another electron, is one way in which x-rays are produced.

The nucleons in nuclei, like the electrons in atoms, can also be thought of as moving about in various orbits, although these are not as well defined and understood as those in atoms. In any case, there is a state of lowest energy, the ground state; except for the very lightest nuclei, all nuclei have excited states as well. These states are shown in Fig. 2.3 for ^{12}C. A comparison of Figs. 2.2 and 2.3 shows that the energies of the excited states and the energies between states are considerably greater for nuclei than for atoms. Although this conclusion is based only on the

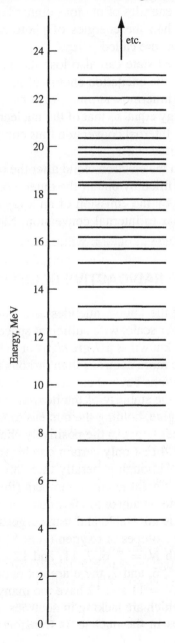

Figure 2.3 The energy levels of carbon 12.

states of hydrogen and ^{12}C, it is found to be true in general. This is due to the fact that the nuclear forces acting between nucleons are much stronger than the electrostatic forces acting between electrons and the nucleus.

Nuclei in excited states may decay to a lower lying state, as do atoms, by emitting a photon with an energy equal to the difference between the energies of the initial and final states. The energies of photons emitted in this way from a nucleus are usually much greater than the energies of photons originating in electronic transitions, and such photons are called *γ-rays*.

A nucleus in an excited state can also lose its excitation energy by internal conversion. In this process, the excitation energy of the nucleus is transferred into kinetic energy of one of the innermost atomic electrons. The electron is then ejected from the atom with an energy equal to that of the nuclear transition less the ionization energy of the electron. Internal conversion thus competes with γ-ray emission in the decay of nuclear-excited states.

The hole remaining in the electron cloud after the departure of the electron in internal conversion is later filled by one of the outer atomic electrons. This transition is accompanied either by the emission of an x-ray or the ejection of another electron in a process similar to internal conversion. Electrons originating in this way are called *Auger electrons*.

2.8 NUCLEAR STABILITY AND RADIOACTIVE DECAY

Figure 2.4 shows a plot of the known nuclides as a function of their atomic and neutron numbers. On a larger scale, with sufficient space provided to tabulate data for each nuclide, Fig. 2.4 is known as a *Segre chart* or the *chart of the nuclides*. The figure depicts that there are more neutrons than protons in nuclides with Z greater than about 20, that is, for atoms beyond calcium in the periodic table. These extra neutrons are necessary for the stability of the heavier nuclei. The excess neutrons act somewhat like nuclear glue, holding the nucleus together by compensating for the repulsive electrical forces between the positively charged protons.

It is clear from Fig. 2.4 that only certain combinations of neutrons and protons lead to stable nuclei. Although generally there are several nuclides with the same atomic number but different neutron numbers (these are the isotopes of the element), if there are either too many or too few neutrons for a given number of protons, the resulting nucleus is not stable and it undergoes *radioactive decay*. Thus, as noted in Section 2.2, the isotopes of oxygen ($A = 8$) with $N = 8, 9$, and 10 are stable, but the isotopes with $N = 5, 6, 7, 11$, and 12 are radioactive. In the case of the isotopes with $N = 5, 6$, and 7, there are not enough neutrons for stability, whereas the isotopes with $N = 11$ and 12 have too many neutrons.

Nuclei such as ^{15}O, which are lacking in neutrons, undergo β^+-*decay*. In this process, one of the protons in the nucleus is transformed into a neutron, and a

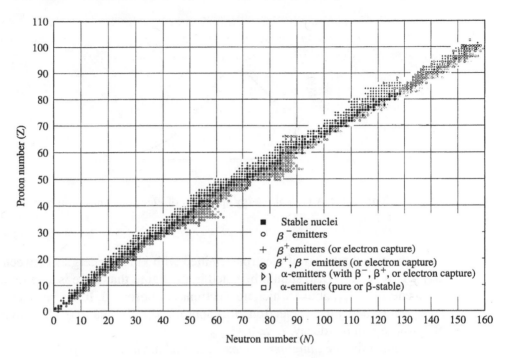

Figure 2.4   The chart of nuclides showing stable and unstable nuclei. (Based on S. E. Liverhant, *Elementary Introduction to Nuclear Reactor Physics*. New York: Wiley, 1960.)

positron and a neutrino are emitted. The number of protons is thus reduced from 8 to 7 so that the resulting nucleus is an isotope of nitrogen, ^{15}N, which is stable. This transformation is written as

$$^{15}\text{O} \xrightarrow{\beta^+} {}^{15}\text{N} + \nu,$$

where β^+ signifies the emitted positron, which in this context is called a β-*ray* and ν denotes the neutrino. By contrast, nuclei like ^{19}O, which are excessively neutron-rich, decay by β^- decay, emitting a negative electron and an antineutrino:

$$^{19}\text{O} \xrightarrow{\beta^-} {}^{19}\text{F} + \bar{\nu},$$

where $\bar{\nu}$ stands for the antineutrino. In this case, a neutron changes into a proton and the atomic number increases by one unit. It should be noted that in both β^+-decay and β^--decay the atomic mass number remains the same.

In both forms of β-decay, the emitted electrons appear with a continuous energy spectrum like that shown in Fig. 2.5. The ordinate in the figure, $N(E)$, is equal to the number of electrons emitted per unit energy, which have a kinetic energy E.

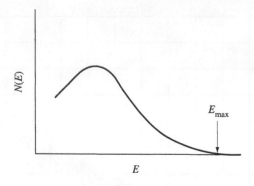

Figure 2.5 A typical energy spectrum of electrons emited in beta decay.

Thus, the actual number of electrons emitted with kinetic energies between E and $E + dE$ is $N(E)\,dE$. It should be noted in the figure that there is a definite maximum energy, $E_{\max}$, above which no electrons are observed. It has been shown that the average energy of the electrons $\overline{E}$ is approximately equal to $0.3E_{\max}$ in the case of β^--decay. In β^+-decay, $\overline{E} \simeq 0.4E_{\max}$.

Frequently, the *daughter nucleus*, the nucleus formed in β-decay, is also unstable and undergoes β-decay. This leads to *decay chains* like the following:

$$^{20}\text{O} \xrightarrow{\beta^-} {}^{20}\text{F} \xrightarrow{\beta^-} {}^{20}\text{Ne (stable)}.$$

A nucleus which is lacking in neutrons can also increase its neutron number by electron capture. In this process, an atomic electron interacts with one of the protons in the nucleus, and a neutron is formed of the union. This leaves a vacancy in the electron cloud which is later filled by another electron, which in turn leads to the emission of γ-rays, which are necessarily characteristic of the daughter element or the emission of an Auger electron. Usually the electron that is captured by the nucleus is the innermost or K-electron, and so this mode of decay is also called *K-capture*. Since the daughter nucleus produced in electron capture is the same as the nucleus formed in β^--decay, these two decay processes often compete with one another.

Another way by which unstable nuclei undergo radioactive decay is by the emission of an α-*particle*. This particle is the highly stable nucleus of the isotope ^{4}He, consisting of two protons and two neutrons. The emission of an α-particle reduces the atomic number by two and the mass number by four. Thus, for instance, the α-decay of $^{238}_{92}\text{U}$ (uranium-238) leads to $^{234}_{90}\text{Th}$ (thorium-234) according to the equation

$$^{238}_{92}\text{U} \rightarrow {}^{234}_{90}\text{Th} + {}^4_2\text{He}.$$

TABLE 2.1   ALPHA-PARTICLE SPECTRUM OF ^{226}Ra

α-particle energy	Relative number of particles (%)
4.782	94.6
4.599	5.4
4.340	0.0051
4.194	7×10^{-4}

Decay by α-particle emission is comparatively rare in nuclides lighter than lead, but it is common for the heavier nuclei. In marked contrast to β-decay, α-particles are emitted in a discrete (line) energy spectrum similar to photon line spectra from excited atoms. This is shown in Table 2.1, where data is given for the four groups of α-particles observed in the decay of ^{226}Ra (Radium-226).

The nucleus formed as the result of β-decay (+ or −), electron capture, or α-decay is often left in an excited state following the transformation. The excited (daughter) nucleus usually decays[5] by the emission of one or more γ-rays in the manner explained in Section 2.7. An example of a situation of this kind is shown in Fig. 2.6 for decay of ^{60}Co—a nuclide widely used in nuclear engineering. A diagram like that shown in the figure is known as a *decay scheme*. It should be especially noted that the major γ-rays are emitted by the daughter nucleus, in this case ^{60}Ni, although they are frequently attributed to (and arise as the result of) the decay of the parent nucleus, ^{60}Co.

Most nuclei in excited states decay by the emission of γ-rays in an immeasurably short time after these states are formed. However, owing to peculiarities in their internal structure, the decay of certain excited states is delayed to a point where the nuclei in these states appear to be semistable. Such long-lived states are called *isomeric states* of the nuclei in question. The decay by γ-ray emission of one of these states is called an *isomeric transition* and is indicated as IT in nuclear data tabulations. In some cases, isomeric states may also undergo β-decay. Figure 2.6 shows the isomeric state found at 58 keV above the ground state of ^{60}Co. As a rule, isomeric states occur at energies very close to the ground state. This state is formed from the β^--decay of ^{60}Fe, not shown in the figure. It is observed that this isomeric state decays in two ways: either to the ground state of ^{60}Co or by β^--decay to the first two excited states of ^{60}Ni. Of these two decay modes, the first is by far the more probable (> 99%) and occurs largely by internal conversion.

---

[5]Strictly speaking, the term *decay* should not be used to describe the emission of γ-rays from nuclei in excited states since only the energy and not the character of the nucleus changes in the process. More properly, γ-ray emission should be referred to as nuclear *dependent-excitation*, not decay. However, the use of the term *decay* is well established in the literature.

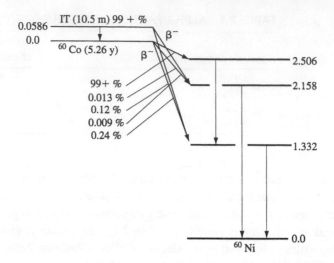

Figure 2.6   Decay scheme of cobalt 60, showing the known radiation emitted. The numbers on the side of the excited states are the energies of these states in MeV above the ground state. The relative occurrence of competing decays is indicated by the various percentages.

In summary, a nucleus without the necessary numbers of protons and neutrons for stability will decay by the emission of α-rays or β-rays or undergo electron capture, all of which may be accompanied by the subsequent emission of γ-rays. It must be emphasized that radioactive nuclei ordinarily do not decay by emitting neutrons or protons.

2.9 RADIOACTIVITY CALCULATIONS

Calculations of the decay of radioactive nuclei are relatively straightforward, owing to the fact that there is only one fundamental law governing all decay processes. This law states that the probability per unit time that a nucleus will decay is a constant independent of time. This constant is called the *decay constant* and is denoted by λ.[6]

Consider the decay of a sample of radioactive material. If at time t there are $n(t)$ atoms that as yet have not decayed, $\lambda n(t)\,dt$ of these will decay, on the average, in the time interval dt between t and $t + dt$. The rate at which atoms decay in the sample is therefore $\lambda n(t)$ disintegrations per unit time. This decay

---

[6]By tradition, the same symbol, λ, is used for both decay constant and the wavelength defined earlier, as well as for mean free path, defined later. Because of their very different uses, no confusion should arise.

rate is called the *activity* of the sample and is denoted by the symbol α. Thus, the activity at time t is given by

$$\alpha(t) = \lambda n(t). \tag{2.23}$$

Activity has traditionally been measured in units of *curies*, where one curie, denoted as Ci, is defined as exactly 3.7×10^{10} disintegrations per second. Units to describe small activities are the *millicurie*, 10^{-3} curie, abbreviated as mCi, the *microcurie*, 10^{-6} curie, denoted by μCi, and the *picocurie*, 10^{-12} curie, which is written as pCi. The SI unit of activity is the *becquerel*, Bq, which is equal to exactly one disintegration per second. One Bq $= 2.703 \times 10^{-11}$ Ci $\simeq 27$ pCi.

Since $\lambda n(t)\, dt$ nuclei decay in the time interval dt, it follows that the decrease in the number of undecayed nuclei in the sample in the time dt is

$$-dn(t) = \lambda n(t)\, dt.$$

This equation can be integrated to give

$$n(t) = n_0 e^{-\lambda t}, \tag{2.24}$$

where n_0 is the number of atoms at $t = 0$. Multiplying both sides of Eq. (2.24) by λ gives the activity of the sample at time t—namely,

$$\alpha(t) = \alpha_0 e^{-\lambda t}, \tag{2.25}$$

where α_0 is the activity at $t = 0$. The activity thus decreases exponentially with time.

The time during which the activity falls by a factor of two is known as the *half-life* and is given the symbol $T_{1/2}$. Introducing this definition

$$\alpha(T_{1/2}) = \alpha_0/2$$

into Eq. (2.25) gives

$$\alpha_0/2 = \alpha_0 e^{-\lambda T_{1/2}}.$$

Then taking the logarithm of both sides of this equation and solving for $T_{1/2}$ gives

$$T_{1/2} = \frac{\ln 2}{\lambda} = \frac{0.693}{\lambda}. \tag{2.26}$$

Solving Eq. (2.26) for λ and substituting into Eq. (2.25) gives

$$\alpha(t) = \alpha_0 e^{-0.693t/T_{1/2}}. \tag{2.27}$$

This equation is often easier to use in computations of radioactive decay than Eq. (2.25), especially with the advent of pocket-size electronic calculators and spread sheet programs, because half-lives are more widely tabulated than decay constants. Equation (2.27) eliminates the need to compute λ.

It is not difficult to show that the average life expectancy or *mean-life*, $\bar{t}$, of a radioactive nucleus is related to the decay constant by the formula

$$\bar{t} = 1/\lambda.$$

From Eq. (2.25), it can be seen that in one mean-life the activity falls to $1/e$ of its initial value. In view of Eq. (2.26), the mean-life and half-life are related by

$$\bar{t} = \frac{T_{1/2}}{0.693} = 1.44 T_{1/2}. \tag{2.28}$$

The exponential decay of a radioactive sample is shown in Fig. 2.7, where the half-life and mean-life are also indicated.

It is frequently necessary to consider problems in which radioactive nuclides are produced in a nuclear reactor or in the target chamber of an accelerator. Let it be assumed for simplicity that the nuclide is produced at the constant rate of R atoms/sec. As soon as it is formed, of course, a radioactive atom may decay. The change in the number of atoms of the nuclide in the time dt is given by a simple rate equation

the time rate of change of the nuclide $=$

the rate of production $-$ the rate of loss

or symbolically

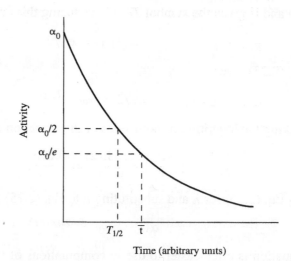

Figure 2.7   The decay of a radioactive sample.

$$dn/dt = -\lambda n + R.$$

This equation can be integrated with the result

$$n = n_0 e^{-\lambda t} + \frac{R}{\lambda}(1 - e^{-\lambda t}), \tag{2.29}$$

where n_0 is again the number of radioactive atoms present at $t = 0$. Multiplying this equation through by λ gives the activity of the nuclide

$$\alpha = \alpha_0 e^{-\lambda t} + R(1 - e^{-\lambda t}). \tag{2.30}$$

If $\alpha_0 = 0$, then, according to this result, α increases steadily from zero and as $t \rightarrow \infty$, α approaches the maximum value $\alpha_{max} = R$. Similarly, n approaches a constant value n_{max} given by R/λ. If $\alpha_0 \neq 0$, then the activity due to decay of the atoms originally present is added to the activity of the newly produced nuclide. In both cases, the activity approaches the value $\alpha_{max} = R$ as $t \rightarrow \infty$.

Example 2.6

Gold-198 ($T_{1/2} = 64.8$ hr) can be produced by bombarding stable ^{197}Au with neutrons in a nuclear reactor. Suppose that a ^{197}Au foil weighing 0.1 g is placed in a certain reactor for 12 hrs and that its activity is 0.90 Ci when removed.

(a) What is the theoretical maximum activity due to ^{198}Au in the foil?

(b) How long does it take for the activity to reach 80 percent of the maximum?

Solution

1. The value of R in Eq. (2.30) can be found from the data at 12 hrs. From Eq. (2.26), $\lambda = 0.693/64.8 = 1.07 \times 10^{-2}$ hr^{-1}. Then substituting into Eq. (2.30) gives

$$0.90 = R\left[1 - e^{1.07 \times 10^{-2} \times 12}\right]$$

or from Eq. (2.26)

$$0.90 = R\left[1 - e^{-0.693 \times 12/64.8}\right].$$

Solving either of these equations yields $R = 7.5$ Ci. (To get R in atoms/sec, which is not required in this problem, it is merely necessary to multiply the prior value of R by 3.7×10^{10}.) According to the previous discussion, the theoretical maximum activity is also 7.5 Ci. [*Ans.*]

2. The time to reach 80% of α_{max} can also be found from Eq. (2.30):

$$0.8R = R(1 - e^{\lambda t}).$$

Solving for t gives $t = 150$ hrs. [*Ans.*]

Another problem that is often encountered is the calculation of the activity of the radioactive nuclide B in the decay chain,

$$A \rightarrow B \rightarrow C \rightarrow \dots .$$

It is clear that an atom of B is formed with each decay of an atom of A. We can write a simple rate equation for this behavior.

the time rate of change of B =

the rate of production, from A − the rate of decay of B to C.

Since $\lambda_A n_A$ is the rate atoms of A decay into atoms of B, the rate atoms of B are produced is $\lambda_A n_A$. The rate of decay of B atoms is $\lambda_B n_B$, so the rate of change of B, dn_B/dt is,

$$\frac{dn_B}{dt} = -\lambda_B n_B + \lambda_A n_A.$$

Substituting Eq. (2.24) for n_A gives the following differential equation for n_B:

$$\frac{dn_B}{dt} = -\lambda_B n_B + \lambda_A n_{A0} e^{-\lambda_A t}, \tag{2.31}$$

where n_{A0} is the number of atoms of A at $t = 0$. Integrating Eq. (2.31) gives

$$n_B = n_{B0} e^{-\lambda_B t} + \frac{n_{A0} \lambda_A}{\lambda_B - \lambda_A} \left(e^{-\lambda_A t} - e^{-\lambda_B t} \right). \tag{2.32}$$

In terms of activity, this equation may be written as

$$\alpha_B = \alpha_{B0} e^{-\lambda_B t} + \frac{\alpha_{A0} \lambda_B}{\lambda_B - \lambda_A} \left(e^{-\lambda_A t} - e^{-\lambda_B t} \right), \tag{2.33}$$

where α_{A0} and α_{B0} are the initial activities of A and B, respectively. Generalizations of Eq. (2.33) for computing the activity of the nth nuclide in a long decay chain have been derived and are found in the references at the end of the chapter.

2.10 NUCLEAR REACTIONS

A *nuclear reaction* is said to have taken place when two nuclear particles—two nuclei or a nucleus and a nucleon—interact to produce two or more nuclear particles or γ-rays. If the initial nuclei are denoted by a and b, and the product nuclei and/or γ-rays (for simplicity it is assumed that there are only two) are denoted by c and d, the reaction can be represented by the equation

$$a + b \rightarrow c + d. \tag{2.34}$$

The detailed theoretical treatment of nuclear reactions is beyond the scope of this book. For present purposes, it is sufficient to note four of the fundamental laws governing these reactions:

1. *Conservation of nucleons.* The total number of nucleons before and after a reaction are the same.
2. *Conservation of charge.* The sum of the charges on all the particles before and after a reaction are the same.
3. *Conservation of momentum.* The total momentum of the interacting particles before and after a reaction are the same.
4. *Conservation of energy.* Energy, including rest-mass energy, is conserved in nuclear reactions.

It is important to note that, as we see, conservation of nucleons and conservation of charge do not imply conservation of protons and neutrons separately.

The principle of the conservation of energy can be used to predict whether a certain reaction is energetically possible. Consider, for example, a reaction of the type given in Eq. (2.34). The total energy before the reaction is the sum of the kinetic energies of the particles a and b plus the rest-mass energy of each particle. Similarly, the energy after the reaction is the sum of the kinetic energies of particles c and d plus their rest-mass energies. By conservation of energy it follows that

$$E_a + E_b + M_a c^2 + M_b c^2 = E_c + E_d + M_c c^2 + M_d c^2, \qquad (2.35)$$

where E_a, E_b, and so on, are the kinetic energies of particles a, b, etc. Equation (2.35) can be rearranged in the form

$$(E_c + E_d) - (E_a + E_b) = [(M_a + M_b) - (M_c + M_d)]c^2. \qquad (2.36)$$

Since the quantities on the left-hand side represents the kinetic energy of the particles, it is evident that the change in the kinetic energies of the particles before and after the reaction is equal to the difference in the rest-mass energies of the particles before and after the reaction.

The right-hand side of Eq. (2.36) is known as the *Q-value* of the reaction; that is,

$$Q = [(M_a + M_b) - (M_c + M_d)]c^2. \qquad (2.37)$$

In all computations and tabulations, Q is always expressed in MeV. Recalling that 1 amu is approximately 931 Mev, we can write the Q value as

$$Q = [(M_a + M_b) - (M_c + M_d)]931 \text{ Mev.} \qquad (2.38)$$

From Eq. (2.36), it is clear that when Q is positive there is a net *increase* in the kinetic energies of the particles. Such reactions are called *exothermic*. When Q is negative there is a net decrease in the energies of the particles, and the reaction is said to be *endothermic*. Since in exothermic reactions there is a net decrease in mass, nuclear mass is converted into kinetic energy, while in endothermic reactions there is a net increase, thus kinetic energy is converted into mass.

Equation (2.37) gives Q in terms of the masses of the nuclei a, b, and so on. However, the Q value can also be in terms of the masses of the neutral atoms containing these nuclei. Thus, in view of the conservation of charge,

$$Z_a + Z_b = Z_c + Z_d, \tag{2.39}$$

where Z_a, Z_b, and so on, are the atomic numbers of a, b, and so on, and Eq. (2.37) can be put in the form

$$Q = \{[(M_a + Z_a m_e) + (M_b + Z_b m_e)]$$
$$- [(M_c + Z_c m_e) + (M_d + Z_d m_e)]\}931 \text{ Mev}, \tag{2.40}$$

where m_e is the electron rest mass in amu. But $M_a + Z_a m_e$ is equal to the mass of the neutral atom of a, $M_b + Z_b m_e$ is the mass of the atom of b, and so on. It follows that Eq. (2.37) is a valid formula for Q, where M_a, M_b, and so on, are interpreted as the masses (in amu) of the neutral atoms in question, although the actual nuclear reaction involves only the atomic nuclei. It is fortunate that Q can be computed from neutral atomic mass data since the masses of most bare nuclei are not accurately known.

Incidentally, in the usual experimental arrangement, one of the particles, say a, is at rest in some sort of target, and the particle b is projected against the target. In this case, Eq. (2.34) is often written in the abbreviated form

$$a(b, c)d$$

or

$$a(b, d)c,$$

whichever is the more appropriate. For example, when oxygen is bombarded by energetic neutrons, one of the reactions that occurs is

$$^{16}O + n \rightarrow {}^{16}N + {}^{1}H.$$

In abbreviated form, this is

$$^{16}O(n, p)^{16}N,$$

where the symbols n and p refer to the incident neutron and emergent proton, respectively.

Example 2.7

Complete the following reaction:

$$^{14}\text{N} + n \rightarrow ? + {}^{1}\text{H}.$$

Solution. The atomic number of ^{14}N is 7, that of the neutron is 0. The sum of the atomic numbers on the left-hand side of the reaction is therefore 7, and the sum on the right must also be 7. Since $Z = 1$ for hydrogen, it follows that Z of the unknown nuclide is $7 - 1 = 6$ (carbon). The total number of nucleons on the left is the sum of the atomic mass numbers—namely, $14 + 1 = 15$. Since the mass number of ^{1}H is 1, the carbon isotope formed in this reaction must be ^{14}C. Thus, the reaction is

$$^{14}\text{N} + n \rightarrow {}^{14}\text{C} + {}^{1}\text{H}.$$

Example 2.8

One of the reactions that occurs when ^{3}H (tritium) is bombarded by deuterons (^{2}H nuclei) is

$$^{3}\text{H}(d, n)^{4}\text{He},$$

where d refers to the bombarding deuteron. Compute the Q value of this reaction.

Solution. The Q value is obtained from the following neutral atomic masses (in amu):

$$
\begin{array}{ll}
M(^{3}\text{H}) = 3.016049 & M(^{4}\text{He}) = 4.002604 \\
M(^{2}\text{H}) = 2.014102 & M(n) = 1.008665 \\
\hline
M(^{3}\text{H}) + M(^{2}\text{H}) = 5.030151 & M(^{4}\text{He}) + M(n) = 5.011269
\end{array}
$$

Thus, from Eq. (2.37), the Q value in amu is $Q = 5.030151 - 5.011269 = 0.018882$ amu. Since 1 amu $= 931.502$ MeV (see Ex. 2.4), $Q = 0.018882 \times 931.502 = 17.588$ MeV, which is positive and so this reaction is exothermic. This means, for instance, that when stationary ^{3}H atoms are bombarded by 1-MeV deuterons, the sum of the kinetic energies of the emergent α-particle (^{4}He) and neutron is $17.588 + 1 = 18.588$ MeV.

2.11 BINDING ENERGY

When a low-energy neutron and a proton combine to form a deuteron, the nucleus of ^{2}H, a 2.23-MeV γ-ray, is emitted and the deuteron recoils slightly with an energy of about 1.3 keV. The reaction in question is

$$p + n \rightarrow d + \gamma,$$

or, in terms of the neutral atoms,

$$^1\text{H} + \text{n} \rightarrow {}^2\text{H} + \gamma.$$

Since the γ-ray escapes from the site of the reaction, leaving the deuteron behind, it follows from the conservation of energy that the mass of the deuteron in energy units is approximately 2.23 MeV less than the sum of the masses of the neutron and proton. This difference in mass between the deuteron and its constituent nucleons is called the *mass defect* of the deuteron.

In a similar way, the masses of all nuclei are somewhat smaller than the sum of the masses of the neutrons and protons contained in them. This mass defect for an arbitrary nucleus is the difference

$$\Delta = ZM_p + NM_n - M_A, \tag{2.41}$$

where M_A is the mass of the nucleus. Equation (2.41) can also be written as

$$\Delta = Z(M_p + m_e) + NM_n - (M_A + Zm_e), \tag{2.42}$$

where m_e is the mass of an electron. The quantity $M_p + m_e$ is equal to the mass of neutral ^{1}H, while $M_A + Zm_e$ is equal to the mass M of the neutral atom. The mass defect of the nucleus is therefore

$$\Delta = ZM(^1H) + NM_n - M, \tag{2.43}$$

which shows that Δ can be computed from the tabulated masses of neutral atoms. Equations (2.41) and (2.43) are not precisely equivalent owing to slight differences in electronic energies, but this is not important for most purposes.

When Δ is expressed in energy units, it is equal to the energy that is necessary to break the nuleus into its constituent nucleons. This energy is known as the *binding energy* of the system since it represents the energy with which the nucleus is held together. However, when a nucleus is produced from A nucleons, Δ is equal to the energy released in the process. Thus, in the case of the deuteron, the binding energy is 2.23 MeV. This is the energy released when the deuteron is formed, and it is also the energy required to split the deuteron into a neutron and proton.

The total binding energy of nuclei is an increasing function of the atomic mass number A. However, it does not increase at a constant rate. This can be seen most conveniently by plotting the average binding energy per nucleon, Δ/A, versus A, as shown in Fig. 2.8. It is noted that there are a number of deviations from the curve at low A, while above $A = 50$ the curve is a smooth but decreasing function of A. This behavior of the binding energy curve is important in determining possible sources of nuclear energy.

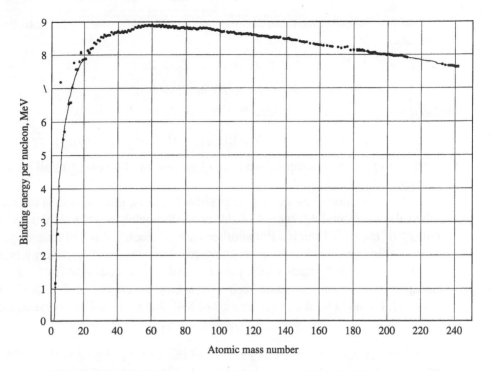

Figure 2.8   Binding energy per nucleon as a function of atomic mass number.

Those nuclei in which the binding energy per nucleon is high are especially stable or tightly bound, and a relatively large amount of energy must be supplied to these systems to break them apart. However, when such nuclei are formed from their constituent nucleons, a relatively large amount of energy is released. By contrast, nuclei with low binding energy per nucleon can be more easily disrupted and they release less energy when formed.

The Q value of a nuclear reaction can be expressed in terms of the binding energies of the reacting particles or nuclei. Consider the reaction given by Eq. (2.34). From Eq. (2.43), the binding energy of a, in units of mass, is

$$\text{BE}(a) = Z_a M(^1\text{H}) + N_a M_n - M_a.$$

The mass of a can then be written as

$$M_a = Z_a M(^1\text{H}) + N_a M_n - \text{BE}(a).$$

Similarly,

$$M_b = Z_b M(^1\text{H}) + N_b M_n - \text{BE}(b),$$

and so on. Substituting these expressions for M_a, M_b, and so on, into Eq. (2.37) and noting that

$$Z_a + Z_b = Z_c + Z_d$$
$$N_a + N_b = N_c + N_d,$$

gives

$$Q = [\text{BE}(c) + \text{BE}(d)] - [\text{BE}(a) - \text{BE}(b)].$$

Here, the c^2 has been dropped because Q and the binding energies all have units of energy.

This equation shows that Q is positive—that is, the reaction is exothermic—when the total binding energy of the product nuclei is greater than the binding energy of the initial nuclei. Put another way, whenever it is possible to produce a more stable configuration by combining two less stable nuclei, energy is released in the process. Such reactions are possible with a great many pairs of nuclides. For instance, when two deuterons, each with a binding energy of 2.23 MeV, react to form ^{3}H, having a total binding energy of 8.48 MeV, according to the equation

$$2{}^2\text{H} \rightarrow {}^3\text{H} + {}^1\text{H}, \tag{2.44}$$

there is a net gain in the binding energy of the system of $8.48 - 2 \times 2.23 = 4.02$ MeV. In this case, the energy appears as kinetic energy of the product nuclei ^{3}H and ^{1}H.

Reactions such as Eq. (2.44), in which at least one heavier, more stable nucleus is produced from two lighter, less stable nuclei, are known as *fusion reactions*. Reactions of this type are responsible for the enormous release of energy in hydrogen bombs and may some day provide a significant source of thermonuclear power.

In the regions of large A in Fig. 2.8, it is seen that a more stable configuration is formed when a heavy nucleus splits into two parts. The binding energy per nucleon in ^{238}U, for instance, is about 7.5 MeV, whereas it is about 8.4 MeV in the neighborhood of $A = 238/2 = 119$. Thus, if a uranium nucleus divides into two lighter nuclei, each with about half the uranium mass, there is a gain in the binding energy of the system of approximately 0.9 MeV per nucleon, which amounts to a total energy release of about $238 \times 0.9 = 214$ MeV. This process is called *nuclear fission*, and is the source of energy in nuclear reactors.

It must be emphasized that the binding energy per nucleon shown in Fig. 2.8 is an average overall of the nucleons in the nucleus and does not refer to any one nucleon. Occasionally it is necessary to know the binding energy of a particular nucleon in the nucleus—that is, the amount of energy required to extract the nucleon from the nucleus. This binding energy is also called the *separation energy* and is entirely analogous to the ionization energy of an electron in an atom. Consider the separation energy E_s of the least bound neutron—sometimes called the *last* neutron—in the nucleus ^AZ. Since the neutron is bound in the nucleus, it follows that the mass of the nucleus (and the neutral atom) ^AZ is less than the sum of the masses of the neutron and the residual nucleus ^{A-1}Z by an amount, in energy Mev, equal to E_s. In symbols, this is

$$E_S = [M_n + M(^{A-1}Z) - M(^AZ)]931 \text{ Mev/amu.} \tag{2.45}$$

The energy E_S is just sufficient to remove a neutron from the nucleus without providing it with any kinetic energy. However, if this procedure is reversed and a neutron with no kinetic energy is absorbed by the nucleus ^{A-1}Z, the energy E_S is released in the process.

Example 2.9

Calculate the binding energy of the last neutron in ^{13}C.

Solution. If the neutron is removed from ^{13}C, the residual nucleus is ^{12}C. The binding energy or separation energy is then computed from Eq. (2.45) as follows:

$$M(^{12}C) = 12.00000$$
$$\underline{M_n \quad\quad = \ \ 1.00866}$$
$$M_n + M(^{12}C) = 13.00866$$
$$\underline{- M(^{13}C) = 13.00335}$$
$$E_S = \ \ 0.00531 \text{ amu} \quad \text{x } 931 \text{ Mev} = 4.95 \text{ MeV } [Ans.]$$

Before leaving the discussion of nuclear binding energy, it should be noted that nuclei containing 2, 6, 8, 14, 20, 28, 50, 82, or 126 neutrons or protons are especially stable. These nuclei are said to be *magic*, and their associated numbers of nucleons are known as *magic numbers*. These correspond to the numbers of neutrons or protons that are required to fill shells (or subshells) of nucleons in the nucleus in much the same way that electron shells are filled in atoms.

The existence of magic nuclei has a number of practical consequences in nuclear engineering. For instance, nuclei with a magic neutron number absorb neutrons to only a very small extent, and materials of this type can be used where neutron absorption must be avoided. For example, Zirconium, whose most abundant isotope contains 50 neutrons, has been widely used as a structural material in reactors for this reason.

2.12 NUCLEAR MODELS

Two models of the nucleus are useful in explaining the various phenomena observed in nuclear physics—the shell model and the liquid drop model. Although neither of these models can completely explain the observed behavior of nuclei, they do provide valuable insight into the nuclear structure and cause many of the nuclear reactions of interest to the nuclear engineer.

Shell Model

The shell model may be thought of as the nuclear analogue to the many electron atom. In this model, the collective interaction of the nucleons in the nucleus generate a potential well. One can then think of a single nucleon as if it is moving in the well created by the average effect of the other nucleons. As in the case of other such potential wells, such a well can have one or more quantized states. These states are then populated in the same way that the atomic orbitals of an atom are populated by electrons. Just as in the atom, there is a maximum number of nucleons that may occupy a shell. When this number is reached, a closed shell results.

Although a detailed discussion of this model is beyond the scope of this text, a few remarks are necessary to understand the stability of certain nuclei and the origin of the magic numbers. The neutrons and protons fill each level in a potential well according to the Pauli exclusion principle. Accounting for the angular momentum for each state, there are $2j + 1$ possible substates for each level with total angular momentum j.

Since we are dealing with two sets of identical particles—neutrons on the one hand and protons on the other—there are really two such wells, one for each. They differ by the coulomb interaction of the protons. The levels are then filled according to the exclusion principle. The differing m_j values will split apart in energy because of the spin orbit interaction. As in the case of the many electron atoms, this may result in reordering of the levels and development of wider gaps in between the energy levels than otherwise would be expected. Since the neutron and proton wells can each have closed shells, the nuclei can be extremely stable when both wells have closed shells and less so when neither do. This phenomenon gives rise to the magic numbers discussed earlier.

Liquid Drop Model

From Section 2.11, the binding energy is the mass defect expressed in energy units. The liquid-drop model of the nucleus seeks to explain the mass defect in terms of a balance between the forces binding the nucleons in the nucleus and the coulombic repulsion between the protons.

The nucleus may be thought of as a drop of nuclear liquid. Just as a water droplet experiences a number of forces acting to hold it together, so does the nuclear droplet. To a first approximation, the mass of a nuclear droplet is just the mass of the components—the neutrons and protons. These are interacting in the nucleus and are bound by the nuclear forces. The binding of each nucleon to its neighbors means that energy and mass must be added to tear the nucleus apart. The mass may then be approximated by:

$$M = NM_n + ZM_p - \alpha A. \tag{2.46}$$

Equation 2.46 overestimates the effect of the bonds between the nucleons since those near the surface cannot have the same number of bonds as those deep inside the nucleus. To correct for this, a surface correction term must be added:

$$M = NM_n + ZM_p - \alpha A + 4\pi R^2 T, \tag{2.47}$$

where T denotes the surface tension. Since the radius R of the nucleus is proportional to $A^{1/3}$, we can rewrite this term:

$$M = NM_n + ZM_p - \alpha A + \beta A^{2/3}. \tag{2.48}$$

The coulombic repulsion tends to increase the energy and hence mass of the nucleus. Using the potential energy associated with the repulsive force, the expression becomes:

$$M = NM_n + ZM_p - \alpha A + \beta A^{2/3} + \gamma Z^2/A^{1/3}. \tag{2.49}$$

There are additional, strictly nuclear effects that must be accounted for in the mass equation. These account for a preference for the nucleons to pair together and for the effect of the Pauli exclusion principle.[7]

In the shell model, the nucleons were thought of as filling two potential wells. The lowest energy system would then be one in which the number of protons would equal the number of neutrons since, in this case, the wells would be filled to the same height, with each level filled according to the Pauli exclusion principle. The nucleus having $N = Z$ should then be more stable than the nucleus with $N \neq Z$. To account for this effect, a correction term must be added to the mass equation:

$$M = NM_n + ZM_p - \alpha A + \beta A^{2/3} + \gamma Z^2/A^{1/3} + \zeta(A - 2Z)^2/A. \tag{2.50}$$

Finally, if one examines the stable nuclides, one finds a preference for nuclei with even numbers of neutrons and protons. The preference reflects that, experimentally, the bond between two neutrons or two protons is stronger than that between a neutron and proton. Nuclei with odd numbers of neutrons and odd numbers of protons would thus be less strongly bound together. When either Z or N is odd and the other even, one would expect the binding to be somewhere in between these two cases. To account for this effect, a pairing term denoted by δ is added to the expression:

$$M = NM_n + ZM_p - \alpha A + \beta A^{2/3} + \gamma Z^2/A^{1/3} + \zeta(A - 2Z)^2/A + \delta. \tag{2.51}$$

---

[7]For a discussion of the Pauli exclusion principle, see references on modern physics.

The term δ is 0 if either N or Z is odd and the other even, positive if both are odd, and negative if both are even. Equation 2.51 is the mass equation.

The coefficients for the mass equation are obtained by fitting the expression to the known nuclei. When this is done, the semi-empirical mass equation is obtained. The values for each of the coefficients are typically taken as:

Mass of neutron	939.573 MeV
Mass of proton	938.280 MeV
α	15.56 MeV
β	17.23 MeV
γ	0.697 MeV
ζ	23.285 MeV
δ	12.0 MeV

The formula can accurately predict the nuclear masses of many nuclides with $Z > 20$. The ability to predict these masses suggests that there is some truth to the way the liquid-drop models the interactions of the nucleons in the nucleus.

For atomic masses, the mass of a hydrogen atom (938.791 MeV) may be substituted for the mass of the proton to account for the mass of the atomic electrons.

Example 2.10

Calculate the mass and binding energy of $^{107}_{47}\text{Ag}$ using the mass equation.

Solution. The mass equation may be used to calculate the binding energy by noting that the negative of the sum of the last five terms represents the binding energy of the constituent nucleons. The atomic mass of the $^{107}_{47}\text{Ag}$ is first obtained by using the mass formula and noting that N is even and Z is odd. The term involving δ is thus taken as zero.

$N \times m_n$	60×939.573 MeV
$Z \times m_H$	47×938.791 MeV
$-\alpha \times A$	-15.56×107 MeV
$+\beta \times A^{2/3}$	$17.23 \times 107^{2/3}$ MeV
$\gamma \times Z^2/A^{1/3}$	$0.697 \times 47^2/107^{1/3}$ MeV
$\zeta \times (A - 2 \times Z)^2/A$	$23.285 \times (107 - 2 \times 47)^2/107$ MeV
Mass (MeV)	99548.1173 MeV
Mass (u)	106.8684 u

The measured mass of $^{107}_{47}\text{Ag}$ is 106.905092 u or within 0.034% of the calculated value. Summing the last four terms gives a total binding energy of 949.44 MeV or 8.9 MeV/nucleon—a value slightly higher than the measured value of approximately 8.6 MeV.

2.13 GASES, LIQUIDS, AND SOLIDS

Before concluding this review of atomic and nuclear physics, it is appropriate to consider the nature of gross physical matter since this is the material encountered in all practical problems. Classically, there are three so-called states of matter: gas, liquid, and solid. The principal characteristics of these are as follows.

Gases   The noble gases—helium, neon, argon, krypton, xenon, and radon —and most metallic vapors are monatomic—that is, they are composed of more or less freely moving, independent atoms. Virtually all other gases consist of equally freely moving diatomic or polyatomic molecules. The random, disordered motion of these particles is one of the characteristic features of all gases.

Solids   Most of the solids used in nuclear systems—namely, metals and ceramics—are crystalline solids. Such solids are composed of large numbers of *microcrystals*, each of which consists of an ordered three-dimensional array or lattice of atoms. Each microcrystal contains an enormous number of individual atoms. Since the regularity in the arrangement of the atoms in the lattice extends over so many atoms (often over the entire microcrystal), such crystals are said to exhibit long-range order. There are a number of other materials that are called *solids* because they are rigid bodies, that do not exhibit long-range order. Examples of such materials are plastics, organic materials, glasses, and various amorphous solids.

Liquids   The microscopic structure of liquids is considerably more complicated than is usually assumed. The atoms and/or molecules in a liquid interact strongly with one another; as a result, they tend to be ordered as they are in a crystal, but not over such long distances. The ordered arrangement breaks down, so to speak, over long distances. For this reason, liquids are said to exhibit short-range order.

The Maxwellian Distribution

In a gas, the energies of the atoms or molecules are distributed according to the Maxwellian distribution function. If $N(E)$ is the density of particles per unit energy, then $N(E)\, dE$ is the number of particles per unit volume having energies between E and $E + dE$. According to the Maxwellian distribution, $N(E)$ is given by the formula

$$N(E) = \frac{2\pi N}{(\pi kT)^{3/2}} E^{1/2} e^{-E/kT}. \tag{2.52}$$

In Eq. (2.52), N is the total number of particles per unit volume; that is, the particle density; k is Boltzmann's constant, which has units of energy per degree Kelvin:

$$k = 1.3806 \times 10^{-23} \text{ joule/}^\circ \text{K}$$
$$= 8.6170 \times 10^{-5} \text{eV/}^\circ \text{K};$$

and T is the absolute temperature of the gas in degrees Kelvin. The function $N(E)$ is plotted in Fig. 2.9.

For solids and liquids, the energy distribution functions are more complicated than the one given in Eq. (2.52). However, it has been shown that, to a first approximation, $N(E)$ for solids and liquids can also be represented by Eq. (2.52), but the parameter T differs somewhat from the actual temperature of the substance. The difference is small for temperatures above about 300° K. At these temperatures, it can often be assumed that Eq. (2.52) applies to solids, liquids, and gases.

The *most probable energy* in a distribution such as the one given in Eq. (2.52) is defined as the energy corresponding to the maximum of the curve. This can be calculated by placing the derivative of $N(E)$ equal to zero. The most probable energy, E_p, in a Maxwellian energy distribution is then easily found to be

$$E_p = \tfrac{1}{2}kT. \tag{2.53}$$

However, the *average energy*, $\overline{E}$, is defined by the integral

$$\overline{E} = \frac{1}{N} \int_0^\infty N(E)E \, dE. \tag{2.54}$$

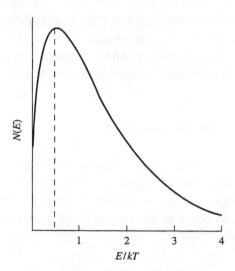

Figure 2.9   The Maxwellian distribution function.

Substituting Eq. (2.52) into Eq. (2.54) and carrying out the integration gives

$$\overline{E} = \tfrac{3}{2}kT. \tag{2.55}$$

The combination of parameters kT in Eqs. (2.53) and (2.55) often appears in the equations of nuclear engineering. Calculations involving these parameters are then expedited by remembering that, for $T_0 = 293.61°K$, kT has the value

$$kT_0 = 0.0253 \text{ eV} \simeq \tfrac{1}{40} \text{ eV}. \tag{2.56}$$

Example 2.11

What are the most probable and average energies of air molecules in a New York City subway in summertime at, say, 38°C (about 100°F)?

Solution. It is first necessary to compute the temperature in degrees Kelvin. From the formula

$$°K = °C + 273.15,$$

it follows that the temperature of the air is 311.15°K. Then using Eqs. (2.53) and (2.55) gives

$$E_p = \frac{1}{2} \times 0.0253 \times \frac{311.15}{293.61} = 0.0134 \text{ eV. } [Ans.]$$

Finally, $\overline{E} = 3E_p$, so that $\overline{E} = 0.0402$ eV. [*Ans.*]

The Gas Law

To a first approximation, gases obey the familiar *ideal gas law*

$$PV = n_M RT, \tag{2.57}$$

where P is the gas pressure, V is the volume, n_M is the number of moles of gas contained in V, R is the gas constant, and T is the absolute temperature. Equation (2.57) can also be written as

$$P = \left(\frac{n_M N_A}{V}\right)\left(\frac{R}{N_A}\right)T,$$

where N_A is Avogadro's number. Since there are N_A atoms per mole, it follows that there are $n_M N_A$ atoms in V. The first factor is therefore equal to N—the total number of atoms or molecules per unit volume. At the same time, the factor R/N_A is the definition of k, Boltzmann's constant. The ideal gas law can thus be put in the convenient form

$$P = NkT. \tag{2.58}$$

From Eq. (2.58) it is seen that gas pressure can be (and, in certain applications, is) expressed in units of energy per unit volume.

2.14 ATOM DENSITY

In nuclear engineering problems, it is often necessary to calculate the number of atoms or molecules contained in 1 cm^3 of a substance. Consider first a material such as sodium, which is composed of only one type of atom. Then if ρ is its physical density in g/cm^3 and M is its gram atomic weight, it follows that there are ρ/M gram moles of the substance in 1 cm^3. Since each gram mole contains N_A atoms, where N_A is Avogadro's number, the atom density N, in atoms per cm^3, is simply

$$N = \frac{\rho N_A}{M}.$$

(2.59)

Example 2.12

The density of sodium is 0.97 g/cm^3. Calculate its atom density.

Solution. The atomic weight of sodium is 22.990. Then from Eq. (2.59),

$$N = \frac{0.97 \times 0.6022 \times 10^{24}}{22.990} = 0.0254 \times 10^{24}. \ [\textit{Ans.}]$$

(It is usual to express the atom densities as a factor times 10^{24}.)

Equation (2.59) also applies to substances composed of individual molecules, except that N is the molecule density (molecules per cm^3) and M is the gram molecular weight. To find the number of atoms of a particular type per cm^3, it is merely necessary to multiply the molecular density by the number, n_i, of those atoms present in the molecule,

$$N_i = n_i \frac{\rho N_A}{M}.$$

(2.60)

The computation of atom density for crystalline solids such as NaCl and for liquids is just as easy as for simple atomic and molecular substances, but the explanation is more complicated. The problem here is that there are no recognizable molecules—an entire microcrystal of NaCl is, so to speak, a molecule. What should be done in this case is to assume that the material consists of hypothetical molecules containing appropriate numbers of the constituent atoms. (These molecules are in fact unit cells of the crystalline solid and contain the appropriate number of atoms.) Then by using the molecular weight for this pseudomolecule in Eq. (2.59), the computed value of N gives the molecular density of this molecule. The atom densities can then be computed from this number in the usual way as illustrated in the following example.

Example 2.13

The density of a NaCl crystal is 2.17 g/cm^3. Compute the atom densities of Na and Cl.

Solution. The atomic weight of Na and Cl are 22.990 and 35.453, respectively. The molecular weight for a pseudomolecule of NaCl is therefore 58.443. Using Eq. (2.59) gives

$$N = \frac{2.17 \times 0.6022 \times 10^{24}}{58.443} = 0.0224 \times 10^{24} \text{ molecules/cm}^3.$$

Since there is one atom each of Na and Cl per molecule, it follows that this is also equal to the atom density of each atom. [*Ans.*]

Frequently it is required to compute the number of atoms of a particular isotope per cm^3. Since, as pointed out in Section 2.2, the abundance of isotopes is always stated in atom percent, the atom density of an isotope is just the total atom density of the element as derived earlier, multiplied by the isotopic abundance expressed as a fraction. Thus, the atom density N_i for the ith isotope is

$$N_i = \frac{\gamma_i \rho N_A}{100M}, \tag{2.61}$$

where γ_i is the isotopic abundance in *atom percent*, abbreviated a/o.

The chemical compositions of mixtures of elements such as metallic alloys are usually given in terms of the percent by weight of the various constituents. If ρ is the physical density of the mixture, then the average density of the ith component is

$$\rho_i = \frac{w_i \rho}{100}, \tag{2.62}$$

where w_i is the *weight percent*, abbreviated w/o, of the component. From Eq. (2.59), it follows that the atom density of this component is

$$N_i = \frac{w_i \rho N_A}{100M_i}, \tag{2.63}$$

where M_i is its gram atomic weight.

With a substance whose composition is specified by a chemical formula, the percent by weight of a particular element is equal to the ratio of its atomic weight in the compound to the total molecular weight of the compound. Thus, with the compound $X_m Y_n$, the molecular weight is $m M_x + n M_y$, where M_x and M_y are the atomic weights of X and Y, respectively, and the percent by weight of the element X is

$$w/o(X) = \frac{m M_x}{m M_x + n M_y} \times 100. \tag{2.64}$$

In some nuclear applications, the isotopic composition of an element must be changed by artificial means. For instance, the uranium used as fuel in many nuclear reactors must be enriched in the isotope ^{235}U (the enrichment process is discussed in Chap. 4). In this case, it is also the practice to specify enrichment in weight percent. The atomic weight of the enriched uranium can be computed as follows. The total number of uranium atoms per cm^3 is given by

$$N = \Sigma N_i,$$

where N_i is the atom density of the ith isotope. Introducing N from Eq. (2.59) and N_i from Eq. (2.63) gives

$$\frac{1}{M} = \frac{1}{100} \Sigma \frac{w_i}{M_i}. \tag{2.65}$$

The application of this formula in computations involving enriched uranium is illustrated in Example 2.16.

Example 2.14

For water of normal (unit) density compute:

(a) the number of H_2O molecules per cm^3,
(b) the atom densities of hydrogen and oxygen,
(c) the atom density of ^2H.

Solution

1. The molecular weight of H_2O is $2 \times 1.00797 + 15.9994 = 18.0153$. The molecular density is therefore,

$$N(H_2O) = \frac{1 \times 0.6022 \times 10^{24}}{18.0153} = 0.03343 \times 10^{24} \text{ molecules/cm}^3. \ [Ans.]$$

2. There are two atoms of hydrogen and one atom of oxygen per H_2O molecule. Thus, the atom density of hydrogen $N(H) = 2 \times 0.03343 \times 10^{24} = 0.06686 \times 10^{24}$ atoms/cm^3, and $N(O) = 0.03343 \times 10^{24}$ atoms/cm^3. [Ans.]

3. The relative abundance of ^2H is 0.015 a/o so that $N(^2H) = 1.5 \times 10^{-4} \times N(H) = 1.0029 \times 10^{-5} \times 10^{24}$ atoms/cm^3. [Ans.]

Example 2.15

A certain nuclear reactor is fueled with 1,500 kg of uranium rods enriched to 20 w/o ^{235}U. The remainder is ^{238}U. The density of the uranium is 19.1 g/cm^3.

(a) How much ^{235}U is in the reactor?

(b) What are the atom densities of ^{235}U and ^{238}U in the rods?

Solution

1. Enrichment to 20 w/o means that 20% of the total uranium mass is ^{235}U. The amount of ^{235}U is therefore

$$0.20 \times 1500 = 300 \text{kg}. \ [Ans.]$$

2. The atomic weights of ^{235}U and ^{238}U are 235.0439 and 238.0508, respectively. From Eq. (2.63),

$$N(^{235}U) = \frac{20 \times 19.1 \times 0.6022 \times 10^{24}}{100 \times 235.0439}$$

$$= 9.79 \times 10^{-3} \times 10^{24} \text{atoms/cm}^3. \ [Ans.]$$

The ^{238}U is present to the extent of 80 w/o, and so

$$N(^{238}U) = \frac{80 \times 19.1 \times 0.6022 \times 10^{24}}{100 \times 238.0508}$$

$$= 3.86 \times 10^{-2} \times 10^{24} \text{atoms/cm}^3. \ [Ans.]$$

Example 2.16

The fuel for a reactor consists of pellets of uranium dioxide (UO_2), which have a density of 10.5 g/cm^3. If the uranium is enriched to 30 w/o in ^{235}U, what is the atom density of the ^{235}U in the fuel?

Solution. It is first necessary to compute the atomic weight of the uranium. From Eq. (2.65),

$$\frac{1}{M} = \frac{1}{100} \left(\frac{30}{235.0439} + \frac{70}{238.0508} \right),$$

which gives $M = 237.141$. The molecular weight of the UO_2 is then $237.141 + 2 \times 15.999 = 269.139$. In view of Eq. (2.64), the percent by weight of uranium in the UO_2 is $(237.141/269.139) \times 100 = 88.1 \ w/o$. The average density of the uranium is therefore $0.881 \times 10.5 = 9.25 \ \text{g/cm}^3$, and the density of ^{235}U is $0.30 \times 9.25 = 2.78 \ \text{g/cm}^3$. The atom density of ^{235}U is finally

$$N(^{235}U) = \frac{2.78 \times 0.6022 \times 10^{24}}{235.0439}$$

$$= 7.11 \times 10^{-3} \times 10^{24} \ \text{atoms/cm}^3. \ [Ans.]$$

REFERENCES

General

Arya, A. P., *Elementary Modern Physics*, Reading, Mass.: Addison-Wesley, 1974.

Beiser, A., *Concepts of Modern Physics*, 5th ed. New York: McGraw-Hill, 1994, Chapters 3, 4, 5, 11, 12, 13, and 14.

Burcham, W. E., *Nuclear Physics: An Introduction*, 2nd ed. Reprint Ann Arbor.

Foster, A. R., and R. L. Wright, Jr., *Basic Nuclear Engineering*, 4th ed. Paramus: Prentice-Hall, 1982, Chapter 2 and 3.

Goble, A. T., and K. K. Baker, *Elements of Modern Physics*, 2nd ed. New York: Ronald Press, 1971, Chapters 8–10.

Kaplan, I., *Nuclear Physics*, 2nd ed. Reading, Mass.: Addison-Wesley Longman, 1962.

Krane, K. S., *Nuclear Physics*, 3rd ed. New York: John Wiley, 1987.

Krane, K. S., *Modern Physics*, 2nd ed. New York: John Wiley, 1995.

Lapp, R. E., and H. L. Andrews, *Nuclear Radiation Physics*, 4th ed. Englewood Cliffs, N.J.: Prentice-Hall, 1972, Chapters 1–7.

Liverhant, S. E., *Elementary Introduction to Nuclear Reactor Physics*. New York: Wiley, 1960.

Meyerhof, W. E., *Elements of Nuclear Physics*, New York: McGraw-Hill, 1967, Chapters 2 and 4.

Oldenberg, O., and N. C. Rasmussen, *Modern Physics for Engineers*, reprint, Marietta, Technical Books, 1992, Chapters 12, 14, and 15.

Semat, H., and J. R. Albright, *Introduction to Atomic and Nuclear Physics*, 5th ed. New York: Holt, Rinehart & Winston, 1972.

Serway, R. A., Moses, C. J., and Moyer, C. A., *Modern Physics*, 3rd ed. Philadelphia: Saunders, 1990.

Tipler, P. A., *Modern Physics*, 2nd ed. New York: Worth, 1977.

Wehr, M. R., Richards, J. A., and Adair, T. W., *Physics of the Atom*, 4th ed. Reading, Mass.: Addison-Wesley, 1984.

Weidner, R. T., and R. L. Sells, *Elementary Modern Physics*, 3rd ed. Boston, Mass.: Allyn & Bacon, 1980.

Williams, W. S. C., *Nuclear and Particle Physics*, Oxford, Eng.: Clarendon Press, 1991.

Nuclear Data

The Chart of the Nuclides. This valuable chart, a must for every nuclear engineer, is available from the Lockheed Martin Distribution Services 10525 Chester Road Cincinnati, OH 45215, USA. This chart is also available in a searchable format from various Internet sites including *http//www.dne.bnl.gov*.

Lederer, C. M., Hollander, J. M., and Perlman, I., *Tables of Isotopes*, 7th ed. New York: Wiley, 1978. This is an extraordinary collection of nuclear data, which includes among other things masses of the nuclides (given in terms of the mass excess, denoted in the tables as Δ; see Problem 2.47), nuclear energy levels, decay schemes, and so on.

The National Nuclear Data Center at Brookhaven National Laboratory, Upton, New York, collects, evaluates, and distributes a wide range of nuclear data. A comprehensive collection of data is available through the National Nuclear Data Center at *http//www.nndc.bnl.gov/nndcscr*.

PROBLEMS

1. How many neutrons and protons are there in the nuclei of the following atoms:
   (a) ^{7}Li,
   (b) ^{24}Mg,
   (c) ^{135}Xe,
   (d) ^{209}Bi,
   (e) ^{222}Rn?

2. The atomic weight of ^{59}Co is 58.93319. How many times heavier is ^{59}Co than ^{12}C?

3. How many atoms are there in 10 g of ^{12}C?

4. Using the data given next and in Example 2.2, compute the molecular weights of
   (a) H_2 gas,
   (b) H_2O,
   (c) H_2O_2.

Isotope	Abundance, a/o	Atomic weight
^{1}H	99.985	1.007825
^{2}H	0.015	2.01410

5. When H_2 gas is formed from naturally occurring hydrogen, what percentages of the molecules have molecular weights of approximately 2, 3, and 4?

6. Natural uranium is composed of three isotopes: ^{234}U, ^{235}U, and ^{238}U. Their abundances and atomic weights are given in the following table. Compute the atomic weight of natural uranium.

Isotope	Abundance, a/o	Atomic weight
^{234}U	0.0057	234.0409
^{235}U	0.72	235.0439
^{235}U	99.27	238.0508

7. A beaker contains 50 g of ordinary (i.e., naturally occurring) water.
    (a) How many moles of water are present?
    (b) How many hydrogen atoms?
    (c) How many deuterium atoms?

8. The glass in Example 2.1 has an inside diameter of 7.5 cm. How high does the water stand in the glass?

9. Compute the mass of a proton in amu.

10. Calculate the mass of a neutral atom of ^{235}U
    (a) in amu;
    (b) in grams.

11. Show that 1 amu is numerically equal to the reciprocal of Avogadro's number.

12. Using Eq. (2.3), estimate the radius of the nucleus of ^{238}U. Roughly what fraction of the ^{238}U atom is taken up by the nucleus?

13. Using Eq. (2.3), estimate the density of nuclear matter in g/cm^3; in Kg/m^3. Take the mass of each nucleon to be approximately 1.5×10^{-24} g.

14. The planet earth has a mass of approximately 6×10^{24} kg. If the density of the earth were equal to that of nuclei, how big would the earth be?

15. The complete combustion of 1 kg of bituminous coal releases about 3×10^7 J in heat energy. The conversion of 1 g of mass into energy is equivalent to the burning of how much coal?

16. The fission of the nucleus of ^{235}U releases approximately 200 MeV. How much energy (in kilowatt-hours and megawatt-days) is released when 1 g of ^{235}U undergoes fission?

17. Compute the neutron–proton mass difference in MeV.

18. An electron starting from rest is accelerated across a potential difference of 5 million volts.
    (a) What is its final kinetic energy?
    (b) What is its total energy?
    (c) What is its final mass?

19. Derive Eq. (2.18). [*Hint:* Square both sides of Eq. (2.5) and solve for mv.]

20. Show that the speed of any particle, relativistic or nonrelativistic, is given by the following formula:

$$v = c\sqrt{1 - \frac{E_{\text{rest}}^2}{E_{\text{total}}^2}},$$

where E_{rest} and E_{total} are its rest-mass energy and total energy, respectively, and c is the speed of light.

21. Using the result derived in Problem 2.20, calculate the speed of a 1-MeV electron, one with a kinetic energy of 1 MeV.

22. Compute the wavelengths of a 1-MeV
    (a) photon,
    (b) neutron.

23. Show that the wavelength of a relativistic particle is given by

$$\lambda = \lambda_C \frac{m_e c^2}{\sqrt{E_{\text{total}}^2 - E_{\text{rest}}^2}},$$

where $\lambda_C = h/m_e c = 2.426 \times 10^{-10}$ cm is called the *Compton wavelength*.

24. Using the formula obtained in Problem 2.23, compute the wavelength of a 1-MeV electron.

25. An electron moves with a kinetic energy equal to its rest-mass energy. Calculate the electron's
    (a) total energy in units of $m_e c^2$;
    (b) mass in units of m_e;
    (c) speed in units of c;
    (d) wavelength in units of the Compton wavelength.

26. According to Eq. (2.20), a photon carries momentum, thus a free atom or nucleus recoils when it emits a photon. The energy of the photon is therefore actually less than the available transition energy (energy between states) by an amount equal to the recoil energy of the radiating system.
    (a) Given that E is the energy between two states and E_γ is the energy of the emitted photon, show that

$$E_\gamma \simeq E\left(1 - \frac{E}{2Mc^2}\right),$$

where M is the mass of the atom or nucleus.
    (b) Compute $E - E_\gamma$ for the transitions from the first excited state of atomic hydrogen at 10.19 eV to ground and the first excited state of ^{12}C at 4.43 MeV to ground (see Figs. 2.2 and 2.3).

27. The first three excited states of the nucleus of ^{199}Hg are at 0.158 MeV, 0.208 MeV, and 0.403 MeV above the ground state. If all transitions between these states and ground occurred, what energy γ-rays would be observed?

28. Using the chart of the nuclides, complete the following reactions. If a daughter nucleus is also radioactive, indicate the complete decay chain.

(a) ^{18}N$\longrightarrow$
(b) ^{83}Y$\longrightarrow$
(c) ^{135}Sb$\longrightarrow$
(d) ^{219}Rn$\longrightarrow$

29. Tritium (^{3}H) decays by negative beta decay with a half-life of 12.26 years. The atomic weight of ^{3}H is 3.016.
    (a) To what nucleus does ^{3}H decay?
    (b) What is the mass in grams of 1 mCi of tritium?

30. Approximately what mass of ^{90}Sr ($T_{1/2} = 28.8$ yr.) has the same activity as 1 g of ^{60}Co ($T_{1/2} = 5.26$ yr.)?

31. Carbon tetrachloride labeled with ^{14}C is sold commercially with an activity of 10 millicuries per millimole (10 mCi/mM). What fraction of the carbon atoms is ^{14}C?

32. Tritiated water (ordinary water containing some ^{1}H^{3}HO) for biological applications can be purchased in 1-cm^3 ampoules having an activity of 5 mCi per cm^3. What fraction of the water molecules contains an ^{3}H atom?

33. After the initial cleanup effort at Three Mile Island, approximately 400,000 gallons of radioactive water remained in the basement of the containment building of the Three Mile Island Unit 2 nuclear plant. The principal sources of this radioactivity were ^{137}Cs at 156 μCi/cm^3 and ^{134}Cs at 26 μCi/cm^3. How many atoms per cm^3 of these radionuclides were in the water at that time?

34. One gram of ^{226}Ra is placed in a sealed, evacuated capsule 1.2 cm^3 in volume.
    (a) At what rate does the helium pressure increase in the capsule, assuming all of the α-particles are neutralized and retained in the free volume of the capsule?
    (b) What is the pressure 10 years after the capsule is sealed?

35. Polonium-210 decays to the ground state of ^{206}Pb by the emission of a 5.305-MeV α-particle with a half-life of 138 days. What mass of ^{210}Po is required to produce 1 MW of thermal energy from its radioactive decay?

36. The radioisotope generator SNAP-9 was fueled with 475 g of ^{238}PuC (plutonium-238 carbide), which has a density of 12.5 g/cm^3. The ^{238}Pu has a half-life of 89 years and emits 5.6 MeV per disintegration, all of which may be assumed to be absorbed in the generator. The thermal to electrical efficiency of the system is 5.4%. Calculate
    (a) the fuel efficiency in curies per watt (thermal);
    (b) the specific power in watts (thermal) per gram of fuel;
    (c) the power density in watts (thermal) per cm^3;
    (d) the total electrical power of the generator.

37. Since the half-life of ^{235}U (7.13×10^8 years) is less than that of ^{238}U (4.51×10^9 years), the isotopic abundance of ^{235}U has been steadily decreasing since the earth was formed about 4.5 billion years ago. How long ago was the isotopic abundance of ^{235}U equal to 3.0 a/o, the enrichment of the uranium used in many nuclear power plants?

38. The radioactive isotope Y is produced at the rate of R atoms/sec by neutron bombardment of X according to the reaction

$$X(n, \gamma)Y.$$

If the neutron bombardment is carried out for a time equal to the half-life of Y, what fraction of the saturation activity of Y will be obtained assuming that there is no Y present at the start of the bombardment?

39. Consider the chain decay

$$A \rightarrow B \rightarrow C \rightarrow,$$

with no atoms of B present at $t = 0$.

(a) Show that the activity of B rises to a maximum value at the time t_m given by

$$t_m = \frac{1}{\lambda_B - \lambda_A} \ln\left(\frac{\lambda_B}{\lambda_A}\right),$$

at which time the activities of A and B are equal.

(b) Show that, for $t < t_m$, the activity of B is less than that of A, whereas the reverse is the case for $t > t_m$.

40. Show that if the half-life of B is much shorter than the half-life of A, then the activities of A and B in Problem 2.39 eventually approach the same value. In this case, A and B are said to be in *secular equilibrium*.

41. Show that the abundance of ^{234}U can be explained by assuming that this isotope originates solely from the decay of ^{238}U.

42. Radon-222, a highly radioactive gas with a half-life of 3.8 days that originates in the decay of ^{234}U (see the chart of nuclides), may be present in uranium mines in dangerous concentrations if the mines are not properly ventilated. Calculate the activity of ^{222}Rn in Bq per metric ton of natural uranium.

43. According to U.S. Nuclear Regulatory Commission regulations, the maximum permissible concentration of radon-222 in air in equilibrium with its short-lived daughters is 3 pCi/liter for nonoccupational exposure. This corresponds to how many atoms of radon-222 per cm^3?

44. Consider again the decay chain in Problem 2.39 in which the nuclide A is produced at the constant rate of R atoms/sec. Derive an expression for the activity of B as a function of time.

45. Complete the following reactions and calculate their Q values. [*Note:* The atomic weight of ^{14}C is 14.003242.]

(a) $^4He(p, d)$
(b) $^9Be(\alpha, n)$
(c) $^{14}N(n, p)$
(d) $^{115}In(d, p)$
(e) $^{207}Pb(\gamma, n)$

46. (a) Compute the recoil energy of the residual, daughter nucleus following the emission of a 4.782-MeV α-particle by ^{226}Ra.

(b) What is the total disintegration energy for this decay process?

47. In some tabulations, atomic masses are given in terms of the mass excess rather than as atomic masses. The mass excess, Δ, is the difference

$$\Delta = M - A,$$

where M is the atomic mass and A is the atomic mass number. For convenience, Δ, which may be positive or negative, is usually given in units of MeV. Show that the Q value for the reaction shown in Eq. (2.38) can be written as

$$Q = (\Delta_a + \Delta_b) - (\Delta_c + \Delta_f).$$

48. According to the tables of Lederer *et al.* (see References), the mass excesses for the (neutral) atoms in the reaction in Example 2.8 are as follows: $\Delta(^3\text{H}) = 14.95$ MeV, $\Delta(^2\text{H}) = 13.14$ MeV, $\Delta(\text{n}) = 8.07$ MeV, and $\Delta(^4\text{He}) = 2.42$ MeV. Calculate the Q value of this reaction using the results of Problem 2.47.

49. The atomic weight of ^{206}Pb is 205.9745. Using the data in Problem 2.35, calculate the atomic weight of ^{210}Po. [*Caution:* See Problem 2.46]

50. Tritium (^3H) can be produced through the absorption of low-energy neutrons by deuterium. The reaction is

$$^2\text{H} + \text{n} \rightarrow ^3\text{H} + \gamma,$$

where the γ-ray has an energy of 6.256 MeV.
(a) Show that the recoil energy of the ^3H nucleus is approximately 7 keV.
(b) What is the Q value of the reaction?
(c) Calculate the separation energy of the last neutron in ^3H.
(d) Using the binding energy for ^2H of 2.23 MeV and the result from part (c), compute the total binding energy of ^3H.

51. Consider the reaction

$$^6Li(\alpha, \text{ p})^9\text{Be}.$$

Using atomic mass data, compute:
(a) the total binding energy of ^6Li, ^9Be, and ^4He;
(b) the Q value of the reaction using the results of part (a).

52. Using atomic mass data, compute the average binding energy per nucleon of the following nuclei:
(a) ^2H
(b) ^4He
(c) ^{12}C
(d) ^{51}V
(e) ^{138}Ba
(f) ^{235}U

53. Using the mass formula, compute the binding energy per nucleon for the nuclei in Problem 2.52. Compare the results with those obtained in that problem.

54. Compute the separation energies of the last neutron in the following nuclei:
   (a) ^{4}He
   (b) ^{7}Li
   (c) ^{17}O
   (d) ^{51}V
   (e) ^{208}Pb
   (f) ^{235}U

55. Derive Eq. (2.53). [*Hint:* Try taking the logarithm of Eq. (2.52) before differentiating.]

56. What is 1 atmosphere pressure in units of eV/cm^3? [*Hint:* At standard temperature and pressure (0°C and 1 atm), 1 mole of gas occupies 22.4 liters.]

57. Calculate the atom density of graphite having density of 1.60 g/cm^3.

58. Calculate the activity of 1 gram of natural uranium.

59. What is the atom density of ^{235}U in uranium enriched to 2.5 *a/o* in this isotope if the physical density of the uranium is 19.0 g/cm^3?

60. Plutonium-239 undergoes α-decay with a half-life of 24,000 years. Compute the activity of 1 gram of plutonium dioxide, ^{239}PuO$_2$. Express the activity in terms of Ci and Bq.

61. It has been proposed to use uranium carbide (UC) for the initial fuel in certain types of breeder reactors, with the uranium enriched to 25 *w/o*. The density of UC is 13.6 g/cm^3.
   (a) What is the atomic weight of the uranium?
   (b) What is the atom density of the ^{235}U?

62. Compute the atom densities of ^{235}U and ^{238}U in UO$_2$ of physical density 10.8 g/cm^3 if the uranium is enriched to 3.5 *w/o* in ^{235}U.

63. The fuel for a certain breeder reactor consists of pellets composed of mixed oxides, UO$_2$ and PuO$_2$, with the PuO$_2$ comprising approximately 30 *w/o* of the mixture. The uranium is essentially all ^{238}U, whereas the plutonium contains the following isotopes: ^{239}Pu (70.5 *w/o*), ^{240}Pu (21.3 *w/o*), ^{241}Pu (5.5 *w/o*), and ^{242}Pu (2.7 *w/o*). Calculate the number of atoms of each isotope per gram of the fuel.

3

Interaction of Radiation with Matter

The design of all nuclear systems—reactors, radiation shields, isotopic generators, and so on—depends fundamentally on the way in which nuclear radiation interacts with matter. In this chapter, these interactions are discussed for neutrons, γ-rays, and various charged particles with energies up to about 20 MeV. Most of the radiation encountered in practical nuclear devices lies in this energy region.

3.1 NEUTRON INTERACTIONS

It is important to recognize at the outset that, since neutrons are electrically neutral, they are not affected by the electrons in an atom or by the positive charge of the nucleus. As a consequence, neutrons pass through the atomic electron cloud and interact directly with the nucleus. In short, neutrons collide with nuclei, not with atoms.

Neutrons may interact with nuclei in one or more of the following ways.

Elastic Scattering　In this process, the neutron strikes the nucleus, which is almost always in its ground state (see Section 2.7), the neutron reappears, and the nucleus is left in its ground state. The neutron in this case is said to have been *elastically scattered* by the nucleus. In the notation of nuclear reactions (see Section 2.10), this interaction is abbreviated by the symbol (n, n).

Inelastic Scattering   This process is identical to elastic scattering except that the nucleus is left in an excited state. Because energy is retained by the nucleus, this is clearly an endothermic interaction. Inelastic scattering is denoted by the symbol (n, n'). The excited nucleus decays, as explained in Section 2.7, by the emission of γ-rays. In this case, since these γ-rays originate in inelastic scattering, they are called *inelastic γ-rays*.

Radiative Capture   Here the neutron is captured by the nucleus, and one or more γ-rays—called *capture γ-rays*—are emitted. This is an exothermic interaction and is denoted by (n, γ). Since the original neutron is absorbed, this process is an example of a class of interactions known as *absorption reactions*.

Charged-Particle Reactions   Neutrons may also disappear as the result of absorption reactions of the type (n, α) and (n, p). Such reactions may be either exothermic or endothermic.

Neutron-Producing Reactions   Reactions of the type (n, 2n) and (n, 3n) occur with energetic neutrons. These reactions are clearly endothermic since in the (n, 2n) reaction one neutron and in the (n, 3n) reaction 2 neutrons are extracted from the struck nucleus. The (n, 2n) reaction is especially important in reactors containing heavy water or beryllium since ^{2}H and ^{9}Be have loosely bound neutrons that can easily be ejected.

Fission   Neutrons colliding with certain nuclei may cause the nucleus to split apart—to undergo *fission*. This reaction, as noted in Chap. 2, is the principal source of nuclear energy for practical applications.

Many of these reactions may be viewed as a two-step process involving the formation of a compound nucleus. For example, the scattering reactions, both elastic and inelastic, may be thought of as a process in which the neutron is first absorbed by the target nucleus to form a new nucleus whose atomic number is unchanged but whose mass number is increased by 1. Then depending on the specific process, the nucleus decays via neutron emission to produce the original nucleus plus a neutron. The product nucleus is left in the ground state or an excited state according to the type of scattering reaction involved. This model is particularly useful in understanding the fission process.

Each of these interactions is discussed in this chapter. However, to describe quantitatively the various interactions, it is necessary to introduce certain parameters.

3.2 CROSS-SECTIONS

The extent to which neutrons interact with nuclei is described in terms of quantities known as *cross-sections*. These are defined by the following type of experiment. Suppose that a beam of monoenergetic (single-energy) neutrons impinges on a thin target of thickness X and area A as shown in Fig. 3.1. If there are n neutrons per cm^3 in the beam and v is the speed of the neutrons, then the quantity

$$I = nv \tag{3.1}$$

is called the *intensity* of the beam. Since the neutrons travel the distance v cm in 1 sec, all of the neutrons in the volume vA in front of the target will hit the target in 1 sec. Thus, $nvA = IA$ neutrons strike the entire target per second, and it follows that $IA/A = I$ is equal to the number of neutrons striking the target per cm^2/sec. Since nuclei are small and the target is assumed to be thin, most of the neutrons striking the target in an experiment like that shown in Fig. 3.1 ordinarily pass through the target without interacting with any of the nuclei. The number that do collide are found to be proportional to the beam intensity, to the atom density N of the target, and to the area and thickness of the target. These observations can be summarized by the equation

$$\text{Number of collisions per second (in entire target)} = \sigma I N A X, \tag{3.2}$$

where σ, the proportionality constant, is called the *cross-section*. The factor NAX in Eq. (3.2) is the total number of nuclei in the target. The number of collisions per second with a single nucleus is therefore just σI. It follows that σ is equal to the number of collisions per second with one nucleus per unit intensity of the beam.

There is another way to view the concept of cross-section. As already noted, IA neutrons strike the target per second. Of these, σI interact with any given nucleus. Therefore, it may be concluded that

$$\frac{\sigma I}{AI} = \frac{\sigma}{A} \tag{3.3}$$

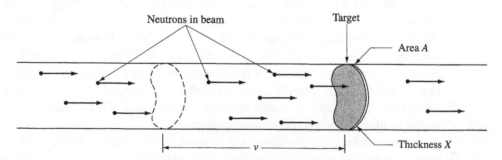

Figure 3.1 Neutron beam striking a target.

is equal to the probability that a neutron in the beam will collide with this nucleus. It is clear from Eq. (3.3) that σ has units of area. In fact, it is not difficult to see that σ is nothing more than the effective cross-sectional area of the nucleus, hence the term *cross-section*.

Neutron cross-sections are expressed in units of *barns*, where 1 barn, abbreviated b, is equal to 10^{-24} cm^2. One thousandth of a barn is called a *millibarn*, denoted as mb.

Up to this point, it has been assumed that the neutron beam strikes the entire target. However, in many experiments, the beam is actually smaller in diameter than the target. In this case, the prior formulas still hold, but now $\mathcal{A}$ refers to the area of the beam instead of the area of the target. The definition of cross-section remains the same, of course.

Each of the processes described in Section 3.1 by which neutrons interact with nuclei is denoted by a characteristic cross-section. Thus, elastic scattering is described by the elastic scattering cross-section, σ_e; inelastic scattering by the inelastic scattering cross-section, σ_i; the (n, γ) reaction (radiative capture) by the capture cross-section, σ_γ; fission by the fission cross-section, σ_f; and so on. The sum of the cross-sections for all possible interactions is known as the *total cross-section* and is denoted by the symbol σ_t; that is

$$\sigma_t = \sigma_e + \sigma_i + \sigma_\gamma + \sigma_f + \cdots . \tag{3.4}$$

The total cross-section measures the probability that an interaction of any type will occur when neutrons strike a target. The sum of the cross-sections of all absorption reactions is known as the *absorption cross-section* and is denoted by σ_a. Thus,

$$\sigma_a = \sigma_\gamma + \sigma_f + \sigma_p + \sigma_\alpha + \cdots , \tag{3.5}$$

where σ_p and σ_α are the cross-sections for the (n, p) and (n, α) reactions. As indicated in Eq. (3.5), fission, by convention, is treated as an absorption process. Similarly, the total scattering cross-section is the sum of the elastic and inelastic scattering cross-section. Thus,

$$\sigma_s = \sigma_e + \sigma_i,$$

and

$$\sigma_t = \sigma_s + \sigma_a.$$

Example 3.1

A beam of 1-Me V neutrons of intensity 5×10^8 neutrons/cm^2-sec strikes a thin ^{12}C target. The area of the target is 0.5 cm^2 and is 0.05 cm thick. The beam has a cross-sectional area of 0.1 cm^2. At 1 Me V, the total cross-section of ^{12}C is 2.6 b. (a) At what rate do interactions take place in the target? (b) What is the probability that a neutron in the beam will have a collision in the target?

Solution

1. According to Table I.3 in Appendix II, $N = 0.080 \times 10^{24}$ for carbon. Then from Eq. (3.2), the total interaction rate is

$$\sigma_t I N \mathcal{A} X = 2.6 \times 10^{-24} \times 5 \times 10^8 \times 0.080 \times 10^{24} \times 0.1 \times 0.05$$

$$= 5.2 \times 10^5 \text{ interactions/sec. } [Ans.]$$

It should be noted that the 10^{-24} in the cross-section cancels with the 10^{24} in atom density. This is the reason for writing atom densities in the form of a number $\times 10^{24}$.

2. In 1 sec, $I \mathcal{A} = 5 \times 10^8 \times 0.1 = 5 \times 10^7$ neutrons strike the target. Of these, 5.2×10^5 interact. The probability that a neutron interacts in the target is therefore $5.2 \times 10^5 / 5 \times 10^7 = 1.04 \times 10^{-2}$. Thus, only about 1 neutron in 100 has a collision while traversing the target.

Example 3.2

There are only two absorption reactions—namely, radiative capture and fission—that can occur when 0.0253-e V neutrons[1] interact with ^{235}U. The cross-sections for these reactions are 99 b and 582 b, respectively. When a 0.0253-e V neutron is absorbed by ^{235}U, what is the relative probability that fission will occur?

Solution. Since σ_γ and σ_f are proportional to the probabilities of radiative capture and fission, it follows that the probability of fission is $\sigma_f / (\sigma_\gamma + \sigma_f) = \sigma_f / \sigma_a = 582/681 = 85.5\%$. [*Ans.*]

To return to Eq. (3.2), this can be written as

$$\text{Number of collisions per second (in entire target)} = I N \sigma_t \times \mathcal{A} X, \qquad (3.6)$$

where σ_t has been introduced because this cross-section measures the probability that a collision of any type may occur. Since $\mathcal{A} X$ is the total volume of the target, it follows from Eq. (3.6) that the number of collisions per cm^3/sec in the target, which is called the *collision density F*, is given by

$$F = I N \sigma_t. \qquad (3.7)$$

The product of the atom density N and a cross-section, as in Eq. (3.7), occurs frequently in the equations of nuclear engineering; it is given the special symbol Σ and is called the *macroscopic cross-section*. In particular, the product $N\sigma_t = \Sigma_t$ is called the *macroscopic total cross-section*, $N\sigma_s = \Sigma_s$ is called the *macroscopic scattering cross-section*, and so on. Since N and σ have units of cm^{-3} and cm^2,

---

[1] For reasons explained in Section 3.6, neutron cross-sections are tabulated at an energy of 0.0253 e V.

respectively, Σ has units of cm^{-1}. In terms of the macroscopic cross-section, the collision in Eq. (3.7) reduces to

$$F = I\Sigma_t. \tag{3.8}$$

Example 3.3

Referring to Example 3.1, calculate the (a) macroscopic total cross-section of ^{12}C at Me V; (b) collision density in the target.

Solution

1. From the definition given previously,

$$\Sigma_t = 0.08 \times 10^{24} \times 2.6 \times 10^{-24} = 0.21 \text{ cm}^{-1}. [Ans.]$$

2. Using Eq. (3.8) gives

$$F = 5 \times 10^8 \times 0.21 = 1.05 \times 10^8 \text{ collisions/cm}^3\text{-sec.} [Ans.]$$

These collisions occur, of course, only in the region of the target that is struck by the beam.

3.3 NEUTRON ATTENUATION

The preceding section refers to experiments involving thin targets. Suppose a thick target of thickness X is placed in a monodirectional beam of intensity I_0 and a neutron detector is located at some distance behind the target as shown in Fig. 3.2. It is assumed that both the target and detector are so small that the detector subtends a small solid angle at the target. For a neutron to reach the detector, it must be traveling directly toward both the target and detector. Consequently, every neutron that has a collision in the target is lost from the beam, and only those neutrons that do not interact enter the detector.

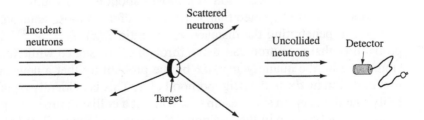

Figure 3.2   Measurement of neutrons that have not collided in a target.

Let $I(x)$ be the intensity of the neutrons that have *not* collided after penetrating the distance x into the target. Then in traversing an additional distance dx, the intensity of the uncollided beam is decreased by the number of neutrons that have collided in the thin sheet of target having an area of 1 cm² and the thickness dx. From Eq. (3.2), this decrease in intensity is given by

$$-dI(x) = N\sigma_t I(x)\,dx = \Sigma_t I(x)\,dx. \tag{3.9}$$

This equation can be integrated with the result

$$I(x) = I_0 e^{-\Sigma_t x}. \tag{3.10}$$

The intensity of the uncollided neutrons thus decreases exponentially with the distance inside the target. The intensity of the beam of uncollided neutrons emerging from the target is then

$$I(X) = I_0 e^{-\Sigma_t X}, \tag{3.11}$$

and this is the intensity measured by the detector.

If the target is very thick, as with a radiation shield, then almost all of the incident neutrons will have at least one collision in the target, so that most of the emerging neutrons will have undergone scattering within the target. Since these neutrons are specifically excluded in the derivation of Eq. (3.11), this equation must not be used to calculate the effectiveness of a shield. To do so would ignore a most important component of the emergent radiation—namely, the scattered neutrons.

When Eq. (3.9) is divided by $I(x)$, the result is

$$-\frac{dI(x)}{I(x)} = \Sigma_t\,dx. \tag{3.12}$$

Since the quantity $dI(x)$ is equal to the number of neutrons out of a total of $I(x)$ that collide in dx, it follows that $dI(x)/I(x)$ is the probability that a neutron which survives up to x without a collision interacts in the next dx. Therefore, from Eq. (3.12), $\Sigma_t\,dx$ is equal to the probability that a neutron will interact in dx, and it may be concluded that Σ_t is the probability per unit path length that a neutron will undergo some sort of a collision as it moves about in a medium.

It should also be noted that, since $I(x)$ refers to those neutrons that have not collided in penetrating the distance x, the ratio $I(x)/I_0 = e^{-\Sigma_t x}$ is equal to the probability that a neutron can move through this distance without having a collision. Now let the quantity $p(x)\,dx$ be the probability that a neutron will have its first collision in dx in the neighborhood of x. This is evidently equal to the probability that the neutron survives up to x without a collision times the probability that it does in fact collide in the additional distance dx. Since Σ_t is the probability of interaction per path length, $p(x)\,dx$ is given by

$$p(x)\, dx = e^{-\Sigma_t x} \times \Sigma_t \, dx$$

$$= \Sigma_t e^{-\Sigma_t x} \, dx.$$

The average distance that a neutron moves between collisions is called the *mean free path*. This quantity, which is designated by the symbol λ, is equal to the average value of x, the distance traversed by a neutron without a collision.

The value of λ is obtained from the probability function $p(x)$ by determining the average distance traveled, that is,

$$\lambda = \int_0^\infty x p(x) \, dx$$

$$= \Sigma_t \int_0^\infty x e^{-\Sigma_t x} \, dx$$

$$= 1/\Sigma_t. \tag{3.13}$$

Example 3.4

Calculate the mean free path of 100-keV neutrons in liquid sodium. At this energy, the total cross-section of sodium is 3.4 b.

Solution. From Table II.3, the atom density of sodium is 0.0254×10^{24}. The macroscopic cross-section is then $\Sigma_t = 0.0254 \times 10^{24} \times 3.4 \times 10^{-24} = 0.0864 \text{ cm}^{-1}$. The mean free path is therefore $\lambda = 1/0.0864 = 11.6 \text{ cm } [Ans.]$

Consider next a homogeneous mixture of 2 nuclear species, X and Y, containing N_X and N_Y atoms/cm^3 of each type, and let σ_X and σ_Y be the cross-sections of the 2 nuclei for some particular interaction. The probability per unit path that a neutron collides with a nucleus of the first type is $\Sigma_X = N_X \sigma_X$ and with a nucleus of the second type is $\Sigma_Y = N_Y \sigma_Y$. The total probability per unit path that a neutron interacts with either nucleus is therefore

$$\Sigma = \Sigma_X + \Sigma_Y = N_X \sigma_X + N_Y \sigma_Y. \tag{3.14}$$

If the nuclei are in atoms that are bound together in a molecule, Eq. (3.14) can be used to define an equivalent cross-section for the molecule. This is done simply by dividing the macroscopic cross-section of the mixture by the number of molecules per unit volume. For example, if there are N molecules $X_m Y_n$ per cm^3, then $N_X = mN$, $N_Y = nN$, and from Eq. (3.14) the cross-section for the molecule is

$$\sigma = \frac{\Sigma}{N} = m\sigma_X + n\sigma_Y. \tag{3.15}$$

Equations (3.14) and (3.15) are based on the assumption that the nuclei X and Y act independently of one another when they interact with neutrons. This is correct for all neutron interactions except elastic scattering by molecules and solids. Low-energy scattering cross-sections for such substances must be obtained from experiment.

Example 3.5

The absorption cross-section of ^{235}U and ^{238}U at 0.0253 e V are 680.8 b and 2.70 b, respectively. Calculate Σ_a for natural uranium at this energy.

Solution. By use of the methods of Section 2.14, the atom densities of ^{235}U and ^{238}U in natural uranium are found to be $3.48 \times 10^{-4} \times 10^{24}$ and 0.0483×10^{24}, respectively. Then from Eq. (3.14) $\Sigma_a = 3.48 \times 10^{-4} \times 680.8 + 0.0483 \times 2.70 = 0.367$ cm^{-1}. [*Ans.*]

Example 3.6

The scattering cross-sections (in barns) of hydrogen and oxygen at 1 Me V and 0.0253 e V are given in the following table. What are the values of σ_s for the water molecule at these energies?

Solution. Equation (3.15) applies at 1 Me V, so that $\sigma_s(H_2O) = 2\sigma_s(H) + \sigma_s(O) = 2 \times 3 + 8 = 14$ b. [*Ans.*] Equation (3.15) *does not apply* at 0.0253, and $\sigma_s(H_2O) \neq 2 \times 21 + 4 = 46$ b. The experimental value of $\sigma_s(H_2O)$ at 0.0253 eV is 103 b. [*Ans.*]

	1 Me V	0.0253 e V
H	3	21
O	8	4

3.4 NEUTRON FLUX

It was shown in Section 3.2 that when a beam of neutrons of intensity I strikes a thin target, the number of collisions per cm^3/sec is given by

$$F = \Sigma_t I, \tag{3.16}$$

where Σ_t is the macroscopic total cross-section.

Consider an experiment of the type shown in Fig. 3.3, in which a small target is exposed simultaneously to several neutron beams. The intensities of the beams are different, but it is assumed that the neutrons in all of the beams have the same energy. In view of the fact that the interaction of neutrons with nuclei is independent of the angle at which the neutrons collide with the nuclei, the total interaction rate is clearly

$$F = \Sigma_t (I_A + I_B + I_C + \cdots). \tag{3.17}$$

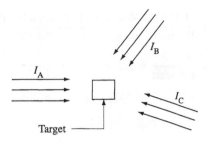

Figure 3.3   Neutron beams striking a target.

The neutrons have been assumed to be monoenergetic. So from Eq. (3.1), this may also be written as

$$F = \Sigma_t(n_A + n_B + n_c + \cdots)v, \qquad (3.18)$$

where n_A, n_B, and so on are the densities of the neutrons in the various beams and v is the neutron speed. Since $n_A + n_B + n_C + \cdots$ is equal to n, the total density of neutrons striking the target, Eq. (3.18) becomes

$$F = \Sigma_t n v. \qquad (3.19)$$

The situation at any point in a reactor is a generalization of this experiment, but with the neutrons moving in *all* directions. It follows that Eq. (3.19) is valid for a reactor, where n is the neutron density at the point where F is computed.

The quantity nv in Eq. (3.19) is called the *neutron flux*, in this case for monoenergetic neutrons, and is given the symbol ϕ; thus

$$\phi = nv. \qquad (3.20)$$

It is evident that the units of neutron flux are the same as the units of beam intensity—namely, neutrons/cm²-sec. In terms of the flux, the collision density is

$$F = \Sigma_t \phi. \qquad (3.21)$$

To understand the importance of the flux and this relationship, consider the following example.

Example 3.7

A certain research reactor has a flux of 1×10^{13} neutrons/cm²-sec and a volume of 64,000 cm³. If the fission cross-section, Σ_f, in the reactor is 0.1 cm⁻¹, what is the power of the reactor?

Solution. The power may be obtained from the fission rate using the relationship between the energy released per fission (200 Me V) and the rate at which fissions are occuring:

$$\text{Power} = \frac{1 \text{ MW}}{10^6 \text{ watt}} \frac{1 \text{ W-sec}}{\text{joules}} \times \frac{1.60 \times 10^{-13} \text{ joule}}{\text{Me V}} \times \frac{200 \text{ Me V}}{\text{fission}} \times \text{fission rate}$$

$$= \frac{1 \text{ MW}}{10^6 \text{ watt}} \times 3.2 \times 10^{-11} \text{ watts/fission/sec} \times \text{fission rate}$$

$$= 3.2 \times 10^{-17} \text{ MW/fission/sec} \times \text{fission rate}$$

From Eq. (3.21), the fission rate is

$$\text{Fission rate} = \Sigma_f \Phi$$

$$= 0.1 \text{ cm}^{-1} \times 1 \times 10^{13} \text{ neutrons/cm}^2\text{-sec.}$$

The reactor power/cm^3 is then

$$\text{Power/cm}^3 = 3.2 \times 10^{-17} \text{ MW/fission/sec} \times 1 \times 10^{12} \text{ fission/sec-cm}^3.$$

The total power is the power/cm^3 times the volume of the active core.

$$\text{Power} = 3.2 \times 10^{-5} \text{ MW/cm}^3 \times 64,000 \text{ cm}^3$$

$$= 2 \text{ MW } [Ans.]$$

3.5 NEUTRON CROSS-SECTION DATA

All neutron cross-sections are functions of the energy of the incident neutrons and the nature of the target nucleus. These factors must be taken into consideration in the choice of materials for use in nuclear devices. Most of the cross-section data needed for such purposes are found in the Brookhaven National Laboratory report BNL-325 and other source, which are discussed in the references at the end of the chapter. Before turning to the data, however, it is of interest to consider the mechanisms by which neutrons interact with nuclei.

Compound Nucleus Formation   Most neutron interactions proceed in 2 steps. The incident neutron, on striking the target nucleus, first coalesces with it to form a compound nucleus. If the target nucleus is $^A Z$, the compound nucleus is $^{A+1} Z$. The compound nucleus may then decay in a number of ways. For example, when 10-Me V neutrons strike an ^{56}Fe target, the compound nucleus is ^{57}Fe, and this nucleus may decay by emitting an elastic or inelastic neutron, a γ-ray, or two neutrons. In symbols, these processes are

$$^{56}\text{Fe} + \text{n} \rightarrow (^{57}\text{Fe})^* \begin{cases} ^{56}\text{Fe} + \text{n (elastic scattering)} \\ ^{56}\text{Fe} + \text{n' (inelastic scattering)} \\ ^{57}\text{Fe} + \gamma \text{ (radiative capture)} \\ ^{55}\text{Fe} + 2\text{n (n, 2n reaction).} \end{cases}$$

One of the striking features of interactions that proceed by way of compound nucleus formation is that their cross-sections exhibit maxima at certain incident neutron energies. Such maxima are called *resonances* and arise in the following way. It is recalled from Section 2.7 that nuclei have various excited states corresponding to different configurations of nucleons in the nucleus. It turns out that the incident neutron and target nucleus are more likely to combine and form a compound nucleus if the energy of the neutron is such that the compound nucleus is produced in one of its excited states. The resonances show up in the cross-section because it is necessary to form the compound nucleus before the interaction can proceed.

It is recalled from Section 2.11 that it takes energy—namely, the neutron separation energy—to remove a neutron from a nucleus. This separation energy reappears, however, when the neutron reenters the nucleus. Therefore, it follows that when a neutron collides with a nucleus, the compound nucleus is formed in an excited state having an energy equal to the kinetic energy of the incident neutron plus the separation energy or binding energy of the neutron in the compound nucleus.[2]

Elastic Scattering   The elastic scattering cross-section as a function of the energy of the incident neutron can be divided into three distinct regions. In the first, low-energy region, σ_e is approximately constant. The scattering in this region does not occur by compound nucleus formation, but merely because of the forces exerted by the target nucleus on the passing neutron. The cross-section for this potential scattering is given by

$$\sigma_e \text{ (potential scattering)} = 4\pi R^2, \tag{3.22}$$

where R is the nuclear radius.

Beyond the potential scattering region, there is a region of resonances that is due to compound nucleus formation. At still higher energies, the resonances crowd together to such an extent that the individual resonances can no longer be resolved; in this region, σ_e is a smooth and slowly varying function of energy.

Figure 3.4 shows these three regions for the target nucleus carbon. Carbon is a relatively light nucleus. With heavier nuclei, the region of resonances is found at lower energies. For example, the resonance region of ^{238}U begins at only 6 e V and ends at roughly 1 ke V.

Example 3.8

Using experimental elastic scattering data, estimate the radius of the C nucleus.

---

[2]This discussion is somewhat simplified, center-of-mass effects having been ignored. For a more complete discussion, see *Introduction to Nuclear Reactor Theory*, noted in the references.

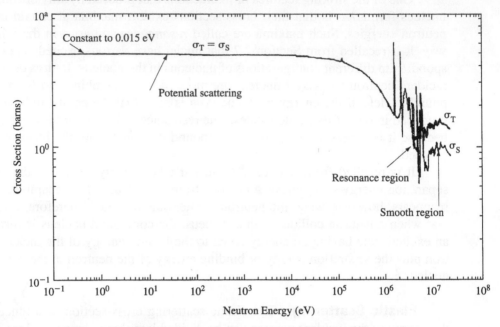

Figure 3.4 The elastic scattering and total cross-section of carbon. (Plotted from data received over the Internet from the Korean Atomic Energy Research Institute using ENDFPLOT and ENDF/B 6.1.)

Solution. From Fig. 3.4, it is observed that σ_e has the constant value of 4.8 b from about 0.02 eV to 0.01 Me V. This is due to potential scattering. Then from Eq. (3.22), $4\pi R^2 = 4.8 \times 10^{-24}$ and $R = 6.2 \times 10^{-13}$ cm. [*Ans.*]

Inelastic Scattering   This process does not occur unless the neutron has sufficient energy to place the target nucleus in its first excited state.[3] As a result, σ_i is zero up to some threshold energy. Generally speaking, the energy at which the first excited state is found decreases with increasing mass number. As a consequence, σ_i is nonzero over a larger energy region for the heavier nuclei than for the lighter nuclei. The threshold for inelastic scattering is 4.80 Me V for C, whereas it is only 44 keV for ^{238}U. At energies well above threshold, σ_i is roughly equal to σ_s.

Radiative Capture   As in the case of elastic scattering, it is convenient to divide the radiative capture cross-section into three regions. In the low-energy

---

[3]Because of center-of-mass effects, the threshold energy for inelastic scattering is actually somewhat higher than the energy of the first excited state. Except for the very light nuclei, however, this can be ignored.

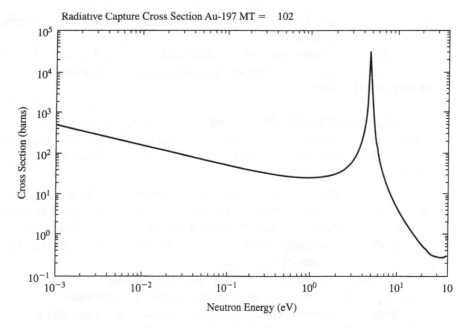

Figure 3.5   The radiative capture cross-section of Au-197 at low energy. (From ENDF/B 6 plotted over the Internet using ENDFPLOT from the Korean Atomic Energy Research Institute.)

region of most nuclei, σ_γ varies at $1/\sqrt{E}$, where E is the neutron energy. Since the neutron speed v is proportional to $\sqrt{E}$, this means that σ_γ varies as $1/v$. The low-energy region of σ_γ is therefore known as the *1/v region*. Neutron cross-sections are often plotted on a log-log scale, and a cross-section that is $1/v$ then appears as a straight line with a slope of $-1/2$. This can be seen in Fig. 3.5, in which the $1/v$ region and the first resonance are shown for ^{197}Au. For a few important nuclei, σ_γ does not show exact $1/v$ behavior at low energy, and such nuclei are called *non-1/v absorbers*.

Above the $1/v$ region, there is a region of resonances that occurs at the same energies as the resonances in σ_s. Near an isolated resonance at the energy E_r, σ_γ is given by the Breit–Wigner one-level formula:

$$\sigma_\gamma = \frac{\gamma_r^2 g}{4\pi} \frac{\Gamma_n \Gamma_g}{(E - E_r)^2 + \Gamma^2/4}. \tag{3.23}$$

In this expression, γ_r is the wavelength of neutrons with energy E_r, g is a constant known as the *statistical factor*, Γ_n and Γ_g are constants called, respectively, the *neutron width* and the *radiation width*, and $\Gamma = \Gamma_n + \Gamma_\gamma$ is called the *total width*. It is easy to show that σ_γ falls to one half of its maximum value at the energies

$E_r \pm \Gamma/2$. In short, Γ is the width of the resonance at one half its height, and this is the origin of the term *width*.

Above the resonances region, which ends at about 1 keV in the heavy nuclei and at increasingly higher energies in lighter nuclei, σ_γ drops rapidly and smoothly to very small values.

Charged-Particle Reactions   As a rule, the (n, p), (n, α), and other charged-particle reactions are endothermic and do not occur below some threshold energy. Their cross-sections also tend to be small, even above threshold, especially for the heavier nuclei.

However, there are some important exothermic reactions in light nuclei. One of these is the reaction ^{10}B(n, α)^{7}Li, the cross-section of which is shown in Fig. 3.6. It is observed that σ_a is very large at low energy; for this reason, ^{10}B is often used to absorb low-energy neutrons. It should also be noted in Fig. 3.6 that σ_a is $1/v$ over several orders of magnitude in energy.

A similar exothermic reaction that also shows a strong $1/v$ behavior is ^{6}Li(n, α)^{3}H. This reaction is used for the production of tritium, ^{3}H.

Some endothermic charged-particle reactions are important in reactors even though their thresholds are high. In water reactors, for example, the ^{16}O(n, p)^{16}N

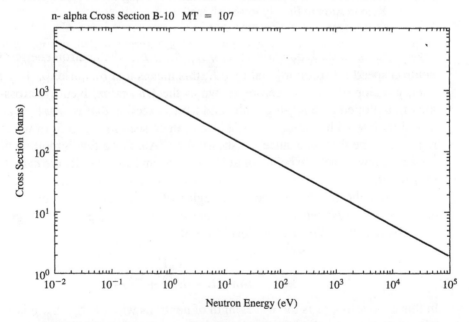

Figure 3.6   The cross-section for B-10, n-alpha reaction from 0.01 eV to 10,000 eV. (From ENDF/B 6 using ENDFPLOT from the Korean Atomic Energy Research Institute over the Internet.)

reaction is the principal source of the radioactivity of the water (the ^{16}N undergoes β-decay, with a half-life of approximately 7 secs, which is accompanied by the emission of 6- to 7-Me V γ-rays), despite that ordinarily only one neutron in several thousand has an energy greater than the 9-Me V threshold for this reaction.

Total Cross-Section   Since σ_t is the sum of all the other cross-sections, the variation of σ_t with energy reflects the behavior of the individual component cross-sections. In particular, at low energy, σ_t behaves as

$$\sigma_t = 4\pi R^2 + \frac{C}{\sqrt{E}}, \tag{3.24}$$

where C is a constant. The first term in this expression is the cross-section for elastic scattering; the second term gives the cross-section for radiative capture or whatever other exothermic reaction is possible at this energy. If the first term in Eq. (3.24) is much larger than the second, then σ_t is a constant at low energy; if the second term dominates, σ_t is $1/v$ in this energy region.

In the resonance region, σ_t exhibits the resonances found in σ_s and σ_i, all of which occur at the same energies in each of these cross-sections. At higher energies above the resonance region, σ_t becomes a smooth and rolling function of energy, as shown in Fig. 3.4.

Hydrogen and Deuterium   The nuclei ^{1}H and ^{2}H, which are present in large amounts in many nuclear reactors, interact with neutrons in a somewhat different manner from other nuclei. For one thing, interactions with ^{1}H and ^{2}H do not involve the formation of a compound nucleus. They also do not have any resonances. The cross-section, σ_s, is constant up to 10 KeV, and σ_γ is $1/v$ at all energies. Furthermore, these nuclei have no excited states (^{1}H is, after all, only a single proton), and so inelastic scattering does not occur.

Example 3.9

The value of σ_γ for ^{1}H at 0.0253 eV is 0.332 b. What is σ_γ at 1 eV?

Solution. Since σ_γ is $1/v$, it can be written as

$$\sigma_\gamma(E) = \sigma_\gamma(E_0) \sqrt{\frac{E_0}{E}},$$

where E_0 is any energy. In this problem, σ_γ is known as 0.0253 eV and so it is reasonable to take this to be the value of E_0. Then

$$\sigma_\gamma(1e\,V) = 0.332 \times \sqrt{\frac{0.0253}{1}}$$

$$= 0.0528 \text{ b. } [Ans.]$$

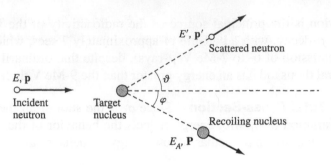

Figure 3.7   Elastic scattering of a neutron by a nucleus.

3.6 ENERGY LOSS IN SCATTERING COLLISIONS

When a neutron is elastically scattered from a nucleus at rest, the nucleus recoils from the site of the collision. The kinetic energy of the scattered neutron is therefore smaller than the energy of the incident neutron by an amount equal to the energy acquired by the recoiling nucleus. In this way, neutrons lose energy in elastic collisions even though the internal energy of the nucleus does not change.

The energy loss in elastic scattering can be found from the laws of conservation of energy and momentum. Let E, $\mathbf{p}$ and E', $\mathbf{p}'$ be the kinetic energy and (vector) momentum of the neutron before and after the collision, respectively, and let E_A and $\mathbf{P}$ be the energy and momentum of the recoiling nucleus. The neutron is scattered at the angle ϑ; the nucleus recoils at the angle φ (see Fig. 3.7).

In view of the fact that the collision is elastic, it follows that

$$E = E' + E_A. \tag{3.25}$$

The conservation of momentum, namely,

$$\mathbf{p} = \mathbf{p}' + \mathbf{P},$$

can be depicted by the vector diagram shown in Fig. 3.8. Then by the law of cosines,

$$P^2 = p^2 + (p')^2 - 2pp' \cos \vartheta. \tag{3.26}$$

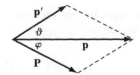

Figure 3.8   Vector diagram for conservation of momentum.

From classical mechanics, $P^2 = 2ME_A$, $p^2 = 2mE$, and $p'^2 = 2mE'$, where M and m are the masses of the nucleus and neutron, respectively. Equation (3.26) can then be written as

$$ME_A = mE + mE' - 2m\sqrt{EE'} \cos \vartheta. \tag{3.27}$$

Since M/m is approximately equal to A, the atomic mass number of the nucleus, Eq. (3.27) is equivalent to

$$AE_A = E + E' - 2\sqrt{EE'} \cos \vartheta.$$

Next, introducing E_A from Eq. (3.25) gives, when arranged,

$$(A + 1)E' - 2\sqrt{EE'} \cos \vartheta - (A - 1)E = 0.$$

This equation is quadratic in $\sqrt{E'}$ and has the solution

$$E' = \frac{E}{(A + 1)^2}[\cos \vartheta + \sqrt{A^2 - \sin^2 \vartheta}]^2 . \tag{3.28}$$

It is of some interest to consider the consequences of Eq. (3.28). In a grazing collision, ϑ is approximately equal to zero, and Eq. (3.28) gives $E' = E$. As would be expected, there is no energy loss in such a collision. Except for the special case of hydrogen, which must be considered separately, it follows from Eq. (3.28) that the minimum value of E', $(E')_{\mathrm{min}}$, occurs when $\vartheta = \pi$. The neutron is then scattered directly backward and suffers the largest possible loss in energy. For $\vartheta = \pi$, Eq. (3.28) gives

$$(E')_{\mathrm{min}} = \left(\frac{A-1}{A+1}\right)^2 E = \alpha E, \tag{3.29}$$

where

$$\alpha = \left(\frac{A-1}{A+1}\right)^2 \tag{3.30}$$

is called the *collision parameter*. Values of α are given in Table 3.1.

The scattering of neutrons by hydrogen is unique because the masses of the neutron and hydrogen nucleus (proton) are essentially equal. It is not difficult to show from classical mechanics, and, indeed, it is a common observation, that a particle striking another particle of the same mass, which is initially at rest, cannot be scattered through an angle greater than 90°. The minimum energy of neutrons scattered from hydrogen must therefore be found by placing $\vartheta = \pi/2$ in Eq. (3.28), and this gives

$$(E')_{\mathrm{min}} = 0.$$

TABLE 3.1   COLLISION PARAMETERS

Nucleus	Mass No.	α	ξ
Hydrogen	1	0	1.000
H_2O		*	0.920†
Deuterium	2	0.111	0.725
D_2O		*	0.509†
Beryllium	9	0.640	0.209
Carbon	12	0.716	0.158
Oxygen	16	0.779	0.120
Sodium	23	0.840	0.0825
Iron	56	0.931	0.0357
Uranium	238	0.983	0.00838

*Not defined.
†An appropriate average value.

Since this result could have also been obtained by placing $A = 1$ in Eq. (3.29), it may be concluded that

$$(E')_{min} = \alpha E \tag{3.31}$$

is valid for *all* nuclei, including hydrogen.

It is also of interest to know the average energy of an elastically scattered neutron. This computation is more difficult than that of maximum and minimum energies and is not given here. It can be shown, however, that for scattering by light nuclei, including hydrogen, and at most of the neutron energies of interest in nuclear reactors, the average energy of the scattered neutron is given approximately by

$$\overline{E'} = \tfrac{1}{2}(1 + \alpha)E.$$

The average energy loss, $\overline{\Delta E}$ then

$$\overline{\Delta E} = E - \overline{E'}$$
$$= \tfrac{1}{2}(1 - \alpha)E,$$

and the average fractional energy loss is

$$\frac{\overline{\Delta E}}{E} = \tfrac{1}{2}(1 - \alpha). \tag{3.32}$$

This equation is also valid for the heavier nuclei, but not for high-energy neutrons. With ^{238}U, for example, Eq. (3.32) is not accurate much above an energy of 100 keV. At higher energies, the energy loss in collisions with the heavier nuclei is less than that predicted by Eq. (3.32).

From Eq. (3.30) and Table 3.1, it is observed that α is zero for $A = 1$ (hydrogen) and increases monotonically to unity with increasing A. In view of Eq. (3.30), it follows that the average fractional energy loss decreases from $\frac{1}{2}$ in the case of hydrogen to almost zero for the heavy nuclei. Thus, on the average, a neutron loses one-half of its energy in a collision with hydrogen. When scattered by carbon, since $\alpha = 0.716$, it loses about 14% of its energy, while in a collision with uranium, $\alpha = 0.983$, a neutron loses less than 1% of its energy. In short, neutrons lose less and less energy the heavier the target nucleus is. It is often necessary to slow down fast neutrons—a process known as *moderation*. From the foregoing discussion, it is clear that materials of low mass number are most effective for this purpose since the neutrons slow down most rapidly in such media.

Neutrons also lose energy through inelastic collisions, as a result of both recoil and internal excitation of the target nucleus. Since the threshold energy for inelastic scattering is so high in light nuclei (usually on the order of several Me V—it does not occur at all in hydrogen), moderation by inelastic scattering is less important than by elastic scattering in these nuclei. With the heavier nuclei, however, the inelastic threshold is much lower, and inelastic scattering is often the principal mechanism for neutron moderation.

In many reactor calculations, especially in connection with neutron moderation, it is convenient to describe neutron collisions in terms of a new variable called *lethargy*. This quantity is denoted by the symbol u and is defined as

$$u = \ln(E_M/E), \tag{3.33}$$

where E_M is an arbitrary energy—usually that of the highest energy neutron in the system. From Eq. (3.33), it should be noted that, at high energy, a neutron's lethargy is low; as it slows down and E decreases, its lethargy increases.

The average change in lethargy in an elastic collision, $\overline{\Delta u}$, like the average fractional energy loss [see Eq. (3.32)], is independent of the energy of the incident neutron. The quantity $\overline{\Delta u}$ appears in many nuclear engineering calculations and is denoted by the symbol ξ. By a derivation that is too lengthy to be given here, it can be shown that ξ is given by

$$\xi = 1 - \frac{(A-1)^2}{2A} \ln\left(\frac{A+1}{A-1}\right), \tag{3.34}$$

where A is the mass number of the target nucleus. Except for small values of A, ξ is well approximated by the simple formula

$$\xi \simeq \frac{2}{A + \frac{2}{3}}. \tag{3.35}$$

Even for $A = 2$, Eq. (3.35) is off by only about 3%. Exact values of ξ are given in Table 3.1.

Example 3.10

A 1-Me V neutron is scattered through an angle of 45° in a collision with a ^{2}H nucleus. (a) What is the energy of the scattered neutron? (b) What is the energy of the recoiling nucleus? (c) How much of a change in lethargy does the neutron undergo in this collision?

Solution

1. Substituting $E = 1$ Me V, $A = 2$, and $\vartheta = 45°$ into Eq. (3.28) gives immediately $E' = 0.738$ Me V. [*Ans.*]

2. Since the collision is elastic (inelastic scattering does not occur with ^{2}H), the recoil energy is $E_A = 1.000 - 0.738 = 0.262$ Me V. [*Ans.*]

3. The lethargy going into the collision is $u = \ln(E_M/E)$ and coming out is $u' = \ln(E_M/E')$. The change in lethargy is therefore $\Delta u = u' - u = \ln(E/E') - \ln(1/0.738) = 0.304$ (a unitless number). [*Ans.*]

Polyenergetic Neutrons   In Sections 3.2 and 3.3, the rate at which neutrons undergo collision in a target was calculated on the assumption that the incident neutrons were monoenergetic. This can easily be generalized to neutrons that are not monoenergetic, but have a distribution in energy.

For this purpose, let $n(E)\,dE$ be the number of neutrons per cm^3 with energies between E and $E + dE$ in a neutron beam incident on a thin target. The intensity of these neutrons is

$$dI(E) = n(E)v(E)\,dE, \tag{3.36}$$

where $v(E)$ is the speed corresponding to the energy E. According to Eq. (3.8), this beam interacts in the target at a rate of

$$dF(E) = n(E)v(E)\Sigma_t(E)\,dE$$

collisions per cm^3/sec, where $\Sigma_t(E)$ is the macroscopic total cross-section. The collision density is then

$$F = \int n(E)v(E)\Sigma_t(E)\,dE, \tag{3.37}$$

in which the integration is carried over all energies in the beam.

To compute the interaction rate for a particular type of interaction, it is merely necessary to replace $\Sigma_t(E)$ in Eq. (3.37) by the appropriate cross-section. An especially important case is the absorption of thermal neutrons—that is, neutrons whose energies are distributed according to the Maxwellian function described in Section 2.13. These neutrons are found in certain types of nuclear reactors called *thermal reactors*, which are discussed in Chapters 4 and 6. The absorption rate in a beam of thermal neutrons is

$$F_a = \int n(E)v(E)\Sigma_a(E)\,dE, \qquad (3.38)$$

where $\Sigma_a(E)$ is the macroscopic absorption cross-section and the integral is evaluated at thermal energies up to about 0.1 eV.

In Section 3.5, it was pointed out that at low energies most nuclei exhibit $1/v$ absorption, either as the result of radiative capture or some other absorption reaction. At these energies, $\Sigma_a(E)$ can be written as

$$\Sigma_a(E) = \Sigma_a(E_0)\frac{v_0}{v(E)}, \qquad (3.39)$$

where E_0 is an arbitrary energy and v_0 is the corresponding speed. When Eq. (3.39) is introduced into Eq. (3.38), the $v(E)$ cancels so that

$$F_a = \Sigma_a(E_0)v_0 \int n(E)\,dE. \qquad (3.40)$$

The remaining integral is equal to the total density of thermal neutron, n, and Eq. (3.40) reduces to

$$F_a = \Sigma_a(E_0)nv_0. \qquad (3.41)$$

Equation (3.41) shows that for a $1/v$ absorber, the absorption rate is independent of the energy distribution of the neutrons and is determined by the cross-section at an arbitrary energy. Furthermore, it may be concluded from Eq. (3.41) that, although the neutrons have a distribution of energies, the absorption rate is the same as that for a monoenergetic beam of neutrons with arbitrary energy E_0 and intensity nv_0.

In view of these results, it has become standard practice to specify all absorption cross-sections $1/v$ or not, at the single energy of $E_0 = 0.0253$ eV. The corresponding speed is $v_0 = 2,200$ meters/sec. Values of cross-sections at 0.0253 eV are loosely referred to as *thermal cross-sections*. These are tabulated in a number of places, including the chart of nuclides; an abridged table is found in Appendix II.

This quantity nv_0 in Eq. (3.41) is called 2,200 meters-per-second flux and is denoted by ϕ_0; that is

$$\phi_0 = nv_0. \qquad (3.42)$$

The absorption rate is then simply

$$F_a = \Sigma_a(E_0)\phi_0. \qquad (3.43)$$

Although only a comparatively few nuclei are non-$1/v$ absorbers, these nuclei are usually important in nuclear systems since their cross-sections tend to be rather high. The absorption rate for such nuclei is again given by Eq. (3.38), but now the integral cannot be simplified as it was in the $1/v$ case. In particular, F_a now depends on the function $n(E)$ as well as $\Sigma_a(E)$. However, by assuming that $n(E)$ is the Maxwellian function, C. H. Westcott computed F_a numerically for all of the important non-$1/v$ absorbers. The resulting value of F_a is a function of the temperature of the neutron distribution and is given in the form

$$F_a = g_a(T)\Sigma_a(E_0)\phi_0, \tag{3.44}$$

where $g_a(T)$, which is called the *non-$1/v$ factor*, is a tabulated function and $\Sigma_a(E_0)$ is again the absorption cross-section at 0.0253 eV. A short table of non-$1/v$ factors is given in Table 3.2.

Although the prior results were derived for a beam of neutrons incident on a thin target, they apply equally well to the more complicated situation found in many nuclear systems in which the neutrons are moving in all directions. In particular, the 2,200 meters-per-second flux is defined at any point where there are n thermal neutrons per cm^3, and Eq. (3.43) or (3.44) can be used to compute the absorption rate at such a point.

Example 3.11

A small indium foil is placed at a point in a reactor where the 2,200 meters-per-second flux is 5×10^{12} neutrons/cm^2-sec. The neutron density can be represented by a Maxwellian function with a temperature of 600°C. At what rate are the neutrons absorbed per cm^3 in the foil?

Solution. From Table II.3 in Appendix II, $N = 0.0383 \times 10^{24}$ and $\sigma_a(E_0) = 194$ b so that $\Sigma_a(E_0) = 0.0383 \times 194 = 7.43$ cm^{-1}. However, indium is non-$1/v$ and from Table 3.2, $g_a(600°C) = 1.15$. From Eq. (3.44), it follows that

$$F_a = 1.15 \times 7.43 \times 5 \times 10^{12} = 4.27 \times 10^{13} \text{neutrons/cm}^3\text{-sec. } [Ans.]$$

3.7 FISSION

It was shown in Section 2.11 that the binding energies of nuclei per nucleon decrease with increasing atomic mass number, for A greater than about 50. This means that a more stable configuration of nucleons is obtained whenever a heavy nucleus splits into two parts—that is, undergoes fission. The heavier, more unstable nuclei might therefore be expected to fission spontaneously without external intervention. Such fissions do occur.

It is interesting to examine the origin of the decrease in the binding energy per nucleon with increasing A. Figure 3.9 shows the terms in the binding energy per

TABLE 3.2 NON-1/V FACTORS*

T,°C	Cd g_a	In g_a	^{135}Xe $g_a^\dagger$	^{149}Sm g_a	^{233}U g_a	^{233}U g_f	^{235}U g_a	^{235}U g_f	^{238}U g_a	^{239}Pu g_a	^{239}Pu g_f
20	1.3203	1.0192	1.1581	1.6170	0.9983	1.0003	0.9780	0.9759	1.0017	1.0723	1.0487
100	1.5990	1.0350	1.2103	1.8874	0.9972	1.0011	0.9610	0.9581	1.0031	1.1611	1.1150
200	1.9631	1.0558	1.2360	2.0903	0.9973	1.0025	0.9457	0.9411	1.0049	1.3388	1.2528
400	2.5589	1.1011	1.1864	2.1854	1.0010	1.0068	0.9294	0.9208	1.0085	1.8905	1.6904
600	2.9031	1.1522	1.0914	2.0852	1.0072	1.0128	0.9229	0.9108	1.0122	2.5321	2.2037
800	3.0455	1.2123	0.9887	1.9246	1.0146	1.0201	0.9182	0.9036	1.0159	3.1006	2.6595
1000	3.0599	1.2915	0.8858	1.7568	1.0226	1.0284	0.9118	0.8956	1.0198	3.5353	3.0079

*Based on C. H. Westcott, "Effective Cross-Section Values for Well-Moderated Thermal Reactor Spectra," Atomic Energy Commission report AECL-1101, January 1962.
†Based on E. C. Smith et al., *Phys. Rev.* **115**, 1693 (1959).

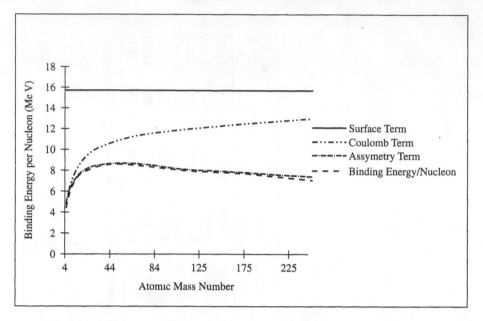

Figure 3.9   Components of the binding energy per nucleon based on the semi-empirical mass formula.

nucleon curve as determined from the liquid drop model of the atom. The decrease in binding energy per nucleon is largely due to the coulomb term overcoming the volume term. The volume term represents the binding due to the strong nuclear forces. It is the coulomb repulsion, then, that is largely responsible for fission.

Although heavy nuclei do spontaneously fission, they do so only rarely. For fission to occur rapidly enough to be useful in nuclear reactors, it is necessary to supply energy to the nucleus. This, in turn, is due to the fact that there are attractive forces acting between the nucleons in a nucleus, and energy is required to deform the nucleus to a point where the system can begin to split in two. This energy is called the *critical energy of fission* and is denoted by E_{crit}. Values of E_{crit} are given in Table 3.3 for several nuclei.

Any method by which energy E_{crit} is introduced into a nucleus, thereby causing it to fission, is said to have *induced* the fission. The most important of these is neutron absorption. It is recalled from Sections 2.11 and 3.5 that when a neutron is absorbed the resulting compound nucleus is formed in an excited state at an energy equal to the kinetic energy of the incident neutron plus the separation energy or binding energy of the neutron in the compound nucleus. If this binding energy alone is greater than the critical energy for fission of the compound nucleus, then fission can occur with neutrons having essentially no kinetic energy. For example, according to Table 3.3, the binding energy of the last neutron in ^{236}U is 6.4 Me V,

TABLE 3.3   CRITICAL ENERGIES FOR FISSION, IN Me V

Fissioning Nucleus $^A Z$	Critical Energy	Binding Energy of Last Neutron in $^A Z$
^{232}Th	5.9	*
^{233}Th	6.5	5.1
^{233}U	5.5	*
^{234}U	4.6	6.6
^{235}U	5.75	*
^{236}U	5.3	6.4
^{238}U	5.85	*
^{239}U	5.5	4.9
^{239}Pu	5.5	*
^{240}Pu	4.0	6.4

*Neutron binding energies are not relevant for these nuclei since they cannot be formed by the absorption of neutrons by the nuclei ^{A-1}Z.

whereas E_{crit} is only 5.3 Me V. Thus, when a neutron of zero kinetic energy is absorbed by ^{235}U, the compound nucleus, ^{236}U, is produced with 1.1 Me V more energy than its critical energy, and fission can immediately occur. Nuclei such as ^{235}U, that lead to fission following the absorption of a zero-energy neutron, are called *fissile*. Note, however, that although it is the ^{235}U that is said to be fissile, the nucleus that actually fissions in this case is the ^{236}U. From Table 3.3 it is evident that ^{233}U and ^{239}Pu (plutonium-239) are also fissile. In addition, ^{241}Pu and several other nuclei not indicated in the table are also fissile.

With most heavy nuclei other than ^{233}U, ^{235}U, ^{239}Pu, and ^{241}Pu, the binding energy of the incident neutron is not sufficient to supply the compound nucleus with the critical energy, and the neutron must have some kinetic energy to induce fission. In particular, this is always the case when the struck nucleus contains an even number of nucleons since the binding energy of the incident neutron to an even-A nucleus is always less than to an odd-A nucleus. (This odd–even variation in binding energy is evident in Table 3.3.) For instance, the binding energy of the last neutron in ^{239}U is only 4.9 Me V, and this is the excitation of the compound nucleus formed when a neutron of zero kinetic energy is absorbed by ^{238}U. Since E_{crit} for ^{239}U is 5.5 Me V, it is clear that fission cannot occur unless the neutron incident on the ^{238}U has an energy greater than about 0.6 Me V. Nuclei such as ^{238}U, which do not fission unless struck by an energetic neutron, are said to be *fissionable but nonfissile*. For reasons that are clear later in this book, the nonfissile isotopes such as ^{238}U cannot alone be used to fuel nuclear reactors, and it is the fissile isotopes, especially ^{235}U and ^{239}Pu, that are the practical fuels of nuclear power.

The semi-empirical mass formula discussed in Chapter 2 may be used to examine the dependence of the critical energy given in Table 3.3. For the case of ^{235}U, the critical energy is obtained by calculating the Q-value for the reaction using the semi-empirical mass formula. The origin of the variation in critical energy is evident by comparing the value of the odd–even term as A is increased from 233 to 238.

Fission Cross-Sections   The cross-sections of fissile nuclei for neutron-induced fission resemble radiative capture cross-sections in their dependence on the energy of the incident neutron. Thus, as seen in Fig. 3.10, where σ_f is given for ^{235}U, there are three distinct regions to the cross-section. At low energy, σ_f is $1/v$ or nearly so; this is followed by a region of resonances; finally, above the resonance region, σ_f is smooth and rolling. It should be noted that σ_f is especially large in the $1/v$ region.

The fission cross-sections of fissionable but nonfissile nuclei, by contrast, are zero up to a threshold energy, which always occurs above the resonance region. As a result, σ_f is relatively smooth at all energies. This is illustrated in Fig. 3.11, where σ_f is shown for ^{238}U.

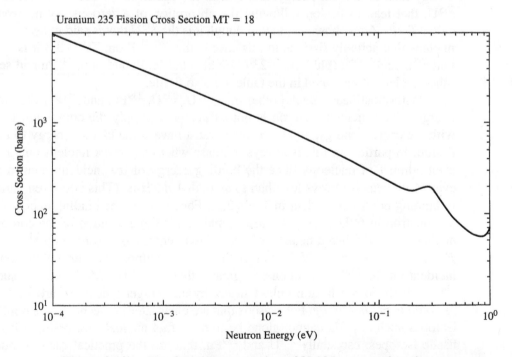

Uranium 235 Fission Cross Section MT = 18

Figure 3.10a   The fission cross-section of U-235; see continuation on next three pages. (Plotted from ENDF/B 6 using ENDFPLOT over the Internet from the Korean Atomic Energy Research Institute.)

Uranium 235 Fission Cross Section MT =18

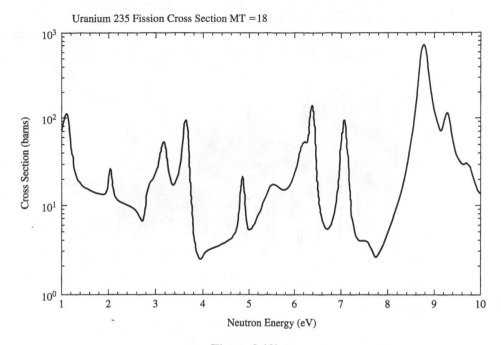

Figure 3.10b

Uranium 235 Fission Cross Section MT = 18

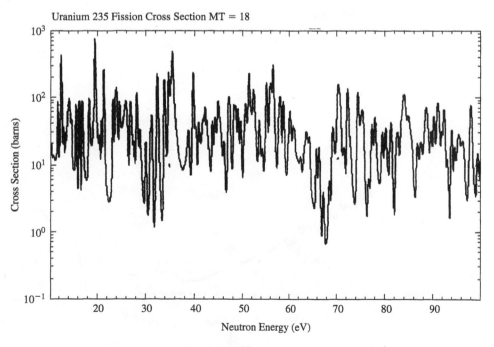

Figure 3.10c

Uranium 235 Fission Cross Section MT = 18

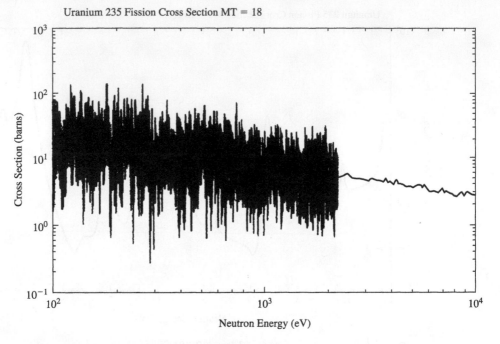

Figure 3.10d

Uranium 235 Fission Cross Section MT = 18

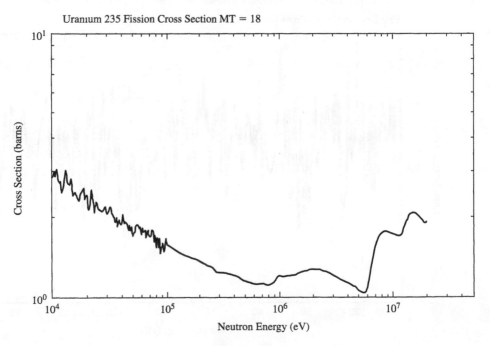

Figure 3.10e

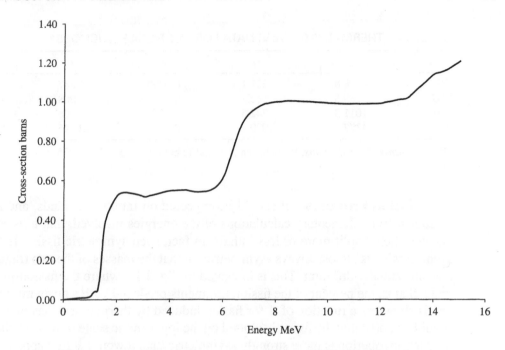

Figure 3.11 The fission cross-section of U-238. (Plotted using ENDF/B 6 and ENDFPLOT over the Internet from Korean Atomic Energy Research Institute.)

It should not be implied from the foregoing discussion that, when a neutron collides with a fissile nucleus, or with a fissionable but nonfissile nucleus above the fission threshold, the result is always fission. This is not the case; neutrons interacting with these nuclei may be scattered, elastically or inelastically, they may be absorbed in radiative capture, and so on. The cross-sections for all of these processes have been measured and are found in ENDF/B 6. However, with fissile nuclei at low energies, only three interactions are possible: elastic scattering, radiative capture, and, of course, fission. The value of σ_s is much smaller than either σ_γ or σ_f, so that radiative capture and fission are by far the more probable events. The ratio of the cross-sections of these two processes is called the *capture-to-fission ratio* and is denoted by the symbol α, that is,

$$\alpha = \frac{\sigma_\gamma}{\sigma_f}. \tag{3.45}$$

This parameter, which is a function of energy, has an important bearing on the design of many reactors. Values of α for the fissile nuclei at 0.0253 e V are given in Table 3.4, along with the cross-sections for these nuclei.

TABLE 3.4   THERMAL (0.0253 e V) DATA FOR THE FISSILE NUCLIDES*

	$\sigma_a{}^\dagger$	σ_f	α	η	ν
^{233}U	578.8	531.1	0.0899	2.287	2.492
^{235}U	680.8	582.2	0.169	2.068	2.418
^{239}Pu	1011.3	742.5	0.362	2.108	2.871
^{241}Pu	1377	1009	0.365	2.145	2.917

*From *Neutron Cross-Sections*, Brookhaven National Laboratory report BNL-325, 3rd ed., 1973.
$^\dagger\sigma_a = \sigma_\gamma + \sigma_f$.

Fission Products   It should be expected on intuitive grounds, and it can be shown from elementary calculations of the energies involved, that a fissioning nucleus should split more or less in half. In fact, such symmetric fission is a rare event. Fission is almost always asymmetric, so that the masses of the two fragments are substantially different. This is indicated in Fig. 3.12, where the fission-product yield, that is, the percent of the fission fragments produced with a given mass number, is shown as a function of A for fission induced by thermal neutrons in ^{235}U. It should be noted that the figure is plotted on the logarithmic scale so that the fission-product distribution is more strongly asymmetric than it would at first appear. With the increasing energy of the incident neutron, fission becomes more symmetric. This is illustrated in Fig. 3.12 by the yield of fission products arising from fission induced by 14 Me V neutrons.

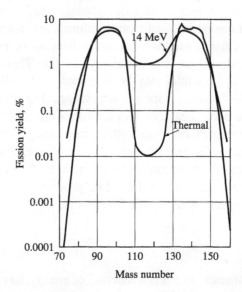

Figure 3.12   Fission-product yields for thermal and 14-MeV fission neutrons in U-235.

When the fission products are initially formed, they are excessively neutron rich; they contain more neutrons than are necessary for their stability. As a result, they decay by emitting a sequence of negative β-rays, which are accompanied by various γ-rays. For example, the isotope ^{115}Pd (palladium-115) is produced directly in fission and decays by the chain

$$^{115}\text{Pd} \xrightarrow{\beta-} {}^{115}\text{Ag} \xrightarrow{\beta-} {}^{115}\text{Cd} \xrightarrow{\beta-} {}^{115}\text{In (stable)}.$$

Many fission-product decay chains of this kind have been identified and can be deduced from the data on the chart of the nuclides.

The radioactivity of the fission products is the cause of a number of problems in the utilization of nuclear energy. For one thing, fission products accumulate in an operating reactor as the fuel undergoes fission, and elaborate precautions must be taken to ensure that they do not escape to the surrounding environment. Furthermore, the heat released by decaying fission products may be so great that a reactor must be cooled after shutdown to prevent damage to the fuel. The continuing emission of radiation from the fission products also tends to make parts of a reactor highly radioactive. When removed from a reactor, these parts must be cooled while being stored prior to disposal or processing.

The quantitative aspects of fission-product decay are complicated by the fact that hundreds of different radioactive nuclides are produced in fission, each with its own characteristic half-life and decay radiation. For many purposes, however, the following expressions may be used to represent approximately the overall decay of the fission products. Thus, the rates at which β-rays and γ-rays are emitted in the time interval from about 10 seconds to several weeks after a single fission are given by

$$\text{Rate of emission of } \beta\text{-rays} \simeq 3.8 \times 10^{-6} t^{-1.2} \beta\text{-rays/sec}, \qquad (3.46)$$

$$\text{Rate of emission of } \gamma\text{-rays} \simeq 1.9 \times 10^{-6} t^{-1.2} \gamma\text{-rays/sec}, \qquad (3.47)$$

where t is the time after fission in days.

To express the prior disintegration rates in units of curies, it is merely necessary to note that each β-ray originates in the decay of a nuclide. Then since 1 Ci $= 3.7 \times 10^{10}$ disintegrations/sec, the fission-product activity t days after 1 fission is

$$\text{Fission product activity} \simeq 3.8 \times 10^{-6} t^{-1.2}/3.7 \times 10^{10}$$

$$= 1.03 \times 10^{-16} t^{-1.2} \text{ Ci}. \qquad (3.48)$$

It is often necessary to calculate the total fission-product activity that accumulates in the fissile fuel of an operating reactor. Suppose, for example, that the reactor has been operating at a constant power of P megawatts (MW) for T days and is then shut down. To determine the activity of the fission products t days after

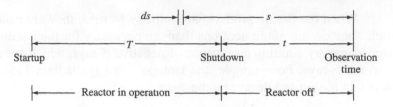

Figure 3.13 Diagram for computing fission-product activity.

shutdown, it is necessary to integrate Eq. (3.48) over the appropriate time span. Let s be the interval in days from the time when a fission occurs to the time that the activity of its fission fragments is measured (see Fig. 3.13). It is shown later in this section that at a power level of P MW the total fission rate is $2.7 \times 10^{21} P$ fissions per day. The number of fissions occurring in time ds is then $2.7 \times 10^{21} P ds$ fissions. From Eq. (3.48), the activity s days later is

$$2.7 \times 10^{21} P ds \times 1.03 \times 10^{-16} s^{-1.2} = 0.28 \times 10^6 P s^{-1.2} ds \text{ Ci.}$$

The total activity at t is then

$$\text{Fission product activity} = 0.28 \times 10^6 P \int_t^{t+T} s^{-1.2} ds$$

$$= 1.4 \times 10^6 P [t^{-0.2} - (t + T)^{-0.2}] \text{ Ci.} \quad (3.49)$$

The prior equations may be used to compute the activity of a single fuel rod that has been left in a reactor for t days and then removed. In this case, P is the power produced by the fuel rod in megawatts and t is the time in days since its removal from the reactor.

It is somewhat more difficult to analytically calculate the total energy released by fission products since, for one thing, the energy spectrum of the emitted radiation changes in time as these nuclides decay. For rough estimates, the average energies of the β- and γ-rays are sometimes taken to be 0.4 Me V and 0.7 Me V, respectively. The rate of energy release from the decaying fission products following one fission is then

$$\text{Decay energy rate} \simeq 2.8 \times 10^{-6} t^{-1.2} \text{Me V/sec,} \quad (3.50)$$

where t is again in days. However, a far better procedure is to use actual experimental data on energy release. This method is discussed in detail in Chapter 8 (see Section 8.1).

Example 3.12

The total initial fuel loading of a particular reactor consists of 120 fuel rods. After the reactor has been operated at a steady power of 100 MW for 1 year, the fuel is removed. Assuming that all rods contribute equally to the total power, estimate the activity of a fuel rod 1 day after removal.

Solution. In Eq. (3.49), $P = 100$, $t = 1$, and $t + T = 1 + 365 = 366$. The activity of all fuel rods 1 day after removal is then

$$1.4 \times 10^6 \times 100 \times [(1)^{-0.2} - (366)^{-0.2}] = 1.4 \times 10^8 (1 - 0.307) = 9.7 \times 10^7 \text{ Ci.}$$

One rod would have an activity of

$$9.7 \times 10^7 / 120 = 8.1 \times 10^5 \text{ Ci. } [Ans.]$$

Fission Neutrons   Most of the neutrons released in fission (usually more than 99%) are emitted essentially at the instant of fission. These are called *prompt neutrons*, in contrast to *delayed neutrons*, which are released comparatively long after the fission event.

The average number of neutrons, both prompt and delayed, released per fission is given the symbol ν. Values of ν for fission induced by 0.0253-e V neutrons are given in Table 3.4. As the energy of the incident neutron is raised, ν increases slowly. One additional neutron is emitted for every 6- to 7-Me V increase in neutron energy.

For later use in reactor calculations, it is convenient to define the parameter η, which is equal to the number of neutrons released in fission per neutron absorbed by a fissile nucleus. Since radiative capture competes with fission, η is always smaller than ν. In particular, η is equal to ν multiplied by the relative probability (σ_f / σ_a) that an absorption leads to fission (see Example 3.2) or

$$\eta = \nu \frac{\sigma_f}{\sigma_a} = \nu \frac{\sigma_f}{\sigma_\gamma + \sigma_f}. \tag{3.51}$$

In terms of α, the capture-to-fission ratio (see Eq. 3.45), this can be written as

$$\eta = \frac{\nu}{1 + \alpha}. \tag{3.52}$$

For a mixture of fissile or fissile and nonfissile nuclides, η is defined as the average number of neutrons emitted per neutron absorbed in the mixture. In this case, η is given by

$$\eta = \frac{1}{\Sigma_a} \sum_i \nu(i) \Sigma_f(i), \tag{3.53}$$

where $\nu(i)$ and $\Sigma_f(i)$ are the value of ν and the macroscopic fission cross-section for the ith nuclide, respectively, and Σ_a is the macroscopic cross-section for the mixture. It should be noted that $\nu(i)$, $\Sigma_f(i)$ and Σ_a in Eq. (3.53) must be computed at the energy of the neutrons inducing the fission. For example, if the fuel is a mixture of ^{235}U and ^{238}U and the fissions are induced by low-energy neutrons, then

$$\eta = \frac{\nu(235)\Sigma_f(235)}{\Sigma_a(235) + \Sigma_a(238)}. \tag{3.54}$$

There are no terms involving ^{238}U in the numerator because this nuclide does not fission with low-energy neutrons. However, if this same fuel were used in a fast reactor (see Section 4.2), in which the fissions are induced by highly energetic neutrons, η would be

$$\eta = \frac{\nu(235)\Sigma_f(235) + \nu(238)\Sigma_f(238)}{\Sigma_a(235) + \Sigma_a(238)}. \tag{3.55}$$

In this expression, all quantities are computed at the elevated energies.

Example 3.13

Calculate the value of η for natural uranium at 0.0253 e V.

Solution. Written out in detail, Eq. (3.54) is

$$\eta = \frac{\nu(235)N(235)\sigma_f(235)}{N(235)\sigma_a(235) + N(238)\sigma_a(238)}.$$

According to Eq. (2.61), the atom density of an isotope is proportional to its isotopic abundance γ, so that

$$\eta = \frac{\nu(235)\gamma(235)\sigma_f(235)}{\gamma(235)\sigma_a(235) + \gamma(238)\sigma_a(238)}.$$

Introducing data from Table 3.4 and Table II.2 of Appendix II gives

$$\eta = \frac{2.418 \times 0.72 \times 582.2}{0.72 \times 680.8 + 99.26 \times 2.70} = 1.34. \ [Ans.]$$

[*Note:* For reasons discussed in Chap. 6, this is not precisely the value used in reactor problems involving natural uranium; see Example 6.11.]

The prompt fission neutrons are emitted with the continuous energy spectrum shown in Fig. 3.14. This spectrum is well described by the function

$$\chi(E) = 0.453e^{-1.036E} \sinh \sqrt{2.29E}, \tag{3.56}$$

where $\chi(E)$ is defined so that $\chi(E)\, dE$ is the fraction of the prompt neutrons with energies between E and $E + dE$ and E is in Me V. The function $\chi(E)$ is normalized so that

$$\int_0^\infty \chi(E)\, dE = 1.$$

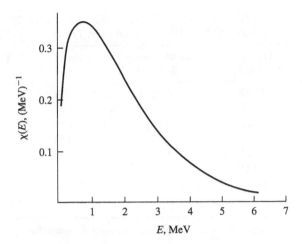

Figure 3.14   The prompt neutron spectrum.

The average energy $\overline{E}$ of the prompt neutrons can be found from the integral

$$\overline{E} = \int_0^\infty E\,\chi(E)\,dE = 1.98\ \mathrm{Me\,V}.$$

The most probable energy, corresponding to the peak of the $\chi(E)$ curve, is 0.73 Me V.

Although delayed neutrons ordinarily comprise less than 1% of the neutrons released in fission, they play an important role in the control of nuclear reactors, as is seen in Chapter 7. These neutrons originate in the decay by neutron emission of nuclei produced in the β-decay of certain fission products. For example, when the fission-product ^{87}Br decays to ^{87}Kr, the latter may be formed in an excited state. In this case, the least bound neutron in the ^{87}Kr is not bound at all and is ejected from the nucleus with an energy of about 0.3 Me V. This neutron is emitted as soon as the excited state is formed. Therefore, it appears to be emitted with the 54.5-sec half-life of the ^{87}Br.

Nuclei such as ^{87}Br are called *delayed-neutron precursors*. There are believed to be about 20 such precursors, most of which have now been positively identified. The precursors can be divided into 6 groups, each with its own characteristic half-life. The group half-lives and decay constants are given in Table 3.5 for low-energy (thermal) fission in ^{235}U. Also included in the table are the observed yields (neutrons per fission) of the delayed neutrons in each group, together with the delayed-neutron fractions β_i. The quantity β_i is defined as the fraction of all of the fission neutrons released in fission that appear as delayed neutrons in the ith group. In other words, β_i is the absolute neutron yield of the ith group divided by ν. The total delayed fraction β is the sum of all the β_i.

TABLE 3.5   DELAYED NEUTRON DATA FOR THERMAL FISSION IN ^{235}U*

Group	Half-Life (sec)	Decay Constant (l_i, sec^{-1})	Energy (ke V)	Yield, Neutrons per Fission	Fraction (β_i)
1	55.72	0.0124	250	0.00052	0.000215
2	22.72	0.0305	560	0.00346	0.001424
3	6.22	0.111	405	0.00310	0.001274
4	2.30	0.301	450	0.00624	0.002568
5	0.610	1.14	—	0.00182	0.000748
6	0.230	3.01	—	0.00066	0.000273

Total yield: 0.0158

Total delayed fraction (β): 0.0065

*Based in part on G. R. Keepin, *Physics of Nuclear Kinetics*, Reading, Mass.: Addison-Wesley, 1965.

Prompt γ-Rays   At the instant of fission, a number of γ-rays are emitted from the fissioning nucleus. These are referred to as *prompt γ-rays* to distinguish them from the fission-product γ-rays. The energy spectrum of the prompt γ-rays is approximately the same as the spectrum of the fission-product γ-rays.

The Energy Released in Fission

In discussing the energy of fission, it is important to distinguish between the total energy released in the process and the energy that can be recovered in a reactor and is therefore available for the production of heat. The recoverable energy and the total energy, in general, are different. This is illustrated in Table 3.6, which gives a breakdown of the component energies as they occur in the neutron-induced fission of ^{235}U.

TABLE 3.6   EMITTED AND RECOVERABLE ENERGIES FOR FISSION OF ^{235}U

Form	Emitted Energy, Me V	Recoverable Energy, Me V
Fission fragments	168	168
Fission-product decay		
β-rays	8	8
γ-rays	7	7
neutrinos	12	—
Prompt γ-rays	7	7
Fission neutrons (kinetic energy)	5	5
Capture γ-rays	—	3–12
Total	207	198–207

As indicated in the table, most (almost 85%) of the energy released in fission appears as the kinetic energy of the fission fragments. These fragments come to rest within about 10^{-3}cm of the fission site so that all of their energy is converted into heat. The energies of the fission-product β-rays and γ-rays, the prompt and delayed neutrons, and the prompt γ-rays are also recoverable since almost none of these radiations ever escape from a nuclear power system. However, the neutrinos that accompany β-decay interact only slightly with matter and escape completely from every nuclear device. Their energy of almost 12 Me V per fission is therefore irrevocably lost for practical purposes.

Because most fission neutrons remain within the confines of a reactor, these neutrons are eventually captured by the nuclei in the system. It is shown in the next chapter, however, that one of the ν-neutrons emitted per fission must be absorbed by a fissionable nucleus and produce another fission in order for a nuclear reactor to remain in operation. Therefore, it follows that the remaining $(\nu - 1)$ neutrons per fission must be absorbed parasitically in the reactor, that is, absorbed in a nonfission reaction. Each absorption usually leads to the production of one or more capture γ-rays, whose energies depend on the binding energy of the neutron to the compound nucleus. Since ν is approximately 2.42 for ^{235}U (its precise value depends on the energy of the neutrons causing the fission), this means that, from about 3 to 12 Me V of capture, γ-radiation is produced per fission depending on the materials used in the reactor. All this γ-ray energy is, of course, recoverable. It is observed in Table 3.6 that the energy of the capture γ-rays compensates to some extent for the energy lost by neutrino emission. In any case, the recoverable energy per fission is approximately 200 Me V. In the absence of more accurate data, this is the value that is normally used at least in preliminary calculations.

Consider now a reactor in which the energy from the fission ^{235}U is released at the rate of P megawatts. In other words, the reactor is operating at a thermal power of P megawatts. With a recoverable energy per fission of 200 Me V, the rate at which fissions occur per second in the entire reactor is

$$
\text{Fission rate} = P \text{ MW} \times \frac{10^6 \text{ joules}}{\text{MW-sec}} \times \frac{\text{fission}}{200 \text{ Me V}}
$$

$$
\times \frac{\text{Me V}}{1.60 \times 10^{-13} \text{ joule}} \times \frac{86,400 \text{ sec}}{\text{day}}
$$

$$
= 2.70 \times 10^{21} P \text{ fissions/day.}
$$

To convert this to grams per day fissioned, which is also called the *burnup rate*, it is merely necessary to divide by Avogadro's number and multiply by 235.0, the gram atomic weight of ^{235}U. This gives simply

$$
\text{Burnup rate} = 1.05P \text{ g/day.} \tag{3.57}
$$

Thus, if the reactor is operating at a power of 1 MW, the ^{235}U undergoes fission at the rate of approximately 1 g/day. To put this another way, the release of 1 megawatt/day of energy requires the fission of 1 g of ^{235}U.

It must be remembered, however, that fissile nuclei are consumed both in fission and radiative capture. Since the total absorption rate is $\sigma_a/\sigma_f = (1 + \alpha)$ times the fission rate, it follows from Eq. (3.57) that ^{235}U is consumed at the rate of

$$\text{Consumption rate} = 1.05(1 + \alpha)P \text{ g/day.} \qquad (3.58)$$

For ^{235}U, the thermal value of α is 0.169. Equation (3.58) shows that this isotope is consumed at the rate of about 1.23 g/day per megawatt of power if the fissions are induced primarily by thermal neutrons.

Example 3.14

The energy released by the fissioning of 1 g of ^{235}U is equivalent to the combustion of how much (a) coal with a heat content of 3×10^7 J/kg (13,000 Btu/lb) and (b) oil at 4.3×10^7 J/kg (6.5×10^6 Btu/barrel)?

Solution. According to the prior discussion, the fissioning of 1 g of ^{235}U releases approximately 1 megawatt/day = 24,000 kWh = 8.64×10^{10} J.

1. This energy is also released by

$$\frac{8.64 \times 10^{10}}{3 \times 10^7} = 2.88 \times 10^3 \text{ kg} = 2.88 \text{ metric tons}$$

$$= 3.17 \text{ ST (short tons) of coal. } [Ans.]$$

2. In terms of oil this is also

$$\frac{8.64 \times 10^{10}}{4.3 \times 10^7} = 2.00 \times 10^3 \text{ kg} = 2.00 \text{ t}$$

$$= 12.6 \text{ barrels. } [Ans.]$$

3.8 γ-RAY INTERACTIONS WITH MATTER

Although the term γ-ray is normally reserved for radiation emitted by nuclei and x-ray refers to radiation originating in transitions of atomic electrons, both forms of radiation are called γ-rays in the present section. There is, of course, no fundamental difference between the two radiations, per se, as they are both electromagnetic radiation.

Gamma rays interact with matter in several ways. Ordinarily, however, only three processes must be taken into account in nuclear engineering problems. These are the photoelectric effect, pair production, and Compton effect.

The Photoelectric Effect   In the photoelectric effect, the incident γ-ray interacts with an entire atom, the γ-ray disappears, and 1 of the atomic electrons is ejected from the atom. The atom recoils in this process, but carries with it very little kinetic energy. The kinetic energy of the ejected photoelectron is therefore equal to the energy of the photon less the binding energy of the electron to the atom—that is, the ionization energy for the electron in question.

If a γ-ray succeeds in ejecting an inner atomic electron, the hole in the electronic structure is later filled by a transition of 1 of the outer electrons into the vacant position. This transition is accompanied by the emission of x-rays characteristic of the atom or by the ejection of an Auger electron (see Section 2.7).

The cross-section per atom for the photoelectric effect is denoted by the symbol σ_{pe}. This cross-section can be used in the same way as the neutron cross-sections discussed in the preceding sections. Thus, if I is the intensity of γ-rays incident on a thin target containing N atoms per cm^3, then $I N \sigma_{pe}$ is the number of photoelectric interactions/cm^3-sec.

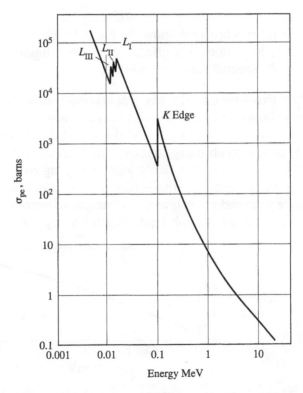

Figure 3.15   The photoelectric cross-section of lead as a function of gamma ray energy.

The cross-section σ_{pe} depends both on the energy E of the incident photon and the atomic number Z of the atom. Figure 3.15 shows σ_{pe} for lead as a function of E. It should be noted that σ_{pe} rises to very large values at low energy—less than 1 Me V. Photons in this energy region obviously do not penetrate far into a lead target (or shield).

As also indicated in Fig. 3.15, there are a number of discontinuities in σ_{pe} at low energy. These are called *absorption edges* and correspond to energies below which it is not possible to eject certain electrons from the lead atom. For instance, below the K-edge, the incident photon does not have sufficient energy to eject a K-electron—the most tightly bound electron. The next most tightly bound electrons after the K-electrons are the L-electrons. For reasons unimportant to the present discussion, these electrons have three slightly different ionization energies. The three edges denoted in the figure as L_I, L_{II}, and L_{III} correspond to the minimum photon energies required to eject the 3 differently bound L-electrons. Above the edges, that is, above the K-edge, σ_{pe} drops off roughly as E^{-3}.

The photoelectric cross-section depends strongly on Z, varying as

$$\sigma_{pe} \sim Z^n, \tag{3.59}$$

where n is the function of E shown in Fig. 3.16. Because of the strong dependence of σ_{pe} on Z, the photoelectric effect is of greatest importance for the heavier atoms, such as lead, especially at lower energies.

Pair Production   In this process, the photon disappears and an electron pair—a positron and a negatron—is created. Since the total rest-mass energy of the 2 electrons is $2m_e c^2 = 1.02$ Me V, this effect does not occur unless the photon has at least this much energy. Above this threshold, the cross-section for a pair production, σ_{pp}, increases steadily with increasing energy, as shown in Fig. 3.17, where the pair production cross-section is shown for lead.

Since pair production is an electromagnetic interaction, it can take place only in the vicinity of a Coulomb field. At most γ-ray energies of interest, this is the

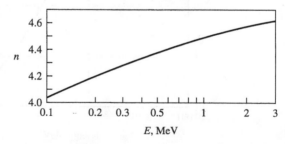

Figure 3.16   The constant n in Eq. (3.53) as a function of gamma ray energy. (From R. D. Evans, The Atomic Nucleus, New York: McGraw-Hill, 1955.)

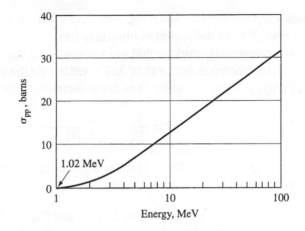

Figure 3.17   The pair production cross-section of lead as a function of
γ-ray energy.

field of the nucleus, not the surrounding electrons. As a result, σ_{pp} is a function of
Z, and, in particular, varies as Z^2, that is,

$$\sigma_{pp} \sim Z^2. \tag{3.60}$$

The total kinetic energy of the negatron–positron pair is equal to the energy
of the photon less 1.02 Me V. Once formed, these electrons move about and lose
energy as a result of collisions with atoms in the surrounding medium. After the
positron has slowed down to very low energies, it combines with a negatron, the
two particles disappear, and two photons are produced (annihilation radiation),
each having an energy of 0.511 Me V.

The Compton Effect   The Compton effect, or *Compton scattering* as it
is sometimes called, is simply the elastic scattering of a photon by an electron, in
which both energy and momentum are conserved. As shown in Fig. 3.18, the incident photon with energy E and wavelength λ is scattered through the angle ϑ and

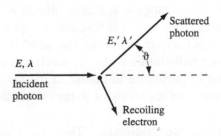

Figure 3.18   The Compton effect.

the struck electron recoils. Since the recoiling electron acquires some kinetic energy, the energy E' of the scattered photon is less than E, and since the wavelength of a photon is inversely proportional to its energy (see Eq. 2.22), the wavelength λ' of the scattered photon is larger than λ. By setting up the equations for the conservation of energy and momentum, it is not difficult to derive the following relation:

$$E' = \frac{E E_e}{E(1 - \cos \vartheta) + E_e},\tag{3.61}$$

where $E_e = m_e c^2 = 0.511$ Me V is the rest-mass energy of the electron. A formula equivalent to Eq. (3.61) is

$$\lambda' - \lambda = \lambda_C (1 - \cos \vartheta),\tag{3.62}$$

where

$$\lambda_C = \frac{h}{m_e c} = 2.426 \times 10^{-10} \text{ cm}\tag{3.63}$$

is called the *Compton wavelength*.

In Compton scattering, the photon interacts with individual electrons, and it is therefore possible to define a Compton cross-section per electron $_e\sigma_C$. This cross-section decreases monotonically with increasing energy from a maximum value 0.665 b (essentially $\frac{2}{3}$ of a barn) at $E = 0$, which is known as the *Thompson cross-section*, σ_T. Figure 3.19 shows $_e\sigma_C$ as a function of photon energy. Incidentally, for $E \gg E_e$, $_e\sigma_C$ behaves roughly as E^{-1}.

The Compton cross-section per atom, σ_C, is equal to the number of electrons in the atom—namely, Z—multiplied by $_e\sigma_C$. Thus,

$$\sigma_C = Z_e \sigma_C.\tag{3.64}$$

From a practical standpoint, the Compton effect is the cause of many difficult problems encountered in the shielding of γ-rays. This is because the photon does not disappear in the interaction as it does in the photoelectric effect and in pair production. The Compton-scattered photon is free to interact again in another part of the system. Although it is true that x-rays and Auger electrons are emitted following the photoelectric effect and that annihilation radiation accompanies pair production, these radiations are always much less energetic than the initial photon and do not tend to propagate in matter to the same extent as Compton-scattered photons. This multiple scattering of γ-rays is considered again in Chapter 10.

Attenuation Coefficients   The total cross-section per atom for γ-ray interaction is the sum of the cross-sections for the photoelectric effect, pair produc-

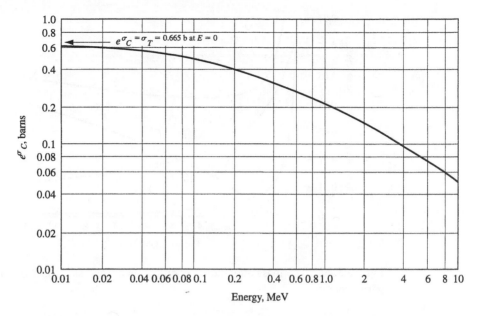

Figure 3.19   The Compton cross-section per electron as a function of gamma ray energy.

tion, and Compton scattering:

$$\sigma = \sigma_{\text{pe}} + \sigma_{\text{pp}} + \sigma_C. \tag{3.65}$$

A macroscopic cross-section can also be defined, like the macroscopic neutron cross-section, by multiplying σ in Eq. (3.65) by the atom density N. By tradition, such macroscopic γ-ray cross-sections are called *attenuation coefficients* and are denoted by the symbol μ. Thus,

$$\mu = N\sigma = \mu_{\text{pe}} + \mu_{\text{pp}} + \mu_C, \tag{3.66}$$

where μ is the total attenuation coefficient and μ_{pe}, μ_{pp}, and μ_C are the attenuation coefficients for the three interaction processes. Like macroscopic cross-sections for neutrons, the various μ's have units of cm^{-1}. It is also convenient to define the quantity μ/ρ, which is called the *mass attenuation coefficient*,[4] where ρ is the physical density. From Eq. (3.66), this is given by

$$\frac{\mu}{\rho} = \frac{\mu_{\text{pe}}}{\rho} + \frac{\mu_{\text{pp}}}{\rho} + \frac{\mu_C}{\rho}. \tag{3.67}$$

---

[4] A new standard is now in use in which μ is used to designate the mass attenuation coefficient and μ^* the attenuation coefficient. In this text, the older convention is used.

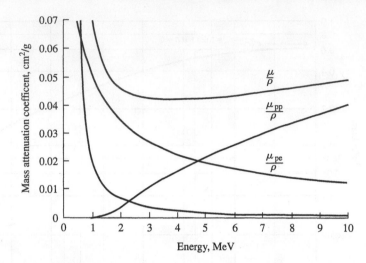

Figure 3.20   The mass attenuation coefficients of lead as a function of γ-ray energy.

Since μ and ρ have units of cm^{-1} and g/cm^3, respectively, it follows that μ/ρ has the units cm^2/g.

Figure 3.20 shows the mass attenuation coefficients, on a linear scale, for lead. There is minimum in μ/ρ at about 3.5 Me V because σ_{pe} and σ_C decrease with increasing γ-ray energy, whereas σ_{pp} increases from its threshold at 1.02 Me V. Also, as shown in the figure, Compton scattering is the dominant mode of interaction from about 0.5 Me V to 5 Me V. Because σ_{pe} and σ_{pp} depend more strongly on Z than does σ_C, the energy range over which Compton scattering is dominant increases with decreasing Z.[5] Thus, in the case of aluminum, for instance, Compton scattering predominates all the way from 0.06 Me V to 20 Me V.

At energies where Compton scattering is the principal mode of interaction,

$$\frac{\mu}{\rho} \simeq \frac{\mu_C}{\rho} = \frac{N\sigma_C}{\rho}.$$

Introducing the usual formula for atom density (see Eq. 2.59),

$$N = \frac{\rho N_A}{M},$$

where N_A is Avogadro's number and M is the gram atomic weight, gives

$$\frac{\mu}{\rho} \simeq \frac{N_A \sigma_C}{M} = N_A \left(\frac{A}{M}\right) {}_e\sigma_C,$$

[5]This is also illustrated in Fig. 3.21.

where use has been made of Eq. (3.64). A check of the chart of the nuclides shows that, except for hydrogen and the very heavy elements, the ratio Z/M is approximately equal to $\frac{1}{2}$. This means that, at those energies where Compton scattering is the dominant process, values of μ/ρ tend to be roughly the same for all elements. This is illustrated in Fig. 3.21, where μ/ρ is shown for a number of elements as a function of γ-ray energy. Numerical values of μ/ρ are given in Table II.4 in Appendix II.

Since attenuation coefficients are essentially macroscopic cross-sections, the value of μ for a mixture of elements is given by the same formula as Eq. (3.14).

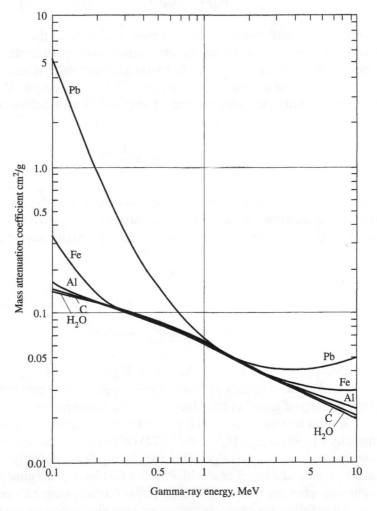

Figure 3.21   The mass attenuation coefficients of several elements. (From S. Glasstone and A. Sesonske, *Nuclear Reactor Engineering.* New York: Van Nostrand, 1967; by permission, US DOE.)

Thus,

$$\mu = \mu_1 + \mu_2 + \cdots, \tag{3.68}$$

where μ_1, μ_2 and so on are the values of μ for the various constituents. Also, it is not difficult to show that the mass absorption coefficient for the mixture is related to the mass absorption coefficients of the constituents by the formula

$$\frac{\mu}{\rho} = \frac{1}{100}\left[\omega_1\left(\frac{\mu}{\rho}\right)_1 + \omega_2\left(\frac{\mu}{\rho}\right)_2 + \cdots\right], \tag{3.69}$$

where ω_1, ω_2, and so on are the percents by weight of the various elements, and $(\mu/\rho)_1$, $(\mu/\rho)_2$, and so on are the mass absorption coefficients of the elements, as given in Table II.4. Equations (3.68) and (3.69) are valid at all energies.

By the same argument that led to Eq. (3.13), it is easy to show that μ is equal to the probability per unit path that a γ-ray will have a collision in a medium and that

$$\lambda = \frac{1}{\mu} \tag{3.70}$$

is the mean free path of the γ-ray. Furthermore, if I_0 is the intensity (γ-rays/cm^2-sec) of the monoenergetic γ-ray beam striking a target of thickness X, then the intensity of the photons that penetrate the target without having a collision is

$$I = I_0 e^{-\mu X}. \tag{3.71}$$

In terms of the mass attenuation coefficient, Eq. (3.71) may be written as

$$I = I_0 e^{-(\mu/\rho)(\rho X)}. \tag{3.72}$$

The quantity ρX in Eq. (3.72) has units of g/cm^2 and is equal to the number of grams contained in an area of 1 cm^2 of the target. Thicknesses of materials are often given in units of g/cm^2 in calculations of γ-ray attenuation.

It must be emphasized, as it was in the earlier discussion of neutrons, that the intensity I given in Eqs. (3.71) and (3.72) refers only to those often very few γ-rays that do not interact in the target. These are by no means the only photons that appear on the far side of a target or shield. Photons that have undergone multiple Compton scattering, photons from annihilation radiation following pair production, and the x-rays that follow the photoelectric effect may also penetrate the target. All of these radiations must be taken into account in shielding calculations. Methods for doing so are given in Chapter 10.

Example 3.15

Calculate the mass attenuation coefficient of UO_2 for 1 MeV γ-rays. What is their mean free path? The density of UO_2 is about 10 g/cm^3.

Solution. The molecular weight of UO_2 is $238 + 2 \times 16 = 270$. The percent by weight that is uranium is then $238/270 = 88.1\%$; the remaining 11.9% is oxygen. From Table II.4, $\mu/\rho = 0.0757$ cm^2/g for uranium and 0.0636 cm^2/g for oxygen. Thus, for UO_2

$$\frac{\mu}{\rho} = 0.881 \times 0.0757 + 0.119 \times 0.0636 = 0.0743 \text{ cm}^2/\text{g. } [Ans.]$$

The value of μ is 10 times this number, that is, $\mu = 0.743$ cm^{-1} since $\rho = 10$. The mean free path is

$$\lambda = \frac{1}{\mu} = \frac{1}{0.743} = 1.35 \text{ cm. } [Ans.]$$

Energy Deposition   In calculations of radiation protection to be considered in Chapter 9, it is necessary to compute the rate at which energy is deposited by a γ-ray beam as it passes through a medium. By analogy with Eq. (3.8), the total collision density at a point where the γ-ray intensity is I is given by

$$F = I\mu, \tag{3.73}$$

where μ is the total attenuation coefficient. If the γ-rays were absorbed at each collision, then the rate at which energy is deposited per unit volume in the medium would be simply $EF = EI\mu$, where E is the energy of the γ-rays.

With both the photoelectric effect and pair production, the incident photon is in fact absorbed. Unless the medium is very thin, most of the secondary radiation emitted subsequent to these interactions—the x-rays, electrons, and annihilation radiation—is also absorbed in the medium. Thus, the total energy of the incident γ-ray can be assumed to be deposited in these processes. In Compton scattering, however, the only energy deposited is the kinetic energy of the recoiling electron. Let $\overline{T}$ be the average energy of this electron. The average energy deposited by Compton scattering is then $\overline{T}I\mu_C$, where μ_C is the Compton attenuation coefficient. It is now convenient to define the *Compton absorption cross-section* σ_{Ca} by the relation

$$E\sigma_{Ca} = \overline{T}\sigma_C. \tag{3.74}$$

The corresponding *Compton absorption coefficient* μ_{Ca} is then given by

$$E\mu_{Ca} = \overline{T}\mu_C. \tag{3.75}$$

In terms of this coefficient, the energy deposition rate per unit volume by Compton scattering is simply $EI\mu_{Ca}$.

The total energy deposition rate W per unit volume from photoelectric effect, pair production, and Compton scattering can now be written as

$$W = EI(\mu_{pe} + \mu_{pp} + \mu_{Ca})$$
$$= EI\mu_a, \qquad (3.76)$$

where

$$\mu_a = \mu_{pe} + \mu_{pp} + \mu_{Ca} \qquad (3.77)$$

is called the *linear absorption coefficient*.[6] It can be seen that, by defining μ_{Ca} according to Eq. (3.75), it is possible to treat all three modes of γ-ray interaction on the same footing.

The quantity μ_a/ρ is called the *mass absorption coefficient*, and representative values are given in Table II.5. It is easy to see from Eq. (3.76) that the rate of energy deposition per unit mass is equal to $EI\mu_a/\rho$.

It should also be mentioned, in concluding this section, that the product EI that appears in Eq. (3.76) is called the *energy intensity* or *energy flux*. This has the units of energy per cm²/sec and is equal to the rate at which energy in a γ-ray beam passes into a medium per cm².

Example 3.16

It is proposed to store liquid radioactive waste in a steel container. If the intensity of γ-rays incident on the interior surface of the tank is estimated to be 3×10^{11} γ-rays/cm²-sec and the average γ-ray energy is 0.8 Me V, at what rate is energy deposited at the surface of the container?

Solution. Steel is a mixture of mostly iron and elements such as nickel and chromium that have about the same atomic number as iron. Therefore, as far as γ-ray absorption is concerned, steel is essentially all iron. From Table II.5, μ_a/ρ for iron at 0.8 Me V is 0.0274 cm²/g. The rate of energy deposition is then

$$0.8 \times 3 \times 10^{11} \times 0.0274 = 6.58 \times 10^9 \text{ Me V/g-sec. } [Ans.]$$

In SI units, this is equivalent to 1.05 J/kg-sec.

3.9 CHARGED PARTICLES

Ordinarily, there are only three varieties of charged particles that must be dealt with in nuclear engineering problems—namely, α-rays, β-rays, and fission fragments. Before considering these particular radiations, it is of interest to discuss the ways in which charged particles interact with matter.

---

[6] μ_a is called the *energy absorption coefficient* by some authors.

Consider a charged particle that is incident on an atom located at some point in bulk matter. A number of different events may occur as this particle nears the atom in question. First of all, because the particle exerts electrical (Coulomb) forces on the atomic electrons, one or more of these electrons may be placed into excited states of the atom, or an electron may be ejected from the atom altogether, leaving the atom ionized. However, the charged particle may penetrate through the cloud of atomic electrons and be elastically scattered from the nucleus. Since momentum and energy are conserved in such a collision, the nucleus necessarily recoils. If the incident particle is sufficiently massive and energetic, the recoiling nucleus may be ejected from its own electron cloud and move into the medium as another charged particle. Also, under certain circumstances, the incident particle, especially if it is an α-particle, may undergo some sort of nuclear reaction when it collides with the nucleus. Finally, the particle may be accelerated by the Coulomb field of the electrons or the nucleus; as a consequence, a photon may be emitted. This type of radiation, which is emitted whenever a charged particle undergoes acceleration, is called *bremsstrahlung*.

It is clear from the foregoing that the interaction of a charged particle with matter is a complicated affair. In any case, it is evident that in its passage through matter such a particle leaves a trail of excitation and ionization along its path. In this context, because they are directly responsible for producing this ionization, charged particles are referred to as *directly ionizing radiation*. By contrast, uncharged particles, such as γ-rays and neutrons, lead to excitation and ionization indirectly only after first interacting in the substance and producing a charged particle. For this reason, γ-rays and neutrons are said to be *indirectly ionizing radiation*. This is not to imply that γ-rays and neutrons do not directly form ions when they interact—they do. For example, a γ-ray produces photoelectric effect. Similarly, a neutron colliding with a nucleus may eject the nucleus from its atom as a highly charged ion. But the ionization arising from these first interactions of γ-rays and neutrons is entirely insignificant compared with the ionization caused by the subsequent interaction of the charged particles.

It is recalled that the interactions of neutrons and γ-rays are described by cross-sections. Although it is also possible to define various interaction cross-sections for charged particles, it is more useful to describe the extent to which charged particles interact with matter in terms of either their *specific ionization* or their *stopping power*. The specific ionization of a particle is defined as the number of ion pairs produced per unit path traveled by the particle. An ion pair is an ionized atom together with its ejected electron. The stopping power is the total energy lost per path length by a charged particle; that is, it is the total rate of decrease in the energy of the particle along its path. If nuclear reactions involving the particle do not occur, then the stopping power, which is denoted by S, can be written as

$$S = \left(\frac{dE}{dx}\right)_{\text{col}} + \left(\frac{dE}{dx}\right)_{\text{rad}}, \tag{3.78}$$

where the first term is the energy loss per unit due to collisions, which give rise to excitation and ionization, and the second term gives the energy loss by radiation.

The first term in Eq. (3.78) is called the *linear energy transfer* (LET), and it is of special interest in connection with the biological effects of radiation. As shown in Chapter 9, these effects depend on the extent to which energy is deposited by radiation as excitation and ionization within biological systems. The magnitude of LET increases rapidly with the mass and charge of a moving particle. Thus, the LET of α-particles is considerably larger than for electrons of the same energy. For example, the LET of a 1-Me V α-particle in water is about 90 keV/μm, whereas it is only 0.19 keV/μm for a 1-Me V electron. For this reason, α-particles and other heavily charged particles are referred to as *high LET radiation*; electrons are called *low LET radiation*.

Although linear energy transfer is due to the interactions of charged particles in matter, it is also possible to refer to the LET of uncharged, indirectly ionizing radiation—that is, γ-rays and neutrons—since, as noted earlier, the bulk of the local deposition of energy by these radiations is due to the ionization and excitation that occurs subsequent to their first interaction in the material. Because γ-rays produce low LET secondary electrons, they are called low LET radiation. However, since neutron interactions lead to the heavy, high LET charged particles, neutrons are known as high LET radiation. The distinction between high and low LET radiation has important biological consequences that are described in Chapter 9.

Alpha Particles   Because α-particles are so massive, they are only slightly deflected when they interact with atomic electrons. Therefore, they move in more or less straight lines as they travel in a medium. However, as an α-particle slows down, it becomes increasingly probable that it will capture an electron to form an He$^+$ ion and then capture a second electron to become a neutral helium atom. When, ultimately, this atom is formed, the specific ionization abruptly drops to zero. This situation is illustrated in Fig. 3.22, where the specific ionization is shown for an energetic α-particle as a function of distance from the end of its track in dry air at 15°C and 1 atm pressure. The maximum value of the specific ionization shown in the figure is 6,600 ion pairs/mm. At this point, the α-particles have an energy of about 0.75 Me V. A curve of the type given in Fig. 3.22 is known as a *Bragg curve*.

The point at which the ionization falls to zero is called the *range* of the α-particle.[7] As would be intuitively expected, the range is a monotonically increasing

---

[7]Because of the statistical nature of the processes involved, all α-particle tracks do not terminate at precisely the same point. The range described earlier is the most probable endpoint.

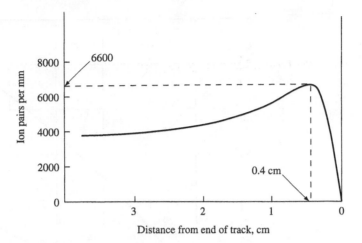

Figure 3.22   Specific ionization of an α-particle in air.

function of the initial energy of the particle. The range as a function of energy is shown in Fig. 3.23 for α-particles in air. Ranges of α-particles in other materials can be found from the range in air by using the *Bragg-Kleeman rule:*

$$R = R_a \left(\frac{\rho_a}{\rho} \right) \sqrt{\frac{M}{M_a}} = 3.2 \times 10^{-4} \frac{\sqrt{M}}{\rho} R_a. \tag{3.79}$$

In this formula, R is the range in a substance of physical density ρ and atomic weight M, and R_a, ρ_a, and M_a are the range, density, and average atomic weight of air, respectively. The numerical constant in Eq. (3.79) is computed for air at 15°C and 1 atm. For compounds or mixtures, $\sqrt{M}$ in Eq. (3.79) is to be replaced by

$$\sqrt{M} = \gamma_1 \sqrt{M_1} + \gamma_2 \sqrt{M_2} + \cdots, \tag{3.80}$$

where γ_1, γ_2, and so on are the fractions of atoms present having atomic weights M_1, M_2, and so on.

The *relative stopping power* of a material is defined as the ratio of the range of α-particles in air to the range of α-particles in the substance in question. From Eq. (3.79), this is given by

$$\text{Relative stopping power} = \frac{R_a}{R} = 3100 \frac{\rho}{\sqrt{M}}. \tag{3.81}$$

It should be noted that the relative stopping power is independent of the initial energy of the particle.

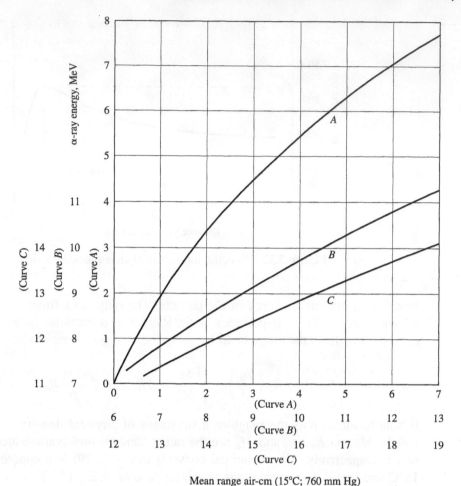

Figure 3.23   Range of α-particles in air as a function of energy.

The stopping powers of most materials are quite high and the ranges of α-particles are consequently very short. For example, the range of a 5-Me V α-particle in aluminum is only 0.0022 cm—about the thickness of thin aluminum foil. Most α-particles are stopped by an ordinary sheet of paper; they are also stopped in the outermost layers of living tissue. Thus, the shielding of α-particles does not ordinarily pose a difficult problem. However, the presence of α-decaying nuclides cannot be ignored in many engineering problems since, as is shown in Chapter 9, these nuclides can lead to serious health hazards when ingested or inhaled.

β-Rays   The attenuation of β-rays in matter in some ways is more complicated than for α-particles. To begin with, β-rays are emitted in a continuous energy spectrum. Furthermore, although they interact with atoms in the same manner as α-particles, β-rays, being less massive particles, are more strongly deflected in each encounter with an atom. As a result, β-rays move in complicated, zigzag paths and not in straight lines as do α-particles.

Nevertheless, it has been found experimentally that the specific ionization of a beam of β-rays varies approximately exponentially with distance into an absorber. This phenomenon appears to be an accident of nature, due in part to the shape of the β-ray spectrum. If $i(x)$ is the specific ionization at the distance x into the absorber, then

$$i(x) = i_0 e^{-\mu x} = i_0 e^{(\mu/\rho)(\rho x)}, \tag{3.82}$$

where i_0 is the specific ionization at $x = 0$ and ρ is the density of the medium. The apparent mass attenuation coefficient μ/ρ is almost independent of the atomic weight of the medium and increases only slowly with the atomic number. An approximate, empirical formula for μ_a/ρ, based on measurements in aluminum, is

$$\frac{\mu}{\rho} = \frac{17}{E_{\max}^{1.14}}, \tag{3.83}$$

where μ/ρ is in cm^2/g and $E_{\max}$ is the maximum β-ray energy in Me V.

A more useful parameter related to the attenuation of β-rays is the maximum range $R_{\max}$. This is defined as the thickness of absorber required to stop the most energetic of the electrons. The product $R_{\max}\rho$, which is the range expressed in units of g/cm^2, is roughly independent of the nature of the absorbing medium. Values of $R_{\max}\rho$ are shown in Fig. 3.24 as a function of the maximum electron energy $E_{\max}$. These data can be represented by the following empirical formulas:

$$R_{\max}\rho = 0.412 E_{\max}^{(1.265 - 0.0954 \ln E_{\max})}, \qquad E_{\max} < 2.5 \text{ Me V} \tag{3.84}$$

and

$$R_{\max}\rho = 0.530 E_{\max} - 0.106, \qquad E_{\max} > 2.5 \text{Me V}. \tag{3.85}$$

In these equations, $R_{\max}\rho$ is in g/cm^2 and $E_{\max}$ is in Me V.

Equations (3.84) and (3.85) give slightly lower values of $R_{\max}\rho$ for air than are actually observed. Figure 3.25 gives the measured ranges of β-rays for air at 15°C and 1 atm pressure. It is evident from a comparison of Figs. 3.23 and 3.25, and also by comparing calculated ranges, that β-rays penetrate considerably further into materials than α-rays of comparable energy. For instance, β-rays with $E_{\max} = 3$ Me V have a range in air of 13 m, whereas the range of 3-Me V α-particles is only 1.7 cm. However, β-rays do not penetrate far into nongaseous materials and, as a result, they are not difficult to shield.

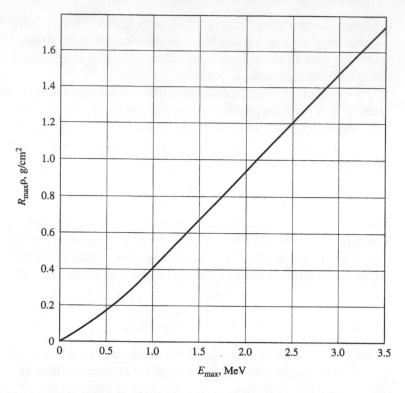

Figure 3.24   Maximum range of β-rays as a function of maximum energy; not valid for air.

Example 3.17

Sodium-24 ($T_{1/2}$ = 15 hr) is often used in medicine as a radioactive tracer. It emits β-rays with a maximum energy of 1.39 Me V. What is the maximum range of the β-rays in animal tissue?

Solution. The density of most animal tissue is approximately unity. From Eq. (3.84), R_{max} is then

$$R_{max} = 0.412 \times 1.39^{(1.265-0.0954 \ln 1.39)}$$

$$= 0.412 \times 1.39^{1.234} = 0.618 \text{ cm } [Ans.]$$

The range of an α-particle of the same energy in tissue is about 9×10^{-4} cm.

Fission Fragments   In Section 3.7, it was pointed out that fissioning nuclei almost always split into two fragments of unequal mass. Since momentum must be conserved in fission, the lighter group of fragments receives somewhat more energy than the heavier group. Therefore, the distribution of fission fragment

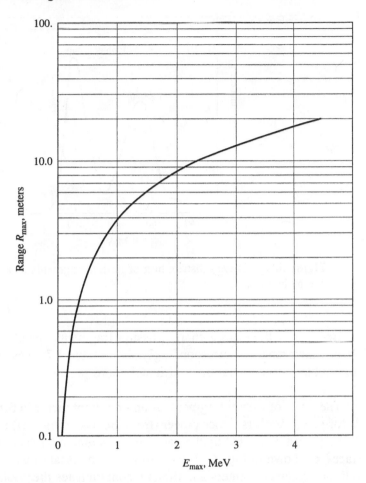

Figure 3.25   Maximum range of β-rays in air at 15 C and 1 atm pressure.

kinetic energies exhibits two peaks, as shown in Fig. 3.26, with the higher energy peak corresponding to the lighter-fission fragments. As seen in the figure, the lighter group with $A \simeq 95$ has an energy of approximately 100 Me V, whereas the heavier group with $A \simeq 140$ receives about 68 Me V.

Once formed, the fission fragments rip through the electron cloud of the orginal fissioning nucleus as they pass into the surrounding medium. In so doing, they usually pick up a number of electrons, although never enough to form a neutral atom. The virgin fission fragment thus appears as a highly energetic and highly ionized atom. The average charge of the lighter group is approximately $+20e$; that of the heavier group is about $+22e$.

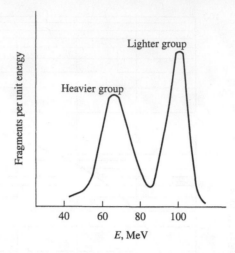

Figure 3.26   Energy distribution of fission fragments as a function of energy in Me V.

Because of their large charge, fission fragments specific ionization is very high, and their range is correspondingly short. Table 3.7 gives approximate ranges of the lighter, more energetic, and hence more penetrating fragments in various materials.

The range of fission fragments is an important factor in the design of the fuel rods for power reactors. Since power (from the fissioning fuel) is produced in these rods, they must be cooled by the passage of a suitable coolant material along their surfaces, as shown in Fig. 3.27. However, it is important that none of the radioactive fission fragments enters and thereby contaminates the coolant. To prevent the escape of such fragments, the fuel is either wrapped in a layer of nonfuel-bearing material, such as stainless steel or an alloy of zirconium, or placed in hollow tubes fabricated from these materials. In view of the short ranges indicated in Table 3.7, this fuel element, *cladding* as it is called, can be quite thin—often no more than 0.05 cm thick.

TABLE 3.7   RANGES OF FISSION FRAGMENTS

Medium	Range, 10^{-3} cm
Aluminum	1.4
Copper	0.59
Silver	0.53
Uranium	0.66
Uranium oxide (U_3O_8)	1.4

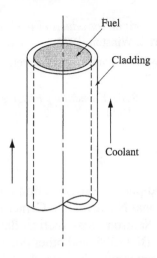

Figure 3.27 A fuel rod with cladding.

REFERENCES

General

Beiser, A., *Concepts of Modern Physics*, 5th ed. New York: McGraw-Hill, 1995, Chapters 3, 4, 5, and 11.

Burcham, W. E., *Nuclear Physics: An Introduction*. reprint, Ann Arbor.

Evans, R. D., *The Atomic Nucleus*. New York: McGraw-Hill, 1955, Chapters 12–14, 18–25.

Foderaro, A., *The Elements of Neutron Interaction Theory*. Cambridge, Mass.: MIT Press, 1971.

Foster, A. R., and R. L. Wright, Jr., *Basic Nuclear Engineering*, 4th ed. Paramus, Prentice-Hall, 1982, Chapters 4 and 8.

Glasstone, S., and A. Sesonske, *Nuclear Reactor Engineering*, 4th ed. New York: Chapman & Hall, 1994, Chapters 1 and 2.

Kaplan, I., *Nuclear Physics*, 2nd ed. Reading, Mass.: Addison-Wesley, 1963, Chapters 15, 16, 18, and 19.

Lamarsh, J. R., *Introduction to Nuclear Reactor Theory*. Reading, Mass.: Addison-Wesley, 1966, Chapters 2 and 3.

Lapp, R. E., and H. L. Andrews, *Nuclear Radiation Physics*, 4th ed. Englewood Cliffs, N.J.: Prentice-Hall, 1972, Chapters 11–13.

Meyerhof, W. E., *Elements of Nuclear Physics*. New York: McGraw-Hill, 1967, Chapters 3 and 5.

Oldenberg, O., and N. C. Rasmussen, *Modern Physics for Engineers*. reprint, Marietta, Technical Books, 1992, Chapters 16 and 17.

Semat, H., and J. R. Albright, *Introduction to Atomic and Nuclear Physics*, 5th ed. New York: Holt, Rinehart & Winston, 1972.

Serway, R. A., Moses, C. J., and C. A. Moyer, *Modern Physics*, 2nd ed. Philadelphia, Saunders, 1997.

Weidner, R. T., and R. L. Sells, *Elementary Physics: Classical and Modern*, 3rd ed. Boston, Mass.: Allyn & Bacon, 1980.

Cross-Section Data

Neutron and charged-particle cross-section data are collected, evaluated, and distributed by the National Nuclear Data Center at Brookhaven National Laboratory, Upton, New York. Neutron cross-section data are published in the report *Neutron Cross-Sections*, BNL-325, and other documents. These cross-sections are also available from the Center on disk in standard format for reactor and shielding computations. Data are also available on the World Wide Web at *http//www.bnl.gov* and a number of other sites. For latest update of availability, the reader should consult the National Nuclear Data Center at *nndc@dne.bnl.gov*.

PROBLEMS

1. Two beams of 1-eV neutrons intersect at an angle of 90°. The density of neutrons in both beams is 2×10^8 neutrons/cm^3.
   (a) Calculate the intensity of each beam.
   (b) What is the neutron flux where the two beams intersect?

2. Two monoenergetic neutron beams of intensities $I_1 = 2 \times 10^{10}$ neutrons/cm^2-sec and $I_2 = 1 \times 10^{10}$ neutrons/cm^2-sec intersect at an angle of 30°. Calculate the neutron flux and current in the region where they intersect.

3. A monoenergetic beam of neutrons, $\phi = 4 \times 10^{10}$ neutrons/cm^2-sec, impinges on a target 1 cm^2 in area and 0.1 cm thick. There are 0.048×10^{24} atoms per cm^3 in the target, and the total cross-section at the energy of the beam is 4.5 b.
   (a) What is the macroscopic total cross-section?
   (b) How many neutron interactions per second occur in the target?
   (c) What is the collision density?

4. The β^--emitter ^{28}Al (half-life 2.30 min) can be produced by the radiative capture of neutrons by ^{27}Al. The 0.0253-eV cross-section for this reaction is 0.23 b. Suppose that a small, 0.01-g aluminum target is placed in a beam of 0.0253-eV neutrons, $\phi = 3 \times 10^8$ neutrons/cm^2-sec, which strikes the entire target. Calculate
   (a) the neutron density in the beam;
   (b) the rate at which ^{28}Al is produced;
   (c) the maximum activity (in curies) that can be produced in this experiment.

5. Calculate the mean free path of 1-eV neutrons in graphite. The total cross-section of carbon at this energy is 4.8 b.

6. A beam of 2-Me V neutrons is incident on a slab of heavy water (D_2O). The total cross-sections of deuterium and oxygen at this energy are 2.6 b and 1.6 b, respectively.
   (a) What is the macroscopic total cross-section of D_2O at 2 Me V?
   (b) How thick must the slab be to reduce the intensity of the uncollided beam by a factor of 10?
   (c) If an incident neutron has a collision in the slab, what is the relative probability that it collides with deuterium?

7. A beam of neutrons is incident from the left on a target that extends from $x = 0$ to $x = a$. Derive an expression for the probability that a neutron in the beam will have its first collision in the second half of the target that is in the region $a/2 < x < a$.

8. *Boral* is a commercial shielding material consisting of approximately equal parts by weight of boron carbide (B_4C) and aluminum compressed to about 95% theoretical density (2.608 g/cm^3) and clad with thin sheets of aluminum 0.25 cm thick. The manufacturer specifies that there are 0.333 g of boron per cm^2 of a boral sheet 0.457 cm in overall thickness. What is the probability that a 0.0253-eV neutron incident normally on such a sheet will succeed in penetrating it?

9. What is the probability that a neutron can move one mean free path without interacting in a medium?

10. A wide beam of neutrons of intensity ϕ_0 is incident on a thick target consisting of material for which $\sigma_a \gg \sigma_s$. The target area is $\mathcal{A}$ and its thickness is X. Derive an expression for the rate at which neutrons are absorbed in this target.

11. Stainless steel, type 304 having a density of 7.86 g/cm^3, has been used in some reactors. The nominal composition by weight of this material is as follows: carbon, 0.08%; chromium, 19%; nickel, 10%; iron, the remainder. Calculate the macroscopic absorption cross-section of SS-304 at 0.0253 eV.

12. Calculate at 0.0253 eV the macroscopic absorption cross-section of uranium dioxide (UO_2), in which the uranium has been enriched to 3 w/o in ^{235}U. The density of UO_2 is approximately 10.5 g/cm^3.

13. The compositions of nuclear reactors are often stated in *volume fractions*, that is, the fractions of the volume of some region that are composed of particular materials. Show that the macroscopic absorption cross-section for the equivalent homogeneous mixture of materials is given by

$$\Sigma_a = f_1 \Sigma_{a1} + f_2 \Sigma_{a2} + \cdots ,$$

where f_i and Σ_{ai} are, respectively, the volume fraction and macroscopic absorption cross-section of the ith constituent at its normal density.

14. Calculate Σ_a and Σ_f (at 0.0253 eV) for the fuel pellets described in Problem 2.63. The pellet density is about 10.6 g/cm^3.

15. Using the fact that the scattering cross-section of ^{209}Bi is approximately 9 b from 0.01 eV to 200 eV, estimate the radius of the ^{209}Bi nucleus and compare with Eq. (2.3).

16. Using the Breit–Wigner formula, compute and plot σ_γ in the vicinity of the first resonance in ^{238}U, which occurs at an energy of 6.67 eV. The parameters of this resonance are: $\Gamma_n = 1.52$ meV (meV = millielectron volts), $\Gamma_\gamma = 26$ meV, and $g = 1$. [*Note:* For reasons given in Section 7.3 (see in particular Fig. 7.12), the values of σ_γ computed from the Breit–Wigner formula do not coincide with those measured with targets at room temperature.]

17. Demonstrate using the Breit–Wigner formula that the width of a resonance at half its height is equal to Γ.

18. There are no resonances in the total cross-section of ^{12}C from 0.01 eV to cover 1 Me V. If the radiative capture cross-section of this nuclide at 0.0253 eV is 3.4 mb, what is the value of σ_γ at 1 eV?

19. The first resonance in the cross-section of aluminum, which is due entirely to scattering, occurs at 5.8 keV. The absorption cross-section at 0.0253 eV is 0.23 b. Calculate for 100 eV:
    (a) σ_a,
    (b) σ_s,
    (c) σ_t.

20. Calculate Σ_a for
    (a) water of unit density at 0.0253 eV;
    (b) water of density 0.7 g/cm^3 at 0.0253 eV;
    (c) water of density 0.7 g/cm^3 at 1 eV.

21. The first resonance in the scattering cross-section of the nuclide $^A Z$ occurs at 1.24 Me V. The separation energies of nuclides $^{A-1} Z, ^A Z$, and $^{A+1} Z$ are 7.00, 7.50, and 8.00 Me V, respectively. Which nucleus and at what energy above the ground state is the level that gives rise to this resonance?

22. There is a prominent resonance in the total cross-section of ^{56}Fe at 646.4 keV. At what energy, measured from the ground state, is the energy level in ^{57}Fe that corresponds to this resonance? [*Hint:* Use the masses of neutral ^{56}Fe and ^{57}Fe to compute the binding energy of the last neutron in ^{57}Fe.]

23. The excited states of ^{17}O occur at the following energies (in Me V) measured from the ground state: 0.871, 3.06, 3.85, 4.55, 5.08, 5.38, 5.70, 5.94, etc. At roughly what energies would resonances be expected to appear in the neutron cross-section of ^{16}O?

24. Using Eq. (3.28), compute and plot E'/E as a function of angle from 0 to π for A = 1, 12, and 238.

25. A 2-Me V neutron traveling in water has a head-on collision with an ^{16}O nucleus.
    (a) What are the energies of the neutron and nucleus after the collision?
    (b) Would you expect the water molecule involved in the collision to remain intact after the event?

26. A 1-Me V neutron strikes a ^{12}C nucleus initially at rest. If the neutron is elastic scattered through an angle of 90°:
    (a) What is the energy of the scattered neutron?
    (b) What is the energy of the recoiling nucleus?
    (c) At what angle does the recoiling nucleus appear?

27. Compute and plot the average fractional energy loss in elastic scattering as a function of the mass number of the target nucleus.

28. Show that the average fractional energy loss in % in elastic scattering for large A is given approximately by

$$\frac{\overline{\Delta E}}{E} \simeq \frac{200}{A}.$$

29. A 1.5-Me V neutron in a heavy water reactor collides with an ^{2}H nucleus. Calculate the maximum and average changes in lethargy in such a collision.

30. Suppose that a fission neutron, emitted with an energy of 2 Me V, slows down to an energy of 1 eV as the result of successive collisions in a moderator. If, on the average, the neutron gains in lethargy the amount ξ in each collision, how many collisions are required if the moderator is
    (a) hydrogen,
    (b) graphite?

31. The 2,200 meters-per-second flux in an ordinary water reactor is 1.5×10^{13} neutrons/cm^2-sec. At what rate are the thermal neutrons absorbed by the water?

32. At one point in a reactor, the density of thermal neutrons is 1.5×10^8 neutrons/cm^3. If the temperature is 450°C, what is the 2,200 meters-per-second flux?

33. A tiny beryllium target located at the center of a three-dimensional Cartesian coordinate system is bombarded by six beams of 0.0253-eV neutrons of intensity 3×10^8 neutrons/cm^2-sec, each incident along a different axis.
    (a) What is the 2,200 meters-per-second flux at the target?
    (b) How many neutrons are absorbed in the target per cm^3/sec?

34. When thermal neutrons interact with ^{14}N, what is the probability that absorption leads to radiative capture?

35. The control rods for a certain reactor are made of an alloy of cadmium (5 w/o), indium (15 w/o) and silver (80 w/o). Calculate the rate at which thermal neutrons are absorbed per gram of this material at a temperature of 400 °C in a 2,200 meters-per-second flux of 5×10^{13} neutrons/cm^2-sec. [*Note:* Silver is a $1/v$ absorber.]

36. From the data in Table 3.3, would you expect ^{232}Th to be fissile? If not, at what neutron energy would you expect fission to be possible?

37. Two hypothetical nuclei, ^AZ and ^{A+1}Z, of atomic weights $M(^AZ) = 241.0600$ and $M(^{A+1}Z) = 242.0621$ have critical fission energies of 5.5 Me V and 6.5 Me V, respectively. Is the nuclide ^AZ fissile?

38. Fission can be induced when γ-rays are absorbed by a heavy nucleus. What energy γ-rays are necessary to induce fission in
    (a) ^{235}U,
    (b) ^{238}U,
    (c) ^{239}Pu?

39. Cross-sections of ^{235}U at 1 Me V are as follows: $\sigma_a = 4.0$ b, $\sigma_i = 1.4$ b, $\sigma_f = 1.2$ b, and $\sigma_a = 1.3$ b. The cross-sections for neutron-producing and charged-particle reactions are all negligible. Compute at this energy

(a) the total cross-section,

(b) the capture-to-fission ratio α.

40. The fission-product ^{131}I has a half-life of 8.05 days and is produced in fission with a yield of 2.9%—that is, 0.029 atoms of ^{131}I are produced per fission. Calculate the equilibrium activity of this radionuclide in a reactor operating at 3,300 MW.

41. Fission-product activity measured at the time t_0 following the burst of a nucleus weapon is found to be α_0. Show that the activity at the time $t = 7^n t_0$ is given approximately by $\alpha = \alpha_0/10^n$. This is known as the 7–10 rule in civil defense.

42. Suppose that radioactive fallout from a nuclear burst arrives in a locality 1 hour after detonation. Use the result of Problem 3.41 to estimate the activity 2 weeks later.

43. The yields of nuclear weapons are measured in kilotons (KT), where 1 KT = 2.6 × 10^{25} Me V. With this in mind,

   (a) How much ^{235}U is fissioned when a 100-KT bomb is exploded?

   (b) What is the total fission-product activity due to this bomb 1 min, 1 hr, and 1 day after detonation? [*Note:* Assume a thermal energy release of 200 Me V per fission.]

44. A research reactor is operated at a power of 250 kilowatts 8 hours a day, 5 days a week, for 2 years. A fuel element, 1 of 24 in the reactor, is then removed for examination. Compute and plot the activity of the fuel element as a function of time up to 2 years after removal.

45. The spontaneous fission rate of ^{238}U is 1 fission per gram per 100 sec. Show that this is equivalent to a half-life for fission of 5.5 × 10^{15} years.

46. Compute and plot the parameter η at 0.0253 eV for uranium enriched in ^{235}U as a function of its enrichment in weight percent ^{235}U.

47. Suppose that 1 kg of ^{235}U undergoes fission by thermal neutrons. Compute the masses (or mass equivalents) in grams for the following, which are produced:

   (a) neutrons,

   (b) β-rays,

   (c) γ-rays,

   (d) neutrinos,

   (e) kinetic energy,

   (f) fission products.

48. The reactor on the nuclear ship *Savannah* operated at a power of 69 MW.

   (a) How much ^{235}U was consumed on a 10,000-nautical-mile voyage at an average speed of 20 knots?

   (b) This is equivalent to how many barrels of 6.5-million-Btu/barrel bunker-C oil?

49. Consolidated Edison's Indian Point No. 2 reactor is designed to operate at a power of 2,758 MW. Assuming that all fissions occur in ^{235}U, calculate in grams per day the rate at which ^{235}U is

   (a) fissioned,

   (b) consumed.

50. Referring to the preceding problem, what is the total accumulated activity of the fission products in the Indian Point No. 2 reactor 1 day after shutdown following 1 year of operation?

51. The photoelectric cross-section of lead at 0.6 Me V is approximately 18 b. Estimate σ_{pe} at this energy for uranium.

52. A 2-Me V photon is Compton scattered through an angle of 30°.
   (a) What is its energy after scattering?
   (b) What is the recoil energy of the struck electron?
   (c) At what angle does the electron appear?

53. Show that Eq. (3.62) follows from Eq. (3.61).

54. Show that the minimum energy of the scattered photon in Compton scattering is given by

$$(E')_{\min} = \frac{E E_e}{2E + E_e},$$

and that for $E \gg E_e$,

$$(E')_{\min} = E_e/2 = 0.255 \text{ Me V}.$$

55. What is the minimum energy of a Compton-scattered photon if its original energy is
   (a) 0.1 Me V,
   (b) 1 Me V,
   (c) 10 Me V?

56. Calculate the mass attenuation coefficient of silica glass (SiO_2, $\rho = 2.21$ g/cm^3) for 3-Me V γ-rays.

57. Derive Eq. (3.69).

58. The mass attenuation coefficient of lead at 0.15 Me V is 1.84 cm^2/g. At this energy, the principal mode of interaction is by the photoelectric effect. What thickness of lead is required to reduce the intensity of a 0.15-Me V γ-ray beam by a factor of 1,000?

59. The density of air at standard temperature and pressure (0°C and 1 atm) is 1.293×10^{-3} g/cm^3. Compute the mean free paths of photons in air under these conditions and compare with the corresponding mean free paths in unit-density water at the following energies:
   (a) 0.1 Me V,
   (b) 1 Me V,
   (c) 10 Me V.

60. At 1 Me V, the Compton cross-section per electron is 0.2112 b, and the Compton energy absorption cross-section per electron is 0.0929 b.
   (a) What is the average energy of the recoiling electron in a Compton interaction at this energy?
   (b) Compute the Compton mass attenuation and mass absorption coefficients at 1 Me V for (i) aluminum, (ii) water.

61. A beam of 0.1-Me V γ-rays with an intensity of 5×10^6 γ-rays/cm^2-sec is incident on thin foils of (i) aluminum, (ii) water. At this energy, the Compton cross-section per electron is 0.4929 b, and the Compton energy absorption cross-section per electron is 0.0685 b. Calculate the energy extracted from the beam per unit volume of the foils due to

(a) Compton scattering,

(b) the photoelectric effect.

62. The absorption of radiation is often measured in units called *rads*, where 1 rad is equal to the absorption of 100 ergs per gram. What intensity of 1 Me V γ-rays incident on a thin slab of water is required to give an absorption rate of 1 rad per second?

63. Determine the range of 5-Me V α-particles in the following media:

(a) air at 15°C, 1 atm;

(b) aluminum;

(c) lead;

(d) unit-density water;

(e) air at 300°C, 10 atm.

64. Determine the relative stopping powers of the media in the preceding problem.

65. Compare the apparent mass attenuation coefficient of 2-Me V (maximum energy) β-rays with the mass attenuation coefficient of 2-Me V γ-rays in aluminum.

66. Compare the maximum ranges of 3-Me V α-rays and β-rays in air at standard temperature and pressure.

67. Near the surface of a flat fuel element in an operating reactor, fissions are occurring at the constant rate of S fissions/cm^3-sec. Given that the average range of the fission fragments is R, show that the rate at which such fragments would escape per cm^2/sec from the surface of the fuel if it were not clad is equal to $SR/2$.

4

Nuclear Reactors and Nuclear Power

Having reviewed in the preceding Chapters the atomic and nuclear physics that form the foundation for nuclear engineering, it is now possible to consider the manner by which nuclear energy is utilized for practical purposes.

4.1 THE FISSION CHAIN REACTION

Nuclear energy is released by way of a fission chain reaction. In this process, which is depicted in Fig. 4.1, neutrons emitted by fissioning nuclei induce fissions in other fissile or fissionable nuclei; the neutrons from these fissions induce fissions in still other fissile or fissionable nuclei; and so on. Such a chain reaction can be described quantitatively in terms of the *multiplication factor*, which is denoted by the symbol k. This is defined as the ratio of the number of fissions (or fission neutrons) in one generation divided by the number of fissions (or fission neutrons) in the preceding generation. In equation form, this is

$$k = \frac{\text{number of fissions in one generation}}{\text{number of fissions in preceding generation}}.$$

(4.1)

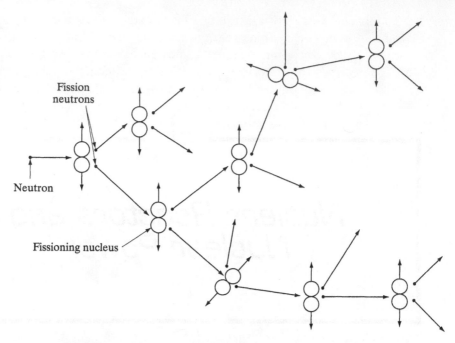

Figure 4.1   Schematic representation of fission chain reaction.

If k is greater than 1, then from Eq. (4.1) the number of fissions increases from generation to generation. In this case, the energy released by the chain reaction increases with time, and the chain reaction, or the system in which it is taking place, is said to be *supercritical*. However, if k is less than 1, the number of fissions decreases with time and the chain reaction is called *subcritical*. Finally, in the special situation where k is equal to 1, the chain reaction proceeds at a constant rate, energy is released at a steady level, and the system is said to be *critical*.

Devices that are designed so that the fission chain reaction can proceed in a controlled manner are called *nuclear reactors*. In a reactor, this control is accomplished by varying the value of k, which can be done by the person operating the system. To increase the power being produced by a reactor, the operator increases k to a value greater than unity so that the reactor becomes supercritical. When the desired power level has been reached, he returns the reactor to critical by adjusting the value of k to be unity, and the reactor then maintains the specified power level. To reduce power or shut the reactor down, the operator merely reduces k, making the reactor subcritical. As a result, the power output of the system decreases. Nuclear bombs and explosives cannot be controlled in this way, and these devices are not referred to as reactors. Several types of nuclear reactors are described later in this Chapter.

To make a reactor critical, or otherwise to adjust the value of k, it is necessary to balance the rate at which neutrons are produced within the reactor with the rate at which they disappear. Neutrons can disappear in two ways as the result of absorption in some type of nuclear reaction, or by escaping from the surface of the reactor. When the sum of the neutron absorption and leakage rates is exactly equal to the neutron production rate, then the reactor is critical. If the production rate is greater than the sum of the absorption and leakage rates, the reactor is supercritical; conversely, if it is smaller, the reactor is subcritical. As would be expected, the production, absorption, and leakage rates depend on the size and composition of a reactor. It is shown in Chapters 5 and 6 that it is possible to derive accurate analytical relationships between these three rates on the one hand and the size and composition on the other. These relationships determine the dimensions and material properties necessary for a reactor to be critical.

4.2 NUCLEAR REACTOR FUELS

Conversion and Breeding

It is recalled from Section 3.7 that η fission neutrons are emitted, on the average, per neutron absorbed by a fissile or fissionable nucleus. To become critical, a reactor obviously must be fueled with a nuclide having a value of η greater than 1. If η were less than 1, a fission in one generation would necessarily lead to less than one fission in succeeding generations, and the reactor could never achieve a critical state. However, η cannot be exactly 1 since, as already noted, some neutrons inevitably are lost either in nonfission absorption reactions or by escaping from the system altogether, and criticality could not be reached.

A check of fission data shows that η is, in fact, substantially greater than 1 for the fissile nuclides at all incident neutron energies, and for the fissionable but nonfissile nuclei such as ^{232}Th and ^{238}U well above their fission thresholds. However, with the nonfissile nuclides, only those neutrons in a reactor with energies above the fission threshold are able to induce fission. For a reactor fueled only with fissionable but nonfissile nuclides, the effective value of η is equal to the actual value of η, multiplied by the fraction of the neutrons absorbed in the reactor with energies above the fission threshold. Regrettably, this number is always less than 1, and it follows that reactors cannot be made critical with nonfissile material alone. Fissile nuclides are therefore essential ingredients in all reactor fuels.

Only one fissile nuclide, ^{235}U, is found in nature, where it occurs with an isotopic abundance of 0.72 $^{a}/_{o}$. The remainder of natural uranium, except for a trace of ^{234}U, is ^{238}U. Thus, only 1 uranium atom out of 139 is ^{235}U. Despite this low concentration of fissile isotope it is possible to fuel certain types of critical reactors with natural uranium, and all of the early reactors were of this type. However, most modern reactors require *enriched uranium* —that is, uranium in which the concentration of ^{235}U has been increased over its natural value. Methods for enriching uranium are discussed in Section 4.7.

As is the case for any natural resource, the supply of ^{235}U is finite. It is shown in Section 4.6 that the world's resources of ^{235}U for nuclear fuel are limited. Indeed, if nuclear power was based on the fission of ^{235}U alone, the era of nuclear energy would be a comparatively short one—probably not more than a century in duration. Fortunately, it is possible to manufacture certain fissile isotopes from abundant nonfissile material, a process known as *conversion*. The two most important fissile isotopes that can be produced by conversion are ^{233}U and ^{239}Pu. The ^{233}U is obtained from thorium by the absorption of neutrons. The reactions involved are as follows:

$$^{232}\text{Th}(n, \gamma)^{233}\text{Th} \xrightarrow{\beta-} {}^{233}\text{Pa} \xrightarrow{\beta-} {}^{233}\text{U}. \tag{4.2}$$

As the result of the reactions in Eq. (4.2), the nonfissile isotope ^{232}Th is converted into fissile ^{233}U. Isotopes like ^{232}Th, which are not fissile but from which fissile isotopes can be produced by neutron absorption, are said to be *fertile*.

It is a relatively simple matter to bring about the reactions given in Eq. (4.2). Naturally occurring thorium is entirely ^{232}Th. Therefore, it is merely necessary to introduce thorium, in one form or another, into a critical reactor where it is exposed to neutrons. After a suitable irradiation time, when the ^{233}U has built up to a desired level, the thorium is withdrawn from the reactor and the ^{233}U is extracted from the thorium. This can be done by chemical means since thorium and uranium are two entirely different chemical elements. With one exception (see Section 4.5), however, comparatively little ^{233}U has been produced from thorium since there is almost no demand for ^{233}U as a reactor fuel. Nevertheless, thorium remains an important potential source of nuclear fuel.

^{239}Pu is obtained from reactions similar to Eq. (4.2)—namely,

$$^{238}\text{U}(n, \gamma)^{239}\text{U} \xrightarrow{\beta-} {}^{239}\text{Np} \xrightarrow{\beta-} {}^{239}\text{Pu}. \tag{4.3}$$

The fertile isotope in this case is the ^{238}U; ^{239}Np is the intermediate nucleus. To realize these reactions in practice, ^{238}U must be irradiated in a reactor. This, however, occurs automatically in most power reactors of current design. Thus, as noted earlier, most present-day reactors are fueled with uranium that is only slightly enriched in ^{235}U. Practically all of the fuel in these reactors is ^{238}U, and the conversion of ^{238}U to ^{239}Pu takes place as a matter of course during the normal operation of these reactors. The plutonium is later extracted chemically from the fuel, uranium and plutonium being different elements. The chemical extraction of fissile material from fertile material is called *fuel reprocessing* and is described in Section 4.8.

After ^{239}Pu has been formed in a reactor, it may absorb a neutron and undergo fission or be transformed into ^{240}Pu. The ^{240}Pu, which is not fissile, may in turn capture another neutron to produce the fissile isotope ^{241}Pu. Finally, the ^{241}Pu may undergo fission or be transformed into ^{242}Pu. Thus, the plutonium extracted from reactor fuel by reprocessing contains, in decreasing amounts, the isotopes ^{239}Pu, ^{240}Pu, ^{241}Pu, and ^{242}Pu. The fractional content of each isotope depends on the burnup of the fuel, defined later in this section, at the time of reprocessing.

The conversion process is described quantitatively in terms of the parameter C, which is called the *conversion ratio* or sometimes the *breeding ratio*. This is defined as the average number of fissile atoms produced in a reactor per fissile fuel atom consumed. Thus, when N atoms of fuel are consumed, NC atoms of fertile material are converted into new fissile atoms. However, if the newly produced fissile isotope is the same as the isotope that fuels the reactor, the new atoms may later be consumed to convert another $NC \times C = NC^2$ atoms of fertile material; these may be consumed to convert NC^3 fertile atoms; and so on. In this way, it is easy to see that the consumption of N fuel atoms results in the conversion of a total of

$$NC + NC^2 + NC^3 + \cdots = \frac{NC}{1 - C}$$

fertile atoms, provided C is less than 1. When $C = 1$, an infinite amount of fertile material can be converted starting with a given amount of fuel.

Example 4.1

In a critical reactor fueled with natural uranium, it is observed that, for every neutron absorbed in ^{235}U, 0.254 neutrons are absorbed in resonances of ^{238}U and 0.640 neutrons are absorbed by ^{238}U at thermal energies. There is essentially no leakage of neutrons from the reactor. (a) What is the conversion ratio for the reactor? (b) How much ^{239}Pu in kilograms is produced when 1 kg of ^{235}U is consumed?

Solution.

1. Each absorption of a neutron by ^{238}U, whether at resonance or thermal energies, produces an atom of ^{239}Pu via reaction Eq. (4.3). Furthermore, each time a neutron is absorbed by a ^{235}U nucleus, that nucleus is consumed. The compound nucleus either undergoes fission or emits a γ-ray to become ^{236}U. Thus, the total number of ^{239}Pu atoms produced per ^{235}U atom consumed is $0.254 + 0.640 = 0.894$. This, by definition, is the value of the conversion factor, so that $C = 0.894$. [*Ans.*]

2. The consumption of 1 kg of ^{235}U involves the disappearance by neutron absorption of $1,000N_A/235$ atoms of ^{235}U, where N_A is Avogadro's number. From the definition of C, this is accompanied by the production of $C \times 1000N_A/235$ atoms of ^{239}Pu, which have a total mass of $C \times 1000 \times 239/235$ g or $C \times 239/235$ kg. Thus, the consumption of 1 kg of ^{235}U leads in this

example to the production of

$$0.894 \times \frac{239}{235} = 0.909 \text{ kg of } ^{239}\text{Pu.}[Ans.]$$

A most important situation occurs when C is greater than 1. In this case, more than one fissile atom is produced for every fissile atom consumed, which is a process described as *breeding*. Reactors that are designed so that breeding will take place are called *breeder reactors* or simply *breeders*. Reactors that convert but do not breed are called *converters;* reactors that neither convert nor breed but simply consume fuel are called *burners*. Breeders are remarkable devices for, in addition to providing power through the energy released in fission, they actually produce more fissile fuel than they consume.

Needless to say, it is more difficult to design a reactor that will breed than one that merely converts. For one thing, while η must be greater than 1 for conversion, it must be greater than 2 for breeding. This is because, in any reactor, one fission neutron must eventually be absorbed in fuel just to keep the reactor critical and maintain the chain reaction. If the reactor is to breed, more than one neutron must be absorbed in fertile material to produce the new fissile isotope. In actual fact, η must be substantially greater than 2 for, as noted earlier, in any reactor some neutrons inevitably are absorbed by nonfuel atoms or lost by leakage.

In this connection, it is important to recognize that η is not a constant, but depends on the energy of the neutron that induces the fission. This variation of η with neutron energy is shown in Fig. 4.2 for the three fissile isotopes ^{233}U, ^{235}U, and ^{239}Pu. At thermal neutron energies (i.e., $E \simeq 0.025$ eV), the value of η for ^{233}U is about 2.29, which is sufficiently in excess of 2 for breeding to be possible. Thus, a properly designed reactor in which most of the fissions are induced by thermal neutrons—that is, a *thermal reactor*—could breed if it were fueled with ^{233}U. By contrast, the thermal neutron values of η for ^{235}U and ^{239}Pu, 2.07 and 2.14, respectively, are not sufficiently greater than 2 to permit breeding.

Returning to Fig. 4.2, it is observed that in the intermediate energy range, from about 1 eV to 100 keV for ^{235}U and from about 10 eV to 20 keV for ^{239}Pu, the value of η falls below 2. Breeding cannot be achieved when these isotopes are used to fuel a reactor in which most of the fissions occur at these energies. It has now been demonstrated that such an intermediate reactor can be made to breed, at least in a limited way, when fueled with ^{233}U. The value of η for ^{233}U is greater than 2 by a sufficient margin at intermediate energies to make breeding possible (see the LWBR discussion in Section 4.5).

Above about 100 keV, as indicated in Fig. 4.2, η rises to values substantially above 2 for all three of the fissile fuels. As far as the value of η is concerned, it should therefore be possible to breed with these fuels provided the reactor is designed in such a way that the bulk of the fissions are induced by neutrons with

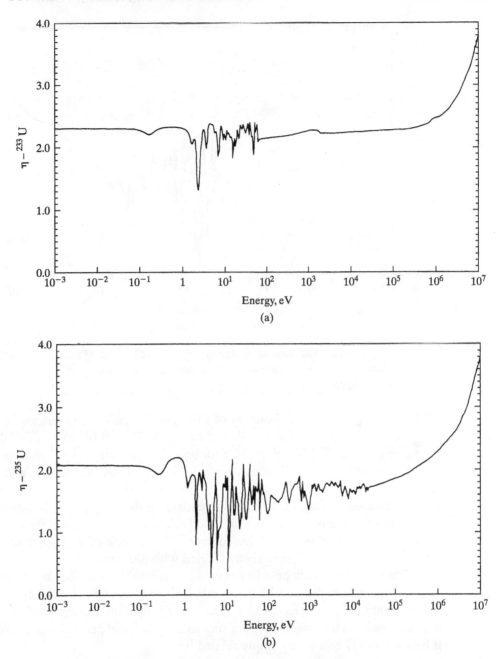

Figure 4.2   Variation of η with energy for (a) ^{233}U and (b) ^{235}U. (Plotted by machine from data on tape at the National Neutron Cross Section Center, Brookhaven National Laboratory.)

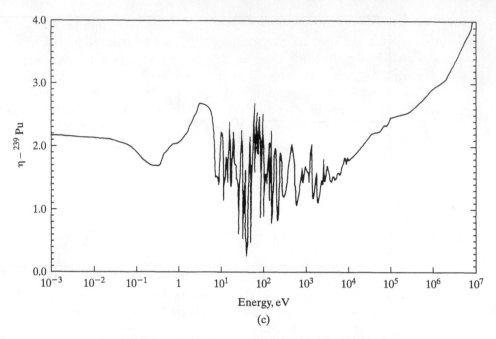

Figure 4.2   Variation of η with energy for (c) ^{239}Pu. (Plotted by machine from data on tape at the National Neutron Cross Section Center, Brookhaven National Laboratory.)

sufficiently high energies. Reactors of this type are called *fast reactors* or, since they are usually designed to breed, *fast breeders*. It must be noted that, although the value of η for ^{235}U is high enough for breeding, unfortunately there is no fertile material in nature that can be converted into ^{235}U. Thus, although a fast reactor fueled with ^{235}U may produce more ^{239}Pu from ^{238}U than the ^{235}U it consumes, the ^{235}U is used up forever. Some fast breeders, in the absence of an initial supply of plutonium, have been started with ^{235}U fuel, but they are ultimately fueled with plutonium. This fuel is also preferred over ^{233}U because of its higher value of η. Thus, fast breeders are almost always fueled with plutonium.

The extent to which breeding occurs in a reactor is described by the *breeding gain*—a parameter denoted by the symbol G. This is defined asthe *net increase* in the number of fissile atoms in a reactor per fuel atom consumed. Since C, the breeding ratio, is the total number of fuel atoms produced per fuel atom consumed, it follows that G and C are simply related by

$$G = C - 1.$$

Suppose that $C = 1.2$. This means that 1.2 new atoms of fuel are produced for each atom consumed. The net increase in the number of fuel atoms per atom consumed is then clearly 0.2, and the breeding gain is 0.2.

Breeding is also described in terms of the *doubling time*. This is defined as the hypothetical time interval during which the amount of fissile material in (or associated with) a reactor doubles. To compute the doubling time, suppose that a reactor is operated at a constant thermal power level of P_0 megawatts. This reactor consumes fissile material at the uniform rate of $w P_0$ grams per day, where w is the fuel consumption rate per unit power. (In the case of a ^{235}U-fueled reactor according to Section 3.7 $w = 1.23$ g per day per thermal megawatt.) This is equivalent to the consumption of $w P_0 N_A / M_f$ atoms of fuel, where N_A is Avogadro's number and M_f is the atomic weight of the fuel. From the earlier definition of G, the consumption of $w P_0 N_A / M_f$ fuel atoms produces $G w P_0 N_A / M_f$ atoms of fuel over and above those consumed, which means that there is a net production rate of $G w P_0$ grams of fuel per day.[1] If none of this new fuel is removed from the reactor, the total amount of fuel in the system will increase linearly with time from the initial fuel inventory m_0. The *linear doubling time* is defined as the time t_{Dl} required for the total amount of fuel in the reactor to reach the value $2 m_0$. Clearly then,

$$G w P_0 \times t_{Dl} = m_0$$

and

$$t_{Dl} = \frac{m_0}{G w P_0}. \tag{4.4}$$

It is not difficult to see that permitting all of the newly produced fuel to accumulate in the reactor is a wasteful procedure. The extra fuel is not required to keep the reactor operating and could better be removed and, together with fuel from other breeders, be used to fuel another breeding reactor. In this way, the total power produced from all the fuel can be increased as the fuel mass increases in both the original and second breeder. This mode of operation also coincides more closely with actual practice since breeder reactors are refueled at regular intervals (normally about once or twice every 2 years) over their lifetime.

It can be shown (see Problem 6.8) that the reactor power which can be produced from a given fuel mass is proportional to the mass—that is,

$$P = \beta m, \tag{4.5}$$

where β is a constant. However, as shown earlier, the rate of increase in mass is given by

$$\frac{dm}{dt} = G w P,$$

---

[1]This assumes that the newly produced fuel and the fuel consumed have the same atomic weight. If they are different, the production rate must be multiplied by the ratio of their atomic weights.

and it follows that

$$\frac{dm}{dt} = Gw\beta m.$$

This equation has the solution

$$m = m_0 e^{Gw\beta t}, \qquad (4.6)$$

where m_0 is again the initial fuel inventory.

The exponential doubling time t_{De}, sometimes called the *compound doubling time,* is now defined as the time in which m reaches $2m_0$ according to Eq. (4.6). This is easily seen to be

$$t_{De} = \frac{\ln 2}{Gw\beta}. \qquad (4.7)$$

However, from Eq. (4.5),

$$\beta = \frac{P_0}{m_0},$$

where P_0 is the initial power. Substituting β into Eq. (4.7) then gives

$$t_{De} = \frac{m_0 \ln 2}{Gw P_0}. \qquad (4.8)$$

Comparing Eqs. (4.4) and (4.8) shows that

$$t_{De} = t_{Dl} \ln 2 = 0.693 t_{Dl}. \qquad (4.9)$$

The growth in total fuel inventory and the two doubling times, t_{Dl} and t_{De}, are shown in Fig. 4.3. In practice, both t_{Dl} and t_{De} have been used. However, because

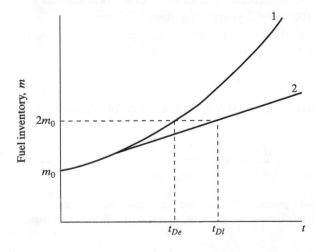

Figure 4.3   Growth in the inventory of fuel by breeding under two conditions: (a) new fuel continually extracted and used for further breeding, (b) new fuel left in reactor.

t_{De} reflects a more usual mode of fuel managment, it is frequently referred to as the *doubling time*.

It should be emphasized that the actual time required by a reactor owner to double his inventory of fissile material is greater than either of the doubling times computed previously, since the time required to remove the newly produced fuel from the reactor and have it chemically separated and fabricated into a form suitable for fueling the reactor was omitted in the derivation. Nevertheless, the doubling time is a reasonable measure of merit for the performance of a breeder. Thus, all other things being equal (which they never are), the best breeder is clearly the one with the lowest value of t_{De}.

Example 4.2

A hypothetical fast breeder reactor is fueled with a mixture of ^{239}Pu and ^{238}U. When operating at full power, the plutonium is consumed at a rate of approximately 1 kg per day. The reactor contains 500 kg of ^{239}Pu at its initial startup, and the breeding gain is 0.15. (a) At what rate is ^{239}Pu being produced? (b) Calculate the linear and exponential doubling times of this reactor.

Solution.

1. From the definition of *breeding gain*, the reactor produces 0.15 extra atoms of ^{239}Pu for every atom of ^{239}Pu consumed. On a per kilogram basis, this means that the reactor produces 0.15 additional kilograms of ^{239}Pu for each kilogram consumed. Since the consumption rate of ^{239}Pu is 1 kg/day, it follows that the instantaneous rate of plutonium production is 0.15 kg/day or about 55 kg/yr. [*Ans.*]

2. The factor $w P_0$ in Eq. (4.4) is the total fuel consumption rate, which is 1 kg/day in this problem. From that equation, it follows that

$$t_{Dl} = \frac{500\text{kg}}{0.15\,\text{kg/day}} = \frac{500\text{ kg}}{55\text{ kg/yr}} = 9.1 \text{ yr.}[\textit{Ans.}]$$

Then from Eq. (4.9),

$$t_{De} = 0.693 \times 9.1 = 6.3 \text{ yr. } [\textit{Ans.}]$$

Nuclear Fuel Performance—Burnup and Specific Burnup

The total energy released in fission by a given amount of nuclear fuel is called the fuel *burnup* and is measured in megawatt days (MWd). The fission energy released per unit mass of the fuel is termed the *specific burnup* of the fuel and is usually expressed in megawatt days per metric ton or per kilogram (that is, MWd/t or MWd/kg) of the heavy metal originally contained in the fuel. According to the discussion in Section 3.7, the fissioning of 1.05 g of ^{235}U yields 1 MWd. Thus, it follows that the maximum theoretical burnup for this fuel is

$$\frac{1 \text{ MWd}}{1.05 \text{ g}} \times \frac{10^6 \text{ g}}{t} = 950,000 \text{ MWd/t}$$

or 950 MWd/kg.

Fuel performance is also described in terms of *fractional burnup* (often expressed in %), denoted as β and defined as the ratio of the number of fissions in a specified mass of fuel to the total number of heavy atoms originally in the fuel. That is,

$$\beta = \frac{\text{number of fissions}}{\text{initial number of heavy atoms}}. \qquad (4.10)$$

Since the fission of all fuel atoms ($\beta = 1$) yields 950,000 MWd/t and the specific burnup at any time is clearly proportional to β, it follows that, for ^{235}U,

$$\text{specific burnup} = 950,000\beta \text{ MWd/t}. \qquad (4.11)$$

The maximum specific burnup that can be obtained from a given reactor fuel depends on a number of factors, which are discussed in Section 7.5.

Example 4.3

A thermal reactor is loaded with 98 metric tons of uranium dioxide (UO_2) fuel in which the uranium is enriched to 3 w/o ^{235}U. The reactor operates at a power level of 3,300 MW for 750 days before it is shut down for refueling. (a) What is the specific burnup of the fuel at shutdown? (b) What is the fractional burnup at shutdown?

Solution.

1. The total burnup of the fuel is $3,300 \times 750 = 2.48 \times 10^6$ MWd. The atomic weight of 3 w/o uranium is approximately 238. Therefore, of the 98 t of UO_2, $98 \times 238/(238 + 2 \times 16) = 86.4$ t is uranium, that is, the heavy metal. The specific burnup is then $2.48 \times 10^6/86.4 = 28,700$ MWd/t. [*Ans.*]
2. From Eq. (4.11), the fractional burnup is

$$\beta = \frac{28,700}{950,000} = 0.0302 = 3.02\%. \text{ [}Ans.\text{]}$$

[*Note.* In any reactor like the one in this example, which is fueled with low-enriched uranium, the bulk of the fuel is ^{238}U. Some of the ^{238}U is converted to the isotopes of plutonium as the reactor operates, and a portion of the plutonium fissions in the reactor. Some ^{238}U also undergoes fission by interactions with fast neutrons. Over 40% of the fissions in such a reactor may actually occur in non-^{235}U atoms. This explains why β in the present case is larger than the fuel enrichment of 3 w/o. In fact, the spent fuel may still have a ^{235}U enrichment on the order of 1 w/o at unloading.]

For convenience, the fissile and fertile nuclides are frequently identified by a simple numerical code that was introduced during the Manhattan Project of World War II. Suppose that the atomic number Z and atomic mass number A of some nuclide are denoted by $Z = ab$ and $A = cde$, where $a, b, \ldots$ and so on are digits. Then according to the code, the nuclide in question is identified by the two-digit number be. Thus, ^{235}U is "25," ^{238}U is "28," ^{239}Pu is "49," ^{241}Pu is "41," and so on. This code often simplifies notation and is used throughout this book.

4.3 NON-NUCLEAR COMPONENTS OF NUCLEAR POWER PLANTS

Since the world's first reactor (an assembly of natural uranium lumps imbedded in graphite) was brought to critical in Chicago on December 2, 1942, a large number of reactors have been designed and built for a variety of purposes: for the conversion of ^{238}U to ^{239}Pu; for the propulsion of ships, aircraft, rockets, and satellites; for medical irradiation; for research; and for the generation of electrical power. It is impossible herein to describe all of these reactors in detail. Instead, attention is focused on reactors in the last category—namely, reactors that are used in stationary power plants for the generation of electricity. Most nuclear engineers today are employed in one way or another with the nuclear electric power industry.

In all nuclear power plants, the fission energy released in the reactor is used to produce steam—either directly in the reactor or in auxiliary heat exchangers called *steam generators*. The reactor or reactor–steam generator combination is called the *nuclear steam supply system* (NSSS). The NSSS serves the same function as the steam boiler in a conventional fossil-fuel plant. The way in which the steam is utilized in a power plant is considered in the present section; reactors and nuclear steam supply systems are discussed in the next two sections.

Figure 4.4 schematically shows the steam system in a nuclear power plant from the point where the steam leaves the reactor or steam generators to the point where it returns as feedwater to be reconverted into steam. As indicated in the figure, the steam is used to drive steam turbines, which are coupled to a generator to produce electricity. Turbines, in principle at least, are simple machines. They consist of a series of bladed wheels affixed to an axle, which rotates at high speed as steam at high temperature and pressure strikes the turbine blades. Steam, of course, is just the gaseous form of water, and it does not contain any liquid water per se. To emphasize this, steam is often referred to as dry steam; steam mixed with liquid water is called wet steam. Steam is always delivered to a turbine as dry as possible, but as it passes through successive turbine stages, its temperature and pressure are reduced, and some of the steam is condensed to droplets of liquid water. It has been found that these droplets impinging on the turbine blades lead to excessive erosion of the blades and hence to reduced turbine lifetime. In addition,

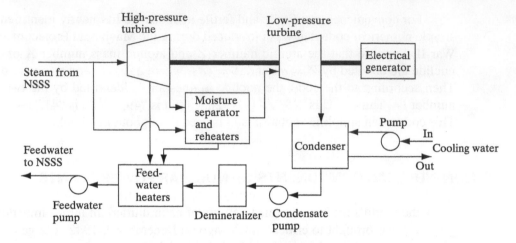

Figure 4.4 Simplified diagram of the power-producing side of a nuclear steam power plant.

the inertia of the liquid in the turbine tends to reduce the turbine efficiency. This problem can be circumvented in a number of ways.

It is recalled from thermodynamics that water is converted to steam by adding heat (the latent heat of vaporization) to water at constant pressure and temperature. The steam so produced is said to be *saturated*. The extraction of any energy from such saturated steam, as occurs when it passes through a turbine, results in the condensation of some of the steam to liquid water. However, if the steam is heated above the temperature at which it was evolved from boiling water, some heat can be extracted without condensation taking place. Steam that has been heated in this way is said to be *superheated*. By superheating the steam before it enters a turbine, the total amount of water produced in the turbine can be significantly reduced. Superheating may be done in a separate unit designed for the purpose, in a portion of the steam generator, or in the reactor.

Another approach to this problem is to remove the wet steam from the turbine when the water content has reached a specified level (separate the steam from the water, which is done in a *moisture separator*), reheat this spent steam using the higher pressure and hotter steam from the reactor or steam generators, and pass the reheated steam (which may be superheated at this point) to one or more lower pressure turbines as shown in Fig. 4.4. Moisture separation and reheating is usually performed in the same device–one for each low-pressure turbine. The wet steam passes over metal blades that draw off the water; then as dry steam, it passes around a series of tubes carrying the hotter NSSS steam. Reheating is utilized in fossil as well as nuclear power plants.

From the low-pressure turbines, the spent steam passes into a condenser, where it is cooled by water from a suitable outside source, and the steam is condensed to water. In this part of the steam cycle, heat is rejected by the system. This rejection of heat is an essential part of the operation of the plant from a thermodynamic point of view. Thus, it is not possible to skip the condenser portion of the cycle and return the spent steam directly to the reactor side of the system. To do so would violate the second law of thermodynamics, which forbids the extraction of heat from any source and the performance of an equivalent amount of work in a cyclical process.

The condensed water, or condensate, is next pumped through a demineralizer and then back to the steam-producing side of the cycle as feedwater. It turns out that the overall efficiency of the system is increased if the feedwater is heated by steam withdrawn from an intermediate stage in the turbine, which is a procedure known as *regeneration*. This heating takes place in the feedwater heaters, shown in Fig. 4.4. These are heat exchangers with turbine steam on the hot side. With the feedwater heated and returned to the reactor or steam generator, the power-producing portion of the steam cycle is complete.

The overall efficiency of a nuclear power plant is defined as the ratio

$$\text{eff} = \frac{W}{Q_R}, \tag{4.12}$$

where W is the rate of electrical energy output in megawatts (denoted by MWe), and Q_R is the rate of thermal energy output from the reactor (written as MWt). The value of W is less than Q_R by an amount equal to the sum of the heat losses in all parts of the system—that is, energy lost in the various heat exchangers, the turbines, pumps, piping, the generator, and also the heat, Q_C, rejected to the coolant in the condenser. In practice, virtually all of the recoverable energy of fission is converted into the steam, which enters the turbine. Relatively little energy is lost in the turbine or piping or is required to operate the various pumps, and the generator performs at high efficiency. It follows that, as a rough approximation,

$$W \simeq Q_R - Q_C. \tag{4.13}$$

Equation (4.12) can then be written as

$$\text{eff} \simeq 1 - \frac{Q_C}{Q_R}. \tag{4.14}$$

It follows from this equation that the smaller the value of Q_C, the greater the efficiency of the plant—or, conversely, the greater the efficiency of the plant, the smaller the amount of heat Q_C rejected to the condenser cooling water for a given plant capacity.

In practice, the efficiency of a steam cycle of the type described earlier depends on many parameters, among which are the temperature T_t of the steam entering the turbine and the temperature T_c of the coolant used in the condenser. In particular, the efficiency increases with increasing T_t and with decreasing T_c. For an actual plant, the value of T_c is determined by the environmental conditions, local air temperature, or the temperature of the condenser cooling water, and only the value of T_t is under the control of the plant designer. To achieve high efficiency, therefore, it is necessary to operate the turbine with steam at the highest possible temperature.

Plant efficiency is an important consideration for both nuclear and fossil fuel plants, but for somewhat different reasons. With fossil fuel plants, the cost of producing electricity is largely determined by the cost of the fuel, which is consumed in large quantity to provide the heat Q_R. It follows from Eq. (4.12) that with higher plant efficiency less fuel must be consumed to produce a given amount of electrical power. Therefore, every effort is made in fossil fuel plants to produce high-temperature steam, and turbines in these plants are operated close to the maximum temperature permitted by the metallurgical properties of the turbine materials.

In contrast, with nuclear power plants, the cost of electricity is largely determined by capital costs (specifically, the interest charges on the money used to build the plant) and not by fuel costs. Thus, electricity can be produced economically in a nuclear plant even if its efficiency is low and fuel consumption is high simply because the cost of electricity is not significantly influenced by fuel costs in the first place. This is a fortunate circumstance for the nuclear industry since it is not possible to produce high-temperature steam, comparable to that used in fossil fuel plants, in many of the nuclear systems currently available. This, in turn, is due to the fact that the temperature of the fuel used in these reactors is restricted to relatively low values—lower, for example, than the temperature of the combustion chamber in a conventional boiler. Such low fuel temperatures are necessary to ensure the integrity of the fuel and make certain that the fission products produced within the fuel remain confined.

However, there is at least one serious consequence of the low efficiency of most present-day nuclear installations—namely, as shown in Eq. (4.14), they necessarily reject more heat to the environment than comparable fossil fuel plants. For example, a nuclear plant operating at an overall efficiency of 33% can be shown (see Problem 4.16) to eject approximately 25% more heat to the environment than a fossil fuel plant with an efficiency of 38%. If this heat is improperly discharged to a body of water or if the body of water is too small for the amount of heat rejected, then the ensuing temperature rise of the water may upset the equilibrium of biological species residing there, and this can lead to the deterioration and eventual stagnation of the water source. This phenomenon is called *thermal pollution*. It is an important consideration both in the siting and in design of all power plants, but especially nuclear power plants. To reduce thermal pollution, many nuclear power plants utilize cooling towers. These are large structures (often over 500 ft high) in which heat is exchanged between the condenser cooling water and the atmosphere. In this way, the heat from the condenser is absorbed by the atmosphere instead of being rejected to a local body of water. Cooling towers must also be used wherever an adequate supply of cooling water cannot be guaranteed year-round.

In addition to efficiency, two other parameters are used to describe the performance of a nuclear power plant (or any type of power plant for that matter). These are the availability and capacity factors. The *availability* is defined as the percent of time, over any given reporting period, that the plant is operational–that is, either in operation or capable of being put into operation if the electrical demand requires it. The *capacity factor* is the percent of the total electrical power that could theoretically be produced during a specified period if the plant were operated at full power 100% of the time. The capacity factor is less than (or possibly equal to) the availability because a plant cannot be operated if it is not available, and because it is not always operated at full power when it is available due to the vagaries of electrical demand and the economics of utility operation.

The availability of American-built light-water nuclear plants (see Section 4.5) is in the neighborhood of 90%, about the same as fossil fired plants. The capacity factors of these nuclear plants generally lie between 80% and 85%. Both the availability and capacity factors of Canadian-built heavy water reactor plants are higher.

Example 4.4

> Over a period of 1 year a nuclear power plant with an output rating of 1,075 MWe actually delivers 255,000 MWd of electric power to the utility grid. During this time, the plant was down—that is, inoperable—28 days for refueling, 45 days for repairs to the nuclear portions of the plant, and 18 days for repairs to the conventional portions of the plant. (a) What was the plant capacity factor during the year? (b) What was the plant availability?
>
> *Solution.*
>
> 1. At full power for the entire year (365 days), the plant could deliver $1075 \times 365 = 392{,}375$ MWd. Since it only delivered 255,000 MWd, the capacity factor is $255{,}000/392{,}375 = 0.65 = 65\%$. [*Ans.*]
> 2. The total number of days the plant was down is $28 + 45 + 18 = 91$. It was therefore operable $365 - 91 = 274$ days. The plant availability is then $274/365 = 0.75 = 75\%$. [*Ans.*]

4.4 COMPONENTS OF NUCLEAR REACTORS

Before discussing specific reactors for the generation of steam, it is helpful to first review the principal component parts or regions of typical reactors. These are shown in Fig. 4.5. It must be emphasized that this drawing is merely schematic and does not depict an actual reactor.

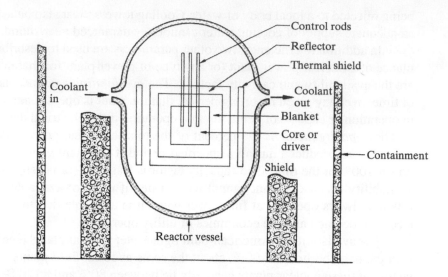

Figure 4.5 Schematic drawing of the principal components of a nuclear reactor together with the radiation shield and containment.

The central region of a reactor shown in Fig. 4.5 is called the *core*.[2] In a thermal reactor, the core contains the fuel, the moderator, and a coolant. In a fast breeder reactor, there is no moderator, only fuel and coolant. The fuel includes the fissile isotope that, as explained earlier, is responsible for both the criticality of the reactor and the release of fission energy. However, the fuel may also contain large amounts of fertile material. Indeed, most modern power reactors (which are thermal) are fueled with uranium enriched to only a few percent ^{235}U, so that most of the fuel is actually ^{238}U.

The moderator, which is only present in thermal reactors, is used to *moderate*—that is, to slow down—the neutrons from fission to thermal energies. It was shown in Section 3.5 that nuclei with low-mass numbers are most effective for this purpose, so the moderator is always a low-mass-number material. Water (two-thirds of atoms of which are hydrogen), heavy water, and graphite (the common form of carbon) are often used as moderators. Beryllium and beryllium oxide (BeO), a white ceramic material, have been used occasionally, but they are very costly.

The coolant, as the name implies, is used to remove the heat from the core and other parts of the reactor where heat may be produced. Water, heavy water, and various gases are the most commonly used coolants for thermal reactors. In the case of water and heavy water, these coolants also frequently serve as the moderator.

---

[2]The core of a breeder reactor is often called the *driver* since this is the part of the reactor that leads to the criticality of the reactor—hence drives the system.

With fast reactors, water and heavy water cannot be used as coolants, at least in liquid form, since these materials would tend to slow down the fission neutrons, which, in this type of reactor, must be kept as energetic as possible. Most fast reactors are cooled by liquid sodium. Sodium has excellent heat transfer properties; with an atomic weight of 23, it is not effective in slowing down neutrons by elastic scattering. (However, some moderation occurs as a result of inelastic scattering.) Gases, of course, can also be used to cool fast reactors.

Surrounding the core of breeder reactors is a region of fertile material called the *blanket.* This region is designed specifically for conversion or breeding. Neutrons that escape from the core are intercepted in the blanket and enter into the various conversion reactions. However, a substantial amount of power may also be produced in the blanket as the result of fissions induced by fast neutrons, so it must be cooled as well as the core.

The region adjacent to the core, or to the blanket if one is present, is called the *reflector.* This is a thickness of moderating material whose function can be understood in the following way. Suppose the core (or blanket) were bare—that is, open and exposed to the air. In this case, all of the neutrons that escaped from the surface of the core would be lost from the reactor—none would return. By placing a region of moderator around the reactor, however, some of these neutrons are returned to the core or blanket after one or more collisions in the reflector. All of the neutrons do not return, of course, but some do, thus saving neutrons for the chain reaction. It is shown later in this book that reflectors significantly reduce the amount of fuel required to make a reactor critical.

The *control rods* shown in Fig. 4.5 are movable pieces of neutron-absorbing material; as their name suggests, they are used to control the reactor. Since they absorb neutrons, any movement of the rods affects the multiplication factor of the system. Withdrawal of the rods increases k, whereas insertion decreases k. Thus, the reactor can be started up or shut down, or its power output can be changed by the appropriate motion of the rods. Control rods must also be adjusted to keep a reactor critical and operating at a specified power level, when, in the course of time, the fuel is consumed and various neutron-absorbing fission products accumulate in the reactor. Several materials have been used for control rods: boron steel, since boron has a high neutron absorption cross-section; hafnium or cadmium, metallic elements that are strong thermal neutron absorbers; silver; and various alloys of these metals. The rods may be cylindrical in shape (rods, in the true sense of the word) or may be sheets, blades, or crossed blades, which are called *cruciform rods.*

The various reactor components just described are all located within the *reactor vessel,* which, if the components are under pressure, is also called the *pressure vessel.* To reduce the thermal stresses in the reactor vessel caused by the absorption of γ-rays emanating from the core, it is necessary in some reactors to place a *thermal shield,* a thick layer of γ-rays absorbing material (usually iron or steel), between the reflector and the inner wall of the vessel. Since the thermal shield absorbs considerable energy, it ordinarily must be cooled along with the core and blanket.

The reactor vessel and all other components of the nuclear steam supply system that contain sources of radiation are surrounded by radiation shielding in varying amounts for the protection of plant personnel during normal operation of the reactor; the amount of shielding varies according to the design of the reactor. To protect the general public from the consequences of a reactor accident—in particular, one involving the release of fission products from the reactor—the entire reactor installation is enclosed in a containment structure. In some plants, this takes the form of a heavily constructed building housing the entire nuclear steam supply system; in other installations, the containment is split, with a portion surrounding the reactor (the primary containment), and the remainder coinciding with the reactor building (the secondary containment).

In addition to the several components illustrated in Fig. 4.5, elaborate safety systems, some of which are described in Chapter 11, must be included in a nuclear power plant for use in emergencies. Provisions must also be made for the loading of fuel and storage of radioactive spent fuel prior to its shipment from the plant. Various sensing devices must be located at several points inside and outside the reactor vessel to monitor the operation of the system. Finally, a considerable amount of structural material is necessary to give support and integrity to the plant. It should be clear that nuclear power plants are complex installations and, by their nature, must be designed with care.

4.5 POWER REACTORS AND NUCLEAR STEAM SUPPLY SYSTEMS

In the relatively short history of nuclear power, many types of reactors have been proposed for the production of steam. The present discussion is confined to those nuclear steam supply systems that are currently in use or that hold promise of being adopted in the future.

Section 4.6 shows that, because of the limited resources of ^{235}U, the nuclear power industry in time will undoubtedly have to shift to the breeder reactor. Present-day reactors that do not breed should be recognized as interim systems, pending the introduction of breeders during the next century.

Power reactor systems consist primarily of five types of reactors: (a) pressurized water reactors and boiling water, which produce steam directly in the core and are the mainstay of the nuclear power industry; (b) evolutionary pressurized water reactors, which have the basic elements of pressurized water reactors but with many significant improvements; (c) evolutionary boiling water reactors; (d) heavy-water moderated reactors; and (e) gas-cooled reactors.

Reactors

The pressurized water and boiling water reactors fall into the general class of *light-water* reactors since they use ordinary or light water for the coolant and moderator. The heavy water reactor, as the name implies, uses D_2O for the moderator, whereas the gas-cooled reactors use a solid such as carbon for the moderator and a gas such as helium for the coolant.

Light-Water Reactors   The most widely used reactor in the world today for producing electric power is the thermal reactor, which is moderated, reflected, and cooled by ordinary (light) water. As noted earlier, water has excellent moderating properties. In addition, its thermodynamic properties are well understood, and it is readily available at little cost. However, water has a high vapor pressure, which means that light-water reactors (LWRs) must be operated at high pressures. Water also absorbs thermal neutrons to such an extent that it is not possible to fuel a LWR with natural uranium—it simply would never become critical. Uranium in water reactors must always be enriched, at least to some extent.

There are basically two types of light-water reactors now in use: the pressurized-water reactor (PWR) and the boiling-water reactor (BWR). Both of these types of reactors are well established in the United States and abroad, and both produce power about as cheaply as comparable coal-fired plants.

The Pressurized-Water Reactor   The PWR was one of the first types of power reactors developed commercially in the United States. This reactor has also become standard on nuclear-powered ships and naval vessels throughout the world.

Figure 4.6 shows a cross-sectional view of a typical PWR. As indicated in the figure, water enters the pressure vessel at a temperature of about 290°C or 554°F, flows down around the outside of the core where it serves as a reflector, passes upward through the core where it is heated, and then exits from the vessel with a temperature of about 325°C or 617°F. The water in a PWR is maintained at a high pressure—approximately 15 MPa or 2250 psi. At this pressure, the water will not boil, at least not to any great extent.

Since the water does not boil in the reactor, the steam for the turbines must be produced external to the reactor. This is done in steam generators, which are heat exchangers with pressurized water on the hot side. A typical steam generator is shown in Fig. 4.7. High-pressure, heated coolant water from the reactor enters at the bottom and passes upward and then downward through several thousand tubes each in the shape of an inverted U. The outer surfaces of these tubes are in contact with lower pressure and cooler feed water returning from the turbine condenser. Heat transferred from the hot water inside the tubes causes the feedwater to boil and produce steam. The lower section of a steam generator where this boiling occurs

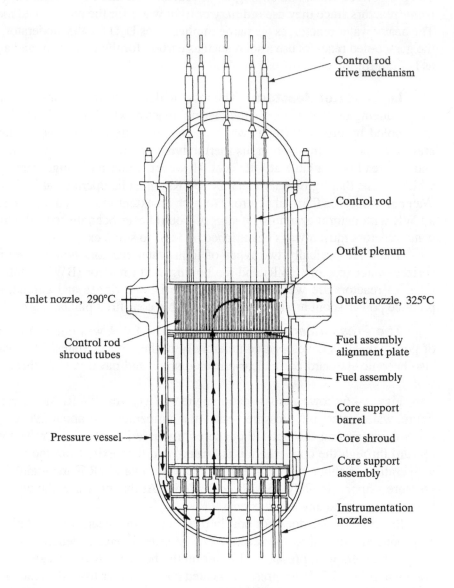

Figure 4.6   Cross-sectional view of a PWR. (Courtesy of Combustion Engineering, Inc.)

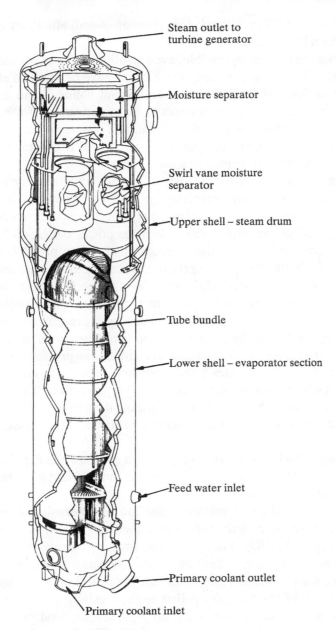

Steam outlet to turbine generator

Moisture separator

Swirl vane moisture separator

Upper shell – steam drum

Tube bundle

Lower shell – evaporator section

Feed water inlet

Primary coolant outlet

Primary coolant inlet

Figure 4.7 A PWR U-tube steam generator. (Courtesy of Westinghouse Electric Corporation.)

is called the *evaporator section*. The wet steam produced in the evaporator passes upward into a portion of the steam generator known as the *steam drum section*. Here the steam is dried in various moisture separators before exiting to the turbines. Steam generators are also manufactured with straight tubes rather than U tubes. Large PWR systems utilize as many as four steam generators, which produce steam at about 293°C or 560°F and 5 MPa or 750 psi. This gives an overall efficiency of between 32% and 33% for a PWR plant.

Because water is essentially incompressible, even small changes in coolant volume could lead to large changes in pressure, which could have deleterious effects on the system. For instance, if for any reason the coolant volume were decreased, the subsequent drop in pressure could result in the vaporization of some of the water in the reactor; this in turn could lead to the burnout of some of the fuel elements. To prevent this from happening, one coolant loop of a PWR is equipped with a pressure-maintaining surge tank known as a *pressurizer*.

As indicated in Fig. 4.8, a pressurizer consists of a tank, containing steam and water in its upper and lower sections, respectively, with a pressure-actuated spray nozzle at the top and pressure-actuated immersion-type heaters at the bottom. The device operates in the following way. Suppose that the power output of the turbine is reduced in response to a drop in the electrical load on the plant. This leads to a temporary increase in the average temperature of the reactor coolant and a corresponding increase in coolant volume. The expansion of the coolant raises the water level in the pressurizer, which raises the pressure of the steam and actuates the valves to the spray nozzle. Water from one of the cold legs of the reactor coolant system sprays into the top portion of the pressurizer and condenses some of the steam. This quenching action reduces the pressure and limits the pressure rise. When the electrical load is increased instead, the attendant decrease in coolant volume drops the water level and reduces the pressure in the pressurizer. This, in turn, causes some of the water to flash to steam, again limiting the pressure change. At the same time, the pressure drop actuates the heater to further limit the pressure reduction.

The major components of a PWR steam supply system are shown in Fig. 4.9. It should be noted that there are four coolant pumps, one for each coolant loop, but only one pressurizer for the entire system.

The fuel in PWR reactors is slightly enriched (from 2 to 5 w/o uranium dioxide, UO_2, which is a black ceramic material with a high melting point of approximately 2,800°C or 5,070°F). The UO_2 is in the form of small cylindrical pellets, about 1 cm in diameter and 2 cm long, which are usually concave on the ends. The pellets are loaded into sealed stainless steel or Zircaloy[3] tubes about 4 m long. At the operating temperature of the fuel, the pellets expand axially and fill the void spaces between the pellets. The result is a solidly packed fuel rod or fuel pin. In addition to providing support for the fuel pellets, the Zircaloy fuel tubes also prevent the escape to the passing coolant of fission products, especially fission product gases, that are released from the pellets during the operation of the reactor.

---

[3]Zircaloy is an alloy of zirconium, a metal with a small neutron absorption cross-section containing small amounts of tin, iron, chromium, and nickel. There are two forms of Zircaloy that have been used in LWRs—Zircaloy-2 and Zircaloy-4; the latter contains no nickel.

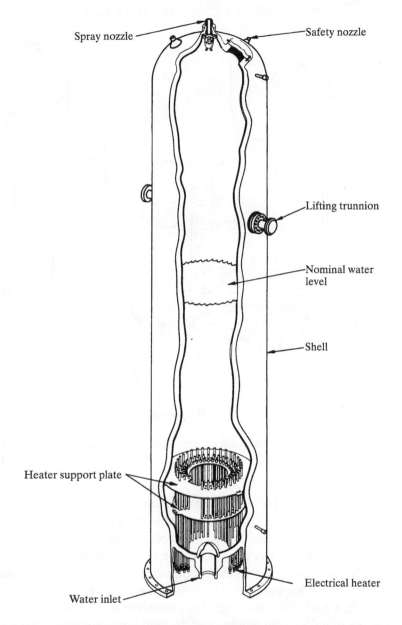

Figure 4.8   A PWR pressurizer. (Courtesy of Westinghouse Electric Corporation.)

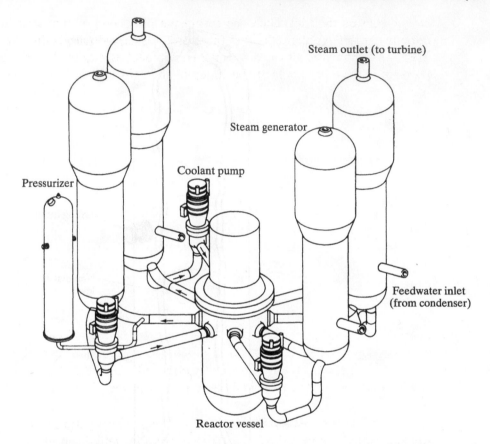

Steam outlet (to turbine)

Steam generator

Coolant pump

Pressurizer

Feedwater inlet
(from condenser)

Reactor vessel

Figure 4.9   Schematic arrangement of the major components of a PWR
steam supply system.

In this context, the fuel tubes are also known as fuel *cladding*. PWR fuel of the
type described here is nominally capable of delivering about 30 MWd per kg of
contained uranium before it must be replaced. Advanced fuels developed in recent
years are capable of extended operation because of higher enrichment and the use
of gadolinium as a control element during initial operation. Such fuel systems have
enrichments as high as 5 *w/o* and can achieve 45 MWd per kg.

On a few occasions, it has been observed that some fuel pellets, although expanding initially as the fuel reaches its operating temperature, later contract due to a
gradual increase in the density of the uranium dioxide and corresponding reduction
in its specific volume (cm^3/g). This *densification* of the fuel, as the phenomenon is
called, is evidently caused by the migration and combination of small voids within
the ceramic. The decrease in the pellet volume leaves void spaces within the fuel
tubes. Because of the high pressure of the moderator coolant, this produces large

stresses across the fuel tubes, increasing the likelihood of fuel-tube rupture. To circumvent this problem, the fuel tubes are pressurized with helium at about 3.4 MPa. As fission product gases accumulate in the tubes over the life of the core, the pressure in the tubes gradually builds up to about 14 MPa near the end of the core life.

The completed, pressurized fuel rods are next arranged in a square lattice structure called a *fuel assembly*, one of which is shown in Fig. 4.10; the fuel assemblies are then arranged in a near-cylindrical array to form the core. The fuel rods in the fuel assemblies are kept apart by various spacers. This is important since rods that come into contact may overheat and release fission products.

Control of the PWR is accomplished by the use of control rods, which normally enter the core from the top (see Fig. 4.6), and a chemical shim system. In this latter system, which is discussed more fully in Chapter 7, the value of the multiplication factor is changed by varying the concentration of a neutron absorber (usually boric acid) dissolved in the coolant water. Fuels capable of higher burnup have gadolinium included to offset the higher enrichment of the fuel during the early stages of core life.

Where the size of the reactor is an important consideration, as it is in a submarine, highly enriched fuel is used. This fuel, enriched to over 90 w/o in ^{235}U and often called fully enriched, makes it possible to reduce the overall dimensions of the core and pressure vessel. The highly enriched uranium is expensive, however, and it cannot be used economically in stationary power plants.

The Boiling-Water Reactor   For a long time, it was thought that if water were permitted to boil within a reactor, dangerous instabilities would result because of uneven formation and movement of the steam bubbles. Experiments carried out in the early 1950s (the famous BORAX experiments) showed that this was indeed the case if the boiling occurred at low pressure. However, when the pressure was raised, the boiling became stable and the reactor was controllable. Since these first demonstrations of the feasibility of boiling-water reactors, the BWR has reached a high state of development. The BWR and PWR are now competing on a neck-and-neck basis for the commercial nuclear power market in light-water reactors.

There are obvious advantages to a BWR. For one thing, the steam is formed in the reactor and goes directly to the turbines—steam generators in separate heat transfer loops are not necessary as they are with the PWR. For this reason, the BWR is said to operate in a *direct cycle*. Furthermore, it is recalled that, for a given amount of water, more heat can be absorbed as *latent heat*—that is, the heat necessary to vaporize a liquid—than as *sensible heat*, which, in the PWR, only changes the temperature of the fluid. Therefore, it follows that less water must be pumped through a BWR per unit time than through a PWR for the same power output. However, the water becomes radioactive in passing through the reactor core

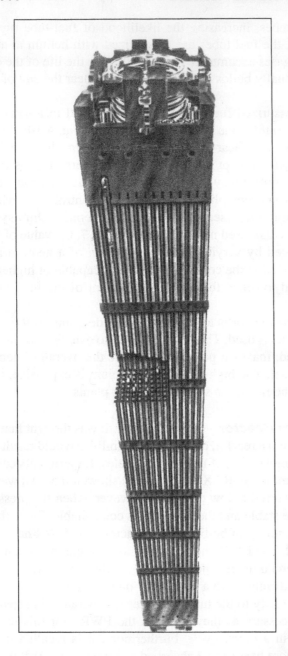

Figure 4.10   Fuel assembly for a PWR. (Courtesy of Babcock & Wilcox Company.)

(see Section 10.12). Since this water is utilized in the electricity-producing side of the plant, all of the components of the steam utilization system—the turbines, condenser, reheaters, pumps, piping, and so on—must be shielded in a BWR plant.

The pressure in a BWR is approximately 7 MPa or 900 psi —about one-half the pressure in a PWR. As a result, the wall of the pressure vessel for a BWR need not be as thick as it is for a PWR. However, it turns out that the power density (watts/cm^3) is smaller in a BWR than in a PWR, and so overall dimensions of a pressure vessel for a BWR must be larger than for a PWR of the same power. As far as the cost of the pressure vessel is concerned, these two effects more or less tend to offset one another.

The internal configuration of a BWR is shown in Fig. 4.11, where the coolant flow is indicated by arrows. Starting with the lower chamber or plenum, the water

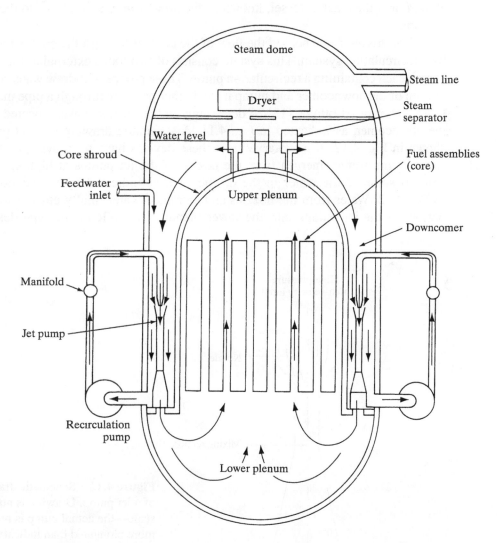

Figure 4.11   Cross-sectional view of a boiling-water reactor. The motion of the water is shown by arrows. (Courtesy of General Electric Company.)

moves upward through the core; as it does, it receives both sensible and latent heat. By the time it reaches the top of the core and enters the upper plenum, a portion of the coolant has been vaporized. This mixture of steam and liquid water next passes through steam separators, which remove most of the water. The steam then goes through a dryer assembly, which removes the remaining water; it then exits from the reactor via a steam line to the turbine. The residual water from the separators and the dryer mixes with feedwater returning from the condenser and passes downward through an annular region external to the core, between the *core shroud* and the reactor vessel, known as the *downcomer,* and returns to the lower plenum.

The driving force behind the flow of the coolant through the core is provided by a recirculation system. This system consists of two loops external to the reactor vessel, each containing a recirculation pump. These pumps withdraw water near the bottom of the downcomer and pump it at a higher pressure through a pipe manifold to a number of jet pumps (18–24 depending on the reactor power) located within the downcomer, as indicated in Fig. 4.11. A schematic drawing of a jet pump is shown in Fig. 4.12. It is observed that these devices have no moving parts. The recirculating water emerges from the nozzle of the jet pumps at high speed and entrains some of the water in the downcomer, as indicated by the suction flow in Fig. 4.12. Water from the nozzles and entrained water finally emerge from the bottom of the jet pumps into the lower plenum. A BWR of the type described

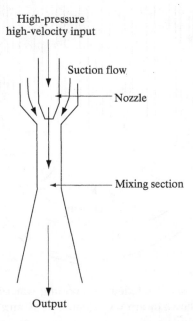

Figure 4.12   Schematic drawing of a jet pump. Drawing is not to scale—the actual pump is much more elongated than indicated. (Courtesy of General Electric Company.)

produces saturated steam at about 290°C or 554°F and 7 MPa or 900 psi. The overall efficiency of a BWR plant is on the order of 33% to 34%.

The fuel for a BWR is essentially the same as for a PWR, that is, slightly enriched UO_2 pellets in sealed tubes—and the core configurations of the two types of reactors are more or less identical. However, BWR control rods are always placed at the bottom of the reactor rather than at the top, as in the case of the PWR. The reason for this is that much of the upper portion of the core of a BWR is normally occupied by steam voids, and movement of the rods in this region does not have as large an effect on the value of k as rod motions in the lower, water-filled part of the core. The rods are thus placed near the part of the reactor where they will do the most good.

The Evolutionary and Advanced Light-Water Reactors

In recent years, economic considerations and concerns over the need for higher levels of safety have prompted an effort to develop new advanced reactor concepts. Both boiling and pressurized water reactor designs of this type were developed. These designs feature the use of simplified primary systems and, in the more advanced designs, passive emergency core-cooling systems. The Advanced Boiling Water Reactor (ABWR) by GE and System 80+ PWR by ABB/Combustion Engineering represent considerable system simplification, whereas the Advanced Passive 600 (AP600) MWe PWR by Westinghouse and the Simplified Boiling Water Reactor (SBWR) by General Electric represent designs with advanced passive cooling features as well.

Construction costs of nuclear power plants as well as other types of central power stations skyrocketted during the late 1970s and early 1980s. The increase in costs was due in part to the general inflationary trends experienced during this time. A major cause of the cost increase for nuclear power plants was the need to backfit modifications needed as a result of the accident at the Three Mile Island Nuclear Generating Station, TMI, discussed in Section 11.7. In an effort to make the use of nuclear energy economically more competitive, the nuclear industry explored ways to both simplify and increase the safety of the nuclear systems. The conclusions from studies made by both the industry and the U.S. Department of Energy (DOE) were the need to decrease the number of components and piping and provide more sources of cooling water to the systems for use during an accident.

Initial designs such as the ABWR and Systems 80+ are based on simplified system designs to reduce costs along with some safety system enhancements. The AP600 and SBWR include both simplifications and the addition of *passive* cooling systems for use during a reactor accident. Passive systems use natural processes driven by gravity and temperature differences. Because these systems do not require electrical power they are thought to be nearly fail-safe and much more reliable. Each of these are discussed in the following sections and comparisons made to conventional reactor designs.

System 80+.   The evolutionary ABB CE System 80+ is an advanced design evolutionary PWR. It is evolutionary in that incremental changes were made in key system components that do not represent radical departures from current designs. The significant features of the System 80+ include the use of a large spherical double-wall concrete and steel containment, a larger water inventory in the primary system, a simplified primary system, part-strength control rods that reduce the need to change levels of boron in the coolant, and a safety depressurization system for reducing system pressure during a small break loss of coolant accident (SBLOCA).

The primary system features two large steam generators and four coolant pumps (Fig. 4.13). The pressurizer size is increased by 33% and the steam generator secondary water volume by 25%. Both of these increases enhance the plant's ability to experience transients without large pressure changes. The increased water volume also provides additional time for operator action in the event of an accident.

Use of a large 61 m (200 ft) diameter spherical steel containment building provides additional volume for steam expansion during a LOCA as well as providing a heat sink. The in-reactor containment refueling water storage tank provides a sump for collection of coolant during an accident.

Advanced Boiling-Water Reactor.   General Electric's Advanced Boiling-Water (ABWR) is another example of an evolutionary designed LWR. Building

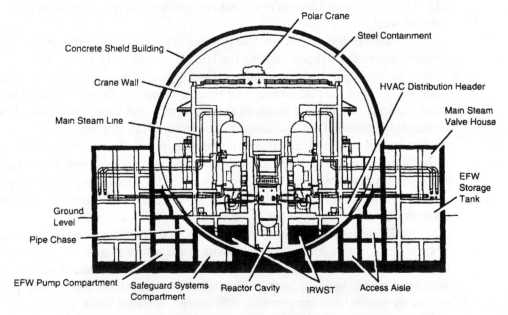

Figure 4.13   Schematic drawing of the System 80+. (Courtesy of ABB Combustion Engineering.)

on current features used in BWRs, this design represents a simplification of the existing technology but without a major departure from current design. Figure 4.14 is a cutaway of the ABWR. The most evident difference is the lack of recirculation pumps. Using centrifugal pumps internal to the reactor vessel, GE's ABWR minimizes the risk of a LOCA by eliminating the complex of recirculation piping and pumps found in conventional BWRs.

The ABWR also uses fine-motion control rods to enable better fuel management and control during operation. Conventional steam separation equipment is used to provide dry steam to the turbines as in current BWRs.

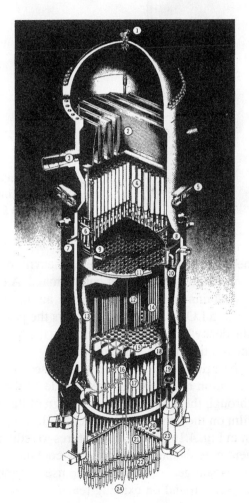

Figure 4.14   Cutaway drawing of the ABWR. (Courtesy of General Electric.)

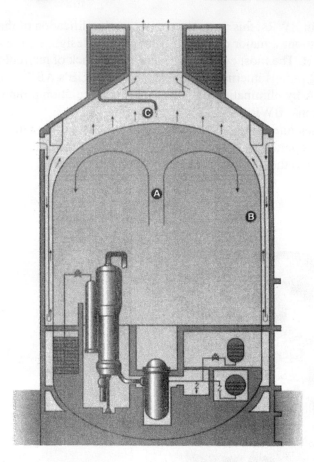

Figure 4.15   Cutaway drawing of the AP600 containment. (Courtesy of Westinghouse Electric.)

AP600 PWR.   The Westinghouse AP 600 relies heavily on the use of passive safety systems to decrease the probability of core damage. A unique feature of the design is the way the containment is used to provide an ultimate heat sink for the decay heat in the event of a LOCA. Figure 4.15 shows the passive containment cooling used to transfer the decay heat from the reactor via the break in the primary system to the containment and then to the atmosphere.

As the steam exits the primary system, it is condensed on the walls of the containment. Natural circulation of air against the outside of the steel-lined containment is encouraged through the unique chimney design of the containment and the presence of a water film on the outside of the steel shell.

The primary system in Fig. 4.16 has several features to reduce the probability of a LOCA and subsequent core damage. The primary coolant pumps are canned rotor pumps located in the steam generator plena. The use of canned rotor pumps eliminates the mechanical seals found on existing reactor circulation pumps. The

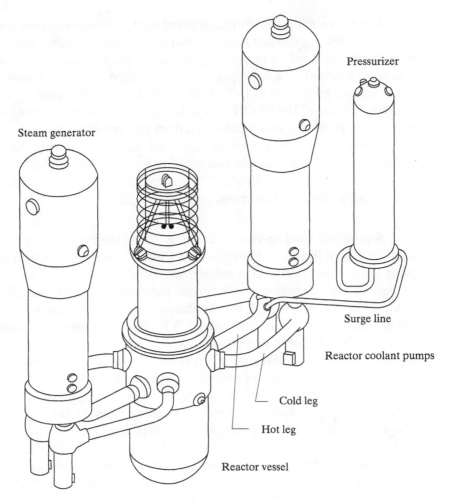

Figure 4.16 AP600 primary system. (Courtesy of Westinghouse Electric.)

pressurizer is also larger by about 30% to reduce the likelihood of opening a safety relief valve during a transient. Core power density is lowered to reduce the probability of core damage.

Passive core cooling is provided during a LOCA by two different means. Two large high-pressure core makeup tank provide water during the early stages of an accident. The design uses an automatic depressurization system to reduce pressure so coolant can flow from an external pool to the reactor under gravity. The use of these two passive systems provides for a high-reliability emergency core-cooling capability.

In addition to the passive containment cooling systems and the use of gravity-fed core-cooling systems, the design also has a passive residual heat removal system that is designed to remove decay heat when the steam generators are not available. Located in a large pool of water called the in-containment refueling water storage tank (IRWST), the heat exchangers provide cooling under normal operating pressure. The IRWST acts as a large source of water during a LOCA and also as a reservoir for the condensation from the containment.

Effort has also been made to reduce the number of components and amount of piping as well. For example, compared to a similar 2 loop plant, the AP600 has 75% less piping in the NSSS and 31% less in the steam systems. There are similar reductions in valves and other components as well.

Simplified Boiling-Water Reactor.   The most significant difference between the SBWR developed by General Electric and a conventional design BWR is the use of natural circulation rather than forced circulation of the reactor coolant. The use of natural circulation reduces the power needed to operate the plant. It also reduces the risk of a LOCA by eliminating the complex piping found in the recirculation system of a conventional BWR. These features both reduce cost and increase safety. Figure 4.17 shows the SBWR in cutaway. An obvious difference in this design is the lack of pumps in either the downcomer or external to the reactor.

In addition to the use of natural circulation, the core and vessel are designed to provide additional coolant inventory. In an accident, the additional coolant is expected to prevent core uncovery, significantly reducing the probability of core damage. Finally, there are several completely passive systems that are designed to replace the conventionally engineered safeguards systems. These systems rely on gravity head to drive water into the core under LOCA conditions and to collect coolant by condensing the steam escaping to the containment from the break in the system. The gravity-driven cooling system (GDCS) is designed to collect the coolant and return it to the vessel. It uses coolant condensed by the passive containment cooling system (PCCS) heat exchangers or heat exchangers located in the PCCS pool above the reactor. Vaporized coolant from the break in the reactor system is condensed in the PCCS and flows to the GDCS tank. There the coolant is injected into the core by gravity.

A key feature of the passive system is the ability to rapidly depressurize the reactor once a serious leak is detected. The process involves the use of the automatic depressurization system (ADS), which is used on conventional reactors today. The ADS opens on decreasing system pressure venting steam from the reactor into the suppression pool. There the steam is condensed. Once the system is depressurized, flow from the GDCS can occur.

Figure 4.18 shows a simulation of the SBWR during a LOCA. Notice that the water level never reaches the top of active fuel (TAF).

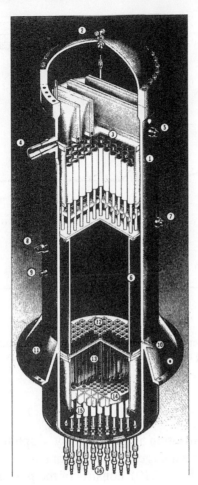

Figure 4.17 Cutaway view of SBWR. (Courtesy of General Electric.)

Russian Reactor Designs   While the United States and Western Europe pursued LWRs of somewhat similar designs, the former Soviet Union and former Eastern Block countries developed their own unique designs. These designs include the RBMK and VVER reactors. Both have found wide application in the former Soviet Union, former Eastern Block countries, and elsewhere. They have several unique features not found in Western reactor designs.

RBMK Reactor   The RBMK or channelized large power reactor is a boiling-water reactor. Unlike Western designed BWRs, it does not have a pressure vessel. Instead, the fuel assemblies are located in separate channelized pressure tubes similar to Canadian CANDU reactors. Each fuel assembly is located in its own pressure tube, which allows the reactor to be refueled online. Unlike Western

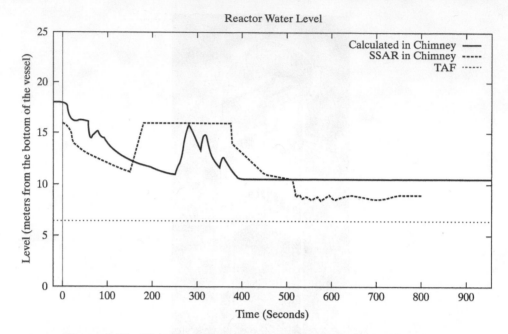

Figure 4.18 Water level prediction in the SBWR from a LOCA simulation using the TRAC computer code. (Courtesy of Penn State Nuclear Engineering Program.)

BWRs, however, the moderation is provided by graphite instead of light water coolant, which makes it much larger than Western reactors.

The primary system (Fig. 4.19), consists of the pressure tubes (usually 1,000 or more), steam and water pipes, steam drums, main circulation pumps, graphite stack, and fuel channels. The main coolant circulation system (MCC) typically consists of two loops. Each loop has two steam water separator drums that provide separation of the steam water mixture. Feedwater enters the steam drums via the feed header and then travels to the suction header via the downcomers. There, three of the four circulation pumps force the coolant into the distribution header and then into the fuel channels where it boils, producing low-quality steam.

The fuel channels are independent of one another and may be isolated from the system, allowing the fuel elements to be removed and replaced. The fuel is low-enrichment (2 w/o) UO_2. Each fuel rod is made up of 1.15 cm by 1.3 cm fuel pellets inside a zirconiun-niobium tube approximately 3.64 m in length. There are 18 fuel rods in each fuel element and they are arranged in a circular pattern approximately 7.9 cm in diameter. A fuel assembly (Fig. 4.20) consists of two fuel elements stacked end to end forming an assembly 7 m long.

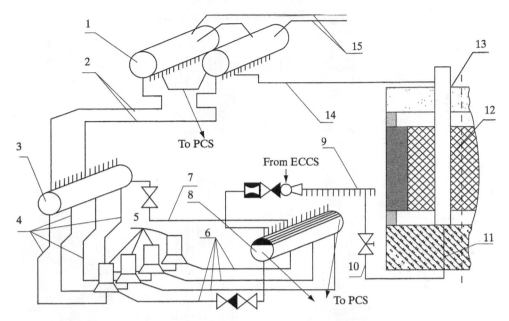

Figure 4.19   One loop of the primary system of a RBMK. 1—separator drum, 2—downcomers, 3—suction header, 4—suction piping of the MCP, 5—MCP tanks, 6—pressure piping of the MCP, 7—bypass headers, 8—pressure header, 9—group distribution header with flow limiter, check valve, and mixer, 10—water piping, 11—channel to core, 12—fuel channel, 13—channel above core, 14—steam-water pipes, 15—steam pipelines. (*Ignalina RBMK-1500, A Source Book*, Lithuanian Energy Institute.)

The fuel assemblies are inserted into the fuel channels, which are housed inside the large graphite stack. The graphite stack consists of columns of graphite blocks of various sizes and shapes. There are other channels in the graphite for control rods, cooling flow, and instrumentation.

The graphite stack is placed inside a hermetically sealed cavity filled with helium and nitrogen. The mixture improves heat transfer from the graphite to the fuel channel and prevents oxidation of the graphite. The entire assembly is then placed within the biological shield. Typically, the reactor is operated inside of a building meeting standard industrial design specifications and not those typical of a Western designed containment building. Typical operating conditions for a 1,500 MWe reactor are given in Table 4.1.

Because of the unique capability to refuel online, the RBMK have relatively high availability. They can also be used to produce plutonium, and hence were limited to the former Soviet Union. As of this writing, there are 16 RBMK reactors

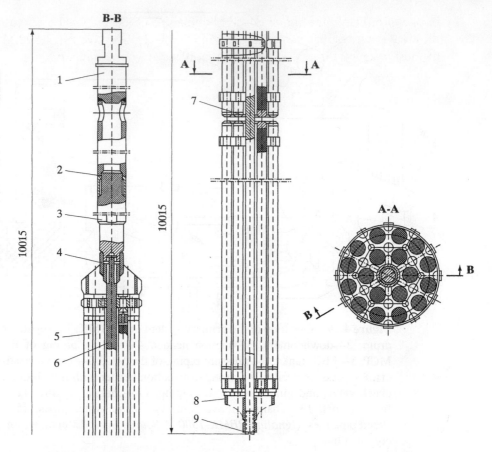

Figure 4.20　Fuel assembly: 1—suspension bracket, 2—top plug, 3—adapter, 4—connecting rod, 5—fuel element, 6—carrier rod, 7—end sleeve, 8—endcap, 9—retaining nut. (*Ignalina RBMK-1500, A Source Book*, Lithuanian Energy Institute.)

TABLE 4.1　RBMK-1500 PARAMETERS

Power	4,800 MWt
Core height	7 m
Core diameter	11.8 m
Number of channels	
Fuel	1,661
Control and shutdown	235
Fuel enrichment	2%
Coolant temperature	
Inlet	260–266 °C
Oulet	284°C
Steam pressure	7.0 MPa

in operation. Only the Ignalina and Ukranian reactors are outside of Russia. Two different power ratings were designed and built—1,000 MWe and 1,500 MWe—by the design agent, the USSR's Ministry of Nuclear Power Industry.

VVER Reactors   Several versions of the VVER series reactor are in use in the former Soviet Union, former Eastern Block, and elsewhere. The basic design of this series of reactors resembles a Western designed PWR. However, these reactors have a number of features that differ significantly from those in use in the United States and other Western countries.

Put into operation in the 1970s, the VVER 440 series reactors are all rated at 440 MWe. All of these design reactors have six loops, isolation valves on each loop and a uniquely designed horizontal tubed steam generator. The NSSS drives two 220 MWe steam turbines. Typical operating conditions for the VVER 440 series is given in Table 4.2.

Design of the reactor core is significantly different from most Western designs since it uses hexagonal fuel assemblies. Each fuel assembly is surrounded by a shroud similar to that found in the U.S. BWRs. The core consists of 312 fuel assemblies and 37 movable control assemblies. The control assemblies have fuel followers that enter the core as the upper portion containing the boron steel control material exits the core. Each assembly consists of 126 fuel rods, each arranged in a triangular pitch. The fuel rod is 9.1 mm in diameter and contains 7.5-mm diameter UO_2 fuel pellets enriched to 2.4–3.6 w/o. The fuel rods are 3.53 m in length. Additional control is accomplished through the use of boric acid as in Western PWRs.

Two series of 440s are in operation—the older VVER 230 and the newer VVER 213. These two differ significantly in the design of the containment and the engineered safety systems. The VVER 230s rely on a unique compartmentalization design to contain fission products in the event of fuel failure. Unlike the later 213s, the 230 series has only limited emergency core-cooling capability. The more recent model VVER 213s have a more advanced emergency core-cooling system and a

TABLE 4.2   VVER 440 PARAMETERS

Power	MWt
Core height	m
Core diameter	m
Fuel enrichment	%
Coolant temperature	
Inlet	°C
Oulet	°C
Steam pressure	MPa
Coolant system pressure	MPa

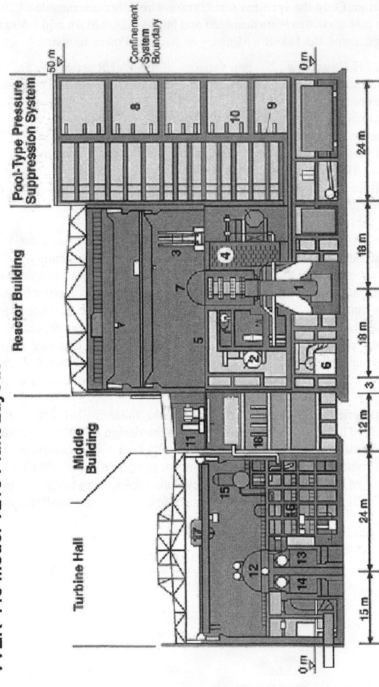

VVER-440 Model V213 Plant Layout

Legend: 1.Reactor pressure vessel, 2.Steam generator, 3.Refueling machine, 4.Spent fuel pit, 5.Confinement system, 6.Make-up feedwater system, 7.Protective cover, 8.Confinement system, 9.Sparging system, 10.Check vales, 11.Intake air unit, 12.Turbine, 13.Condenser, 14.Turbine block, 15. Feedwater tank with degasifier, 16. Preheater, 17.Turbine hall crane, 18.Electrical instrumentation and control compartments.

Figure 4.21 Layout of a VVER 213 showing the containment. (From the Web, www.insp.anl.gov.)

pressure suppression system for minimizing containment pressure during LOCAs. Figure 4.21 shows the layout of the reactor cooling system and containment for a VVER 213.

The larger VVER 1000-series reactor is a four-loop PWR closely resembling a standard Western PWR. Rated at 1,000 MWe, these reactors came into service in the mid-1980s. They differ from the 440 design series in that they use a containment similar to Western PWRs and have four loops with a steam generator in each loop. Like the VVER 440 reactors, the VVER 1000s use horizontal steam generators and two turbines. Figure 4.22 shows the layout of a typical VVER 1000.

Like the VVER 440, the VVER 1000 uses hexagonal fuel elements. The 1000-series fuel assemblies are, however, much larger than those of the VVER 440 design and do not have fuel assembly shrouds. The fuel also uses a zoning of the enrichment with fuel pins of lower enrichment located at the periphery of

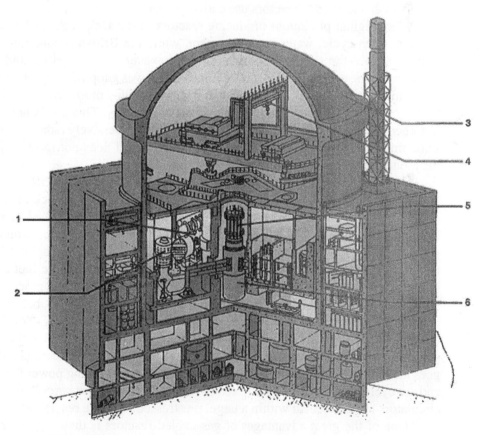

Figure 4.22   Layout of a VVER 1000 showing the containment. (From the Web, www.insp.anl.gov.)

the fuel assemblies. The control rod designs are nearly identical to those of Western reactors and consist of identical control pins connected at the top into a spider web unit. The unit can be inserted into corresponding water channels in the fuel assembly.

Gas Cooled Thermal Reactor   Natural uranium, graphite-moderated reactors were developed in the United States during World War II for the conversion of ^{238}U to ^{239}Pu for military purposes. Following the war, this type of reactor formed the basis for the nuclear weapons programs of the United States and several other nations. It is not surprising, therefore, that natural uranium-fueled reactors became the starting point for the nuclear power industry, especially in nations such as Great Britain and France, which at the time lacked the facilities for producing the enriched uranium necessary to fuel reactors of the light-water type. Both of these countries have now constructed diffusion plants, however, and recent versions of both British and French reactors use enriched fuel.

The original plutonium-producing reactors in the United States had a once-through, open-cycle, water coolant system, while the British-production reactors utilized a once-through air-cooling system. However, for the British and French power reactors, a closed-cycle gas-cooling system was adopted early on. This provides containment and control over radioactive nuclides produced in or absorbed by the coolant. In these reactors, the coolant gas is CO_2. This gas is not a strong absorber of thermal neutrons and it does not become excessively radioactive. At the same time, CO_2 is chemically stable below 540°C and does not react with either the moderator or fuel.

A drawing of the Hinkley Point B advanced gas reactor (AGR) in western England is shown in Fig. 4.23. The reactor core is a 16-sided stack of interconnected graphite bricks, which contains vertical fuel channels arranged in a square lattice. The fuel consists of slightly enriched (1.4%–2.6% in different zones) UO_2 pellets in stainless steel tubes, which in turn are located within a graphite sleeve to form a single fuel element. These fuel elements are arranged in the fuel channels in such a way that the CO_2 can pass smoothly between the fuel and moderator.

The actual flow of gas in the reactor is somewhat complicated, but the details are unimportant for the present discussion. In effect, however, the CO_2 flows up through the channels in the core into a plenum and then through one or the other of several steam generators surrounding the core. From the steam generators, the CO_2 goes through gas circulators (blowers) that provide the pumping power for the gas, and the cycle is complete. As indicated in Fig. 4.23, the core, steam generators, and circulators are all located within a large, prestressed concrete reactor vessel.

One of the great advantages of gas-cooled reactors is their high thermal efficiency. The plant described previously produces superheated steam at approximately 540°C and 16 MPa; it operates at an overall efficiency of about 40%, as

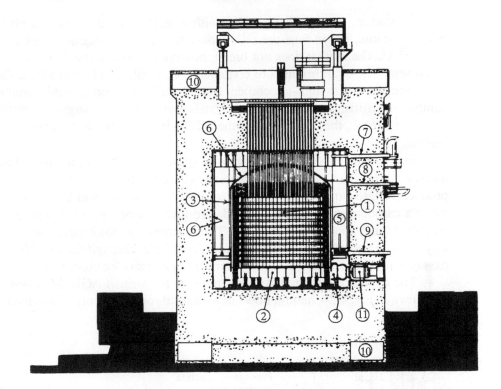

Figure 4.23 The Hinkley Point B advanced gas reactor: (1) reactor core, (2) supporting grid, (3) gas baffle, (4) circulator outlet gas duct, (5) steam generator, (6) thermal insulation, (7,8) steam duct penetrations, (9) steam generator feedwater inlet penetrations, (10) gallery for access to cable stressing concrete reactor vessel, (11) gas circulators. (Courtesy of the Nuclear Power Group, Limited.)

high as the most efficient fossil fuel plant available today. Nevertheless, the British AGR evidently does not produce electricity as cheaply as a PWR, and in 1979, the British government announced that in the future its nuclear program would be based on the construction of PWRs.

In the United States, gas-cooled reactor technology has been centered at the General Atomic Company, where the high-temperature gas-cooled reactor (HTGR) was developed. This is a graphite-moderated, helium-cooled, thermal reactor. As a coolant, helium is excellent. It is far more inert than CO_2, does not absorb neutrons, and therefore does not become radioactive. However, the helium picks up small amounts of radioactive gases that escape from the fuel along with radioactive particles from the walls of the cooling channels, so as a practical matter the coolant in the HTGR is somewhat radioactive.

At startup, the HTGR is fueled with a mixture of thorium and highly enriched uranium, but in time the ^{233}U, converted from the thorium, replaces some of the ^{235}U. The reactor does not breed, however, so that some ^{235}U must always be present. The equilibrium core contains ^{235}U, fertile ^{232}Th, and all the recycled ^{233}U. Because it uses high-enrichment fuel, the HTGR is considerably smaller than comparable British gas-cooled reactors fueled with natural or slightly enriched uranium. As a result, the capital cost of the HTGR is much less than that of the British reactor.

Fuel for the HTGR is in the form of small particles of uranium and thorium dicarbides (U, Th) C_2, which are specially coated to prevent the release of fission products. These particles are cast into rods about 5 cm long and 1.3 cm in diameter using a carbonaceous binder. The rods are then inserted into holes in hexagonal graphite blocks about 80 cm high and 35 cm across, and the blocks are arranged in a cylindrical array to form the core. Additional holes through the graphite provide passages for the coolant gas while others hold channels for control rods.

The operation of the HTGR is similar to the British AGR described earlier. As shown in Fig. 4.24, helium flows downward through the core, flows through the

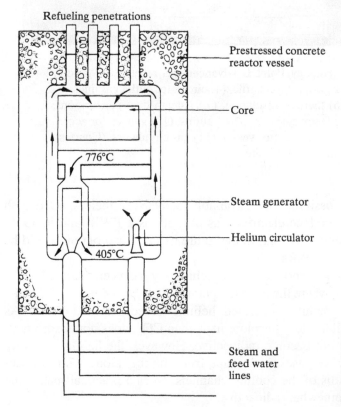

Figure 4.24   Schematic drawing of the Fort St. Vrain 330-MWe HTGR showing the flow of the helium coolant. (Courtesy of General Atomic Company.)

steam generators, and is pumped back to the core through the circulation blowers. As in the British reactors, all of these components are located within a prestressed concrete reactor vessel. Steam is obtained from the HTGR at approximately 540°C and 16 MPa, which gives an overall plant efficiency of about 40%. The most recent U.S. reactor of this type was the Ft. St. Vrain reactor. This unit was a 330 MWe HTGR and operated until its shutdown in 1989.

A unique feature of the HTGR is the very high temperature of the circulating helium—between 815°C and 870°C . Such helium can be used directly in a gas turbine to drive an electrical generator, thus eliminating the need for an intermediate steam cycle. There are many advantages to such a system. For one thing, gas turbines and their associated cycle components are considerably more compact than comparable steam cycle equipment. Furthermore, the temperature of the reject (second-law) heat is so high that this energy can be used in a number of practical applications, leading to an overall efficiency of the HTGR system as high as 50%. Besides producing electricity, the HTGR can provide high-temperature heat required in many chemical processes, such as the gasification of coal and desalination of sea water, among others.

The Heavy-Water Reactor   The heavy-water moderated and cooled reactor (HWR) has been under development in several countries, especially Canada. The interest in heavy-water reactors in Canada dates back to World War II, when the country was deeply involved in HWR research for military purposes. After the war, the Canadians immediately recognized the unique advantages of heavy-water reactors. Since an HWR can be fueled with natural uranium, the large resources of uranium known to exist in Canada could be used directly, thus making it unnecessary for that country to construct costly uranium enrichment plants or to depend on the United States or any other nation for enriched fuel. The Canadians also recognized that the gas-cooled, graphite-moderated reactors of the type being built in Great Britain had to be quite large if fueled with natural uranium; in the Canadian economy, constructing such a plant would have proved too expensive. The Canadian heavy-water reactor, known as CANDU, the acronym for *CAN*ada *D*euterium *U*ranium, has been successful in Canada and has been exported abroad.

A heavy-water reactor can operate on natural uranium because the absorption cross-section of deuterium (D $= ^2$H) for thermal neutrons is very small—much smaller, for example, than the cross-section of ordinary hydrogen (H $= ^1$H). However, since the deuterium in D_2O is twice as heavy as the hydrogen in H_2O, D_2O is not as effective in moderating neutrons as H_2O. Neutrons, on the average, lose less energy per collision in D_2O than they do in H_2O, and they require more collisions and travel greater distances before reaching thermal energies than in H_2O. The core of an HWR is therefore considerably larger than that of an LWR, and if an HWR is built like an LWR, with the fuel immersed in a D_2O-filled pressure vessel, the vessel must be very large.

To avoid the use of a large pressure vessel, the CANDU utilizes the *pressure tube* concept. In this design, the reactor consists of a large horizontal cylindrical tank, called the *calandria,* which contains the D_2O moderator under essentially no pressure. This tank is penetrated by several hundred horizontal tubes or channels, which contain the fuel. The D_2O coolant flows through these tubes at a pressure of about 10 MPa and does not boil. Thus by pressurizing the coolant rather than the whole reactor, a large pressure vessel is not required.

A simplified flow diagram for the coolant in a CANDU reactor is shown in Fig. 4.25. As indicated, a complete coolant circuit involves two fuel tubes and two circulation loops. The D_2O enters the reactor at a temperature of about 266°C and exits at 310°C. It passes from the reactor to a *header*—that is, a junction chamber for the coolant tubes—and then to an inverted U-tube steam generator where steam is produced and carried to the turbines. The coolant then returns to the reactor, passing in the opposite direction through an adjacent fuel tube, where it is heated again before flowing to a second steam generator. The pressurizer shown in the figure performs the same function in the CANDU system as it does in the PWR. Although most of the heat from the fuel is carried away by the D_2O coolant, some energy is deposited in the D_2O moderator. This energy is removed by the moderator coolant loop shown at the bottom of the figure.

Unfortunately, the pressurized D_2O coolant in a CANDU reactor cannot be raised to a temperature high enough to produce steam at a pressure and temperature comparable to that produced in light-water reactors. Therefore, the thermodynamic efficiency of CANDU plants is only between 28% and 30%, and CANDU power stations must use larger turbines with specially designed blades to prevent erosion by wet steam. To circumvent these problems, the Canadians have experimented with other coolants (but the same moderator)—in particular, boiling light water or an organic fluid. The organic coolant is capable of providing an especially high-outlet temperature from the reactor. However, both light water and organic fluids increase the total amount of neutron absorption in the reactor. Thus, neither of these coolants provide sufficient advantage over the D_2O coolant to be adopted in CANDUs, as long as these are fueled with natural uranium.

CANDU fuel is arranged in what Canadians call *bundles*, one of which is shown in Fig. 4.26. It is 49.5 cm long and 10.2 cm in diameter, and consists of 37 hollow Zircaloy tubes filled with natural uranium UO_2 pellets. The fuel tubes are welded to end plates and kept apart by spacers. This fuel is capable of delivering a burnup of about 7.5 MWd/kg.

A unique feature of CANDU reactors is that they are refueled *online*—that is, while the reactor is in operation. As shown in Fig. 4.27, this is accomplished by two fueling machines operated together at opposite ends of a single fuel tube. A fresh fuel bundle is introduced into one of the machines, which then pushes the bundle into the end of the fuel tube. This simultaneously pushes a bundle out at the far end

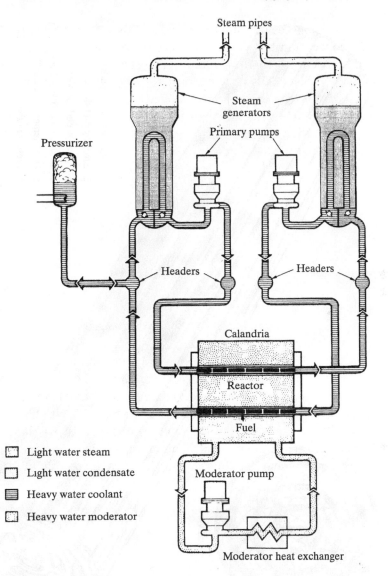

Steam pipes

Steam generators

Primary pumps

Pressurizer

Headers

Headers

Calandria

Reactor

Fuel

☐ Light water steam

☐ Light water condensate

▤ Heavy water coolant

▥ Heavy water moderator

Moderator pump

Moderator heat exchanger

Figure 4.25 Schematic diagram of a CANDU nuclear steam supply system.

of the tube, where it is received by the second refueling machine. The latter bundle is then transferred to an onsite underwater spent fuel repository. Online refueling contributes to very high availability of CANDU plants (about 90%), since, unlike LWR plants, they need not be shut down to refuel.

The overall size and hence the D_2O inventory of a CANDU reactor can be substantially reduced by using slightly enriched rather than natural uranium. This

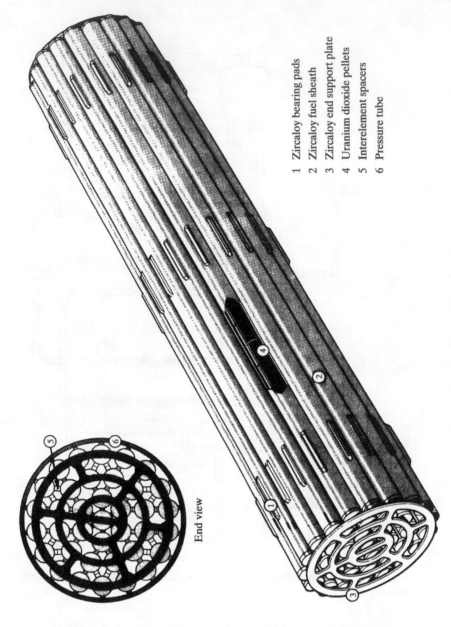

End view

1 Zircaloy bearing pads
2 Zircaloy fuel sheath
3 Zircaloy end support plate
4 Uranium dioxide pellets
5 Interelement spacers
6 Pressure tube

Figure 4.26 An element fuel bundle.

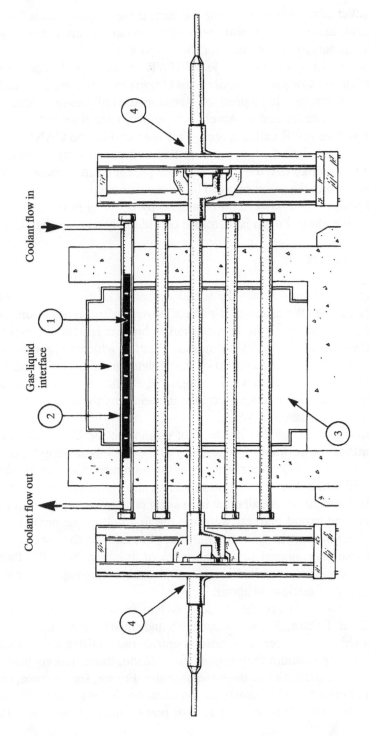

Figure 4.27 Schematic drawing of CANDU reactor showing: (1) fuel bundles, (2) pressure tubes, (3) heavy-water moderator, (4) refueling machines.

has the further advantage of reducing the annual fuel requirements for the reactor and thus tends to conserve uranium resources. An enrichment in the neighborhood of 1.2 w/o appears to be optimum for this purpose.

CANDU is not the only type of HWR in the world. The German firm Kraftwerk Union (now part of the Siemens Group) has designed a natural uranium-fueled HWR that uses a large pressure vessel instead of pressure tubes. Two reactors of this type are located in Atucha, Argentina, near Buenos Aires. Japan has also developed an HWR called Fugen. This reactor, like the CANDU, consists of a tank of D_2O at relatively low pressure that is penetrated by a large number of pressure tubes. Unlike CANDU, however, the pressure tubes contain ordinary water that boils within the reactor core. The reactor is fueled with a mixture of oxides of natural uranium and plutonium (obtained by conversion in LWRs) instead of natural uranium alone. Fugen is therefore considerably smaller than CANDU.

Breeder Reactors

It is shown in Section 4.6 that the world's reserves of ^{235}U are not adequate to indefinitely support the needs of a nuclear power industry based only on burner or converter reactors. With the introduction of breeder reactors, however, the fuel base switches from ^{235}U to ^{238}U or thorium, both of which are considerably more plentiful than ^{235}U. Furthermore, all of the depleted uranium—that is, the residual uranium, mostly ^{238}U, remaining after the isotope enrichment process—can be utilized as breeder fuel. Breeders are capable of satisfying the electrical energy needs of the world for thousands of years.

There is a considerable difference of opinion in the United States and elsewhere regarding the desirability or urgency of the need to develop breeders, owing to a divergence of views on the adequacy of uranium resources and the potential for nuclear proliferation. Some authorities feel that sufficient amounts of uranium will be discovered and produced, albeit at increasing cost, to satisfy the requirements of the domestic and export markets well into the first half of the next century. Other experts are less optimistic about uranium supplies and see the development of the breeder as insurance against future shortages of uranium. Yet still others express deep concern over the possible diversion of plutonium to weapons use resulting in the proliferation of nuclear weapons.

Abroad, the need for the breeder is less in doubt. The major industrial nations—Great Britain, France, Germany, and Japan—have essentially no indigenous uranium resources, and these countries must utilize to the fullest extent possible all of the uranium they import. In addition, these nations possess or can lay claim to large quantities of depleted uranium. France, for instance, has enough depleted uranium on hand to satisfy its electrical needs for the next 100 years if the uranium is used for fuel in breeder reactor power plants. France and Japan have active breeder programs. The former Soviet Union also made a major commitment to development of the breeder.

Four types of breeder reactors have been developed to date: the liquid-metal cooled fast breeder reactor (LMFBR), the gas-cooled fast breeder reactor (GCFR), the molten salt breeder reactor (MSBR), and the light-water breeder reactor (LWBR). Each of these is considered in turn.

The LMFBR   The fundamental principles underlying the fast breeder reactor concept were discovered before the end of World War II, and the potential impact of breeder reactors on future energy supplies was immediately recognized. The first experimental breeder reactor was a small plutonium-fueled, mercury-cooled device, operating at a power level of 25 kW, that first went critical in 1946 in Los Alamos, New Mexico. A 1.3 MW breeder, cooled with a mixture of sodium and potassium, was placed in operation in 1951 at the Argonne National Laboratory in Idaho. This reactor, the Experimental Breeder Reactor-I (EBR-I), produced steam in a secondary loop that drove a turbine generator. The system produced 200 kW of electricity, the world's first nuclear-generated electricity—and it came from an LMFBR! Since these early experiments, dozens of LMFBRs have been constructed around the world. As of this writing, the LMFBR is the only breeder that has reached a stage of significant commercialization anywhere in the world.

The LMFBR operates on the uranium–plutonium fuel cycle (for further discussion of reactor fuel cycles, see Section 4.6). This means that the reactor is fueled with bred isotopes of plutonium in the core or driver, and the blanket is natural or depleted uranium. It is recalled from Fig. 4.2 that η, the number of fission neutrons emitted per neutron absorbed by ^{239}Pu, increases monotonically with increasing neutron energy for energies above 100 keV. It follows that the breeder ratio and breeding gain increase with the average energy of the neutrons inducing fission in the system. Therefore, insofar as possible, every effort must be made to prevent the fission neutrons in a fast reactor from slowing down. This means, in particular, that lightweight nuclei must largely be excluded from the core. There is no moderator, of course, so the core and blanket contain only fuel rods and coolant.

Sodium has been universally chosen as the coolant for the modern LMFBR. With an atomic weight of 23, sodium does not appreciably slow down neutrons by elastic scattering, although, as noted earlier, it does moderate neutrons to some extent by inelastic scattering. Sodium is also an excellent heat transfer agent, so that an LMFBR can be operated at high power density. This, in turn, means that the LMFBR core can be comparatively small. Furthermore, because sodium has such a high boiling point (882°C at 1 atm), reactor coolant loops can be operated at high temperature and at essentially atmospheric pressure without boiling, and no heavy pressure vessel is required. The high coolant temperature also leads to high-temperature, high-pressure steam, and high plant efficiency. Finally, sodium, unlike water, is not corrosive to many structural materials. Reactor components immersed in liquid sodium for years appear like new after the excess sodium has been washed off.

Sodium also has some undesirable characteristics. Its melting point, 98°C, is much higher than room temperature, so the entire coolant system must be kept heated at all times to prevent the sodium from solidifying. This is accomplished by winding a spiral of insulated heating wire called *tracing* along coolant piping, valves, and so forth. Sodium is also highly reactive chemically. Hot sodium reacts violently with water and catches fire when it comes in contact with air, emitting dense clouds of white sodium peroxide smoke. This latter behavior is actually an advantage for sodium cooling since it represents a kind of built-in leak detector. Thus any sodium oozing from an incipient crack in a cooling pipe[4] or from the packing of a valve can immediately be identified by the presence of white smoke. For this reason, moreover, LMFBRs are inherently very tight systems and emit far less radiation to the environment than comparable LWRs. Operation of these reactors has yet to show that they can be operated with high availability on a large commercial scale, however. The operational problems center on the use of Na and how to properly design the reactor internals, valves, and pumps to reliably operate the system.

Unfortunately, sodium absorbs neutrons, even fast neutrons, leading to the formation of the beta-gamma emitter ^{24}Na, with a half-life of 15 hours. Therefore, sodium that passes through the reactor core becomes radioactive. LMFBR plants operate on the steam cycle—that is, the heat from the reactor is ultimately utilized to produce steam in steam generators. Because of the radioactivity of the sodium and because sodium reacts so violently with water, it is not considered sound engineering practice to carry the sodium coolant directly from the reactor to the steam generators. Leaks have often occurred in steam generators between the sodium on one side and the water on the other, and such leaks could lead to the release of radioactivity. Therefore, all LMFBRs utilize two sodium loops: the primary reactor loop carrying radioactive sodium, and an intermediate sodium loop containing nonradioactive sodium, which carries the heat from the primary loop via an intermediate heat exchanger (symbolized by IHX) to the steam generator.

The detailed manner in which the intermediate sodium loop is arranged divides LMFBRs into two categories: the loop-type LMFBR and the pool-type LMFBR. A schematic diagram of a loop-type LMFBR plant is shown in Fig. 4.28. As indicated in the figure, the core and blanket of the reactor and control rods, not shown) are located in a reactor vessel not unlike that of an LWR, except that the vessel need not withstand high pressure. To catch any sodium that might leak

---

[4]Note that, because LMFBRs operate at close to atmospheric pressure, there is no tendency for the sodium coolant to burst out of the system like the water in an LWR. This is one of the distinct safety features of an LMFBR.

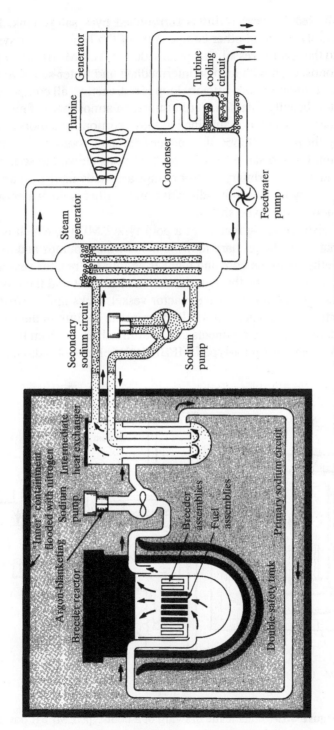

Figure 4.28 Schematic drawing of a loop-type LMFBR power plant.

Generator

Turbine

Turbine
cooling
circuit

Condenser

Steam
generator

Feedwater
pump

Secondary
sodium circuit

Sodium
pump

Intermediate
heat exchanger

"Inner" containment
flooded with nitrogen

Sodium
pump

Argon-blanketing

Breeder reactor

Breeder
assemblies

Fuel
assemblies

Primary sodium circuit

Double-safety tank

171

through the reactor vessel, this is surrounded by a safety tank. In the loop-type system, the IHX and all other components of the heat transfer system are located external to the reactor vessel. A typical 1,000 MWe LMFBR plant has three or four primary loops, each with its own intermediate and water-steam loop.

Since the sodium in a primary loop is radioactive, all components in a primary loop must be heavily shielded. Furthermore, an atmosphere of nitrogen, with which sodium does not react, is maintained in those portions of a loop-type LMFBR plant containing the primary loops; this reduces the likelihood of fires involving radioactive sodium. For the same reason, empty regions above the sodium within the reactor vessel and the primary reactor pumps are made inert with argon. The sodium in the secondary loops is not radioactive and neither is the water or steam, so none of these loops needs to be shielded.

The distinguishing feature of a pool-type LMFBR, which is pictured in Fig. 4.29, is that all of the primary loops—including the pumps and the IHXs—are located, together with the core and blanket, within the reactor vessel. The sodium is circulated by the pumps through the core and blanket and then through the IHXs. The secondary sodium enters the reactor vessel, picks up heat from the IHXs, and then exits to the steam generators. To prevent the sodium in the IHXs from becoming radioactive, these components are carefully shielded from the reactor neutrons. The sodium leaving a pool-type LMFBR is therefore not radioactive.

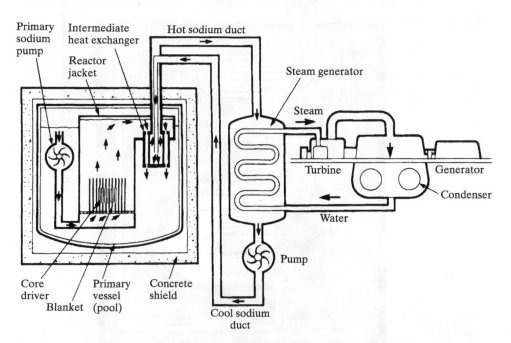

Figure 4.29   Schematic drawing of a pool-type LMFBR power plant.

Of the two types of LMFBRs, the loop-type appears based on the simpler concept. Except for the presence of the intermediate loop, it is not much different in design from an ordinary PWR. All primary loop components—the reactor, pumps, heat exchangers, and so on—are separate and independent. This makes inspection, maintenance, and repairs easier than when these components are immersed in hot, radioactive, and opaque sodium, as they are in pool-type systems. However, substantial amounts of shielding are required around all the primary loops in a loop-type plant, which makes these plants resemble large, heavily built fortresses.

With a pool-type LMFBR, by contrast, no radioactivity leaves the reactor vessel, so no other component of the plant must be shielded.[5] Furthermore, the usual practice is to locate pool-type reactor vessels at least partially underground, so that only the uppermost portion of the vessel requires heavy shielding. It is possible, therefore, to walk into the reactor room where a pool-type reactor is operating and even walk across the top of the reactor without receiving a significant radiation dose—so tight and compact is this type of LMFBR. The two operating commercial LMFBRs are the French Super Phénix and the Japanese Monju reactor. Super Phénix is a pool type whereas Monju is a loop type LMFBR.

The core of an LMFBR consists of an array of fuel assemblies, which are hexagonal stainless steel[6] cans between 10 and 15 cm across and 3 or 4 m long that contain the fuel and fertile material in the form of long pins. An assembly for the central region of the reactor contains fuel pins at its center and blanket pins at either end. Assemblies for the outer part of the reactor contain only blanket pins. When these assemblies are placed together, the effect is to create a central cylindrical driver surrounded on all sides by the blanket.

The fuel pins are stainless steel tubes 6 or 7 mm in diameter, containing pellets composed of a mixture of oxides of plutonium (PuO_2) and uranium (UO_2). The equivalent enrichment of the fuel—that is, the percent of the fuel that is plutonium—ranges between 15 and 35 w/o depending on the reactor in question. The fuel pins are kept apart by spacers, or in some cases, wire wound helically along each pin. The pins in the blanket, which contain only UO_2, are larger in diameter, about 1.5 cm, because they require less cooling than the fuel pins. Both fuel and blanket pins are more tightly packed (closer together) in an LMFBR than in an LWR or HWR because the heat transfer properties of sodium are so much better than those of water. The liquid sodium coolant enters through holes near the bottom of each assembly and passes upward around the pins, removing heat as it goes and then exiting at the top of the core.

---

[5]In both loop-type and pool-type LMFBRs, shielding must also be provided for sodium dump tanks and spent fuel areas.

[6]Zirconium and its alloys cannot be used in an LMFBR because zirconium reacts chemically with sodium.

Although virtually all present-day LMFBRs operate with uranium–plutonium oxide fuel, there is considerable interest in the future use of fuel composed of uranium–plutonium carbide since larger breeding ratios are possible with this kind of fuel. This, in turn, is due to the fact that, although there are two atoms of oxygen per atom of uranium in the oxide (chemical formula UO_2), there is only one atom of carbon per uranium atom in the carbide (whose formula is UC). Light atoms such as carbon and oxygen tend to moderate fission neutrons. Since there are fractionally (one-half) fewer of the atoms in the carbide than in the oxide, it follows that the energy distribution of neutrons in a carbide-fueled LMFBR is shifted to higher energies than in a comparable oxide-fueled LMFBR. The breeding ratio, therefore, goes up because η increases with neutron energy. At the same time, the more energetic neutron spectrum leads to additional fissions in the fertile material. This effect is shown in Table 4.3, where breeding ratios and compound doubling times are given for the different types of fuel.

LMFBRs can also operate with thorium–^{233}U oxide fuel. In this case, the driver portions of the fuel assemblies contain ^{233}U instead of plutonium, and thorium takes the place of natural or depleted uranium. The blanket assemblies are all thorium. However, as indicated in Table 4.3, the substitution of thorium–^{233}U for uranium–plutonium has a devastating effect on both breeding ratio and doubling time. An LMFBR under these circumstances effectively does not breed.

The control rods for LMFBRs are usually stainless steel tubes filled with boron carbide, although other materials have also been used. At one time, it was thought that a fast reactor like the LMFBR would be more difficult to control than most thermal reactors because the fission neutrons do not spend as much time slowing down and diffusing before inducing further fissions. However, years of operating experience have shown that fast reactors, in fact, are highly stable and easily controlled.

Steam from an LMFBR plant is delivered superheated to the turbines at about 500°C and between 16 and 18 MPa. The overall plant efficiency is in the neighborhood of 40%.

TABLE 4.3 BREEDING RATIOS AND COMPOUND DOUBLING TIMES IN YEARS FOR LMFBRS*

Compound		U–Pu fuel	Th–^{233}U fuel
Oxide	Breeding Ratio	1.277	1.041
	Doubling Time	16	112
Carbide	Breeding Ratio	1.421	1.044
	Doubling Time	9	91

*From Chang, Y. I., et al., *Alternative Fuel Cycle Options*. Argonne National Laboratory, 77–70, September 1977.

Other Reactor Types

The GCFR   This reactor concept is a logical extrapolation from HTGR technology. It is a helium-cooled reactor fueled with a mixture of plutonium and uranium. The core of the GCFR is similar to that of an LMFBR, with mixed PuO_2 and UO_2 pellets in stainless steel pins, except that the pins are not as close together as they are in the LMFBR. Also, the pins in the GCFR have a roughened outer surface to enhance heat transfer to the passing coolant. Figure 4.30 shows a drawing of a 360 MWe (1090 MWt) GCFR nuclear steam supply system. The helium, at a pressure of 10.5 MPa, enters the core from the bottom at a temperature of 298°C and leaves from the top of the core at 520°C. The heated gas then passes down through three steam generators under the action of electrically driven circulators (blowers). The GCFR, unlike the LMFBR, requires no intermediate heat exchangers. The reactor is provided with auxiliary circulators and heat exchangers for use in case of a failure of the main cooling loops. All of the components except the

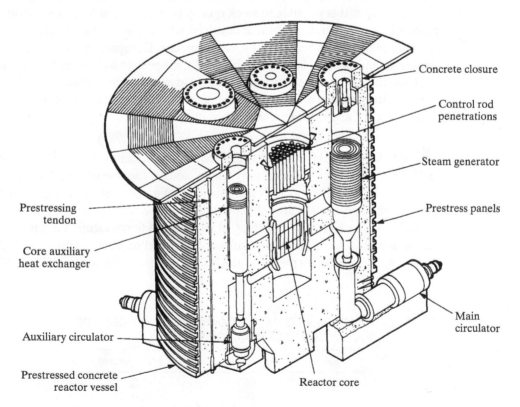

Figure 4.30   Cross-section of a gas-cooled fast reactor. (Courtesy of General Atomic Company.)

main circulator motors are located within a prestressed concrete reactor vessel as shown in the figure. The steam generators produce superheated steam at 485°C and 10.5 MPa. In principle, the entire steam cycle can be circumvented in a GCFR, as it can in an HTGR, and the helium coolant used directly in a gas turbine, with all the advantages this offers.

Being a gas, helium is essentially a void in the reactor and it has almost no effect on the neutrons in the reactor. The neutron spectrum is therefore harder (more energetic) than in the LMFBR, where moderation occurs as the result of inelastic collisions with the sodium. It follows that the breeding ratio is higher for the GCFR than for the LMFBR since the average value of η is higher and the doubling time is correspondingly shorter.

One additional advantage of helium cooling should be mentioned. The economics of a power reactor are determined by many factors, including the availability of the plant—the fraction of time during which the plant can, in fact, be operated. It is important that the time required for routine maintenance and corrective repairs be kept as low as possible. Since the coolant in a GCFR does not become overly radioactive, it is possible to work on any part of the coolant loops soon after the reactor is shut down. If a problem develops within the reactor, the source of the trouble can be quickly determined by visual means. Comparable maintenance and repairs are much more difficult with the LMFBR because sodium is both radioactive and opaque, and because it must be heated at all times to prevent solidification. Some authorities believe that the GCFR is potentially capable of providing the lowest power costs of any reactor conceived to date.

The MSBR   This is a thermal breeder that operates on the ^{233}U–thorium cycle. It is recalled that ^{233}U is the only fissile isotope capable of breeding in a thermal reactor. The MSBR concept is a unique design among reactors in that the fuel, fertile material, and coolant are mixed together in one homogeneous fluid. This is composed of various fluoride salts that, at an elevated temperature, melt to become a clear, nonviscous fluid. The composition of a typical molten salt mixture is given in Table 4.4. All of the elements in the salt mixture, with the exception of the

TABLE 4.4 COMPOSITION IN MOLE PERCENT OF A MOLTEN SALT REACTOR FUEL

Salt	Mole percent
^{7}LiF	72
BeF_2	16
ThF_4	12
$^{233}UF_4$	0.3

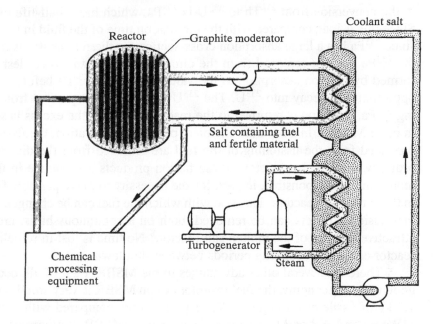

Figure 4.31   Schematic drawing of a molten salt breeder reactor. (Courtesy of Oak Ridge National Laboratory.)

thorium and uranium, have very small thermal neutron absorption cross-sections.[7] This is a vital consideration in a thermal breeder since the value of η is low, and neutrons must be carefully conserved. In addition to their excellent neutronic properties, the fluoride salts have a high solubility for uranium, they are among the most stable of all chemical compounds, they have very low vapor pressure at high temperature, and they have reasonably good heat transfer properties. Furthermore, these salts are not damaged by radiation, do not react violently with air or water, and are inert to some common structural materials and graphite.

Figure 4.31 shows a diagram of the principal components in an MSBR plant. The reactor core consists of an assembly of graphite moderator elements, which are provided with channels for the passage of the molten salt mixture. When this fluid passes through the core, the system becomes critical and the fission energy is absorbed directly in the fluid. The heated fluid then flows through a heat exchanger in an intermediate coolant loop and returns to the reactor, a portion being diverged for chemical processing.

Several important operations are performed in the chemical processing part of an MSBR plant. It is remembered from Section 4.2 that the intermediate nuclide

[7]At 0.0253 eV, $\sigma_a(^7\text{Li}) = 0.037$ b, $\sigma_a(\text{Be}) = 0.0092$ b, $\sigma_a(\text{F}) = 0.0095$ b.

in the conversion from ^{232}Th to ^{233}U is ^{233}Pa, which has a half-life of 27.4 days. Since this is long compared with the circulation time of the fluid in the reactor and since ^{233}Pa has a large absorption cross-section for thermal neutrons (about 41 b), the ^{233}Pa must be removed from the circulating fluid and stored, lest it be transformed by neutron absorption to the valueless isotope ^{234}Pa before it has had an opportunity to decay into ^{233}U. The ^{233}U is eventually separated from the decaying ^{233}Pa, a portion is returned to the reactor fluid, and the excess is sold for fuel in other MSBRs. In a separate chemical processing operation, the fission products produced from the fissioning of the fuel are removed from the diverted fluid. In ordinary, solid-fueled reactors, these fission products accumulate in the fuel elements and are responsible, in part, for the necessity to make periodic fuel changes in these types of reactors. The ease with which the fuel can be changed and the fact that fission products can be removed, both on a continuous basis, are especially attractive features of the molten salt reactor.[8] No time is lost in refueling, and the reactor can operate for long periods between shutdowns.

There are several other advantages to the MSBR. First of all, because of the good neutron economy, the fuel inventory of an MSBR is very small—about 1.0 to 1.2 kg of fissile material per MWe of plant output compared with about 3 kg per MWe for an LWR or 3 to 4 kg per MWe for the LMFBR. Furthermore, because of the low vapor pressure of the molten salts, the MSBR operates at just a little above atmospheric pressure and thus no expensive pressure vessel is required. Finally, since high temperatures are possible with the molten salts, the MSBR can produce superheated steam at 24 MPa and 540°C, which leads to a very high overall plant efficiency of about 44%.

There are also drawbacks to the MSBR. For one thing, as the fuel flows out of the reactor, it carries the delayed neutron precursors with it; when the neutrons are emitted, they activate the entire fuel-containing loop. Maintenance and repairs to any component in the system may therefore require extensive reactor downtime and extensive automatic, remotely operated equipment. The breeding ratio of the MSBR is in the 1.05 to 1.07 range, much smaller than for the LMFBR or GCFR. Doubling times are expected to be from 13 to 20 years. Finally, the need to handle and process the radioactive mixture limits the commercial application of this technology.

The LWBR   For many years, it was the accepted view that it is not possible to build an ordinary light-water reactor that will breed even if it is fueled with

---

[8]However, the fact that solid fuel elements are not used in the MSBR may actually hinder its development since reactor manufacturers are naturally reluctant to pursue reactor systems that do not carry with them the promise of long-term fuel-fabrication contracts.

^{233}U. For one thing, it was thought that too many neutrons would be lost at thermal energies owing to the large absorption cross-section of water. It was recognized that this problem could be avoided by reducing the amount of water relative to fuel in the core. In so doing, the energy spectrum of the neutrons would be shifted to higher energies due to the decreased moderation from the water. This shift would, in turn, mean that a larger fraction of the neutrons would be absorbed in the ^{233}U at intermediate energies ($1\,\mathrm{eV} \leq E \leq 10\,\mathrm{keV}$), where the early data showed that η fell to a value only slightly greater than 2. The overall result would therefore be the same—namely, that the reactor would not breed. During the early 1960s, however, new experiments showed that η is, in fact, sufficiently larger than 2 in the intermediate energy range to make breeding possible, provided losses of neutrons are strictly controlled.

Even when a special effort is made in the design of the LWBR to reduce neutron losses, its overall breeding gain will be very small—too small to make the reactor a net producer of ^{233}U for other reactors of this type. However, enough excess ^{233}U (namely, from 1%–2%) would be obtained over the life of a core to compensate for the loss of ^{233}U accompanying the chemical reprocessing of the fuel. Thus, once an LWBR is put into operation, it presumably could be fueled indefinitely with ^{232}Th, of which there are abundant resources.

To see whether breeding can actually be achieved in a light-water reactor, the U.S. Department of Energy developed an LWBR core that was installed in the government-owned pressurized water reactor at Shippingport, Pennsylvania. The system operated well, producing a breeding ratio between 1.01 and 1.02 as designed. This experiment confirmed the technical feasibility of installing such breeder cores in existing and future pressurized water reactors. Such a conversion from the burning of relatively scarce ^{235}U to more plentiful thorium would significantly improve the overall utilization of nuclear fuel in these reactors and could extend indefinitely the light-water reactor component of the nuclear power industry.

A cross-section through the LWBR core is given in Fig. 4.32. As indicated in the figure, the core consists of hexagonal modules arranged in a symmetrical array surrounded by a reflector-blanket region. Each module contains an axially movable seed region–that is, a region having a multiplication factor greater than unity and a stationary, annular hexagonal blanket with $k < 1$. Each of these regions, in turn, consists of arrays of tightly packed, but not touching, fuel rods containing pellets of thorium dioxide (ThO_2) and $^{233}UC_2$, the latter in varying amounts from 0 to 6 w/o in the seed and from 0 to 3 w/o in the blanket region. A module cross-section showing the fuel rods, which are smaller in the seed than in the blanket, is shown in Fig. 4.33.

Control of the reactor is accomplished by the movement of the seed region within each module. This changes the leakage of neutrons from the core and hence

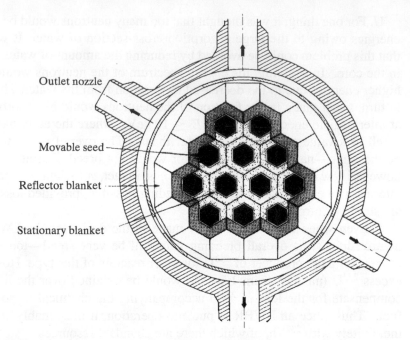

Outlet nozzle

Movable seed

Reflector blanket

Stationary blanket

Figure 4.32 LWBR core cross-section. (Courtesy of the U.S. Department of Energy.)

the value of the multiplication factor. As indicated in Fig. 4.34, the leakage is small when the seed is inserted, but increases as the seed is withdrawn. This novel method of control is necessary for the LWBR to conserve neutrons. Ordinary control rods cannot be used because they would absorb too many neutrons to permit breeding. Figure 4.34 also shows the rather complex distribution of ThO_2 and $^{233}UO_2$ in the fuel modules, which is supposed to provide the reactor with good control and power characteristics. The fuel distribution shown is obtained by differential loading of the fuel tubes, which extend the full length of each module.

Mobile Power Reactors

One of the widest applications of nuclear power is in its use for propulsion of ships. This idea was fostered by the legendary Admiral Hyman G. Rickover, who also played a key role in development of the commercial nuclear power program. Starting with the *Nautilus*, the U.S. Navy and later the British, French, and former Soviet Union all adopted nuclear propulsion as the primary power source for submarines. Both the U.S. and former Soviet Union Navies also use nuclear power for the propulsion of surface ships.

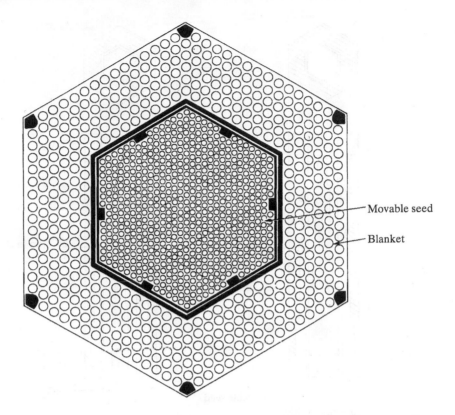

Figure 4.33 Typical LWBR fuel module cross-section. (Courtesy of the U.S. Department of Energy.)

The requirements of a mobile power reactor differ somewhat from those of a commercial central power station, in that the reactors and associated systems must be compact and long lived. As a result, the designs are different from those of the larger commercial reactors. Initially, two competing concepts were developed— a PWR and a liquid metal-cooled design. The PWR reactor concept was used in the *Nautilus* and the liquid metal-cooled reactor in the *Seawolf*. The *Nautilus* went operational in January 1955 and demonstrated its capabilities by sailing underneath the North Pole in August 1958.

Because of maintenance and reliability problems, the liquid metal-cooled design was abandoned in favor of the PWR concept, and the *Seawolf* reactor was removed and replaced with a reactor similar to that of the *Nautilus*. Later reactors were designed for the propulsion of surface ships. The success of nuclear power is clear. All submarines are nuclear powered, and eventually all U.S. aircraft carriers will be as well.

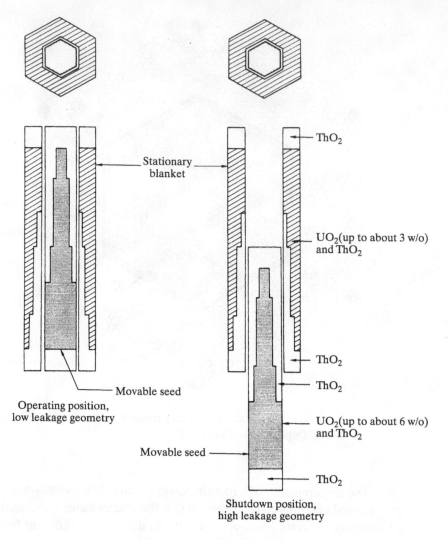

Figure 4.34   Variable geometry control concept. (Courtesy of the U.S. Department of Energy.)

The clear advantages of nuclear power over fossil fuel in a submarine is obvious. With its nearly unlimited undersea endurance and its ability to provide a compact high power source without the need for oxygen, the nuclear reactor enables submarines to stay submerged for months at a time. The designs are similar to those found in commercial reactors, but they use a much higher classified level of enrichment. Based on a typical PWR reactor primary secondary system, the re-

TABLE 4.5   NUCLEAR SUBMARINES AUTHORIZED, UNDER CONSTRUCTION, COMMISSIONED, AND DECOMMISSIONED IN THE UNITED STATES (1997)

	Authorized by Congress	Under Construction	Commissioned	Decommissioned
SEAWOLF (SSN 21) Class	3	2	0	0
USS LOS ANGELES (SSN 688) Class	62	0	54	8
Other Fast Attack Submarines	68	0	11*	59
Total Fast Attack Submarines	133	2	66	67
TRIDENT/USS OHIO (SSNB 726) Class	18	0	18	0
POLARIS/POSEIDON Submarines	41	0	0	39*
Total Ballistic Missile Submarines	59	0	18*	39
Total Submarines	192	2	84	106
Research Vessels (NR-1)	1	0	1	0
Nuclear-powered Aircraft Carriers	10	1	9	0
Nuclear-Powered Guided Missile Cruisers	9	0	2	7
Total Nuclear-Powered Surface Ships	19	1	11	7
Total Nuclear-Powered Ships	212	3	96	113

*Two ships originally authorized by Congress as Fleet Ballistic Missile submarines are currently operating as fast attack submarines.

actor provides steam for both a propulsion turbine and for electrical generation as well.

Table 4.5 lists the number of nuclear submarines and surface ships in operation in the U.S. Navy. A submarine of the newer *Seawolf* class is shown in Fig. 4.35.

Figure 4.35 USS Seawolf, SSN 21, transiting the Thames River in Connecticut. (U.S. NAVY photo.)

Nuclear Power Around the World

As of this writing, nuclear power is employed as a source of power throughout the world. The United States, Europe, and the former Eastern Block countries derive significant amounts of electricity from nuclear power plants. Over 30 countries rely on nuclear energy for a portion of their electrical needs. In 1996, 442 nuclear power plants generated 2,300 billion kilowatt hours of electricity. In some countries, nearly all of the electricity is generated by nuclear energy. Nearly 42% of the electricity in Europe is generated by nuclear, 17% in eastern Asia, and 21% in the United States. By country, the percentages are even higher. In France, for example, over 75% of the electricity comes from nuclear energy. In the United States, 105 nuclear power plants contribute 674 billion kilowatt hours of electricity—or 20% of the nation's electricity supply. Figure 4.36 shows the distribution of electricity generated by nuclear energy in the top 19 producer countries.

An additional 40 nuclear power plants are under construction and will add 95 million kilowatts of capacity by the year 2010. Table 4.6 lists many of the major users of nuclear energy and the types of reactors employed.

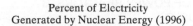

Percent of Electricity
Generated by Nuclear Energy (1996)

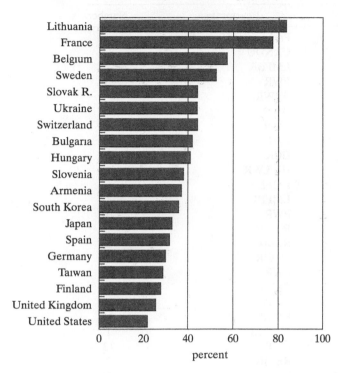

Figure 4.36 Electricity generated from nuclear energy. (Courtesy of the International Atomic Energy Agency.)

4.6 NUCLEAR CYCLES

The procurement, preparation, utilization, and ultimate disposition of the fuel for a reactor is called the reactor *fuel cycle*. Those parts of a fuel cycle that precede the use of the fuel in a reactor are collectively called the *front end* of the fuel cycle. That portion related to the fuel after it has been withdrawn from a reactor is called the *back end* of the fuel cycle.

Figure 4.37 shows the *once-through* fuel cycle for a PWR plant. The mass flows indicated in the figure are in kilograms for a nominal 1,000 MWe (1 GWe) plant operated for 1 year at a capacity factor of 0.75—that is, the masses correspond to 0.75 GWe-year. The fuel cycle for a BWR plant is entirely similar except that the mass flows are somewhat different.

The cycle begins with the mining of uranium ore. This ore contains uranium in the form of a number of complex oxides and is reduced to the oxide U_3O_8, which is then converted to uranium hexafluoride, UF_6, the form in which uranium is accepted at current isotope enrichment plants. Approximately 0.5% of the uranium is lost in the conversion to UF_6. Following enrichment to about 3 w/o in ^{235}U,

TABLE 4.6   NUMBER OF POWER REACTORS, BY
COUNTRY, IN THE WORLD

Country	Reactor type	Number
Argentina	Pressurized Heavy Water	3
Canada	CANDU	22
France	LMFBR	2
	PWR	58
India	PHWR	14
	BWR	2
Japan	BWR	28
	PWR	23
	GCR	1
	HWLWR	1
	LMFBR	1
Kazakhstan	LMFBR	1
Korea	PWR	10
	PHWR	2
Lithuania	RBMK	2
Pakistan	PHWR	1
	PWR	1
Romania	PHWR	2
Russia	PWR	14
	RBMK	11
	LMFBR	1
Ukraine	PWR	13
	RBMK	2
United Kingdom	GCR	20
	AGR	14
	PWR	1
United States	PWR	69
	BWR	36
Taiwan	BWR	4
	PWR	2
China	PWR	3
Germany	PWR	14
	BWR	6
Sweden	BWR	9
	PWR	3
Spain	PWR	7
	BWR	2

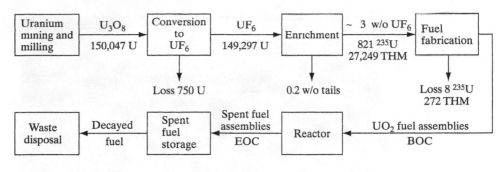

Fuel composition		
	BOC	EOC
^{235}U	813	220
U	26,977	25,858
Pu fissile	—	178
Pu	—	246
THM	26,977	26,104
FP	—	873

Notes:
1. Mass flows in kg's per 0.75 GWe-yr.
2. Abbreviations:
   BOC = beginning of refueling cycle
   EOC = end of refueling cycle
   FP = fission products
   Pu = total plutonium
   Pu fissile = ^{239}Pu + ^{241}Pu
   THM = total heavy metal = U + Pu
   U = total uranium

Figure 4.37 LWR once-through fuel cycle. (Based on *Nuclear Proliferation and Civilian Nuclear Power,* U.S. Department of Energy report DOE/NE-0001/9, Volume 9, 1980.)

which occurs essentially without loss of uranium but that leaves residual tails, the UF_6 is converted to UO_2. Finally, the UO_2 is fabricated into PWR or BWR fuel assemblies and, in time, loaded into the reactor. Fuel fabrication is accompanied by about 1% loss of uranium.

Approximately once every 2 years, the reactors are shut down and a portion of the fuel (one third of the core of a PWR, one quarter of the BWR) is removed and placed in a spent fuel pool adjacent to the reactor containment building. In the once-through cycle, the spent fuel, after much of its radioactivity has died away, is removed from the pool and disposed of as radioactive waste. Because of the limited capacity of reactor spent fuel pools, it is necessary to transfer the spent fuel to special storage facilities. There the fuel is stored in dry storage casks prior to its ultimate disposal. Figure 4.38 is an example of a dry storage facility used for the interim storage of spent fuel.

The once-through cycle is quite wasteful of nuclear energy resources. Both the PWR and BWR have conversion ratios of about 0.6. Since most of the fuel in these reactors is fertile ^{238}U, non-negligible amounts of the fissile plutonium isotopes ^{239}Pu and ^{241}Pu are produced while the reactors are in operation. Thus, as

Figure 4.38 Dry cask storage facility located at the Calvert Cliffs nuclear power station. (Courtesy of Baltimore Gas and Electric.)

indicated in Fig. 4.37, a LWR in a nominal 1,000 MWe nuclear plant discharges about 180 kg of fissile plutonium plus 220 kg of ^{235}U at each refueling. This fissile material, if consumed in a reactor, would release the energy equivalent to that of about 1 million tons of coal. For this reason, the once-through cycle is sometimes called a *throw-away* fuel cycle.

The plutonium and uranium in spent fuel can be utilized if it is recycled as shown in Fig. 4.39. In this arrangement, the spent fuel is reprocessed (reprocessing is discussed in Section 4.8); that is, the plutonium and uranium are chemically extracted from the fuel. The plutonium, in the form of PuO_2, is then mixed with UO_2,[9] fabricated into what is called *mixed-oxide fuel,* and returned to the reactor. The residual, slightly enriched uranium produced in the reprocessing plant is converted to UF_6 for reenrichment. It is estimated that the adoption of a uranium–plutonium recycle would reduce the cumulative U_3O_8 requirements of LWRs by about 40% over the 30-year lifetimes of these reactors. Nevertheless, as of this writing (2000), it is still not clear whether the additional costs of reprocessing and mixed-oxide fuel fabrication can in fact render uranium–plutonium recycling an economically attractive fuel option. However, economics notwithstanding, it is reasonable for some nations such as Japan to adopt recycling in an effort to reduce their dependence on foreign uranium suppliers and ensure independence.

The fuel cycle for a breeder is similar to that of an LWR with recycling because, of course, reprocessing and refabrication play key roles in any breeder system. Figure 4.40 shows the fuel cycle for an LMFBR. Either natural or depleted

---

[9]Natural uranium, depleted uranium, or the slightly enriched uranium from the reprocessing plant can be used in mixed-oxide fuel. The proportions of fissile plutonium, ^{235}U, and ^{238}U in the fuel would amount to between 3 and 3.5 w/o.

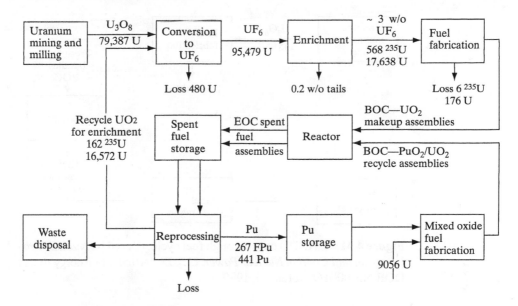

Figure 4.39 LWR with plutonium and uranium recycling. (Based on *Nuclear Proliferation and Civilian Nuclear Power*, U.S. Department of Energy report DOE/NE-0001/9, Volume 9, 1980.)

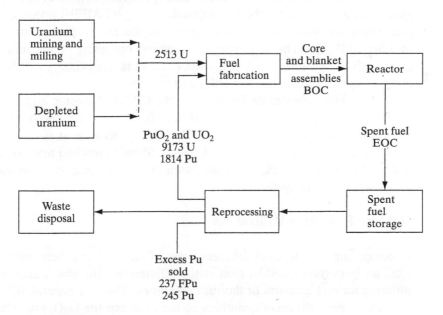

Figure 4.40 Fuel cycle for an LMFBR. (Based on *Nuclear Proliferation and Civilian Nuclear Power*, U.S. Department of Energy report DOE/NE-0001/9, Volume 9, 1980.)

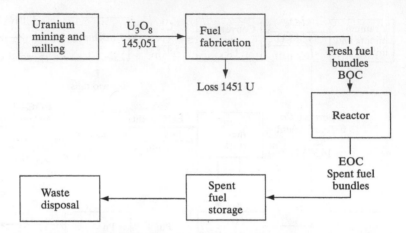

Figure 4.41   CANDU once-through fuel cycle. (Based on *Nuclear Proliferation and Civilian Nuclear Power,* U.S. Department of Energy report DOE/NE-0001/9, Volume 9, 1980.)

uranium can be used in the blanket. This uranium, together with the uranium and plutonium from the reprocessing plant, is fabricated into core and blanket assemblies and introduced into the reactor. The spent fuel, after a cooling period in the spent fuel pool, is sent to be reprocessed. Since an LMFBR produces more fissile plutonium than it needs for its own operation, the excess plutonium is sold for use in other LMFBRs. The fuel cycles for breeders operating on thorium and ^{233}U are essentially the same as that in Fig. 4.40, except that the primary fuel is now thorium and the reprocessed product is ThO_2 and $^{233}UO_2$.

The extremely simple fuel cycle of the CANDU reactor is given in Fig. 4.41. In this cycle, there is no enrichment necessary since the reactor is fueled with natural uranium. The Canadians also do not reprocess their spent fuel in view of the abundant uranium resources in Canada. Should enriched uranium ever be used in CANDU-like reactors, UF_6 conversion and enrichment steps would have to be introduced into the cycle.

Nuclear Resource Utilization

In comparing the merits of different types of reactors and their associated fuel cycles, an important consideration is the efficiency with which each is capable of utilizing natural uranium or thorium resources. Thus, in general, different reactor systems require different quantities of uranium ore (or U_3O_8) or thorium to produce the same amount of electrical energy. Since most countries lack abundant indigenous uranium resources, uranium utilization is often a key factor in a national decision to adopt a particular type of reactor and fuel cycle.

The *nuclear resource utilization, U*, is defined quantitatively as the ratio of the amount of fuel that fissions in a given nuclear system to the amount of natural uranium or thorium input required to provide those fissions—that is,

$$U = \frac{\text{fuel fissioned}}{\text{resource input}}. \tag{4.15}$$

In short, U is the fraction of the naturally occurring resource that can be utilized in a reactor fuel cycle for the purpose of generating electric power.

As a first example, consider the CANDU reactor operating on a once-through cycle. This reactor is fueled directly with natural uranium dioxide, and the maximum specific burnup of the fuel is about 7,500 MWd per metric ton of uranium. Let F be the mass of fuel that fissions out of a total uranium fuel load, L. Then $U = F/L$, which is also equal to the burnup, B, of the fuel. According to Eq. (3.57), the fissioning of approximately 1 gram of fissile material releases 1 MWd. Thus, it follows that

$$U = 7{,}500 \text{ g/t} = 7{,}500 \text{ g}/10^6 \text{ g} = 0.0075.$$

Thus, only .75% of the natural uranium fuel introduced into a CANDU reactor eventually undergoes fission on the once-through cycle.

Consider next an LWR, also on a once-through cycle. Since the fuel is now partially enriched, the amount of natural uranium M_U required to produce a fuel load L is considerably larger than L due to the nature of the enrichment process. Thus, for 3 w/o-enriched fuel, it is shown in the next section that $M_U = 5.48L$. The uranium utilization is then

$$U = \frac{F}{M_U} = \frac{F}{5.48L}.$$

At a nominal maximum specific burnup of 30,000 MWd/t,

$$U = 30{,}000/5.48 \times 10^6 = 0.0055,$$

somewhat less than for a CANDU.

With an LWR or CANDU or other converter reactors operated with recycling, the calculation of fuel utilization is much more complicated. A simplified diagram depicting the material flow in such a cycle is shown in Fig. 4.42. In this illustration, the Ls represent the masses of fuel, either fissile or fertile, loaded into the reactor per cycle; the Fs and Rs, respectively, the masses of fuel fissioned or having undergone radiative capture; the Ds the masses of fuel discharged from the reactor; and the Ws the masses of fuel lost in reprocessing and fuel fabrication, taken together.

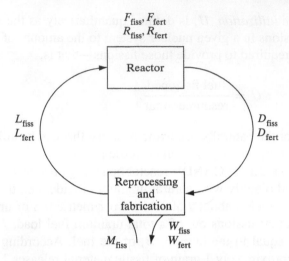

Figure 4.42 Diagram for computing resource utilization in a converter reactor.

Finally, M_{fiss} is the mass of fissile material—not the total amount of fuel—added as makeup per cycle.

The amount of natural uranium M_U required to produce M_{fiss} can be written as

$$M_U = \zeta M_{\text{fiss}}, \qquad (4.16)$$

where the parameter ζ is determined by the nature of the enrichment process. It is shown in the next section that ζ is numerically equal to about 200 and is almost independent of the enrichment of the fuel.

The uranium utilization is then given by

$$U = \frac{F}{M_U} = \frac{F}{\zeta M_{\text{fiss}}}, \qquad (4.17)$$

where

$$F = F_{\text{fissions}} + F_{\text{fert}} \qquad (4.18)$$

is the total mass of fissions in the fissile and fertile material. To obtain an expression for U, it is necessary to introduce three parameters—α, β, and γ. The first,

$$\alpha = \frac{R_{\text{fiss}}}{F_{\text{fiss}}}, \qquad (4.19)$$

is an application of the capture-to-fission ratio, from Eq. (3.45), to a mixture of fissile materials;[10] the second,

---

[10]Note that in converters—for instance, those operating on the uranium–plutonium cycle—the fuel contains a mixture of ^{235}U and the fissile isotopes of plutonium.

$$\beta = \frac{F_{\text{fert}}}{F}, \qquad (4.20)$$

is the fraction of fissions (all fast fissions) occurring in fertile material; and the third,

$$\gamma = \frac{W_{\text{fiss}}}{D_{\text{fiss}}} = \frac{W_{\text{fert}}}{D_{\text{fert}}}, \qquad (4.21)$$

is the fraction of isotope lost in reprocessing and fabrication. This fractional loss is presumed to be the same for fissile and fertile materials.

From Fig. 4.41, it is easy to demonstrate that the fissile makeup M_{fiss} is given by

$$M_{\text{fiss}} = L_{\text{fiss}} - D_{\text{fiss}} + W_{\text{fiss}}, \qquad (4.22)$$

so that, by incorporating Eq. (4.21),

$$M_{\text{fiss}} = (L_{\text{fiss}} - D_{\text{fiss}})(1 - \gamma) + \gamma L_{\text{fiss}}, \qquad (4.23)$$

where the difference $L_{\text{fiss}} - D_{\text{fiss}}$ is the net decrease in fissile material resulting from the operation of the reactor. Since neutron capture in fertile material produces fissile material, it follows that

$$L_{\text{fiss}} - D_{\text{fiss}} = F_{\text{fiss}} + R_{\text{fiss}} - R_{\text{fert}}. \qquad (4.24)$$

But from the definition of the conversion ratio C,

$$R_{\text{fert}} = (F_{\text{fiss}} + R_{\text{fiss}})C,$$

so that

$$L_{\text{fiss}} - D_{\text{fiss}} = (F_{\text{fiss}} + R_{\text{fiss}})(1 - C).$$

Next, using the definition of α and β, this expression becomes

$$L_{\text{fiss}} - D_{\text{fiss}} = F(1 + \alpha)(1 - \beta)(1 - C)$$

and

$$M_{\text{fiss}} = F(1 + \alpha)(1 - \beta)(1 - \gamma)(1 - C) + \gamma L_{\text{fiss}}. \qquad (4.25)$$

Introducing Eq. (4.25) into Eq. (4.17) gives

$$U = \frac{F}{\zeta[F(1 + \alpha)(1 - \beta)(1 - \gamma)(1 - C) + \gamma L_{\text{fiss}}]}. \qquad (4.26)$$

This formula can be simplified by dividing numerator and denominator by L, the total fuel load, and observing that F/L is equal to B, the specific burnup, and

L_{fiss}/L is the fuel enrichment e. The final expression for U then becomes

$$U = \frac{B}{\zeta[B(1+\alpha)(1-\beta)(1-\gamma)(1-C)+\gamma e]}. \tag{4.27}$$

It was implicitly assumed in the derivation of Eq. (4.27) that, in supplying the reactor with the necessary fissile material to maintain the cycle, sufficient fertile material would automatically be included to make up for fertile material lost in fission, conversion, reprocessing, and fabrication. This is the case for reactors such as LWRs that use low-enriched or natural uranium fuel having relatively low conversion factors. However, for reactors with high values of C—that is, near breeders, provision must be made for separate introduction of fertile material aside from that which accompanies the other fuel. The total makeup is then

$$M = M_{\text{fert}} + M_{\text{fiss}}.$$

From Fig. 4.42, the total material balance is

$$M = L - D + W$$
$$= F(1-\gamma) + \gamma L,$$

and so

$$M_{\text{fert}} = F(1-\gamma) + \gamma L - M_{\text{fiss}}. \tag{4.28}$$

The definition of U is now

$$U = \frac{F}{M_U + M_{\text{fert}}} = \frac{F}{\zeta M_{\text{fiss}} + M_{\text{fert}}}, \tag{4.29}$$

where the fertile material is treated as additional resource material. Inserting Eqs. (4.25) and (4.28) and again using the definitions for burnup and enrichment gives finally

$$U = \frac{B}{B(1-\gamma) + \gamma + (\zeta-1)[B(1+\alpha)(1-\beta)(1-\gamma)(1-C)+\gamma e]}. \tag{4.30}$$

Figure 4.43 shows a plot of U as a function of conversion ratio for nominal values of $B = 0.04$ (4% burnup), $(1+\alpha)(1-\beta) \simeq 1$,[11] $\gamma = 0.02$ (2% losses) and $e = 0.08$ (8% average fuel enrichment). More appropriate numbers can be used for each reactor type and for its corresponding fuel enrichment and conversion factor; these are simply compromise values lying between high- and low-efficiency

---

[11]Detailed calculations show that α and β are numerically small and within a few percentages of one another.

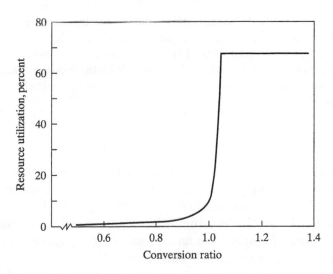

Figure 4.43   Nuclear resource utilization as a function of conversion ratio.

converters. It is observed in the figure that the resource utilization is extremely low, although much higher than in the once-through cycle even up to high values of C. Thus, even at $C = 0.95$, U is about 0.05, which means that only 5% of uranium or thorium resources are utilized in such a high converter with recycling.

When the term corresponding to M_{fiss} in Eq. (4.30) is equal to zero, the cycle is self-sustaining on fertile material—that is, no outside fissile material must be added to maintain the reactor in operation. In the prior example, this occurs at a conversion ratio of 1.0484. At this point, U is given simply by

$$U = \frac{B}{B(1 - \gamma) + \gamma} \tag{4.31}$$

and in the previous example it has the value of 0.676. Any increase in C causes the reactor to breed. However, this does not increase the resource utilization because any additional fertile material added to the reactor over and above M_{fert} computed earlier is merely converted to excess fissile material. It eventually fissions, but not in the reactor in question. Note that breeders do not consume plutonium; they consume uranium. Thus, as shown in Fig. 4.43, U is a constant in the breeding region and is not a function of C. Incidentally, values of U in the range from 0.60 to 0.70, corresponding to a resource utilization of between 60% and 70%, are typical for true breeders such as the LMFBR.

Example 4.5

Suppose a given nation (such as the United States) has sufficient natural uranium to meet its nuclear electric power needs for 30 years using only LWRs on the once-through cycle. How long would this same uranium last if used to fuel LMFBRs at the same electrical capacity?

Solution. According to the earlier discussion, $U = 0.0055$ for the once-through LWR and about 0.67 for an LMFBR. Therefore, the resource would last approximately $30 \times 0.67/0.0055 = 3,655$ years. [*Ans.*] However, note that the 30-year resource supply is based on conventional sources of uranium. With the breeder, costly, nonconventional uranium resources can be utilized since uranium costs are irrelevant to breeder power costs.

Resource Limitations

Table 4.7 summarizes the uranium resource requirements for various nuclear plants and fuel cycles over a 30-year operating period—the assumed lifetime of the plants. In each case, the plant is assumed to produce 1,000 MWe at a capacity factor of 70%. The tails assay (see Section 4.7) was taken as 0.2 w/o. The table demonstrates again the savings in natural uranium resources resulting from recycle and the enormously reduced need for uranium with breeder reactors.[12]

The total, world-wide requirements for uranium over the years depend on the growth of installed nuclear capacity and the mix of reactor types and fuel cycles that is adopted with the passage of time. It is impossible, of course, to accurately predict future nuclear electric capacity because so many uncertain factors enter into decisions to construct new nuclear plants—the growth in the world population,

TABLE 4.7 LIFETIME NATURAL URANIUM
REQUIREMENTS FOR DIFFERENT TYPES OF NUCLEAR
POWER PLANTS AND FUEL CYCLE OPTIONS*

Plant (Reactor) Types and Cycle	Natural Uranium Requirements (t)
LWR	
Once-through	4,260
U-Pu recycling	2,665
HWR	
Natural uranium, once-through	3,655
Natural uranium, Pu recycling	1,820
Low-enriched uranium	2,505
LMFBR	
U-Pu recycling	36[†]

*From *International Nuclear Fuel Cycle Evaluation,* Summary
Volume, International Atomic Energy Agency, 1980.
[†]Natural or depleted uranium.

[12]Values of all fuel cycle parameters, those in Table 4.7, and in Figs. 4.37 and 4.39 through 4.41, are sensitive to the assumptions used in their derivation. Such calculated values should therefore be treated as nominal only.

TABLE 4.8   PROJECTIONS OF NUCLEAR ELECTRIC
GENERATING CAPACITY (GWE)

Year	United States*		World*	
	Low	High	Low	High
1994	99		340	
1995	100		345	
2000	100		345	
2005	100		371	
2010	91	91	415	487

*From *Uranium 1995-Resources, Production, and Demand,*
Nuclear Energy Agency, OECD and IAEA (1996).

the state of world economics, the climate of international politics (wars and other political confrontations), the availability of competing energy sources, and so on. Predictions of installed nuclear capacity made today can appear ludicrous only a few years later.

Since precise prognostication of nuclear capacity cannot be made, it has become a common practice to give projections in terms of high- and low-capacity projections. Future capacity then lies somewhere in between, although there is by no means any guarantee that this will be so. Table 4.8 gives such projections for the United States and the world as prepared by the U.S. Department of Energy and the International Atomic Energy Agency (IAEA). These predictions do not provide the breakdown by reactor type or fuel cycle; that will be determined over the long term by decisions that have not as yet been made. However, some things are evident now. For instance, because of the long lead time required to build a nuclear plant in the United States, especially a new type of plant, and because breeder development has been terminated by the government, it is reasonably safe to presume that none of the U.S. nuclear capacity will originate in breeder plants until well into this century. For other countries, such as Japan, who have active breeder programs, this may not be the case.

Whether sufficient uranium will be available to fuel expanding world nuclear capacity is an issue of continuing controversy even among experts. Uranium is not an especially rare element. It is present in the earth's crust at a concentration by weight of about four parts per million, which makes uranium more abundant than such common substances as silver, mercury, and iodine. There are an estimated 10^{14} tons of uranium located at a depth of less than 12 or 13 miles, but most of this is at such low concentrations it will probably never be recovered.

Most of the uranium used in nuclear plants over the period up to the year 2025 will undoubtedly come from deposits that are now or could be exploited at

TABLE 4.9   ESTIMATED WORLD* URANIUM
RESOURCES AS OF JANUARY, 1995, IN 1,000 t OF
URANIUM†

Resource	Up to $80/kg U	$80 to $130/kg U
Conventional		
Reasonably assured (RAR)	2,124	2,951
Estimated additional reserves Category I	637	1,015
Undiscovered conventional Estimated additional (EAR-II) and speculative		11,000

*U.S. Totals: RAR 366, EAR I & II 1273, RAR + EAR = 1639
and speculative of 1,339,000.
†From *Uranium 1995-Resources, Production, and Demand,
op. cit.*

a forward cost,[13] in U.S. dollars, of up to $130 per kg U or about $50 per lb of
U_3O_8. Estimates of these so-called *conventional resources* are shown in Table 4.9
under two categories: *reasonably assured resources* (RAR) and *estimated additional resources-category I* (EAR–I), which reflect different levels of confidence in
their actual existence. It should be noted that almost 80% of the total RAR occur in
North America, Africa, and Australia. Resources recoverable at a cost of less than
$40 per kg of U, denoted as *known resources*, amount to 886,000 tons.

In addition to the RAR and EAR-I categories, the more nebulous categories
of *estimated additional resources–category II* (EAR–II) and *speculative resources*
are also defined. EAR–II refers to uranium that is expected to occur in deposits for
which the evidence is inferred from like mineral deposits. By comparison, speculative resources are those based on indirect evidence and extrapolation and that
consist of the quantity of uranium that conceivably could be discovered in areas
geologically favorable for the location of uranium deposits. It is estimated that 11
million tons of uranium may be in the speculative category worldwide. According to the U.S. Department of Energy, the United States may have as much as 1.4

[13]*Forward cost* is a specialized term used to describe the economic availability of uranium
resources. It includes the estimated costs of developing, building, and operating uranium mines and
mills at the sites of established uranium resources. It does not include the cost of exploration, the cost
of money (interest charges), marketing costs, profits, and so forth, which contribute to the *price* of
uranium or U_3O_8. As a rule of thumb, the actual market price of uranium is about twice its forward
cost.

million tons in this category. However, it is doubtful that much, if any, of these speculative resources will actually be recovered before the year 2025.

Besides the conventional uranium resources, there are abundant *unconventional resources* of low-grade materials. In the United States, for example, there is a geological formation called Chattanooga shale that extends westward from eastern Tennessee and underlies at shallow depths much of the areas of half a dozen midwestern states. Much of this shale is rich in uranium (up to about 66 ppm[14]), but not to an extent that it can be extracted for less than $130 per kg. Nevertheless, this shale contains an estimated total of 5 million tons of uranium. Indeed, an area of this shale only 7 miles square contains the energy equivalent to all the petroleum in the world. Similar deposits in alum shale in Sweden were actually mined at one time and produced significant amounts of uranium. The mining was halted, however, because of environmental concerns.

More recently, additional fissile material from the former Soviet Union and the U.S. weapons programs have become available. In the form of highly enriched uranium and plutonium, these materials are now being considered as possible fuels for commercial reactors. The consumption of these materials in a reactor has been cited as the best way to dispose of the weapons grade material.

Sea water also contains a virtually unlimited amount of uranium but at a very low concentration—about 0.003 ppm. No economically feasible method for its extraction has yet been devised.

Estimates of uranium resources, proven and speculative, are not directly related to the availability of uranium in the marketplace for consumption in nuclear plants. Uranium in the ground, especially uranium in the speculative category, obviously cannot be used to fuel reactors. Mines must first be developed, mills must be constructed and placed into operation, transportation must be arranged, and so on. All of this requires the investment of effort and capital long before the demand for fuel has been firmly established. Needless to say, if uranium production and stockpiles fail to meet the demand, reactors somewhere in the world have to be shut down or reduced in power, and the price of uranium would go up for the others.

Several studies of uranium supply and demand have been carried out, especially in connection with the International Nuclear Fuel Cycle Evaluation exercise and the periodic uranium resources, production, and demand assessment by the OECD and IAEA. These studies show that, until about 1990, an oversupply of uranium existed. The oversupply was caused mainly by an overly optimistic estimate of uranium utilization. For example, studies cited in the second edition of this book predicted between 255 and 395 GWe of nuclear-generating capacity in the United States alone. These estimates should be compared with the actual level of

---

[14]The average grade of ore processed in the United States in 1978 contained about 1,400 ppm.

100 GWe of nuclear-generating capacity. Similar high estimates were also made for the nuclear-generating capacity worldwide.

Since 1990, reductions have occurred in uranium production and exploration. While uranium requirements have continued to increase, production has contracted. The difference between production and requirements since 1990 was met by a drawdown of existing stockpiles. The imbalance between production and requirements has also led to a rebound in market price for uranium since 1995.

Although it is possible to predict short-term requirements for uranium, a number of uncertainties in the long-term uranium market have recently been introduced. These include the decision to convert high-enrichment uranium from warheads to low-enriched uranium by blending and the availability of uranium from the former Soviet Union and Eastern Block countries. After the breakup of the Soviet Union and independence of the former Eastern Block countries, production in these areas declined drastically. Prior to the breakup, these countries had considerable excess production. Because of facility closures and environmental concerns, it is uncertain whether these countries' production capabilities will ever be regained.

At present, insufficient uranium production capability exists to meet present and future requirements. The demand for uranium is being met by a combination of production and utilization of the excess inventory built prior to 1991. This could eventually lead to a destabilization of the uranium market, resulting in significantly higher prices for uranium.

The HEU from both the United States and former Soviet Union nuclear weapons may help alleviate the shortfall. In 1997, the U.S. and Russian governments signed an agreement under which Russian HEU would be blended down to LEU for commercial reactor application. It is expected that similar efforts will occur in the United States as well. The first shipments of LEU converted from HEU were delivered to the U.S. Enrichment Corporation in 1995. Eventually plutonium from nuclear weapons will also be available for use in mixed-oxide (MOX) fuel. Some European reactors are already licensed to use MOX fuel derived from recycling of fissile reactor material.

Recycled material may also play an important role in meeting demand. Although at current uranium prices, recycled material is too costly, future price increases may eventually negate this difference. It is estimated that over 150,000 tons of heavy metal are stored in spent fuel. This material represents a significant resource that could be used. Further, the removal from the waste stream of the unburnt plutonium and fissile uranium will help the concerns over long-term waste disposal.

Changes on the demand side may also impact the imbalance. In recent years, there has been a trend toward improved fuel utilization through better, more efficient fuel design as well as improved fuel management. A negative impact on the demand side may arise from higher than expected use of nuclear power for the generation of electricity. Recently, there has been a growing realization that nuclear energy is environmentally friendly, producing no acid rain, green house gasses, or ozone-depleting emissions. As Third World countries accelerate their economies, it is expected that the per capita consumption of electricity will rise. Many such countries are looking toward nuclear-based generation to meet these energy needs.

Only if the breeder option were adopted would the total capacity no longer be resource limited. Initially, the growth in breeder capacity is determined by the growth in electrical demand, the construction capability of industry, and the availability of plutonium from LWRs for fueling new breeders. Eventually, the breeders would provide sufficient plutonium for fueling new breeders. Beyond this point in time, indicated by the sharp break in the breeder curves, the rate of increase in capacity is very rapid, determined only by the doubling time of the reactors. As discussed in Section 4.5, this is shorter for carbide than for oxide fuel.

4.7 ISOTOPE SEPARATION

The problem of enriching uranium is a formidable one because it requires the separation of two isotopes, ^{235}U and ^{238}U, which have nearly the same atomic weight. Ordinary chemical separation is out of the question, of course, since both isotopes are the same element—uranium. Separation must be accomplished, therefore, by physical means, and several processes have been developed for this purpose. Before describing some of these processes, it is useful to consider some of the basic relationships that underlie isotope separation.

Material Balance and Separative Work

The amount of starting or input material (i.e., feed material) that must be furnished to an enrichment plant to obtain a given amount of enriched product depends on the original enrichment of the feed, the desired enrichment of the product, and the enrichment of the residual, depleted uranium, or *tails*. Specifically, suppose it is required to produce M_P kg of uranium enriched to x_P weight fraction[15] in ^{235}U from M_F kg of feed with enrichment x_F, leaving behind M_T kg of tails of enrichment x_T. The overall material flow is depicted in Fig. 4.44. Since there is little or no loss of uranium in the plant, it follows that

$$M_F = M_P + M_T. \tag{4.32}$$

---

[15] Although isotope abundance is always given in atom percent, uranium enrichment is specified in weight percent or weight fraction. The isotope abundance of ^{235}U in natural uranium is 0.72%. This corresponds to an enrichment of 0.711 *w/o* or a weight fraction of 0.00711.

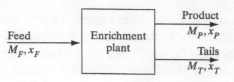

Figure 4.44 Flow diagram for enrichment plant.

The total amount of ^{235}U is also the same before and after enrichment, so that

$$x_F M_F = x_P M_P + x_T M_T.$$

Eliminating M_T from the last two equations gives

$$M_F = \left(\frac{x_P - x_T}{x_F - x_T}\right) M_P. \tag{4.33}$$

If the feed is natural uranium, then the value of x_F is fixed by nature at 0.00711. The value of x_T is set by the operator of the enrichment plant. At this writing, the enrichment plants in the United States are operating with $x_T = 0.002$— that is, 0.2 w/o. The denominator in Eq. (4.33) is therefore constant, and it follows that the amount of feed needed to produce a specified amount of product increases linearly with the enrichment. Except for very low enrichments, M_F is directly proportional to x_P. The variation of M_F/M_P with enrichment is shown in Fig. 4.45.

At the enrichment x_P, the mass of ^{235}U in the product mass M_P is

$$M_{25} = x_P M_P.$$

Substituting for M_P in Eq. (4.33) then gives

$$M_F = \left(\frac{x_P - x_T}{x_F - x_T}\right) \frac{M_{25}}{x_P}. \tag{4.34}$$

Except for low enrichments, $x_P - x_T \simeq x_P$, so that

$$M_F \simeq \frac{1}{x_F - x_T} M_{25}.$$

With $x_F = 0.00711$ and $x_T = 0.002$, this gives

$$M_F \simeq 196 M_{25}.$$

Thus, as stated earlier in Section 4.6, the order of 200 kg (or 200 atoms) of natural uranium feed are required to produce enriched fuel containing 1 kg (or 1 atom) of ^{235}U, almost independent of the fuel's enrichment. It is shown in Section 4.6 that this circumstance has considerable bearing on the utilization of natural uranium resources in nonbreeding reactors.

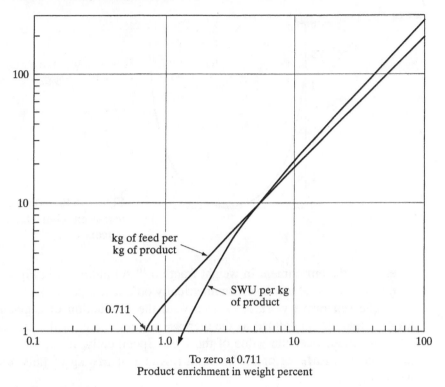

Figure 4.45   Feed and separative work, in kg, as a function of product enrichment.

The cost of enriching uranium is described in terms of a special unit called the *separative work unit,* (SWU). This concept is based on the observation that isotope separation is equivalent to the unmixing of gases that had been previously irreversibly mixed. Since separated isotopes represent a more ordered situation than unseparated isotopes, the entropy of the former is clearly smaller than that of the latter. According to the laws of thermodynamics, work must be performed on a system to decrease its entropy in an isothermal process. Therefore, it follows that work must be performed to separate isotopes, and this work is measured in SWU.

Separative work can be expressed in terms of a function $V(x)$ known as the *value function,* given by

$$V(x) = (1 - 2x) \ln \left(\frac{1-x}{x} \right), \tag{4.35}$$

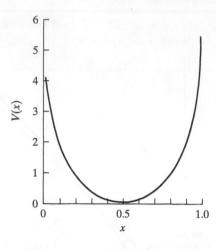

Figure 4.46 The value function versus enrichment x in weight percent.

where x is the enrichment in weight fraction.[16] As indicated in Fig. 4.46, $V(x)$ is zero at $x = 0.5$ and increases symmetrically on either side of this point.

The separative work associated with the production of a given amount of product material is defined as the increase in total value of the product and tails, taken together, over the value of the feed. Specifically, if M_P kg of product are produced from M_F kg of feed, with a residual of M_T kg of tails, the separative work is

$$\text{SWU} = M_P V(x_P) + M_T V(x_T) - M_F V(x_F), \qquad (4.36)$$

where x_P, x_T, and x_F are the respective enrichments. However, since

$$M_T = M_F - M_P,$$

Eq. (4.36) can be written as

$$\text{SWU} = M_P[V(x_P) - V(x_T)] - M_F[V(x_F) - V(x_T)]. \qquad (4.37)$$

It should be noted that SWU has units of mass (kilograms). For fixed values of x_F and x_T, the separative work necessary to produce a given amount of product increases monotonically with enrichment. This is illustrated in Fig. 4.45, where SWU/M_P is plotted as a function of x_P.

Example 4.6

One third of the core of a PWR is removed and replaced with fresh fuel approximately once a year. For a nominal 1,000 MWe PWR plant, each reload requires about 33,000 kg of 3.2 w/o-enriched UO_2. If the feed to the enrichment plant is

---

[16]It can be shown that $V(x)$ is proportional to the total amount of fluid that has to be pumped in all stages of a separation system per unit amount of material processed.

natural uranium, and the tails assay is 0.20 w/o: (a) How much feed is required per reload? (b) How much would the enrichment cost at the 1981 U.S. price of $130.75 per kg SWU?

Solution.

1. There are

$$\frac{238}{238 + 2 \times 16} \times 33,000 = 29,089 \text{ kg}$$

of enriched uranium in 33,000 kg of UO_2. Using $x_P = 0.032$, $x_F = 0.00711$, and $x_T = 0.002$ in Eq. (4.33) gives

$$M_F = \frac{0.032 - 0.002}{0.00711 - 0.002} \times 29,089 = 170,777 \text{ kg } [Ans.]$$

2. The value of the 3.2 w/o uranium is

$$V(0.032) = (1 - 2 \times 0.032) \ln\left(\frac{1 - 0.032}{0.032}\right) = 3.191.$$

Similarly, $V(0.00711) = 4.869$ and $V(0.002) = 6.188$. From Eq. (4.36), the separative work is

$$SWU = 29,089(3.191 - 6.188 - 170,777(4.869 - 6.188)$$

$$= 138,075 \text{ kg}.$$

At $130.75 per SWU, the charge for enrichment would be

$$138,075 \times 130.75 = \$18,053,306. [Ans.]$$

It is of some interest to consider the variation of feed and separative work requirements with the residual enrichment of the tails. As shown in Fig. 4.47, the feed requirements decrease with decreasing tails enrichment for fixed values of x_F and x_P. This would be expected since decreasing tails enrichment means that more ^{235}U is extracted per kg of feed. However, the SWU required to produce a given product goes up with decreasing tails enrichment simply because it takes more work to extract more ^{235}U out of a given mass of feed. Whether it makes sense to increase or decrease tails enrichment at any time depends on the relative costs of separative work and uranium feed, on the one hand, and the desire to conserve natural uranium resources, on the other. For instance, in producing ~ 3 w/o uranium for LWRs, a reduction in tails enrichment from 0.20 w/o to 0.10 w/o would lead to a saving of about 15% in feed, but this would increase the SWU per kg of product by 43%. Therefore, a decrease in tails assay would not be indicated until the ratio of the cost of uranium to separative work increases substantially or a shortfall in uranium supply appears imminent.

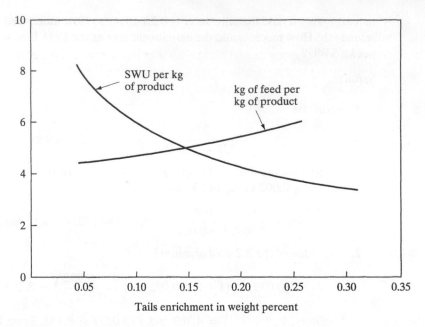

Figure 4.47 Feed and separative work, in kg, as a function of tails enrichment.

The rate at which an enrichment facility can produce enriched uranium is determined by its ability to deliver separative work in a unit of time. Accordingly, enrichment plant capacities are measured in SWU/yr or, more appropriately, in millions of SWU/yr. Clearly, the greater the separative work capacity, the more reactors can be serviced by a given facility or facilities. For example, in 1983, the U.S. government had 27.3 million SWU/yr of separative capacity in operation. Approximately 116,000 SWU/yr are needed to fuel a 1,000 MWe PWR (aside from the initial loading; see Problem 4.33). It follows that U.S. enrichment plants will be capable of handling the fuel needs of $27.3 \times 10^6/116,000 = 235$ one thousand MWe PWRs for a total electrical capacity of 235 GWe. Today, this represents an excess in capacity despite the sale of fuel on the international market and is therefore being reduced.

Uranium Enrichment Processes

At least a dozen methods have been devised for the separation of isotopes, but only a few have been shown to be economically practical or hold promise of becoming so. Some of the more prominent of these methods are considered briefly.

Gaseous Diffusion   The first economic enrichment process to be successfully developed was the gaseous diffusion method. This process is based on the
observation that, in a mixture of two gases, on average the lighter molecules travel
faster than the heavier molecules. As a result, the lighter molecules strike the walls
of a container more frequently than their heavier counterparts. If a portion of the
container consists of a porous material having holes large enough to permit the
passage of individual gas molecules, but not so large that a mass flow-through of
the gas can occur, then more light molecules than heavy molecules will pass out of
the container. The gas leaving the container is therefore somewhat enriched in the
lighter molecules, whereas the residual gas is correspondingly depleted. This phenomenon is illustrated in Fig. 4.48, which shows the actual type of container, barrier configuration, and piping connections used in some gaseous diffusion plants.
A single container like that shown in the figure is called a *diffuser.*

Uranium, of course, is not a gas, and the diffusion enrichment of uranium is
carried out using uranium hexaflouride, UF_6. Although this substance is a solid at
room temperature, it is easily vaporized. However, this requires that all components
of a diffusion plant be maintained at an appropriate temperature to ensure that the
UF_6 remains in gaseous form. Although UF_6 is a stable compound, it is highly reactive with water and corrosive to most common metals. As a consequence, internal
gaseous pathways must be fabricated from nickel, or austenitic stainless steel, and
the entire system must be leak tight. Despite its unpleasant characteristics, UF_6
is the only compound of uranium sufficiently volatile to be used in the gaseous
diffusion process.

Fortunately, fluorine only consists of the single isotope ^{19}F, so that the difference in molecular weights of different molecules of UF_6 is only due to the differ-

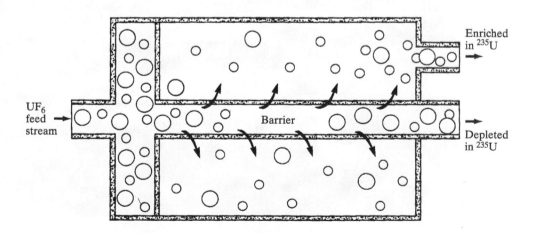

Figure 4.48   Operation of a diffuser in a gaseous diffusion process.

ence in weights of the uranium isotopes. However, because the molecular weights of $^{235}UF_6$ and $^{238}UF_6$ are so nearly equal, very little separation of the $^{235}UF_6$ and $^{238}UF_6$ is affected by a single pass through a barrier (i.e., in one diffuser). It is necessary, therefore, to connect a great many diffusers together in a sequence of stages, using the outputs of each stage as the inputs for two adjoining stages, as shown in Fig. 4.49. Such a sequence of stages is called a *cascade*. In practice, diffusion cascades require thousands of stages depending on the desired product enrichment.

As indicated in Fig. 4.49, the gas must be compressed at each stage to make up for a loss in pressure across the diffuser. This leads to compression heating of the gas, which then must be cooled before entering the diffuser. The requirements for pumping and cooling make diffusion plants enormous consumers of electricity. For instance, a plant capacity of 10 million SWU/yr requires about 2,700 MW of

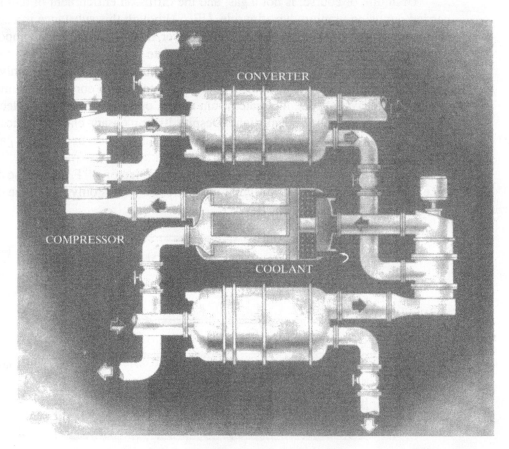

Figure 4.49   Three stages of a gaseous diffusion cascade. (U.S. Department of Energy.)

electrical power. In the United States, the diffusion plants of the Department of Energy make it the nation's single largest user of electricity.

Although this is the principle method used today in the United States, other countries have developed this technology. China, France, and Argentina have all built prototype or large-scale plants using this method.

Gas Centrifuge   The gas centrifuge process for separating isotopes is a sophisticated version of the common method used for many years in biology and medicine for fractionating blood and other biological specimens. Historically, centrifuging was one of the first processes considered for enriching uranium for military purposes in World War II, and it was carried to the pilot plant stage before being abandoned in favor of gaseous diffusion. The separation of the isotopes of chlorine had already been successfully demonstrated in the 1930s by J. W. Beams at the University of Virginia.

In the gas centrifuge method, the gases to be separated—namely, $^{235}UF_6$ and $^{238}UF_6$—are placed in an appropriate container or rotor and rotated at high speed. The rotation creates what is, in effect, a strong gravitational field. As a result, the heavier gas tends to move to the periphery of the vessel, whereas the lighter gas accumulates nearer the center—buoyed by the heavier gas. The amount of separation that can be attained in a single centrifuge machine depends on the mass difference between the two gas components, the length of the rotor, and, most important, its speed of rotation. The basic principles of the centrifuge are shown in Fig. 4.50, provided by the U.S. Department of Energy. The electromagnetically driven rotor is located in an evacuated chamber. UF_6 is fed into the rotor at the axis, and the enriched and depleted fractions are withdrawn near the top and bottom of the rotor, as shown. An important feature in the operation of an isotope-separating centrifuge is that an axial countercurrent of the gas is maintained within the rotor, as indicated in the figure, by the establishment of a temperature difference between the ends of the rotor and/or the appropriate mechanical design of the 2 output scoops. As the gas moves upward near the rotor axis, the $^{238}UF_6$ diffuses toward the outer wall. The gas arriving at the upper scoop is therefore slightly enriched in $^{235}UF_6$; similarly, the stream reaching the lower scoop is depleted in $^{235}UF_6$.

Since the rate of gas flow through one centrifuge is not large, a great many machines must be connected in parallel in a commercial-size enrichment plant. Also, because the necessary degree of enrichment cannot be carried out in a single centrifuge, the machines must be connected together in cascade as in a gaseous diffusion plant.

The principle advantage of the gas centrifuge method over gaseous diffusion is that, according to the U.S. Department of Energy, a gas centrifuge plant uses only 96% less electric power than a gaseous diffuson plant of the same separative work capacity. Thus, although a nominal 10 million SWU/yr diffusion plant requires

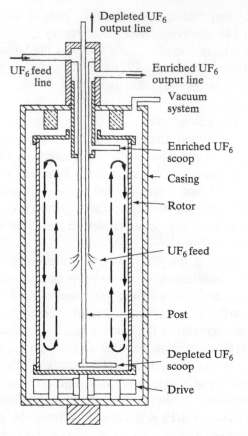

Figure 4.50 Schematic representation of a gas centrifuge. (U.S. Department of Energy.)

2,700 MW of electrical capacity, the same enrichment plant with centrifuges would need only 109 MW of electricity.

Several countries have developed enrichment facilities based on the centrifuge process. These include the United Kingdom, Russia, China, Pakistan, India, Brazil, Germany, The Netherlands, Japan, and possibly Iran and Israel.

Aerodynamic Processes   Aerodynamic processes exploit the small differences in mass between ^{238}U and ^{235}U bearing molecules in much the same way as gaseous diffusion plants operate. To date, two methods have been tried at least in a prototype manner—the Becker nozzle and the aerodynamic process.

The Becker nozzle was devised by E. W. Becker in the Federal Republic of Germany. In this process, a mixture of UF_6 and hydrogen flows at supersonic speeds through a specially designed curved nozzle as shown in Fig. 4.51. Due to the centrifugal pressure gradient created in the gas, the heavier molecules tend to

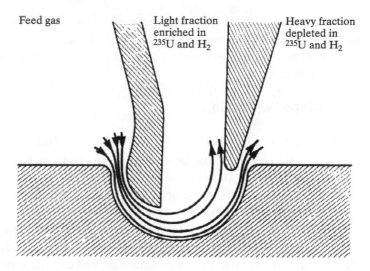

Figure 4.51   Principle features of a Becker nozzle process.

diffuse preferentially toward the outer wall, as they do in a centrifuge machine. The emerging gas can therefore be separated into lighter and heavier fractions by placing a sharp divider at an appropriate location on the far side of the nozzle. It has been shown that the nozzle process provides a much larger degree of separation per stage than the gaseous diffusion process. Unfortunately, the process was not found competitive with other methods.

The Republic of South Africa has developed enrichment facilities based on a similar concept to the Becker nozzle. In the aerodynamic process, uranium hexaflouride gas and hydrogen enters a tube at high speed through holes in the side. The gas spirals down the tube. The tube narrows along the axis, and the heavier fraction, rich in ^{238}U, exits to the side. The lighter fraction exits through the bottom. The plant successfully operated for several years, but proved uneconomical.

Electromagnetic Separation   This method is an extension on a large scale of the principle of the mass spectrometer. Ionized uranium gas is accelerated by an electrical potential into a magnetic field, where the more massive particles travel in arcs of larger radii than the lighter particles. The separated beams are then collected in properly spaced graphite receivers. The uranium ions react with the graphite and form uranium carbide, which is then chemically treated to obtain the separated isotopes.

One of the obvious advantages of the electromagnetic process is that virtually complete separation of the uranium isotopes can be obtained in one pass through the machine. A large pilot plant was placed in operation in Oak Ridge, Tennessee, during World War II. The Iraqi government also chose this method for their clandestine nuclear program. However, the method is less economical than gaseous diffusion plants for enriching uranium. Today it yields small laboratory amounts of elements enriched in particular isotopes but was actively pursued by Iraq for its weapons program.

Laser Isotope Separation

The newest and most ingenious method for separating isotopes relies on laser technology and the light-absorbing properties of atoms or molecules. The science of laser optics lies outside the scope of this book; suffice it to say that lasers have the unique ability to produce intense beams of light of well-defined frequencies—that is, beams of essentially monochromatic light.

As explained in Chapter 2, atoms normally exist in their ground states, but they can be placed in excited states by absorption of appropriate amounts of energy. Molecules, like atoms, also have a ground state and excited states. Whereas the excited states of atoms are due to different configurations of electrons, the states in molecules may also arise because of different vibrational or rotational energies of the molecular structure. In either case, the absorption of electromagnetic radiation occurs in the same way. If the energy of the incident radiation does not correspond to a possible transition between states, very little absorption occurs; when the incident energy matches a possible transition, considerable energy absorption may result. This phenomenon of resonance absorption is shown in Fig. 4.52 for excitation resonances in ^{235}U and ^{238}U at approximately 5,027 angstroms (1 angstrom=1

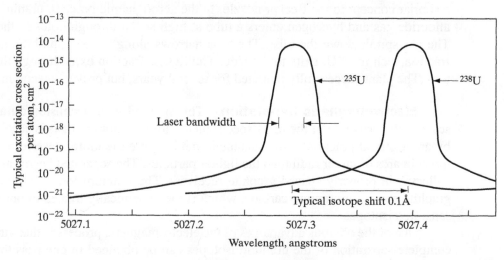

Figure 4.52 Excitation cross-section as a function of wavelength. (Courtesy of Jersey Nuclear-Avco Isotopes, Inc.)

Å $= 10^{-8}$ cm), the wavelength of green light, which corresponds to annihilation incident photon energy of about 2.4 eV.

Laser isotope separation (LIS) is based on the fact that the energies of the excited states of atoms and molecules depend to some degree on their atomic masses. This is reflected in a shift, known as the *isotope shift,* in the energies of the excitation resonances. In the case of uranium atoms, as indicated in Fig. 4.52, the resonances shown in ^{235}U and ^{238}U are separated by about 0.1 Å. By exposing uranium vapor to a laser beam having a bandwidth (i.e., a variation in wavelength) less than 0.1 Å, it is possible to place the ^{235}U in an excited state while leaving the ^{238}U in its ground state. Subsequent exposure of the vapor to other laser beams can then raise the energy of the excited ^{235}U to a point where the atom becomes ionized. In principle, the charged ion can then be extracted by electron fields from the system without disturbing the neutral ^{238}U.

The mechanism just described relates to separation effected with individual atoms, and it is sometimes referred to as *atomic vapor* LIS (AVLIS). A similar method based on the selective excitation and decomposition of UF_6 gas is called *molecular* LIS (MLIS). Both techniques are being rapidly developed in major laboratories around the world.

The main components of one proposed AVLIS chamber are indicated in Fig. 4.53. Molten metallic uranium contained in the water-cooled crucible at the bottom of the unit is vaporized by exposure to an intense beam of high-energy electrons. The vapor expands rapidly into the chamber and enters the extraction structure, consisting of a number of parallel plates. Here it is exposed to four intense laser beams—the choice of four different beams is based on the detailed physics of the optical excitation of uranium. The lasers selectively ionize a portion of the ^{235}U atoms, which are drawn to the product collector plates by electric and magnetic fields. The neutral ^{238}U atoms pass through the product collectors to the tails collector. However, all the atoms falling on the product collectors are not solely ^{235}U. Much of the material that accumulates on the collectors is simply uranium vapor, which has flowed directly to the collecting plates from the crucible source. The uranium collected is therefore enriched in ^{235}U, but is not pure ^{235}U.

For the sake of clarity, only one laser beam and mirror are shown in Fig. 4.53; in fact, several beams, reflected repeatedly within the chamber, are used to increase the utilization of laser energy. To further increase laser utilization, several chambers are connected in series so the beams traverse each in turn. The overall energy consumption of an AVLIS plant is estimated to be less than one tenth that of a gaseous diffusion plant of the same capacity.

For complex reasons, the AVLIS system operates most satisfactorily at low-feed enrichments, with either natural uranium or depleted uranium. Thus, the proposed pilot plant could use depleted uranium (enrichment 0.2–0.3 w/o) to produce 3 w/o-enriched uranium for use in LWRs, leaving a tails enrichment of 0.08 w/o.

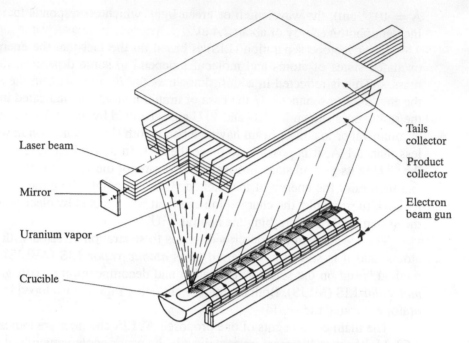

Figure 4.53   Schematic representation of one proposed laser isotope separation process. (Courtesy of Jersey Nuclear-Avco Isotopes, Inc.)

Operating on natural uranium, the AVLIS system would increase the availability of ^{235}U for LWRs by about 18% (see Problem 4.39).

Other Processes   There are two other competing processes that may prove more economical than current techniques. Although still under development, these methods are considered technically feasible and potentially commercially viable. The methods are chemical exchange and plasma separation.

The chemical exchange process relies on the behavior of molecules bearing the heavier ^{238}U atom to be different from that of the lighter ^{235}U-bearing molecules. Thus, in certain reactions, one isotope tends to concentrate in one molecule over another. The process is similar to the enrichment of water in deuterium discussed later.

The second process relies on the cyclotron frequency of ^{235}U being different from that of ^{238}U. This difference allows the ions of one species to be preferentially collected over the other.

Production of Heavy Water

Up to this point, the discussion of isotope separation has centered on the problem of producing enriched uranium to fuel reactors moderated or cooled with ordinary water. As explained earlier, such reactors cannot be made critical with natural uranium. However, natural uranium can be used as fuel in heavy-water moderated reactors such as CANDU and the Argentine reactors at Atucha (see Section 4.5). With these reactors, the burden of isotope separation lies in producing the heavy water, rather than in enriching the fuel. If heavy water is used both as the moderator and coolant, approximately one metric ton is required per MWe of nuclear plant capacity. Heavy water has been produced commercially by a variety of methods, including electrolysis, distillation of water or liquid hydrogen, and a method of chemical exchange described later. This last method has come to be preferred because it uses an abundant feed material (water) and requires little energy.

Because they possess the same number of protons and electrons, isotopes of the same element are normally assumed to be chemically identical. However, such isotopes may not undergo chemical reactions at precisely the same rate, owing to differences in their atomic mass. Naturally the difference in reaction rates increases with the disparity among the isotopic masses, and it is most pronounced in the case of ordinary hydrogen and deuterium, whose masses differ by a factor of two.

Consider, for example, the reaction between deuterated hydrogen sulfide and water:

$$H_2O + HDS \rightleftharpoons HDO + H_2S.$$

If H and D were chemically the same, the equilibrium constant for the reaction

$$K = \frac{[HDO][H_2S]}{[H_2O][HDS]}$$

where the brackets denote concentration, would be equal to unity. Actually, K is substantially greater than unity and is also a function of temperature. Thus, at $25°C$, K is 2.37; at $125°C$, K is 1.84. By writing K in the form

$$K = \frac{[HDO][H_2S]}{[HDS][H_2O]},$$

it is clear that, in equilibrium, the concentration of HDO in H_2O is greater than that of HDS in H_2S. What is more, the relative concentration of HDS in H_2O increases with increasing temperature. These circumstances make it possible to separate D from H and underlie the so-called *GS dual-temperature exchange process*[17] by which most heavy water is produced commercially.

The GS process takes place in a chemical exchange tower where H_2S gas is bubbled upward through a sequence of perforated trays, along and around which

---

[17]Although the GS process was invented by K. Geib and J. S. Spevack, the initials GS evidently refer to Girdler–Sulfide, Girdler being the name of the company that designed the first GS plant in the United States. (See Benedict *et al.* in the references.)

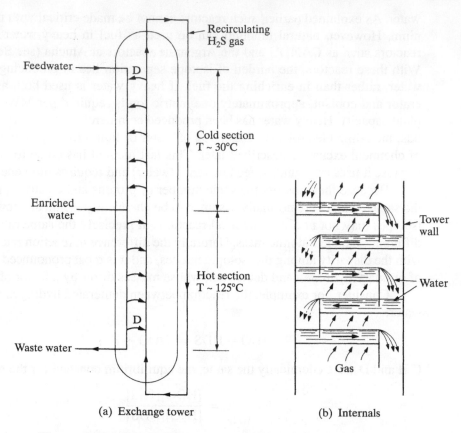

(a) Exchange tower   (b) Internals

Figure 4.54 Schematic drawings showing (a) water, gas, and deuterium flows in exchange tower; and (b) internal structure of tower.

water flows downward as shown in Fig. 4.54. The tower is divided into two parts: a hot section (T $\simeq$ 125°C) and a cold (cooler) section (T$\simeq$ 32°C). Since the equilibrium concentration of deuterium in the H_2S leaving the hot section is greater than it is leaving the cold section, it follows that there is a net transfer of D from the HDO to the H_2O in the cold section. The overall effect is that D is carried around the tower and enriches the feedwater in the cold section. In short, the water is enriched in the cold section; the H_2S is reconstituted in the hot section. Some of the enriched water is extracted at the junction of the two sections and passes into additional separation towers. By adjusting the water and H_2S flow rates, the H_2S can be entirely reconstituted within the tower. In some versions of the GS process, especially those used in Canada, the enriched gas rather than the enriched water is carried forward to additional towers until at the last tower the enriched water is finally withdrawn.

In practice, only a few stages (sets of exchange towers) are required to produce water enriched to between 20 and 30 w/o D_2O. Further enrichment to 99.75 w/o required in CANDUs is carried out using vacuum distillation.

4.8 FUEL REPROCESSING

In an LWR fuel cycle with recycle or in any breeder reactor cycle, it is necessary to reprocess the spent fuel to recover the plutonium or uranium. Several methods have been developed over the years to do so. However, virtually all reprocessing plants that have been built since the 1950s utilize the Purex solvent extraction process.

The Purex process is based on the following experimental facts. Uranium and plutonium can exist in a number of valence (oxidation) states. Because of differences in their oxidation and reduction potentials, it is possible to oxidize or reduce one of these elements without disturbing the other. Furthermore, compounds of uranium and plutonium in different states have different solubilities in organic solvents. For instance, in the 4^+ and 6^+ states, the nitrates of both uranium and plutonium are soluble in certain solvents, while in the 3^+ state their compounds are virtually insoluble in the same solvents.

Therefore, solvent extraction involves three critical steps: (a) separating the uranium and plutonium from the fission products by absorbing the first two in the appropriate solvent, leaving the latter in solution; (b) reducing the oxidation state of the plutonium to 3^+ so that it is no longer soluble in the solvent; and (c) back-extracting the plutonium into aqueous solution.

A simplified flow diagram for a Purex reprocessing plant is shown in Fig. 4.55. The fuel rods are first cut into short lengths—3 to 5 cm long. This occurs in the so-called *mechanical head-end* of the plant. The chopped pieces are next heated to remove radioactive gases, mostly tritium and the fission product gas ^{85}Kr, trapped within the fuel. These gases are collected and stored.

The batch of fuel is then dissolved in a concentrated solution of nitric acid (HNO_3). This aqueous solution of uranium and plutonium, both of which at this point are in high states of oxidation, together with the fission products and dissolved remnants of the fuel assemblies, next passes through a filter to remove undissolved components of the assemblies and enters at the middle of the first extraction column.

In this column, the organic solvent tributylphosphate (TBP), diluted in the kerosenelike substance dodecane, flows up the column extracting uranium and plutonium from the aqueous solution. At the same time, more nitric acid enters from the top of the column to scrub the rising solvent of any fission products that it may have picked up. The organic solution that leaves the top of the column contains essentially all the uranium and plutonium and only a trace of fission products,

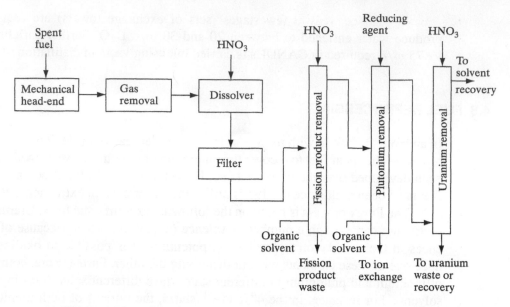

Figure 4.55 Simplified flow diagram of a PUREX reprocessing plant.

whereas the aqueous solution exiting at the bottom holds most of the fission products and very little uranium or plutonium.

The organic solution passes next into a second column, where it counterflows against a dilute solution of a chemical-reducing agent (a ferrous compound is often used), which reduces the plutonium to the 3^+ state, but leaves the uranium in the 6^+ state. Since the plutonium is no longer soluble in the TBP, it passes into the aqueous solution before leaving the column.

The uranium is stripped from the organic solvent in a third column, where it passes into a counterflowing stream of dilute nitric acid. The solvent leaving the top of the column, from which most of the plutonium, uranium, and fission products have now been removed, is piped to a recovery plant for purification and reuse. The uranium exits the column in aqueous solution.

To further purify the uranium and plutonium fractions, their respective solutions can be processed through additional extraction columns. However, the plutonium is often purified and concentrated by ion exchange. This process involves passing the plutonium solution into an ion exchange resin and then eluting the plutonium with dilute acid. The concentration of the purified plutonium can then be increased by partially evaporating the solution, taking care not to approach criticality. This is the usual form of the plutonium output from a fuel reprocessing plant—a highly purified solution of plutonium nitrate. It is an easy matter to transform the plutonium to the oxide PuO_2.

4.9 RADIOACTIVE WASTE DISPOSAL

Radioactive wastes in several different forms are produced at various points in the fuel cycle of a nuclear power plant: during the mining of uranium, the manufacture of the fuel, the operation of the reactor, and the processing and recycling of the fuel (if this is part of the cycle). Wastes are also produced when the plant is ultimately decommissioned and dismantled. The nuclear power industry is not the sole purveyor of radioactive wastes, however. Hospitals have become a major source of such wastes due to the widespread use of radiopharmaceuticals in medicine. Radioactive wastes are also a significant by-product of nuclear weapons programs.

It is usual to classify waste in four categories. *High-level waste* consists of spent fuel, if this is discarded as waste, and any wastes generated in the first stages of a fuel reprocessing plant (since this waste contains the bulk of the fission products). *Transuranic (TRU) waste* consists mostly of the isotopes of plutonium at concentrations in excess of 10^{-9} Ci/g; TRU wastes are generated by fuel reprocessing, plutonium fuel fabrication, and manufacturing of nuclear weapons. *Low-level waste* contains less than 10^{-9} Ci/g of TRU nuclides. It also includes material that is free of TRU and requires little or no shielding, but it is still potentially dangerous. *Mine and mill tailings* consist of residues from uranium mining and milling operations; such residues contain low concentrations of naturally occurring radionuclides.

A perusal of the chart of nuclides reveals that the vast majority of the several hundred fission products are very short-lived. Only five have half-lives between 1 and 5 years; two–namely, ^{90}Sr and ^{137}Cs–have half-lives of about 30 years, and three—^{93}Zr, ^{129}I, and ^{135}Cs–have half-lives in excess of a million years, and hence are effectively stable. In 100 years, the activity of a 5-year nuclide decreases by a factor of 10^6, and the shorter lived nuclides disappear altogether. Over the long-term, therefore, the fission product activity of high-level waste is due only to ^{90}Sr and ^{137}Cs. The latter radionuclide decays into stable ^{137}Ba. However, ^{90}Sr decays to ^{90}Y, which decays with a 64-hr half-life to stable ^{90}Zr. Thus, high-level waste ultimately contains three fission products, ^{90}Sr, ^{90}Y and ^{137}Cs.

The half-lives of many of the TRU nuclides tend to be considerably longer. For instance, the half-life $T_{1/2}$ of ^{239}Pu is 24,000 years. Therefore, the activity of these nuclides dies off more slowly than that of the fission products. In the spent fuel from a typical LWR, the TRU activity exceeds the fission product activity after approximately 700 years.

The total high-level activity from spent fuel, including both the fission products and the TRU nuclides, depends on the nature of the reactor fuel cycle. With the once-through cycle, the activity of the fuel persists for hundreds of thousands of years owing to the presence of TRU material. However, with a closed cycle,

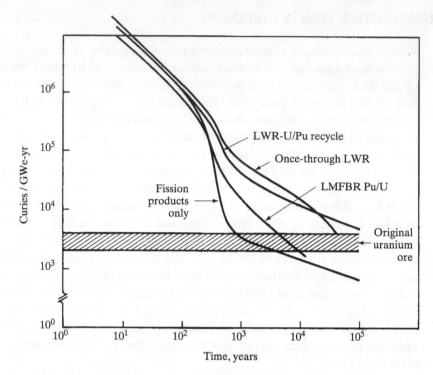

Figure 4.56 High-level waste activities from different reactor fuel cycles. (Based on *Nuclear Proliferation and Civilian Nuclear Power,* U.S. Department of Energy report DOE/NE-0001/9, Volume 9, 1980.)

the plutonium isotopes are returned to the reactor where, on fissioning, they are transformed into short-lived fission products.

This situation is illustrated in Fig. 4.56, where the activity of high-level waste generated in 1 GWE year is shown for the LWR with and without recycling; for the LMFBR (with recycling, of course) and for a special case in which LWR spent fuel is reprocessed, the fission products are treated as waste and the plutonium is merely stored. Except in the early and late years, the activity in this last cycle is due almost entirely to ^{90}Sr, ^{90}Y, and ^{137}Cs. Also shown in the figure is the range of activities of the uranium ore required to produce the original fuel. It is observed that the high-level activity associated with the fuel cycles approaches within an order of magnitude of the ore in about 1,000 years.[18]

[18]A more fair comparison of ore and high-level waste should take into account the relative toxicity of the heavy elements, which decay in large part by α-emission, with that of fission products, which emit β-rays. The toxicity of α-emitters is about 10 times greater than that of β-rays (see Section 9.2).

Most of the high-level waste in the world today is a by-product of nuclear weapons programs. For the most part, pending permanent disposal, this waste is either stored in liquid form in large tanks or is being solidified to reduce the possibility of leakage. Increasing amounts of high-level waste are also accumulating in the spent-fuel pools of nuclear power plants. Commercial reprocessing of spent fuel is occurring in a number of foreign countries. However, because of proliferation concerns, the United States is not participating.

How commercial high-level waste is disposed of clearly depends on whether or not spent fuel is reprocessed. With no reprocessing, the fuel may be disposed of intact. Presumably this would be done by placing the spent fuel assemblies in suitable containers and burying these containers in some stable geological setting. Historically, stable rock formations have been considered for this purpose.

If the fuel is reprocessed and the plutonium is recycled (or stored), the disposal problem becomes more manageable and susceptible to unique technological solution. Since fission products represent a small fraction of the mass of the fuel, reprocessing substantially reduces the volume of the waste. The waste, in liquid form, can then be calcinated—that is, dried at high temperature; mixed with *frit,* the substance from which glass is made; and then vitrified—that is, made into glass. Another newer technique involves surrounding beads of waste with layers of ceramic as in the fabrication of HTGR fuel. These and similar methods for solidifying the waste effectively immobilize the radioactive particles, something that cannot be done as easily with unprocessed spent fuel. The glass or ceramic complexes are finally placed in cannisters and deposited in stable geological formations.

If a canister holding either a whole fuel assembly or solidified waste should disintegrate, even soon after its emplacement in a repository, there is good reason to believe that the fission products and TRU nuclides would not diffuse far into the environment. Strong support for this contention is furnished by what has become known as the *Oklo phenomenon.* Oklo is the name of a uranium mine in the African nation of Gabon, where France obtains much of the uranium for its nuclear program. When uranium from this mine was introduced into a French gaseous diffusion plant, it was discovered that the feed uranium was already depleted below the 0.711 *w/o* of ordinary natural uranium. It was as if the uranium had already been used to fuel some unknown reactor.

And so it had. French scientists found traces of fission products and TRU waste at various locations within the mine. These observations were puzzling at first because it is not possible to make a reactor go critical with natural uranium, except under very special circumstances with a graphite or heavy-water moderator, neither of which could reasonably be expected to have ever been present in the vicinity of Oklo. The explanation of the phenomenon is to be found in the fact that the half-life of ^{235}U, 7.13×10^8 years, is considerably shorter than the half-life of ^{238}U, 4.51×10^9 years. Since the original formation of the earth, more ^{235}U has therefore decayed than ^{238}U. This, in turn, means that the enrichment of natural uranium was greater years ago than it is today. Indeed, it is easy to show (see Problem 2.37) that about 3 billion years ago this enrichment was in the neighborhood of 3 w/o—sufficiently high to form a critical assembly with ordinary water, which is known to have been present near Oklo at that time. The relevance of the Oklo phenomenon to present-day disposal of radioactive wastes is that neither the fission products (identified by their stable daughters) nor the plutonium migrated from the Oklo site in the billions of years since the reactor was critical.

Compared with high-level waste, low-level waste represents more of a nuisance than a hazard. However, it comprises a much larger volume. A nominal 1,000 MWe LWR power plant produces the order of 100 m^3 of solid low-level waste, mostly contaminated laundry wastes, protective clothing, glassware, tools, containers, and so on, and somewhat higher level waste in the form of spent resins from reactor coolant demineralizers, air filters, and so on. Such wastes are normally placed in drums and shipped off-site to waste depositories, where the drums, several hundred per GWe-year, are buried. Liquid low-level wastes are usually solidified or retained for decay, diluted, and discharged to the environment.

In terms of total activity, a typical large hospital generates more low-level waste than a nuclear power plant. The bulk of this activity is due to ^{3}H and ^{14}C; the remainder consists of short-lived nuclides that quickly decay. Fortunately, the total amount of ^{3}H and ^{14}C originating in medical institutions is trivial compared with their natural production rates in the atmosphere from cosmic rays (see Section 9.7). At least in principle, these wastes can safely be incinerated, although, for political reasons, this procedure has not been generally adopted.

The major source of concern from uranium mining and mill tailings is the increased release of the radioactive gas radon—in particular, the isotope ^{222}Rn, which has a half-life of 3.8 days. This nuclide is one of the products in the long decay chain beginning with ^{238}U and is the immediate daughter of the decay of ^{226}Ra. Chemically, radon is a noble gas, and therefore it readily diffuses out of solid materials containing uranium or radium. Although radon does not present a health hazard, its longer lived daughters do, especially ^{210}Pb ($T_{1/2} = 19.4$ years). When these daughter products, formed by the decay of radon in the atmosphere, are inhaled, they may become attached to the tissues at the base of the bronchial network. Their subsequent decay can lead to lung cancer (see Chap. 9 for further discussion of this process). Disposal of such tailings is either by placement underground, the preferred but more costly method, or by covering the tailings with no less than 3 m of earth and then planting vegetation to prevent erosion.

REFERENCES

General

Bodansky, D., *Nuclear Energy Principles, Practices, and Prospects*. Woodbury: AIP Press, 1996.

El-Wakil, M. M., *Nuclear Energy Conversion*, Revised. LaGrange Park: American Nuclear Society, 1982, Chapters 1 and 3–12.

Foster, A. R., and R. L. Wright, Jr., *Basic Nuclear Engineering*, 4th ed. Paramus: Prentice-Hall, 1982, Chapter 1.

Glasstone, S., and A. Sesonske, *Nuclear Reactor Engineering*, 4th ed. New York: Chapman & Hall.

Lamarsh, J. R., *Introduction to Nuclear Reactor Theory*. Reading, Mass.: Addison-Wesley Longman, 1966, Chapter 4.

Oldenburg, O., and N. C. Rasmussen, *Modern Physics for Engineers*. New York: McGraw-Hill, 1966, Chapters 1, 2, 12, 14, and 15.

Wills, J. G., *Nuclear Power Plant Technology*, Reprint. Marietta: Technical Books, 1992, Chapters 1–4.

Fuel Enrichment

Benedict, M., and T. H. Pigford, *Nuclear Chemical Engineering*, 2nd ed. New York: McGraw-Hill, 1981.

Cohen, K., "The Theory of Isotope Separation," Chapter 6 in National Nuclear Energy Series, div. iii, vol. 1B. New York: McGraw-Hill, 1951.

Villani, S., *Uranium Enrichment*. New York: Springer-Verlag, 1979.

Von Halle, E., R. L. Hoglund, and J. Shacter, "Diffusion Separation Methods" in *Encyclopedia of Chemical Technology*, 2nd ed., vol. 7. New York: Interscience, 1965.

AEC Gaseous Diffusion Plant Operations, U.S. Atomic Energy Commission report ORO-684-1972.

Nuclear Fuel Resources

Nuclear Fuel Supply, U.S. Atomic Energy Commission report WASH-1242, 1973.

Nuclear Energy and Its Fuel Cycle: Prospects to 2025, Paris: OECD, Nuclear Energy Agency, 1982.

Uranium, 1995 Resources, Production, and Demand, periodic reports of the OECD, Nuclear Energy Agency, Paris, and the International Atomic Energy Agency, Vienna.

Nuclear Fuel Resources

Cochran, R. G. and N. Tsoulfandis, *The Nuclear Fuel Cycle: Analysis and Management*. Lagrange Park: American Nuclear Society, 1990.

Nuclear Reactor Types

Almenas, K., A. Kaliatka, and E. Uspuras, *IGNALINA RBMK-1500 - A Source Book*. Lithuanian Energy Institute, 1994.

Freeman, L. B., et al., "Physics Experiments and Lifetime Performance of a Light Water Breeder Reactor," Nucl. Sci. and Eng., 102, 341–364, 1989.

Glasstone, S., and A. Sesonske, *Nuclear Reactor Engineering*, 4th ed. New York: Chapman & Hall, 1994, Chapters 13 and 15.

Judd, A. M., *Fast Breeder Reactors: An Engineering Introduction*. New York: Pergamon Press Reprint, 1981.

Lahey, R.T., and F. J. Moody, *The Thermal Hydraulics of a Boiling Water Nuclear Reactor*, 2nd ed. LaGrange Park: American Nuclear Society, 1993, Chapter 1.

See also *Nuclear Engineering International* and *Nuclear News* for periodic updates on reactor developments, uranium supply, and enrichment technologies.

PROBLEMS

1. If 57.5% of the fission neutrons escape from a bare sphere of ^{235}U, what is the multiplication factor of the sphere? In this system, the average value of η is 2.31.

2. Measurements on an experimental thermal reactor show that, for every 100 neutrons emitted in fission, 10 escape while slowing down and 15 escape after having slowed down to thermal energies. No neutrons are absorbed within the reactor while slowing down. Of those neutrons absorbed at thermal energies, 60% are absorbed in fission material. (a) What is the multiplication factor of the reactor at the time these observations are made? (b) Suppose the thermal leakage is reduced by one third. How would this change the value of k? [*Note:* The values of η and ν for the reactor fuel are 2.07 and 2.42, respectively.]

3. (a) Show that the energy released in the nth generation of a fission chain reaction, initiated by one fission, is given by

$$E_n = k^n E_R,$$

where k is the multiplication factor and E_R is the recoverable energy per fission. (b) Show that the total energy released up to and including the nth generation is given by

$$E_n = \frac{k^{n+1} - 1}{k - 1} E_R$$

4. Show that the fraction, F, of the energy released from a supercritical chain reaction that originates in the final m generations of the chain is given approximately by

$$F = 1 - k^{-m},$$

provided the total number of generations is large.

5. (a) Most of the energy from a nuclear explosion is released during the final moments of the detonation. Using the result of the previous problem, compute the number of fission generations required to release 99% of the total explosive yield. Use the nominal value $k = 2$. (b) If the mean time between generations is the order of 10^{-8} sec over what period of time is energy released during a nuclear explosion?

6. A burst of 1×10^9 neutrons from a pulsed accelerator is introduced into a subcritical assembly consisting of an array of natural uranium rods in water. The system has a multiplication factor of 0.968. Approximately 80% of the incident neutrons are absorbed in uranium. (a) How many first-generation fissions do the neutrons produce in the assembly? (b) What is the total fission energy in joules released in the assembly by the neutron burst?

7. It is found that, in a certain thermal reactor, fueled with partially enriched uranium, 13% of the fission neutrons are absorbed in resonances of ^{238}U and 3% leak out of the reactor, both while these neutrons are slowing down; 5% of the neutrons that slow down in the reactor subsequently leak out; of those slow neutrons that do not leak out, 82% are absorbed in fuel, 74% of these in ^{235}U. (a) What is the multiplication factor of this reactor? (b) What is its conversion ratio?

8. A natural uranium-fueled converter operates at a power of 250 MWt with a conversion ratio of 0.88. At what rate is ^{239}Pu being produced in this reactor in kg/year?

9. Assuming a recoverable energy per fission of 200 MeV, calculate the fuel burnup and consumption rates in g/MWd for (a) thermal reactors fueled with ^{233}U or ^{239}Pu; (b) fast reactors fueled with ^{239}Pu. [*Note:* In part (b), take the capture-to-fission ratio to be 0.065.]

10. Because of an error in design, a thermal reactor that was supposed to breed on the ^{232}Th–^{233}U cycle unfortunately has a breeding ratio of only 0.96. If the reactor operates at a thermal power level of 500 megawatts, how much ^{232}Th does it convert in 1 year?

11. What value of the breeding gain is necessary for a fast breeder operating on the ^{238}U–^{239}Pu cycle to have an exponential doubling time of 10 years if the specific power for this type of reactor is 0.6 megawatts per kilogram of ^{239}Pu?

12. Per GWe-year, a typical LMFBR produces 558 kg and consumes 789 kg of fissile plutonium in the core while it produces 455 kg and consumes 34 kg of fissile plutonium in the blanket. What is the breeding ratio for this reactor? [*Note:* As would be expected, there is a net consumption of plutonium in the core and a net production in the blanket, with a positive output from the reactor as a whole. By adjusting the properties of the blanket, it is easy to make breeders into net consumers of plutonium if dangerously large stockpiles of this material should ever accumulate in the world.]

13. A certain fossil fueled generating station operates at a power of 1,000 MWe at an overall efficiency of 38% and an average capacity factor of 0.70. (a) How many tons of 13,000 Btu per pound of coal does the plant consume in 1 year?(b) If an average coal-

carrying railroad car carries 100 tons of coal, how many car loads must be delivered to the plant on an average day? (c) If the coal contains 1.5% by weight sulfur and in the combustion process this all goes up the stack as SO_2, how much SO_2 does the plant produce in 1 year?

14. Had the plant described in Problem 4.13 been fueled with 6.5 million Btu per barrel bunker-C fuel oil containing .37% sulfur, (a) how many barrels and tons of oil would the plant consume in 1 year? (b) how much SO_2 would the plant release in 1 year? [*Note:* 1 U.S. petroleum barrel = 5.61 cubic feet = 42 U.S. gallons; the density of bunker-C oil is approximately the same as water.]

15. Rochester Gas and Electric's Robert Emmett Ginna nuclear power plant operates at a net electric output of 470 megawatts. The overall efficiency of the plant is 32.3%. Approximately 60% of the plant's power comes from fissions in ^{235}U, the remainder from fissions in converted plutonium, mostly ^{239}Pu. If the plant were operated at full power for 1 year, how many kilograms of ^{235}U and ^{239}Pu would be (a) fissioned? (b) consumed?

16. Consider two electrical generating stations—one a fossil fuel plant and the other nuclear—both producing the same electrical power. The efficiency of the fossil fuel plant is $(eff)_f$, whereas that of the nuclear plant is $(eff)_n$.
(a) Show that the ratio of the heat rejected to the environment by the two plants is given by

$$\frac{Q_{cn}}{Q_{cf}} = \frac{1 - (eff)_n}{1 - (eff)_f} \times \frac{(eff)_f}{(eff)_n}.$$

(b) Evaluate Q_{cn}/Q_{Cf} for the case where $(eff)_f = 38\%$ and $(eff)_n = 33\%$.

17. An MSBR power plant produces 1,000 MWe at an overall efficiency of 40%. The breeding ratio for the reactor is 1.06, and the specific power is 2.5 MWt per kilogram of ^{233}U. (a) Calculate the linear and exponential doubling times for this reactor. (b) What is the net production rate of ^{233}U in kg/year?

18. A 3,000 MW reactor operates for 1 year. How much does the mass of the fuel change during this time as the result of the energy release?

19. Referring to the nominal LWR cycle described in Fig. 4.37, compute (a) the specific burnup of the fuel in MWd/t; (b) the fractional burnup of the fuel; (c) the enrichment of the fresh fuel; (d) the enrichment (^{235}U) of the spent fuel; (e) the fraction of the power originating in fissions in ^{235}U and plutonium, respectively; (f) the 30-year requirement for uranium for this system. The plant operates at an overall efficiency of 33.4%.

20. Show that the fuel cycle data on the LMFBR in Fig. 4.40 are consistent with the reactor operating at a thermal efficiency of 36.5%.

21. Referring again to the LMFBR in Fig. 4.40, (a) what fraction of the total power is produced by the core and blanket, respectively? (b) What is the average specific burnup of the fuel in the core? (c) What is the breeding ratio of the reactor?

22. Using the data in Fig. 4.41, compute (a) the enrichment of the fresh fuel; (b) the residual enrichment of the spent fuel; (c) the fraction of power originating in fissions

in ^{235}U; (d) the burnup of the fuel. [*Note:* The reactor operates at an efficiency of 30%.]

23. Reliable nuclear weapons cannot be made using plutonium containing much more than 7 w/o^{240}Pu because this isotope has a high spontaneous fission rate that tends to preinitiate—that is, fizzle—the device. Using Fig. 4.37 and Figs. 4.39 through 4.40, determine whether any of these commercial power systems produce weapons-grade plutonium ($< 7\ w/o\ ^{240}$Pu). Assume that all nonfissioning plutonium is ^{240}Pu (some of it is actually ^{242}Pu).

24. Compute and plot annual and cumulative uranium requirements through the year 2015 for the projected nuclear electric capacities given in Table 4.8 for each of the following assumptions: (a) all reactors are PWRs with no recycling; (b) all reactors are PWRs, and all operate with recycling after 2000; (c) all reactors are PWRs until 2000 after which 20% of new reactors are LMFBRs. [*Note:* For simplicity, ignore the extra fuel required for startup; that is, take the annual uranium needs to be the values in Table 4.7 divided by 30.]

25. A nuclear fuel-fabricating company needs 10,000 kg of 3 w/o uranium and furnished natural uranium (as UF$_6$) as feed. Assuming a tails assay of 0.2 w/o, (a) how much feed is required? (b) How much will the enrichment cost?

26. Suppose that the tails enrichment in the preceding problem were decreased to 0.15 w/o. (a) How much uranium feed would this save? (b) How much more would the enriched fuel cost? (c) What is the cost per kg of feed saved?

27. If a customer supplies DOE as part of his feed—uranium that is partially enriched above (or depleted below) the natural uranium level—the total amount of natural uranium feed that he must furnish to obtain a given amount of product is reduced (or increased) by an amount equal to the natural uranium feed needed to provide the partially enriched (or depleted) feed. The customer receives a credit (or debit) for the separative work represented by his enriched (or depleted) feed.

   Suppose that the fuel-fabricating company in Problem 4.25 offers to supply as part of its feed 10,000 kg of 1% assay uranium. How much natural uranium feed is required in addition, and how much will the total job cost?

28. Repeat Problem 4.27 for the case in which the customer furnishes 10,000 kg of .6% feed.

29. Referring to Problem 4.28, suppose exactly 33,000 kg of 3.2 w/o UO$_2$ is required per reload. (a) How much UF$_6$ must be given to the enrichment plant, assuming 1% loss in fabrication? (b) How much yellow cake must be used in the conversion to UF$_6$ assuming .5% loss in conversion?

30. It is proposed to produce 25 kg of 90 w/o for a nuclear weapon by enriching 20 w/o fuel from a research reactor. (a) How much fresh reactor fuel would be required? (b) Compute the total SWU required. (c) Compare the SWU/kg required to produce 20 w/o fuel starting from natural uranium with the SWU/kg for 90 w/o material beginning at 20 w/o. [*Note:* Assume the tails enrichment to be 0.2 w/o.]

31. (a) Derive an explicit expression for the mass of tails produced in a specified enrichment requiring a given amount of separative work. (b) Show that in enriching natu-

ral uranium to 3 w/o, approximately 1 kg of 0.2 w/o tails are produced per SWU expended. (c) In 1981, the total enrichment capacity in operation in the world was 28.4×10^6 SWU/yr. At what rate was depleted uranium being produced, assuming all the capacity was used to produce 3 w/o fuel?

32. In 1980, the United States had approximately 300,000 tons of depleted uranium in storage at its gaseous diffusion plants. If the entire 1980 electrical capacity of 600 GWe were furnished by LMFBRs fueled with this uranium, how long would the 1980 depleted uranium stock (which is continuing to grow; see preceding problem) last? Use LMFBR data from Fig. 4.40.

33. Show that approximately 116,000 SWU of separative work is required annually to maintain the nominal 1,000 MWe LWR described in Fig. 4.37.

34. The enrichment consortium EURODIF operates gaseous diffusion plants in France with a capacity of 10.8 million SWU/year. Using the result of the preceding problem, how much LWR capacity can this company service?

35. Compute and plot the relative cost in SWU of enriched uranium, per unit of contained ^{235}U, as a function of enrichment to 93 w/o. Compare with Fig. 4.45.

36. Based on the projections of future installed nuclear electric generating capacity given in Table 4.8, how large an industry (dollars per year) can isotope enrichment be expected to be in 2010 based on current U.S. SWU prices if most of the capacity is in LWRs?

37. The principle of enrichment by gaseous diffusion can be seen in the following way. Consider a chamber (diffuser) split into two volumes, A and B, by a porous barrier with openings having a total area S as shown in Fig. 4.48. Volume A contains two isotopic species of mass M_1 and M_2 at atom densities N_{10} and N_{20}, respectively. Volume B is at low pressure and is pumped out at the rate of F m³/sec. According to kinetic theory, the number of atoms or molecules in a gas striking the wall of a container per m²/sec is

$$J = \frac{1}{4}N\bar{v},$$

where N is the atom density and $\bar{v}$ is the average speed given by

$$\bar{v} = 2\sqrt{\frac{2kT}{\pi M}},$$

where M is the mass of the atom and T is the absolute temperature. (a) Show that in equilibrium the atom concentration in B is

$$\frac{N}{V_B} = (J_1 + J_2)\frac{S}{F},$$

where J_1 and J_2 refer to the two isotopic species. (b) Show that the ratio of equilibrium concentrations in B is given by

$$\frac{N_1}{N_2} = \frac{N_{1o}}{N_{2o}} \sqrt{\frac{M_2}{M_1}}.$$

38. The quantity

$$\frac{N_1/N_{1o}}{N_2/N_{2o}} = \sqrt{\frac{M_2}{M_1}}$$

in the preceding problem is called the *ideal separation factor* α_o. Compute α_o for a gas consisting of molecules of $^{235}UF_6$ and $^{238}UF_6$. (The small value of α_o in this case means that a great many stages must be used to enrich uranium to useful levels.)

39. If, using LIS, it is possible to enrich diffusion plant tails from 0.2 *w/o* to 3 *w/o* with residual tails of 0.08 *w/o*, verify that the utilization of natural uranium for 3 *w/o* fuel would be increased by 18%.

40. To how much energy in eV does the isotopic shift of 0.1 Å shown in Fig. 4.52 correspond?

41. (a) Show that the specific activity of fuel irradiated for T days up to a burnup of B MWd/kg, t days after removal from a reactor, is given by

$$\alpha = 1.4 \times 10^6 \frac{B}{T}[t^{-0.2} - (t + T)^{-0.2}]Ci/kg.$$

(b) Compute and plot the specific activity for fuel irradiated for 3 years to a burnup of 33 MWd/kg from 1 day to 1 year after removal from the reactor.

42. The fission yields of ^{90}Sr and ^{137}Cs are 0.0593 atoms per fission, respectively. Calculate the specific activity due to these nuclides and to ^{90}Y in spent fuel irradiated to 33 MWd/kg from the time the fuel is removed from the reactor up to 1,000 years.

43. Carbon-14, with a half-life of 5,730 years, is produced in LWRs by way of an (n,p) reaction with nitrogen impurities (in both the fuel and the coolant water) and via an (n,α) reaction with ^{17}O. About 60 to 70 Ci of ^{14}C are generated this way per GWe-year. (a) What is the total ^{14}C activity in the fuel unloaded each year from a 1,000 MWe reactor that has operated with a 0.70 capacity factor? Assume that one third of the core is unloaded each year. (b) How much ^{14}C will be produced per year in the year 2010 by all the reactors in the world if high or low projections of Table 4.8 come to pass and all reactors are LWRs?

5

Neutron Diffusion and Moderation

To design a nuclear reactor properly, it is necessary to predict how the neutrons will be distributed throughout the system. The importance of a knowledge of the neutron distribution is illustrated in an example in the following section. Unfortunately, determining the neutron distribution is a difficult problem in general. The neutrons in a reactor move about in complicated paths as the result of repeated nuclear collisions. To a first approximation, however, the overall effect of these collisions is that the neutrons undergo a kind of diffusion in the reactor medium, much like the diffusion of one gas in another. The approximate value of the neutron distribution can then be found by solving the diffusion equation—essentially the same equation used to describe diffusion phenomena in other branches of engineering such as molecular transport. This procedure, which is sometimes called the *diffusion approximation,* was used for the design of most of the early reactors. Although more sophisticated methods have now been developed, it is still widely used to provide first estimates of reactor properties.

5.1 NEUTRON FLUX

It was shown in Section 3.4 that, for monoenergetic neutrons, the reaction rate F is related to the macroscopic cross-section Σ_i and the flux ϕ by:

$$F = \Sigma_t \phi.$$

or

$$F = \Sigma_t n v \tag{5.1}$$

It is easy to enlarge these results to include neutrons that have a distribution of energies. Thus, let $n(E)$ be defined as the neutron density per unit energy; that is, $n(E)dE$ is the number of neutrons per cm^3 with energies between E and $E + dE$. From Eq. (5.1), the interaction rate for these essentially monoenergetic neutrons is

$$dF = \Sigma(E) \times n(E)dE \times v(E), \tag{5.2}$$

where the energy dependence of all parameters is noted explicitly. The total interaction rate is then given by the integral

$$F = \int_0^\infty \Sigma_t(E)n(E)v(E)dE = \int_0^\infty \Sigma_t(E)\phi(E)dE, \tag{5.3}$$

where

$$\phi(E) = n(E)v(E) \tag{5.4}$$

is called the *energy-dependent flux* or the *flux per unit energy*. The limits on the integrals in Eq. (5.3) are written as 0 and ∞ to indicate that the integration is to be carried out over all neutron energies.

Equation (5.3) refers to the total interaction rate. Rates of particular interactions can be found from similar expressions. Thus, the number of scattering collisions per cm^3/sec is

$$F_s = \int_0^\infty \Sigma_s(E)\phi(E)dE; \tag{5.5}$$

the number of neutrons absorbed per cm^3/sec is

$$F_a = \int_0^\infty \Sigma_a(E)\phi(E)dE; \tag{5.6}$$

and so on.

In the next few sections, it is assumed that the neutrons are monoenergetic. The diffusion of nonmonoenergetic neutrons is considered in Section 5.8.

5.2 FICK'S LAW

Diffusion theory is based on *Fick's law,* which was originally used to account for chemical diffusion. It was shown early in chemistry that if the concentration of a

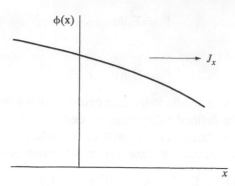

Figure 5.1   Neutron flux and current.

solute is greater in one region of a solution than in another, the solute diffuses from the region of higher concentration to the region of lower concentration. Furthermore, it was found that the rate of solute flow is proportional to the negative of the gradient of the solute concentration. This is the original statement of Fick's law.

To a good approximation, neutrons in a reactor behave in much the same way as a solute in a solution. Thus, if the density (or flux) of neutrons is higher in one part of a reactor than in another, there is a net flow of neutrons into the region of lower neutron density. For example, suppose that the flux[1] varies along the x-direction as shown in Fig. 5.1. Then Fick's law is written as

$$J_x = -D\frac{d\phi}{dx}. \qquad (5.7)$$

In this expression, J_x is equal to the *net* number of neutrons that pass per unit time through a unit area perpendicular to the x-direction. J_x has the same units as flux—namely, neutrons/cm^2-sec. The parameter D in Eq. (5.7) is called the *diffusion coefficient* and has units of cm.

Equation (5.7) shows that if, as in Fig. 5.1, there is a negative flux gradient, then there is a net flow of neutrons along the positive x-direction as indicated in the figure. To understand the origin of this flow, consider the neutrons passing through the plane at $x = 0$. These neutrons pass through the plane from left to right as the result of collisions to the left of the plane; conversely, they flow from right to left as the result of collisions to the right of the plane. However, since the concentration of neutrons and the flux is larger for negative values of x, there are more collisions per cm^3/sec on the left than on the right. Therefore, more neutrons are scattered from left to right than the other way around, with the result that there is a net flow of neutrons in the positive x-direction through the plane, just as predicted by Eq. (5.7). It is important to recognize that neutrons do not flow from regions of high

---

[1]It is usual in nuclear engineering to make calculations with the flux, which is proportional to neutron density, rather than with the density.

flux to low flux because they are in any sense pushed that way. There are simply more neutrons scattered, or moving, in one direction than in the other.

The flux is generally a function of three spatial variables, and in this case Fick's law is

$$\mathbf{J} = -D \text{ grad } \phi = -D\nabla\phi. \tag{5.8}$$

Here $\mathbf{J}$ is known as the *neutron current density vector* or simply *the current,* and grad $= \nabla$ is the gradient operator. We have assumed that the diffusion coefficient, D, is not a function of the spatial variables.[2] The physical significance of the vector $\mathbf{J}$ may be seen by taking the dot product of $\mathbf{J}$ with a unit vector in the x-direction $\mathbf{a}_x$. This gives the x-component of $\mathbf{J}$, namely J_x:

$$\mathbf{J} \cdot \mathbf{a}_x = J_x,$$

which, as already noted, is equal to the net flow of neutrons per second per unit area normal to the x-direction. It follows that if $\mathbf{n}$ is a unit vector pointing in an *arbitrary* direction, then

$$\mathbf{J} \cdot \mathbf{n} = J_n \tag{5.9}$$

is equal to the net flow of neutrons per second per unit area normal to the direction of $\mathbf{n}$.

Example 5.1

In Section 5.6, it is shown that the flux at the distance r from a point source emitting S neutrons per second in an infinite moderator is given by the formula

$$\phi(r) = \frac{Se^{-r/L}}{4\pi Dr},$$

where L is a constant. Find expressions for (a) the neutron current in the medium, (b) the net number of neutrons flowing out through a sphere of radius r surrounding the source.

Solution.

1. Because of the geometry of the problem, the neutron current density vector clearly must point outward in the radial direction. From Appendix III, the r-component of the gradient, ∇_r, in spherical coordinates is

$$\nabla_r = \mathbf{a}_r \frac{d}{dr},$$

   where $\mathbf{a}_r$ is a unit radial vector. Then from Fick's law, Eq. (5.8),

$$\mathbf{J}(r) = -D\mathbf{a}_r \frac{d}{dr}\left(\frac{Se^{-r/L}}{4\pi Dr}\right) = \mathbf{a}_r \frac{S}{4\pi}\left(\frac{1}{r^2} + \frac{1}{rL}\right)e^{-r/L}. \text{ [Ans.]}$$

---

[2]Vectors and vector operators are discussed in Appendix III.

2. Since $\mathbf{J}$ is everywhere normal to the surface of the sphere, the net number of neutrons crossing per unit area of the sphere is just the magnitude of $\mathbf{J}$, that is, $\mathbf{J} \cdot \mathbf{a}_r = J$. The net flow through the whole sphere of area $4\pi r^2$ is then

$$4\pi r^2 J(r) = S\left(1 + \frac{r}{L}\right)e^{-r/L}. \ [Ans.]$$

Returning to Eqs. (5.7) and (5.8), it may be noted that, since J_x and $\mathbf{J}$ have the same units as ϕ, D has units of length.[3] It can be shown by arguments,[4] which are too lengthy to be reproduced here, that D is given *approximately* by the following formula:

$$D = \frac{\lambda_{tr}}{3}, \tag{5.10}$$

where λ_{tr} is called the *transport mean free path* and is given in turn by

$$\lambda_{tr} = \frac{1}{\Sigma_{tr}} = \frac{1}{\Sigma_s(1 - \overline{\mu})}. \tag{5.11}$$

In this equation, Σ_{tr} is called the *macroscopic transport cross-section*. Σ_s is the macroscopic scattering cross section of the medium, and $\overline{\mu}$ is the average value of the cosine of the angle at which neutrons are scattered in the medium. The value of $\overline{\mu}$ at most of the neutron energies of interest in reactor calculations can be computed from the simple formula

$$\overline{\mu} = \frac{2}{3A}, \tag{5.12}$$

Example 5.2

The scattering cross-section of carbon at 1 eV is 4.8 b. Estimate the diffusion coefficient of graphite at this energy.

Solution. Using Eq. (5.12) with $A = 12$ gives $\overline{\mu} = \dfrac{2}{36} = 0.555$. From Table II.3, the atom density of graphite is 0.08023×10^{24}. Introducing these values into Eqs. (5.11) and (5.12) gives

$$D = \frac{1}{3\Sigma_s(1 - \overline{\mu})} = \frac{1}{3 \times 0.08023 \times 4.8(1 - 0.055)}$$
$$= 0.916 \text{ cm.} \ [Ans.]$$

It must be emphasized that Fick's law is not an exact relation. Rather, it is an approximation that in particular is not valid under the following conditions:

---

[3]In chemistry and chemical engineering, Fick's law is written in terms of concentration rather than flux. The units of the diffusion coefficient in that case are cm²/sec.
[4]See particularly the first two references by Glasstone et al.

1. In a medium that strongly absorbs neutrons;
2. Within about three mean free paths of either a neutron source or the surface of a medium; and
3. When the scattering of neutrons is strongly anisotropic.

To some extent, these limitations are present in every practical reactor problem. Nevertheless, as noted earlier, Fick's law and diffusion theory are often used to estimate reactor properties. Higher order methods are available for cases near sources or boundaries and in strongly absorbing media.

5.3 THE EQUATION OF CONTINUITY

Consider an arbitrary volume V within a medium containing neutrons. As time goes on, the number of neutrons in V may change if there is a net flow of neutrons out of or into V, if some of the neutrons are absorbed within V, or if sources are present that emit neutrons within V. The *equation of continuity* is the mathematical statement of the obvious fact that, since neutrons do not disappear unaccountably,[5] the time rate of change in the number of neutrons in V must be accounted for in terms of these processes. In particular, it follows that

$$\begin{bmatrix} \text{Rate of change in} \\ \text{number of neutrons in } V \end{bmatrix} = \begin{bmatrix} \text{rate of production} \\ \text{of neutrons in } V \end{bmatrix}$$

$$- \begin{bmatrix} \text{rate of absorption} \\ \text{of neutrons in } V \end{bmatrix} - \begin{bmatrix} \text{rate of leakage of} \\ \text{neutrons from } V \end{bmatrix} \qquad (5.13)$$

Each of these terms is considered in turn.

Let n be the density of neutrons at any point and time in V. The total number of neutrons in V is then

$$\int_V n\, dV,$$

where the subscript on the integral indicates that the integration is to be performed throughout V. The rate of change in the number of neutrons is

$$\frac{d}{dt} \int_V n\, dV,$$

---

[5]Neutrons do disappear, of course, when they undergo β-decay. However, this process has a comparatively long half-life and need not be taken into consideration.

which can also be written as

$$\int_V \frac{\partial n}{\partial t} dV.$$

In moving the time derivative inside the integral, it is necessary to change to partial derivative notation because n may be a function of space variables as well as time.

Next, let s be the rate at which neutrons are emitted from sources per cm^3 in V. The rate at which neutrons are produced throughout V is given by

$$\text{Production rate} = \int_V s\,dV.$$

The rate at which neutrons are lost by absorption per cm^3/sec is equal to $\Sigma_a\phi$, where Σ_a is the macroscopic absorption cross-section (which may be a function of position) and ϕ is the neutron flux. Throughout the volume V, the total loss of neutrons per second due to absorption is then

$$\text{Absorption rate} = \int_V \Sigma_a\phi\,dV.$$

Consider next the flow of neutrons into and out of V. If $\mathbf{J}$ is the neutron current density vector on the surface of V and $\mathbf{n}$ is a unit normal pointing outward from the surface, then, according to the results of the preceding section,

$$\mathbf{J}\cdot\mathbf{n}$$

is the net number of neutrons passing outward through the surface per cm^2/sec. It follows that the total rate of leakage of neutrons (which may be positive or negative) through the surface A of the volume is

$$\text{Leakage rate} = \int_A \mathbf{J}\cdot\mathbf{n}\,dA.$$

This surface integral can be transformed into a volume integral by using the divergence theorem (see Appendix III). Thus,

$$\int_A \mathbf{J}\cdot\mathbf{n}\,dA = \int_V \text{div } \mathbf{J}\,dV,$$

and so

$$\text{Leakage rate} = \int_V \text{div } \mathbf{J}\,dV.$$

The equation of continuity can now be obtained by introducing the prior results into Eq. (5.13). This gives

$$\int_V \frac{\partial n}{\partial t} dV = \int_V s\, dV - \int_V \Sigma_a \phi\, dV - \int_V \text{div } \mathbf{J}\, dV.$$

All of the previous integrals are to be carried out over the same volume, and so their integrands must also be equal. The equation must hold for any arbitrary volume. Therefore, the integrands on the right when summed must equal to the integrand on the left. Thus,

$$\frac{\partial n}{\partial t} = s - \Sigma_a \phi - \text{div } \mathbf{J}. \tag{5.14}$$

Equation (5.14) is the general form of the equation of continuity. If the neutron density is not a function of time, this equation reduces to

$$\text{div } \mathbf{J} + \Sigma_a \phi - s = 0, \tag{5.15}$$

which is known as the *steady-state equation of continuity*.

5.4 THE DIFFUSION EQUATION

Unfortunately, the continuity equation has two unknowns—the neutron density, n, and the neutron current density vector, $\mathbf{J}$. To eliminate one of these requires a relationship between them. The relationship is based on the approximation that the current and flux are related by Fick's law (Eq. 5.8). On substitution of Fick's law into the equation of continuity (Eq. 5.14), one obtains the neutron diffusion equation. Assuming that D is not a function of position, this gives

$$D\nabla^2 \phi - \Sigma_a \phi + s = \frac{\partial n}{\partial t}, \tag{5.16}$$

where the symbol $\nabla^2 = \text{div grad}$ is called the *Laplacian*. Formulas for the Laplacian in various coordinate systems are given in Appendix III. Since $\phi = nv$, where v is the neutron speed, Eq. (5.16) can also be written as

$$D\nabla^2 \phi - \Sigma_a \phi + s = \frac{1}{v} \frac{\partial \phi}{\partial t}. \tag{5.17}$$

In the remainder of this chapter, only time-independent problems are considered. In this case, Eq. (5.17) becomes

$$D\nabla^2 \phi - \Sigma_a \phi + s = 0. \tag{5.18}$$

This is the *steady-state diffusion equation*.

It is often convenient to divide Eq. (5.18) by D, which gives

$$\nabla^2 \phi - \frac{1}{L^2} \phi = \frac{s}{D}, \tag{5.19}$$

where the parameter L^2 is defined as

$$L^2 = \frac{D}{\Sigma_a}. \tag{5.20}$$

The quantity L appears frequently in nuclear engineering problems and is called the *diffusion length*; L^2 is called the *diffusion area*. Since D and Σ_a have units of cm and cm^{-1}, respectively, it follows from Eq. (5.20) that L^2 has units of cm^2 and L has units of cm. A physical interpretation of L and L^2 is given later in this chapter.

5.5 BOUNDARY CONDITIONS

The neutron flux can be found by solving the diffusion equation. Since the diffusion equation is a partial differential equation, it is necessary to specify certain boundary conditions that must be satisfied by the solution. Some of these are determined from obvious requirements for a physically reasonable flux. For example, since a negative or imaginary flux has no meaning, it follows that ϕ must be a real, non-negative function. The flux must also be finite, except perhaps at artificial singular points of a source distribution.

In many problems, neutrons diffuse in a medium that has an outer surface—that is, a surface between the medium and the atmosphere. It was pointed out in Section 5.2 that Fick's law is not valid in the immediate vicinity of such a surface, and it follows that the diffusion equation is not valid there either. Higher order methods show, however, that if the flux calculated from the diffusion equation is assumed to vanish at a small distance d beyond the surface, then the flux determined from the diffusion equation is very nearly equal to the exact flux in the interior of the medium. The assumption that the flux vanishes a small distance d beyond the surface is clearly nonphysical. Rather, it is a convenient mathematical approximation that provides a high degree of accuracy for estimates of the flux inside the medium. This state of affairs is illustrated in Fig. 5.2.

The parameter d is known as the *extrapolation distance*, and for most cases of interest it is given by the simple formula

$$d = 0.71\lambda_{\mathrm{tr}}, \tag{5.21}$$

where λ_{tr} is the transport mean free path of the medium. From Eq. (5.10),

$$\lambda_{\mathrm{tr}} = 3D$$

and so d becomes

$$d = 2.13D. \tag{5.22}$$

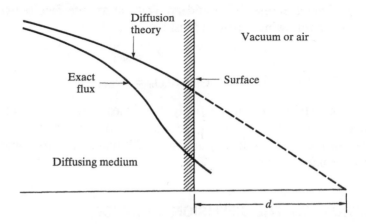

Figure 5.2   The extrapolation distance at a surface.

Measured values of D for nongaseous materials (some of which are given in Section 5.9) are usually less than 1 cm and frequently much less so. Thus, from Eq. (5.22), it is seen that d is usually small compared with most reactor dimensions. It is often possible, therefore, when solving the diffusion equation to assume that the flux vanishes at the actual surface of the system.

In those cases where d is not negligible, we must establish a mathematical boundary for the problem. This boundary is referred to as the *extrapolated boundary*. It is located a distance d from the real physical boundary. This boundary condition is referred to as the *vacuum boundary condition*. It is typically imposed in problems where there is a media having a low density or in a region at the edge of the problem where there are few neutrons that reenter the regions of interest. A typical case is in the air-filled annulus outside the reactor vessel. In such cases, the dimensions of the problem are increased by d. For our case, if the distance from the center of the diffusing media in Fig. 5.2 is a, then the distance to the extrapolated boundary is $a + d$, and we define the quantity $\tilde{a}$ as

$$\tilde{a} = a + d.$$

It is on the boundary at $\tilde{a}$ that the flux is assumed to vanish.[6]

It is also necessary to specify boundary conditions at an interface between two different diffusing media, such as the interface between the reactor core and the reflector. Since the neutrons cross an interface without hindrance, it is not difficult to see that both the flux and component of the current normal to the surface must

---

[6]It must be emphasized that the assumption that the flux vanishes is merely a mathematical convenience. The flux is nonzero beyond the boundary and varies in a way defined by the geometry and material properties in that region.

be continuous across the boundary. Thus, at an interface between two regions A and B, the following relations must be satisfied:

$$\phi_A = \phi_B \tag{5.23}$$

$$(J_A)_n = (J_B)_n, \tag{5.24}$$

where ϕ_A and ϕ_B are, respectively, the fluxes in region A and B evaluated at the interface, and $(J_A)_n$ and $(J_B)_n$ are the normal components of the neutron current evaluated at the interface. Equations (5.23) and (5.24) are sometimes called *interface boundary conditions*.

5.6 SOLUTIONS OF THE DIFFUSION EQUATION

Some simple diffusion problems are now considered to illustrate how the diffusion equation may be solved subject to the prior boundary conditions. These methods are applied in Chapter 6 to calculations of reactor properties.

Infinite Planar Source

Consider first an infinite planar source emitting S neutrons per cm^2/sec in an infinite diffusing medium. Examining Fig. 5.3, one sees that there is no variation in the y or z direction that could cause the flux to change. The flux in this case can only be a function of x—the distance from the plane. Also on further examination, one sees that the problem has symmetry about the $x = 0$ plane. The solution can then be divided into one for $x > 0$ and one for $x < 0$. Since there are no neutron sources present except at $x = 0$, the diffusion equation (Eq. 5.19) for $x \neq 0$ becomes

$$\frac{d^2\phi}{dx^2} - \frac{1}{L^2}\phi = 0, \quad x \neq 0. \tag{5.25}$$

Because of symmetry, we need only solve the equation in one-half of the plane. Then, by an appropriate transformation, one may obtain the solution for the other

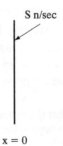

S n/sec

x = 0

Figure 5.3   Planar source at origin $x = 0$.

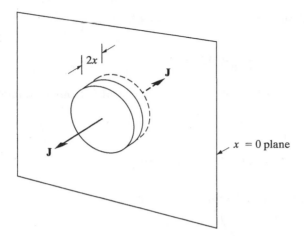

Figure 5.4   Pillbox in the $x = 0$ plane.

half plane. As discussed earlier, this is a second order differential equation that requires two boundary conditions.

The boundary conditions may be determined by recalling that the flux must be everywhere finite and positive definite. For this problem, the flux must remain finite as x goes to either positive or negative infinity. This condition may be used here.

Equation (5.25) has the following general solution:

$$\phi = Ae^{-x/L} + Ce^{x/L}, \tag{5.26}$$

where A and C are constants[7] to be determined by the boundary conditions. For the moment, consider the right-half plane where we have only positive values of x. Since the second term in Eq. (5.26) increases without limit with increasing x, it follows that C must be taken to be zero. Equation (5.26) then reduces to

$$\phi = Ae^{-x/L}. \tag{5.27}$$

The constant A in Eq. (5.27) is found in the following way. Suppose that a small pillbox of unit surface area and thickness $2x$ is constructed at the source plane as shown in Fig. 5.4. There clearly is no net flow of neutrons parallel to the source plane through the sides of the pillbox since the source plane and medium are infinite in this direction. Therefore, in view of the symmetry of the problem in the x direction, the net flow of neutrons out of the pillbox is simply $2J(x)$, where $J(x)$ is the neutron current density on the surface of the pillbox located at x. In the limit as x goes to zero, the net flow out of the box must approach S, the source

---

[7]The symbol B has a special meaning in nuclear engineering and is not used as a constant in calculations of the present type.

density of the plane. It follows that

$$\lim_{x \to 0} J(x) = \frac{S}{2}. \tag{5.28}$$

This relation is known as a *source condition* and is useful for other situations as well.[8]

From Fick's law,

$$\mathbf{J} = -D\frac{d\phi}{dx} = \frac{DA}{L}e^{-x/L}.$$

Inserting this into Eq. (5.28) and taking the limit gives

$$A = \frac{SL}{2D}.$$

From Eq. (5.27), the flux is then

$$\phi = \frac{SL}{2D}e^{-x/L}.$$

This solution is valid only for the right-half plane where x is positive. However, again because of the symmetry of the problem, the flux must be the same at $-x$ as at $+x$. Thus, a solution valid for all x can be obtained by replacing x by its absolute values $|x|$.

$$\phi = \frac{SL}{2D}e^{-|x|/L}. \tag{5.29}$$

Point Source

Consider next a point source emitting S neutrons per second isotropically in an infinite medium. If the source point is taken to be at the center of a spherical coordinate system, the flux obviously only depends on r. Then, with the Laplacian expressed in spherical coordinates (see Appendix III), the diffusion equation becomes for $r \neq 0$

$$\frac{1}{r^2}\frac{d}{dr}r^2\frac{d\phi}{dr} - \frac{1}{L^2}\phi = 0. \tag{5.30}$$

The same approach may be applied here as for the planar source. The appropriate source condition for this problem is obtained by drawing a small sphere around the

---

[8]It must be emphasized that this source condition is only valid for a planar source in a symmetric problem. For the treatment of an asymmetric case, see Problem 5.13.

source and computing the net number of neutrons that pass through its surface per second. If the sphere has the radius r, this number is just $4\pi r^2 J(r)$, so that in the limit, as r goes to zero, the following condition is obtained:

$$\lim_{r \to 0} r^2 J(r) = \frac{S}{4\pi}. \tag{5.31}$$

To solve Eq. (5.30), it is convenient to introduce a new variable, w, defined by

$$w = r\phi. \tag{5.32}$$

When Eq. (5.32) is substituted into Eq. (5.30), the following equation is found for w:

$$\frac{d^2 w}{dr^2} - \frac{1}{L^2} w = 0.$$

The general solution to this equation is obviously the same as for the planar source

$$w = ae^{-r/L} + Ce^{r/L},$$

and ϕ is therefore

$$\phi = A\frac{e^{-r/L}}{r} + C\frac{e^{r/L}}{r},$$

where A and C are unknown constants. Again, as in the preceding example, ϕ must remain finite as r becomes infinite, so that C must be placed equal to zero. The constant A is found from the source condition, Eq. (5.31). Thus,

$$J = -D\frac{d\phi}{dr} = DA\left(\frac{1}{rL} + \frac{1}{r^2}\right)e^{-r/L},$$

and so

$$A = \frac{S}{4\pi D}.$$

The flux is therefore given by

$$\phi = \frac{Se^{-r/L}}{4\pi Dr}. \tag{5.33}$$

It is important to note that this solution is quite different than would be obtained for a point source in a vacuum (see Problem 5.1). This difference occurs since the diffusing media both absorbs and scatters neutrons, whereas a vacuum does not.

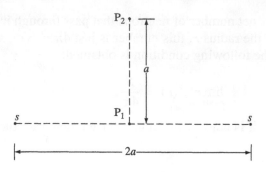

Figure 5.5 Two point sources.

Example 5.3

Two point sources, each emitting S neutrons/sec, are located $2a$ cm apart in an infinite diffusing medium as shown in Fig. 5.5. Derive expressions for the flux and current at the point P_1 midway between the sources.

Solution. Since the flux is a scalar quantity and the diffusion equation is a linear equation, the total flux at P_1 is the sum of the contributions from each source.[9] Thus,

$$\phi(P_1) = 2 \times \frac{Se^{-a/L}}{4\pi Da} = \frac{Se^{-a/L}}{2\pi Da}. \quad [Ans.]$$

The neutron current at P_1 in Fig. 5.5 must clearly be zero because the current vectors from the sources are equal and oppositely directed at this point. (In Problem 5.6 at the end of the chapter, it is required to find the flux and current at the point P_2. The individual current vectors do not cancel in this case.)

Bare Slab

Finally, consider an infinite slab of thickness $2a$ that has an infinite planar source at its center emitting S neutrons per cm²/sec as in Fig. 5.6. If the zero of the coordinate system is at the center of the slab, the diffusion equation for $x \neq 0$ and

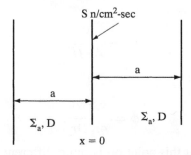

Figure 5.6 Infinite slab with planar source at $x = 0$.

---

[9]Since the diffusion equation is a linear equation, the ability to add the flux and current from different sources and obtain the total flux or current is a general property.

$|x| \le a$ is again Eq. (5.25). We may again use symmetry and the source condition given by Eq. (5.28). Now, however, the condition on the flux as $|x| \to \infty$ must be different from that used before. Here the flux is required to vanish at the extrapolated surfaces of the slab—that is, at $x = a + d$ for the right-half plane and at $x = -a - d$ for the left, where d is the extrapolation distance. These boundary conditions are,

$$\phi(a + d) = \phi(-a - d) = 0.$$

For the right-half plane, the general solution to Eq. (5.25) is

$$\phi = Ae^{-x/L} + Ce^{x/L}. \tag{5.34}$$

Then, in view of the boundary condition at $a + d$,

$$\phi(a + d) = Ae^{-(a+d)/L} + Ce^{(a+d)/L} = 0,$$

so that

$$C = -Ae^{-2(a+d)/L}.$$

Substituting this result into Eq. (5.34) gives

$$\phi = A[e^{-x/L} - e^{x/L - 2(a+d)/L}].$$

The constant A is found from the source condition (Eq. 5.28) in the usual way, and is

$$A = \frac{SL}{2D}(1 + e^{-2(a+d)/L})^{-1}.$$

For positive x, therefore, ϕ is given by

$$\phi = \frac{SL}{2D} \frac{e^{-x/L} - e^{x/L - 2(a+d)/L}}{1 + e^{-2(a+d)/L}}.$$

In view of the symmetry of the problem, a solution valid for all x is obtained by substituting $|x|$ for x; thus

$$\phi = \frac{SL}{2D} \frac{e^{-|x|/L} - e^{|x|/L - 2(a+d)/L}}{1 + e^{-2(a+d)/L}}. \tag{5.35}$$

This solution can be written in more convenient form if the numerator and denominator are multiplied by $e^{(a+d)/L}$. This gives

$$\phi = \frac{SL}{2D} \frac{e^{(a+d-|x|)/L} - e^{-(a+d-|x|)/L}}{e^{(a+d)/L} + e^{-(a+d)/L}}$$

$$= \frac{SL}{2D} \frac{\sinh[(a + d - |x|)/L]}{\cosh[(a + d)/L]}. \tag{5.36}$$

where sinh and cosh are hyperbolic functions defined as

$$\sinh x = \frac{e^x - e^{-x}}{2}, \quad \cosh x = \frac{e^x + e^{-x}}{2}.$$

Example 5.4

(a) Derive an expression for the number of neutrons that leak per second from 1 cm^2 on both sides of the slab discussed earlier. (b) What is the probability that a source neutron will leak from the slab?

Solution.

1. Since the current represents the net number of neutrons passing through a surface per cm^2/sec, the number of neutrons leaking per cm^2/sec can be found by computing the neutron current at the surface of the slab. Consider first the surface located at $x = a$. By use of Fick's law,

$$J(a) = -D \frac{d}{dx} \frac{SL}{2D} \frac{\sinh[(a + d - x)/L]}{\cosh[(a + d)/L]}\bigg|_{x=a}$$

$$= \frac{S}{2} \frac{\cosh(d/L)}{\cosh[(a + d)/L]}.$$

2. The probability that a neutron escapes is equal to the number that escape per cm^2/sec divided by the number emitted per cm^2/sec by the source. For example, if 100 neutrons are emitted per cm^2/sec and 10 escape from the surface, then the probability that a source neutron will ultimately escape is $10/100 = 100 = 0.10$. The leakage probability is thus

$$\frac{2J(a)}{S} = \frac{\cosh(d/L)}{\cosh[(a + d)/L]}. \ [Ans.]$$

5.7 THE DIFFUSION LENGTH

It is of interest at this point to examine the physical interpretation of the diffusion length, which appears in the diffusion equation and in so many of its solutions. To this end, consider a monoenergetic point source emitting S neutrons per second in an infinite homogeneous moderator. As these neutrons diffuse about in the medium, individual neutrons move in complicated, zigzag paths due to successive collisions as indicated in Fig. 5.7. Eventually, however, every neutron is absorbed in the medium—none can escape since the medium is infinite.

The number of neutrons, dn, that are absorbed per second at a distance from the source between r and $r + dr$ is given by

$$dn = \Sigma_a \phi(r) dV,$$

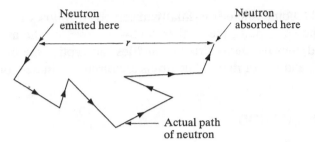

Neutron
emitted here

Neutron
absorbed here

r

Actual path
of neutron

Figure 5.7   Trajectory of a
neutron in a moderating medium.

where $\phi(r)$ is the flux from the point source and $dV = 4\pi r^2 dr$ is the volume of a spherical shell of radius r and thickness dr. Introducing $\phi(r)$ from Eq. (5.33) gives

$$dn = \frac{S\Sigma_a}{D}re^{-r/L}dr = \frac{S}{L^2}re^{-r/L}dr,$$

where use has been made of the definition of L^2 from Eq. (5.20). Since S neutrons per second are emitted by the source and dn are absorbed per second between r and $r + dr$, it follows that the probability $p(r)dr$ that a source neutron is absorbed in dr is

$$p(r)dr = \frac{1}{L^2}re^{-r/L}dr.$$

It is now possible to compute the average distance from the source at which a neutron is absorbed by averaging r over the probability distribution $p(r)dr$. For somewhat obscure reasons, however, it is more usual in nuclear engineering to compute the average of the square of this distance, rather than the average of the distance itself. Thus,

$$\overline{r^2} = \int_0^\infty r^2 p(r)dr$$

$$= \frac{1}{L^2}\int_0^\infty r^3 e^{-r/L}dr = 6L^2.$$

Solving for L^2 gives

$$L^2 = \frac{1}{6}\overline{r^2}. \tag{5.37}$$

In words, Eq. (5.37) states that L^2 is equal to one-sixth the average of the square of the vector (crow-flight) distance that a neutron travels from the point where it is emitted to the point where it is finally absorbed. It follows from this result that the greater the value of L, the further neutrons move, on the average, before they are absorbed, thus the more diffusive and less absorptive the medium is. Measured values of L and L^2 for thermal neutrons are discussed in Section 5.9.

5.8 THE GROUP-DIFFUSION METHOD

In Chap. 3, the energy dependence of the flux was discussed and a method introduced for approximating the dependence for the thermal flux. Since neutrons in a nuclear reactor actually have a distribution in energy, this distribution must be accounted for in the diffusion equation. To begin with, neutrons are emitted in fission with a continuous energy spectrum $\chi(E)$, and this distribution broadens as the neutrons are scattered in the medium and diffuse about the system, losing energy in elastic and inelastic collisions. In thermal reactors, it is recalled, most of the fission neutrons succeed in slowing down all the way to thermal energies before they are absorbed in the fuel. In fast reactors, they slow down much less before inducing the fissions required to maintain the chain reaction.

One of the most effective ways to calculate the slowing down and diffusion of neutrons is by the group-diffusion method. In this method, the entire range of neutron energy is divided into N energy intervals as indicated in Fig. 5.8. All of the neutrons within each energy interval are then lumped together, and their diffusion, scattering, and absorption are described in terms of suitably averaged diffusion coefficients and cross-sections.

Consider, for instance, the neutrons in the gth energy interval—these are called the gth group neutrons. (Note that, by convention, the most energetic group is denoted by $g = 1$; the least energetic group has $g = N$.) To obtain an equation

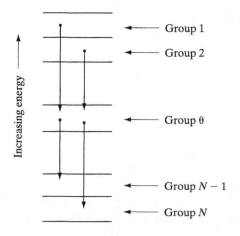

Figure 5.8   Energy groups for a group-diffusion calculation.

describing these neutrons, it is first necessary to define the flux of neutrons in this group as

$$\phi_g = \int_g \phi(E)dE, \tag{5.38}$$

where $\phi(E)$ is the energy-dependent flux defined in Section 5.1, and the subscript on the integral means that the integration is carried out over all energies within the group. Clearly all we have done is sum the flux within the energy range defined by the gth energy group.

Neutrons disappear from the gth group both in absorption reactions and as the result of scattering collisions that cause them to change their energy such that they no longer belong to the gth energy group but instead to another group. The total absorption rate per cm^3 in the gth group is given by the integral

$$\text{Absorption rate} = \int_g \Sigma_a(E)\phi(E)dE.$$

Thus, we have again summed simply all the events that occur in the gth group. By defining the macroscopic *group* absorption cross-section Σ_{ag} as

$$\Sigma_{ag} = \frac{1}{\phi_g} \int_g \Sigma_a(E)\phi(E)dE, \tag{5.39}$$

the absorption rate can be written as a product of the group cross-section and the group flux

$$\text{Absorption rate} = \Sigma_{ag}\phi_g. \tag{5.40}$$

The rate at which neutrons in the gth group are scattered into the hth group is written in a form analogous to Eq. (5.40)—namely,

$$\text{Transfer rate, } g \text{ to } h = \Sigma_{g \to h}\phi_g. \tag{5.41}$$

The quantities $\Sigma_{g \to h}$ are called *group transfer cross-sections;* their derivation is somewhat complicated and is not given here. In any case, they may be viewed as known numbers that depend on the scattering properties of the medium. The total rate at which neutrons are scattered out of the gth group per cm^3/sec is then[10]

---

[10]It is assumed here that neutrons can only lose energy in a scattering collision, which is true for energetic neutrons. However, at low thermal energies, a neutron can gain energy in a collision with the nucleus of an atom in thermal motion. Such upscattering must be taken into account in group-diffusion calculations involving low-energy neutrons.

$$\text{Total transfer rate out of } g = \sum_{h=g+1}^{N} \Sigma_{g \to h} \phi_g. \qquad (5.42)$$

Note that here we have assumed that neutrons lose energy when scattered, which is true except for very slow neutrons. In this case, we are assuming only down scattering to occur and no upscattering.

Neutrons enter the gth group either from sources that emit neutrons directly into the group or as the result of scattering from groups at higher energy. The number per cm^3/sec entering g from the hth group is

$$\text{Transfer rate } h \text{ to } g = \Sigma_{h \to g} \phi_h, \qquad (5.43)$$

and the total number scattered into g is obtained by summing over all groups that can scatter into the gth group–that is, all higher energy groups:

$$\text{Total transfer rate into } g = \sum_{h=1}^{g-1} \Sigma_{h \to g} \phi_h. \qquad (5.44)$$

Again, we have assumed that no upscattering takes place.

Combining the prior terms in the obvious way gives the following steady-state diffusion equation for the gth group neutrons:

$$D_g \nabla^2 \phi_g - \Sigma_{ag} \phi_g - \sum_{h=g+1}^{N} \Sigma_{g \to h} \phi_g + \sum_{h=1}^{g-1} \Sigma_{h \to g} \phi_h = -s_g. \qquad (5.45)$$

In this equation, D_g, the group-diffusion coefficient, is given by

$$D_g = \frac{1}{\phi_g} \int_g D(E) \phi(E) dE = \frac{1}{3\phi_g} \int_g \frac{1}{\Sigma_{\text{tr}}(E)} \phi(E) dE, \qquad (5.46)$$

where Σ_{tr} is the transport cross-section, and s_g is the total number of neutrons emitted per cm^3/sec into the group from sources.

The multigroup approximation forms the basis for many reactor analysis codes used today. Computer codes such as DIFF3D and SIMULATE use this approach.

Example 5.5

Microscopic three-group cross-sections of sodium for calculations of sodium-cooled fast reactors are given in Table 5.1. Suppose that at some point in such a reactor the three group fluxes are $\phi_1 = 6 \times 10^{14}$, $\phi_2 = 1 \times 10^{15}$, and $\phi_3 = 3 \times 10^{15}$. Calculate for this point the number of neutrons (a) absorbed per cm^3/sec in sodium; (b) scattered per cm^3/sec from the first group to the second as the result of collisions with sodium. The sodium is at normal density.

TABLE 5.1   NOMINAL THREE-GROUP CROSS-SECTIONS, IN BARNS, FOR SODIUM*

g	Energy range, MeV	σ_γ	σ_{tr}	$\sigma_{g \to g+1}$	$\sigma_{g \to g+2}$
1	1.35–∞	0.0005	2.0	0.24	0.06
2	0.4–1.35	0.001	3.2	0.18	—
3	0–0.4	0.001	3.7	—	—

*From *Reactor Physics Constants,* 2nd ed., Argonne National Laboratory report ANL-5800, 1963.

Solution.

1. The total absorption rate F_a is the sum of the absorption rates in three groups,

$$F_a = \sum_{g=1}^{3} \Sigma_{ag} \phi_g = N \sum_{g=1}^{3} \sigma_{\gamma g} \phi_g,$$

where N is the atom density of sodium. From Table II.3, $N = 0.02541 \times 10^{24}$. Then using the values of σ_γ in Table 5.1 gives

$$F_a = 0.02541(0.0005 \times 6 \times 10^{14} + 0.001 \times 1 \times 10^{15} + 0.001 \times 3 + 10^{15})$$

$$= 1.09 \times 10^{11} \text{ neutrons/cm}^3\text{-sec. } [Ans.]$$

2. The rate at which neutrons are scattered from the first to the second group is

$$F_{1 \to 2} = \Sigma_{1 \to 2} \phi_1$$

$$= N \sigma_{1 \to 2} \phi_1$$

$$= 0.02541 \times 0.24 \times 6 \times 10^{14}$$

$$= 3.66 \times 10^{12} \text{ neutrons/cm}^3\text{-sec } [Ans.]$$

Because of the complexity of the group-diffusion equations, Eq. (5.45), or the *multigroup equations* as they are sometimes called, it is common practice to use a computer program to evaluate the group fluxes. The techniques by which this is done involve approximating the derivatives by numerical methods and then requiring the equations hold only at a series of discrete points in space. The equations are then reduced to algebraic equations valid for only those points. The exact approaches vary, and such methods lie beyond the scope of this text. It suffices here to note that many computer programs such as those mentioned earlier have been written by which the equations represented by Eq. (5.45) can be evaluated.

In the special case of a one-group calculation, the group-transfer terms are missing and Eq. (5.45) reduces to

$$D\nabla^2 \phi - \Sigma_a \phi = -s, \tag{5.47}$$

where the subscript denoting the group number has been omitted. It is observed that Eq. (5.47) is precisely the same as the diffusion equation for monoenergetic neutrons, as would be expected from the derivation of the group equations. The one-group and multigroup methods are utilized in Chapter 6 for calculations of reactor criticality.

5.9 THERMAL NEUTRON DIFFUSION

An important application of the one-group method is in the diffusion of thermal neutrons. It is recalled that these are neutrons that have slowed down to thermal energies and whose energy distribution is given by the Maxwellian function (see Section 2.12),

$$n(E) = \frac{2\pi n}{(\pi k T)^{3/2}} E^{1/2} e^{-E/kT}. \tag{5.48}$$

In this formula, $n(E)$ is defined so that $n(E)dE$ is the number of neutrons per cm^3 with energies between E and $E + dE$, and n is the total neutron density (i.e. neutrons per cm^3).

From Eq. (5.4), the energy-dependent flux for thermal neutrons is

$$\phi(E) = n(E)v(E)$$

$$= \frac{2\pi n}{(\pi k T)^{3/2}} \left(\frac{2}{m}\right)^{1/2} e^{-E/kT}. \tag{5.49}$$

Using the usual formula for speed,

$$v(E) = \left(\frac{2E}{m}\right)^{1/2},$$

where m is the neutron mass, has been introduced. The one-group thermal flux, which is denoted as ϕ_T, is then given by Eq. (5.38) as

$$\phi_T = \int_T \phi(E)dE, \tag{5.50}$$

where the subscript T on the integral signifies that the integration is to be carried out over thermal energies, which are normally taken to extend up to about $5kT \simeq 0.1$ eV. However, since the exponential in $\phi(E)$ drops to such small values above this energy, very little error is made by carrying the integral to infinity. Substituting Eq. (5.49) then gives

$$\phi_T = \frac{2\pi n}{(\pi k T)^{3/2}} \left(\frac{2}{m}\right)^{1/2} \int_0^\infty e^{-E/kT} dE = \frac{2n}{\sqrt{\pi}} \left(\frac{2kT}{m}\right)^{1/2}. \tag{5.51}$$

At this point, it is convenient to denote by E_T the neutron energy corresponding to kT and to let v_T be the corresponding speed—that is,

$$E_T = kT \tag{5.52}$$

and

$$\frac{1}{2}mv_T^2 = E_T. \tag{5.53}$$

The numerical values of E_T and v_T can be computed from the following formulas:

$$E_T = 8.617T \times 10^{-5} \text{ eV} \tag{5.54}$$

and

$$v_T = 1.284T^{1/2} \times 10^4 \text{ cm/sec}, \tag{5.55}$$

where T is the temperature in degrees Kelvin. Equation (5.51) can now be written as

$$\phi_T = \frac{2}{\sqrt{\pi}}nv_T. \tag{5.56}$$

It is important to recognize that the thermal flux ϕ_T and the 2,200 meters-per-second flux ϕ_0, which was defined in Section 3.6, are quite different concepts. Thus, although ϕ_T represents the flux of all thermal neutrons lumped together and computed by the integral in Eq. (5.50), ϕ_0 is not a real flux, but a pseudoflux calculated by assuming that all the thermal neutrons have only one energy—namely, $E_0 = 0.0253$ eV. That is ϕ_0 is a number created to yield the same reaction rate obtained in the actual case using the value of the cross-section at E_0. In contrast, the flux ϕ_T is the sum over energy of all the flux of particles at each energy in the thermal range. The thermal flux ϕ_T is appropriate for calculations involving the diffusion of thermal neutrons, whereas the flux ϕ_0 is more useful in computations of neutron absorption rates in targets exposed to thermal neutrons. To put it simply, ϕ_T is used in the design of a reactor, whereas ϕ_0 is appropriate when using a reactor.

The relationship between the two fluxes can be found by dividing ϕ_0 by ϕ_T, where ϕ_0 is given by

$$\phi_0 = nv_0,$$

where $v_0 = 2,200$ meters/sec. This gives

$$\frac{\phi_0}{\phi_T} = \frac{\sqrt{\pi}}{2}\frac{v_0}{v_T}. \tag{5.57}$$

However, v_0 can be computed from Eq. (5.55) as

$$v_0 = 1.284 T_0^{1/2} \times 10^4 \text{ cm/sec},$$

using the value of $T_0 = 293.61$ K. Then Eq. (5.57) becomes

$$\frac{\phi_0}{\phi_T} = \frac{\sqrt{\pi}}{2} \left(\frac{T_0}{T} \right)^{1/2}. \tag{5.58}$$

which is the desired relation.

The diffusion coefficient in a one-group calculation of thermal neutron diffusion is denoted as D and is obtained by evaluating the integral in Eq. (5.46) or from measurements of thermal neutrons. Values of D are given in Table 5.2 for several moderators.

The one-group thermal absorption cross-section, denoted by Σ_a, is found from Eq. (5.39):

$$\overline{\Sigma}_a = \frac{1}{\phi_T} \int_T \Sigma_a(E) \phi(E) dE.$$

The integral is just the total absorption rate, and from Eq. (3.38) this is equal to

$$g_a(T) \Sigma_a(E_0) \phi_0,$$

where $g_a(T)$ is the non-$1/v$ factor, $\Sigma_a(E_0)$ is the macroscopic absorption cross-section at 0.0253 eV, and ϕ_0 is the 2,200 meters-per-second flux. Thus, $\overline{\Sigma}_a$ becomes

$$\overline{\Sigma}_a = g_a(T) \Sigma_a(E_0) \phi_0 / \phi_T$$

$$= \frac{\sqrt{\pi}}{2} g_a(T) \Sigma_a(E_0) \left(\frac{T_0}{T} \right)^{1/2}, \tag{5.59}$$

TABLE 5.2 THERMAL NEUTRON DIFFUSION PARAMETERS OF COMMON MODERATORS AT 20°C*

Moderator	Density, g/cm³	$\overline{D}$, cm	$\overline{\Sigma}_a$, cm⁻¹	L_T^2, cm²	L_T, cm
H_2O	1.00	0.16	0.0197	8.1	2.85
D_2O†	1.10	0.87	9.3×10^{-5}	9.4×10^3	97
Be	1.85	0.50	1.04×10^{-3}	480	21
Graphite	1.60	0.84	2.4×10^{-4}	3500	59

*Based on *Reactor Physics Constants,* 2nd ed., Argonne National Laboratory report ANL-5800, 1963, Section 3.3.
†D_2O containing 0.25 weight/percent H_2O. These values are very sensitive to the amount of H_2O impurity (see Problem 5.28).

where use has been made of Eq. (5.58). Numerical values of $\overline{\Sigma}_a$ at 20°C for the common moderators, for which $g_a(T)$ is unity, are given in Table 5.2.

With the thermal flux defined and the parameters D and $\overline{\Sigma}_a$ known, it is now possible to write down the one-group-diffusion equation for thermal neutrons. From Eq. (5.47), this is

$$\overline{D}\nabla^2\phi_T - \overline{\Sigma}_a\phi_T = -s_T, \qquad (5.60)$$

in which s_T is the density of thermal neutron sources. Dividing Eq. (5.60) by $\overline{D}$ gives

$$\nabla^2\phi_T - \frac{1}{L_T^2}\phi_T = -\frac{s_T}{\overline{D}}, \qquad (5.61)$$

where

$$L_T^2 = \frac{\overline{D}}{\overline{\Sigma}_a} \qquad (5.62)$$

is called the *thermal diffusion area*; L_T is the *thermal diffusion length*. Table 5.2 also gives values of these parameters.

It is observed that Eq. (5.61) is identical in form to the diffusion equation for monoenergetic neutrons (see Eq. 5.19), and it follows that the solutions to this equation are valid for thermal neutrons provided D, L^2, and ϕ are replaced by $\overline{D}$, L_T^2, and ϕ_T. This is illustrated by the next example.

Example 5.6

A point source emitting 10^7 thermal neutrons per second is located in an infinite body of unit-density water at room temperature. What is the thermal neutron flux 15 cm from the source?

Solution. The flux is given by Eq. (5.33) with appropriate changes in designations of the parameters—that is,

$$\phi_T = \frac{Se^{-r/L_t}}{4\pi\overline{D}r}.$$

From Table 5.2, $L_T = 2.85$ cm and $\overline{D} = 0.16$ cm. Also, $S = 10^7$ and $r = 15$ cm. Inserting these values gives

$$\phi_T = \frac{10^7 \times e^{-15/2.85}}{4\pi \times 0.16 \times 15} = 1.72 \times 10^3 \text{ neutrons/cm}^2\text{-sec. } [Ans.]$$

The values of the thermal diffusion parameters given in Table 5.2 are for normal density and room temperature. It is often necessary to know these parameters at other densities and temperatures, and these can be found by appropriately modifying the tabulated values. Consider first $\overline{\Sigma}_a$. Being a macroscopic cross-section,

$\overline{\Sigma}_a$ is proportional to N, the atom density, which in turn is proportional to ρ, the physical density. It follows that

$$\overline{\Sigma}_a \sim \rho.$$

The dependence of $\overline{\Sigma}_a$ on temperature is given by Eq. (5.59).

According to Eqs. (5.10) and (5.11), the diffusion coefficient is inversely proportional to Σ_s. Since this is also proportional to ρ,

$$\overline{D} \sim \frac{1}{\rho}.$$

The situation with regard to temperature is somewhat more complicated. For moderators other than H_2O and D_2O, $\overline{D}$ is essentially independent of temperature. However, for H_2O, $\overline{D}$ varies approximately as

$$\overline{D} \sim T^{0.470},$$

whereas for D_2O, it behaves as

$$\overline{D} \sim T^{0.112},$$

where T is the absolute temperature. The dependence of $\overline{D}$ on ρ and T may then be summarized by

$$\overline{D}(\rho, T) = \overline{D}(\rho_0, T_0) \left(\frac{\rho_0}{\rho}\right) \left(\frac{T}{T_0}\right)^m, \tag{5.63}$$

where $\overline{D}(\rho_0, T_0)$ is the value of $\overline{D}$ at density ρ_0 and temperature T_0, and $m = 0.470$ for H_2O and 0.112 for D_2O and zero otherwise.

Finally, by combining the previous results with the definition of L_T^2 given in Eq. (5.62), it is not difficult to see that

$$L_T^2(\rho, T) = L_T^2(\rho_0, T_0) \left(\frac{\rho_0}{\rho}\right)^2 \left(\frac{T}{T_0}\right)^{m+1/2}. \tag{5.64}$$

The utility of these expressions is shown by the following example.

Example 5.7

Calculate the diffusion coefficient and diffusion area of ordinary water at 500°F and at a pressure of 2,000 psi. The density of water at this temperature and pressure is 49.6 lb/ft³.

Solution. From Table 5.2, $\overline{D}(\rho_0, T_0) = 0.16$ cm at $\rho_0 = 1$ g/cm³ $= 62.4$ lb/ft³ and $T_0 = 293°$K. Using the formula relating temperature on the Celsius and Fahrenheit scales, namely,

$$°C = \frac{5}{9}(°F - 32),$$

it is found that 500°F is equivalent to 260°C. The absolute temperature is then $T = 260 + 273 = 533°K$. Substituting these values into Eq. (5.64) then gives

$$\overline{D} = 0.16 \times \left(\frac{62.4}{49.6}\right) \times \left(\frac{533}{293}\right)^{0.470} = 0.267 \text{ cm. } [Ans.]$$

Finally, introducing $L_T^2(\rho_0, T_0) = 8.1 \text{ cm}^2$ from Table 5.2 into Eq. (5.64) gives

$$L_T^2 = 8.1 \times \left(\frac{62.4}{49.6}\right)^2 \times \left(\frac{533}{293}\right)^{0.970} = 22.9 \text{ cm}^2. [Ans.]$$

5.10 TWO-GROUP CALCULATION OF NEUTRON MODERATION

In a number of reactor calculations, especially those involving the criticality of thermal reactors to be discussed in Chapter 6, at least two groups of neutrons must be used to obtain reasonably accurate results. One group is necessary to describe the thermal neutrons in the manner explained in the preceding section. The second or fast group includes all the neutrons having energies above thermal—that is, the neutrons that are slowing down from fission energies to about $5kT$.

As preparation for the computations to be presented in Chapter 6, it is convenient at this point to illustrate the use of two groups in a simple example. Consider, therefore, the problem of computing the two-group fluxes as a function of distance from a point source emitting S fission (fast) neutrons per second in an infinite uniform moderator. The group-diffusion equation for the fast neutrons can be obtained from the general group-diffusion equation, Eq. (5.45), with $g = 1$. However, the absorption cross-sections of all moderating materials are small, especially above the thermal energy region, and so Σ_{a_1} can be taken to be zero. Also, since there are only 2 groups in the calculation, neutrons scattered out of the fast group necessarily must enter the thermal group. Thus, only $\Sigma_{1\rightarrow 2}$ is nonzero in the third term of Eq. (5.45). Furthermore, the last term in this equation is zero because no neutrons are scattered into the fast group. The diffusion equation for the fast neutrons is thus simply approximated by

$$D_1 \nabla^2 \phi_1 - \Sigma_1 \phi_1 = 0. \tag{5.65}$$

In this equation, Σ_1 has been written for the transfer cross-section $\Sigma_{1\rightarrow 2}$. The source term is zero since there are no sources except at the point source.

In nuclear reactor theory, the number of neutrons that slow down to thermal energies per cm^3/sec is known as the *slowing-down density* and is usually denoted by the symbol q_T. The term $\Sigma_1\phi_1$ in Eq. (5.65) is equal to the number of neutrons scattered per cm^3/sec from the fast group to the thermal group. Since the neutrons entering the thermal group have obviously just slowed down, it follows that $\Sigma_1\phi_1$ is also equal to the slowing-down density—that is,

$$q_T = \Sigma_1\phi_1. \tag{5.66}$$

To turn next to the thermal neutrons, these are described by the thermal diffusion equation, Eq. (5.61). However, the only source of thermal neutrons in the problem are those that slow down out of the fast group. Since these appear at the rate of q_T neutrons per cm^3/sec, the source term in Eq. (5.61) is $\Sigma_1\phi_1$. The thermal group equation is then

$$\nabla^2\phi_T - \frac{1}{L_T^2}\phi_T = \frac{\Sigma_1\phi_1}{D}. \tag{5.67}$$

To find ϕ_T from Eq. (5.67), it is necessary first to solve Eq. (5.65) for ϕ_1. In view of the symmetry of the problem, both ϕ_1 and ϕ_T clearly depend only on r, the distance from the source. Dividing Eq. (5.65) through by D_1 and writing the Laplacian in spherical coordinates then gives

$$\frac{1}{r^2}\frac{d}{dr}r^2\frac{d\phi_1}{dr} - \frac{1}{\tau_T}\phi_1 = 0. \tag{5.68}$$

where τ_T is defined as

$$\tau_T = \frac{D_1}{\Sigma_1}. \tag{5.69}$$

The parameter τ_T occurs in many types of reactor problems and is called the *neutron age*. It is given this odd name because it can be shown to be a function of how long it takes a neutron to slow down. The age does not have units of time, however. As seen from Eq. (5.69), τ_T actually has units of cm^2.

A comparison of Eq. (5.68) with Eq. (5.30), which describes the diffusion of neutrons from a point source, shows that these two equations are identical except that τ_T in Eq. (5.68) replaces L^2 in Eq. (5.30). As a consequence, the solution to Eq. (5.68) is given by Eq. (5.33), but with $\sqrt{\tau_T}$ written for L and D_1, for D

$$\phi_1 = \frac{Se^{-r/\sqrt{\tau_T}}}{4\pi D_1 r}. \tag{5.70}$$

Furthermore, it is recalled from Section 5.7 that L^2 is equal to one sixth the average of the square of the crow-flight distance that a neutron travels from the source to the point where it is finally absorbed. The same kind of interpretation can be made for τ_T, except that neutrons are not absorbed in the fast group—they are scattered out of that group and into the thermal group. It is not difficult to see, therefore, that the parameter τ_T is equal to one-sixth the average of the square of the crow-flight distance from the point where a neutron is emitted as a fast neutron to the point where it slows down to thermal energies. In symbols,

$$\tau_T = \frac{1}{6}\overline{r^2}. \tag{5.71}$$

The age of neutrons as defined by Eq. (5.71) can be determined experimentally. Measured values of τ_T for fission neutrons in various moderators are given in Table 5.3.

When Eq. (5.70) is substituted into Eq. (5.67), the result is an inhomogeneous differential equation. The details of solving this equation are not given here. It can be readily verified by direct substitution, however, that the solution is

$$\phi_T = \frac{SL_T^2}{4\pi r \overline{D}(L_T^2 - \tau_T)}(e^{-r/L_T} - e^{-r/\sqrt{\tau_T}}). \tag{5.72}$$

This expression gives the thermal flux arising from a point source of fast neutrons according to two-group theory. It must be emphasized that Eq. (5.72) is by no means exact. Indeed, a two-group model is not particularly appropriate for computing the slowing down of neutrons in most media—there are far better and more accurate (and more complicated) methods for calculating the slowing down of neutrons from point sources. Nevertheless, as shown in Chapter 6, the two-group model can be used for rough, first-order calculations of thermal reactors.

Incidentally, the experimental values of τ_T given in Table 5.3 are for moderators at room temperature and normal density. Neither D_1 nor Σ_1 in Eq. (5.69) is particularly sensitive to temperature, but both depend on density. To compute τ_T at other than normal density, it may be noted that τ_T, like L_T^2, varies inversely as the square of the density. Thus, the value of τ_T at density ρ is given by

$$\tau_T(\rho) = \tau_T(\rho_0)\left(\frac{\rho_0}{\rho}\right)^2, \tag{5.73}$$

where $\tau_T(\rho_0)$ is the age at density ρ_0.

TABLE 5.3 FAST-GROUP CONSTANTS FOR VARIOUS MODERATORS

Moderator	D_1, cm	Σ_1, cm^{-1}	τ_T, cm^2
H_2O	1.13	0.0419	~27
D_2O	1.29	0.00985	131
Be	0.562	0.00551	102
Graphite	1.016	0.00276	368

REFERENCES

Duderstadt, J., and L. Hamilton, *Nuclear Reactor Analysis.* New York: John Wiley, 1975.

Glasstone, S., and M. C. Edlund, *The Elements of Nuclear Reactor Theory.* Princeton, N.J.: Van Nostrand, 1952, Chapter 5.

Glasstone, S., and A. Sesonske, *Nuclear Reactor Engineering,* 4th ed. New York: Chapman & Hall, 1994, Chapter 3.

Isbin, H. S., *Introductory Nuclear Reactor Theory.* New York: Reinhold, 1963, Chapter 4.

Lamarsh, J. R., *Introduction to Nuclear Reactor Theory.* Reading, Mass.: Addison-Wesley, 1966, Chapter 5.

Liverhant, S. E., *Elementary Introduction to Nuclear Reactor Physics.* New York: Wiley, 1960.

Murray, R. L., *Nuclear Reactor Physics.* Englewood Cliffs, N.J.: Prentice-Hall, 1957, Chapter 2.

Zweifel, P. F., *Reactor Physics.* New York: McGraw-Hill, 1973, Chapter 2.

PROBLEMS

1. A point source emits S neutrons/sec isotropically in an infinite vacuum. (a) Show that the neutron flux at the distance r from the source is given by

$$\phi = \frac{S}{4\pi r^2}.$$

(b) What is the neutron current density vector at the same point? [*Note:* Neutrons do not diffuse in a vacuum.]

2. Three isotropic neutron sources, each emitting S neutrons/sec, are located in an infinite vacuum at the three corners of an equilateral triangle of side a. Find the flux and current at the midpoint of one side.

3. Using Eqs. (5.10) and (5.11), estimate the diffusion coefficients of (a) beryllium, (b) graphite, for monoenergetic 0.0253 eV neutrons.

4. The neutron flux in a bare spherical reactor of radius 50 cm is given by

$$\phi = 5 \times 10^{13} \frac{\sin 0.0628r}{r} \text{neutrons/cm}^2\text{-sec},$$

where r is measured from the center of the reactor. The diffusion coefficient for the system is 0.80 cm. (a) What is the maximum value of the flux in the reactor? (b) Calculate the neutron current density as a function of position in the reactor. (c) How many neutrons escape from the reactor per second?

5. Isotropic point sources each emitting S neutrons/sec are placed in an infinite moderator at the four corners of a square of side a. Compute the flux and current at the midpoint of any side of the square and at its center.

6. Find expressions for the flux and current at the point P_2 in Fig. 5.5

7. An isotropic point source emits S neutrons/sec in an infinite moderator. (a) Compute the net number of neutrons passing per second through a spherical surface of radius r centered on the source. (b) Compute the number of neutrons absorbed per second within the sphere. (c) Verify the equation of continuity for the volume within the sphere.

8. Two infinite planar sources each emitting S neutrons/cm^2 are placed parallel to one another in an infinite moderator at the distance a apart. Calculate the flux and current as a function of distance from a plane midway between the two.

9. Suppose the two planar sources in the preceding problem are placed at right angles to one another. Derive expressions for the flux and current as a function of distance from the line of intersection of the sources in a plane bisecting the angle between the sources.

10. An infinite moderator contains uniformly distributed isotropic sources emitting S neutrons/cm^3-sec. Determine the steady-state flux and current at any point in the medium.

11. An infinite bare slab of moderator of thickness $2a$ contains uniformly distributed sources emitting S neutrons/cm^3-sec. (a) Show that the flux in the slab is given by

$$\phi = \frac{S}{\Sigma_a}\left(1 - \frac{\cosh x/L}{\cosh\left(\dfrac{a+d}{L}\right)}\right),$$

where x is measured from the center of the slab. (b) Verify the equation of continuity by computing per unit area of the slab the total number of neutrons (i) produced per sec within the slab; (ii) absorbed per second within the slab; and (iii) escaping per second from the slab. [*Hint:* The solution to an inhomogeneous differential equation is the sum of solutions to the homogeneous equation plus a particular solution. Try a constant for the particular solution.]

12. A point source emitting S neutrons/sec is placed at the center of a sphere of moderator of radius R. (a) Show that the flux in the sphere is given by

$$\phi = \frac{S}{4\pi D \sinh\left(\dfrac{R+d}{L}\right)} \frac{\sinh\dfrac{1}{L}(R+d-r)}{r},$$

where r is the distance from the source. (b) Show that the number of neutrons leaking per second from the surface of the sphere is given by

$$\text{No. leaking/sec} = \frac{(R+d)S}{L \sinh\left(\dfrac{R+d}{L}\right)}.$$

(c) What is the probability that a neutron emitted by the source escapes from the surface?

13. An infinite planar source emitting S neutrons/cm²-sec is placed between infinite slabs of beryllium and graphite of thickness a and b, respectively, as shown in Fig. 5.9. Derive an expression for the neutron flux in the system. [*Note:* Since the media are different on opposite sides of the source, this problem is not symmetric and the source condition Eq. (5.28) is not valid. The appropriate boundary conditions for this problem are

$$\text{(i)} \quad \lim_{x \to 0}[\phi(x > 0) - \phi(x < 0)] = 0$$
$$\text{(ii)} \quad \lim_{x \to 0}[J(x > 0) - J(x < 0)] = S.$$

Condition (i) states, in effect, that ϕ is continuous at the source, whereas (ii) accounts for the neutrons emitted from the source.]

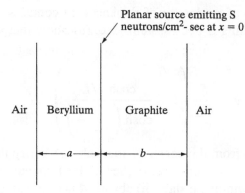

Planar source emitting S
neutrons/cm²- sec at $x = 0$

Air | Beryllium | Graphite | Air

|←——a——→|←——b——→|

Figure 5.9   Infinite planar source between two finite slabs of material.

14. A sphere of moderator of radius R contains uniformly distributed sources emitting S neutrons/cm³-sec. (a) Show that the flux in the sphere is given by

$$\phi = \frac{S}{\Sigma_a}\left[1 - \frac{R+d}{r}\frac{\sinh r/L}{\sinh\left(\dfrac{R+d}{L}\right)}\right].$$

(b) Derive an expression for the current density at any point in the sphere.

(c) How many neutrons leak from the sphere per second? (d) What is the average probability that a source neutron will escape from the sphere?

15. The three-group fluxes for a bare spherical fast reactor of radius $R = 50$ cm are given by the following expressions:

$$\phi_1(r) = \frac{3 \times 10^{15}}{r} \sin\left(\frac{\pi r}{R}\right)$$

$$\phi_2(r) = \frac{2 \times 10^{16}}{r} \sin\left(\frac{\pi r}{R}\right)$$

$$\phi_3(r) = \frac{1 \times 10^{16}}{r} \sin\left(\frac{\pi r}{R}\right).$$

The group-diffusion coefficients are $D_1 = 2.2$ cm, $D_2 = 1.7$ cm, and $D_3 = 1.05$ cm. Calculate the total leakage of neutrons from the reactor in all three groups. [*Note:* Ignore the extrapolation distance.]

16. The thermal flux in the center of a beam tube of a certain reactor is 2×10^{13} neutrons/cm^2-sec. The temperature in this region is 150°C. Calculate (a) the thermal neutron density; (b) the energy E_T; (c) the 2,200 meters-per-second flux.

17. The thermal flux at the center of a graphite research reactor is 5×10^{12} neutrons/cm^2-sec. The temperature of the system at this point is 120°C. Compare the neutron density at this point with the atom density of the graphite.

18. The thermal flux in a bare cubical reactor is given approximately by the function

$$\phi_T(x, y, z) = A \cos\left(\frac{\pi x}{\tilde{a}}\right) \cos\left(\frac{\pi y}{\tilde{a}}\right) \cos\left(\frac{\pi z}{\tilde{a}}\right),$$

where A is a constant, a is the length of a side of the cube, $\tilde{a}$ is a plus $2d$, d is the extrapolation distance, and x, y, and z are measured from the center of the reactor. Derive expressions for the (a) thermal neutron current as a function of position in the reactor; (b) number of thermal neutrons leaking per second from each side of the reactor; and (c) total number of thermal neutrons leaking per second from the reactor.

19. A planar source at the center of an infinite slab of graphite 2 meters thick emits 10^8 thermal neutrons per cm^2/sec. Given that the system is at room temperature, calculate the: (a) total number of thermal neutrons in the slab per cm^2 at any time; (b) number of thermal neutrons absorbed per cm^2/sec of the slab; (c) neutron current as a function of position in the slab; (d) total number of neutrons leaking per cm^2/sec from the two surfaces of the slab; (e) probability that a source neutron does not leak from the slab.

20. The thermal flux in a bare cubical reactor of side $a = 800$ cm is given by the expression

$$\phi_T(x, y, z) = 2 \times 10^{12} \cos\left(\frac{\pi x}{\tilde{a}}\right) \cos\left(\frac{\pi y}{\tilde{a}}\right) \cos\left(\frac{\pi z}{\tilde{a}}\right),$$

where x, y, and z are measured from the center of the reactor and $\tilde{a} = a + 2d$. The temperature is 400°C, and the measured values of the thermal diffusion coefficient and diffusion length are 0.84 cm and 17.5 cm, respectively. (a) How many moles of thermal neutrons are there in the entire reactor? (b) Calculate the neutron current density vector as a function of position in the reactor. (c) How many thermal neutrons leak from the reactor per second? (d) How many thermal neutrons are absorbed in the reactor per second? (e) What is the relative probability that a thermal neutron will leak from the reactor?

21. The thermal (0.0253 eV) cross-section for the production of a 54-minute isomer of ^{116}In via the (n, γ) reaction with ^{115}In is 157 b. A thin indium foil weighing 0.15 g is placed in the beam tube described in Problem 5.16. (a) At what rate are thermal neutrons absorbed by the foil? (b) After 1 hour in the beam tube, what is the 54-minute activity of ^{116}In? [*Note:* ^{115}In has an isotopic abundance of 95.7% and is a non-$1/v$ absorber.]

22. The thermal flux in a bare spherical reactor 1 m in diameter is given approximately by

$$\phi_T(r) = 2.29 \times 10^{14} \frac{\sin 0.0628r}{r} \text{ neutrons/cm}^2\text{-sec.}$$

If the reactor is moderated and cooled by unit density water that takes up to one-third of the reactor volume, how many grams of ^{2}H are produced per year in the reactor? Assume that the water is only slightly warmed by the heat of the reactor.

23. Tritium (^{3}H) is produced in nuclear reactors by the absorption of thermal neutrons by ^{6}Li via the reaction ^{6}Li(n, α)^{3}H. The cross-section for this reaction at 0.0253 eV is 940 b and it is $1/v$. (a) Show that the annual production of ^{3}H in a thermal flux ϕ_T, per gram of ^{6}Li, is given by

$$\text{production rate/g} = 2.28 \times 10^9 \phi_T \text{ atoms/g-yr.}$$

(b) Compute the annual production in curies of ^{3}H per gram of ^{6}Li in a flux of $\phi_T = 1 \times 10^{14}$ neutrons/cm^2-sec.

24. A radioactive sample with half-life $T_{1/2}$ is placed in a thermal reactor at a point where the thermal flux is ϕ_T. Show that the sample disappears as the result of its own decay and by neutron absorption with an effective half-life given by

$$\left(\frac{1}{T_{1/2}}\right)_{\text{eff}} = \frac{1}{T_{1/2}} + \frac{\overline{\sigma}_a \phi_T}{\ln 2},$$

where $\overline{\sigma}_a$ is the average thermal absorption cross-section of the sample.

25. Calculate the thermal diffusion coefficient and diffusion length of water near the outlet of a pressurized-water reactor—that is, at about 300°C and density of 0.68 g/cm³.

26. Calculate for natural uranium at room temperature and 350°C the value of $\overline{D}$, Σ_a, and L_T. The measured value of L_T at room temperature is 1.55 cm. [*Note:* The value of σ_s is constant at low neutron energies and is approximately the same for ^{235}U and ^{238}U. The densities of uranium at 350°C and room temperature are essentially the same.]

27. Repeat the calculation of Problem 5.26 for uranium enriched to 2 weight/percent in ^{235}U.

28. Compute and plot for mixtures of D_2O and H_2O, at H_2O concentrations up to 5 weight/percent: (a) Σ_a at 0.0253 eV; (b) L_T at room temperature.

29. Calculate the thermal neutron diffusion length at room temperature in water solutions of boric acid (H_3BO_3) at the following concentrations: (a) 10 g/liter, (b) 1 g/liter, and (c) 0.1 g/liter. [*Hint:* Because of the small concentration of the boric acid, the diffusion coefficient for the mixture is essentially the same as that of pure water.]

30. An infinite slab of ordinary water 16 cm thick contains a planar source at its center emitting 10^8 thermal neutrons per cm²/sec. Compute and plot the thermal flux within the slab.

31. Repeat the calculations of Problem 5.30 and plot the thermal flux for boric acid solutions having the concentrations given in Problem 5.29.

32. Verify that Eq. (5.72) is in fact the solution to Eq. (5.68). Discuss this solution in the cases where $\tau_T \to 0$ and $\tau_T \to L_T^2$.

33. Calculate the neutron age of fission neutrons in water for the conditions given in Problem 5.25.

34. An infinite slab of moderator of extrapolated thickness $\widetilde{a}$ contains sources of fast neutrons distributed according to the function

$$s(x) = S \cos\left(\frac{\pi x}{\widetilde{a}}\right).$$

Using the two-group method, derive an expression for the thermal flux in the slab.

6

Nuclear Reactor Theory

In a critical reactor, as explained in Section 4.1, there is a balance between the number of neutrons produced in fission and the number lost, either by absorption in the reactor or by leaking from its surface. One of the central problems in the design of a reactor is the calculation of the size and composition of the system required to maintain this balance. This problem is the subject of the present Chapter.

Calculations of the conditions necessary for criticality are normally carried out using the group-diffusion method introduced at the end of the last Chapter. It is reasonable to begin the present discussion with a one-group calculation. Although this type of calculation is most appropriate for fast reactors, it is shown later in the Chapter that the one-group method can also be used in a modified form for computations of some thermal reactors.

6.1 ONE-GROUP REACTOR EQUATION

Consider a critical fast reactor containing a homogeneous mixture of fuel and coolant. It is assumed that the reactor consists of only one region and has neither a blanket nor a reflector. Such a system is said to be a *bare reactor*.

This reactor is described in a one-group calculation by the one-group time-dependent diffusion equation (see Eq. [5.17])

$$D\nabla^2\phi - \Sigma_a\phi + s = -\frac{1}{v}\frac{\partial\phi}{\partial t}. \tag{6.1}$$

Here ϕ is the one-group flux, D and Σ_a are the one-group diffusion coefficient and macroscopic absorption cross-section for the fuel-coolant mixture, s is the source density (i.e., the number of neutrons emitted per cm^3/sec), and v the neutron speed. Note, we are making no assumptions as to the state of the reactor so that there may be time dependence involved. If this were not the case, then there would be no time derivative term, and a balance between the source and the absorption and leakage would exist.

Nominal constants and one-group cross-sections for a fast reactor are given in Table 6.1.

In a reactor at a measurable power, the source neutrons are emitted in fission. To determine s, let Σ_f be the fission cross-section for the fuel. If there are v neutrons produced per fission, then the source is

$$s = v\Sigma_f\phi.$$

Here we are assuming that there is no nonfission source of neutrons present in the reactor—a good approximation for a reactor except at very low power levels.

If the fission source does not balance the leakage and absorption terms, then the right-hand side of Eq. (6.1) is nonzero. To balance the equation, we multiply the source term by a constant $1/k$, where k an unknown constant. If the source is too small, then k is less than 1. If it is too large, then k is greater than 1. Eq. (6.1) may now be written as

$$D\nabla^2\phi - \Sigma_a\phi + \frac{1}{k}v\Sigma_f\phi = 0. \tag{6.2}$$

TABLE 6.1   NOMINAL ONE-GROUP CONSTANTS FOR A FAST REACTOR*

Element or Isotope	σ_γ	σ_f	σ_a	σ_{tr}	v	η
Na	0.0008	0	0.0008	3.3	—	—
Al	0.002	0	0.002	3.1	—	—
Fe	0.006	0	0.006	2.7	—	—
^{235}U	0.25	1.4	1.65	6.8	2.6	2.2
^{238}U	0.16	0.095	0.255	6.9	2.6	0.97
^{239}P	0.26	1.85	2.11	6.8	2.98	2.61

*From *Reactor Physics Constants*, 2nd ed., Argonne National Laboratory report ANL-5800, 1963.

The equation may be rewritten as an eigenvalue equation by letting

$$B^2 = \frac{1}{D}\left(\frac{1}{k}\Sigma_f - \Sigma_a\right),$$

where B^2 is defined as the geometric buckling. Then Eq. (6.2) may be written as

$$\nabla^2\phi = -B^2\phi, \tag{6.3}$$

Finally, this expression for the leakage term may be substituted into Eq. (6.2) and the diffusion equation rewritten as

$$-DB^2\phi - \Sigma_a\phi + \frac{1}{k}\nu\Sigma_f\phi = 0 \tag{6.4}$$

or

$$\nabla^2\phi + B^2\phi = 0. \tag{6.5}$$

This important result is known as the *one-group reactor equation*. The one-group equation may be solved for the constant k:

$$k = \frac{\nu\Sigma_f\phi}{DB^2\phi + \Sigma_a\phi} = \frac{\nu\Sigma_f}{DB^2 + \Sigma_a}. \tag{6.6}$$

Note that this equation does not give a value for k since B^2 is still unknown.

Physically, Eq. (6.6) may be interpreted as follows. The numerator is the number of neutrons that are born in fission in the current generation, whereas the denominator represents those that were lost from the previous generation. Since all neutrons in a generation are either absorbed or leak from the reactor, then the numerator must also equal the number born in the previous generation. But this is the definition of the multiplication factor from Section 4.1 for a finite reactor. We may now define the multiplication factor for a reactor as the birth rate, $\nu\Sigma_f\phi$, divided by the leakage rate plus the absorption rate, $DB^2\phi + \Sigma_a\phi$, of the neutrons.

The source term in the one-group equation may be written in terms of the fuel absorption cross-section using the following. Let Σ_{aF} be the one-group absorption cross-section of the fuel, and let η be the average number of fission neutrons emitted per neutron absorbed in the fuel.[1] The source term is then given by,

$$s = \eta\Sigma_{aF}\phi. \tag{6.7}$$

---

[1]If the fuel consists of more than one species of fissile nuclide or a mixture of fissile and fissionable nuclides, then η should be computed as in Section 3.7 using Eq. (3.55).

This may also be written as,

$$s = \eta \frac{\Sigma_{aF}}{\Sigma_a} \Sigma_a \phi = \eta f \Sigma_a \phi, \qquad (6.8)$$

where

$$f = \frac{\Sigma_{aF}}{\Sigma_a} \qquad (6.9)$$

is called the *fuel utilization*. Since Σ_a is the cross-section for the mixture of fuel and coolant, whereas Σ_{aF} is the cross-section of the fuel only, it follows that f is equal to the fraction of the neutrons absorbed in the reactor that are absorbed by the fuel.

The source term in Eq. (6.8) can also be written in terms of the multiplication factor for an infinite reactor. For this purpose, consider an infinite reactor having the same composition as the bare reactor under discussion. With such a reactor, there can be no escape of neutrons as there is from the surface of a bare reactor. All neutrons eventually are absorbed, either in the fuel or the coolant. Furthermore, the neutron flux must be a constant, independent of position. The number of fissions in one generation is simply related to the number of neutrons born in one generation by ν, the number of neutrons produced per fission. Since all neutrons born must eventually be absorbed in the system, this must be equal to $\Sigma_a \phi$, the number of neutrons absorbed per cm^3/sec everywhere in the system. Of these neutrons, the number $f \Sigma_a \phi$ are absorbed in fuel and release $\eta f \Sigma_a \phi$ fission neutrons in the next generation. The number of neutrons born in this new generation can again be related to the fission rate by ν. Sooner or later, all of these neutrons must in turn be absorbed in the reactor. Thus, the absorption of $\Sigma_a \phi$ neutrons in one generation leads to the absorption of $\eta f \Sigma_a \phi$ in the next. Now according to the discussion in Section 4.1, the multiplication factor is defined as the number of fissions in one generation divided by the number in the preceding generation. Since in an infinite medium these are related to the absorption rate, it follows that,

$$k_\infty = \frac{\eta f \Sigma_a \phi}{\Sigma_a \phi} = \eta f, \qquad (6.10)$$

where the subscript on k_∞ signifies that this result is only valid for the infinite reactor.

Example 6.1

Calculate f and k_∞ for a mixture of ^{235}U and sodium in which the uranium is present to $1 w/o$.

Solution. From Eq. (6.9), f is

$$f = \frac{\Sigma_{aF}}{\Sigma_a} = \frac{\Sigma_{aF}}{\Sigma_{aF} + \Sigma_{aS}},$$

where Σ_{aF} and Σ_{aS} are the macroscopic absorption cross-sections of the uranium and sodium, respectively. Dividing numerator and denominator by Σ_{aF} gives

$$f = \frac{1}{1 + \Sigma_{aS}/\Sigma_{aF}} = \frac{1}{1 + N_S \sigma_{aS}/N_F - \sigma_{aF}},$$

where N_F and N_S are the atom densities of the uranium and sodium. Next let ρ_F and ρ_S be the number of grams of uranium and sodium per cm^3 in the mixture. Then

$$\frac{N_S}{N_F} = \frac{\rho_S}{\rho_F} \frac{M_F}{M_S},$$

where M_F and M_S are the gram atomic weights of uranium and sodium. Since $1 w/o$ of the mixture is fuel, this means that,

$$\frac{\rho_F}{\rho_F + \rho_S} = 0.01$$

or

$$\frac{\rho_S}{\rho_F} = 99.$$

Using the values of σ_a given in Table 6.1, the value of f is given by

$$f^{-1} = 1 + 99 \times \frac{235}{23} \times \frac{0.0008}{1.65} = 1.49$$

and

$$f = 0.671. [Ans.]$$

The value of k_∞ is

$$k_\infty = \eta f = 2.2 \times 0.671 = 1.48.$$

Since this is greater than unity, an infinite reactor with this composition would be supercritical. [*Ans.*]

Since η and f are constants that only depend on the material properties of the reactor, the value of k_∞ is the same for a bare reactor as for an infinite reactor of the same composition. The source term in Eq. (6.8) can therefore be written as,

$$s = k_\infty \Sigma_a \phi. \tag{6.11}$$

Introducing Eq. (6.11) and Eq. (6.3) into the one-group reactor equation, Eq. (6.4) gives,

$$-DB^2\phi - \Sigma_a\phi + \frac{k_\infty}{k}\Sigma_a\phi = -\frac{1}{v}\frac{\partial\phi}{\partial t}.$$

If the reactor is just critical (i.e., $k = 1$), then the right-hand side is zero and

$$-DB^2\phi + (k_\infty - 1)\Sigma_a\phi = 0. \tag{6.12}$$

Dividing by D yields

$$-B^2\phi + \frac{k_\infty - 1}{L^2}\phi = 0, \tag{6.13}$$

where,

$$L^2 = \frac{D}{\Sigma_a} \tag{6.14}$$

is the one-group diffusion area. Equation (6.13) may be solved for the buckling, B^2, for a critical reactor

$$B^2 = \frac{k_\infty - 1}{L^2}. \tag{6.15}$$

It is shown in the next sections that the one-group diffusion equation, together with the usual boundary conditions on ϕ, not only determines the shape of the flux in the reactor, but also leads to a condition that must be satisfied for the reactor to be critical.

6.2 THE SLAB REACTOR

As the first example of a bare reactor, consider a critical system consisting of an infinite bare slab of thickness a as shown in Fig. 6.1. The reactor equation in this

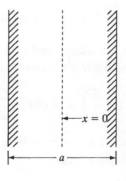

Figure 6.1   The infinite slab reactor.

case is

$$\frac{d^2\phi}{dx^2} + B^2\phi = 0, \tag{6.16}$$

where x is measured from the center of the slab.

To determine the flux within the reactor, Eq. (6.16) must be solved subject to the boundary condition that ϕ vanishes at the extrapolated faces of the slab—that is, at $x = \tilde{a}/2$ and at $x = -\tilde{a}/2$ where $\tilde{a} = a + 2d$. Then the boundary conditions become

$$\phi\left(\frac{\tilde{a}}{2}\right) = \phi\left(-\frac{\tilde{a}}{2}\right) = 0. \tag{6.17}$$

It may also be noted that, because of the symmetry of the problem, there can be no net flow of neutrons at the center of the slab. Since the neutron current density is proportional to the derivative of ϕ, this means that

$$\frac{d\phi}{dx} = 0 \tag{6.18}$$

at $x = 0$. The condition given by Eq. (6.18) is equivalent to requiring that ϕ be an even function—that is,

$$\phi(-x) = \phi(x), \tag{6.19}$$

and has a continuous derivative within the reactor. [In any problem where it is clear that the flux is a well-behaved function, Eq. (6.19) is often easier to apply than Eq. (6.18).]

In any case, the general solution to Eq. (6.16) is

$$\phi(x) = A \cos Bx + C \sin Bx, \tag{6.20}$$

where A and C are constants to be determined. Placing the derivative of Eq. (6.20) equal to zero at $x = 0$ gives immediately $C = 0$, so that ϕ reduces to

$$\phi(x) = A \cos Bx.$$

Next, introducing the boundary condition given by Eq. (6.17) at $\tilde{a}/2$ or $-\tilde{a}/2$ (it makes no difference which since the cosine is an even function) gives

$$\phi\left(\frac{\tilde{a}}{2}\right) = A \cos\left(\frac{B\tilde{a}}{2}\right) = 0. \tag{6.21}$$

This equation can be satisfied either by taking $A = 0$, which leads to the trivial solution $\phi(x) = 0$, or by requiring that

$$\cos\left(\frac{B\widetilde{a}}{2}\right) = 0. \tag{6.22}$$

This, in turn, is satisfied if B assumes any of the values B_n, where

$$B_n = \frac{n\pi}{\widetilde{a}} \tag{6.23}$$

and n is an odd integer, as can readily be seen by direct substitution into Eq. (6.22).

The various constants B_n are known as *eigenvalues*, and the corresponding functions $\cos B_n x$ are called *eigenfunctions*. It can be shown that if the reactor under consideration is not critical, the flux is the sum of all such eigenfunctions, each multiplied by a function that depends on the time. However, if the reactor is critical, all of these functions except the first die out in time, and the flux assumes the steady-state shape of the first eigenfunction or *fundamental*—namely,

$$\phi(x) = A \cos B_1 x = A \cos\left(\frac{\pi x}{\widetilde{a}}\right). \tag{6.24}$$

This is the flux in a critical slab reactor.

The square of the lowest eigenvalue B_1^2 is called the *buckling* of the reactor. The origin of this term can be seen by solving the equation satisfied by the flux—namely,

$$\frac{d^2\phi}{dx^2} + B_1^2\phi = 0, \tag{6.25}$$

for B_1^2. The result is

$$B_1^2 = -\frac{1}{\phi}\frac{d^2\phi}{dx^2}.$$

The right-hand side of this expression is proportional to the curvature of the flux in the reactor, which, in turn, is a measure of the extent to which the flux curves or buckles. Since in the slab reactor

$$B_1^2 = \left(\frac{\pi}{\widetilde{a}}\right)^2, \tag{6.26}$$

the buckling decreases as a increases. In the limit, as a becomes infinite, $B_1^2 = 0$, ϕ is a constant and has no buckle.

It should be observed that the value of the constant A in Eq. (6.24), which determines the magnitude of ϕ, has not been established in the prior analysis. Mathematically, this is because the reactor equation (Eq. [6.5] or [6.16]) is homogeneous, and ϕ multiplied by any constant is still a solution to the equation. Physically the reason that the value of A has not been established is that the magnitude of the flux in a reactor is determined by the power at which the system is operating, and not by its material properties.

To find an expression for A, it is necessary to make a separate calculation of the reactor power. In particular, there are $\Sigma_f \phi(x)$ fissions per cm³/sec at the point x, where Σ_f is the macroscopic fission cross-section. If the recoverable energy is E_R joules per fission[2] (with a recoverable energy of 200 MeV, $E_R = 3.2 \times 10^{-11}$ joules), then the total power per unit area of the slab, in watts/cm², is

$$P = E_R \Sigma_f \int_{-a/2}^{a/2} \phi(x)\,dx. \tag{6.27}$$

Note that the integration is carried out over the physical dimensions of the reactor and not to the extrapolated boundary. Inserting the expression for $\phi(x)$ from Eq. (6.20) and performing the integration gives

$$P = \frac{2\tilde{a}E_R\Sigma_f A \sin(\frac{\pi a}{2\tilde{a}})}{\pi}. \tag{6.28}$$

The final formula for the thermal flux in a slab reactor is then

$$\phi(x) = \frac{\pi P}{2\tilde{a}E_R\Sigma_f \sin(\frac{\pi a}{2\tilde{a}})} \cos\left(\frac{\pi x}{\tilde{a}}\right),$$

which, if d is small compared to a, reduces to

$$\phi(x) = \frac{\pi P}{2aE_R\Sigma_f} \cos\left(\frac{\pi x}{a}\right). \tag{6.29}$$

6.3 OTHER REACTOR SHAPES

It is not possible, of course, to construct a reactor in the form of an infinite slab. Therefore, it is necessary to generalize the results of the preceding section for reactors of more realistic shapes. This can easily be done for the following reactors: sphere, infinite cylinder, rectangular parallelepiped, and finite cylinder. These reactors, together with the coordinate systems used to describe them, are shown in Fig. 6.2. It should be noted that each of these reactors is bare; reflected reactors are discussed in Section 6.6.

Sphere

Consider first a critical spherical reactor of radius R. The flux in this reactor is a function only of r, and the reactor equation is

---

[2]In Lamarsh's book on reactor theory, E_R, when expressed in joules, was denoted by γ.

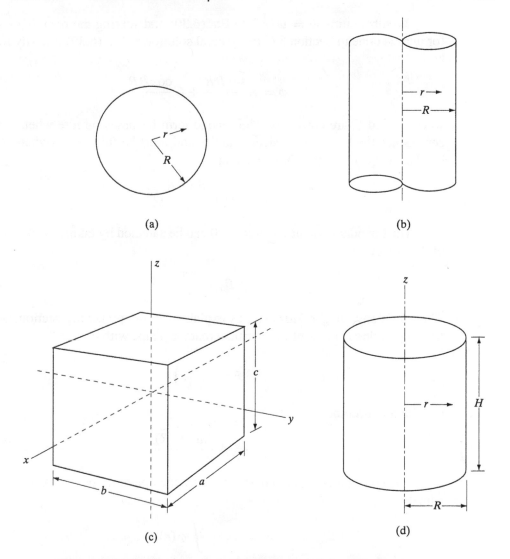

(a)   (b)

(c)   (d)

Figure 6.2   (a) The spherical reactor; (b) the infinite cylindrical reactor; (c) the parallelepiped reactor; (d) the finite cylindrical reactor.

$$\frac{1}{r^2}\frac{d}{dr}r^2\frac{d\phi}{dr} + B^2\phi = 0, \tag{6.30}$$

where use has been made of the Laplacian in spherical coordinates. The flux must satisfy the boundary condition $\phi(\widetilde{R}) = 0$ and remain finite throughout the interior of the reactor.

By substituting $\phi = w/r$ into Eq. (6.30) and solving the resulting equation for w as is done in Section 5.6, the general solution to Eq. (6.30) is easily found to be

$$\phi = A \frac{\sin BR}{r} + C \frac{\cos BR}{r},$$

where A and C are constants. The second term becomes infinite when r goes to zero. Since the flux is a physical quantity and must be finite everywhere, C must be placed equal to zero. Thus, ϕ becomes

$$\phi = A \frac{\sin BR}{r}. \tag{6.31}$$

The boundary condition $\phi(\widetilde{R}) = 0$ can be satisfied by taking B to be any one of the eigenvalues

$$B_n = \frac{n\pi}{\widetilde{R}},$$

where n is any integer. However, as explained in the preceding section, only the first eigenvalue is relevant for a critical reactor. Thus, with $n = 1$, the buckling is

$$B_1^2 = \left(\frac{\pi}{\widetilde{R}} \right)^2 \tag{6.32}$$

and the flux becomes

$$\phi = A \frac{\sin(\pi r / \widetilde{R})}{r}. \tag{6.33}$$

The constant A is again determined by the operating power of the reactor—namely,

$$P = E_R \Sigma_f \int \phi(r) dV, \tag{6.34}$$

where dV is the differential volume element. In view of the geometry of the problem, dV is given by

$$dV = 4\pi r^2 \, dr$$

and Eq. (6.34) becomes

$$P = 4\pi E_R \Sigma_f \int_0^R r^2 \phi(r) \, dr.$$

Introducing the flux from Eq. (6.33) and carrying out the integration gives

$$P = 4\pi E_R \Sigma_f A \frac{\widetilde{R}}{\pi} \left[\frac{\widetilde{R}}{\pi} \sin\left(\frac{\pi R}{\widetilde{R}} \right) - R \cos\left(\frac{\pi R}{\widetilde{R}} \right) \right].$$

If d is small, the flux in the sphere may therefore be written as

$$\phi = \frac{P}{4 E_R \Sigma_f R^2} \frac{\sin(\pi r / \widetilde{R})}{r}. \tag{6.35}$$

Infinite Cylinder

Next consider an infinite critical cylindrical reactor of radius R, which is just critical. In this reactor, the flux depends only on the distance r from the axis. With the Laplacian approximation to cylindrical coordinates described in Appendix III, the reactor equation becomes

$$\frac{1}{r} \frac{d}{dr} r \frac{d\phi}{dr} + B^2 \phi = 0,$$

or, when the differentiation in the first term is carried out,

$$\frac{d^2\phi}{dr^2} + \frac{1}{r} \frac{d\phi}{dr} + B^2 \phi = 0. \tag{6.36}$$

Besides satisfying this equation, ϕ must also satisfy the usual boundary conditions including $\phi(\widetilde{R}) = 0$.

Equation (6.36) is a special case of *Bessel's equation*,

$$\frac{d^2\phi}{dr^2} + \frac{1}{r} \frac{d\phi}{dr} + \left(B^2 - \frac{m^2}{r^2} \right) \phi = 0, \tag{6.37}$$

in which m is a constant. Since Eq. (6.37) is a second-order differential equation, it has two independent solutions. These are denoted as $J_m(Br)$ and $Y_m(Br)$ and are called *ordinary Bessel functions of the first and second kind*, respectively.[3] These functions appear in many engineering and physics problems and are widely tabulated.

Comparing Eq. (6.36) with Eq. (6.37) shows that, in the present problem, m is equal to zero. The general solution to the reactor equation can therefore be written as

$$\phi = A J_0(Br) + C Y_0(Br),$$

where A and C are again constants. The functions $J_0(x)$ and $Y_0(x)$ are plotted in Fig. 6.3. It is observed that $Y_0(x)$ is infinite at $x = 0$, while $J_0(0) = 1$. Therefore,

---

[3] A short table of Bessel functions is given in Appendix V.

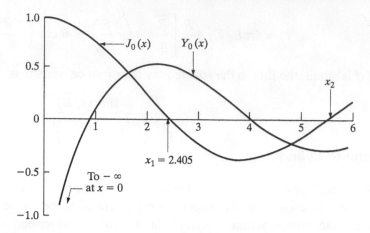

Figure 6.3   The Bessel functions $J_0(x)$ and $Y_0(x)$.

since ϕ must remain finite within the reactor, C must be taken to be zero. Thus, ϕ reduces to

$$\phi = A J_0(Br). \tag{6.38}$$

The boundary condition $\phi(\widetilde{R}) = 0$ now becomes

$$\phi(\widetilde{R}) = A J_0(Br) = 0. \tag{6.39}$$

As shown in Fig. 6.3, the function $J_0(x)$ is equal to zero at a number of values of x, labeled $x_1, x_2, \ldots$, so that $J_0(x_n) = 0$. This means that Eq. (6.39) is satisfied provided B is any one of the values

$$B_n = \frac{x_n}{\widetilde{R}},$$

which are the eigenvalues of the problem. However, since in a critical reactor only the lowest eigenvalue is important, it follows that the buckling is

$$B_1^2 = \left(\frac{x_1}{\widetilde{R}}\right)^2 = \left(\frac{2.405}{\widetilde{R}}\right)^2. \tag{6.40}$$

This one-group flux is then

$$\phi = A J_0\left(\frac{2.405r}{\widetilde{R}}\right). \tag{6.41}$$

The constant A is again determined by Eq. (6.34), where, for the infinite cylinder, $dV = 2\pi r\,dr$. Thus, the power per unit length of the cylinder is

$$P = 2\pi E_R \Sigma_f \int_0^R \phi(r) r \, dr$$

$$= 2\pi E_R \Sigma_f A \int_0^R J_0\left(\frac{2.405r}{R}\right) r \, dr.$$

The integral can be evaluated using the formula

$$\int J_0(x') x' dx' = x J_1(x),$$

which gives for small d

$$P = 2\pi E_R \Sigma_f R^2 A J_1(2.405)/2.405 = 1.35 E_R \Sigma_f R^2 A.$$

The final expression for the flux is then

$$\phi = \frac{0.738P}{E_R \Sigma_f R^2} J_0\left(\frac{2.405r}{R}\right). \tag{6.42}$$

Finite Cylinder

An interesting flux distribution is obtained for the case of a finite cylinder of height H and radius R. In this reactor, the flux depends on the distance r from the axis and the distance z from the midpoint of the cylinder. With the Laplacian appropriate to cylindrical coordinates described in Appendix III, the reactor equation becomes

$$\frac{1}{r}\frac{\partial}{\partial r} r \frac{\partial \phi}{\partial r} + \frac{\partial^2 \phi}{\partial z^2} + B^2 \phi = 0,$$

or, when the differentiation in the first term is carried out,

$$\frac{d^2\phi}{\partial r^2} + \frac{1}{r}\frac{\partial \phi}{\partial r} + \frac{\partial^2 \phi}{\partial z^2} + B^2 \phi = 0. \tag{6.43}$$

Besides satisfying this equation, ϕ must also satisfy the usual boundary conditions including $\phi(\tilde{R}, z) = 0$ and $\phi(r, \tilde{H}/2) = 0$. The solution is obtained by assuming separation of variables with

$$\phi(r, z) = R(r)Z(z).$$

Upon substitution into Eq. (6.43), one obtains

$$\frac{1}{R}\frac{1}{r}\frac{\partial}{\partial r} r \frac{\partial R}{\partial r} + \frac{1}{Z}\frac{\partial^2 Z}{\partial z^2} = -B^2.$$

Since this equation must be satisfied for any r or z combination, the first and second terms must be constants. One can then write

$$\frac{d^2 R}{dr^2} + \frac{1}{r}\frac{dR}{dr} + B_r^2 R = 0,$$

which is Bessel's equation of the first-kind order zero as before. Further,

$$\frac{d^2 Z}{dz^2} + B^2 Z = 0,$$

The buckling B^2 is

$$B^2 = B_r^2 + B_z^2$$

whose solution is

$$Z(z) = A \cos B_z z + C \sin_z Bz,$$

is the same as for the infinite slab reactor. The solution must satisfy the boundary conditions and be positive definite. The solution obtained is then the product of the slab and infinite cylinder solutions with the required boundary conditions. Thus,

$$\phi(r, z) = A J_0 \left(\frac{2.405r}{\tilde{R}} \right) \cos \frac{\pi z}{\tilde{H}},$$

where $\tilde{H} = H + 2d$ and $\tilde{R} = R + d$. The constant A may be determined as before using Eq. (6.34) if the reactor power is known.

Rectangular Parallelepiped

The derivation of the flux distribution in the parallelepiped reactors is somewhat lengthy, but may be obtained in a manner similar to the finite cylinder. Values of the buckling and flux for this reactor geometry are given in Table 6.2, which also summarizes the results obtained earlier for the slab, sphere, infinite cylinder, and finite cylinder.

TABLE 6.2   BUCKLINGS, B^2, AND FLUXES FOR CRITICAL BARE REACTORS (ASSUMING d IS SMALL)

Geometry	Dimensions	Buckling	Flux	A	Ω
Infinite slab	Thickness a	$\left(\frac{\pi}{a}\right)^2$	$A \cos\left(\frac{\pi x}{a}\right)$	$1.57 P / a E_R \Sigma_f$	1.57
Rectangular parallelepiped	$a \times b \times c$	$\left(\frac{\pi}{a}\right)^2 + \left(\frac{\pi}{b}\right)^2 + \left(\frac{\pi}{c}\right)^2$	$A \cos\left(\frac{\pi x}{a}\right) \cos\left(\frac{\pi y}{b}\right) \cos\left(\frac{\pi z}{c}\right)$	$3.87 P / V E_R \Sigma_f$	3.88
Infinite cylinder	Radius R	$\left(\frac{2.405}{R}\right)^2$	$A J_0 \left(\frac{2.405r}{R}\right)$	$0.738 P / R^2 E_R \Sigma_f$	2.32
Finite cylinder	Radius R				
	Height H	$\left(\frac{2.405}{R}\right)^2 + \left(\frac{\pi}{H}\right)^2$	$A J_0 \left(\frac{2.405r}{R}\right) \cos\left(\frac{\pi z}{H}\right)$	$3.63 P / V E_R \Sigma_f$	3.64
Sphere	Radius R	$\left(\frac{\pi}{R}\right)^2$	$A \frac{1}{r} \sin\left(\frac{\pi r}{R}\right)$	$P / 4 R^2 E_R \Sigma_f$	3.29

Maximum-to-Average Flux and Power

The maximum value of the flux, ϕ_{max}, in a uniform bare reactor is always found at the center of the reactor. Since the power density is also highest at the center, it is of some interest to compute the ratio of the maximum flux to its average value throughout the reactor. This ratio, which is denoted by Ω, is a measure of the overall variation of the flux within a reactor and is an indication of the extent to which the power density at the center of the system exceeds the average power density.

Consider the case of a bare spherical reactor. The value of ϕ_{max} is obtained by taking the limit of Eq. (6.35) as r goes to zero. Thus,

$$\phi_{max} = \frac{P}{4E_R \Sigma_f R^2} \lim_{r \to 0} \frac{\sin(\pi r/R)}{r} = \frac{\pi P}{4E_R \Sigma_f R^3}. \tag{6.44}$$

The average value of ϕ is given by

$$\phi_{av} = \frac{1}{V} \int \phi dV, \tag{6.45}$$

where the integral is carried out over the volume of the sphere. However, this integral is proportional to the reactor power—that is,

$$P = E_R \Sigma_f \int \phi dV,$$

and so

$$\phi_{av} = \frac{P}{E_R \Sigma_f V}, \tag{6.46}$$

which is a valid result for all geometries. Dividing Eq. (6.44) by Eq. (6.46) gives, for the particular case of a sphere,

$$\Omega = \frac{\phi_{max}}{\phi_{av}} = \frac{\pi^2}{3} = 3.29. \tag{6.47}$$

Values of Ω for other geometries are given in Table 6.2.

For reasons to be discussed in the next two Chapters, it is important that the flux distribution in an actual reactor be as uniform—that is, as *flat*—as possible. The values of Ω in Table 6.2 would be quite unacceptable for a real reactor. However, it must be borne in mind that these values were derived for bare reactors. In such systems, the flux undergoes large changes from the reactor center to its surface. As is shown in Section 6.6, with reflected reactors, the flux does not fall to low values at the core-reflector interface, and the flux in these reactors (and all power reactors are reflected) is considerably flatter. The flux is made still flatter in most modern reactors by nonuniformly distributing the fuel in the core and by the addition of absorbers (see Section 7.6).

Example 6.2

A bare spherical reactor of radius 50 cm operates at a power level of 100 megawatts $= 10^8$ joules/sec. $\Sigma_f = 0.0047$ cm^{-1}. What are the maximum and average values of the flux in the reactor?

Solution. The maximum value of ϕ can be found directly from Eq. (6.44). Introducing numerical values gives

$$\phi_{\text{max}} = \frac{\pi \times 10^8}{4 \times 3.2 \times 10^{-11} \times 0.0047 \times (50)^3}$$

$$= 4.18 \times 10^{15} \text{neutrons/cm}^2\text{-sec. [\textit{Ans.}]}$$

The average value of the flux follows from Eq. (6.47):

$$\phi_{\text{av}} = \frac{\phi_{\text{max}}}{\Omega} = \frac{4.18 \times 10^{15}}{3.29} = 1.27 \times 10^{15} \text{neutrons/cm}^2\text{-sec. [\textit{Ans.}]}$$

6.4 THE ONE-GROUP CRITICAL EQUATION

It was seen earlier that a necessary condition for each of the reactors to be critical is that B^2 must equal the first eigenvalue B_1^2. But for a critical reactor, Eq. (6.6) gives k as

$$k = \frac{\nu \Sigma_f}{\Sigma_a + D B^2}.$$

Solving for B^2 when $k = 1$ gives

$$B^2 = \frac{\nu \Sigma_f - \Sigma_a}{D}, \tag{6.48}$$

or the critical buckling, B_c^2, must be

$$B_c^2 = \frac{\dfrac{\nu \Sigma_f}{\Sigma_a} - 1}{\dfrac{D}{\Sigma_a}}. \tag{6.49}$$

The right-hand side is a function of the material properties of the system. The right side must be equal to the square of the first eigenvalue B_1^2, which depends only on the dimensions and geometry of the reactor. Using the definition for k_∞ and L^2, this becomes

$$\frac{k_\infty - 1}{L^2} = B_1^2. \tag{6.50}$$

This equation determines the conditions under which a given bare reactor is critical. For instance, if the physical properties of the reactor are specified, then the left-hand side of Eq. (6.50) is known and the dimensions of the reactor must be adjusted so that B_1^2 on the right satisfies the equation. However, if the dimensions are specified, then B_1^2 is known and the properties of the reactor must be adjusted so that the left-hand side of the equation is equal to the right. In this way, the critical size or critical mass of fuel may be determined. An example of this procedure follows.

Example 6.3

A fast reactor assembly consisting of a homogeneous mixture of ^{239}Pu and sodium is to be made in the form of a bare sphere. The atom densities of these constituents are $N_F = 0.00395 \times 10^{24}$ for the ^{239}Pu and $N_S = 0.0234 \times 10^{24}$ for the sodium. Estimate the critical radius R_c of the assembly.

Solution. Introducing $B_1^2 = (\pi/\widetilde{R})^2$ from Eq. (6.32) into Eq. (6.50) and solving for $\widetilde{R}$ gives $\widetilde{R}_c$, the critical radius to the extrapolated boundary:

$$\widetilde{R}_c = \pi \sqrt{\frac{L^2}{k_\infty - 1}}.$$

Therefore, it is necessary to compute k_∞ and L^2 to find $\widetilde{R}$.

Using the cross-sections in Table 6.1 gives

$$\Sigma_{aF} = 0.00395 \times 2.11 = 0.00833 \text{ cm}^{-1},$$

$$\Sigma_{aS} = 0.0234 \times 0.0008 = 0.000019 \text{ cm}^{-1},$$

$$\Sigma_a = \Sigma_{aF} + \Sigma_{aS} = 0.00835.$$

Then from Eq. (6.9)

$$f = \frac{0.00833}{0.00835} \simeq 1$$

and

$$k_\infty = \eta f \simeq 2.61.$$

To compute $L^2 = D/\Sigma_a$ requires the value of D. This in turn is given by (see Eq. [5.10])

$$D = \frac{1}{3\Sigma_{tr}},$$

where Σ_{tr} is the macroscopic transport cross-section. From the values of σ_{tr} given in Table 6.1,

$$\Sigma_{tr} = 0.00395 \times 6.8 + 0.0234 \times 3.3 = 0.104 \text{ cm}^{-1},$$

so that

$$D = \frac{1}{3 \times 0.104} = 3.21 \text{ cm}$$

and

$$d = 2.13D = 6.84 \text{ cm}.$$

Finally,

$$L^2 = \frac{3.21}{0.00835} = 384 \text{ cm}^2.$$

Inserting k_∞ and L^2 into the prior expression for R then gives

$$\widetilde{R}_c = \pi \sqrt{\frac{384}{2.61 - 1}} = 49.5 \text{ cm}$$

$$R_c = \widetilde{R}_c - d = 42.7 \text{ cm. } [Ans.]$$

The reverse of this problem—namely, the calculation of the critical composition or mass of a reactor whose size is given—is somewhat more complicated, at least for a fast reactor, and is delayed until the discussion of thermal reactors in the next section.

To return to Eq. (6.50), this may be rearranged and put into the following form,

$$\frac{k_\infty}{1 + B^2 L^2} = 1, \tag{6.51}$$

where the subscript denoting the first or critical eigenvalue has been omitted, and where it is understood that, from this point on, B^2 refers to the buckling for a critical reactor. Equation (6.51) is known as the one-group *critical equation* for a bare reactor.

It is instructive to examine the physical meaning of the critical equation. For this purpose, consider a critical bare reactor of arbitrary geometry. The number of neutrons absorbed in this reactor per second is $\Sigma_a \int_V \phi dV$. However, the number that leak from the surface of the reactor per second is given by $\int_A \mathbf{J} \cdot \mathbf{n} \, dA$, where $\mathbf{J}$ is the neutron current density at the surface, $\mathbf{n}$ is a unit vector normal to the surface, and the integral is evaluated over the entire surface. From Fick's law and the divergence theorem,

$$\int_A \mathbf{J} \cdot \mathbf{n} dA = \int_V \text{div } \mathbf{J} dV = -D \int_V \nabla^2 \phi \, dV. \tag{6.52}$$

However, from the reactor equation (see Eq. [6.5]), this can be written as

$$-D \int_V \nabla^2 \phi \, dV = DB^2 \int_V \phi \, dV. \tag{6.53}$$

Neutrons may either leak from the reactor or be absorbed within its interior—there is no other alternative. Therefore, the relative probability, P_L, that a neutron is absorbed—that is, will not leak—is equal to the number absorbed in the reactor per second divided by the sum of that number and the number of neutrons that leak. From the results of the preceding paragraph, it follows that

$$P_L = \frac{\Sigma_a \int_V \phi \, dV}{\Sigma_a \int_V \phi \, dV + DB^2 \int_V \phi \, dV} = \frac{\Sigma_a}{\Sigma_a + DB^2}.$$

Dividing numerator and denominator by Σ_a gives

$$P_L = \frac{1}{1 + B^2 L^2}, \tag{6.54}$$

where the definition of L^2 has been used. Equation (6.54) is the one-group formula for the nonleakage probability for a bare reactor.

Comparing Eqs. (6.51) and (6.54) shows that the critical equation can be written as

$$k_\infty P_L = 1. \tag{6.55}$$

This result has the following interpretation. A total of $\Sigma_a \int_V \phi \, dV$ neutrons are absorbed in the reactor per second, and this leads to the release of

$$\eta f \Sigma_a \int_V \phi \, dV = k_\infty \Sigma_a \phi \, dV$$

fission neutrons. Owing to leakage, however, only $P_L k_\infty \Sigma_a \int_V \phi \, dV$ of these are absorbed in the system to initiate a new generation of neutrons. From the definition of k, the multiplication factor of the reactor, it follows that

$$k = \frac{P_L k_\infty \Sigma_a \int_V \phi \, dV}{\Sigma_a \int_V \phi \, dV} = k_\infty P_L = \eta f P_L. \tag{6.56}$$

Thus, the left-hand side of the critical equation is actually the multiplication factor for the reactor, and the critical equation follows by merely placing $k = 1$.

Example 6.4

What is the average probability that a fission neutron in the assembly described in Example 6.3 is absorbed within the system?

Solution. The probability in question is the nonleakage probability given in Eq. (6.54). Using the values $\tilde{R} = 48.5$ cm and $L^2 = 384$ cm^2 from the example gives

$$P_L = \frac{1}{1 + \left(\frac{\pi}{48.5}\right)^2 \times 384} = 0.38. \; [Ans.]$$

Thus, there is a 38% chance that a neutron will be absorbed and a 62% chance that it will escape.

6.5 THERMAL REACTORS

Thermal reactors are sufficiently distinctive to warrant separate discussion. It is recalled that these systems contain fuel, coolant, various structural materials, and a moderator to slow down the fission neutrons to thermal energies. For convenience, in this section, all the materials in the reactor other than fuel are called moderator. In short, the reactor is assumed to consist only of fuel and moderator.

The Four-Factor Formula

To begin with, consider an infinite reactor composed of a homogeneous fuel–moderator mixture. If $\overline{\Sigma}_a$ is the macroscopic thermal absorption cross-section of the mixture, that is,

$$\overline{\Sigma}_a = \overline{\Sigma}_{aF} + \overline{\Sigma}_{aM} \tag{6.57}$$

where $\overline{\Sigma}_{aF}$ and $\overline{\Sigma}_{aM}$ are the cross-sections of the fuel and moderator, respectively, then there is a total of $\overline{\Sigma}_a \phi_T$ neutrons absorbed per cm^3/sec everywhere in the reactor, where ϕ_T is the thermal flux (see Section 5.9). Of this number, the fraction

$$f = \frac{\overline{\Sigma}_{aF}}{\overline{\Sigma}_a} = \frac{\overline{\Sigma}_{aF}}{\overline{\Sigma}_{aF} + \overline{\Sigma}_{aM}} \tag{6.58}$$

is absorbed by the fuel. The parameter, f, which was called the fuel utilization in Section 6.1, is known as the *thermal utilization* in thermal reactors.

There are $f\overline{\Sigma}_a \phi_T$ neutrons absorbed per cm^3/sec in fuel. As a result, $\eta_T f \overline{\Sigma}_a \phi_T$ fission neutrons are emitted per cm^3/sec, where η_T is the average number of neutrons emitted per thermal neutron absorbed in fuel. The parameter η_T is computed from the integral

$$\eta_T = \frac{\int \eta(E)\sigma_{aF}(E)\phi(E)\,dE}{\int \sigma_{aF}(E)\phi(E)\,dE}, \tag{6.59}$$

where $\phi(E)$ is the Maxwellian flux given by Eq. (5.49). Values of η_T are given in Table 6.3, where it is observed that η_T is a slowly varying function of temperature.

In thermal reactors containing large amounts of fissionable but nonfissile materials such as ^{238}U, a small fraction of the fissions is induced by fast neutrons

TABLE 6.3 VALUES OF η_T, THE AVERAGE NUMBER OF
FISSION NEUTRONS EMITTED PER NEUTRON
ABSORBED IN A THERMAL FLUX, AT THE
TEMPERATURE T

T, °C	^{233}U	^{235}U	^{239}Pu
20	2.284	2.065	2.035
100	2.288	2.063	1.998
200	2.291	2.060	1.947
400	2.292	2.050	1.860
600	2.292	2.042	1.811
800	2.292	2.037	1.785
1000	2.292	2.033	1.770

interacting with the nonfissile nuclides. These fast fissions can be taken into account by introducing a factor ϵ, called the *fast fission factor*, which is defined as the ratio of the total number of fission neutrons produced by both fast and thermal fission to the number produced by thermal fission alone. It follows from this definition that the total number of fission neutrons produced per cm^3/sec in an infinite thermal reactor is equal to $\epsilon \eta_T f \overline{\Sigma}_a \phi_T$. The value of ϵ for reactors fueled with natural or slightly enriched uranium ranges from about 1.02 to 1.08.

As already noted, there can be no leakage of neutrons from an infinite reactor; all of the fission neutrons must eventually be absorbed somewhere in the reactor. In a thermal reactor, most of the neutrons are absorbed after they have slowed down to thermal energies. However, some neutrons may be absorbed while slowing down by nuclei having absorption resonances at energies above the thermal region. If p is the probability that a fission neutron is not absorbed in any of these resonances, then, of the $\epsilon \eta_T f \overline{\Sigma}_a \phi_T$ fission neutrons produced per cm^3/sec, only $p \epsilon \eta_T f \overline{\Sigma}_a \phi_T$ actually succeed in slowing down to thermal energies. The parameter p is known as the *resonance escape probability* and is one of the most important factors in the design of a thermal reactor.

From the foregoing discussion, it follows that the absorption of $\overline{\Sigma}_a \phi_T$ thermal neutrons leads to the production of $p \epsilon \eta_T f \overline{\Sigma}_a \phi_T$ new thermal neutrons, all of which must eventually be absorbed in an infinite reactor. The absorption of this generation of thermal neutrons leads to another generation of thermal neutrons, and so on. The multiplication factor of the reactor is therefore

$$k_\infty = \frac{p \epsilon \eta_T f \overline{\Sigma}_a \phi_T}{\overline{\Sigma}_a \phi_T} = \eta_T f p \epsilon, \qquad (6.60)$$

where the subscript on k_∞ again signifies that the equation is valid only for infinite systems. Since k_∞ is the product of four factors—η_T, f, p, and ϵ—Eq. (6.60) is called the *four-factor formula*. It may be noted that the order of the factors in the right-hand side of Eq. (6.60) has been rearranged to conform with normal usage.

Criticality Calculations

The one-group method, at least in the form given earlier in this Chapter, gives only rough estimates of the critical size or composition of a thermal reactor. This is due to the fact that, although most of the fission neutrons are ultimately absorbed at thermal energies, they may diffuse about over considerable distances while slowing down. Such fast-neutron diffusion must be taken into consideration. To this end, as discussed in Section 5.10, it is usual to describe a thermal reactor by at least two neutron groups: one group for the fast neutrons—those with energies above the thermal region—and a second group for the thermal neutrons.

In such a two-group calculation, it can usually be assumed that there is no absorption of neutrons in the fast group; resonance absorption is taken into account by the introduction of the resonance escape probability. Neutrons are then lost from the fast group only as a result of scattering into the thermal group. In particular, following the discussion in Section 5.10, there are $\Sigma_1\phi_1$ neutrons scattered per cm^3/sec out of the fast group, where ϕ_1 is the fast flux. Also, in a thermal reactor, it can be assumed that the bulk of the fissions is induced by thermal neutrons; the few fissions induced by fast neutrons are accounted for in the fast fission factor. It follows that $\eta_T f \epsilon \overline{\Sigma}_a \phi_T = (k_\infty/p)\overline{\Sigma}_a\phi_T$ fission neutrons are emitted per cm^3/sec, and these neutrons necessarily appear as source neutrons in the fast group. The source density for the fast group is therefore

$$s_1 = \frac{k_\infty}{p}\overline{\Sigma}_a\phi_T.$$

Substituting this expression into the group diffusion equation (see Eq. [5.45]) gives the following equation for the fast group:

$$D_1\nabla^2\phi_1 - \Sigma_1\phi_1 + \frac{k_\infty}{p}\overline{\Sigma}_a\phi_T = 0. \qquad (6.61)$$

In the absence of resonance absorption, all of the $\Sigma_1\phi_1$ neutrons per cm^3/sec scattered out of the fast group would appear as source neutrons in the thermal flux equation. With resonance absorption present, however, only $p\Sigma_1\phi_1$ neutrons per cm^3/sec actually succeed in entering the thermal group. Thus, the thermal source term is

$$s_T = p\Sigma_1\phi_1.$$

Introducing this expression into the thermal diffusion equation then gives

$$\overline{D}\nabla^2\phi_T - \overline{\Sigma}_a\phi_T + p\Sigma_1\phi_1 = 0. \qquad (6.62)$$

Equations (6.61) and (6.62) are the two-group equations describing a bare thermal reactor.

It is not difficult to show that, in a bare reactor, all group fluxes have the same spatial dependence, which is determined by the one-group reactor equation. Thus, for a bare thermal reactor, the two-group fluxes may be written as

$$\phi_1 = A_1\phi, \tag{6.63}$$

$$\phi_T = A_2\phi, \tag{6.64}$$

where A_1 and A_2 are constants and ϕ satisfies the equation

$$\nabla^2\phi + B^2\phi = 0. \tag{6.65}$$

Substituting the last three equations into Eqs. (6.61) and (6.62) yields

$$-(D_1 B^2 + \Sigma_1)A_1 + \frac{k_\infty}{p}\overline{\Sigma}_a A_2 = 0 \tag{6.66}$$

and

$$p\Sigma_1 A_1 - (\overline{D}B^2 + \overline{\Sigma}_a)A_2 = 0. \tag{6.67}$$

Equations (6.66) and (6.67) are a set of homogeneous linear algebraic equations in the two unknowns, A_1 and A_2.

According to Cramer's rule,[4] Eqs. (6.66) and (6.67) have nontrivial solutions only if the determinant of the coefficients multiplying A_1 and A_2 vanishes—that is, if

$$\begin{vmatrix} -(D_1 B^2 + \Sigma_1) & \frac{k_\infty}{p}\overline{\Sigma}_a \\ p\Sigma_1 & -(\overline{D}B^2 + \overline{\Sigma}_a) \end{vmatrix} = 0.$$

---

[4]Consider the set of homogeneous linear equations:

$$\begin{array}{llll} a_{11}x_1 & a_{12}x_2 & \cdots & a_{1N}x_N = 0, \\ a_{21}x_1 & a_{22}x_2 & \cdots & a_{2N}x_N = 0, \\ \cdots & \cdots & \cdots & \cdots \\ a_{N1}x_1 & a_{N2}x_2 & \cdots & a_{NN}x_N = 0, \end{array}$$

where a_{mn} are constants. Cramer's rule states that there is no solution other than $x_1 = x_2 = \cdots = x_N = 0$ unless the determinant of the coefficients a_{mn} vanishes—that is,

$$\begin{vmatrix} a_{11} & a_{12} & \cdots & a_{1N} \\ a_{21} & a_{22} & \cdots & a_{2N} \\ \cdots & \cdots & \cdots & \cdots \\ a_{N1} & a_{N2} & \cdots & a_{NN} \end{vmatrix} = 0.$$

When this equation is satisfied, the equations in the original set are not independent. That is, although there appear to be N distinct equations in the set, at least one of these equations can be written as a combination of other equations in the set. Since there are N unknowns, but only $N - 1$ independent equations, it is possible to determine only $N - 1$ of the xs in terms of the remaining x.

Multiplying out the determinant gives

$$k_\infty \overline{\Sigma}_a \Sigma_1 - (\overline{D}B^2 + \overline{\Sigma}_a)(D_1 B^2 + \Sigma_1) = 0,$$

or, on rearranging,

$$\frac{k_\infty \Sigma_1 \overline{\Sigma}_a}{(D_1 B^2 + \Sigma_1)(\overline{D}B^2 + \overline{\Sigma}_a)} = 1.$$

Dividing numerator and denominator by $\Sigma_1 \overline{\Sigma}_a$ yields finally

$$\frac{k_\infty}{(1 + B^2 L_T^2)(1 + B^2 \tau_T)} = 1. \tag{6.68}$$

In this expression,

$$L_T^2 = \frac{\overline{D}}{\overline{\Sigma}_a} \tag{6.69}$$

is the thermal diffusion area discussed in Section 5.9, and τ_T is the parameter called *neutron age*, which was defined in Section 5.10 as

$$\tau_T = \frac{D_1}{\Sigma_1}. \tag{6.70}$$

Equation (6.68) is the two-group critical equation for a bare thermal reactor. The factor

$$P_T = \frac{1}{1 + B^2 L_T^2} \tag{6.71}$$

in this equation was shown in Section 6.4 to be the probability that a thermal neutron will not leak from the reactor. Also, it is not difficult to show that

$$P_F = \frac{1}{1 + B^2 \tau_T} \tag{6.72}$$

is the probability that a fission neutron will not escape from the reactor while slowing down. Therefore, by the argument given in Section 6.4, it follows that the left-hand side of Eq. (6.68) is the multiplication factor of the reactor—that is,

$$k = k_\infty P_T P_F, \tag{6.73}$$

and the critical equation follows by writing $k = 1$.

Since all reactors are designed in such a way that there is as little leakage of neutrons as possible, both P_T and P_F are ordinarily very close to unity. The quantities $B^2 L_T^2$ and $B^2 \tau_T$ are therefore small. When the denominator of the critical equation, Eq. (6.68), is multiplied out, the term $B^4 L_T^2 \tau_T$ can often be ignored. The resulting expression is

$$\frac{k_\infty}{1 + B^2(L_T^2 + \tau_T)} = 1$$

or

$$\frac{k_\infty}{1 + B^2 M_T^2} = 1, \tag{6.74}$$

where

$$M_T^2 = L_T^2 + \tau_T \tag{6.75}$$

is called the *thermal migration area*.

It is observed that Eq. (6.74) is identical in form to the critical equation (Eq. [6.51]) in a one-group calculation. For this reason, Eq. (6.74) is known as the *modified one-group* critical equation—modified in the sense that L^2 in the one-group equation has been replaced by M_T^2. Furthermore, from Eqs. (6.64) and (6.65), it follows that the thermal flux is given by the same equation as in a one-group calculation—namely,

$$\nabla^2 \phi_T + B^2 \phi_T = 0, \tag{6.76}$$

where B^2 is the buckling given in Table 6.2. Thus, the only difference between an ordinary one-group calculation and a modified one-group calculation of a bare thermal reactor is that, in the latter, L_T^2 is replaced by M_T^2; the flux and the buckling are the same as before.

It may be noted that if τ_T is much less than L_T^2, Eq. (6.74) reduces to the one-group critical equation, Eq. (6.51). In this case, the reactor is well described by one-group theory. A comparison of the values of τ_T and L_T^2 in Tables 5.2 and 5.3 shows that, with the exception of water, τ_T is in fact less than L_T^2 for the common moderators; it is much less than L_T^2 for D_2O and graphite. Therefore, it would be quite wrong to use an ordinary one-group calculation, omitting τ_T, for a water-moderated reactor. Somewhat less error would be involved in such a computation of a D_2O or graphite-moderated system, provided the value of L_T^2, with fuel mixed in with the moderator, is still substantially larger than τ_T.

Applications

The prior results can now be applied to the practical problem of determining the critical composition or critical dimensions of a bare thermal reactor. For simplicity, the discussion is limited to reactors that do not contain resonance absorbers or nuclei that undergo fast fission. In short, the reactor is assumed to consist of a homogeneous mixture of a fissile isotope such as ^{235}U and moderator. In this case, $p = \epsilon = 1$. Then according to Eq. (6.60), k_∞ reduces to

$$k_\infty = \eta_T f. \qquad (6.77)$$

Reactors with resonance absorption and fast fission are treated in Section 6.8.

It is recalled that there are two situations to be considered: (a) the physical size of the reactor is specified and its critical composition must be determined, or (b) the composition of the reactor is specified and its critical size must be determined. These problems are each taken up in turn.

Case 1. Size Specified With size given, B^2 can be immediately computed from the appropriate formula in Table 6.2. The composition must then be adjusted so that k_∞ and M_T^2 have the values necessary to satisfy the critical equation, Eq. (6.74).

It is first convenient to introduce the parameter Z defined by

$$Z = \frac{\overline{\Sigma}_{AF}}{\overline{\Sigma}_{AM}} = \frac{N_F \overline{\sigma}_{aF}}{N_M \overline{\sigma}_{aM}}, \qquad (6.78)$$

where, as usual, the subscripts F and M denote the fuel and moderator, respectively. Then from Eq. (6.58), the thermal utilization can be written as

$$f = \frac{Z}{Z + 1}. \qquad (6.79)$$

In view of Eq. (6.77), k_∞ takes the form

$$k_\infty = \frac{\eta_T Z}{Z + 1}. \qquad (6.80)$$

Consider next the thermal diffusion area, which is one of the two factors entering into M_T^2 (see Eq. [6.75]). From Eq. (6.69),

$$L_T^2 = \frac{\overline{D}}{\overline{\Sigma}_a},$$

where $\overline{D}$ and $\overline{\Sigma}_a$ refer to the homogeneous mixture of fuel and moderator. However, $\overline{D}$ is essentially equal to $\overline{D}_M$, the diffusion coefficient for the moderator, since the concentration of the fuel in the moderator is ordinarily small for homogeneous thermal reactors. Thus, L_T^2 becomes

$$L_T^2 = \frac{\overline{D}_M}{\overline{\Sigma}_a} = \frac{\overline{D}_M}{\overline{\Sigma}_{aF} + \overline{\Sigma}_{aM}}.$$

Dividing numerator and denominator by $\overline{\Sigma}_{aM}$ and using the definition of Z from Eq. (6.78) gives

$$L_T^2 = \frac{L_{TM}^2}{Z+1},\qquad(6.81)$$

where L_{TM}^2 is the thermal diffusion area of the moderator. On solving Eq. (6.79) for Z and substituting into Eq. (6.81), the result is

$$L_T^2 = (1-f)L_{TM}^2.\qquad(6.82)$$

The age τ_T, like $\overline{D}$, depends primarily on the scattering properties of the medium. In the homogeneous type of reactor under consideration, very little fissile material is necessary to reach criticality. Furthermore, the scattering cross-sections of fissile atoms are not significantly larger than those of the ordinary moderators. It is possible, therefore, to ignore the presence of the fuel altogether and use for τ_T the value of the age for the moderator alone, τ_{TM}, which is given in Table 5.3.

Introducing Eqs. (6.80) and (6.81) into the critical equation (Eq. 6.74) then gives

$$\frac{\eta_T Z}{Z + 1 + B^2(L_{TM}^2 + Z\tau_{TM} + \tau_{TM})} = 1.$$

When this equation is solved for Z, the result is found to be

$$Z = \frac{1 + B^2(L_{TM}^2 + \tau_{TM})}{\eta_T - 1 - B^2\tau_{TM}}.\qquad(6.83)$$

This is the value of Z, which leads to a critical reactor with the specified value of B^2.

To find the total mass of fuel required for criticality using this value of Z— that is, the *critical mass*—the procedure is as follows. From Eq. (6.78), the atom density of the fuel is

$$N_{\mathrm{F}} = Z \frac{\overline{\sigma}_{aM}}{\overline{\sigma}_{aF}} N_{\mathrm{M}}.\qquad(6.84)$$

The total number of fuel atoms in the reactor is $N_{\mathrm{F}}V$, where V is the reactor volume, and the total number of moles of fuel is $N_{\mathrm{F}}V/N_A$, where N_A is Avogadro's number. If M_{F} is the gram atomic weight of the fuel, it follows that m_{F}, the fuel mass, is given by

$$m_{\mathrm{F}} = \frac{N_{\mathrm{F}}V M_{\mathrm{F}}}{N_A}\qquad(6.85)$$

$$= Z\frac{\overline{\sigma}_{aM}V M_{\mathrm{F}}}{\overline{\sigma}_{aF}N_A}.\qquad(6.86)$$

However, the total mass of the moderator is

$$m_{\mathrm{M}} = \frac{N_{\mathrm{M}} V M_{\mathrm{M}}}{N_A},$$

so that Eq. (6.86) can also be written as

$$m_{\mathrm{F}} = Z \frac{\overline{\sigma}_{a\mathrm{M}} M_{\mathrm{F}}}{\overline{\sigma}_{a\mathrm{F}} M_{\mathrm{M}}} m_{\mathrm{M}}. \tag{6.87}$$

Introducing the formulas given in Chapter 5 for $\overline{\sigma}_{a\mathrm{M}}$ and $\overline{\sigma}_{a\mathrm{F}}$ (see Eq. [5.59]) gives finally

$$m_{\mathrm{F}} = Z \frac{\sigma_{a\mathrm{M}}(E_0) M_{\mathrm{F}}}{g_{a\mathrm{F}}(T) \sigma_{a\mathrm{F}}(E_0) M_{\mathrm{M}}} m_{\mathrm{M}}, \tag{6.88}$$

where the non-$1/v$ factor of the moderator has been taken to be unity and the cross-sections are evaluated at the energy $E_0 = 0.0253$ eV. Because the fuel concentration is normally so small, the moderator mass can be computed using its ordinary density as shown in the following example.

Example 6.5

A bare spherical thermal reactor, 100 cm in radius, consists of a homogeneous mixture of ^{235}U and graphite. The reactor is critical and operates at a power level of 100 thermal kilowatts. Using modified one-group theory, calculate (a) the buckling; (b) the critical mass; (c) k_∞; (d) L_T^2; (e) the thermal flux. For simplicity, make all computations at room temperature.

Solution.

1. From Table 6.2, $B^2 = (\pi/R)^2 = (\pi/100)^2 = 9.88 \times 10^{-4}$ cm^{-2}, and $B = 3.14 \times 10^{-2}$. [*Ans.*]

2. According to Tables 5.2, 5.3, and 6.3, $L_{T\mathrm{M}}^2 = 3500$ cm^2, $\eta_T = 2.065$, and $\tau_{T\mathrm{M}} = 368$ cm^2. Then from Eq. (6.83),

$$Z = \frac{1 + 9.88 \times 10^{-4}(3500 + 368)}{2.065 - 1 - 9.88 \times 10^{-4} \times 368} = 6.87.$$

Using the values $\sigma_{a\mathrm{M}}(E_0) = 0.0034$ b, $\sigma_{a\mathrm{F}}(T) = 681$ b, $g_a(T) = 0.978$ in Eq. (6.88) gives

$$m_{\mathrm{F}} = \frac{6.87 \times 0.0034 \times 235}{0.978 \times 681 \times 12} m_{\mathrm{M}} = 6.87 \times 10^{-4} m_{\mathrm{M}}.$$

The density of graphite is approximately 1.60 g/cm^3, so the total mass of graphite in the reactor is $m_{\mathrm{M}} = \frac{4}{3} \pi R^3 \times 1.60 = 6.70 \times 10^6$ g $= 6,700$ kg. It follows that the critical mass is $m_{\mathrm{F}} = 6.87 \times 10^{-4} \times 6,700 = 4.60$ kg. [*Ans.*]

3. From Eq. (6.79),

$$f = \frac{Z}{Z+1} = \frac{6.87}{7.87} = 0.873,$$

and

$$k_\infty = \eta_T f = 2.065 \times 0.873 = 1.803. \ [Ans.]$$

4. From Eq. (6.82),

$$L_T^2 = (1 - f)L_{TM}^2 = (1 - 0.873) \times 3500 = 444 \ \text{cm.}^2 \ [Ans.]$$

5. The thermal flux is given by

$$\phi_T = A\frac{\sin Br}{r},$$

where, according to Table 6.2, $A = P/4R^2 E_R \overline{\Sigma}_f$. Here $P = 100 \ \text{kW} = 10^5$ joules/sec and $E_R = 3.2 \times 10^{-11}$ joule. The value of $\phi\Sigma_f = N_F \phi \sigma_f$ can be found by noting from Eq. (6.85) that

$$N_F = \frac{m_F N_A}{V M_F},$$

so that

$$\phi\Sigma_f = \frac{m_F N_A \phi \sigma_f}{V M_F}$$

$$= \frac{m_F N_A}{V M_F} \times 0.886 g_{fF}(T)\sigma_f(E_0).$$

Using the values $g_{fF}(T) = 0.976$ and $\sigma_f(E_0) = 582$ b gives $\phi\Sigma_f = 1.41 \times 10^{-3} \ \text{cm}^{-1}$. The constant A is then

$$A = \frac{10^5}{4 \times 10^4 \times 3.2 \times 10^{-11} \times 1.41 \times 10^{-3}} = 5.54 \times 10^{13}$$

and the flux is for d small

$$\phi_T(r) = 5.54 \times 10^{13}\frac{\sin Br}{r}. \ [Ans.]$$

The maximum value of ϕ_T occurs at $r = 0$ and is equal to $\phi_T(0) = 5.54 \times 10^{13} B = 1.74 \times 10^{12}$ neutrons/cm²-sec.

Case 2. Composition Specified   When the composition is given and the critical dimensions must be found, the parameters k_∞ and M_T^2 can be computed directly. The value of B^2 can then be obtained from Eq. (6.74)—namely,

$$B^2 = \frac{k_\infty - 1}{M_T^2}. \tag{6.89}$$

If the geometry of the reactor is specified, the dimensions can then be determined from the appropriate formula for B^2 in Table 6.2. Thus, for a cubical reactor of side a and d small, $a = \pi\sqrt{3}/B$; for a sphere, the critical radius is $R = \pi/B$.

However, if it is only specified that the reactor is, say, a finite cylinder, then the buckling formula in the table only provides a relationship between its height H and radius R, which must be satisfied if the reactor is critical. Each combination of H and R leads to a different reactor volume and hence a different critical mass. It is not difficult to show that this mass is smallest when H and R satisfy the relationship $H = 1.82R$. The cross-section through the axis of such a cylinder is almost square. The parallelepiped with least volume, as might be expected, is a cube.

Example 6.6

A 5-watt experimental thermal reactor is constructed in the form of a cylinder. The reactor is fueled with a homogeneous mixture of ^{235}U and ordinary water with a fuel concentration of 0.0145 g/cm^3. Because of its low power, the system operates at essentially room temperature and atmospheric pressure. (a) Calculate the dimensions of the cylinder that has the smallest critical mass. (b) Determine the critical mass.

Solution.

1. The ratio of the number of atoms of ^{235}U per cm^3 to water molecules per cm^3 is

$$\frac{N_F}{N_M} = \frac{\rho_F M_M}{\rho_M M_F},$$

where ρ_F and ρ_M are the densities of ^{235}U and water in the mixture, respectively, and M_F and M_M are their gram atomic and molecular weights. Introducing these values gives

$$\frac{N_F}{N_M} = \frac{0.0145 \times 18}{1 \times 235} = 1.11 \times 10^{-3}.$$

The average thermal absorption cross-sections of fuel and moderator are

$$\phi\sigma_{aF} = 0.886 g_a(20°C)\sigma_{aF}(E_0) = 0.886 \times 0.978 \times 681 = 590 \text{ b}$$

and

$$\phi\sigma_{aM} = 0.886 \times 2 \times \sigma_{aH}(E_0) = 0.886 \times 2 \times 0.332 \doteq 0.588 \text{ b}.$$

Then from Eq. (6.78),

$$Z = \frac{N_F\phi\sigma_{aF}}{N_M\phi\sigma_{aM}} = 1.11 \times 10^{-3} \times \frac{590}{0.588} = 1.11.$$

The thermal utilization is

$$f = \frac{Z}{Z+1} = \frac{1.11}{2.11} = 0.526$$

and

$$k_\infty = \eta_T f = 2.065 \times 0.526 = 1.0862.$$

According to Table 5.2, L_{TM}^2 for water is 8.1 cm^2, so that from Eq. (6.82), L_T^2 for the fuel–moderator mixture is

$$L_T^2 = (1 - f)L_{TM}^2 = (1 - 0.526) \times 8.1 = 3.84 \text{ cm}^2.$$

The neutron age from Table 5.3 is 27 cm^2, and hence

$$M_T^2 = L_T^2 + \tau_T = 3.84 + 27 = 30.8 \text{ cm}^2.$$

Then from Eq. (6.89),

$$B^2 = \frac{k_\infty - 1}{M_T^2} = \frac{1.0862 - 1}{30.8} = 2.80 \times 10^{-3} \text{ cm}^{-2}.$$

From Table 6.2, $B^2 = (2.405/R)^2 + (\pi/H)^2$. However, for minimum critical mass, it is necessary that $H = 1.82R$. This means that

$$B^2 = \left(\frac{2.405}{R}\right)^2 + \left(\frac{\pi}{1.82R}\right)^2 = \frac{8.763}{R^2}.$$

Solving for R^2 gives

$$R^2 = \frac{8.763}{B^2} = \frac{8.763}{2.80 \times 10^{-3}} = 3.13 \times 10^3.$$

Thus, $R = 55.9$ cm and $H = 101.7$ cm. [*Ans.*]

2. The reactor volume is $\pi R^2 H = \pi \times (55.9)^2 \times 101.7 = 9.98 \times 10^5$ cm^3. Since there are 0.0145 g/cm^3 of ^{235}U, the fuel mass is

$$m_F = 0.0145 \times 9.88 \times 10^5 = 1.45 \times 10^4 \text{ g} = 14.5 \text{ kg. [*Ans.*]}$$

6.6 REFLECTED REACTORS

It is pointed out in Chapter 4 that the neutron economy is improved when the core of a reactor is surrounded by a reflector—that is, by a thick, unfueled region of moderator. The neutrons that otherwise would leak from the bare core now pass into the reflector, and some of these diffuse back into the core. The net result is that the critical size, and hence mass of the system is reduced.

Criticality calculations for reflected reactors are now considered within the framework of the one-group diffusion theory. It is recalled that this method is applicable to calculations of fast reactors and thermal reactors, such as those moderated by D_2O or graphite, for which $\tau_T \ll L_T^2$. Reflected water reactors for which $\tau_T \gg L_T^2$ are treated separately later in this section.

As a specific example, consider a spherical reactor consisting of a core of radius R surrounded by an infinite reflector. In the following analysis, parameters that refer to the core and reflector are denoted by the subscripts c and r, respectively.

According to the one-group theory, the flux in the core ϕ_c satisfies the equation (see Eq. [6.5])

$$\nabla^2 \phi_c + B^2 \phi_c = 0, \tag{6.90}$$

where for a critical reactor

$$B^2 = \frac{k_\infty - 1}{L_c^2}. \tag{6.91}$$

Since there is no fuel in the reflector, the flux in this region satisfies the one-group diffusion equation

$$\nabla^2 \phi_r - \frac{1}{L_r^2} \phi_r = 0. \tag{6.92}$$

To find the flux throughout the reactor and the conditions for criticality, it is necessary to solve Eqs. (6.90) and (6.92) for ϕ_c and ϕ_r subject to boundary conditions.

The general solution to Eq. (6.90) is derived in Section 6.3 and is

$$\phi_c = A \frac{\sin Br}{r} + C \frac{\cos Br}{r},$$

where A and C are constants. Since ϕ_c must be finite everywhere, even at the center of the reactor (at $r = 0$), it is necessary to place $C = 0$ so that ϕ_c reduces to

$$\phi_c = A \frac{\sin Br}{r}. \tag{6.93}$$

The general solution to Eq. (6.92) is

$$\phi_r = A' \frac{e^{-r/L_r}}{r} + C' \frac{e^{r/L_r}}{r},$$

where A' and C' are constants. However, ϕ_r must remain finite as r goes to infinity, and C' must therefore be taken to be zero. The flux in the reflector is then

$$\phi_r = A' \frac{e^{-r/L_r}}{r}. \tag{6.94}$$

The functions ϕ_c and ϕ_r must also satisfy the interface boundary conditions (see Section 5.5)—namely, continuity of neutron flux. Similarly, the radial component of the current at the core-reflector interface—that is, at $r = R$, must also be continuous. Written out in detail, these conditions are

$$\phi_c(R) = \phi_r(R) \tag{6.95}$$

and

$$\mathbf{J}_c(R) \cdot \mathbf{n} = \mathbf{J}_r(R) \cdot \mathbf{n}$$

or

$$D_c \phi_c'(R) = D_r \phi_r'(R), \tag{6.96}$$

where the prime indicates differentiation with respect to r.

Introducing Eqs. (6.93) and (6.94) into Eq. (6.95) gives

$$A \frac{\sin BR}{R} = A' \frac{e^{-R/L_r}}{R}. \tag{6.97}$$

Next, differentiating Eqs. (6.93) and (6.94) and inserting the results into Eq. (6.96) yields

$$A D_c \left(\frac{B \cos BR}{R} - \frac{\sin BR}{R^2} \right) = -A' D_r \left(\frac{1}{RL_r} + \frac{1}{R} \right) e^{-R/L_r}. \tag{6.98}$$

Equations (6.97) and (6.98) are homogeneous linear equations in the unknowns A and A'; they have nontrivial solutions only if the determinant of the coefficients vanishes. Multiplying out this determinant (which is equivalent to dividing one equation by the other) gives

$$D_c \left(B \cot BR - \frac{1}{R} \right) = -D_r \left(\frac{1}{L_r} + \frac{1}{R} \right).$$

For computational purposes, it is convenient to rearrange this equation in the following form:

$$BR \cot BR - 1 = -\frac{D_r}{D_c} \left(\frac{R}{L_r} + 1 \right). \tag{6.99}$$

Equation (6.99) must be satisfied for the reactor to be critical. For example, if the composition of the core is given, then B^2 is known from Eq. (6.91), and R can be calculated from Eq. (6.99). However, if R is specified, B^2 must be computed from Eq. (6.99), and the critical composition of the core can be determined by using Eq. (6.91). The only complicating feature of these calculations that is not present with bare reactors is that Eq. (6.99) is a transcendental equation.

It is instructive to consider the solution of Eq. (6.99) graphically. For instance, suppose that the composition of the core, and hence B, is specified. The critical

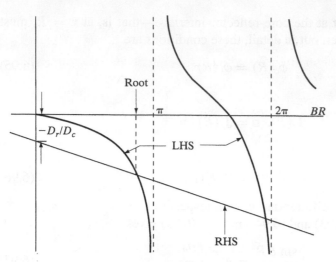

Figure 6.4 A plot of the two sides of Eq. (6.99).

core radius can then be found by plotting the left-hand side (LHS) and the right-hand side (RHS) of Eq. (6.99) separately as functions of BR. This is shown in Fig. 6.4. The LHS has an infinite number of branches, each of width π, while the RHS is a straight line of slope $-D_r/D_c BL_r$ [5] that intersects the vertical axis at $-D_r/D_c$. Every value of BR corresponding to an intersection of the LHS and RHS is a root or solution to Eq. (6.99), and since there are an infinite number of intersections, there are also an infinite number of solutions. However, as explained earlier in this Chapter, only the first solution is relevant to a critical reactor, the higher roots corresponding to solutions that die out in time. With the value of BR obtained from the first intersection and with B known, the critical radius can be immediately determined.

It should be noted in Fig. 6.4 that the first intersection of LHS and RHS occurs at a value of BR that is less than π. If the reactor is bare instead of reflected, BR is exactly equal to π. Therefore, it follows that the critical core radius is less for the reflected reactor than for the bare reactor of the same composition—a conclusion that is anticipated on physical grounds.

Incidentally, it may be noted that, in the special case in which the moderator in the core and reflector are the same, $D_r = D_c$, and Eq. (6.99) reduces to

$$B \cot BR = -\frac{1}{L_r}. \qquad (6.100)$$

[5] As a function of BR,

$$\text{RHS} = -\frac{D_r}{D_c}\left(\frac{BR}{BL_r}+1\right).$$

This equation is not transcendental in R. Thus, if B is known, R can be calculated directly. This case is illustrated in Example 6.8.

Having obtained the condition for criticality, it is now necessary to return to Eqs. (6.97) and (6.98) and find the values of the constants A and A' that determine the flux. These equations are not independent, however, but are related through Eq. (6.99). It is possible, therefore, to find A' in terms of A or A in terms of A', but both constants cannot be found independently. For example, from Eq. (6.97), A' is

$$A' = Ae^{R/L_r} \sin BR, \tag{6.101}$$

where BR is the known solution to Eq. (6.99). The constant A, which fixes the magnitude of the flux throughout the reactor, is determined by the reactor power. As in Section 6.3, this is given by

$$P = E_R \Sigma_f \int \phi_c dV, \tag{6.102}$$

where the integral extends over the reactor core—the power-producing region. Inserting ϕ_c from Eq. (6.93) and using the volume element $dV = 4\pi r^2\,dr$ gives

$$P = 4\pi E_R \Sigma_f A \int_0^R r \sin Br\,dr$$

$$= \frac{4\pi E_R \Sigma_f A}{B^2}(\sin BR - BR \cos BR).$$

When this is solved for A, the result is

$$A = \frac{PB^2}{4\pi E_R \Sigma_f(\sin BR - BR \cos BR)}. \tag{6.103}$$

The prior procedure—solving the core and reflector equations, obtaining an equation that must be satisfied for criticality, and evaluating the flux constants in terms of the reactor power—can be carried through analytically for only a few reactors—namely, the fully reflected parallelepiped and cylinder. In practice, the properties of these reactors are often estimated from calculations of spherical reactors of the same composition and volume.

It may be noted that the analysis of reflected reactors is carried out in this section without regard for whether the reactors are fast or thermal. The various formulas derived earlier are therefore valid for both types of systems. To use these formulas in calculations of thermal reactors, it is merely necessary to replace each reactor parameter with its appropriate thermal averaged value. The one-group flux is then the thermal flux provided the one-group approximation holds. This is illustrated in the next example.

Example 6.7

The core of a thermal reactor is a sphere 300 cm in radius that contains a homogeneous mixture of ^{235}U and D_2O. This is surrounded by an infinite graphite reflector. Calculate for room temperature (a) the critical concentration of ^{235}U in grams per liter; (b) the critical mass.

Solution.

1. To find the fuel concentration, the value of B must first be found from Eq. (6.99). Introducing the values from Table 5.2, $\overline{D}_r = 0.84$ cm, $\overline{D}_c = 0.87$ cm, $R = 300$ cm, and $L_r = L_{Tr} = 59$ cm, gives

$$BR \cot BR = 1 - \frac{0.84}{0.87}\left(\frac{300}{59} + 1\right) = -4.88.$$

Solving, graphically or otherwise, for BR yields $BR = 2.64$ and

$$B = \frac{2.64}{300} = 9.80 \times 10^{-3} \text{ cm}^{-1}.$$

Next, letting

$$Z = \frac{N_F \overline{\sigma}_{aF}}{N_M \overline{\sigma}_{aM}},$$

it follows as in Section 6.5 that

$$k_\infty = \frac{\eta_T Z}{Z + 1}$$

and

$$L_T^2 = (1 - f)L_{TM}^2 = L_{TM}^2/(Z + 1).$$

Inserting these expressions into the definition of B^2, specifically

$$B^2 = \frac{k_\infty - 1}{L_T^2}$$

and solving for Z gives

$$Z = \frac{1 + B^2 L_{TM}^2}{\eta_T - 1}.$$

Introducing the values $B = 8.80 \times 10^{-3}$cm^{-1}, $L_{TM}^2 = 9.4 \times 10^3$ cm^2, and $\eta_T = 2.065$ then gives

$$Z = \frac{1 + (8.80 \times 10^{-3})^2 \times 9.4 \times 10^3}{2.065 - 1} = 1.622$$

From the definition of Z and the usual formula for computing atom density, it is easy to see that the ratio of the physical densities of fuel to moderator is given by

$$\frac{\rho_F}{\rho_M} = Z\frac{M_F\bar{\sigma}_{aM}}{M_M\bar{\sigma}_{aF}} = Z\frac{M_F\sigma_{aM}(E_0)}{M_M g_{aF}(20°C)\sigma_{aF}(E_0)},$$

where M_M is the molecular weight of D_2 and the absorption cross-sections are taken at $E_0 = 0.0253$ eV. Using the values $M_F = 235$, $\sigma_{aM}(E_0) = (\bar{\Sigma}_{aM}/N) \times \sqrt{4/\pi} = 9.3 \times 10^{-5} \times 1.128/.03323 = 3.16 \times 10^{-3}$ (taking into account the 0.25 w/o H_2 in D_2O), $M_M = 20$, $g_{aF}(20°C) = 0.978$, and $\sigma_{aF}(E_0) = 681$ b, gives

$$\frac{\rho_F}{\rho_M} = 1.622 \times \frac{235 \times 3.16 \times 10^{-3}}{20 \times 0.978 \times 681} = 9.04 \times 10^{-5}.$$

The density of D_2O is 1.1 g/cm^3 so that

$$\rho_F = 9.04 \times 10^{-5} \times 1.1 = 9.94 \times 10^{-5} \text{ g/cm}^3 = 0.0994 \text{ g/li. [}Ans.\text{]}$$

2. The critical mass is

$$m_F = V\rho_F = \frac{4}{3} \times \pi \times (300)^3 \times 9.94 \times 10^{-5}$$

$$= 1.124 \times 10^4 \text{ g} = 11.24 \text{ kg. [}Ans.\text{]}$$

Example 6.8

A large spherical thermal reactor, moderated and reflected by an infinite reflector of graphite, is fueled with ^{235}U at a concentration of 2×10^{-4} g/cm^3. (a) Calculate the critical radius of the core. (b) Compute what the critical radius is if the reactor is bare. Make calculations at room temperature.

Solution.

1. Since the moderator and reflector are the same, the radius of the reactor is determined by Eq. (6.100),

$$B \cot BR = -\frac{1}{L_{Tr}},$$

where L_{Tr} is the thermal diffusion length of the reflector and B is found from

$$B^2 = \frac{k_\infty - 1}{L_{Tc}^2}.$$

First, it is convenient to compute Z:

$$Z = \frac{N_F\bar{\sigma}_{aF}}{N_M\bar{\sigma}_{aM}} = \frac{\rho_F M_M\bar{\sigma}_{aF}}{\rho_M M_F\bar{\sigma}_{aM}} = \frac{\rho_F M_M g_{aF}(20°C)\sigma_{aF}(E_0)}{\rho_M M_F\sigma_{aM}(e_0)}.$$

Substituting $\rho_F = 2 \times 10^{-4}$ g/cm^3, $\rho_M = 1.6$ g/cm^3, $M_F = 235$, $M_M = 12$, $g_{aF}(20°C) = 0.978$, $\sigma_{aF}(E_0) = 681$ b, $\sigma_{aM} = 3.4 \times 10^{-3}$ b, gives

$$Z = \frac{2 \times 10^{-4} \times 12 \times 0.978 \times 681}{1.6 \times 235 \times 3.4 \times 10^{-3}} = 1.25.$$

Then

$$
\begin{aligned}
f &= \frac{Z}{Z+1} = 0.556, \\
k_\infty &= 2.065 \times 0.556 = 1.148, \\
L_{Tc}^2 &= (1-f)L_{TM}^2 = (1 - 0.556) \times 3500 = 1554 \text{ cm}^2,
\end{aligned}
$$

and

$$B^2 = \frac{1.148 - 1}{1554} = 9.52 \times 10^{-5},$$

so that $B = 9.76 \times 10^{-3}$ cm^{-1}.

From Eq. (6.100),

$$
\begin{aligned}
\cot BR &= -\frac{1}{BL_{Tr}} = -\frac{1}{9.76 \times 10^{-3} \times 59} = -1.74, \\
BR &= 2.62,
\end{aligned}
$$

and finally

$$R = \frac{2.62}{B} = \frac{2.62}{9.76 \times 10^{-3}} = 268 \text{ cm. } [Ans.]$$

2. If the reactor is unreflected and since d is small, the radius R_0 is given by

$$B = \frac{\pi}{R_0}$$

or

$$R_0 = \frac{\pi}{B} = \frac{\pi}{9.76 \times 10^{-3}} = 322 \text{ cm. } [Ans.]$$

Flux in a Reflected Thermal Reactor

Although the one-group method may provide reasonable values for the critical mass, this method does not accurately predict the flux throughout the reactor, especially in the case of a thermal reactor. To do so requires a two-group or multi-group calculation. Although such computations for a reflected reactor are beyond the scope of this book, the results are interesting and have an important bearing on the design of a thermal reactor.

  To be specific, consider a two-group calculation of a reflected system. As usual, the fast flux, ϕ_1, describes the behavior of the fast neutrons; the thermal

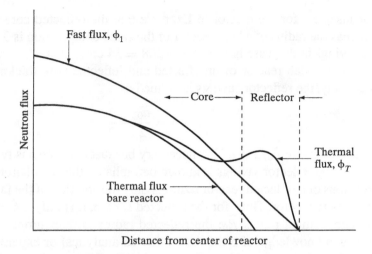

Figure 6.5 Fast and thermal fluxes in a reflected thermal reactor and the thermal flux in the equivalent reactor.

neutrons are included in the thermal flux, ϕ_T. One of the striking results of such a computation is that the thermal flux is found to rise near the core–reflector interface and exhibits a peak in the reflector as shown in Fig. 6.5. The behavior of ϕ_T, which is not predicted in one-group theory, is due to the thermalization in the reflector of fast neutrons escaping from the core. The thermalized neutrons are not absorbed as quickly in the reflector as neutrons thermalized in the core since the reflector, being unfueled, has a much smaller absorption cross-section. The thermal neutrons tend to accumulate in the reflector until they leak back into the core, escape from the outer surface of the reflector, or are absorbed.

An important consequence of the rise in the thermal flux in the reflector is that it tends to flatten the thermal flux distribution in the core. This effect can be seen in Fig. 6.5, which also shows the flux for the bare reactor. The flux in the core near the core–reflector interface is built up by the flow of thermal neutrons out of the reflector. Thus, a reflector not only reduces the critical size and mass of a reactor, it also reduces the maximum-to-average flux ratio.

Reflector Savings

The decrease in the critical dimensions of a reactor core as the result of the use of a reflector is called the *reflector savings*. Thus, for a spherical reactor, if R_0 is the radius of the bare core and R is the core radius with the reflector present, the reflector savings is defined as

$$\delta = \widetilde{R}_0 - R. \tag{6.104}$$

For instance, for the reactor in Example 6.8, the reflected core radius is 268 cm, whereas the radius of a bare reactor of the same composition is 322 cm. The reflector savings in this case is $\delta = 322 - 268 = 54$ cm.

For a slab reactor of unreflected and reflected core thickness, $\widetilde{a}_0$ and a, respectively, the reflector savings is defined as

$$\delta = \frac{\widetilde{a}_0}{2} - \frac{a}{2}, \tag{6.105}$$

where the factor $\widetilde{a}_0/2 - a/2$ is necessary because the slab has two sides.

If the reflector savings is known beforehand, the calculation of the critical dimensions of a reflected reactor requires only the solution of the far simpler problem of the bare reactor. Thus, for the reflected sphere, it is only necessary to determine the bare critical radius $\widetilde{R}_0$; the reflected radius is then simply $R = \widetilde{R}_0 - \delta$. Of course, a knowledge of δ requires either an analytical or experimental solution to the reflected reactor problem. The practical importance of the reflector savings lies in the fact that δ is relatively insensitive to changes in the composition of the reactor. This means that if δ is determined for one reactor, then the same value of δ can be used for a different reactor of similar composition.

For two- and three-dimensional reactor configurations, the reflector savings must be accounted for in each of the terms contributing to the buckling. Consider the case of a fully reflected finite cylindrical reactor with reflectors on the top, bottom, and sides of the cylinder. From Table 6.2, the buckling for the bare reactor is given by

$$B^2 = \left(\frac{2.405}{R}\right)^2 + \left(\frac{\pi}{H}\right)^2 .$$

Accounting for the extrapolation distance d,

$$B^2 = \left(\frac{2.405}{\widetilde{R}}\right)^2 + \left(\frac{\pi}{\widetilde{H}}\right)^2 ,$$

where $\widetilde{R}$ and $\widetilde{H}$ are given by

$$\widetilde{R} = R + d$$

and

$$\widetilde{H} = H + 2d,$$

respectively.

For the fully reflected case, the reflector savings must be accounted for and these become

$$\widetilde{R} = R + \delta$$

and

$$\tilde{H} = H + 2\delta.$$

If the cylinder is reflected on only the bottom and not the top, then $\tilde{H}$ is given by

$$\tilde{H} = H + d + \delta.$$

For purposes of making rough calculations of reflected dimensions and/or critical masses of thermal reactors, except those moderated or reflected with water, the following simple formula can be used to estimate δ:

$$\delta \simeq \frac{\overline{D}_c}{\overline{D}_r} L_{Tr}, \tag{6.106}$$

where the symbols have their usual meanings. Equation (6.106) is only valid provided the reflector is several diffusion lengths thick. In this case, the reflector is effectively infinite; that is, an increase in reflector thickness does not reduce the critical size of the core. In practice, virtually all reactors have reflectors that are many diffusion lengths thick, and hence can be thought of as infinite.

As mentioned earlier, the one-group method, employed in this section to treat reflected reactors, cannot be accurately applied to water-moderated and water-reflected systems, because τ_T is much larger than L_T^2. However, to handle these types of reactors, the following empirical formula, developed by R. W. Deutsch, may be used to obtain the reflector savings:

$$\delta = 7.2 + 0.10(M_T^2 - 40.0), \tag{6.107}$$

where δ is in cm, and M_T^2, the migration area of the core, is in cm^2. The use of Eq. (6.107) is illustrated by the following example.

Example 6.9

Calculate the critical core radius and critical mass of a spherical reactor, moderated and reflected by unit-density water. The core contains ^{235}U at a concentration of 0.0145 g/cm^3 (as in Example 6.6).

Solution. From Example 6.6, $M_T^2 = 30.8$ cm^2 and so

$$\delta = 7.2 + 0.10(30.8 - 40.0) = 6.3 \text{ cm}.$$

Also from that example, $B^2 = 2.80 \times 10^{-3}$ cm^{-2} so that $B = 0.0529$. The bare radius is given by

$$\tilde{R}_0 = \frac{\pi}{B} = \frac{\pi}{0.0529} = 59.4 \text{ cm},$$

and the reflected radius is

$$R = \tilde{R}_0 - \delta = 59.4 - 6.3 = 53.1 \text{ cm. } [Ans.]$$

The critical mass is then

$$m_F = 0.0145 \times \frac{4}{3} \times \pi \times (53.1)^3 = 9.09 \times 10^3 \text{ g}$$
$$= 9.09 \text{ kg. } [Ans.]$$

This is considerably less than the 14.5 kg for the unreflected reactor of Example 6.6.

6.7 MULTIGROUP CALCULATIONS

The one-group method gives only the roughest estimates of the properties of a critical reactor. To obtain more accurate results, it is necessary to perform multigroup calculations of the type described in Section 5.8. To set up the multigroup equations for a critical reactor, three new group constants must first be defined:

Σ_{fg} = the group-averaged macroscopic fission cross-section;

ν_g = the average number of fission neutrons released as the result of fissions induced by neutrons in the gth group; and

χ_g = the fraction of fission neutrons that are emitted with energies in the gth group.

The number of fissions per cm^3/sec in the hth group can now be written as $\Sigma_{fh}\phi_h$, where ϕ_h is the flux in the hth group. As a result of these fissions, $\nu_h \Sigma_{fh}\phi_h$ neutrons are released. The total number of neutrons emitted as the result of fissions in all groups is then $\sum \nu_h \Sigma_{fh}\phi_h$, where the sum is carried out over all groups. If the fraction χ_g of these neutrons appears in the gth group, it follows that the source term s_g for the gth multigroup equation, Eq. (5.45), is

$$s_g = \chi_g \sum_{h=1}^{N} \nu_h \Sigma_{fh}\phi_h. \tag{6.108}$$

Introducing this expression into Eq. (5.45) then gives

$$D_g \nabla^2 \phi_g - \Sigma_{ag}\phi_g - \sum_{h=g+1}^{N} \Sigma_{g \to h}\phi_g + \sum_{h=1}^{g-1} \Sigma_{h \to g}\phi_h + \chi_g \sum_{h=1}^{N} \nu_h \Sigma_{fh}\phi_h = 0. \tag{6.109}$$

This is the gth group equation in a set of N multigroup equations.

For the usual reactor, which has several regions of differing material properties, there is a set of equations like those represented by Eq. (6.109) for each region. In this case, it is necessary to solve the equations in each region and satisfy boundary conditions at every interface as well as at the reactor surface. This procedure can only be carried out in a practical way with a high-speed computer, and many computer programs have been written for solving the multigroup equations (e.g. DIF3D).

For an infinite reactor, the fluxes are independent of position, $\phi_g = $ constant, $\nabla^2 \phi_g = 0$, and the multigroup equations reduce to a set of linear algebraic equations with constant coefficients in the unknowns $\phi_1, \phi_2, \ldots, \phi_N$. Since these are homogeneous equations, according to Cramer's rule, there can be no solutions other than $\phi_1 = \phi_2 = \cdots = \phi_N = 0$ unless the determinant of the coefficients of the fluxes is equal to zero, which is actually the requirement for the criticality of the infinite system. By changing the concentration of the fuel, the value of the determinant also changes. The concentration for which the determinant is zero is the composition required for the system to be critical. The relative magnitudes of the fluxes $\phi_1, \phi_2, \ldots$ can be found by solving the multigroup linear equations using values for the coefficients computed at the critical composition. Absolute values of the fluxes are not obtained, however, because these depend on the reactor power level. Nevertheless, the relative values of the flux are sufficient to estimate the energy dependence of the flux or neutron spectrum in the reactor.

Such calculations of neutron spectra in infinite reactors are used in reactor design in the following way. Because of the complexity of the multigroup equations for a finite, multiregion reactor, it is often too costly in computer time to utilize a large number of groups in multigroup computations, especially in the early stages of the design of a reactor when these computations must be carried out repeatedly with varying design parameters. Therefore, it is more practical to perform the initial design calculations for the finite reactor using only a few groups, but with multigroup calculations carried out with a large number of groups for the infinite reactor. Only when the design of a reactor has been more or less decided on, are large-scale, space-dependent multigroup calculations undertaken. These provide accurate estimates of the critical mass, the flux and power distributions, and other properties of the system.

6.8 HETEROGENEOUS REACTORS

Up to this point, it has been assumed that the reactor core consists of a homogeneous mixture of fissile material, coolant, and, if the reactor is thermal, moderator, whereas in most reactors, the fuel is actually contained in fuel rods of one sort or another. Such nonhomogeneous reactors are divided into two classes. If the neutron mean free path at all neutron energies is large compared with the thickness of the fuel rods, then it is unlikely that a neutron will have more than one successive colli-sion within any one fuel rod. In this case, the reactor core is homogeneous as far as the neutrons are concerned and the reactor is called *quasi-homogeneous*. However, if the neutron mean free path at some energy is comparable to or smaller than the thickness of a fuel rod, then the neutrons may undergo several collisions within a fuel rod, the fuel and moderator must be treated as separate regions, and the reactor is said to be *heterogeneous*. Reactors with fuel rods in the form of thin plates are often of the quasi-homogeneous type, especially if the fuel is highly enriched and therefore present in the rod at low concentration. Most modern (thermal) power reactors, however, are fueled with only slightly enriched (a few weight percent) uranium, the fuel rods are not thin, and these reactors are heterogeneous.

From the standpoint of the neutrons, a quasi-homogeneous reactor is in fact homogeneous. Thus, the formulas developed in the preceding sections for homo-geneous systems can be used with quasi-homogeneous reactors. It is merely neces-sary in the computations to use the atom densities for the equivalent homogeneous reactors as shown in the following example.

Example 6.10

The fuel plates of a mobile thermal reactor are thin fuel sandwiches as shown in Fig. 6.6. The cladding is zirconium, and the meat is a mixture of ^{235}U and zirconium, containing approximately 150 atoms of zirconium for every atom of uranium. Both the cladding and meat are 0.05 cm thick. The plates are immersed in a water moder-ator coolant in such a way that the total volumes of the plates and the water are the same. At room temperature: (a) What is the mean free path of thermal neutrons in the zirconium cladding and the meat? (b) What is the value of k_∞ for the system?

Solution.

1. According to Table II.3, the atom density of zirconium, N_Z, in units of 10^{24}, is 0.0429, and its total cross-section at thermal energies is approximately 6.6 b.

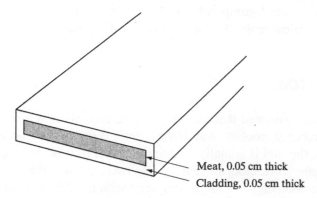

Meat, 0.05 cm thick
Cladding, 0.05 cm thick   **Figure 6.6**   A fuel plate sandwich.

The mean free path in the cladding is therefore

$$\lambda = 1/\Sigma_t = 1/0.0429 \times 6.6 = 3.5 \text{ cm. } [Ans.]$$

To find the mean free path in the meat, it is necessary to compute Σ_t for the ^{235}U–zirconium mixture. Since the uranium is only present as an impurity, the atom concentration of the zirconium in the meat is the same as in the cladding—that is, $N_Z = 0.0429$. The atom concentration of the ^{235}U is then $N_{25} = 0.0429/150 = 2.86 \times 10^{-4}$. For ^{235}U,

$$\sigma_t = \sigma_a + \sigma_s = 681 + 8.9 \simeq 690 \text{ b.}$$

Thus, for the meat[6]

$$\Sigma_t = 2.86 \times 10^{-4} \times 690 + 0.0429 \times 6.6 = 0.480 \text{ cm}^{-1}.$$

The mean free path is then

$$\lambda = 1/\Sigma_t = 2.1 \text{ cm. } [Ans.]$$

The mean free path in the cladding is $3.5/0.05 = 70$ times larger than the thickness of the cladding, whereas it is $2.1/0.05 = 42$ times larger in the meat. The reactor core is clearly quasi-homogeneous.

2. As usual, $k_\infty = \eta_T f$, where $\eta_T = 2.065$ for ^{235}U at room temperature. To compute f, the atom densities of water, zirconium, and ^{235}U of the equivalent homogenized mixture must be used. Since the water and zirconium each occupy one-half the volume, their respective atom densities are just one-half their normal values. In units of 10^{24}:

$$N_W = 0.5 \times 0.0334 = 0.0167,$$

where 0.0334 is the molecular density of unit-density water, and

$$N_Z = 0.5 \times 0.0429 = 0.0215.$$

The uranium is present in a volume equal to one-third of the sandwich or one-sixth of the total volume, and its atom density there is 2.86×10^{-4} so that for the mixture

$$N_{25} = 2.86 \times 10^{-4}/6 = 4.77 \times 10^{-5}.$$

At 0.0253 eV, the microscopic absorption cross-sections of ^{235}U, zirconium, and water (per molecule) are 681 b, 0.185 b, and 0.664 b, respectively. At 20°C, the non-$1/v$ factor for ^{235}U is 0.978. The value of f is then

---

[6]This calculation would appear to violate the provisions of Section 3.3—that thermal scattering cross-sections must not be added. However, most of the contribution from the fuel to Σ is from absorption, and this can be added to scattering.

$$f = \frac{4.77 \times 10^{-5} \times 0.978 \times 681}{4.77 \times 10^{-5} \times 681 \times 0.978 + 0.0215 \times 0.185 + 0.0167 \times 0.664} = 0.6783.$$

Thus,

$$k_\infty = 2.065 \times 0.6783 = 1.401. \; [Ans.]$$

Calculations for heterogeneous thermal reactors are considerably more diffi-cult than those for homogeneous reactors. It is still possible to write $k_\infty = \eta_T f p \epsilon$, and to use the equations for criticality derived earlier, but the factors in these for-mulas must be computed with some care.

The Value of η_T

Most heterogeneous reactors are fueled with natural or partially enriched uranium in metallic or oxide form. The term fuel then refers to a mixture of ^{235}U, ^{238}U, and oxygen. In either case, η_T can be found from the formula

$$\eta_T = \frac{\nu_{25} \overline{\Sigma}_{f25}}{\overline{\Sigma}_{a25} + \overline{\Sigma}_{a28}}, \tag{6.110}$$

since the absorption cross-section of oxygen is essentially zero.

Example 6.11

Compute the value of η_T for natural uranium at room temperature.

Solution. It is convenient first to rewrite Eq. (6.110) in the form

$$\eta_T = \frac{\nu_{25} \overline{\sigma}_{f25}}{\overline{\sigma}_{a25} + \frac{N_{28}}{N_{25}} \overline{\sigma}_{a28}}.$$

The ratio $N_{28}/N_{25} = 138$ for natural uranium. Then using the values $\nu_{25} = 2.42$, $\sigma_{f25} = 582 \, b$, $g_{f25} \, (20°C) = 0.976$, $\sigma_{a25} = 681 \, b$, $g_{a25} \, (20°C) = 0.978$, $\sigma_{a28} = 2.70$ b, $g_{a28} \, (20°C) = 1.0017$ gives

$$\eta_T = \frac{2.42 \times 582 \times 0.976}{681 \times 0.978 + 138 \times 2.70 \times 1.0017} = 1.32. \; [Ans.]$$

Thermal Utilization

Thermal utilization is defined as the probability that a thermal neutron, if it is ultimately absorbed in the core, is in fact absorbed in fuel. This, in turn, is equal to the ratio of the number of neutrons that are absorbed in fuel per second to the number absorbed per second in both fuel and moderator.[7]

[7]It is assumed here that the reactor core consists of only fuel and moderator. In real reactors, cladding, structural materials, control rods, and other materials are always present.

The number of neutrons absorbed in the fuel per second is given by

$$\int_{V_F} \overline{\Sigma}_{aF} \phi_T \, dV,$$

where $\overline{\Sigma}_{aF}$ is the macroscopic thermal absorption cross-section of the fuel, and the integration is carried out over the volume of the fuel V_F. This number can also be written as

$$\overline{\Sigma}_{aF} \overline{\phi}_T V_F,$$

where $\overline{\phi}_{tF}$ is the average thermal flux in the fuel. The number of neutrons absorbed per second in the moderator is given by a similar expression, so that f becomes

$$f = \frac{\overline{\Sigma}_{aF} \overline{\phi}_{TF} V_F}{\overline{\Sigma}_{aF} \overline{\phi}_{TF} V_F + \overline{\Sigma}_{aM} \overline{\phi}_{TM} V_M}. \tag{6.111}$$

By dividing the numerator and denominator of Eq. (6.111) by $\overline{\phi}_{TF}$, f can be put in the form

$$f = \frac{\overline{\Sigma}_{aF} V_F}{\overline{\Sigma}_{aF} V_F + \overline{\Sigma}_{aM} V_M \zeta}, \tag{6.112}$$

where

$$\zeta = \overline{\phi}_{TM} / \overline{\phi}_{TF} \tag{6.113}$$

is called, for reasons discussed later, the *thermal disadvantage factor*.

The values of ϕ_{TF} and ϕ_{TM} cannot be calculated exactly except by numerical methods. However, the qualitative behavior of the flux in the fuel and moderator is shown in Fig. 6.7. It is seen that the flux is smaller in the fuel than in the moderator. This is to be expected on physical grounds since the absorption cross-section of the fuel is so much higher than that of the moderator. It also follows that ϕ_{TF} is smaller than ϕ_{TM}, so that from Eq. (1.113), ζ is greater than unity. This fact has important consequences that are considered later in this section.

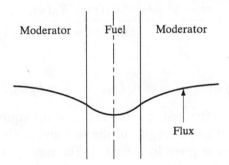

Moderator   Fuel   Moderator

Flux

Figure 6.7   Neutron flux in and near a fuel rod.

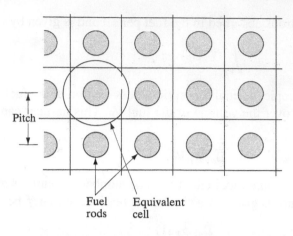

Pitch

Fuel   Equivalent
rods   cell

Figure 6.8 A square lattice of fuel rods showing an equivalent cell.

Although f cannot be calculated analytically, it is possible to derive approximate formulas for this parameter. In these approximations, an infinite array or lattice of fuel rods is divided into unit cells as shown in Fig. 6.8. Calculations of the rate at which neutrons are absorbed in the fuel and moderator are then carried out in the equivalent cylindrical cell—that is, the cell whose volume is equal to that of the noncylindrical cell. This procedure is called the *Wigner–Seitz method*.

The results of these calculations are usually presented in the following form:

$$\frac{1}{f} = \frac{\overline{\Sigma}_{aM} V_M}{\overline{\Sigma}_{aF} V_F} F + E, \tag{6.114}$$

where F and E are called *lattice functions*. In the simplest calculations of these functions, in which diffusion theory is assumed to be valid in both the fuel and moderator, F and E are given by

$$F(x) = \frac{x I_0(x)}{2 I_1(x)} \tag{6.115}$$

and

$$E(y, z) = \frac{z^2 - y^2}{2y} \left[\frac{I_0(y) K_1(z) + K_0(y) I_1(z)}{I_1(z) K_1(y) - K_1(z) I_1(y)} \right]. \tag{6.116}$$

In these formulas,

$$x = a/L_F, \qquad y = a/L_M, \qquad z = b/L_M,$$

where a is the radius of the fuel rod, b is the radius of the equivalent cell, and L_F and L_M are the thermal diffusion lengths of the fuel and moderator, respectively. A short list of values of L_F is given in Table 6.4. The functions I_n and K_n, where

TABLE 6.4 THERMAL DIFFUSION LENGTHS
OF FUEL

Fuel	L_F, cm
Natural uranium	1.55
U_3O_8	3.7

$n = 0$ and 1, are called *modified Bessel functions*. A short table of these functions is given in Appendix V.

Equations (6.115) and (6.116) are reasonably accurate for reactors in which $a \ll b$—that is, reactors in which the moderator volume is much larger than the fuel volume. This is usually the case in large gas-cooled reactors. However, these equations provide only rough estimates of f for reactors with more tightly packed lattices. Most modern power reactors are of this latter type, and more complicated formulas must be used for these systems.

It may be mentioned that, when x, y, and z are less than about 0.75, the following series expansions can be used with good accuracy to compute F and E for cylindrical rods:

$$F(x) = 1 + \frac{1}{2}\left(\frac{x}{2}\right)^2 - \frac{1}{12}\left(\frac{x}{2}\right)^4 + \frac{1}{48}\left(\frac{x}{2}\right)^6 - \cdots , \tag{6.117}$$

$$E(y, z) = 1 + \frac{z^2}{2}\left[\frac{z^2}{z^2 - y^2}\ln\left(\frac{z}{y}\right) - \frac{3}{4} + \frac{y^2}{4z^2} + \cdots\right]. \tag{6.118}$$

Example 6.12

The core of an experimental reactor consists of a square lattice of natural uranium rods imbedded in graphite. The rods are 1.02 cm in radius and 25.4 cm apart.[8] Calculate the value of f for this reactor.

Solution. It is first necessary to determine the radius b of the cylindrical cell having the same volume as the unit cell. Both cells are shown in Fig. 6.9. These cells are of equal length, so for equal volumes it is necessary that

$$\pi b^2 = (25.4)^2$$

and so

$$b = 25.4/\sqrt{\pi} = 14.3 \text{ cm.}$$

---

[8]The distance between the centers of nearest rods, 25.4 cm in the example, is called the *lattice spacing* or *lattice pitch*.

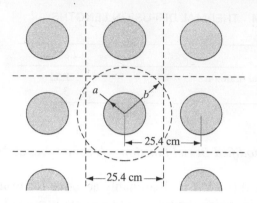

Figure 6.9 A square lattice with 25.4-cm pitch.

For natural uranium, $L_T = 1.55$ cm; for graphite, $L_T = 59$ cm. Thus,

$$x = 1.02/1.55 = 0.658,$$

$$y = 1.02/59 = 0.0173,$$

$$z = 14.3/59 = 0.242.$$

Introducing these values into Eqs. (6.117) and (6.118) gives $F = 1.0532$ and $E = 1.0557$.

From Table II.3, at 0.0253 eV, $\Sigma_{aM} = 0.0002728$ cm^{-1} and $\Sigma_{aF} = 0.3668$ cm^{-1}. Also, $V_M/V_F = (b^2 - a^2)/a^2 = 195.6$. Substituting into Eq. (6.114) then gives

$$\frac{1}{f} = \frac{0.0002728 \times 195.6}{0.3668} \times 1.0532 + 1.0557 = 1.2089,$$

so that

$$f = 0.8272. \ [Ans.]$$

Resonance Escape Probability

It is recalled from Section 6.5 that the *resonance escape probability*, as the name implies, is the probability that a fission neutron will escape capture in resonances as it slows down to thermal energies. Analytical-numerical methods have been devised for computing p from nuclear cross-section data, but these are beyond the scope of this text. However, many measurements of p have been carried out, and these have shown that p can be expressed approximately by the formula

$$p = \exp\left[-\frac{N_F V_F I}{\zeta_M \Sigma_{sM} V_M}\right]. \tag{6.119}$$

TABLE 6.5   CONSTANTS FOR COMPUTING I*

Fuel	A	C
^{238}U (metal)	2.8	38.3
^{238}UO$_2$	3.0	39.6
^{232}Th (metal)	3.9	20.9
^{232}ThO$_2$	3.4	24.5

*Based on W.G. Pettus and M. N. Baldwin, Babcock and Wilcox Company reports BAW-1244, 1962 and BAW-1286, 1963.

TABLE 6.6   VALUES OF $\zeta_M \Sigma_{sM}$

Moderator	$\zeta_M \Sigma_{sM}$
Water	1.46
Heavy water	0.178
Beryllium	0.155
Graphite	0.0608

Here N_F is the atom density of the fuel lump, in units of 10^{24}, V_F and V_M are the volumes of fuel and moderator, respectively, ζ_M is the average increase in lethargy per collision in the moderator (see Section 3.6), Σ_{sM} is the macroscopic scattering cross-section of the moderator at resonance energies, and I is a parameter known as the *resonance integral*. Values of I for cylindrical fuel rods are well represented by the following empirical expression:

$$I = A + C/\sqrt{a\rho}, \tag{6.120}$$

in which A and C are measured constants given in Table 6.5, a is the fuel rod radius, and ρ is the density of the fuel. For convenience in making calculations, values of $\zeta_M \Sigma_{sM}$ are given in Table 6.6.

It should be noted that, according to Eq. (6.119), I has dimensions of cross-section. The constants in Table 6.5, when introduced into Eq. (6.120), give I in barns. However, a must be given in cm and ρ given in g/cm^3. Also, it should be pointed out that, although the values of the constants A and C in Table 6.5 are given for the pure resonance absorbers ^{238}U and ^{232}Th, these constants are also valid for slightly enriched uranium and for thorium containing a small quantity of converted ^{233}U. Small amounts of ^{235}U or ^{233}U in the fuel do not significantly alter the extent of resonance absorption in the ^{238}U or the ^{232}Th—the fuel is still essentially all ^{238}U or ^{232}Th, as the case may be.

Example 6.13

Calculate p for the lattice described in Example 6.12.

Solution. The rod radius is $a = 1.02$ cm, and the uranium density is 19.1 g/cm³. From Eq. (6.120), the resonance integral is therefore

$$I = 2.8 + 38.3/\sqrt{1.02 \times 19.1}$$

$$= 11.5 \text{ b}.$$

From Table II.3, $N_F = 0.0483$ in units of 10^{24}. From Table 6.6, $\zeta_M \Sigma_{sM} = 0.0608$; from Example 6.12, $V_M/V_F = 195.6$. Introducing these values into Eq. (6.119) gives

$$p = \exp\left[-\frac{0.0483 \times 11.5}{0.0608 \times 195.6}\right]$$

$$= \exp(-0.04670) = 0.9544. \text{ [Ans.]}$$

Fast Fission

The fast fission factor, ϵ, the ratio of the total number of neutrons emitted from both thermal and fast fission to the number emitted in thermal fission alone, has been calculated and measured for a large number of heterogeneous reactors. Results of experiments with 1.50 cm diameter, slightly enriched uranium rods[9] in ordinary water are shown in Fig. 6.10. It is observed that ϵ increases as the ratio of the uranium volume to water increases. This would be expected physically. Thus with more uranium metal present in the system, the probability is greater that a fission neutron will strike a uranium nucleus before its energy falls below the fast fission

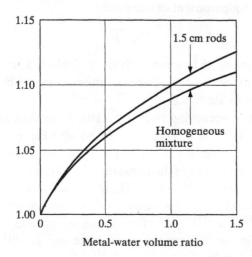

Figure 6.10 The fast fission factor as a function of metal–water volume ratio.

---

[9]As noted earlier, most modern power reactors, at least those in the United States, use slightly (2–3 *w/o*) enriched uranium. Since at low enrichments the ^{238}U content of the fuel is essentially independent of enrichment, the value of ϵ does not depend strongly on enrichment.

threshold in a collision with the moderator. Also shown in Fig. 6.10 are computed values of ϵ for a homogeneous mixture of uranium and water. It is evident from the figure that ϵ is not a sensitive function of the radius of the fuel rods.

Considerable additional data on ϵ, including data on moderators other than water, are found in the report ANL-5800 (see References).

The Value of k_∞

It is of interest to compare the values of the parameters in the four-factor formula for a heterogeneous thermal reactor with those of a homogeneous mixture of the same materials. Consider first the thermal utilization, which for the heterogeneous system is given by Eq. (6.111) or (6.112):

$$f_{\text{hetero}} = \frac{\overline{\Sigma}_{a\text{F}}\overline{\phi}_{T\text{F}}V_{\text{F}}}{\overline{\Sigma}_{a\text{F}}\overline{\phi}_{T\text{F}}V_{\text{F}} + \overline{\Sigma}_{a\text{M}}\overline{\phi}_{T\text{M}}V_{\text{M}}} = \frac{\overline{\Sigma}_{a\text{F}}V_{\text{F}}}{\overline{\Sigma}_{a\text{F}}V_{\text{F}} + \overline{\Sigma}_{a\text{M}}V_{\text{M}}\zeta},$$

where ζ is the disadvantage factor (cf. Eq. (6.113). If the system is homogenized, the terms $\overline{\Sigma}_{a\text{F}}\overline{\phi}_{T\text{F}}V_{\text{F}}$ and $\overline{\Sigma}_{a\text{M}}\overline{\phi}_{T\text{M}}V_{\text{M}}$ give the total absorption in the fuel and moderator. However, in a homogeneous system, the fluxes in the fuel and moderator are the same—that is, $\overline{\phi}_{T\text{F}} = \overline{\phi}_{T\text{M}}$—so that $\zeta = 1$, and therefore

$$f_{\text{homo}} = \frac{\overline{\Sigma}_{a\text{F}}V_{\text{F}}}{\overline{\Sigma}_{a\text{F}}V_{\text{F}} + \overline{\Sigma}_{a\text{M}}V_{\text{M}}}. \tag{6.121}$$

It was shown earlier that ζ is greater than unity for a heterogeneous reactor. From a comparison of Eqs. (6.112) and (6.121), it follows that

$$f_{\text{hetero}} < f_{\text{homo}}. \tag{6.122}$$

The physical reason that f is smaller for the heterogeneous system than for its homogenized equivalent is that, in the heterogeneous case, as noted earlier, the flux is lower in the fuel than it is in the moderator. This depression in the flux is caused by the fact that some of the neutrons entering the fuel from the moderator are absorbed near the surface of the fuel, and they simply do not survive to contribute to the flux in its interior. The outer portion of the fuel thus shields the interior, and this state of affairs is known as *self-shielding*.

What is a disadvantage for the thermal utilization, however, is necessarily an advantage as far as the resonance escape probability is concerned. Thus, although self-shielding decreases the absorption of thermal neutrons in the fuel, it also reduces the number of neutrons that are absorbed parasitically by resonance absorbers in the fuel. Therefore, it follows that the resonance escape probability for a heterogeneous system is larger than for the equivalent homogeneous mixture. Furthermore, since the average absorption cross-section of the fuel at resonance energies is much larger than the absorption cross-section at thermal energies, the depression in the flux of resonance neutrons is greater than it is for thermal neutrons. As a consequence, the decrease in f for the heterogeneous system is more than offset by the increase in p.

The fast fission factor for a heterogeneous reactor is also larger than for the homogeneous system, as may be seen in Fig. 6.10. This is because in the heterogeneous case the fission neutrons pass through a region of pure fuel, where they are likely to induce fast fission before they encounter the moderator. The effect of heterogeneity is not as pronounced with ϵ, however, as it is with f and p.

In view of the fact that

$$(fp)_{\text{hetero}} > (fp)_{\text{homo}}$$

and

$$\epsilon_{\text{hetero}} > \epsilon_{\text{homo}},$$

it follows that k_∞ for a heterogeneous reactor is greater than for its homogeneous equivalent.[10] This result is of considerable importance in the design of certain reactors. In particular, the maximum value of k_∞ for a homogeneous mixture of natural uranium and graphite is only 0.85, and such a reactor cannot become critical. By lumping the fuel into a heterogeneous lattice, k_∞ can be made sufficiently greater than unity that a critical reactor can be constructed using these materials.

REFERENCES

General

Duderstadt, J. J. & L. J. Hamilton, *Nuclear Reactor Analysis*. New York: John Wiley & Sons, 1976, Chapters 5 & 7.

Glasstone, S., and M. C. Edlund, *The Elements of Nuclear Reactor Theory*. Princeton, N.J.: Van Nostrand, 1952, Chapters 7–9.

Glasstone, S., and A. Sesonske, *Nuclear Reactor Engineering*, 4th ed. New York: Chapman & Hall, 1994, Chapters 3 & 4.

Isbin, H. S., *Introductory Nuclear Reactor Theory*. New York: Reinhold, 1963, Chapters 6–8.

Lamarsh, J. R., *Introduction to Nuclear Reactor Theory*. Reading, Mass.: Addison-Wesley, 1966, Chapters 6–11.

Liverhant, S. E., *Elementary Introduction to Nuclear Reactor Physics*. New York: Wiley, 1960.

---

[10]This conclusion is valid for heterogeneous reactors fueled with uranium enriched to about 5% in ^{235}U. Above this enrichment, p is of less overall importance in determining the value of k_∞, and lumping the fuel tends to reduce the overall absorption of neutrons by the fissile material.

Meem, J. L., *Two Group Reactor Theory*. New York: Gordon & Breach, 1964, Chapters 3, 7, and 9.

Murray, R. L., *Nuclear Reactor Physics*. Englewood Cliffs, N.J.: Prentice-Hall, 1957, Chapters 3–5.

Sesonske, A., *Nuclear Power Plant Analysis*. Oak Ridge: U.S. DOE report TID-26241, 1973, Chapter 5.

Zweifel, P. F., *Reactor Physics*. New York: McGraw-Hill, 1973, Chapter 3.

Computer Codes

The Radiation Shielding Information Center (RSIC) at the Oakridge National Laboratory, Oak Ridge, Tennessee, collects, maintains, and distributes a library of computer programs or codes for the solution of problems in nuclear reactor design and engineering. The center also provides the documentation necessary to understand and use the various codes. Information is available on the web at `http://epicws.epm.ornl.gov/rsic.html`.

Reactor Data

Considerable useful data are found in the Argonne National Laboratory report *Reactor Physics Constants*, ANL-5800, L. J. Templin, Editor, 1963.

Bessel Functions

These functions are tabulated in many places including:
Abramowitz, M., and I. A. Stegun, Editors, *Handbook of Mathematical Tables*, 10th ed. Washington, D.C.: U.S. Government Printing Office, 1972.

Etherington, H., Editor, *Nuclear Engineering Handbook*. New York: McGraw-Hill, 1958.

PROBLEMS

Note: Make all computations at room temperature, unless otherwise stated, and take the recoverable energy per fission to be 200 MeV.

1. Calculate the fuel utilization and infinite multiplication factor for a fast reactor consisting of a mixture of liquid sodium and plutonium, in which the plutonium is present to 3.0 w/o. The density of the mixture is approximately 1.0 g/cm^3.

2. The core of a certain fast reactor consists of an array of uranium fuel elements immersed in liquid sodium. The uranium is enriched to 25.6 $^w/_o$ in ^{235}U and comprises 37% of the core volume. Calculate for this core (a) the average atom densities of sodium, ^{235}U, and ^{238}U; (b) the fuel utilization; (c) the value of η; (d) the infinite multiplication factor.

3. A bare-cylinder reactor of height 100 cm and diameter 100 cm is operating at a steady-state power of 20 MW. If the origin is taken at the center of the reactor, what is the power density at the point $r = 7$ cm, $z = -22.7$ cm?

4. In a spherical reactor of radius 45 cm, the fission rate density is measured as 2.5×10^{11} fissions/cm^3-sec at a point 35 cm from the center of the reactor. (a) At what steady-state power is the reactor operating? (b) What is the fission rate density at the center of the reactor?

5. Using the flux function given in Table 6.2 for a critical finite cylindrical reactor, derive the value of the constant A.

6. The core of a certain reflected reactor consists of a cylinder 10-ft high and 10 ft in diameter. The measured maximum-to-average flux is 1.5. When the reactor is operated at a power level of 825 MW, what is the maximum power density in the reactor in kW/liter?

7. Suppose the reactor described in Example 6.3 was operated at a thermal power level of 1 kilowatt. How many neutrons would escape from the reactor per second? [*Hint:* See Example 6.4.]

8. Show that in a one-group model, the power produced by a reactor per unit mass of fissile material is given by

$$\frac{\text{watts}}{\text{g}} = \frac{\text{kW}}{\text{kg}} = \frac{3.2 \times 10^{11} \sigma_f \bar{\phi} N_A}{M_F},$$

where σ_f is the one-group fission cross-section, $\bar{\phi}$ is the average one-group flux, N_A is Avogadro's number, and M_F is the gram atomic weight of the fuel.

9. (a) Estimate the critical radius of a hypothetical bare spherical reactor having the same composition as the reactor in Problem 6.1. (b) If the reactor operates at a thermal power level of 500 MW, what is the maximum value of the flux? (c) What is the probability that a fission neutron will escape from the reactor?

10. An infinite slab of moderator of thickness $2a$ contains at its center a thin sheet of ^{235}U of thickness t. Show that in one-group theory the condition for criticality of this system can be written approximately as

$$\frac{2D}{(\eta - 1)L\Sigma_{aF}t} \coth \frac{a}{L} = 1,$$

where D and L are the diffusion parameters of the moderator and Σ_{aF} is the macroscopic absorption cross-section of the ^{235}U.

11. A large research reactor consists of a cubical array of natural uranium rods in a graphite moderator. The reactor is 25 ft on a side and operates at a power of 20 MW. The average value of $\bar{\Sigma}_f$ is 2.5×10^{-3} cm^{-1}. (a) Calculate the buckling. (b) What is the maximum value of the thermal flux? (c) What is the average value of the thermal flux? (d) At what rate is ^{235}U being consumed in the reactor?

12. Show with a recoverable energy per fission of 200 MeV the power of a ^{235}U-fueled reactor operating at the temperature T given by either of the following expressions:

$$P = 4.73 m_F g_F(T)\overline{\phi}_0 \times 10^{-14} \text{ MW}$$

or

$$P = 7.19 m_F g_F(T)^{-1/2}\overline{\phi}_T \times 10^{-13} \text{ MW},$$

where m_F is the total amount in kg of ^{235}U in the reactor, $g_F(T)$ is the non-1/v factor for fission, $\overline{\phi}_0$ is the average 2,200 meters-per-second flux, and $\overline{\phi}_T$ is the average thermal flux.

13. Solve the following equations:

$$3x + 4y + 7z = 16,$$

$$x - 6y + z = 2,$$

$$2x + 3y + 3z = 12.$$

14. Solve the following equations:

$$3.1x + 4.0y + 7.2z = 0,$$

$$x - 5.0y - 9.0z = 0,$$

$$7x + 4.5y + 8.1z + 8.1z = 0.$$

15. A homogeneous solution of ^{235}U and H_2O contains 10 grams of ^{235}U per liter of solution. Compute (a) the atom density of ^{235}U and the molecular density of H_2O; (b) the thermal utilization; (c) the thermal diffusion area and length; (d) the infinite multiplication factor.

16. Compute the thermal diffusion length for homogeneous mixtures of ^{235}U and the following moderators at the given fuel concentrations and temperatures. Graphite: $N(25)/N(C) = 4.7 \times 10^{-6}$; $T = 200°C$. Beryllium: $N(25)/N(Be) = 1.3 \times 10^{-5}$; $T = 100°C$. D_2O: $N(25)/N(D_2O) = 1.4 \times 10^{-6}$; $T = 20°C$. H_2O: $N(25)/N(H_2O) = 9.2 \times 10^{-4}$; $T = 20°C$.

17. Consider a critical bare slab reactor 200 cm thick consisting of a homogeneous mixture of ^{235}U and graphite. The maximum thermal flux is 5×10^{12} neutrons/cm^2-sec. Using modified one-group theory, calculate: (a) the buckling of the reactor; (b) the critical atomic concentration of uranium; (c) the thermal diffusion area; (d) the value of k_∞; (e) the thermal flux and current throughout the slab; (f) the thermal power produced per cm^2 of this slab.

18. The binding energy of the last neutron in ^{13}C is 4.95 MeV. Estimate the recoverable energy per fission in a large graphite-moderated, ^{235}U-fueled reactor from which there is little or no leakage of neutrons or γ-rays.

19. Calculate the concentrations in grams per liter of (1) ^{235}U, (2) ^{233}U, and (3) ^{239}Pu required for criticality of infinite homogeneous mixtures of these fuels and the following moderators: (a) H_2O, (b) D_2O, (c) Be, (d) graphite.

20. A bare-spherical reactor 50 cm in radius is composed of a homogeneous mixture of ^{235}U and beryllium. The reactor operates at a power level of 50 thermal kilowatts.

Using modified one-group theory, compute: (a) the critical mass of ^{235}U; (b) the thermal flux throughout the reactor; (c) the leakage of neutrons from the reactor; (d) the rate of consumption of ^{235}U.

21. The flux in a bare-finite cylindrical reactor of radius r and height H is given by

$$\phi = A J_0 \left(\frac{2.405r}{\tilde{R}} \right) \cos \left(\frac{\pi z}{\tilde{H}} \right).$$

Find A if the reactor is operated at a power of P watts.

22. It is proposed to store H_2O solutions of fully enriched uranyl (= u'rá-nil) sulfate ($^{235}UO_2SO_4$) with a concentration of 30 g of this chemical per liter. Is this a safe procedure when using a tank of unspecified size?

23. Show that the flux in a bare-cubical reactor of side a is

$$\phi = A \left(\cos \frac{\pi}{\tilde{a}} x \right)^3.$$

24. If the reactor in Problem 6.23 is operating at P watts, show that the constant A is

$$A = \frac{P}{E_R \Sigma_f \dfrac{8}{\pi^3} \tilde{a}^3 \left(\sin \dfrac{\pi}{2} \dfrac{a}{\tilde{a}} \right)^3}.$$

25. A bare-thermal reactor in the shape of a cube consists of a homogeneous mixture of ^{235}U and graphite. The ratio of atom densities is $N_F/N_M = 1.0 \times 10^{-5}$ and the fuel temperature is 250°C. Using modified one-group theory, calculate: (a) the critical dimensions; (b) the critical mass; (c) the maximum thermal flux when the reactor operates at a power of 1 kW.

26. The original version of the Brookhaven Research Reactor consisted of a cube of graphite that contained a regular array of natural uranium rods, each of which was located in an air channel through the graphite. When the reactor was operated at a thermal power level of 22 MW, the average fuel temperature was approximately 300°C and the maximum thermal flux was 5×10^{12} neutrons/cm²-sec. The average values of L_T^2 and τ_T were 325 cm² and 396 cm², respectively, and $k_\infty = 1.0735$. (a) Calculate the critical dimensions of the reactor; (b) What was the total amount of natural uranium in the reactor?

27. Using one-group theory, derive expressions for the flux and the condition for criticality for the following reactors: (a) an infinite slab of thickness a, infinite reflector on both sides; (b) an infinite slab of thickness a, reflectors of thickness b on both sides; (c) a sphere of radius R, reflector of thickness b.

28. The core of a spherical reactor consists of a homogeneous mixture of ^{235}U and graphite with a fuel–moderator atom ratio $N_F/N_M = 6.8 \times 10^{-6}$. The core is surrounded by an infinite graphite reflector. The reactor operates at a thermal power of 100 kW. Calculate the: (a) value of k_∞; (b) critical core radius; (c) critical mass; (d) reflector savings; (e) thermal flux throughout the reactor; (f) maximum-to-average flux ratio.

29. Estimate the new critical radius and the critical mass of a reactor with the same composition as described in Problem 6.20 when the core is surrounded by an infinite beryllium reflector.

30. Show that the solution to the one-speed diffusion equation in an infinite cylinder made of a nonmultiplying media is

$$\phi = AK_0 + CI_0,$$

where K_0 and I_0 are modified bessel functions.

31. An infinite-cylindrical reactor core of radius R is surrounded by an infinitely thick reflector. Using one-group theory, (a) find expressions for the fluxes in the core and reflector; (b) show that the condition for criticality is

$$D_c B \frac{J_1(BR)}{J_0(BR)} = \frac{D_r}{L_r} \frac{K_1(R/L_r)}{K_0(R/L_r)},$$

where J_0 and J_1 are ordinary Bessel functions and K_0 and K_1 are modified Bessel functions.

32. The core of an infinite planar thermal reactor consists of a solution of ^{239}Pu and H_2O with a plutonium concentration of 8.5 g/liter. The core is reflected on both faces by infinitely thick H_2 reflectors. Calculate the: (a) reflector savings, (b) critical thickness of the core, (c) critical mass in g/cm^2.

33. The core of a thermal reactor consists of a sphere, 50 cm in radius, that contains a homogeneous mixture of ^{235}U and ordinary H_2O. This core is surrounded by an infinite H_2O reflector. (a) What is the reflector savings? (b) What is the critical mass? (c) If the maximum thermal flux is 1×10^{13} neutrons/cm^2-sec, at what power is the reactor operating? [*Hint:* Compute M_T^2 for core assuming the reactor is bare and of radius 50 cm. Use this value of M_T^2 to estimate δ and then compute new M_T^2 assuming reactor is bare and of radius $50 + \delta$. Iterate until convergence is obtained.]

34. A spherical-breeder reactor consists of a core of radius R surrounded by a breeding blanket of thickness b. The infinite multiplication factor for the core k_∞ is greater than unity, whereas that for the blanket k_∞ is less than unity. Using one-group theory, derive the critical conditions and expressions for the flux throughout the reactor.

35. Plate-type fuel elements for an experimental reactor consist of sandwiches of uranium and aluminum. Each sandwich is 7.25 cm wide and 0.16 cm thick. The cladding is aluminum that is 0.050 cm thick. The meat is an alloy of fully enriched uranium and aluminum which has 20 $^w/_o$ uranium and a density of approximately 3.4 g/cm^3. (a) Estimate the mean free path of thermal neutrons in the meat and cladding in this fuel element. (b) Is a reactor fueled with these elements quasi-homogeneous or heterogeneous?

36. The fuel sandwiches in the reactor in the preceding problem were braised into aluminum holders and placed in a uniform array in an ordinary water moderator. The (total) metal–water volume ratio is 0.73, and there are 120 atoms of aluminum per atom of uranium. Calculate: (a) the thermal utilization; (b) k_∞; (c) the thermal diffusion length. [*Note:* In part (c), compute the value of $\overline{D}$ for the homogeneous mixture by adding together the macroscopic transport cross-sections of the aluminum and water. For Σ_{tr} of water, use $\frac{1}{3}\overline{D}$, where $\overline{D}$ is the experimental value of $\overline{D}$ for water, corrected of course for density.]

37. A heterogeneous uranium–water lattice consists of a square array of natural uranium rods 1.50 cm in diameter with a pitch of 2.80 cm. Calculate the: (a) radius of the equivalent cell; (b) uranium-water volume ratio V_F/V_M; (c) thermal utilization; (d) resonance escape probability; (e) fast fission factor (from Fig. 6.10); (f) k_∞.

38. Repeat the calculations of Problem 6.37 for the case in which the uranium is enriched to 2.5 w/o in ^{235}U.

39. In a hexagonal lattice (also called a *triangular lattice*), each fuel rod is surrounded by six nearest neighbor rods, equally spaced at a distance equal to the pitch s of the lattice. Show that the radius of the equivalent cell in this lattice is given by

$$b = 0.525\ s.$$

40. Calculate k_∞ for a hexagonal lattice of 1.4-cm-radius natural uranium rods and graphite if the lattice pitch is 20 cm. [*Note:* The fast fission factor for this lattice is 1.03.]

7

The Time-Dependent Reactor

The previous Chapter was concerned with critical nuclear reactors operating at constant power. Needless to say, reactors are not always critical. It is necessary, for instance, for a reactor to be supercritical to start it up or raise its power level, whereas it must be subcritical to shut it down or reduce power. The study of the behavior of the neutron population in a noncritical reactor is called *reactor kinetics*. This subject is discussed in Section 7.2.

The degree of criticality of a reactor is usually regulated by the use of *control rods* or *chemical shim*. Control rods are pieces or assemblies of neutron-absorbing material whose motions have an effect on the multiplication factor of the system. Thus, if a control rod is withdrawn from a critical reactor, the reactor tends to become supercritical; if a rod is inserted, the system falls subcritical. With chemical shim, control is accomplished by varying the concentration of a neutron-absorbing chemical, usually boric acid (H_3BO_3), in a water moderator or coolant. Some elementary calculations of the effectiveness of control rods and chemical shim in reactor control are given in Section 7.3.

One of the most important factors affecting criticality is the reactor temperature. Several of the parameters entering into the value of k are temperature dependent, and changes in T necessarily lead to changes in k. This circumstance is especially relevant to the question of reactor safety and is discussed in Section 7.4. A related matter, the effect on k of the formation of voids within a reactor, is also considered in Section 7.4.

Besides changing reactor power level and responding to changes in temperature, control rods and/or chemical shim systems are used to compensate for the burnup of the fuel over the lifetime of a reactor core. In this connection, it should be noted that a reactor is initially fueled with more than the minimum amount of fuel necessary for criticality—if this were not the case and it were fueled with only its minimum critical mass, the reactor would fall subcritical after the first fission. Control rods, boric acid, gadolinium, and other neutron absorbers are introduced into the core to compensate for the excess fuel. Then as fissions occur, the fuel is consumed, the gadolinium is depleted, and the rods are slowly withdrawn or the H_3BO_3 concentration is reduced to keep the reactor critical.

Two fission-product nuclides are formed with each fission, so there is an accumulation of fission products as the fuel is consumed. Certain of these nuclides—in particular, ^{135}Xe and ^{149}Sm—have very large absorption cross-sections, and their presence in a reactor can have a profound effect on the value of k. Most of the other fission products have much smaller σ_a and tend to reduce k more gradually over the life of the core. The generation and accumulation of *fission-product poisons* is discussed in Section 7.5.

In reactors containing fertile material, the production by conversion of new fissile material compensates to some extent for fuel burnup and fission product formation. These changes in the properties of a reactor core with time have an obvious effect on determining the useful life of the total energy output which can be expected from a given core, and therefore have a bearing on overall power costs. This interplay between fuel properties and the economics of power production, together with the management decisions it makes necessary, form the subject of *nuclear fuel management* which concludes this chapter.

7.1 CLASSIFICATION OF TIME PROBLEMS

The time dependence of the neutron population may be divided into three classes of problems. These classes are the short, intermediate, and long time problems. The first class of short time problems usually occurs over seconds to tens of minutes. In this class of problems, the depletion of the fuel is small and can be ignored. Such problems arise during reactor transient and safety analysis calculations.

The second class of problems involves time scales that are on the order of hours to at most a day or two. Here again, the depletion of the fuel is ignored. However, the changing characteristics of the fission products must be accounted for.

Finally, the long-term problems involve time periods of days to months. This class of problems requires a detailed knowledge of the fuel depletion and the distribution of both the flux and fuel in the reactor. This problem is of major concern for fuel-management calculations.

Long Time or Depletion Problems   These problems are concerned with the variation of the neutron flux over long periods of time. However, they can be treated in a relatively simple manner by assuming that the changes in the flux occur slowly over time, and hence the reactor may be treated as if it is in a series of stationary states. The problems are solved using an adiabatic approximation where the reactor is assumed to go through several stationary configurations. These configurations are determined by solving the steady-state time-independent case.

This type of problem is solved by assuming that the neutron production and loss terms are balanced. The diffusion equation is written as

$$D\nabla^2\phi - \Sigma_a\phi = \lambda\upsilon\Sigma_f\phi, \tag{7.1}$$

where λ is an eigenvalue of the problem. It is assumed to be adjustable and is varied to give the solution. For a time-independent solution, λ must be equal to 1. If $\lambda \neq 1$, then something must be changed, such as the buckling so that the leakage is changed, or the absorption or the fission cross-section so that the reaction rates are altered.

The design problem is then to obtain an eigenvalue of $\lambda = 1$ throughout the life of the reactor. Once the condition is obtained, the reactor power will be constant and time independent. The multiplication factor of the core will be equal to one. One can identify λ as equal to $1/k$ where k is the multiplication factor.

The process in reality has introduced a parameter λ that is equal to one when the problem is time independent. In doing the analysis in this manner, we have solved an artificial problem that corresponds to the case of $k = 1$. Values of λ other than one are not real since the problem is then time dependent. But the value of λ does give us an understanding of what state the reactor is in. The management of the fuel in a reactor is largely based on this type of an approach.

Short Time Problems   These problems are typically encountered when there is an action that causes the conditions in the reactor system to be perturbed. For example, the steam demand might suddenly be altered due to a change in turbine load. In a BWR, this is seen as a change in pressure in the reactor vessel. For a PWR, the change manifests itself as an alteration in the primary system temperature. These changes lead to changes in the moderation rate of the neutrons, causing the multiplication factor, k, to change.

The method most often used to solve short-term problems is based on the assumption that the shape of the reactor flux does not change. The reactor is assumed to act as a point, hence the name *point kinetics*. Such an approach is the basis for most analyses of reactor systems except where flux shapes are known to vary with time.

Intermediate Time Problems   This class of problems is dominated by the changing concentration of the fission products due to their radioactive decay. Since several of the fission products are produced in significant quantities and have large enough thermal neutron absorption cross-sections, they become of concern due to their absorption of thermal neutrons. Since their concentrations are varying with time, the variation will affect the absorption term in the multiplication factor. This change in absorption rate must then be accounted for in the calculation if the reactor is to stay critical.

Solution methods vary depending on the need to account for the spatial variation of the fission products. If dependence of the concentration is ignored, then the point kinetics equation may be used to solve for the time-dependent power by properly accounting for the time variation of k.

7.2 REACTOR KINETICS

It is recalled from Chapter 3 that most of the neutrons emitted in fission appear virtually at the instant of fission; these are the prompt neutrons. A small fraction of the fission neutrons appears long after the fission event; these fissions are delayed neutrons. The time behavior of a reactor depends on the various properties of these two types of neutrons. The prompt neutrons are considered first.

Prompt Neutron Lifetime

Following their emission, the prompt fission neutrons slow down as the result of elastic and inelastic collisions with nuclei in the system. In a fast reactor, they do not slow down very much—only to the order of tens or hundreds of keV—before they are absorbed or leak out of the core. In thermal reactors, however, most of them succeed in reaching thermal energies without being absorbed or escaping from the system. As thermal neutrons, they diffuse about in the reactor; some eventually are absorbed and some leak out. The average time between the emission of the prompt neutrons and their absorption in a reactor is called the *prompt neutron lifetime* and is denoted by l_p.

Consider first the value of l_p in an infinite thermal reactor. It can be shown theoretically, and it has been found experimentally, that the time required for a neutron to slow down to thermal energies is small compared to the time that the neutron spends as a thermal neutron before it is finally absorbed. The average lifetime of a thermal neutron in an infinite system is called the *mean diffusion time* and is given the symbol t_d. It follows, therefore, that

$$l_p \simeq t_d \tag{7.2}$$

for an infinite thermal reactor.

It is an easy matter to calculate t_d. A thermal neutron of energy E travels, on the average, an absorption mean free path $\lambda_a(E)$ before being absorbed. It survives, therefore, the time

$$t(E) = \lambda_a(E)/v(E), \tag{7.3}$$

where $v(E)$ is the neutron speed corresponding to the energy E. The mean diffusion time is then the average value of $t(E)$—that is,

$$t_d = \overline{t(E)}. \tag{7.4}$$

Since $\lambda_a(E) = 1/\Sigma_a(E)$, where $\Sigma_a(E)$ is the macroscopic absorption cross-section at the energy E, Eq. (7.3) can be written as

$$t(E) = \frac{1}{\Sigma_a(E)v(E)}. \tag{7.5}$$

If the absorption is $1/v$, as it is at least approximately in thermal reactors, $\Sigma_a(E) = \Sigma_a(E_0)v_0(E)/v(E)$, where $E_0 = 0.0253$ eV and $v_0 = 2,200$ m/sec, and Eq. (7.5) becomes

$$t(E) = \frac{1}{\Sigma_a(E_0)v_0}. \tag{7.6}$$

Thus, $t(E)$ is a constant, independent of E, and

$$t_d = \frac{1}{\Sigma_a(E_0)v_0} = \frac{\sqrt{\pi}}{2\overline{\Sigma}_a v_T}, \tag{7.7}$$

where use has been made of Eqs. (5.55) and (5.59) with $g_a(T) = 1$. Values of t_d for several moderators are given in Table 7.1.

TABLE 7.1 APPROXIMATE DIFFUSION
TIMES FOR SEVERAL MODERATORS

Moderator	t_d, sec
H_2O	2.1×10^{-4}
D_2O^*	4.3×10^{-2}
Be	3.9×10^{-3}
Graphite	0.017

*With 0.25% H_2O impurity.

If the reactor consists of a mixture of fuel and moderator, then $\overline{\Sigma}_a = \overline{\Sigma}_{aF} + \overline{\Sigma}_{aM}$, and Eq. (7.7) becomes

$$t_d = \frac{\sqrt{\pi}}{2v_T(\overline{\Sigma}_{aF} + \overline{\Sigma}_{aM})}, \tag{7.8}$$

which can also be written as

$$t_d = \frac{\sqrt{\pi}}{2v_T\overline{\Sigma}_{aM}} \frac{\overline{\Sigma}_{aM}}{\overline{\Sigma}_{aF} + \overline{\Sigma}_{aM}}.$$

The first factor in this equation is the mean diffusion time for the moderator, t_{dM}; the second factor is equal to $1 - f$, where f is the thermal utilization (see Eq. 6.58). Thus, for the mixture,

$$t_d = t_{dM}(1 - f). \tag{7.9}$$

Example 7.1

Calculate the prompt neutron lifetime in an infinite, critical thermal reactor consisting of a homogeneous mixture of ^{235}U and unit density H_2O at room temperature.

Solution. Since the reactor is critical,

$$k_\infty = \eta_T f = 1,$$

so that $f = 1/\eta_T$. From Table 6.3, $\eta_T = 2.065$ and therefore $f = 0.484$. According to Table 7.1, t_d for water is 2.1×10^{-4} sec. Using Eqs. (7.2) and (7.8),

$$l_p \simeq t_d = 2.1 \times 10^{-4}(1 - 0.484) = 1.08 \times 10^{-4} \text{ sec. [Ans.]}$$

The prior results only pertain to thermal reactors. Prompt neutron lifetimes are considerably shorter in fast reactors than in thermal reactors since the neutrons never have an opportunity to reach thermal energies. In a fast reactor, the value of l_p is on the order of 10^{-7} sec.

Reactor with No Delayed Neutrons

As noted at the beginning of this section, the delayed neutrons play an important role in reactor kinetics. This is a remarkable fact since so few fission neutrons are delayed—less than 1% for thermal fission in ^{235}U. To understand the importance of the delayed neutrons, it is helpful to consider first the kinetics of a reactor in the absence of delayed neutrons—that is, assuming that all neutrons are emitted promptly in fission. For the moment, the discussion is restricted to the infinite thermal reactor.

Next we note that the absorption of a fission neutron in the fission process begins a new generation of fission neutrons. The *mean generation time*, Λ, is defined as the time between the birth of a neutron and subsequent absorption-inducing fission. In the typical case of $k \simeq 1$ and in the absence of delayed neutrons, l_p is approximately equal to Λ—the time between successive generations of neutrons in the chain reaction. It follows from the definition of k_∞ that the absorption of a neutron from one generation leads to the absorption, l_p sec later, of k_∞ neutrons in the next generation. Thus, if $N_F(t)$ is the number of fissions (which, of course, is proportional to the number of neutron absorptions) occurring per cm³/sec at the time t, then the fission rate l_p seconds later will be

$$N_F(t + l_p) = k_\infty N_F(t). \qquad (7.10)$$

The first term in this equation can be expanded as

$$N_F(t + l_p) \simeq N_F(t) + l_p \frac{dN_F(t)}{dt},$$

and when this is substituted back into Eq. (7.10), the result is

$$\frac{dN_F(t)}{dt} \simeq \frac{k_\infty - 1}{l_p} N_F(t). \qquad (7.11)$$

The solution to this equation is

$$N_F(t) = N_F(0) \exp\left(\frac{k_\infty - 1}{l_p}\right) t, \qquad (7.12)$$

where $N_F(0)$ is the fission rate at $t = 0$. Equation (7.12) can also be written as

$$N_F(t) = N_F(0)e^{t/T}, \qquad (7.13)$$

in which

$$T = \frac{l_p}{k_\infty - 1} \qquad (7.14)$$

is called the *reactor period*—in the absence of delayed neutrons.

Example 7.2

Suppose that the reactor described in Example 7.1 is critical up to time $t = 0$, and then k_∞ is increased from 1.000 to 1.001. Compute the response of the reactor to this change in k_∞.

Solution. From Example 7.1, l_p is approximately 10^{-4} sec, so that from Eq. (7.14),

$$T = \frac{10^{-4}}{1.001 - 1.000} = 0.1 \text{ sec.}$$

The flux (and power) would therefore increase as e^{10t}, where t is in seconds. [*Ans.*]

The period computed in Example 7.2 is very short. Thus with a period of 0.1 sec, the reactor would pass through 10 periods in only 1 second, and the fission rate (and power) would increase by a factor of $e^{10} = 22,000$. Had the reactor originally been operating at a power of 1 megawatt, the system would reach a power of 22,000 megawatts in 1 sec if it did not destroy itself first—as it undoubtedly would.

Fortunately, this analysis, in which delayed neutrons have been omitted, does not describe the kinetics of an actual reactor. As is shown presently, the delayed neutrons considerably increase the reactor period. As a consequence, reactors can be controlled rather easily.

Reactor with Delayed Neutrons

In an accurate reactor kinetics calculation, it is necessary to consider in detail the production and decay of each of the six groups of delayed neutron precursors. The resulting mathematical analysis is necessarily rather complicated. To simplify the discussion, it is assumed for the moment that there is only one group of delayed neutrons that appear from the decay of a single hypothetical precursor.[1]

Consider an infinite homogeneous thermal reactor that may or may not be critical. Since the thermal flux must be independent of position, the time-dependent diffusion equation for the thermal neutrons is (see Eq. [5.16]):

$$s_T - \overline{\Sigma}_a \phi_T = \frac{dn}{dt}, \tag{7.15}$$

where s_T is the source density of neutrons slowing down into the thermal energy region, and n is the density of thermal neutrons. From Eq. (5.56),

$$\phi_T = \frac{2}{\sqrt{\pi}} n v_T,$$

and Eq. (7.15) becomes

$$s_T - \overline{\Sigma}_a \phi_T = \frac{\sqrt{\pi}}{2 v_T} \frac{d\phi_T}{dt}. \tag{7.16}$$

---

[1]The one delayed neutron group may be thought of as an effective delayed neutron group when an average over all six groups is taken.

Dividing this equation through by $\overline{\Sigma}_a$ and making use of Eq. (7.7) gives

$$\frac{s_T}{\overline{\Sigma}_a} - \phi_T = t_d \frac{d\phi_T}{dt}. \tag{7.17}$$

Finally, since $t_d \simeq l_p$, Eq. (7.17) can also be written as

$$\frac{s_T}{\overline{\Sigma}_a} - \phi_T = l_p \frac{d\phi_T}{dt}. \tag{7.18}$$

If all of the fission neutrons were prompt, then, in view of the definition of k_∞, the source density would be (see Eq. [6.11]):

$$s_T = k_\infty \overline{\Sigma}_a \phi_T.$$

However, the fraction β of the fission neutrons is delayed, and so only the fraction $(1 - \beta)$ is prompt. The contribution to s_T due to the prompt neutrons is then

$$s_T \text{ (from prompt neutrons)} = (1 - \beta)k_\infty \overline{\Sigma}_a \phi_T.$$

A delayed neutron slows down very rapidly after it is emitted by its precursor. It follows that the contribution to the thermal source density from the delayed neutrons is equal to the rate of decay of the precursor multiplied by the probability p that the delayed neutron escapes resonance capture while slowing down. Thus,

$$s_T \text{ (from delayed neutrons)} = p\lambda C,$$

where p is the resonance escape probability, λ is the decay constant of the precursor, and C is the precursor concentration in atoms/cm^3. Combining the prior results gives

$$s_T = (1 - \beta)k_\infty \overline{\Sigma}_a \phi_T + p\lambda C. \tag{7.19}$$

Introducing this expression into Eq. (7.18) yields the following equation for the thermal flux:

$$(1 - \beta)k_\infty \phi_T + \frac{p\lambda C}{\overline{\Sigma}_a} - \phi_T = l_p \frac{d\phi_T}{dt}. \tag{7.20}$$

The equation determining the precursor concentration can be obtained from the following argument. In Section 6.5, it was shown that the rate at which fission neutrons, prompt and delayed, are produced is $\eta_T \epsilon f \overline{\Sigma}_a \phi_T / p$. The rate at which delayed neutrons are produced is then $\beta k_\infty \overline{\Sigma}_a \phi_T / p$. Now since each delayed neutron appears as a result of the decay of a precursor, it follows that the rate of production of the precursor is also equal to $\beta k_\infty \overline{\Sigma}_a \phi_T / p$. The precursor decays, of course, at the usual rate λC, so that C is governed by the equation

$$\frac{dC}{dt} = \frac{\beta k_\infty \overline{\Sigma}_a \phi_T}{p} - \lambda C. \tag{7.21}$$

Equations (7.20) and (7.21) are coupled differential equations that must be solved simultaneously to determine ϕ_T or C.

Consider now a specific problem. Suppose that, up to the time $t = 0$, the reactor is critical and $k_\infty = 1$. A step change is then made in k_∞ so that the reactor becomes either supercritical or subcritical. It is required to determine ϕ_T as a function of time after $t = 0$.

This problem can be solved by assuming solutions of the form

$$\phi = Ae^{\omega t} \tag{7.22}$$

and

$$C = C_0 e^{\omega t}, \tag{7.23}$$

where A and C_0 are constants and ω is a parameter to be determined. Inserting these functions into Eq. (7.21) gives

$$C_0 = \frac{\beta k_\infty \overline{\Sigma}_a A}{p(\omega + \lambda)}.$$

When C_0 and Eqs. (7.22) and (7.23) are introduced into Eq. (7.20), it is found that the constant A cancels, which leaves

$$(1 - \beta)k_\infty + \frac{\lambda \beta k_\infty}{\omega + \lambda} - 1 = \omega l_p.$$

It is convenient to rewrite this equation in the following form:

$$\frac{k_\infty - 1}{k_\infty} = \frac{\omega l_p}{1 + \omega l_p} + \frac{\omega}{1 + \omega l_p}\frac{\beta}{\omega + \lambda}. \tag{7.24}$$

The left-hand side (LHS) of Eq. (7.24) is known as the *reactivity*, in this case of the infinite reactor, and is denoted by the symbol ρ; thus

$$\rho = \frac{k_\infty - 1}{k_\infty}. \tag{7.25}$$

For a finite reactor, the reactivity is defined as

$$\rho = \frac{k - 1}{k}. \tag{7.26}$$

In terms of the reactivity, Eq. (7.24) may be written as

$$\rho = \frac{\omega l_p}{1 + \omega l_p} + \frac{\omega}{1 + \omega l_p} \frac{\beta}{\omega + \lambda}. \tag{7.27}$$

This result is known as the *reactivity equation* for one group of delayed neutrons.

Before considering the significance of the reactivity equation, it should be noted that if a reactor is supercritical, then $k > 1$ and ρ is positive. In this situation, the reactor is said to have *positive reactivity*. However, when the reactor is subcritical, $k < 1$, ρ is negative, and the reactor is said to have *negative reactivity*. It should be noted that, from Eq. (7.26), ρ is restricted to values in the range $-\infty < \rho < 1$.

Example 7.3

What is the reactivity corresponding to a change in k_∞ from 1.000 to 1.001?

Solution. From Eq. (7.26)

$$\rho = \frac{1.001 - 1}{1.001} \simeq 10^{-3}.$$

Since the numerator is equal to the change in k, ρ is equal to the fractional change in k. In this case, ρ is often expressed in percent—specifically, $\rho = 100 \times 10^{-3} = 0.1$ percent. [*Ans.*]

The reactivity equation, Eq. (7.27), gives a relationship between the reactivity of a reactor and those values of ω for which the functions given in Eqs. (7.22) and (7.23) are solutions to the differential equations, Eqs. (7.20) and (7.21). In short, the problem of solving the differential equations has been reduced to the problem of finding the roots of Eq. (7.27) for a specified value of ρ. This can most conveniently be done by plotting the right-hand side (RHS) of the equation as a function of ω as shown in Fig. 7.1. As indicated in the figure, this consists of three distinct branches. The roots of the equation are then located at the intersections, where these curves cross the horizontal lines corresponding to given values of ρ, the other for negative ρ. It is seen that there are two roots—ω_1 and ω_2—for either $+\rho$ and $-\rho$. Thus, it follows that the flux behaves as

$$\phi_T = A_1 e^{\omega_1 t} + A_2 e^{\omega_2 t}, \tag{7.28}$$

where A_1 and A_2 are constants.

It is observed from Fig. 7.1 that when ρ is positive, ω_1 is positive and ω_2 is negative. Thus, as time goes on, the second term in Eq. (7.28) dies out and the flux ultimately *increases* as $e^{\omega_1 t}$. In contrast, when ρ is negative, both ω_1 and ω_2 are negative, but ω_2 is more negative than ω_1. Thus, the second term in Eq. (7.28)

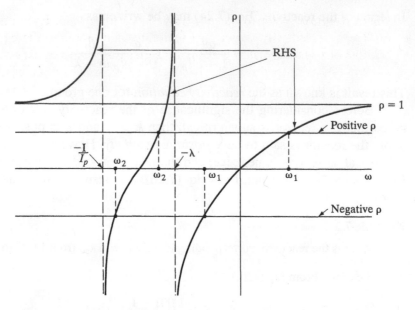

Figure 7.1 Plot of the reactivity equation for one group of delayed neutrons (not to scale).

again dies out more rapidly than the first, and the flux now *decreases* as $e^{\omega_1 t}$. In either case, positive ρ or negative ρ, the flux approaches $e^{\omega_1 t}$—that is,

$$\phi_T \to e^{\omega_1 t}. \tag{7.29}$$

The reciprocal of ω_1 is called the *reactor period* or sometimes the *stable period*, and it is denoted as T—that is,

$$T = \frac{1}{\omega_1}. \tag{7.30}$$

The ultimate behavior of the flux may therefore be written as

$$\phi_T \to e^{t/T}. \tag{7.31}$$

Had the prior derivation included all six delayed neutron groups instead of only one, the resulting reactivity equation would have been of the same form as Eq. (7.27), but necessarily more complicated. This general reactivity equation is

$$\rho = \frac{\omega l_p}{1 + \omega l_p} + \frac{\omega}{1 + \omega l_p} \sum_{i=1}^{6} \frac{\beta_i}{\omega + \lambda_i}, \tag{7.32}$$

where β_i and λ_i refer to the ith delayed neutron group. By plotting the RHS of Eq. (7.32), it is easily shown that there are seven roots to the equation for either positive or negative ρ. The flux again is given by a sum of exponentials, but there are now seven instead of two. However, just as in the simple case of one delayed group, with increasing time, ϕ once more approaches $e^{\omega_1 t}$, where ω_1 is the first root

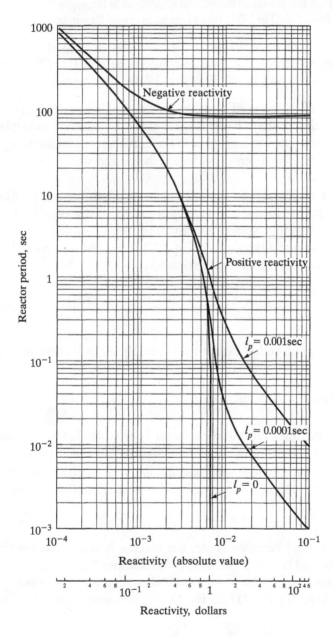

Figure 7.2 Reactor period as a function of positive and negative reactivity for a ^{235}U-fueled reactor.

of the equation, whether ρ is positive or negative. The reactor period is again the reciprocal of ω_1. Values of the period for a ^{235}U-fueled reactor computed from Eq. (7.32) are shown in Fig. 7.2 as a function of reactivity.

Example 7.4

Using Fig. 7.2, determine the period of the reactor described in Example 7.1 following the change in k_∞ from 1.000 to 1.001 (the same as given in Example 7.2).

Solution. It was shown in Example 7.1 that $l_p \simeq 10^{-4}$ sec. In Example 7.3, the reactivity equivalent to this change in k_∞ was computed to be $\rho = 10^{-3}$. Thus, the reactor period, which may be read directly from Fig. 7.2, is 57 sec. [*Ans.*]

The remarkable effect of the delayed neutrons on the time behavior of a reactor is evident by comparing the results of Examples 7.2 and 7.4. In the first example, the delayed neutrons were not taken into account in the calculation, and the period was only 0.1 sec. With the delayed neutrons included, the period was increased to 57 secs—a factor of over 500.

To turn now to the kinetics of a fast reactor, it is recalled that the prompt neutron lifetime in this type of reactor is very short—on the order of 10^{-7} sec. Therefore, except for very large insertions of reactivity resulting in large values of ω (short reactor periods), the quantity of ωl_p in Eq. (7.32) can be neglected. The reactivity equation for fast reactors is then approximately

$$\rho = \omega \sum_{i=1}^{6} \frac{\beta_i}{\omega + \lambda_i}. \tag{7.33}$$

For ^{235}U-fueled fast reactors, the period can be read from the $l_p = 0$ curve in Fig. 7.2.

The Prompt Critical State

It is shown in Section 6.1 that the multiplication factor is proportional to the total number of neutrons, prompt and delayed, emitted per fission. However, since only the fraction $(1 - \beta)$ of the fission neutrons are prompt, the multiplication factor as far as the prompt neutrons are concerned is actually $(1 - \beta)k$. Therefore, when

$$(1 - \beta)k = 1, \tag{7.34}$$

the reactor is critical on prompt neutrons alone, the reactor is said to be *prompt critical*. In this case, the period is very short, as it is in Example 7.2.

The reactivity corresponding to the prompt critical condition can be found by inserting $k = 1/(1 - \beta)$ from Eq. (7.34) into Eq. (7.26). It is easy to see that this gives

$$\rho = \beta \qquad (7.35)$$

as the condition for prompt criticality. Thus, since $\beta = 0.0065$ for fissions induced by thermal neutrons in ^{235}U, a ^{235}U-fueled thermal reactor becomes prompt critical with the addition of only about 0.0065, or .65% in reactivity.

Because reactors have such short periods when they are prompt critical, it is normal practice to restrict (positive) reactivity additions to less than β. There are some research reactors, such as the TRIGA reactor produced by General Atomics, with intrinsic properties that cause them to shut down rapidly when they become supercritical. When these reactors are suddenly made prompt critical, they deliver a sharp pulse of neutrons. The Pennsylvania State University's Breazeale Reactor, for example, may be pulsed to a power several thousand times that of its maximum steady-state value. Figure 7.3 shows a power pulse from this reactor. The pulse is generated by ejecting a control rod from the reactor, which increases the reactivity by 2.72β.

The amount of reactivity necessary to make a reactor prompt critical—namely, $\rho = \beta$—is used to define a unit of reactivity known as the *dollar*. Since the value of β varies from fuel to fuel, the dollar is not an absolute unit. Values of the delayed neutron fractions for the most important fissile and fertile isotopes are given in Table 7.2. For example, a dollar is worth 0.0065 in reactivity for a ^{235}U-fueled reactor, but it is worth only $\beta = 0.0026$ for a ^{233}U-fueled reactor. The reactivity equivalent to one one-hundredth of a dollar is called a *cent*.

Example 7.5

The reactivity of the reactor in Example 7.3 is 0.001. If the reactor is fueled with ^{235}U, what is its reactivity in dollars?

Solution. The value of β is 0.0065. Thus, ρ is

$$\rho = \frac{0.001}{0.0065} = 0.154 \text{ dollars}$$

$$= 15.4 \text{ cents. } [Ans.]$$

The Prompt Jump (Drop)

Following a sudden change in multiplication factor, a previously critical reactor exhibits a behavior given by Eq. (7.28)—namely,

$$\phi = A_1 e^{\omega_1 t} + A_2 e^{\omega_2 t}$$

in the one delayed-group model and by a sum of seven exponentials in the more realistic case of six delayed groups,

$$\phi = A_1 e^{\omega_1 t} + A_2 e^{\omega_2 t} + \cdots + A_7 e^{\omega_7 t}.$$

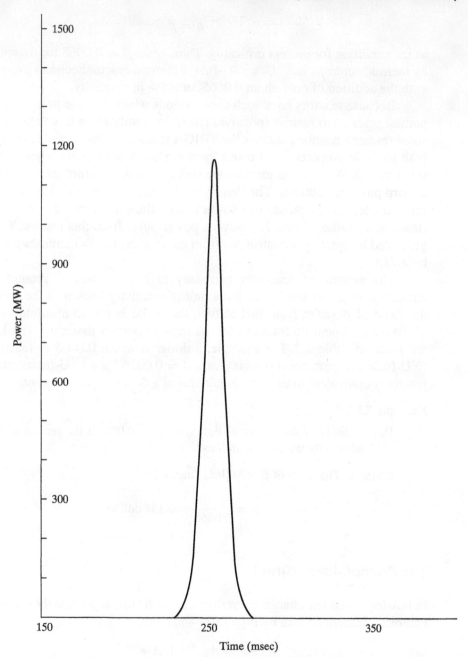

Figure 7.3 Power pulse obtained from Penn State's Breazeale nuclear reactor by ejecting a control rod with a worth of 2.72β. The maximum power is 1164 MW.

TABLE 7.2   DELAYED NEUTRON FRACTIONS

Nuclide	β (thermal neutron fission)	β (fast fission*)
^{232}Th	—	0.0203
^{233}U	0.0026	0.0026
^{235}U	0.0065	0.0064
^{238}U	—	0.0148
^{239}Pu	0.0021	0.0020

*Fission induced by prompt neutron spectrum.

As explained earlier, all but the first term dies away quickly, and the flux then reduces to the first term, rising or falling with reactor period $T = 1/\omega_1$. Exact calculations show that, in the case of the one delayed-group model, A_2 is negative for positive ρ and positive for negative ρ. The effect of the rapid die away of a negative term is to give a sudden rise in ϕ following the insertion of positive reactivity as shown in Fig. 7.4. Likewise, the die away of a positive term gives a sudden drop in ϕ for negative reactivity. To determine the overall time behavior of a reactor, it is necessary to know the level to which the flux first rises or drops before assuming a stable period.

The exact computations of the early response to changes in reactivity are rather complicated, especially when all seven terms of the flux are included. A simple, approximate result can be obtained, however, by merely assuming that the concentrations of the delayed neutron precursors do not change over the time of the sudden rise or drop in flux. This is the basis for the *prompt jump* approximation.

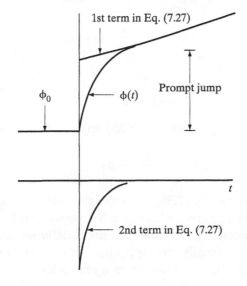

Figure 7.4   Time behavior of a reactor after step insertion of reactivity showing origin of prompt jump.

In the discussion that follows, we consider two situations. The first is when k_∞ is initially equal to unity, and the second is when $k \neq 1$.

k_∞ **Equal to One**   In the first case, the precursor concentration is taken to be constant at its value in the critical reactor. Then from Eq. (7.21), $dC/dt = 0$, thus

$$C = \frac{\beta \overline{\Sigma}_a \phi_{T0}}{p\lambda}, \qquad (7.36)$$

where k_∞ is placed equal to unity since the (infinite) reactor is originally critical. Also in Eq. (7.36), ϕ_{T0} is the flux prior to the change in reactivity. Introducing Eq. (7.36) into Eq. (7.20) gives

$$l_p \frac{d\phi_T}{dt} = [(1-\beta)k_\infty - 1]\phi_T + \beta\phi_{T0},$$

where k_∞ is the multiplication factor after the change in reactivity. The solution to this equation subject to the condition $\phi_T(0) = \phi_{T0}$ is easily found to be

$$\phi_T = \phi_{T0} \exp\frac{[(1-\beta)k_\infty - 1]t}{l_p} + \frac{\beta\phi_{T0}}{1-(1-\beta)k_\infty}\left[1 - \exp\frac{[(1-\beta)k_\infty - 1]t}{l_p}\right].$$

$$(7.37)$$

Now for $(1-\beta)k_\infty < 1$, that is, for reactivity less than prompt critical, the two exponentials in Eq. (7.37) die out with a period

$$T = \frac{l_p}{(1-\beta)k_\infty - 1} \simeq \frac{l_p}{k_\infty - 1},$$

which is the very short period given by Eq. (7.14) in the absence of delayed neutrons. Thus, from Eq. (7.37), it is seen that ϕ_T quickly assumes the value

$$\phi_T = \frac{\beta}{1-(1-\beta)k_\infty}\phi_{T0}. \qquad (7.38)$$

On introducing $k_\infty = 1/(1-\rho)$ from Eq. (7.25), Eq. (7.38) becomes finally

$$\phi_T = \frac{\beta(1-\rho)}{\beta - \rho}\phi_{T0}. \qquad (7.39)$$

As a first application of Eq. (7.39), consider the case of a positive reactivity change, such as might be required to increase the power level of a reactor. For reasons of safety, such reactivity insertions are not usually so large as to cause a very short reactor period. Normally, power is raised at a period no less than about 2 minutes. From Fig. 7.2, this corresponds to a reactivity of $\rho \simeq 0.0006$ for a

^{235}U-fueled reactor. When this value of ρ is introduced into Eq. (7.39), it is seen that $\phi_T \simeq \phi_{T0}$. It follows that for positive reactivity insertions, the prompt jump in the flux is usually negligible, and the flux can be assumed to increase from its initial value on a stable exponential with a constant period.

With negative reactivity, the situation is different, for now very large (negative) reactivities can be introduced. Thus, when a reactor is *scrammed*, several or all of the control rods may be quickly inserted into the system. Suppose, for instance, that 20% in negative reactivity is suddenly inserted in a reactor fueled with ^{235}U. Then substituting $\beta = 0.0065$ and $\rho = -0.2$ into Eq. (7.39) gives

$$\phi_T = 0.038\phi_{T0}.$$

Thus, the reactor flux (and the power since this is proportional to the flux) drops quickly to about 4% of its initial value before the reactor goes on its stable period.[2]

Incidentally, in the one delayed-group approximation, it should be observed in Fig. 7.1 that, as ρ becomes increasingly negative, the first root of the reactivity equation approaches the value $-\lambda$, the decay constant for a hypothetical one-group of delayed neutrons. When the reactivity equation for the actual case of six delayed neutron groups is plotted, the value of ω_1 approaches λ_1, the decay constant for the *longest lived* precursor—namely, the precursor with a mean life of about 80 sec. Thus, with large negative reactivity insertions, the (negative) period has the value

$$T = \frac{1}{\omega_1} = \frac{1}{\lambda_1} = 80 \text{ sec.}$$

This result is seen from Fig. 7.2. As shown in the figure a reactor goes on this 80-sec period with negative reactivity in excess of about .4%.

Example 7.6

A ^{235}U-fueled reactor operating at a constant power of 500 MW is scrammed by the sudden insertion of control rods worth 10% in reactivity. To what (fission) power level has the reactor fallen 10 minutes later?

Solution. The power drops suddenly to the level P_1, where

$$P_1 = \frac{\beta(1 - \rho)}{\beta - \rho} P_0$$

$$= \frac{0.0065(1 + 0.10)}{0.0065 + 0.10} \times 500 = 33.5 \text{ MW.}$$

---

[2]This result and the preceding discussion only pertain to the power produced as the result of fission within the reactor. Power produced by the decay of radioactive fission products is not included. It is shown in Chapter 8 that fission product decay power amounts to about 7% of the reactor power when the reactor has been operated long enough for the fission products to reach equilibrium levels. If this is the case, scramming the reactor can at most reduce the power to 7% of its initial value.

It continues to drop according to the relation

$$P = P_1 e^{-t/T},$$

where $T = 80$ sec. After 10 minutes, the power falls to

$$P = 33.5 e^{-(10 \times 60)/80} = 33.5 e^{-7.5}$$

$$= 0.0185 \text{ MW. } [Ans.]$$

k_∞ Not Equal to One   In the second case, $k_\infty \neq 1$, we assume that the prompt neutron lifetime l_p is very small, such that the neutron population jumps quickly to its new value, but there is little change in the precursor concentration. Using Eq. (7.20),

$$(1 - \beta)k_\infty \phi_T + \frac{p\lambda C}{\overline{\Sigma}_a} - \phi_T = l_p \frac{d\phi_T}{dt} \qquad (7.40)$$

and dividing by k_∞ to obtain

$$\frac{(1 - \beta)k_\infty \phi_T}{k_\infty} + \frac{1}{k_\infty} \frac{p\lambda C}{\overline{\Sigma}_a} - \frac{1}{k_\infty}\phi_T = l_p \frac{1}{k_\infty} \frac{d\phi_T}{dt}. \qquad (7.41)$$

On rearranging

$$\frac{(k_\infty - 1)\phi_T}{k_\infty} + \frac{1}{k_\infty} \frac{p\lambda C}{\overline{\Sigma}_a} - \beta\phi_T = l_p \frac{1}{k_\infty} \frac{d\phi_T}{dt} \qquad (7.42)$$

and defining the reactivity in $ units as

$$\rho' = \frac{1}{\beta} \frac{(k_\infty - 1)}{k_\infty} \qquad (7.43)$$

gives

$$\rho' \beta \phi_T + \frac{1}{k_\infty} \frac{p\lambda C}{\overline{\Sigma}_a} - \beta\phi_T = l_p \frac{1}{k_\infty} \frac{d\phi_T}{dt}$$

On solving for ϕ_T, this yields

$$\phi_T = \frac{1}{(1 - \rho')\beta} \left[\frac{1}{k_\infty} \frac{p\lambda C}{\overline{\Sigma}_a} - l_p \frac{1}{k_\infty} \frac{d\phi_T}{dt} \right].$$

If the prompt neutron lifetime is small, one can show that the last term is much less than the precursor term for $\rho' \leq 0.50\$$, and thus may be neglected so that

$$\phi_T \simeq \frac{1}{(1 - \rho')\beta} \times \frac{p\lambda C}{\nu \overline{\Sigma}_f \epsilon p}. \qquad (7.44)$$

Equation (7.44) is valid both before and after a reactivity change is made. Denoting the reactivity after the insertion as ρ_+, then Eq. (7.44) for ϕ_T^+ the flux after becomes

$$\phi_T^+ \simeq \frac{1}{(1 - \rho_+')\beta} \times \frac{p\lambda C}{v\overline{\Sigma}_f}. \tag{7.45}$$

A similar expression is obtained for the flux ϕ_T^- immediately before the change in reactivity. The ratio of ϕ_T^+ to ϕ_T^- is then

$$\frac{\phi_T^+}{\phi_T^-} = \frac{1 - \rho_-'}{1 - \rho_+'}. \tag{7.46}$$

The power ratio is

$$\frac{P^+}{P^-} = \frac{1 - \rho_-'}{1 - \rho_+'}. \tag{7.47}$$

This equation is a more general form of the prompt jump approximation and is valid for reactivities less than 0.50\$. Use of this expression is given in the problems.

Small Reactivities

As shown in Fig. 7.2, the curve $\rho(\omega)$—that is, the plot of the reactivity equation—passes through the origin. This means that, for small reactivities, the first root of the equation is also small. In this case, the terms containing ω in the denominator of Eq. (7.27) or (7.32) can be ignored with the result

$$\rho = \omega_1 \left(l_p + \sum_i \frac{\beta_i}{\lambda_i} \right) = \omega_1 \left(l_p + \sum_i \beta_i \bar{t}_i \right), \tag{7.48}$$

where $\bar{t}_i = 1/\lambda_i$ is the mean life of the ith precursor group. The reactor period can now be written as

$$T = \frac{1}{\omega_1} = \frac{1}{\rho} \left(l_p + \sum_i \beta_i \bar{t}_i \right). \tag{7.49}$$

Values of the sum on the Right-Hand Side of Eq. (7.49) are given in Table 7.3 for the fissile nuclides. In all cases, this sum is much larger than the usual value of l_p, and Eq. (7.49) reduces to

$$T = \frac{1}{\rho} \sum_i \beta_i \bar{t}_i. \tag{7.50}$$

Examples of the use of this formula are given in the problems.

TABLE 7.3   VALUES OF THE SUM $\Sigma \beta_i \bar{t}_i$

Nuclide	$\Sigma \beta_i \bar{t}_i$, sec
^{233}U	0.0479
^{235}U	0.0848
^{239}Pu	0.0324

7.3 CONTROL RODS AND CHEMICAL SHIM

As explained in the introduction to the chapter, control rods are used in two ways: (a) to change the degree of reactor criticality for the purpose of raising or lowering the power level, and (b) to keep a reactor critical by compensating for the changes in the properties of the system that take place over its lifetime.

When a rod is used in the first manner, the effect is to place a reactor on a stable period. In this case, the *rod worth* is defined as the magnitude of the reactivity required to give the observed period. However, the worth of a rod used to keep a reactor critical is measured in terms of the change in multiplication factor of the system for which the rod can compensate. Despite the fundamental difference in these two definitions of *rod worth*, it can be shown[3] that these definitions of rod worth are essentially identical. In the present section, the worths of rods in a number of important situations are derived.

Central Control Rod

As a first example, consider the problem of calculating the worth of a rod of radius a inserted along the axis of a bare cylindrical thermal reactor of extrapolated radius R and height H. The calculations are considerably simplified if it is assumed that when the rod is withdrawn from the reactor, the hole left by the rod is filled with core material and, contrariwise, when the rod is inserted, a cylinder of core material is extruded.

Suppose that the reactor is just critical when the rod is out of the reactor. The neutron flux then has the shape shown in Fig. 7.5, which is determined by the equation

$$\nabla^2 \phi_T + B_0^2 \phi_T = 0, \tag{7.51}$$

where B_0^2 is the usual buckling of a bare cylindrical reactor. If the rod is now fully inserted into the reactor, the flux takes on the form indicated in Fig. 7.5, owing to the absorption of neutrons in the rod. The shape of the flux can be found from the

---

[3]Lamarsh, J. R. *Nuclear Reactor Theory*, p. 499 (see references).

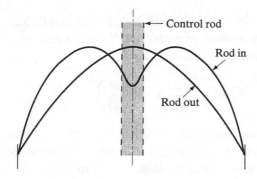

Figure 7.5 Flux in a bare reactor with and without a central control rod.

equation

$$\nabla^2 \phi_T + B^2 \phi_T = 0, \tag{7.52}$$

where B^2 is the buckling of the system with the rod present.

From modified one-group theory, the multiplication factor of the initially critical reactor is given by

$$k_0 = \frac{k_\infty}{1 + B_0^2 M_T^2} = 1,$$

while the multiplication factor of the reactor with the rod inserted is

$$k = \frac{k_\infty}{1 + B^2 M_T^2}.$$

According to Eq. (7.26), this change in the multiplication factor is equivalent to a (negative) reactivity of

$$\rho = \frac{k - 1}{k} = \frac{k - k_0}{k}.$$

Since, by definition, the rod worth ρ_ω is equal to the magnitude of ρ, it follows that

$$\rho_\omega = |\rho| = \frac{k_0 - k}{k} = \frac{(B^2 - B_0^2)M_T^2}{1 + B_0^2 M_T^2}. \tag{7.53}$$

In the event that $B^2 \simeq B_0^2$, Eq. (7.53) can be written as

$$\rho_\omega = \frac{2M^2 B_0 \Delta B}{1 + B_0^2 M_T^2}, \tag{7.54}$$

where

$$\Delta B = B - B_0.$$

To compute ρ_ω, it is necessary to solve Eqs. (7.51) and (7.52) and determine B_0^2 and B^2, the lowest eigenvalues in each case. With the rod out, B_0^2 is given by the formula derived in Chapter 6—namely,

$$B_0^2 = \left(\frac{2.405}{\widetilde{R}}\right)^2 + \left(\frac{\pi}{\widetilde{H}}\right)^2.$$

The calculation of B^2 is somewhat more difficult for two reasons. First, the geometry of the problem is more complicated with the rod present. Second, the rod, especially if it is a strong absorber of neutrons, tends to distort the flux to such an extent that diffusion theory is not valid in its vicinity. This second difficulty can be circumvented by requiring the flux to satisfy the following boundary condition at the surface of the rod:

$$\frac{1}{\phi_T}\frac{d\phi_T}{dr} = \frac{1}{d}, \tag{7.55}$$

where d is called the *extrapolation distance*. This is the same procedure as was used in Chapter 5 in connection with the diffusion of neutrons near a bare surface, where diffusion theory also was not valid.

Values of d have been computed using advanced theoretical methods. The results of these calculations can be expressed approximately by the following formula[4]

$$d = 2.131\overline{D}\frac{a\Sigma_t + 0.9354}{a\Sigma_t + 0.5098}, \tag{7.56}$$

where $\overline{D}$ is the thermal neutron diffusion coefficient and Σ_t is the macroscopic total cross-section, both for the reactor material surrounding the rod. Equation (7.56) is valid for strongly absorbing rods (i.e., *black rods*).[5] Other formulas for d for less strongly absorbing rods (i.e., *grey rods*) are found in the literature.

The solving of Eq. (7.52) subject to the boundary condition Eq. (7.55), as well as the usual conditions at the surface of the reactor, is somewhat tedious and is not given here. In particular, the calculation of B^2 requires the solution of a complicated transcendental equation. However, in the case where B^2 is not too different from B_0^2, this equation can be solved for ΔB. When this is substituted into Eq. (7.54), the following result is obtained:

$$\rho_\omega = \frac{7.43 M_T^2}{(1 + B_0^2 M_T^2)R^2}\left[0.116 + \ln\left(\frac{R}{2.405a}\right) + \frac{d}{a}\right]^{-1}. \tag{7.57}$$

[4]B. Pellaud, *Nuclear Science and Engineering* **33**, 169 (1968).

[5]A *black* substance is defined as one that absorbs all thermal neutrons incident on it; a *grey* substance absorbs some but not all such neutrons.

Equation (7.57) gives the worth of a black control rod located at the center of a bare cylindrical thermal reactor as computed by modified one-group theory.

Example 7.7

A small bare reactor is a square cylinder (i.e., $H = 2R$) of extrapolated height (i.e., $\tilde{H} = H + 2d$) 70 cm. The reactor consists of a mixture of ordinary water and ^{235}U at room temperature. What is the worth of a black control rod 1.90 cm in radius inserted at the center of the reactor?

Solution. It is first necessary to compute M_T^2 for the critical reactor. In terms of the thermal utilization, the modified one-group critical equation is (see Section 6.5)

$$\frac{k_\infty}{1 + B_0^2 M_T^2} = \frac{\eta_T f}{1 + B_0^2[(1 - f)L_{TM}^2 + \tau_{TM}]} = 1,$$

where L_{TM}^2 and τ_{TM} are the diffusion area and age for water. Solving for f gives

$$f = \frac{1 + B_0^2(L_{TM}^2 + \tau_{TM})}{\eta_T + B_0^2 L_{TM}^2}. \tag{7.58}$$

The value of B_0^2 is

$$B_0^2 = \left(\frac{2.405}{35}\right)^2 + \left(\frac{\pi}{70}\right)^2 = 6.74 \times 10^{-3}.$$

Using this value together with $L_{TM}^2 = 8.1 \text{ cm}^2$, $\tau_{TM} = 27 \text{ cm}^2$, and $\eta_T = 2.065$, in Eq. (7.58) gives

$$f = \frac{1 + 6.74 \times 10^{-3}(8.1 + 27)}{2.065 + 6.74 \times 10^{-3} \times 8.1} = 0.583.$$

This, in turn, gives for M_T^2 the value

$$M_T^2 = (1 - 0.583) \times 8.1 + 27 = 30.4 \text{ cm}^2.$$

In the present problem, the reactor consists of a very dilute solution of ^{235}U and water. Thus, both $\overline{D}$ and Σ_t in Eq. (7.56) have the values for water alone—namely, $\overline{D} = 0.16$ cm (from Table 5.2) and $\Sigma_t = 3.443 \text{ cm}^{-1}$ (from Table II.3). Introducing these values together with $a = 1.90$ cm into Eq. (7.56) gives

$$d = 2.131 \times 0.16 \times \frac{1.90 \times 3.443 + 0.9354}{1.90 \times 3.443 + 0.5098} = 0.361 \text{ cm}.$$

Inserting the prior results for M_T^2, B_0^2, and d into Eq. (7.57) then gives

$$\rho_\omega = \frac{7.43 \times 30.4}{(1 + 6.74 \times 10^{-3} \times 30.4) \times (35)^2}\left[0.116 + \ln\left(\frac{35}{2.405 \times 1.90}\right) + \frac{0.361}{1.90}\right]^{-1}$$

$$= 0.065 = 6.5\%. \ [Ans.]$$

Cluster Control Rods

In practice, almost no reactors are controlled with only one rod as described in the preceding discussion. This is because the presence of a single, strongly absorbing rod would lead to the large distortion of the flux indicated in Fig. 7.5, and this in turn would give rise to an undesirable power and temperature distribution in the reactor core. At the same time, reliance on only one rod to control a reactor is not wise from the standpoint of safety since this provides no reserve control capability in case the rod does not function properly.

Many research reactors, which operate at relatively low power—on the order of a few megawatts or less—are controlled with a small number of rods usually arranged in a symmetric fashion about the center of the core. Formulas for the worth of such rings of rods have been derived, but are too complicated to be reproduced here. They are found in the references at the end of the chapter.

However, with power reactors operating as they do at high power densities and high temperatures, it is extremely important that the neutron flux remain as uniform as possible throughout the core. For this reason, power reactors always contain a great many rods. There are two different types of control rod systems in general use today. In one of these, a number of small cylindrical rods, about the size of the fuel rods, are attached at one end by a metal plate called a *spider* and move together through individual sheaths in a fuel assembly as shown in Fig. 7.6. Each group of rods so connected is known as a *cluster control rod*. A typical power reactor may have 50 or 60 such clusters, each containing on the order of 20 rods, giving about 1,000 individual control rods. The rods are composed of a strong neutron absorber. Various alloys of cadmium, indium, and silver, hafnium, or steel containing boron have been used in some reactors—more recently, hollow rods filled with boron carbide (B_4C) powder (sometimes the boron is enriched with the isotope ^{10}B) have been employed.

In many power reactors, the control rod drives (i.e., the mechanisms that insert or withdraw the rods) are interconnected electrically or *ganged* so that several rods move simultaneously in response to a signal from the reactor operator. Consolidated Edison's Indian Point Reactors Nos. 2 and 3, for example, each contain 61 cluster control rods, which are divided into four independently moving groups or *banks*. As far as the reactor operator is concerned, therefore, each reactor has only four independent control rods.

Although the reactor operator obviously must have the final control over the operating conditions, power level, and so on, of a reactor, the ordinary movement of the rods in most power reactors is controlled, subject to operator intervention, by an on line computer. This computer accepts as input data the values of a number of reactor parameters, such as the pressure, temperature, power density, and so on, which are continuously measured by various monitors located throughout the

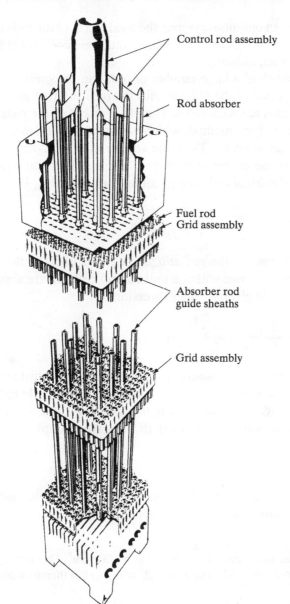

Control rod assembly

Rod absorber

Fuel rod
Grid assembly

Absorber rod
guide sheaths

Grid assembly

Figure 7.6 A cluster control rod assembly. (Courtesy of Westinghouse Electric Corporation.)

system. Using these data, the computer determines what motions of the control rods are necessary to maintain a desired pressure or temperature or to obtain the optimum power distribution within the reactor. The results of these computations are automatically transmitted into the required movement of the rods. Such computer control over rod motion leads to longer core life. At the same time, because the control rods are used in an optimum manner, the total worth of the rods required for the control of the reactor is reduced over what would be necessary if the rods were moved all together or haphazardly.

An estimate of the worth of a large number of rods can be obtained by comparing the multiplication factor of the reactor with the rods inserted to the multiplication factor when the rods are withdrawn. The value of k with the rods present can be found by the Wigner–Seitz method, which was used in Section 6.8 for the calculation of heterogeneous reactors. Thus, the reactor core is divided into unit cells, with a control rod at the center of each cell, and the diffusion equation is solved in the equivalent cylindrical cell. The resulting formula for ρ_ω is

$$\rho_\omega = \frac{f_R}{1 - f_R}, \tag{7.59}$$

where f_R is a parameter known as the *rod utilization* and is equal to the fraction of the fission neutrons slowing down within a unit control cell that are absorbed in the control rod. This, in turn, is given by the expression

$$\frac{1}{f_R} = \frac{(z^2 - y^2)d}{2a} + E(y, z). \tag{7.60}$$

In this equation, a is the control rod radius; d is the extrapolation distance at the surface of the rod; $y = a/L_T$, where L_T is the thermal diffusion length in the reactor core material; $z = R_c/L_T$, where R_c is the radius of the equivalent cell; and $E(y, z)$ is the lattice function defined by Eq. (6.116). An example of the use of these formulas follows.

Example 7.8

Suppose that the reactor described in Example 7.7 were to be controlled by 100 evenly spaced black control rods 0.508 cm in radius. Estimate the total worth of the rods.

Solution. The cross-sectional area of the reactor is $\pi(35)^2$ cm^2. One one-hundredth of this is the area of one control cell. The radius R_c of this cell is therefore determined from

$$\pi R_c^2 = \pi(35)^2/100,$$

so that $R_c = 3.5$ cm. The thermal diffusion area is given by

$$L_T^2 = (1 - f)L_{TM}^2,$$

and inserting $f = 0.583$ from Example 7.7 and $L_{TM}^2 = 8.1$ cm^2 gives $L_T^2 = 3.38$ cm^2. Thus, $L_T = 1.85$ cm, $a/L_T = 0.275$, and $R_c/L_T = 1.89$.

Using the following values of the Bessel functions:

$$I_0(0.275) = 1.019, \quad I_1(0.275) = 0.1389, \quad I_1(1.89) = 1.435;$$

$$K_0(0.275) = 1.453, \quad K_1(0.275) = 3.371, \quad K_1(1.89) = 0.1618;$$

in Eq. (6.116) gives

$$E(0.275, 1.89) = \frac{(1.89)^2 - (0.275)^2}{2 \times 0.275} \left[\frac{1.019 \times 0.1618 + 1.453 \times 1.435}{1.435 \times 3.371 - 0.1618 \times 0.1389} \right]$$

$$= 2.971.$$

From Eq. (7.56), the value of d is found to be

$$d = 2.131 \times 0.16 \times \frac{0.508 \times 3.443 + 0.9354}{0.508 \times 3.443 + 0.5098} = 0.405 \text{ cm}.$$

Introducing this into Eq. (7.60) then yields

$$\frac{1}{f_R} = \frac{[(1.89)^2 - (0.275)^2] \times 0.405}{2 \times 0.508} + 2.971 = 4.365.$$

Thus, $f_R = 0.229$ and

$$\rho_\omega = \frac{0.229}{1 - 0.229} = 0.297 = 29.7\%. \, [Ans.]$$

Cruciform Rods

Another type of control rod that has been widely used in power reactors is the *cruciform* rod, which consists of two crossed blades as shown in Fig. 7.7. These rods have the advantage of fitting snugly into the corners and along the edges of the fuel assemblies as indicated in Fig. 7.8, and they have good physical integrity. Figure 7.8 also shows the array of 137 cruciform rods used in a 780-MWe boiling water reactor plant. Cruciform rods are made of either solid neutron-absorbing metal or hollow, cross-like forms filled with rods of neutron absorber—often these rods are hollow tubes filled with B_4C powder as in the cluster control rod. This latter type of construction is used for the rod illustrated in Fig. 7.7.

Because of the complicated geometry of cruciform rods, it is not possible to determine their worth exactly by analytic calculations. In the actual design of reactors containing such rods, their worth is computed on fast digital computers using two- or three-dimensional multigroup codes. However, a rough estimate of their worth can be found from the following simple method.

To begin with, the reactor is again divided into cells, but this time the unit cell is bounded by the blades of two rods as shown in Fig. 7.9. Next, this cell is replaced by the simple slab geometry of Fig. 7.10, obtained by bending the blades

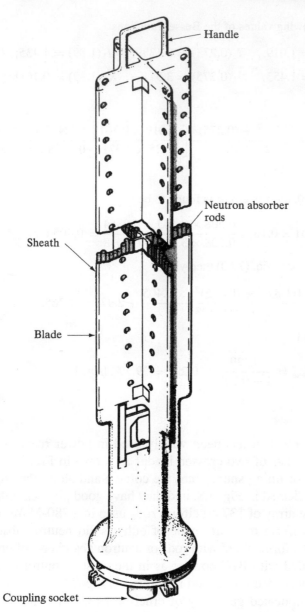

Handle

Neutron absorber rods

Sheath

Blade

Coupling socket

Figure 7.7 A cruciform control rod. (Courtesy of General Electric Company.)

at the top and bottom of the cell through 90°. The value of f_R can then be found as follows. First, the neutron diffusion equation in slab geometry is solved to determine the current density of neutrons flowing into the rod. Then this current density is multiplied by the actual surface area of the blades to give the number of neutrons absorbed per second by the rod. Finally, this number, divided by the number

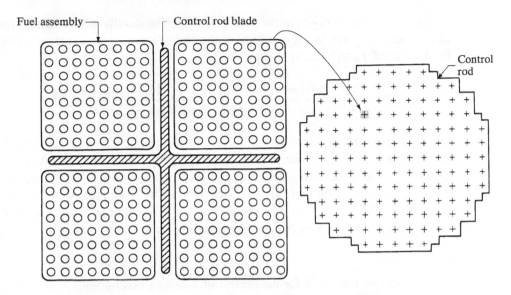

Figure 7.8 A cruciform rod in place between four fuel assemblies and the complete array of rods for a BWR core. (Courtesy of General Electric Company.)

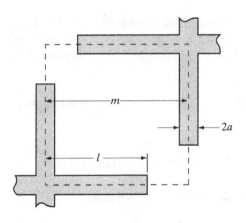

Figure 7.9   Cell for calculation of worth of closely spaced cruciform rods.

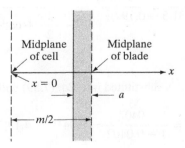

Figure 7.10   Slab geometry for solving diffusion equation.

of neutrons slowing down in the cell, gives f_R. The details of this straightforward calculation are given in the problems at the end of the chapter. As shown there, the final value of f_R to be used in Eq. (7.59) is

$$f_R = \frac{4(l-a)L_T}{(m-2a)^2} \frac{1}{d/L_T + \coth[(m-2a)/2L_T]},$$

where the parameters l, a, and m are those shown in Fig. 7.9, d is the extrapolation distance at a planar surface, and L_T is the thermal diffusion length in the core material. If the blades are black to thermal neutrons, as they usually are, then $d = 2.131\overline{D}$—a conclusion that can be reached either by placing $a = \infty$ in Eq. (7.56) or by using Eq. (5.22) for a bare planar surface. The latter is equivalent to a black planar surface since, in both cases, neutrons are irretrievably lost.

Example 7.9

Cruciform rods used for the control of a certain boiling-water reactor consist of blades 9.75 in wide and 0.312 in thick. The rods are placed in a rectangular array, with 44.5 cm between the centers of the nearest rods. The parameters of the core material are $L_T = 1.2$ cm and $\overline{\Sigma}_a = 0.20$ cm. The rods are black to thermal neutrons. Calculate the worth of the rods.

Solution. The dimensions of the rod in cm are $l = 9.75 \times 2.54/2 = 12.38$ cm and $a = 0.312 \times 2.54/2 = 0.396$ cm. In view of Fig. 7.9, a closest distance between rods of 44.5 cm means that $m = 44.5/\sqrt{2} = 31.5$ cm. The value of $\overline{D}$ for computing d is found from

$$\overline{D} = \overline{\Sigma}_a L_T^2 = 0.20 \times (1.2)^2 = 0.288 \text{ cm},$$

so that

$$d = 2.131\overline{D} = 0.614 \text{ cm}.$$

Introducing these values into Eq. (7.60) gives

$$f_R = \frac{4 \times (12.38 - 0.396) \times 1.2}{(31.5 - 0.792)^2} \frac{1}{\dfrac{0.614}{1.2} + \coth\left(\dfrac{31.5 - 0.792}{2 \times 1.2}\right)}$$

$$= 0.0402.$$

To find ρ_ω, this number is substituted into Eq. (7.59) with the result

$$\rho_\omega = \frac{0.0402}{1 - 0.0402} = 0.0419 = 4.19\%. \quad [\text{Ans.}]$$

Control Rods in Fast Reactors

Only a few fast reactors have been designed or built as of this writing. Thus, it is not possible to state with certainty what types of control rods will come into general use as fast reactor design becomes standardized. However, rods filled with B_4C, enriched in the isotopes ^{10}B, are used in the French-designed Super Phénix and appear to hold considerable promise for this type of reactor. This is because the absorption cross-section of boron is still significant at high neutron energies, whereas σ_a drops rapidly to very low values with increasing energy for most other substances.

In any case, the absorption cross-sections of all materials, including boron, are relatively small at the energies of importance in fast reactors. Thus, σ_a for normal boron in the energy range of about 0.1 MeV to 0.4 MeV is only 0.27 b. The absorption mean free path, λ_a, of neutrons of this energy in media containing boron is therefore relatively large. With B_4C, for instance, at a density of 2 g/cm^3 (this is 80% of theoretical density), the atom density of boron $N_B = 0.087 \times 10^{24}$ atoms/cm^3, and so $\lambda_a = 1/0.087 \times 0.27 = 42.6$ cm. This is considerably larger than the thickness of any control rod likely to be used in a fast reactor.

The fact that at the relevant energies λ_a is much larger than control rod dimensions means that the neutron flux inside the rod is more or less the same as it is in the surrounding medium. Therefore, for calculational purposes, the boron contained in the rod can be assumed to be uniformly distributed in the reactor (or in a region thereof). This makes it possible to estimate the worth of control rods in fast reactors by elementary means. By contrast, the situation in a thermal reactor is just the opposite. There is a marked drop in the flux near a control rod in a thermal reactor, making it much more difficult to analyze.

The only effect of a uniformly distributed poison on the multiplication factor of a fast reactor is to change the value of f—the fuel utilization (see Section 6.1). The transport properties of the reactor are little affected by the presence of the boron. According to Eq. (6.56), the multiplication factor for a fast reactor is given by

$$k = k_\infty P_L = \eta f P_L,$$

where P_L is the nonleakage probability. Thus, it follows from the definition of rod worth in Eq. (7.53) that

$$\rho_\omega = \frac{k_0 - k}{k} = \frac{f_0 - f}{f}, \tag{7.61}$$

where the symbols without the subscript refer to the reactor with the rods inserted.

If σ_{aF}, σ_{aC}, and σ_{aB} are, respectively, the one-group macroscopic absorption cross-sections of the fuel, the coolant (including structural material in the core), and the boron, f is

$$f = \frac{\Sigma_{aF}}{\Sigma_{aF} + \Sigma_{aC} + \Sigma_{aB}}$$

while

$$f_0 = \frac{\Sigma_{aF}}{\Sigma_{aF} + \Sigma_{aC}}.$$

Substituting these expressions into Eq. (7.61) gives

$$\rho_\omega = \frac{\Sigma_{aB}}{\Sigma_{aF} + \Sigma_{aC}}. \tag{7.62}$$

This is the total worth of the control rods in a fast reactor.

Example 7.10

The fast reactor assembly described in Example 6.3 is to be controlled with five rods each consisting of a hollow stainless steel tube 5 cm in inner diameter and 76 cm long, filled with normal enrichment B_4C at a density of 2.0 g/cm^3. Estimate the worth of the rods using one-group theory.

Solution. First it is necessary to find the total number of boron atoms in all the rods. The total volume of the rods is $5 \times \pi (2.5)^2 \times 76 = 7,460$ cm^3. The mass of B_4C in the rods is therefore $2.0 \times 7,460 = 14,920$ g. The molecular weight of B_4C is $4 \times 10.8 + 12.0 = 55.2$, and so the total number of atoms of boron in the reactor is

$$\frac{4 \times 14,920 \times 0.6 \times 10^{24}}{55.2} = 649 \times 10^{24}.$$

The atom density of boron averaged over the whole volume of the reactor is then

$$N_B = \frac{649 \times 10^{24}}{\frac{4}{3}\pi (41.7)^3} = 2.14 \times 10^{-3} \times 10^{24} \text{ atoms/cm}^3.$$

Using the value $\sigma_a = 0.27$ b gives

$$\Sigma_{aB} = 2.14 \times 10^{-3} \times 0.27 = 0.000578 \text{ cm}^{-1}.$$

Substituting this value of Σ_{aB}, together with $\Sigma_{aF} = 0.00833$ and $\Sigma_{aC} = 0.000019$ from Example 6.3, into Eq. (7.62) gives finally

$$\rho_\omega = \frac{0.000578}{0.00833 + 0.000019}$$

$$= 0.0692 = 6.97\%. \text{ [Ans.]}$$

In actual fast reactor design, the worth of the control rods is computed using multigroup codes and a computer. For such accurate calculations, the rods are homogenized only over those regions of the reactor where the rods are located. Also, because the absorption cross-section increases rapidly with decreasing energy, it is necessary to take into account the dip in the group fluxes inside the rods in the lower energy groups.

Partially Inserted Rods

Up to this point, it has been assumed that the rods are either fully inserted or fully withdrawn from the reactor core. In practice, some or all of the rods are inserted at the time of startup of a reactor and are slowly withdrawn to keep the reactor critical as the fuel is consumed and as fission product poisons accumulate. So it is necessary to know the worth of the rods as a function of their distance of insertion. This must also be known for rods that are more quickly inserted or withdrawn for the purpose of changing the power level.

Let $\rho_\omega(x)$ be the worth of one or more rods that are inserted the distance x, parallel to the axis of a cylindrical reactor core of total height H, and let $\rho_\omega(H)$ be the worth of these rods when fully inserted. It has been found from experiments and can be also shown on theoretical grounds that $\rho_\omega(x)$ and $\rho_\omega(H)$ are approximately related by the formula

$$\rho_\omega(x) = \rho_\omega(H)\left[\frac{x}{H} - \frac{1}{2\pi}\sin\left(\frac{2\pi x}{H}\right)\right]. \tag{7.63}$$

This function $\rho_\omega(x)$ is shown in Fig. 7.11, where it is observed that $\rho_\omega(0) = 0$ while $x = H$ and $\rho_\omega(x) = \rho_\omega(H)$.

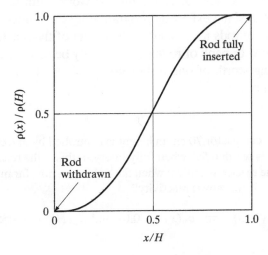

Figure 7.11 The worth of a partially inserted control rod as a function of the distance of insertion.

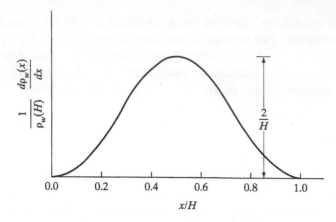

Figure 7.12 Fractional rate of change of rod worth as a function of the fraction of rod insertion.

It is of some interest to compute the derivative of $\rho_\omega(x)/\rho_\omega(H)$:

$$\frac{1}{\rho_\omega(H)}\frac{d\rho_\omega(x)}{dx} = \frac{1}{H}\left[1 - \cos\left(\frac{2\pi x}{H}\right)\right].   \tag{7.64}$$

This expression, which gives the fractional rate of change in reactivity per unit distance of movement of the rods, is plotted in Fig. 7.12. As shown in the figure, the reactivity changes very little per unit distance when the rod is first introduced into the reactor and also just as it is fully inserted. Physically, this is because, in these two instances, the end of the rod is moving through regions where the flux is very small and so few additional neutrons are absorbed in the rod as a result of its motion. However, in the region near the center of the core, the change in $\rho_\omega(x)$ is the greatest because the flux is also greatest there.

In many reactors, the problem of changing rod worth with distance of rod insertion is circumvented to some extent by programming the motions of the rods so that, as one rod or bank of rods moves out of the center of the reactor where the flux is highest, a second row or bank of rods automatically begins to be withdrawn. In this way, the decreasing worth of one rod or group of rods is compensated by the increasing worth of another.

Example 7.11

A bare square cylindrical reactor 70 cm in height is controlled by one control rod located on the axis that is worth 6.5% when fully inserted (this is the reactor described in Example 7.7). If the reactor is critical when the rod is out, how far must the rod be inserted to introduce 1% (negative) reactivity?

Solution. Using Eq. (7.63), with $\rho_\omega(x) = 0.01$ and $\rho_\omega(H) = 0.065$, and letting $y = 2\pi x/H$, gives

$$0.01 = 0.065 \left[\frac{y}{2\pi} - \frac{1}{2\pi} \sin y \right]$$

or

$$0.966 = y - \sin y.$$

This is a transcendental equation that can be solved either graphically or numerically. The solution is $y = 1.91$. Then, with $H = 70$ cm, the final result is

$$x = \frac{yH}{2\pi} = \frac{1.91 \times 70}{2\pi} = 21.3 \text{ cm. } [Ans.]$$

Chemical Shim

It was pointed out in the introduction to this chapter that certain water-moderated and water-cooled reactors can be controlled in part by varying the concentration of boric acid (H_3BO_3) in the water.[6] Such chemical shim control cannot be used alone to control a reactor since the process of changing H_3BO_3 concentration, although done remotely and automatically at the behest of the reactor operator, cannot be made to respond as quickly as control rods to sudden control requirements. Therefore, chemical shim is always used in conjunction with and as a supplement to mechanical control rods. When these two types of control mechanisms are present in a reactor, the control rods provide reactivity control for fast shutdown and for compensating for reactivity changes due to temperature changes[7] that accompany changes in power. The chemical shim is used to keep the reactor critical during xenon transients[8] and to compensate for depletion of fuel and buildup of fission products over the life of the core.

There are a number of reasons for using chemical shim. For one thing, it substantially reduces the number of control rods required in a reactor. Since control rods and their drive mechanisms are expensive, this results in substantial cost savings. Furthermore, because the H_3BO_3 is more or less uniformly distributed throughout the reactor core, changes in the reactivity of the system resulting from changes in the concentration of the H_3BO_3 can be made without disturbing the power distribution in the core.

Chemical shim affects the reactivity of a thermal reactor primarily by changing the value of the thermal utilization. The boron has comparatively little effect on the transport properties of the core, especially in a reactor having a fairly uniform flux distribution. The worth of a given concentration of boric acid can therefore be computed from Eq. (7.62), which was derived from uniformly distributed boron in

---

[6] The solubility of H_3BO_3 is 63.5 g/liter at room temperature, 267 g/liter at 100°C.
[7] This effect is discussed in Section 7.4.
[8] These are discussed in Section 7.5.

a fast reactor. For a thermal reactor, this equation takes the form

$$\rho_\omega = \frac{\overline{\Sigma}_{aB}}{\overline{\Sigma}_{aF} + \overline{\Sigma}_{aM}}, \tag{7.65}$$

where $\overline{\Sigma}_{aB}$, $\overline{\Sigma}_{aF}$, and $\overline{\Sigma}_{aM}$ are, respectively, the macroscopic thermal average cross-sections of the boron, fuel, and moderator (including coolant and structural material).

Dividing the numerator and denominator of Eq. (7.65) by $\overline{\Sigma}_{aM}$ gives

$$\rho_\omega = \frac{\overline{\Sigma}_{aB}/\overline{\Sigma}_{aM}}{\overline{\Sigma}_{aF}/\overline{\Sigma}_{aM} + 1}. \tag{7.66}$$

It is easily shown that the term $\overline{\Sigma}_{aF}/\overline{\Sigma}_{aM}$ is equal to $f_0/(1 - f_0)$, where f_0 is the thermal utilization in the absence of the boron. Substituting this result into Eq. (7.66) then gives

$$\rho_\omega = (1 - f_0)\frac{\overline{\Sigma}_{aB}}{\overline{\Sigma}_{aM}}. \tag{7.67}$$

The concentration of H_3BO_3 is usually specified in parts per million (ppm) of water—one ppm meaning one gram of boron per 10^6 grams of H_2O. If C is the concentration in ppm, then the ratio of the mass of boron to the mass of water is

$$\frac{m_B}{m_W} = C \times 10^{-6}.$$

The atomic weight of boron is 10.8 and the molecular weight of water is 18.0. The ratio of the atom density of boron to the molecular density of water is therefore

$$\frac{N_B}{N_W} = \frac{18.0}{10.8} \times C \times 10^{-6},$$

and the term $\overline{\Sigma}_{aB}/\overline{\Sigma}_{aW}$ in Eq. (7.67) becomes

$$\frac{\overline{\Sigma}_{aB}}{\overline{\Sigma}_{aM}} = \frac{N_B\overline{\sigma}_{aB}}{N_W\overline{\sigma}_{aW}} = \frac{18.0 \times 759}{10.8 \times 0.66} \times C \times 10^{-6} = 1.92C \times 10^{-3}.$$

Introducing this expression into Eq. (7.67) gives the final formula for shim worth,

$$\rho_\omega = 1.92C \times 10^{-3}(1 - f_0). \tag{7.68}$$

Example 7.12

The thermal utilization of a certain rod-and-shim controlled PWR is 0.930 at startup (clean and cold). At this time, the total excess reactivity in the reactor is 20.5%. If the control rods are worth 8.5%, what is the minimum H_3BO_3 concentration in ppm and g/liter of unit-density water that is necessary to *hold down* the reactor—that is, to keep it from becoming critical?

Solution. The shim must be worth $20.5 - 8.5 = 12\%$. Then using Eq. (7.68),

$$C = \frac{\rho_\omega \times 10^3}{1.92(1 - f_0)} = \frac{0.12 \times 10^3}{1.92(1 - 0.930)} = 893 \text{ ppm. } [Ans.]$$

The molecular weight of H_3BO_3 is 61.8. Therefore, the shim system must contain $(61.8/10.8) \times 893 = 5110$ g of H_3BO_3 per 10^6 g of water. At unit density, 10^6 g of water occupies 10^3 liters. The concentration of the boric acid is thus 5,110 g per 10^3 liters $= 5.11$ g/liter. [*Ans.*]

It should be noted that, in this example, the control rods alone do not have sufficient reactivity to keep the reactor from going critical at the beginning of core life. This is actually the case in many PWRs.

7.4 TEMPERATURE EFFECTS ON REACTIVITY

As noted at the beginning of this chapter, many of the parameters that determine the multiplication factor of a reactor depend on temperature. As a result, a change in the temperature leads to a change in k and alters the reactivity of the system. It is now shown that this effect has an important bearing on the operation of a reactor and, ultimately, the safety of the system.

Temperature Coefficients

The extent to which the reactivity is affected by changes in temperature is described in terms of the *temperature coefficient of reactivity*, denoted as α_T. This is defined by the relation

$$\alpha_T = \frac{d\rho}{dT}, \tag{7.69}$$

where ρ is the reactivity and T is the temperature. Introducing the definition of ρ from Eq. (7.26)—namely,

$$\rho = \frac{k - 1}{k} = 1 - \frac{1}{k}$$

and differentiating gives

$$\alpha_T = \frac{1}{k^2} \frac{dk}{dT}. \tag{7.70}$$

In all cases of interest, k is close to unity and so Eq. (7.70) can be written approximately as

$$\alpha_T = \frac{1}{k}\frac{dk}{dT}.\tag{7.71}$$

This equation is more convenient than Eq. (7.69) for calculational purposes and is often taken as the definition of α_T. According to Eq. (7.71), α_T is equal to the fractional change in k per unit change in temperature and has units of (degrees)$^{-1}$.

The response of a reactor to a change in temperature depends on the algebraic sign of α_T. Consider first the situation when α_T is positive. Since k is always positive, dk/dT is also positive, which means that an increase in T leads to an increase in k. Now suppose that for some reason the temperature of the reactor increases. This increases the value of k, which, in turn, leads to an increase in the power level of the reactor, giving rise to a further increase in the temperature, another increase in k, and so on. Thus, with α_T positive, an increase in temperature leads to ever-increasing temperature and power until the reactor is either shut down by outside intervention or it melts down.

However, suppose the reactor temperature were decreased. If α_T is positive, a decrease in T leads to a decrease in k. This reduces the reactor power, which reduces the temperature, giving a further decrease in k, and so on, until the reactor eventually shuts down. Thus, if $\alpha_T > 0$, an increase in T leads to meltdown, a decrease in T to shutdown—in the absence of external intervention.

The situation is quite different when α_T is negative. In this case, dk/dT is negative, and an increase in T gives a decrease in k. Now an increase in reactor temperature leads to a decrease in power, which tends to decrease the temperature and return the reactor to its original state. Furthermore, a decrease in T results in an increase in k, so that if T goes down the power goes up and again the reactor tends to return to its original state. Clearly, a reactor with positive α_T is inherently *unstable* to changes in its temperature, whereas a reactor having a negative α_T is inherently *stable*.

It must be recognized that the temperature ordinarily does not change uniformly throughout a reactor. For instance, an increase in reactor power, is reflected first by a rise in the temperature of the fuel since this is the region where the power is generated. In a thermal reactor, the coolant and moderator temperatures do not increase until heat has been transferred from the fuel to these regions. It is necessary in discussing temperature coefficients to specify the component whose temperature is undergoing a change. Thus, the *fuel temperature coefficient* is defined as the fractional change in k per unit change in fuel temperature, the *moderator temperature coefficient* is the fractional change in k per unit change in moderator temperature, and so on.

Because the temperature of the fuel reacts immediately to changes in reactor power, the fuel temperature coefficient is also called the *prompt temperature coefficient*, denoted as α_{prompt}. The value of α_{prompt} determines the first response of a reactor to changes in either fuel temperature or reactor power. For this reason, α_{prompt} is the most important temperature coefficient insofar as reactor safety is concerned. In the United States, the Nuclear Regulatory Commission (NRC) will not license a reactor unless α_{prompt} is negative; all licensed reactors[9] are thereby assured of being inherently stable.

Nuclear Doppler Effect

The prompt temperature coefficient of most reactors is negative owing to a phenomenon known as the *nuclear Doppler effect*. This effect, which admittedly is somewhat complicated, can be explained in the following way. For the moment, the discussion is restricted to thermal reactors.

It is recalled from Chapter 3 that neutron cross-sections exhibit resonances at certain energies and that in the heavier nuclei these resonances are due almost entirely to absorption, not scattering. Throughout that chapter, it is tacitly assumed that the nuclei with which the neutrons interact are at rest in the laboratory reference frame. In particular, the Breit–Wigner formula,

$$\sigma_\gamma(E) = \frac{\lambda_r^2 g}{4\pi} \frac{\Gamma_n \Gamma_\gamma}{(E - E_r)^2 + \Gamma^2/4},$$

describing a resonance of width Γ at the energy E_r, is based on this assumption.

In fact, however, nuclei are located in atoms that are in continual motion owing to their thermal energy. As a result of these thermal motions, even a beam of strictly monoenergetic neutrons impinging on a target appears to the nuclei in the target to have a continuous spread in energy. This, in turn, has an effect on the observed shape of a resonance. Thus, it can be shown that when the cross-section is averaged over the motions of the nuclei, the resonance becomes shorter and wider than when the nuclei are at rest. Furthermore, the effect becomes more pronounced as the temperature of the target goes up, as indicated in Fig. 7.13. The change in shape of the resonance with temperature is called *Doppler broadening*. In Fig. 7.13, the unbroadened resonance for 0°K was computed from the Breit–Wigner formula, since at this temperature there is no thermal motion. An important aspect of the Doppler effect is that, although the shape of a resonance changes with temperature, the total area under the resonance remains essentially constant.

---

[9]All reactors in the United States must be licensed by the Nuclear Regulatory Commission (NRC) except those owned by the Department of Defense, Department of Energy, or the NRC itself.

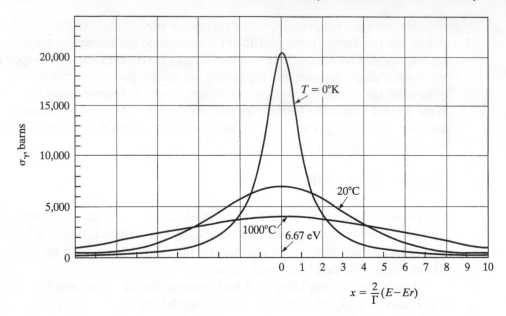

Figure 7.13 Doppler broadening of the capture cross-section of ^{238}U at 6.67-eV resonance.

Consider now neutrons slowing down in a thermal reactor past a particular resonance. The number of neutrons absorbed per cm^3/sec in this resonance is given by

$$F_a = \int \phi(E) \Sigma_a(E) dE, \qquad (7.72)$$

where $\phi(E)$ is the energy-dependent flux, $\Sigma_a(E)$ is the macroscopic absorption cross-section, and the integration is carried out over the resonance. Equation (7.72) can also be written as

$$F_a = \phi_{av} \int \Sigma_a(E) dE,$$

in which ϕ_{av} is the average value of the flux in the resonance. The remaining integral is simply the total area under the resonance and, as just noted, this is a constant, independent of temperature. Therefore, it follows that the number of neutrons absorbed in the resonance is proportional to the average flux in the resonance.

It is not difficult to see that the value of ϕ_{av} depends on temperature. Thus, suppose the temperature increases. Then, as explained earlier, the magnitude of the absorption cross-section decreases due to the Doppler effect, and this, in turn, causes ϕ_{av} to increase. The situation is analogous to the placing of a neutron absorber, such as a control rod or an absorbing foil, in a medium in which neutrons are diffusing. If the cross-section of the absorber is increased, the neutron flux in its vicinity goes down; but if the cross-section is reduced, the flux obviously goes up.

Finally, since ϕ_{av} increases with temperature and since the number of neutrons absorbed in the resonance is proportional to ϕ_{av}, it must be concluded that *resonance absorption increases with increasing temperature*. This, of course, decreases k and accounts for the negative value of the prompt temperature coefficient.

To calculate α_{prompt} for a thermal reactor, the multiplication factor may be written as

$$k = k_\infty P = \eta_T f p \epsilon P, \tag{7.73}$$

where P is the overall nonleakage probability (see Section 6.5). Next, the resonance escape probability, p, is singled out by taking logarithms:

$$\ln k = \ln \eta_T f \epsilon P + \ln p.$$

Differentiating with respect to the temperature and holding all parameters constant except p then gives,

$$\frac{d}{dT}(\ln k) = \frac{1}{k}\frac{dk}{dT}(\ln p) = \frac{1}{p}\frac{dp}{dT}$$

or

$$\alpha_{prompt} = \alpha_{prompt}(p). \tag{7.74}$$

From Eq. (6.119), p is given by the formula

$$p = \exp\left[-\frac{N_F V_F I}{\zeta_M \Sigma_{sM} V_M}\right], \tag{7.75}$$

where I is the resonance integral, V_F and V_M are the volumes in a unit cell of fuel and moderator, respectively, N_F is the atom density of the fuel, Σ_{sM} is the macroscopic scattering cross-section of the moderator, and ζ_M is a constant.[10] The prompt coefficient of p can be computed by assuming that the moderator temperature remains constant while the temperature of the fuel changes. Then both σ_{sM} and V_M are constant. Furthermore, the product $N_F V_F$, which is equal to the total number of fuel atoms in a fuel rod, does not change with temperature. The entire temperature dependence of p is therefore contained in I.

Many measurements of resonance absorption have been carried out at different temperatures, and it has been found that I can be represented by the formula

---

[10]Values of $\zeta_M \Sigma_{sM}$ are given in Table 6.6.

TABLE 7.4   CONSTANTS FOR COMPUTING β_i

Fuel	$A' \times 10^4$	$C' \times 10^2$
^{238}U (metal)	48	1.28
^{238}UO$_2$	61	0.94
Th (metal)	85	2.68
ThO$_2$	97	2.40

$$I(T) = I(300\text{K})[1 + \beta_I(\sqrt{T} - \sqrt{300\text{K}})], \tag{7.76}$$

where the temperature is in Kelvin. The temperature dependence of I is due to the Doppler effect as explained earlier. The parameter β_I is a function of the properties of the fuel and is given approximately by

$$\beta_I = A' + C'/a\rho. \tag{7.77}$$

Here, A' and C' are constants given in Table 7.4, a is the rod radius in cm, and ρ is the fuel density in g/cm^3.

Returning to Eq. (7.75) and differentiating $\ln p$ gives

$$\alpha_{\text{prompt}}(p) = \alpha_{\text{prompt}} = \frac{N_F V_F}{\zeta_M \Sigma_{sM} V_M} \frac{dI}{dT}.$$

From Eq. (7.76), however,

$$\frac{dI}{dT} = \frac{I(300\text{K})\beta_I}{2\sqrt{T}},$$

and so

$$\alpha_{\text{prompt}} = \frac{N_F V_F I(300\text{K})}{\zeta_M \Sigma_{sM} V_M} \frac{\beta_I}{2\sqrt{T}}.$$

This may also be written as

$$\alpha_{\text{prompt}} = \frac{\beta_I}{2\sqrt{T}} \ln\left[\frac{1}{p(300\text{K})}\right]. \tag{7.78}$$

Equation (7.78) gives the prompt temperature coefficient of a thermal reactor.

Example 7.13

The resonance escape probability for a certain natural uranium ($\rho = 19.1$ g/cm^3)-fueled reactor is 0.878 at 300K. The fuel rods are 2.8 cm in diameter. Compute the prompt temperature coefficient of the reactor at an operating temperature of 350°C.

Solution. The rod radius is $2.8/2 = 1.4$ cm. From Eq. (7.77) and Table 7.4, β_I is then,

$$\beta_I = 48 \times 10^{-4} + 1.28 \times 10^{-2}/1.4 \times 19.1$$
$$= 53 \times 10^{-4}.$$

The temperature in degrees Kelvin is

$$T = 273 + 350 = 623 \text{K}.$$

Equation (7.78) then gives

$$\alpha_{\text{prompt}} = -\frac{53 \times 10^{-4}}{2\sqrt{623}} \ln\left(\frac{1}{0.878}\right)$$
$$= -1.39 \times 10^{-5} \text{ per} °C \text{ or K.}[11]$$

The prompt temperature coefficient due to the Doppler effect is more difficult to determine for a fast reactor. This is because, for one thing, at the elevated energies of interest in this type of reactor, the resonances are so close together in energy that they overlap, and their properties are less well known. In addition, the amount of resonance absorption depends in detail on the neutron spectrum, and this varies from reactor to reactor. Nevertheless, methods have been developed for computing this important temperature coefficient for fast reactors.

One additional aspect of this problem should be mentioned. In a fast reactor, fissions as well as radiative capture occur following the absorption of neutrons in high-energy, overlapping resonances. Thus, in a fast reactor with a fuel mixture of, say, ^{239}Pu and ^{238}U, an increase in temperature not only increases the parasitic capture of neutrons by the ^{238}U, but also tends to increase the fission rate of the ^{239}Pu. The Doppler effect in this case is, so to speak, a mixed blessing. Therefore, it is necessary to ensure the safe operation of the reactor to include enough ^{238}U in the system so that, on balance, the Doppler effect provides a negative value of α_{prompt}. This is not a serious restriction, however, since substantial amounts of ^{238}U are normally present in this type of reactor to maximize the breeding gain.

The Moderator Coefficient

Although of less immediate importance than α_{prompt}, the moderator temperature coefficient, α_{mod}, determines the ultimate behavior of a reactor in response to changes in fuel temperature. It also determines the effect on the reactor of changes in the temperature of the coolant entering the system. It is desirable for α_{mod} to be negative to ensure stability during normal operation and during accident conditions.

---

[11]A Celsius (centigrade) degree is the same size as a Kelvin degree.

In a thermal reactor, an increase in the moderator temperature affects the multiplication factor in two ways: (a) the temperature at which thermal cross-sections are computed is changed, and (b) the physical density of the moderator changes owing to thermal expansion. Of these effects, the second is ordinarily the more significant.

To compute α_{mod}, k is first written as in Eq. (7.73):

$$k = \eta_T f p \epsilon P.$$

The value of η_T is not a function of moderator temperature. Although ϵ does depend to some extent on the moderator density, it can also be taken to be independent of T. Taking logarithms of both sides of Eq. (7.73) and differentiating with respect to T gives,

$$\alpha_{\mathrm{mod}} = \alpha_T(f) + \alpha_T(p) + \alpha_T(P). \tag{7.79}$$

The moderator coefficient is therefore the sum of three other temperature coefficients.

Consider first $\alpha_T(f)$. It is recalled that f is equal to the probability that a thermal neutron eventually is absorbed in fuel, and is given by the ratio of the macroscopic cross-section of the fuel to that of the fuel, moderator, and coolant taken together. If the fuel and moderator are both solids and the reactor is cooled by a weakly absorbing gas, then a change in moderator temperature does not alter the total number of absorbers in the system and f remains more or less constant. In this case, $\alpha_T(f)$ is essentially zero.

The situation is quite different if, as in pressurized-water reactors, the fuel is a solid, but the coolant and/or moderator are liquids. Now because the coefficients of expansion for liquids are greater than for solids, and since the liquid in a reactor is always held in a metallic vessel of some sort, an increase in the moderator temperature results in the expulsion of some liquid from the reactor. The amount of parasitic absorption of thermal neutrons in the moderator and coolant is thereby reduced relative to the absorption in the fuel, with the result that f increases. It may be concluded that, in this case, $\alpha_T(f)$ is positive. This effect is especially pronounced when a chemical shim is present because some of the boron is expelled along with the moderator coolant.

Consider next $\alpha_T(p)$. In Eq. (7.75) the only factor that depends on the moderator temperature is $\Sigma_{s\mathrm{M}} V_{\mathrm{M}}$, which is proportional to the total number of moderator-coolant atoms in the reactor. If the moderator is a solid, this number does not change with temperature, and $\alpha_T(p)$ is zero. However, with a liquid moderator, as explained earlier, an increase in temperature leads to the ejection of some of the moderator. This decreases the value of $\Sigma_{s\mathrm{M}} V_{\mathrm{M}}$ and leads to a decrease in p. Thus, for a liquid moderator coolant, $\alpha_T(p)$ is negative.

The temperature coefficient of the nonleakage probability is also negative. This may be seen from Eqs. (6.71) and (6.72), from which the overall nonleakage probability (which includes the leakage of both fast and thermal neutrons) is given by

$$P = \frac{1}{(1 + B^2 L_T^2)(1 + B^2 \tau_T)},$$ (7.80)

where B^2 is the buckling, L_T^2 is the thermal diffusion area, and τ_T is the age of fission neutrons to thermal. The value of B^2 is determined by the dimensions of the system, which do not change significantly when the moderator temperature changes. From the definition of $L_T^2 = \overline{D}/\overline{\Sigma}_a$ (see Eq. 5.62) and $\tau_T = D_1/\Sigma_1$ (see Eq. 5.69), it is easy to see that both of these parameters vary inversely with the square of the physical density ρ_d[12] of the medium provided the medium is uniform and homogeneous. For a solid fuel and solid moderator, the values of L_T^2 and τ_T depend in a complex way on the densities of these two media. However, a change in temperature of a solid moderator has little effect on its density; P is therefore constant and $\alpha_T(P) = 0$. With solid fuel and liquid moderator coolant, L_T^2 and τ_T vary as ρ_d^{-n}, where n is a number less than 2. In this case,

$$P = \frac{1}{\left(1 + \dfrac{C_1}{\rho_d^n}\right)\left(1 + \dfrac{C_2}{\rho_d^n}\right)},$$ (7.81)

where C_1 and C_2 are constants. Now if T increases, ρ_d decreases because of the expulsion of liquid, and P goes down. It follows that $\alpha_T(P)$ is negative.

A negative value of $\alpha_T(P)$ means that the leakage of neutrons increases with increasing moderator temperature. This, in turn, is simply due to the fact that, other things being equal, neutrons diffuse more readily through less dense media. However, there is another way to view these conclusions. Clearly a fast neutron that does not escape from a reactor core (and is not absorbed in a resonance) must slow down in the core. When the moderator density decreases and faster neutrons escape, the number of neutrons moderated to thermal energies also decreases. Thus, a decrease in moderator density leads to (a) a decrease in thermal neutron absorption in the moderator—a positive effect on k; and (b) a decrease in the moderation of fast neutrons—a negative effect on k. Whether α_{mod} is positive or negative depends (aside from the effect on the resonance escape probability, which is always negative) on whether a given moderator "absorbs more than it moderates" or "moderates more than it absorbs." In the important case of water, it can be shown that this material moderates more than it absorbs, so that α_{mod} is negative. However,

---

[12]The symbol ρ_d is used in the present discussion for density rather than ρ to avoid confusion with the reactivity.

this situation can be reversed if there is too much boron present in the water as a chemical shim leading to a positive α_{mod}.[13]

The negative moderator coefficient has an important effect on the control of a pressurized-water reactor following changes in the demand for power from the turbine. For instance, suppose that the throttle (control valve) on the turbine is closed somewhat in response to a reduction in load. This ultimately leads to an increase in the temperature of the coolant water entering the reactor since less energy is being extracted from the steam passing through the turbine. Because α_{mod} is negative, this leads, in turn, to a decrease in reactor power. However, had the load been increased, the effect would have been the reverse, and the reactor power would have increased. In either case, the reactor automatically responds to changes in load in the desired manner. This behavior of a reactor is called *automatic load following*.

The Void Coefficient

Consider a liquid-moderated and/or liquid-cooled reactor. At a point where the liquid boils, the volume occupied by the vapor is essentially a *void* since the density of the vapor is so much less than that of the liquid. The void fraction, x, is defined as the fraction of a given volume occupied by voids. For instance, if 25% of a volume is occupied by vapor and the remainder is liquid, then $x = 0.25$.

Changing the void fraction in a reactor normally affects the reactivity of the system. This phenomenon is described in terms of the *void coefficient*, α_V, which is defined as,

$$\alpha_V = \frac{d\rho}{dx},$$

(7.82)

where ρ is the reactivity and x is the void fraction. Clearly, α_V is a dimensionless number.

In view of the previous definition, it follows that, if α_V is positive, an increase in the void fraction leads to an increase in reactivity. This is obviously not a desirable state of affairs since an increase in reactivity gives an increase in power, which would gives rise to additional boiling, more voids, a further increase in reactivity, and so on, until much of the liquid is boiled away and the core melts down. By contrast, with α_V negative, an increase in x gives a decrease in reactivity and in power, and the reactor tends to be restored to its initial condition. Therefore, it is clearly desirable for α_V to be negative.

The presence of voids alters the average density of the moderator or coolant. In particular, if the void fraction is x, the average density of the mixture of liquid

---

[13]Some reactors operate very early in life with a small positive temperature coefficient due to the need for high boron concentration with the new fuel designs.

and vapor is given by

$$\rho_d = (1 - x)\rho_l + x\rho_g, \tag{7.83}$$

where ρ_l and ρ_g are the densities of the liquid and vapor, respectively. However, ρ_g is usually much less than ρ_l (except near the critical point of the substance in question), and so Eq. (7.83) can be written as

$$\rho_d = (1 - x)\rho_l. \tag{7.84}$$

As would have been expected, ρ_d decreases as x increases.

Since changing x changes the average density of the moderator or coolant, the value of α_V is related, in the case of thermal reactors, to the value of α_{mod}. It is recalled that, with a moderator coolant such as water, which moderates more than it absorbs, a decrease in density has a negative effect on the reactivity. It follows that this is also the case with the void coefficient. With water reactors, provided the thermal absorption cross-section is not made too high by the addition of a chemical shim, an increase in void fraction decreases the reactivity and α_V is negative.

The negative void coefficient is used to provide boiling-water reactors with automatic load following. This is accomplished by coupling the load demand on the turbine to the recirculation flow system of the reactor (see Fig. 4.4). An increase in turbine load leads to increased recirculation flow, whereas a reduction in load results in decreased recirculation flow. In the case of increased turbine load, the increase in the recirculation flow through the reactor initially sweeps out a portion of the voids in the core. This reduction in void content increases the reactivity of the system because α_V is negative, and this causes the reactor to produce more power. Eventually, a larger void content is reestablished at a higher power level. In equilibrium, the core produces the exact power required by the turbine. Had the turbine load been decreased, the recirculation flow would have been reduced, the void content would have gone up, and the reactor power would have decreased, again following turbine demand.

The situation regarding the void coefficient of sodium-cooled fast reactors is more complex. Sodium does not absorb neutrons to any great extent, so that the expulsion of coolant from the reactor due to the formation of voids has only a negligible effect on the overall absorption of neutrons in the system. However, the sodium slows down neutrons by inelastic scattering at high energy and by elastic scattering at the lower end of the energy spectrum. The removal of sodium due to void formation reduces these moderating effects. As a result, the neutron spectrum in the reactor becomes harder—that is, more energetic. It is remembered from Chapter 4, however (see Fig. 4.2), that η, the average number of fission neutrons released per neutron absorbed in fuel, increases rapidly with incident neutron energy for all of the fissile nuclides at fast reactor energies. Thus, as a consequence of the formation of voids, the average value of η goes up and the reactivity increases. This tends to make α_V positive.

Besides hardening the neutron spectrum, an increase in the void fraction increases the neutron leakage because of the reduced density of the sodium. This tends to make α_V negative. Therefore, the sign of α_V depends on which effect is greater—the increase in η or the increase in neutron leakage.

In practice, the sign of α_V is a function of the location where a void forms in the reactor. Thus, if a void occurs near the center of a large reactor, the increase in leakage from this region is of little consequence since the neutrons are merely absorbed in another part of the system. In this region, α_V is positive. In contrast, if a void forms near the outside of the core, the additional neutrons that escape are lost from the core (at least some of them are), and in this case α_V is negative. Precisely how large sodium-cooled fast reactors can be built to provide a negative α_V in all parts of the system or how such reactors can be controlled with positive values of α_V in some regions has not been settled as of this writing since so few fast reactors have been built.

7.5 FISSION PRODUCT POISONING

All fission products absorb neutrons to some extent, and their accumulation in a reactor tends to reduce its multiplication factor. Since absorption cross-sections decrease rapidly with increasing neutron energy, such fission product poisons are of the greatest importance in thermal reactors. This section considers only reactors of this type.

To a good approximation, the only effect of fission product poisons on the multiplication factor is on thermal utilization. Thus, the reactivity equivalent of poisons in a previously critical reactor can be written as,

$$\rho = \frac{k - k_0}{k} = \frac{f - f_0}{f}, \tag{7.85}$$

where the parameters without a subscript refer to the poisoned reactor. In the absence of poisons, f_0 is given by

$$f_0 = \frac{\overline{\Sigma}_{aF}}{\overline{\Sigma}_{aF} + \overline{\Sigma}_{aM}},$$

where $\overline{\Sigma}_{aF}$ and $\overline{\Sigma}_{aM}$ are the macroscopic thermal absorption cross-sections of the fuel and everything else except the fuel, respectively. With poisons present, f becomes,

$$f = \frac{\overline{\Sigma}_{aF}}{\overline{\Sigma}_{aF} + \overline{\Sigma}_{aM} + \overline{\Sigma}_{aP}},$$

where $\overline{\Sigma}_{aP}$ is the macroscopic cross-section of the poison. From Eq. (7.85), the reactivity due to the poisons is then,

$$\rho = \frac{f - f_0}{f} = -\frac{\overline{\Sigma}_{aP}}{\overline{\Sigma}_{aF} + \overline{\Sigma}_{aM}}. \tag{7.86}$$

Not surprisingly, Eq. (7.86) is essentially the same (except for sign) as the expression for the worth of a chemical shim, which is a uniformly distributed poison.

Equation (7.86) can be put in a more convenient form by writing the multiplication factor of the unpoisoned reactor as[14]

$$k = 1 = \eta_T f p \epsilon = \frac{\eta_T p \epsilon \overline{\Sigma}_{aF}}{\overline{\Sigma}_{aF} + \overline{\Sigma}_{aM}} \tag{7.87}$$

$$= \frac{v p \epsilon \overline{\Sigma}_f}{\overline{\Sigma}_{aF} + \overline{\Sigma}_{aM}}, \tag{7.88}$$

where $\overline{\Sigma}_f$ is the macroscopic fission cross-section. Solving Eq. (7.88) for $\overline{\Sigma}_{aF} + \overline{\Sigma}_{aM}$ and inserting this into Eq. (7.86) gives

$$\rho = -\frac{\overline{\Sigma}_{aP}/\overline{\Sigma}_f}{v p \epsilon}. \tag{7.89}$$

Equation (7.89) is in a form suitable for calculation of fission product poisoning.

Xenon-135

The most important fission product poison is ^{135}Xe, whose thermal (2,200 m/sec) absorption cross-section is 2.65×10^6 b and is non-1/v (the non-1/v factor is given in Table 3.2). This isotope is formed as the result of the decay of ^{135}I and is also produced directly in fission. The ^{135}I is formed in fission and by the decay of ^{135}Te (tellurium-135). These processes and their half-lives are summarized next:

$$^{135}\text{Te} \xrightarrow[11\text{sec}]{\beta-} {}^{135}\text{I} \xrightarrow[6.7\text{hr}]{\beta-} {}^{135}\text{Xe} \xrightarrow[9.2\text{hr}]{\beta-} {}^{135}\text{Cs} \xrightarrow[2.3 \times 10^6\text{yr}]{\beta-} {}^{135}\text{Ba(stable)}$$

$$\uparrow \qquad\qquad \uparrow \qquad\qquad \uparrow$$

$$\text{Fission} \qquad \text{Fission} \qquad \text{Fission}$$

---

[14]Leakage of neutrons from the reactor core is ignored in the present treatment. Such leakage has little effect on fission product poisoning.

TABLE 7.5   FISSION PRODUCT YIELDS (ATOMS PER FISSION) FROM THERMAL FISSION*

Isotope	^{233}U	^{235}U	^{239}Pu
^{135}I	0.0475	0.0639	0.0604
^{135}Xe	0.0107	0.00237	0.0105
^{149}Pm	0.00795	0.01071	0.0121

*From M. E. Meek and B. F. Rider, "Compilation of Fission Product Yields," General Electric Company Report NEDO-12154, 1972.

Because ^{135}Te decays so rapidly to ^{135}I, it is possible to assume that all ^{135}I is produced directly in fission. The effective yields of ^{135}I and ^{135}Xe for the three fissile nuclei are given in Table 7.5; the decay constants of these isotopes are shown in Table 7.6.

Because xenon is produced in part by the decay of iodine, the xenon concentration at any time depends on the iodine concentration. This, in turn, is determined by the rate equation

$$\frac{dI}{dt} = \gamma_I \overline{\Sigma}_f \phi_T - \lambda_I I, \tag{7.90}$$

where I is the number of ^{135}I atoms/cm^3, γ_I is the effective yield (atoms per fission) of this isotope, and $\overline{\Sigma}_f$ is the thermal fission cross-section.

One atom of ^{135}Xe is formed by the decay of each atom of ^{135}I so that the total rate of formation of ^{135}Xe is $\lambda_I I + \gamma_X \overline{\Sigma}_f \phi_T$, where γ_X is the fission yield of the xenon. The ^{135}Xe disappears as the result of its natural radioactive decay and because of neutron absorption. Therefore, the xenon rate equation is

$$\frac{dX}{dt} = \lambda_I I + \gamma_X \overline{\Sigma}_f \phi_T - \lambda_X X - \overline{\sigma}_{aX} \phi_T X, \tag{7.91}$$

where X is the ^{135}Xe concentration in atoms/cm^3 and $\overline{\sigma}_{aX}$ is the thermal absorption cross-section of ^{135}Xe. From Eq. (7.91), it may be noted that the effective decay

TABLE 7.6   DECAY CONSTANTS FOR FISSION PRODUCT POISONING CALCULATIONS

Isotope	λ, sec^{-1}	λ, hr^{-1}
^{135}I	2.87×10^{-5}	0.1035
^{135}Xe	2.09×10^{-5}	0.0753
^{149}Pm	3.63×10^{-6}	0.0131

constant of ^{135}Xe is $\lambda_X + \overline{\sigma}_{aX}\phi_T$ and its effective half-life is

$$(T_{1/2})_{\text{eff}} = \frac{0.693}{\lambda_X + \overline{\sigma}_{aX}\phi_T}.$$

In an operating reactor, the quantities $\overline{\Sigma}_f$ and ϕ_T may be functions of time, and the solutions to Eqs. (7.90) and (7.91) depend on the nature of these functions. However, several special solutions to these equations are now considered.

Equilibrium Xenon

Because the half-lives of ^{135}Xe and ^{135}I are so short and the absorption cross-section of xenon is so large, the concentrations of these isotopes, in all reactors except those operating at very low flux, quickly rise to their saturation or equilibrium values, I_∞ and X_∞. These concentrations can be found by placing the time derivatives in Eqs. (7.90) and (7.91) equal to zero. Thus, from Eq. (7.90),

$$I_\infty = \frac{\gamma_I \overline{\Sigma}_f \phi_T}{\lambda_I}, \tag{7.92}$$

and from Eq. (7.91),

$$X_\infty = \frac{\lambda_I I_\infty + \gamma_X \overline{\Sigma}_f \phi_T}{\lambda_X + \overline{\sigma}_{aX}\phi_T} \tag{7.93}$$

$$= \frac{(\gamma_I + \gamma_X)\overline{\Sigma}_f \phi_T}{\lambda_X + \overline{\sigma}_{aX}\phi_T}. \tag{7.94}$$

The macroscopic absorption cross-section of the xenon is then

$$\overline{\Sigma}_{aX} = X_\infty \overline{\sigma}_{aX} = \frac{(\gamma_I + \gamma_X)\overline{\Sigma}_f \phi_T \overline{\sigma}_{aX}}{\lambda_X + \overline{\sigma}_{aX}\phi_T}. \tag{7.95}$$

The form of Eq. (7.95) can be improved by dividing the numerator and denominator of the Right-Hand Side by $\overline{\sigma}_{aX}$. This gives

$$\overline{\Sigma}_{aX} = \frac{(\gamma_I + \gamma_X)\overline{\Sigma}_f \phi_T}{\phi_X + \phi_T}, \tag{7.96}$$

where ϕ_X is a temperature-dependent parameter having the dimensions of flux:

$$\phi_X = \frac{\lambda_X}{\overline{\sigma}_{aX}} = 0.770 \times 10^{13} \text{ cm}^{-2} \text{ sec}^{-1} \tag{7.97}$$

at $T = 20°$C. From Eq. (7.91), it can be seen that ϕ_X is the thermal flux at which the disappearance rates of ^{135}Xe by neutron absorption and natural decay are equal.

On inserting Eq. (7.96) into Eq. (7.89), the reactivity equivalent of equilibrium xenon is found to be

$$\rho = -\frac{\gamma_I + \gamma_X}{\nu p \epsilon} \frac{\phi_T}{\phi_X + \phi_T}. \tag{7.98}$$

There are now two situations to be considered. First, if $\phi_T \ll \phi_X$, Eq. (7.98) reduces to

$$\rho = -\frac{(\gamma_I + \gamma_X)\phi_T}{\nu p \epsilon \phi_X}, \tag{7.99}$$

and it is seen that, in this low-flux case, ρ increases linearly with ϕ_T. However, if $\phi_T \gg \phi_X$, ρ takes on its maximum value

$$\rho = -\frac{\gamma_I + \gamma_X}{\nu p \epsilon}. \tag{7.100}$$

To get some idea of the magnitude of the equilibrium xenon poisoning effect, suppose a reactor is fueled with ^{235}U and contains no resonance absorbers or fissionable material other than ^{235}U. In this case, $p = \epsilon = 1$, and Eq. (7.100) gives

$$\rho = -\frac{\gamma_I + \gamma_X}{\nu} = -\frac{0.066}{2.42} = -2.73\%,$$

or about four dollars. This is the maximum reactivity due to equilibrium xenon in a ^{235}U-fueled reactor. If the reactor contains resonance absorbers, the maximum reactivity is somewhat higher.

Xenon after Shutdown—Reactor Deadtime

Although the fission production of ^{135}Xe ceases when a reactor is shut down, this isotope continues to be produced as the result of the decay of the ^{135}I present in the system. The xenon concentration initially increases after shutdown, although it eventually disappears by its own decay.

If the iodine concentration at shutdown is I_0, its concentration at a time t later is given by,

$$I(t) = I_0 e^{-\lambda_I t}. \tag{7.101}$$

Inserting this function into Eq. (7.91) and noting that $\phi_T = 0$ after shutdown yields for the concentration of ^{135}Xe,

$$X(t) = X_0 e^{-\lambda_X t} + \frac{\lambda_I I_0}{\lambda_I - \lambda_X}(e^{-\lambda_X t} - e^{-\lambda_I t}). \tag{7.102}$$

If the xenon and iodine had reached equilibrium prior to shutdown, then I_0 and X_0 would be given by Eqs. (7.92) and (7.93), and the reactivity equivalent of the xenon becomes

$$\rho = -\frac{1}{\nu p \epsilon} \left[\frac{(\gamma_I + \gamma_X)\phi_T}{\phi_X + \phi_T} e^{-\lambda_X t} + \frac{\gamma_I \phi_T}{\phi_I - \phi_X}(e^{-\lambda_X t} - e^{-\lambda_I t}) \right]. \qquad (7.103)$$

Here, ϕ_I is the temperature-dependent parameter

$$\phi_I = \frac{\lambda_I}{\sigma_{aX}} = 1.055 \times 10^{13} \text{ cm}^{-2} \text{ sec}^{-1} \qquad (7.104)$$

at $T = 20°C$, and ϕ_T is evaluated, of course, prior to shutdown.

Figure 7.14 shows the (negative) reactivity due to xenon buildup after shutdown in a ^{235}U-fueled reactor for four values of the flux prior to shutdown. As

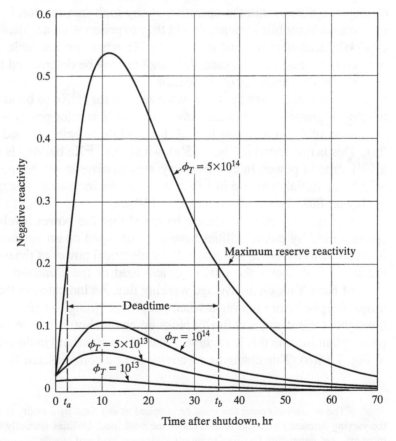

Figure 7.14   Xenon-135 buildup after shutdown for several values of the operating flux before shutdown.

shown in the figure, the reactivity rises to a maximum, which occurs at about 10 hours after shutdown, and then decreases to zero. It should be noted that the (negative) reactivity rise is greatest in reactors that have been operating at the highest flux before shutdown. This is true simply because the accumulated concentration of ^{135}I at shutdown is greatest in these reactors.

The postshutdown buildup of xenon is of little importance in low-flux reactors, but may be troublesome in reactors designed to operate at high flux. In particular, if at any time after the shutdown of a reactor the positive reactivity available by removing all control rods is less than the negative reactivity due to the xenon, the reactor cannot be restarted until the xenon has decayed. This situation is indicated in Fig. 7.14, where the horizontal line represents a hypothetical reserve reactivity of 20%. It is clear from the figure that, during the time interval from t_a to t_b, the reactor that previously operated at a flux of 5×10^{14} cannot be restarted. This period is known as the *reactor deadtime*. The existence of a deadtime can be of major significance in the operation of any high-flux reactor, but it is of greatest importance in mobile systems should they experience an accidental scram. This is especially true near the end of the core life when the available excess reactivity may be very small. In this case, the core life may be determined by the maximum xenon override capability of the reactor.

It is not necessary to shut down a reactor for ^{135}Xe to build up, although the buildup is greatest in this case. Any reduction in reactor power increases the concentration of ^{135}Xe because less of this nuclide is being burned up in the lower flux. This is illustrated in Fig. 7.15(a), where the ^{135}Xe buildup is shown following a 50% drop in power. In a similar way, any increase in reactor power is accompanied by an initial decrease in ^{135}Xe owing to the increased burnup of the ^{135}Xe in the higher flux. This is shown in Fig. 7.15(b).

In this connection, it should be noted that the power levels of some of the reactors used by public utilities have to be changed on an almost continuing basis in response to the changing demands for electrical power of the consuming public. Figure 7.16(a) shows the total electrical load of the Consolidated Edison Company of New York on an average working day. As indicated in the figure, the load drops sharply before midnight when people are going to bed, it continues to drop throughout the night, and finally picks up again shortly before dawn. If a nuclear power plant follows this load curve,[15] the xenon reactivity in the reactor is as shown in Fig. 7.16(b). This changing reactivity can be compensated for by adjusting the

---

[15]The minimum power that must be supplied at any time by a utility is called its *base load;* the varying demand above this level is called the *peak load*. Utilities normally use their newest and most efficient generating facilities to supply their base load and handle the variable peak load with their older and less efficient equipment. A reactor that is base loaded operates at full power for long periods of time with a more or less constant inventory of xenon.

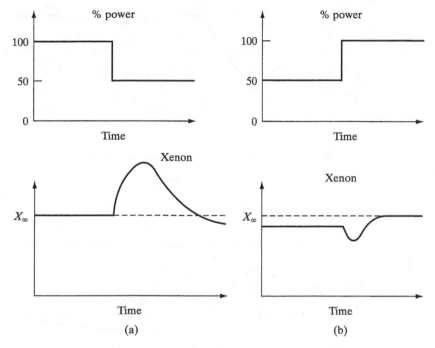

Figure 7.15   Xenon buildup and burnout following step changes in power: (a) from full to half power; (b) from half to full power.

control rods or, if the reactor has a chemical shim system, by varying the boron concentration.

Samarium-149

The thermal (2,200 m/sec) cross-section of ^{149}Sm is 41,000 b and is non-1/v (see Table 3.2; the average thermal cross-section at 20°C is 58,700 b). Although this isotope is of much less concern in a reactor than ^{135}Xe, it is often included separately in reactor calculations. ^{149}Sm is not formed directly in fission but appears as the result of the decay of ^{149}Nd (neodymium-149) as follows:

$$^{149}\text{Nd} \xrightarrow[1.7\text{hr}]{\beta-} {}^{149}\text{Pm} \xrightarrow[53\text{hr}]{\beta-} {}^{149}\text{Sm(stable)}$$

$$\uparrow$$
$$\text{Fission}$$

Because ^{149}Nd decays comparatively rapidly to ^{149}Pm (promethium-149), ^{149}Pm may be assumed to be produced directly in fission with the yield of γ_P atoms per fission (see Table 7.5). The concentration P of the promethium in atoms/cm^3

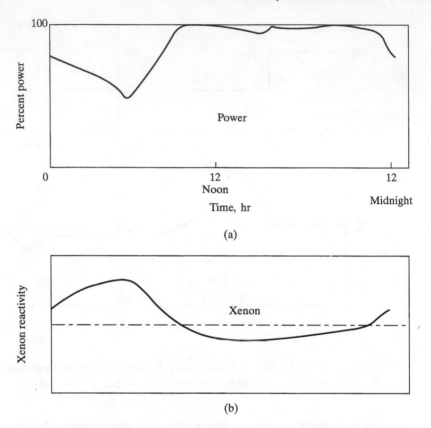

Figure 7.16 (a) Nominal weekday load of Consolidated Edison Company of New York; (b) the buildup and burnout of xenon in a reactor responding to this load.

is then determined by the equation

$$\frac{dP}{dt} = \gamma_P \overline{\Sigma}_f \phi_T - \lambda_P P, \tag{7.105}$$

where λ_P is given in Table 7.6. Since the samarium is stable, it disappears only as the result of neutron capture. The relevant equation is then

$$\frac{dS}{dt} = \lambda_P P - \overline{\Sigma}_{aS} \phi_T S, \tag{7.106}$$

where S is the atom density of ^{149}Sm, and $\overline{\Sigma}_{aS}$ is its average thermal absorption cross-section.

Equilibrium Samarium

Since the absorption cross-section of ^{149}Sm is much less than that of ^{135}Xe and the half-life of ^{149}Pm is longer than those of ^{135}I and ^{135}Xe, it follows that it takes somewhat longer for the promethium and samarium concentrations to reach their equilibrium values than it does for xenon. Nevertheless, in all reactors except those operating at very low flux, these isotopes come into equilibrium in, at most, a few days' time.

When the time derivatives in Eqs. (7.105) and (7.106) are placed equal to zero, these equilibrium concentrations are found to be

$$P_\infty = \frac{\gamma_P \overline{\Sigma}_f \phi_T}{\lambda_P} \tag{7.107}$$

and

$$S_\infty = \frac{\gamma_P \overline{\Sigma}_f}{\overline{\sigma}_{aS}}. \tag{7.108}$$

The macroscopic absorption cross-section of equilibrium ^{149}Sm is then

$$\overline{\Sigma}_{aS} = \gamma_P \overline{\Sigma}_f. \tag{7.109}$$

Substituting in Eq. (7.89), this is equivalent to the reactivity

$$\rho = -\frac{\gamma_P}{\nu p \epsilon}. \tag{7.110}$$

It should be particularly noted that this reactivity is independent of the flux or power of the reactor. By contrast, it is recalled that the reactivity due to ^{135}Xe increases with flux. With the value of γ_P given in Table 7.5 for ^{235}U, Eq. (7.110) gives

$$\rho = -\frac{0.0107}{2.42} = -0.442\%,$$

or about 68 cents, for a reactor in which $p = \epsilon = 1$.

Samarium after Shutdown

After shutdown, ^{149}Sm builds up as the accumulated ^{149}Pm decays. However, unlike ^{135}Xe, which undergoes β-decay, ^{149}Sm is stable and remains in a reactor until the system is brought back to critical, whereon it is removed by neutron absorption.

With the concentrations of promethium and samarium at shutdown denoted by P_0 and S_0, respectively, the samarium concentration at the time t later is

$$S(t) = S_0 + P_0(1 - e^{\lambda_P t}). \tag{7.111}$$

At $t \to \infty$, the samarium concentration approaches $S_0 + P_0$, which it clearly must as eventually all the promethium decays to samarium. If the promethium and samarium are at their equilibrium concentrations prior to shutdown, P_0 and S_0 are given by Eqs. (7.107) and (7.108). The postshutdown reactivity due to ^{149}Sm can then be written as

$$\rho = -\frac{\gamma_P}{\nu p \epsilon} \left[1 + \frac{\phi_T}{\phi_S}(1 - e^{-\lambda_P t}) \right], \tag{7.112}$$

where ϕ_S is defined as

$$\phi_S = \frac{\lambda_P}{\overline{\sigma}_{aS}} = 6.180 \times 10^{13} \text{ cm}^{-2} \text{ sec}^{-1} \tag{7.113}$$

at 20°C.

Equation (7.112) is shown in Fig. 7.17 for a ^{235}U-fueled reactor operated at four different values of ϕ_T before shutdown. It is observed that, although the equilibrium samarium is independent of the flux, the postshutdown samarium increases

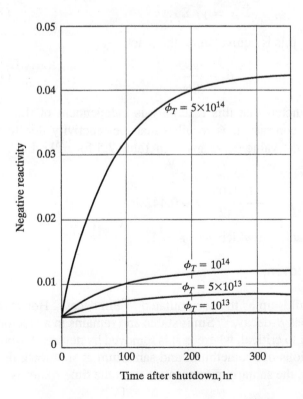

Figure 7.17 Samarium-149 buildup after shutdown for several values of flux.

with increasing flux. Thus, according to Eq. (7.112), the maximum value of $-\rho$ is

$$\rho = -\frac{\gamma_P}{\nu p \epsilon}\left(1 + \frac{\phi_T}{\phi_S}\right),\tag{7.114}$$

which increases linearly with ϕ_T.

Other Fission Product Poisons

In addition to ^{135}Xe and ^{149}Sm, many fission products are formed in a reactor. Some of these are radioactive and have half-lives that are short compared with the average life of a reactor core (1–2 years). Others have absorption cross-sections that are so large that their effective half-life, $0.693/(\lambda + \overline{\sigma}_a\phi_T)$, is also short compared with core life. Such nuclides, as in the case of ^{135}Xe and ^{149}Sm, quickly reach saturation concentrations in the reactor, and these change only when the reactor power is changed or as the flux and fuel mass change over the life of the core. The reactivity associated with these fission products, which include among others ^{151}Sm, ^{155}Eu, ^{113}Cd, and ^{157}Gd, can be calculated in the same way as ^{135}Xe and ^{149}Sm.

At the other extreme are the fission products that are either stable or whose half-lives or effective half-lives are long compared with core life. The formation of these nuclides permanently alters the neutronic properties of a reactor, and such fission products are called *permanent poisons*. Studies have shown that these poisons accumulate at an average rate of approximately 50 b/fission. If all this cross-section is assumed to be associated with a single hypothetical nuclide having the cross-section $\overline{\sigma}_p \simeq 50$ b, then the macroscopic cross-section of the permanent poisons builds up according to the relation

$$\frac{d\overline{\Sigma}_p}{dt} = \overline{\sigma}_p\overline{\Sigma}_f\phi_T.\tag{7.115}$$

The total accumulation during a time t is then

$$\overline{\Sigma}_p = \overline{\sigma}_p\int_0^t \overline{\Sigma}_f\phi_T dt.\tag{7.116}$$

Once $\overline{\Sigma}_f$ and ϕ_T are known as a function of time and space, Eq. (7.116) can be evaluated.

It is somewhat more tedious to compute the poisoning due to nuclides whose effective half-lives are comparable to the core lifetime—too long for the nuclides to saturate and too short for them to be considered permanent. This involves the calculation of many decay chains over the life of the core, and the computations

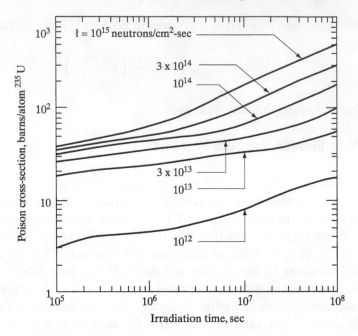

Figure 7.18 Buildup of fission product poisons in barns of absorption cross-section per original atom of ^{235}U. (From ANL-5800, 2nd ed., 1963.)

are usually performed on a computer. Figure 7.18 shows the results of one such set of calculations, which were carried out assuming that the thermal flux remains constant throughout core life—admittedly not a realistic assumption for most power reactors. Permanent poisons and ^{135}Xe and ^{149}Sm were included in the calculation that led to Fig. 7.18.

Example 7.14

Using Fig. 7.18, estimate the total reactivity due to fission product poisons in a ^{235}U-fueled research reactor operated at a constant thermal flux of 10^{12} neutrons/cm^2-sec, 1 year after startup. There is essentially no depletion of the ^{235}U at this flux over this period.

Solution. According to the figure, at 1 year ($t = 3.16 \times 10^7$ sec), the poisoning amounts to about 12 b per original ^{235}U atom. If there were N_{25} atoms of ^{235}U per cm^3 at startup, the macroscopic thermal poison cross-section 1 year later is then

$$\overline{\Sigma}_p = 0.886 \times 12 N_{25} = 10.6 N_{25}.$$

The reactivity associated with this value of $\overline{\Sigma}_p$ is given by Eq. (7.89),

$$\rho = \frac{\overline{\Sigma}_p}{\nu\Sigma_f} = -\frac{10.6N_{25}}{\nu\overline{\sigma}_{f25}N_{25}} = -\frac{10.6}{2.42 \times 0.886 \times 582}$$

$$= -0.00849 = -0.849\%. \ [Ans.]$$

7.6 CORE PROPERTIES DURING LIFETIME

One of the most important and difficult problems in reactor design is the prediction of the properties of a reactor core throughout its life. How these properties change in time must be known to ensure that the reactor will operate safely throughout the lifetime of the core. The NRC specifies safety limits for the operation of every reactor at the time it is licensed. The reactor designer and operator must be certain that conditions do not develop in the system that can lead to a violation of these limits.

The changing reactor properties also determine the length of time that a given reactor core will remain critical, producing power at the desired level. This is an important practical consideration. It obviously makes little sense for a public utility to purchase a reactor that must be shut down every month for refueling, if it can obtain one that needs to be refueled every 2 years. However, it makes equally little sense to buy a core that has fuel enough to remain critical for 20 years when radiation damage is known to compromise the fuel elements in one-tenth of this time.

Burnup–Lifetime Calculations

Computations of changing reactor properties are known as *burnup calculations* or *lifetime calculations*. Consider a problem of this type for a thermal reactor whose core consists of a uniform distribution of fuel assemblies, each containing the same number of identical fuel rods. The fuel is slightly enriched uranium dioxide. For simplicity, it is assumed that the reactor is controlled by a chemical shim system, so there are no control rod motions. At startup, the reactor is necessarily fueled with more fissile material than needed to become critical, and the excess reactivity of the fuel is held down by the chemical shim.

Suppose that the reactor is first brought to critical from a *clean* (no fission product poisons), *cold* [16] (room temperature) condition. Then as the reactor heats

---

[16]Some reactors, including most PWRs, are brought to critical after the reactor has reached its normal operating temperature. This is accomplished by pumping the water coolant through the primary coolant system using the primary coolant pumps. The friction of the water flowing through the system raises the temperature to the operating level. Once this temperature has been reached, the control rods (or most of them) are withdrawn, and the reactor is brought to critical by boron dilution.

up to its operating temperature—a process that takes several hours—the reactivity of the system tends to drop owing to the negative temperature coefficient. This drop in reactivity between cold and hot conditions is known as the *temperature defect* of the reactor. To compensate for the temperature defect, the concentration of the boron shim must be reduced during the startup period to keep the reactor critical. The buildup of equilibrium ^{135}Xe also requires dilution of the boron.

Once the reactor is operating in a steady state at the desired power level, the neutron flux takes on the spatial distribution of the type described in Chapter 6 and shown in Fig. 6.5. As indicated in that figure, the thermal flux (and the fluxes of higher energy neutrons) is greatest at the center of the reactor. This means that fuel is consumed, fertile material is converted, and fission product poisons are produced more rapidly in this region than in other parts of the core. This, in turn, has the effect of reducing the flux in the center of the reactor relative to that on the outside. It follows that, although the reactor initially had uniform properties, it becomes nonuniform as time goes on. Since the analytical methods discussed in Chapter 6 do not apply to nonuniform systems, the reactor properties must be determined by numerical methods.

In practice, burnup computations begin by dividing the time scale into a number of intervals of length Δt. The multigroup diffusion equations are then solved at $t = 0$, the beginning of the first interval—that is, at beginning of the life of the core (BOL). This calculation gives the fluxes at BOL and the boron concentration required for the reactor to be critical. It is now assumed that the multigroup fluxes computed for t_0 remain constant through the first interval to the time $t_1 = t_0 + \Delta t$, and these fluxes are used to compute at t_1 the spatial distributions of fuel, converted material, and poisons as follows.

Fuel   If it is assumed that all of the ^{235}U is consumed as the result of thermal neutron absorption, the atom density at the point $\mathbf{r}$ and time t is determined by the equation

$$\frac{dN_{25}(\mathbf{r}, t)}{dt} = -N_{25}(\mathbf{r}, t)\overline{\sigma}_{a25}\phi_T(\mathbf{r}, t). \tag{7.117}$$

Since $\phi_T(\mathbf{r}, t) = \phi_T(\mathbf{r}, t_0)$ and is constant during the first interval, it follows from Eq. (7.117) that

$$N_{25}(\mathbf{r}, t) = N_{25}(\mathbf{r}, t_0)e^{-\sigma_{a25}\phi_T(\mathbf{r}, t_0)\Delta t}. \tag{7.118}$$

Converted Material   Plutonium-239 is formed from ^{238}U according to the reaction

$$^{238}\text{U}(n, \gamma)^{239}\text{U} \xrightarrow[\beta^-]{239} \text{Np} \xrightarrow[\beta^-]{239} \text{Pu}.$$

Some ^{241}Pu may also be formed as a result of the successive capture by ^{239}Pu of two neutrons, especially if the ^{239}Pu is left in the reactor for a long time. The production of ^{241}Pu will be ignored in the present treatment.

Neutrons are absorbed by ^{238}U both at resonance and thermal energies. The number absorbed in resonance is $(1 - p)$ multiplied by the number of neutrons that slow down per second in the reactor, where p is the resonance escape probability. This latter number is

$$\epsilon P_F [\nu_{25} N_{25}(\mathbf{r}, t) \overline{\sigma}_{f25} + \nu_{49} N_{49}(\mathbf{r}, t) \overline{\sigma}_{f49}] \phi_T(\mathbf{r}, t),$$

in which ϵ is the fast fission factor, P_F is the fast nonleakage probability and the subscript 49 refers to the ^{239}Pu. The concentration of ^{239}Pu is then given by the rate equation

$$\frac{dN_{49}(\mathbf{r}, t)}{dt} = N_{28} \overline{\sigma}_{a28} \phi_T(\mathbf{r}, t) \tag{7.119}$$

$$+ (1 - p) \epsilon P_F [\nu_{25} N_{25}(\mathbf{r}, t) \overline{\sigma}_{f25} + \nu_{49} N_{49}(\mathbf{r}, t) \overline{\sigma}_{f49}] \phi_T(\mathbf{r}, t)$$

$$- N_{49}(\mathbf{r}, t) \overline{\sigma}_{a49} \phi_T(\mathbf{r}, t)$$

The first and last terms in this equation are the rates of formation and disappearance of ^{239}Pu due to thermal neutron absorption; the concentration of the ^{238}U is taken to be constant.

Solving Eq. (7.119) gives at t_1

$$N_{49}(\mathbf{r}, t_1) = \frac{N_{28} \overline{\sigma}_{a28} + (1 - p) \epsilon P_F \nu_{25} N_{25}(\mathbf{r}, t_0) \overline{\sigma}_{f25}}{\overline{\sigma}_{a49} - (1 - p) \epsilon P_F \nu_{49} \overline{\sigma}_{f49}} \tag{7.120}$$

$$\times \{1 - \exp -[\overline{\sigma}_{a49} - (1 - p) \epsilon P_F \nu_{49} \overline{\sigma}_{f49}] \phi_T(\mathbf{r}, t_0) \Delta t\}.$$

For simplicity, the concentration of the ^{235}U is assumed to remain constant relative to the buildup of ^{239}Pu.

Poisons   The concentrations of the fission product poisons can be calculated in a straightforward way, at least in the case of the saturating and permanent poisons. For instance, the ^{135}Xe concentration is easily obtained by solving the rate equations, Eqs. (7.90) and (7.91), with $\phi_T(\mathbf{r}, t) = \phi_T(\mathbf{r}, t_0)$. However, the resulting formulas are complicated and are not given here. From Eq. (7.116), the permanent poison cross-section at t_1 is

$$\overline{\Sigma}_{pp}(\mathbf{r}, t_1) = \overline{\sigma}_{pp} \overline{\Sigma}_{f25}(\mathbf{r}, t_0) \phi_T(\mathbf{r}, t_0) \Delta t. \tag{7.121}$$

For simplicity, the concentration of the ^{235}U is assumed to remain constant relative to the buildup of ^{239}Pu. The nonsaturating nuclides must be calculated using an appropriate computer code, as was done to obtain Fig. 7.18.

At this point, all of the properties of the reactor at t_1 are known as a function of position, and the multigroup equations are solved numerically using these new properties. If the reactor is assumed to be operated at constant power, the usual case in problems of this kind, the magnitude of the fluxes is obtained by requiring that the integral

$$P = E_R \int_V \overline{\Sigma}_f(\mathbf{r}, t_1)\phi_T(\mathbf{r}, t_1)dV$$

be satisfied, where P is the reactor power, V is the volume of the core, and $\overline{\Sigma}_f(\mathbf{r}, t_1)$ includes the fission cross-sections of both ^{235}U and ^{239}Pu.

The fluxes obtained in this way for $t = t_1$ are next used to compute new core properties at $t_2 = t_1 + \Delta t$; the multigroup calculations are repeated, the fluxes again being normalized to the reactor power; the fluxes at t_2 are used to compute the properties at $t_3 = t_2 + \Delta t$; and so on.

At each stage of these computations, the boron concentration required for criticality is determined. Besides providing a description of how the reactor properties change with time, the calculations give a curve of critical boron concentration as a function of time after startup. Such a curve, the result of calculations of the type just described, is shown in Fig. 7.19 for one of the PWR cores used by the Consolidated Edison Company of New York.

It is observed in Fig. 7.19 that, as expected, the critical boron concentration, after an initial drop due to temperature and xenon, decreases slowly and steadily

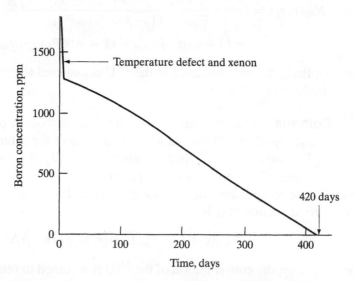

Figure 7.19   Boron concentration in a PWR required for the criticality as function of time after startup.

with time and after 420 days falls to zero. At this point, the reactor can no longer be kept critical by diluting boron. Unless rods are present in the core and can be removed, the reactor cannot remain critical at its original power level. However, this is not necessarily the end of life for the core. The reactor can be kept critical somewhat longer by gradually reducing power as the necessary positive reactivity is supplied by the negative temperature coefficient accompanying the decreasing reactor temperature and by the reduction in the xenon concentration due to the decreasing neutron flux. This process is referred to as *coastdown*. For many practical purposes, however, the time at which the boron concentration drops to zero in Fig. 7.19 is often used to define the end of core life.

Optimum Design and Operation—Fuel Management

Having once performed calculations of the type just described, which at every stage, except at $t = t_0$, involves the computation of the properties of a nonuniform reactor, it is reasonable to ask whether the original, clean core, with its uniform distribution of fuel, represents the optimum design for the system. Would it not be better perhaps to start with a core already having some sort of nonuniform distribution of fuel?

This question has been studied in detail, and the uniform core is not the optimum design. To operate a reactor safely and efficiently, it is important that the power density be as flat—that is, uniform—as possible across the core. The uniformly loaded core, however, has severe power peaks near the center. By nonuniformly distributing the fuel the power density can be significantly flattened. At the same time, this leads to a longer core life for a given initial fuel loading. In addition, it is desirable to limit the flux reaching the reactor vessel so as to minimize radiation damage to the vessel and extend its life.

Figure 7.20 shows a typical loading pattern used in the current generation of PWR cores. Each square represents one fuel assembly of the type shown in Fig. 4.10. With this arrangement, the enrichment of the fuel decreases from the outside of the core toward the center. At the time of refueling, the central assemblies with the lowest enrichment are replaced by fuel from the outer region with a higher enrichment, and fresh fuel is added to the core periphery. By loading and refueling the core in this way, the power distribution is much flatter across the core than it would be with a uniform fuel loading. To reduce the leakage of neutrons out of the core, and thus minimize the flux at the reactor vessel, some cores now replace the outer one or two rows of fuel pins by stainless steel rods. This lowers the radiation damage in the reactor vessel.

The power distribution can also be flattened and the lifetime of the core extended by appropriate movement of the control rods. This procedure is used, in particular, in boiling water reactors. In addition, the current BWRs have a

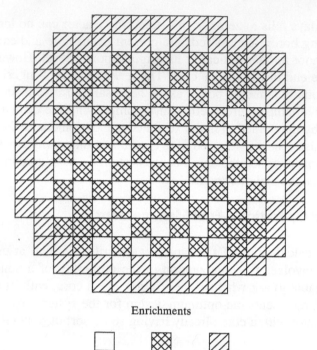

Enrichments

2.25 $^w/_o$ 2.80 $^w/_o$ 3.30 $^w/_o$

Figure 7.20 Fuel loading pattern at the beginning of a PWR's life. (Courtesy of Westinghouse Electric Corporation.)

nonuniform distribution of fuel within each fuel assembly to flatten the power across the assembly. This is shown in Fig. 7.21, where the numbers 1, 2, 3, and 4 represent, in that order, fuel rods of decreasing enrichment. The rods indicated by a zero contain no fuel.

The problems discussed earlier—how fuel should be loaded and reloaded, how the control rods should be moved, and, in general, how the controllable parameters of a reactor should be varied to optimize its performance and ensure the safe operation of the system—are matters referred to as *fuel management*.[17] Because of the essentially infinite variety of possible fuel loading and reloading patterns, control rod movements (combined control rod movements and shim control, if both are present), and so on, fuel-management problems are often formidable. To handle such problems, multigroup lifetime computations of the type described earlier must be performed repeatedly for a variety of different operating strategies, at considerable cost in computer time. Modern computer codes such as SIMULATE

---

[17]More properly, these problems form the subject of *incore* fuel management. Problems concerned with the purchase or leasing of fuel, the reprocessing of spent fuel elements, and other matters related to the management of fuel outside of the reactor are usually called *out-of-core* fuel management.

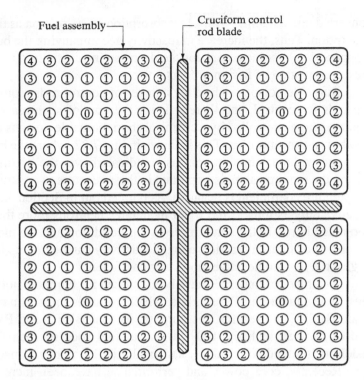

Figure 7.21   Fuel loading pattern within fuel assemblies of a BWR. (Courtesy of General Electric Company.)

enable the core design to be quickly evaluated for life and flux shape. Based on the results of the study of a particular core design, the fuel assemblies can be shuffled to maximize core life and flatten the flux.

Since it is not always possible to accurately compute the day-to-day behavior of a reactor before it has gone into operation, many modern power reactors are equipped with online computers that continually analyze data from instruments monitoring the behavior of the system. The output from these computers is used for updating fuel-management decisions.

Burnable Poisons

Considerably more control reactivity is required before the startup of a new clean, cold core, that is loaded with fresh fuel, and during the core's early operating stages than is needed toward the end of the core life. To reduce these control requirements, it is common practice to place a *burnable poison* at selected locations in the core. This is a nuclide that has a large absorption cross-section which is converted into a nuclide with a low absorption cross-section as the result of neutron absorption. Thus, the increase in reactivity accompanying the burnup of the poison compensates to some extent for the decrease in reactivity due to fuel burnup and the accumulation of fission product poisons.

Burnable poisons reduce the number of control rods required in reactors that use rods for control, and thereby lower the cost of such reactors. In contrast, in water reactors controlled by a chemical shim, burnable poisons decrease the necessary boron concentration. This is shown in Fig. 7.22, where the boron concentration with and without burnable poisons is indicated over the lifetime of the core. Reducing the required shim is important in these reactors to ensure that the moderator temperature coefficient is negative. As pointed out in Section 7.3, this coefficient is negative only as long as the water moderates neutrons more than it absorbs them. Too much boron in the water reverses this condition and the moderator coefficient becomes positive. In practice, enough reactivity can be tied up in burnable poisons to give the desired negative moderator coefficient.

A number of materials have been utilized for burnable poisons. Boron in various forms, including Pyrex glass, which contains approximately 12 w/o boron oxide (B_2O_3), is used for this purpose. Recent BWRs and PWRs use gadolinia (Gd_2O_3) mixed with the UO_2 in several fuel rods of each fuel assembly. The gadolinia is zoned both radially and axially in the more advanced fuel designs.

Recently, core power and performance have been increased by the use of even more sophisticated core designs. These designs rely on the use of higher enrichment fuel, usually 5 w/o enrichment, and extensive use of burnable poisons,

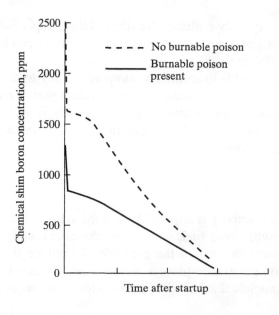

Figure 7.22 Boron shim with and without burnable poison. (Courtesy of Westinghouse Electric Corporation.)

TABLE 7.7 TYPICAL REACTIVITY WORTHS FOR
CONTROL ELEMENTS 3,000 MWT LIGHT-WATER
REACTOR

	PWR	BWR
Excess reactivity at 20°C	$45	$38
No Xe or Sm	$k = 1.41$	$k = 1.33$
Total control rod worth	$11	$26
	~60 clusters	140–185 rods
Fixed burnable poisons	$13	$18
Chemical shim worth	$26	–
Net reactivity	–$5	–$6

usually gadolinia. The unique feature of the fuel is the zoning of the gadolinia and the uranium to produce very uniform burnup and power. The use of these advanced fuels in both BWRs and PWRs has resulted in cycle lengths of 24 months and power upratings of as much as 20% for some reactors.

Table 7.7 summarizes the reactivity worth of the various components found in current-generation reactor designs.

Note that the excess reactivity in a BWR is less than in a PWR. But the number of control rods in a PWR is much less than in a BWR due to the presence of chemical shim in a PWR.

REFERENCES

Duderstadt, J. J., and L. J. Hamilton, *Nuclear Reactor Analysis*. New York: Wiley, 1976. Chapter 6.

Glasstone, S., and A. Sesonske, *Nuclear Reactor Engineering*, 4th ed. New York: Chapman & Hall 1994, Chapter 5.

Hetrick, D. L., *Dynamics of Nuclear Reactors*. LaGrange Park: American Nuclear Society, 1993.

Isbin, H. S., *Introductory Nuclear Reactor Theory*. New York: Reinhold, 1963, Chapters 9 and 16.

Lamarsh, J. R., *Introduction to Nuclear Reactor Theory*. Reading, Mass.: Addison-Wesley, 1966, Chapters 12–14.

Meem, J. L., *Two Group Reactor Theory*. New York: Gordon & Breach, 1964, Chapters 5 and 8.

Murray, R. L., *Nuclear Reactor Physics*. Englewood Cliffs, N.J.: Prentice-Hall, 1957, Chapters 6–8.

Schultz, M. A., *Control of Nuclear Reactors and Power Plants*, 2nd ed. New York: McGraw-Hill, 1961.

Sesonske, A., *Nuclear Power Plant Analysis*. U.S. DOE report TID-26241, 1973, Chapters 5–9.

Zweifel, P. F., *Reactor Physics*. New York: McGraw-Hill, 1973, Chapters 4, 7, and 10.

PROBLEMS

1. Compute the prompt neutron lifetime for an infinite critical thermal reactor consisting of a homogeneous mixture of ^{235}U and (a) D_2O, (b) Be, (c) graphite.

2. Calculate the thermal neutron diffusion time in water having a density of 44 lb/ft^3.

3. Express the following reactivities of a ^{239}Pu-fueled thermal reactor in dollars: (a) 0.001, (b) 4%, (c) −0.01.

4. Express the following reactivities of a ^{235}U-fueled thermal reactor in percent: (a) 0.001, (b) $2, (c) −50 cents.

5. Plot the reactivity equation for one-delayed-neutron group for a ^{235}U-fueled thermal reactor and prompt neutron lifetimes of (a) 0 sec, (b) 10^{-4} sec, (c) 10^{-3} sec. Take $\beta = 0.0065$ and $\lambda = 0.1$ sec^{-1}.

6. From the plot in Problem 7.5, determine the periods of ^{235}U-fueled thermal reactors with $l_p = 10^{-3}$ sec and reactivities of (a) +0.1%, (b) −10 cents, (c) +$.50, (d) +$1.00.

7. Compare the answers to Problem 7.6, which are based on the one-delayed-group model, with periods determined from Fig. 7.2.

8. A ^{235}U-fueled reactor originally operating at a constant power of 1 milliwatt is placed on a positive 10-minute period. At what time will the reactor power level reach 1 megawatt?

9. Calculate the ratio of the concentration of the 22.7-sec precursor to the thermal neutron density in a critical infinite thermal reactor consisting of a mixture of ^{235}U and ordinary water, given that the thermal flux is 10^{13} neutrons/cm^2-sec. [*Hint:* Place $dC/dt = 0$ in Eq. (7.21).]

10. The first term on the Right-Hand Side of Eq. (7.19) gives the number of prompt neutrons slowing down per cm^3/sec while the second term gives this number for the delayed neutrons. Compare the magnitude of these two terms in a critical reactor.

11. Fifty cents in reactivity is suddenly introduced into a critical fast reactor fueled with ^{235}U. What is the period of the reactor?

12. The reactor in Problem 7.8 is scrammed by the instantaneous insertion of 5 dollars of negative reactivity after having reached a constant power level of 1 megawatt. Approximately how long does it take the power level to drop to 1 milliwatt?

13. When a certain research reactor operating at a constant power of 2.7 megawatts is scrammed, it is observed that the power drops to a level of 1 watt in 15 minutes. How much reactivity was inserted when the reactor was scrammed?

14. An infinite reactor consists of a homogeneous mixture of ^{235}U and H_2O. The fuel concentration is 5% smaller than that required for criticality. What is the reactivity of the system?

15. One cent worth of either positive or negative reactivity places a ^{235}U-fueled reactor on what period?

16. It is desired to double the power level of a ^{235}U-fueled reactor in 20 minutes. If $l_p = 10^{-4}$ sec, (a) on what period should the reactor be placed? (b) how much reactivity should be introduced?

17. During test-out procedures, a ^{239}Pu-fueled thermal reactor is operated for a time at a power of 1 megawatt. The power is then to be increased to 100 megawatts in 8 hours. (a) On what stable period should the reactor be placed? (b) What reactivity insertion is required?

18. A ^{235}U-fueled research reactor is started up at a power level of 10 watts. Then reactivity according to the following schedule is inserted into the reactor:

t (sec)	ρ' ($)	Power (watts)	T (sec)
0 – 100	0	10	
100 – 300	0.099		100
300 → ∞	−8.00		

Complete the table.

19. An experimental reactor facility is a bare square cylinder 100 cm high, composed of small beryllium blocks with thin foils of ^{235}U placed in between, so that the system can be considered to be a homogeneous mixture of Be and ^{235}U. The reactor is to be controlled with a single black control rod 2.5 cm in radius and located along the axis of the system. (a) If the reactor is critical with the rod fully withdrawn, how much negative reactivity is introduced into the system when the rod is fully inserted? (b) Assuming that the rod moves into the reactor instantaneously, on what period does the reactor go?

20. Suppose it is desired to control the reactor described in Example 7.7 with one central rod having a worth of 10%. How big should the rod be? [*Hint:* Plot ρ_ω versus a.]

21. If the reactor in Problem 7.19 is controlled by an array of 25 more or less uniformly distributed black rods 0.3 in. in diameter, what is the total worth of the rods?

22. A certain pressurized-water reactor is to be controlled by 61 cluster control assemblies, each assembly containing 20 black rods 1.15 cm in diameter. The reactor core is a cylinder 320 cm in diameter. The average thermal diffusion length in the core is 1.38 cm, $D = 0.21$ cm, and Σ_t in the core material is approximately 2.6 cm^{-1}. Calculate the total worth of the rods.

23. Consider the problem of determining the worth of a cruciform control rod in the one-dimensional geometry shown in Fig. 7.10. First suppose that q_T fission neutrons slow down to thermal energies per cm^3/sec. The quantity q_T can be treated as a source term in the thermal diffusion equation, which becomes

$$-D\frac{d^2\phi_T}{dx^2} - \Sigma_a\phi_T = -q_T.$$

(a) Show that the solution to this equation, subject to the boundary conditions,

$$\frac{1}{\phi_T}\frac{d\phi_T}{dx}T = 0$$

at $x = 0$ and

$$\frac{1}{\phi_T}\frac{d\phi_T}{dx} = -\frac{1}{d}$$

at $x = (m/2) - a$, is

$$\phi_T(x) = \frac{q_T}{\Sigma_a}\left[1 - \frac{\cosh(x/L_T)}{(d/L_T)\sinh[(m - 2a)/2L_T] + \cosh[(m - 2a)/2L_T]}\right].$$

(b) Compute f_R by multiplying the neutron current density at the blade surface by the area of the blades in the cell and then dividing by the total number of neutrons thermalizing per second in the cell.

24. It is proposed to use the cruciform rods described in Example 7.9 to control the PWR described in Problem 7.22. How far apart should these rods be placed to provide the same total worth as the cluster control rods in that problem?

25. Suppose the fast reactor described in Example 6.3 is controlled with 50 rods, each rod containing approximately 500 g of natural boron. Estimate the total worth of the rods.

26. The core of a fast reactor is a square cylinder 77.5 cm in diameter. The composition of the core *by volume* is as follows: 25% fuel; 25% stainless steel cladding and structural material; 50% liquid sodium. The fuel consists of a mixture of ^{238}U and ^{239}Pu having a density of 19.1 g/cm^3, the plutonium making up 20 w/o of the mixture. It is desired to provide 5.5% reactivity control in 10 control rods containing B$_4$C. How much B$_4$C is required per rod?

27. A control rod 100 cm long is worth 50 cents when totally inserted. (a) How much reactivity is introduced into the reactor when the rod is pulled one-quarter of the way out? (b) At what rate is reactivity introduced at this point per cm motion of the rod?

28. Suppose that, at some time during its operating history, the reactor described in Example 7.7 is critical with the rod withdrawn one-half of its full length. If the rod is now suddenly withdrawn another 10 cm, (a) how much reactivity is introduced? (b) on what period does the reactor power rise?

29. What concentration of boron in ppm will give the same worth as the single control rod in Example 7.7?

30. An infinite ^{235}U-fueled, water-moderated reactor contains 20% more ^{235}U than required to become critical. What concentration of (a) boron in ppm or (b) boric acid in g/liter is required to hold down the excess reactivity of the system?

31. A reactor is operating at constant power. Suddenly the temperature of the incoming coolant drops below its previous value. Discuss qualitatively the subsequent behavior of the reactor in the two cases: (a) α_T is positive, (b) α_T is negative. Show, in particular, that the reactor is unstable in (a) and stable in (b).

32. Calculate the prompt temperature coefficient at room temperature of a reactor lattice consisting of an assembly of 1-in diameter natural uranium rods in a heavy-water moderator, in which the moderator volume-to-fuel volume ratio is equal to 30.

33. A CO_2-cooled, graphite-moderated reactor fueled with rods of slightly enriched uranium dioxide 1.5 cm in diameter has a resonance escape probability of 0.912 at 300K. What is the value of p at the fuel operating temperature of 665°C?

34. The overall temperature coefficient of a ^{235}U-fueled reactor is -2×10^{-5} per °C and is independent of temperature. By how much does the reactivity of the system drop when its temperature is increased from room temperature (about 70°F) to the operating temperature of 550°F? Give your answer in percent and in dollars. [*Note:* The decrease in reactivity calculated in this problem is called the *temperature defect* of the reactor.]

35. Show that the temperature coefficient of B^2 is given by

$$\alpha_T(B^2) = -\frac{2}{3}\beta,$$

where β is the coefficient of volume expansion of the reactor structure. [*Hint:* Recall that β is equal to three times the coefficient of linear expansion.]

36. A pressurized water reactor fueled with stainless steel-clad fuel elements is contained in a stainless steel vessel that is a cylinder 6 ft in diameter and 8 ft high. Water having an average temperature of 575°F occupies approximately one-half of the reactor volume. How much water is expelled from the reactor vessel if the average temperature of the system is increased by 10°F? [*Note:* The volume coefficients of expansion of water and stainless steel at 300°C are 3×10^{-3} per °C and 4.5×10^{-5} per °C, respectively.]

37. The thermal utilization of the reactor in Problem 7.36 is 0.682 at 575°F. Using the results of that problem and ignoring the presence of structural material in the core, estimate $\alpha_T(f)$ at approximately 575°F.

38. What is the effective half-life of ^{135}Xe in a thermal flux of 10^{14} neutrons/cm^2-sec at a temperature of 800°C?

39. Compute and plot the equilibrium xenon reactivity as a function of thermal flux from $\phi_T = 5 \times 10^{12}$ to $\phi_T = 5 \times 10^{14}$.

40. Using Fig. 7.14, plot the maximum post-shutdown xenon reactivity as a function of thermal flux from $\phi_T = 10^{13}$ to $\phi_T = 5 \times 10^{14}$.

41. A ^{235}U-fueled reactor operating at a thermal flux of 5×10^{13} neutrons/cm^2-sec is scrammed at a time when the reactor has 5% in reserve reactivity. Compute the time to the onset of the deadtime and its duration.

42. Calculate the equilibrium concentration, in atoms/cm^3, of ^{135}Xe and ^{149}Sm in an infinite critical ^{235}U-fueled, water-moderated thermal reactor operating at a temperature of 200°C and a thermal flux of 10^{13} neutrons/cm^2-sec.

43. How much reactivity is tied up in ^{135}Xe and ^{149}Sm in the reactor described in Problem 7.42?

44. Gadolinium-157 is a stable nuclide having an absorption cross-section at 0.0253 eV of 240,000 b. It is formed from the decay of the fission product ^{157}Sm according to the following chain:

$$^{157}\text{Sm} \quad \xrightarrow[0.5m]{\beta^-} \quad ^{157}\text{Eu} \xrightarrow[15.2h]{\beta^-} \quad ^{157}\text{Gd}$$
$$\uparrow$$
$$\text{Fission}$$

Neither ^{157}Sm nor ^{157}Eu absorbs neutrons to a significant extent. The ^{235}U fission yield of ^{157}Sm is 7×10^{-5} atoms per fission. (a) What is the equilibrium reactivity tied up in ^{157}Gd in a reactor having an average thermal flux of 2.5×10^{13} neutrons/cm^2-sec? (b) What is the maximum reactivity due to this nuclide after the shutdown of the reactor in part (a)?

45. A nonradioactive fission product has an absorption cross-section of 75 b. Should this be considered a permanent poison in a reactor having a thermal flux of 3×10^{13} neutrons/cm^2-sec?

46. What fraction of the poisoning in Example 7.14 is due to ^{135}Xe?

47. An infinite reactor containing no fertile material operates at constant power over its lifetime. (a) Show that the atom density of the fuel decreases according to the relation

$$N_F(t) = N_F(0)[1 - \overline{\sigma}_{aF}\phi_T(0)t],$$

where $N_F(0)$ and $\phi_T(0)$ are, respectively, the fuel atom density and thermal flux at startup. (b) Find an expression for the flux as a function of time.

48. An infinite ^{139}Pu-fueled fast breeder has the breeding gain G and is operated at constant power before refueling. Derive expressions for the concentration of ^{239}Pu and the (one-group) fast flux as a function of time after startup.

8

Heat Removal from Nuclear Reactors

For a reactor to operate in a steady state, with an internal temperature distribution that is independent of time, all of the heat released in the system must be removed as fast as it is produced. This is accomplished, in all reactors except those operating at very low power levels, by passing a liquid or gaseous coolant through the core and other regions where heat is generated. The nature and operation of this coolant system is one of the most important considerations in the design of a nuclear reactor.

The temperature in an operating reactor varies from point to point within the system. As a consequence, there is always one fuel rod, usually one of the rods near the center of the reactor, that at some point along its length is hotter than all the rest. This maximum fuel temperature is determined by the power level of the reactor, the design of the coolant system, and the nature of the fuel. However, metallurgical considerations place an upper limit on the temperature to which a fuel rod can safely be raised. Above this temperature, there is a danger that the fuel may melt, which can lead to the rupture of the cladding and release of fission products. One of the major objectives in the design of a reactor coolant system is to provide for the removal of the heat produced at the desired power level while ensuring that the maximum fuel temperature is always below this predetermined value.

It should be noted that, from a strictly nuclear standpoint, there is no theoretical upper limit to the power level that can be attained by any critical reactor having sufficient excess reactivity to overcome its negative temperature coefficient. Thus, by removing control rods and placing a reactor on a positive period, its power could be increased indefinitely were it not for the fact that eventually a point would be reached where the coolant system is no longer able to remove all of the heat being produced. Beyond this point, the fuel would heat up and eventually a portion of the core would melt down. This situation is avoided in the actual operation of a reactor by reinserting some of the control rods or adding boron to the shim system to return the reactor to critical when a desired power level has been reached. Throughout the present chapter, it is assumed that the reactors under consideration are critical and operating at a constant power.

Before beginning the discussion of the design of reactor coolant systems, a word is in order concerning units. As pointed out in Appendix I, where the matter of units is considered in some detail, the SI system of units has been adopted by most of the nations of the world, with the exception of the United States. All U.S. scientists and many U.S. engineers and engineering societies also use this system. However, there is still strong resistance to the adoption of SI units. As a result, most American manufacturers of nuclear power equipment base their designs on the English system. Since this state of affairs is likely to persist for several years to come, English units are used throughout the present chapter as the primary unit. SI units are provided where appropriate. Tables for conversion from English to SI units are given in Appendix I.

8.1 GENERAL THERMODYNAMIC CONSIDERATIONS

From a thermodynamic point of view, a nuclear reactor is a device in which energy is produced and transferred to a moving fluid. Thus, as indicated in Fig. 8.1, heat is released in a reactor at the rate of q BTU/hr or watts and absorbed by the coolant, which enters the reactor at the temperature T_{in} and exits from the reactor at the temperature T_{out}, passing through the system at the rate of w lb/hr or kg/hr.

With all power reactors except the BWR and RBMK, there is no (net) change in phase of the coolant as it passes through the reactor; that is, the coolant does not boil. In these reactors, the heat from the reactor merely increases the temperature of the coolant, a process that occurs at essentially constant pressure—namely, the reactor coolant pressure. The heat in BTUs or joules required to raise the temperature of a unit mass of coolant from the temperature T_{in} to T_{out} is

$$\int_{in}^{out} c_p(T)dT,$$

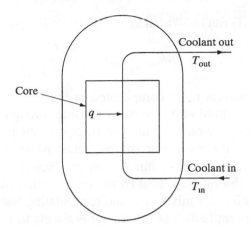

Coolant out

T_{out}

Core

q

Coolant in

T_{in}

Figure 8.1   Schematic drawing of coolant flow through a reactor.

where $c_p(T)$ is the specific heat at constant pressure per unit mass of the coolant. Since w lb or kg of coolant flows through the reactor per hour, the rate at which heat produced in the reactor is absorbed by the coolant is given by

$$q = w \int_{T_{\text{in}}}^{T_{\text{out}}} c_p(T)dT. \tag{8.1}$$

Equation (8.1) can also be written in terms of the thermodynamic function *enthalpy*. This is defined for a substance by the relation

$$h = u + Pv, \tag{8.2}$$

where h is the enthalpy per unit mass of the substance (or *specific enthalpy*), u is its internal energy per unit mass, P is the pressure, and v is the *specific volume* of the substance (i.e., the volume per unit mass ft^3/lb or m^3/kg). The units of h are Btu/lb or joules/kg. In thermodynamics, it is shown that when heat is added to a substance at constant pressure, essentially all of the heat is used to increase its enthalpy.[1] Thus, if h_{in} and h_{out} are the specific enthalpies of the coolant entering and leaving the reactor, respectively, it follows that

$$h_{\text{out}} = h_{\text{in}} + \int_{T_{\text{in}}}^{T_{\text{out}}} c_p(T)dT. \tag{8.3}$$

---

[1] A very small amount of heat may also be used to increase the kinetic energy of the substance or change its gravitational potential energy by virtue of raising the substance from a lower to a higher level in the earth's gravitational field. However, both of these contributions are negligible for reactor coolants.

It also follows that Eq. (8.1) can be written as

$$q = w(h_{\text{out}} - h_{\text{in}}).$$ (8.4)

The situation is somewhat more complicated in the case of a BWR or an RBMK. In these reactors, a portion of the water passing through the core is vaporized to steam, exits from the reactor via a steam pipe to the turbine, and later returns as feed water from the condenser and reheaters. Most of the water that goes through the core is recirculated within the reactor. However, in steady-state operation, there is no net absorption of heat by the recirculating water. Therefore, although the incoming feed water mixes with the recirculating water before passing through the core, the overall effect of the reactor is simply to vaporize the feed water, as is indicated schematically in Fig. 8.2.

The change in enthalpy from the point where the feed water enters a BWR to the point where the steam exits from the reactor consists of two stages. The water is first heated from its entering temperature T_{in} to the temperature at which it boils— that is, the *saturation temperature* T_{sat} for the given system pressure. The water temperature does not rise above this value. The associated increase in enthalpy is

$$h_f = h_{\text{in}} + \int_{T_{\text{in}}}^{T_{\text{sat}}} c_p(T)dT.$$

This is the enthalpy of the saturated water. With the onset of boiling, the water absorbs an amount of heat equal to the *heat of vaporization,* denoted by h_{fg}, per unit mass of water that changes phase. Thus, the specific enthalpy of the steam is

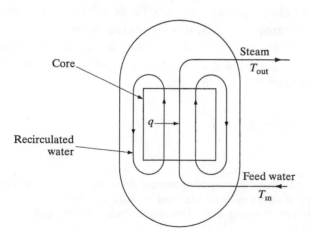

Figure 8.2   Schematic drawing of coolant flow through a BWR.

$$h_{\text{out}} = h_f + h_{fg} \tag{8.5}$$

$$= h_{\text{in}} + \int_{T_{\text{in}}}^{T_{\text{sat}}} c_p(T)dT + h_{fg}.$$

Again, the rate at which heat is absorbed by the coolant is given by Eq. (8.4),

$$q = w(h_{\text{out}} - h_{\text{in}}),$$

where w is the steam or feed water flow rate (the two are equal) in lb/hr or kg/hr.

The advantage of expressing q in terms of enthalpy is that h has been computed using Eq. (8.3) or Eq. (8.5) and is available in tabular form for a number of substances. Abridged tables of h (these are called *steam tables*) are given in Appendix IV for water, helium, and sodium. It should be noted that only the *difference* in enthalpy between two different temperatures is meaningful and useful; the absolute value of enthalpy has no meaning. Thus, the zero of enthalpy can be set arbitrarily, and the values of enthalpy computed from this starting point. In the case of water, for example, in the English system of units, h is taken to be zero for the saturated liquid at 32°F or in SI units as 0°C.

Again, for water, it is observed in Appendix IV that the specific enthalpy is tabulated for saturated water and steam as a function of the temperature, and associated pressure in Table IV.1, and as a function of pressure and the associated temperature in Table IV.2. (It must be remembered that there is only one pressure for a given temperature; conversely, there is one temperature for a given pressure at which water and steam are in equilibrium—i.e., at which water will boil.) However, the enthalpy of a liquid is not a sensitive function of pressure. Thus, the enthalpy of water in the pressurized water reactor is nearly equal to the value of h indicated in the tables for saturated liquid at the actual water temperature independent of its pressure.[2] This procedure is used in the first of the following two examples, which illustrate the use of Eq. (8.4) for PWR and BWR systems.

Example 8.1

A PWR operates at a thermal power of 3,025 MW. Water enters the reactor at 542.6°F and flows through the reactor at a rate of 136.3×10^6 lb/hr. The system pressure is 2,250 psia.[3] At what temperature does the coolant leave the reactor?

---

[2]Data for correcting saturation values of h for higher pressure are given in Table 4 of J. H. Keenan and F. G. Keyes, *Thermodynamic Properties of Steam Including Data for the Liquid and Solid Phases,* New York: Wiley, 1959. In the temperature region from 400°F to 600°F and up to pressures of 300 psi, the correction for h amounts to less than 1%.

[3]Psia means *absolute pressure* in lb/in². The difference between the reactor pressure and atmospheric pressure is called *gauge pressure* and is abbreviated psig.

Solution. According to Table I.9, 1 kW = 3,412 Btu/hr. Then

$$q = 3{,}205 \times 10^3 \times 3{,}412 = 10{,}321 \times 10^6 \text{ Btu/hr.}$$

From Eq. (8.4), the rise in enthalpy through the core is

$$h_\text{out} - h_\text{in} = \frac{q}{w} = \frac{10{,}321 \times 10^6}{136.3 \times 10^6} = 75.7 \text{ Btu/lb.}$$

Interpolating from the values of enthalpy in Table IV.1 for saturated water at 540°F and 550°F gives the value of h at 542.6°F as 539.7 Btu/lb. The enthalpy of the emerging water is then

$$h_\text{out} = 539.7 + 75.7 = 615.4 \text{ Btu/lb.}$$

A second interpolation from Table IV.1 shows that this corresponds to water at a temperature of 599.1°F. [*Ans.*]

Example 8.2

A BWR operating at pressure of 6.895 MPa produces 2.93×10^6 kg of steam per hour. Feed water enters the reactor at 190.6°C. (a) What is the temperature of the steam? (b) At what power is the reactor operating?

Solution.

1. The steam emerges at essentially the temperature at which the water boiled. According to Table IV.2, at a pressure of 6.895 MPa, this is 284.86°C. [*Ans.*]
2. From Table IV.2, h of the steam is 2,773.2 kJ/kg, while from Table IV.1, at 190.6°C, h of the feed water is 807.8 kJ/kg. Then Eq. (8.4) gives

$$q = 2.93 \times 10^6(2{,}773.2 - 807.8) = 5{,}758 \times 10^6 \text{ J/hr} = 1{,}599 \text{ MW. } [Ans.]$$

8.2 HEAT GENERATION IN REACTORS

The starting point in the design of a reactor cooling system is the determination of the spatial distribution of the heat produced within the reactor. This problem is discussed in this section.

It was pointed out in Chapter 3 that the energy released in fission appears in several forms—as kinetic energy of the fission neutrons, as prompt fission γ-rays, as γ-rays and β-rays from the decay of fission products, and in the emission of neutrinos. With the exception of the neutrinos, virtually all of this energy is ultimately absorbed somewhere in the reactor. However, because these various radiations are attenuated in different ways by matter, their energy tends to be deposited in different locations. In the following discussion of the deposition of energy, it is assumed that the reactor fuel is in the form of individual fuel rods. This is the case for virtually all power reactor concepts except the molten salt breeder reactor (see Section 4.5).

Heat Production in Fuel Elements (Fuel Rods)

As indicated in Table 3.6, the fission fragments have a kinetic energy of about 168 MeV per fission. These highly charged particles have an extremely short range (see Section 3.9), and therefore their energy is deposited near the site of the fission within the fuel. Similarly, most of the 8 MeV of the fission product β-rays is also deposited in the fuel. However, many of the γ-rays from the decaying fission products and those emitted directly in fission pass out of the fuel since they are less strongly attenuated than charged particles. Some of these γ-rays are absorbed in the surrounding coolant and/or moderator, in the thermal shield, or in the radiation shielding that surrounds the reactor. However, because of the proximity of fuel rods in most reactors, many of the γ-rays are intercepted by and absorbed in neighboring rods.

The prompt neutrons are emitted with a total kinetic energy of about 5 MeV per fission. In a thermal reactor, the bulk of this energy is deposited in the moderator as the neutrons slow down. The capture γ-rays emitted following the absorption of these neutrons in nonfission reactions are therefore produced and absorbed throughout the reactor. In a fast reactor, the fission neutrons slow down very little before they are absorbed, and their kinetic energy appears as an addition to the energy of the capture γ-rays. The delayed neutrons contribute negligibly to the energy of a reactor. As already noted, none of the energy of the neutrinos is retained within a reactor.

It should be clear from the foregoing remarks that the spatial deposition of fission energy depends on the details of the reactor's structure. Nevertheless, for preliminary calculations, it may be assumed that approximately one-third of the total γ-ray energy—about 5 MeV—is absorbed in the fuel. This, together with the 168 MeV from the fission fragments and 8 MeV from the β-rays, gives 181 MeV per fission (about 90% of the recoverable fission energy), which is deposited in the fuel, most in the immediate vicinity of the fission site. The remainder (about 20 MeV) of the recoverable energy is deposited in the coolant and/or moderator, in various structural materials, and in the blanket, reflector, and shield.

The rate at which fission occurs in the fuel, and hence the rate of production of heat, varies from fuel rod to fuel rod, and it is also a function of position within any given rod. In particular, if E_d is the energy deposited locally in the fuel per fission, then the rate of heat production per unit volume at the point $\mathbf{r}$ is given by the expression[4]

$$q'''(\mathbf{r}) = E_d \int_0^\infty \Sigma_{fr}(E)\phi(\mathbf{r}, E)dE, \tag{8.6}$$

where $\Sigma_{fr}(E)$ is the macroscopic fission cross-section of the fuel and $\phi(\mathbf{r}, E)$ is the energy-dependent flux as a function of position. The natural units of q''' are MeV/sec-cm^3. However, for engineering calculations, q''' should be converted to units of Btu/hr-ft^3 or kW/lit using the conversion factors in Appendix I.

In the thermal reactor, most of the fissions are induced by thermal neutrons. In this case, according to the results of Section 5.9, the integral in Eq. (8.6) can be written as

$$q'''(\mathbf{r}) = E_d \overline{\Sigma}_{fr}\phi_T(\mathbf{r}), \tag{8.7}$$

where $\overline{\Sigma}_{fr}$ is the thermal fission cross-section of the fuel and $\phi_T(\mathbf{r})$ is the thermal flux. In the framework of a multigroup calculation, Eq. (8.6) can also be expressed in the form

$$q'''(\mathbf{r}) = E_d \sum_g \Sigma_{fg}\phi_g(\mathbf{r}), \tag{8.8}$$

where the sum is carried out over all the groups and the notation is the same as in Section 5.8.

The spatial dependence of the flux depends on the geometry and structure of the reactor. However, many heat removal calculations are carried out for the theoretical case of the bare, finite cylinder. For a thermal reactor, the thermal flux is then (see Table 6.2)

$$\phi_T(r, z) = \frac{3.63P}{E_R\overline{\Sigma}_f V} J_0\left(\frac{2.405r}{\tilde{R}}\right)\cos\left(\frac{\pi z}{\tilde{H}}\right), \tag{8.9}$$

where P is the total power of the reactor in joules, E_R is the recoverable energy per fission in joules, V is the reactor volume in cm^3, and $\tilde{R}$ and $\tilde{H}$ are its outer dimensions in centimeters to the extrapolated boundaries. In obtaining Eq. (8.9), it was assumed that the fuel is homogeneously distributed throughout the reactor,

---

[4]The following notation for heat production and flow is well established in heat transfer literature and is used in this chapter:

q''' :   power density, rate of energy production per unit volume; BTU/hr-ft^3 or kW/lit,
q'' :   heat flux; BTU/hr-ft^2 or W/cm^2,
q' :   linear power; Btu/hr-ft or kW/m,
q :   heat transfer rate; Btu/hr or MW.

and $\overline{\Sigma}_f$ is the value of the macroscopic fission cross-section of this mixture in units of cm^{-1}.

Equation (8.9) can also be used to approximate the flux in a reactor where the fuel is contained in separate fuel rods, provided the value of $\overline{\Sigma}_f$ is computed for the equivalent homogeneous mixture. For instance, suppose that there are n fuel rods of radius a and length H, the height of the core. Then if $\overline{\Sigma}_{fr}$ is the macroscopic fission cross-section of the rod, the total fission cross-section in the entire core[5] is $\overline{\Sigma}_{fr} \times n\pi a^2 H$. Therefore, the average value of $\overline{\Sigma}_f$ in the core is

$$\overline{\Sigma}_f = \frac{\overline{\Sigma}_{fr} n\pi a^2 H}{\pi R^2 H} = \frac{\overline{\Sigma}_{fr} na^2}{R^2}, \tag{8.10}$$

and the flux is

$$\phi_T(r, z) = \frac{3.63 P R^2}{E_R \overline{\Sigma}_{fr} V a^2 n} J_0\left(\frac{2.405r}{\widetilde{R}}\right) \cos\left(\frac{\pi z}{\widetilde{H}}\right)$$

$$= \frac{1.16 P}{E_R \overline{\Sigma}_{fr} H a^2 n} J_0\left(\frac{2.405r}{\widetilde{R}}\right) \cos\left(\frac{\pi z}{\widetilde{H}}\right), \tag{8.11}$$

where $V = \pi R^2 H$ has been substituted.

When the expression for ϕ_T in Eq. (8.11) is introduced into Eq. (8.7), the rate of heat production per unit volume of a fuel rod becomes

$$q'''(r, z) = \frac{1.16 P E_d}{H a^2 n E_R} J_0\left(\frac{2.405r}{\widetilde{R}}\right) \cos\left(\frac{\pi z}{\widetilde{H}}\right). \tag{8.12}$$

It should be noted that, in this procedure for obtaining q''', any variation of the flux across the diameter of the fuel rods has been ignored. The dependence of q''' on r in Eq. (8.12) gives the change in the flux from rod to rod across the diameter of the core—not across any individual fuel rod. Indeed, given the derivation of Eq. (8.12), q''' must be taken to be constant across individual rods. This does not introduce significant errors in heat transfer calculations, especially for the small-diameter, weakly absorbing (low-enrichment) fuel rods currently used in most power reactors.

From Eq. (8.12), it is evident that the maximum rate of heat production occurs in the middle ($z = 0$) of the central rod ($R = 0$). In this case, both functions in Eq. (8.12) are unity, and the maximum value of q''' is

$$q'''_{max} = \frac{1.16 P E_d}{H a^2 n E_R}. \tag{8.13}$$

---

[5]Recall that macroscopic cross-section is cross-section per unit volume.

The maximum rate of heat production in a noncentral rod located at $r \neq 0$ is

$$q_{max}'''(r) = q_{max}''' J\left(\frac{2.405r}{\widetilde{R}}\right). \tag{8.14}$$

The total rate at which heat is produced in any fuel rod is given by the integral

$$q_r(r) = \pi a^2 \int_{-H/2}^{H/2} q'''(r, z) dz.$$

Introducing $q'''(r, z)$ from Eq. (8.12) gives

$$
\begin{aligned}
q_r(r) &= \frac{1.16\pi P E_d}{H n E_R} J_0\left(\frac{2.405r}{R}\right) \int_{-H/2}^{H/2} \cos\left(\frac{\pi z}{H}\right) dz \\
&= \frac{2.32 P E_d}{n E_R} J_0\left(\frac{2.405r}{R}\right).
\end{aligned}
\tag{8.15}
$$

The formulas derived previously for the bare cylindrical reactor should not be taken too seriously for calculations of the heat production in a real reactor. In particular, Eq. (8.13) considerably overestimates the value of q_{max}''' for a reflected and/or nonuniformly fueled reactor, which has a smaller maximum-to-average flux ratio than a bare reactor (see Section 6.3). To see this effect, it is first noted that

$$q_{max}''' = \overline{\Sigma}_{fr} E_d \phi_{max}, \tag{8.16}$$

where ϕ_{max} is the maximum value of the thermal flux. The total reactor power is given by

$$P = \overline{\Sigma}_f E_R \phi_{av} V, \tag{8.17}$$

where ϕ_{av} is the average thermal flux and $\overline{\Sigma}_f$ is again the macroscopic fission cross-section averaged over the entire core volume V. Dividing Eq. (8.16) by Eq. (8.17) and rearranging gives

$$q_{max}''' = \frac{P \overline{\Sigma}_{fr} E_d \phi_{max}}{\overline{\Sigma}_f E_R \phi_{av} V} = \frac{P \overline{\Sigma}_{fr} E_d \Omega}{\overline{\Sigma}_f E_R V}, \tag{8.18}$$

where Ω is the maximum-to-average flux ratio. Finally, introducing Eq. (8.10) for $\overline{\Sigma}_f$ yields

$$q_{max}''' = \frac{P E_d R^2 \Omega}{a^2 n V E_R} = \frac{P E_d \Omega}{\pi H a^2 n E_R}. \tag{8.19}$$

Suppose that $\Omega = 2.4$, a reasonable value for an actual reactor. Then comparing Eqs. (8.13) and (8.19) shows that

$$(q'''_{max})_{actual} = \frac{\Omega}{1.16\pi}(q'''_{max})_{bare} \simeq \frac{2}{3}(q'''_{max})_{bare}.$$

Example 8.3

The extrapolated dimensions of a certain pressurized water reactor are $\tilde{R} = 67$ in and $\tilde{H} = 144$ in. The reactor operates at the thermal power of 1,893 MW. It contains 193 fuel assemblies, each consisting of 204 UO_2 fuel rods 0.42 in. in diameter. Assuming that the assemblies are uniformly distributed throughout the reactor, calculate the total energy production rate and the maximum energy production rate per ft^3 in rods located (a) at the axis of the reactor, (b) 20 in from the axis.

Solution.

1. There are $n = 193 \times 204 = 39,372$ rods in the reactor. Assuming that $E_d = 180$ MeV and $E_R = 200$ MeV, Eq. (8.15) gives for the rod at $r = 0$

$$q_r(0) = \frac{2.32 \times 1893 \times 180}{39,372 \times 200} = 0.100\text{MW}$$

$$= 3.43 \times 10^5 \text{ Btu/hr. } [Ans.]$$

Comparing Eqs. (8.13), (8.14), and (8.15) shows that

$$q'''_{max} = \frac{1}{2Ha^2}q_r(0).$$

Therefore,

$$q'''_{max} = \frac{3.43 \times 10^5}{2 \times \left(\dfrac{144}{12}\right) \times \left(\dfrac{0.21}{12}\right)^2}$$

$$= 4.66 \times 10^7 \text{ Btu/hr-ft}^3 \text{ (482.3 kW/lit). } [Ans.]$$

2. To obtain similar results for a rod located at $r = 20$ in, it is merely necessary to multiply the prior answers by the factor appearing in Eqs. (8.14) and (8.15), yielding,

$$J_0\left(\frac{2.405r}{\tilde{R}}\right) = J_0\left(\frac{1.405 \times 20}{67}\right) = J_0(0.718) = 0.875.$$

Thus, for this rod,

$$q_r(20 \text{ in}) = 3.43 \times 10^5 \times 0.875$$

$$= 3.00 \times 10^5 \text{ Btu/hr (87.9 kW) } [Ans.]$$

and

$$q'''_{\text{max}}(20 \text{ in}) = 4.66 \times 10^7 \times 0.875$$

$$= 4.08 \times 10^5 \text{ Btu/hr-ft}^3 (4.2 \text{ kW/lit}). \text{ [}Ans.\text{]}$$

Radiation Heating

As noted already, roughly 10% of the recoverable energy of fission is absorbed outside the fuel. In thermal reactors, the kinetic energy of the fission neutrons is deposited in the surrounding moderator and coolant in more or less the same spatial distribution as the fissions from which these neutrons originate. However, only 2% to 3% of the fission energy appears in this form, and it is often assumed that this kinetic energy is deposited uniformly throughout the core.

The calculation of the energy deposition from the longer range γ-rays is a more difficult problem. In principle, this can be determined by evaluating the integral

$$q'''(\mathbf{r}) = \int \phi_\gamma(\mathbf{r}, E_\gamma) E_\gamma \mu_a(E_\gamma) dE_\gamma, \tag{8.20}$$

where $\phi_\gamma(\mathbf{r}, E_\gamma)$ is the γ-ray flux as a function of position and energy, and $\mu_a(E_\gamma)$ is the linear absorption coefficient. The computation $\phi_\gamma(\mathbf{r}, E_\gamma)$ is complicated by the fact that γ-rays undergo multiple Compton scattering—a problem discussed in Chapter 10. However, in source-free regions outside the core, such as the thermal shield or reactor vessel, the γ-ray flux may be presumed to fall off approximately exponentially, provided the region in question is not too many mean-free paths thick. In this case,

$$q'''(r) = \int -e^{-\mu_a(E_\gamma)r} \phi_{\gamma 0}(E_\gamma) E_\gamma \mu_a(E_\gamma) dE_\gamma. \tag{8.21}$$

Here, $\phi_{\gamma 0}(E_\gamma)$ is the γ-ray flux as a function of energy incident on the region. Equation (8.21) is usually evaluated by dividing the γ-ray spectrum into a number of energy groups. This procedure is illustrated in Section 10.11.

Fission Product Decay Heating

After a few days of reactor operation, the β- and γ-radiation emitted from decaying fission products amounts to about 7% of the total thermal power output of the reactor. When the reactor is shut down, the accumulated fission products continue to decay and release energy within the reactor. This fission product decay energy can be quite sizable in absolute terms, and a means for cooling the reactor core after shutdown must be provided in all reactors except those operating at very low power levels. If this is not done, the temperature of the fuel may rise to a point where the integrity of the fuel is compromised and fission products are released.

Consider a reactor that has been operating at a constant thermal power P_0 long enough for the concentrations of the radioactive fission products to come to equilibrium. Since the rate of production of the fission products is proportional to the reactor power, it follows that the activity of the fission products at any time after the reactor has been shut down is also proportional to P_0. The ratio $P(t_s)/P_0$, where $P(t_s)$ is the power (rate of energy release) emanating from the fission products at the time t_s after shutdown, is therefore independent of P_0. Figure 8.3 shows this ratio as a function of t_s, in seconds, for a ^{235}U-fueled thermal reactor. The figure is based on a recoverable energy per fission of 200 MeV.

Figure 8.3 can also be used to give values of the fission product decay power for a reactor that is operated for the finite time t_0 and then shut down. The ratio $P(t_0, t_s)/P_0$, where t_s is again the cooling time (reactor off), is then obtained from the expression

$$\frac{P(t_0, t_s)}{P_0} = \frac{P(t_s)}{P_0} - \frac{P(t_0 + t_s)}{P_0}. \tag{8.22}$$

The two terms on the right of this equation are to be found from Fig. 8.3.

If a ^{235}U-fueled reactor contains substantial quantities of ^{238}U, as many of these reactors do, the decay of ^{239}U and ^{239}Np, formed by the absorption of neu-

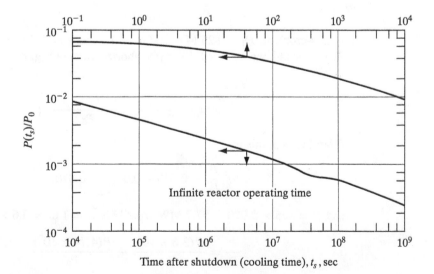

Figure 8.3   The ratio $P(t_s)/P_0$ of the fission product decay power to the reactor operating power as a function of time t_s after shutdown. (Subcommittee ANS-5 of the American Nuclear Society, 1968.)

trons in the ^{238}U, must also be taken into account. Using the equations of radioactive decay derived in Section 2.9, it is easy to show that

$$\frac{P_{29}}{P_0} = 2.28 \times 10^{-3} C \left(\frac{\overline{\sigma}_{a25}}{\overline{\sigma}_{f25}}\right)[1 - e^{-4.91 \times 10^{-4} t_0}]e^{-4.91 \times 10^{-4} t_s} \tag{8.23}$$

and

$$\frac{P_{39}}{P_0} = 2.17 \times 10^{-3} C \left(\frac{\overline{\sigma}_{a25}}{\overline{\sigma}_{f25}}\right)[(1 - e^{-3.41 \times 10^{-6} t_0})e^{-3.41 \times 10^{-6} t_s}$$

$$- 7.0 \times 10^{-3}(1 - e^{-4.91 \times 10^{-4} t_0})e^{-4.91 \times 10^{-4} t_s}]. \tag{8.24}$$

In these equations, P_{29} and P_{39} are the decay powers of ^{239}U and ^{239}Np, respectively, C is the conversion factor for the reactor, and $\overline{\sigma}_{a25}$ and $\overline{\sigma}_{f25}$ are the effective thermal cross-sections of ^{235}U. The times t_0 and t_s are again in seconds.

Example 8.4

A certain ^{235}U-fueled reactor operates at a thermal power of 825 MW for 1.5 years and is then shut down. (a) Using Fig. 8.3, compute the decay energy at the following times: at shutdown, 1 hour after shutdown, 1 year after shutdown. (b) If the conversion factor for the reactor is $C = 0.88$, what are the contributions to the decay energy at the prior times due to ^{239}U and ^{239}Np?

Solution.

1. In this problem, $P_0 = 825$ MW and $t_0 = 1.5$ years $= 1.5 \times 3.16 \times 10^7$ sec. Then at shutdown, $t_s = 10^{-1}$ sec (the shortest time in Fig. 8.3), and from Eq. (8.22),

$$\frac{P}{P_0} = \frac{P(10^{-1})}{P_0} - \frac{P(4.74 \times 10^7)}{P_0}.$$

From Fig. 8.3, this is

$$\frac{P}{P_0} = 0.070 - 0.0007 \simeq 0.070,$$

and $P = 825 \times 0.070 = 57.7$ MW. [*Ans.*] For $t_s = 1$ hr $= 3.6 \times 10^3$ sec:

$$\frac{P}{P_0} = \frac{P(3.6 \times 10^3)}{P_0} - \frac{P(4.74 \times 10^7)}{P_0}$$

$$= 0.014 - 0.0007 \simeq 0.014,$$

so that $P = 825 \times 0.014 = 11.5$ MW. [*Ans.*] Finally, for $t_s = 1$ yr $= 3.16 \times 10^7$ sec, $t_0 + t_s = 7.90 \times 10^7$ sec:

$$\frac{P}{P_0} = \frac{P(3.16 \times 10^7)}{P_0} - \frac{P(7.90 \times 10^7)}{P_0}$$

$$= 0.00079 - 0.00063 = 0.00016,$$

$$P = 825 \times 0.00016 = 0.132 \text{ MW. } [Ans.]$$

2. If the difference between the non-$1/v$ factors for $\overline{\sigma}_{\alpha 25}$ and $\overline{\sigma}_{f25}$ is ignored, $\overline{\sigma}_{a25}/\overline{\sigma}_{f25} = 681/582$. Then with $t_0 = 4.74 \times 10^7$ sec, the exponential in the brackets of Eq. (8.23) is negligible, and with $t_s = 0$ this equation gives

$$\frac{P_{29}}{P_0} = 2.28 \times 10^{-3} \times 0.88 \times \frac{681}{582} = 2.35 \times 10^{-3}.$$

Then $P_{29} = 825 \times 2.35 \times 10^{-3} = 1.95$ MW. [Ans.]
For $t_s = 1$ hr,

$$P_{29} = 1.95 \exp[-4.9 \times 10^{-4} \times 3,600] = 0.33 \text{ MW. } [Ans.]$$

With $t_s = 1$ year, $P_{29} \simeq 0$, as the half-life of ^{239}U is only 23.5 minutes. [Ans.]
For P_{39}, using Eq. (8.24) in a straightforward way, the results are: at shutdown, 1.84 MW; after 1 hour, $\simeq 1.84$ MW; after 1 year, $\simeq$ zero. [Ans.]

8.3 HEAT FLOW BY CONDUCTION

Energy is removed from a reactor by two fundamentally different heat transfer processes—*conduction* and *convection*. In conduction, heat is transmitted from one location in a body to another as a result of a temperature difference existing in the body—there is no macroscopic movement of any portion of the body. It is by this mechanism, as shown in this section, that heat produced in a fuel rod is transferred to the surface of the rod. Heat convection involves the transfer of heat to a moving liquid or gas, again as the result of a temperature difference and the later rejection of this heat at another location. Thus, the heat conducted to the surface of a fuel rod is carried into the coolant and out of the system by convection. Such convective heat transfer is discussed in Section 8.4.

For completeness, it should be mentioned that heat can also be transferred as thermal radiation across a vacuum or other rarefied space between a hotter body and a colder one. However, this process is of relatively little importance except in some gas-cooled reactors and is not considered further.

The Equations of Heat Conduction

The fundamental relationship governing heat conduction is *Fourier's law,* which for an isotropic medium is written as

$$\mathbf{q}'' = -k \text{ grad } T. \tag{8.25}$$

Here $\mathbf{q}''$, which is called the *heat flux*, is defined so that $\mathbf{q}'' \cdot \mathbf{n}$ is equal to the rate of heat flow across a unit area with unit outward normal $\mathbf{n}$. Thus, $\mathbf{q}''$ is entirely analogous to the neutron current density vector defined in Chapter 5. In the English system, $\mathbf{q}''$ has units of Btu/hr-ft^2 and W/m^2 in SI units. The parameter k in Eq. (8.25) is called the *thermal conductivity* and has units of Btu/hr-ft-°F in the English system and W/m-K in SI units. Values of k for a number of important substances are given in Appendix IV. In general, k is a function of temperature. The function T in Eq. (8.25) is the temperature in °F or K as appropriate. The similarity between Eq. (8.25) and Fick's law of diffusion should be especially noted.

Consider an arbitrary volume V of material throughout a portion of which heat is being produced. From the conservation of energy, the net rate at which heat flows out of the surface of V in the steady state, must be equal to the total rate at which heat is produced within V. If this were not the case, the substance would change temperature and therefore would not be in a steady state. In equation form,

$$\begin{bmatrix} \text{Net rate of flow} \\ \text{of heat out of } V \end{bmatrix} - \begin{bmatrix} \text{rate of heat production} \\ \text{within } V \end{bmatrix} = 0. \qquad (8.26)$$

The net rate at which heat flows out of the surface of V is

$$\text{Heat flow} = \int_A \mathbf{q}'' \cdot \mathbf{n} \, dA, \qquad (8.27)$$

where the vector $\mathbf{q}''$ is as defined earlier, $\mathbf{n}$ is a unit vector normal to the surface, and the integral is taken over the entire surface. From the divergence theorem, Eq. (8.27) can also be written as

$$\text{Heat flow} = \int_V \text{div } \mathbf{q}'' dV, \qquad (8.28)$$

The total rate of heat production within V is equal to

$$\text{Heat production} = \int_V q''' dV, \qquad (8.29)$$

where q''' is the rate at which heat is produced per unit volume.

Equations (8.28) and (8.29) can now be introduced into Eq. (8.26). Since the integrals are over the same arbitrarily selected volumes, their integrands must be equal and the following expression is obtained:

$$\text{div } \mathbf{q}'' - q''' = 0. \qquad (8.30)$$

This result is the *steady-state equation of conductivity* for heat transfer and is analogous to the equation of continuity (Eq. 5.15) discussed in Section 5.3 in connection with neutron diffusion. There is no term in Eq. (8.30) equivalent to the absorption term appearing in the neutron case; heat simply does not vanish within a medium as neutrons are inclined to do.

When Fourier's law, Eq. (8.25), is substituted into Eq. (8.30) and the resulting equation is divided through by k, which is assumed to be constant, the result is

$$\nabla^2 T = \frac{q'''}{k} = 0. \tag{8.31}$$

This is called the *steady-state heat conduction equation* and is of a form known as *Poisson's equation*. In a region where there are no heat sources, $q''' = 0$ and Eq. (8.31) reduces to

$$\nabla^2 T = 0, \tag{8.32}$$

which is called *Laplace's equation.*

These results are now applied to some problems of interest in nuclear reactors. One of the central problems, as is seen, is the calculation of the heat that can be transferred out of a fuel rod and ultimately into a coolant for a given maximum temperature in the fuel. The maximum fuel temperature is a preset condition that must not be exceeded for reasons of safety.

Plate-Type Fuel Elements

Consider first a plate-type fuel element or rod like that shown in Fig. 8.4, consisting of a fueled central strip (the "meat") of thickness $2a$ surrounded on all sides by cladding of thickness b. It is assumed that heat is generated uniformly within the fuel at the rate of q''' Btu/hr-ft^3 or kW/lit, and that the temperature has reached a steady-state distribution throughout the element.

Ordinarily, the total thickness of such an element is small compared with either its width or length. It is reasonable to ignore the negligible amount of heat flowing out through the edges or ends of the element. In short, the heat flows only in the x direction, where x is the distance from the center of the element normal to the surface (see Fig. 8.4). The temperature distribution in the fuel is then determined

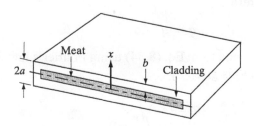

Figure 8.4 A plate-type fuel element.

by Poisson's equation in one dimension,

$$\frac{d^2T}{dx^2} + \frac{q'''}{k_f} = 0,$$ (8.33)

where k_f is the thermal conductivity of the fuel.

Two boundary conditions must be specified with a second-order differential equation like Eq. (8.33). In the present problem, these are

(i) $T(0) = T_m$,

where T_m is the maximum (central temperature) in the fuel, and

(ii) $\dfrac{dT}{dx} = 0$

at $x = 0$. The second condition follows from the symmetry of the problem, which permits no heat flow at the center of the fuel.

Integrating Eq. (8.33) twice gives the general solution

$$T = \frac{q'''}{2k_f}x^2 = C_1x + C_2,$$

where C_1 and C_2 are constants. Placing $x = 0$ immediately gives $C_2 = T_m$. Also, in view of condition (ii), C_1 must be taken to be zero. The temperature within the fuel is therefore

$$T = T_m - \frac{q'''}{2k_f}x^2.$$ (8.34)

Using this equation, the temperature T_s at the surface of the fuel (at the fuel-cladding interface) may be evaluated. Thus, writing $x = a$ in Eq. (8.34) gives

$$T_s = T_m - \frac{q'''a^2}{2k_f}.$$ (8.35)

The total rate of heat production in the fuel is equal to q''' multiplied by the fuel volume $2Aa$, where A is the area of one face of the fuel. In the steady state, all of the heat produced within the fuel flows out of the fuel. The heat flowing through *one face* of the fuel is therefore

$$q = q'''Aa.$$ (8.36)

This result can also be obtained from Eq. (8.34) using Fourier's law. Thus, the heat flux (heat flow per unit area) is given by

$$q'' = k_f\frac{dT}{dx},$$

where the derivative is to be evaluated at $x = a$. Inserting Eq. (8.34) and performing the differentiation yields

$$q'' = q'''a.$$

The total rate of heat flowing out of one side of the fuel is then

$$q = q''A = q'''Aa,$$

which is the same as Eq. (8.36).

It is sometimes convenient to rewrite Eq. (8.36) in a form in which q''' has been eliminated through Eq. (8.35). Thus, solving Eq. (8.35) for q''' and inserting the result into Eq. (8.36) gives

$$q = \frac{T_m - T_s}{a/2k_f A}. \tag{8.37}$$

This expression may be viewed as the heat transfer analogue of Ohm's law in electricity—namely,

$$I = \frac{V}{R}, \tag{8.38}$$

where I is the current, V is the potential difference, and R is the resistance. In the present case, q corresponds to I, $T_m - T_s$ is analogous to the potential difference, and $a/2k_f A$ is called the *thermal resistance*. The value of this analogy is clear shortly.

Turning next to the temperature distribution in the cladding of the plate-type fuel element shown in Fig. 8.4, it is first observed that, since there is little or no heat generated in this region, $q''' = 0$ and the heat conduction equation reduces to

$$\frac{d^2T}{dx^2} = 0. \tag{8.39}$$

This is the simplest form of Laplace's equation (Eq. 8.32). The boundary conditions are now

$$\text{(i)} \quad T(a) = T_s$$
$$\text{(ii)} \quad T(a+b) = T_c,$$

where T_c is the temperature at the outer surface of the cladding.

Integrating Eq. (8.39) twice gives

$$T = C_1 x + C_2.$$

The constants C_1 and C_2 are easily found from conditions (i) and (ii), and the final expression for T is

$$T = T_s - \frac{x-a}{b}(T_s - T_c). \tag{8.40}$$

It should be observed that T is a linear function of position in the source-free cladding, whereas it is quadratic in the region containing fuel. The temperature distribution in the fuel element is as shown in Fig. 8.5.

Since there are no sources or sinks for heat within the cladding, all of the heat passing into the cladding from the fuel is conducted to the outer surface of the cladding. This rate of heat flow can be found by multiplying the heat flux, which is obtained by applying Fourier's law to Eq. (8.40), by the area of the element. This gives

$$q = \frac{k_c A}{b}(T_s - T_c), \tag{8.41}$$

where k_c is the conductivity of the cladding. If Eq. (8.41) is put in the form

$$T_s - T_c = q\left(\frac{b}{k_c A}\right)$$

and Eq. (8.37) is rewritten as

$$(T_m - T_s) = q\left(\frac{a}{2k_f A}\right),$$

the temperature T_s, which is not a conveniently measured parameter, can be eliminated by adding the last two equations. The result is

$$(T_m - T_c) = q\left(\frac{a}{2k_f A} + \frac{b}{k_c A}\right)$$

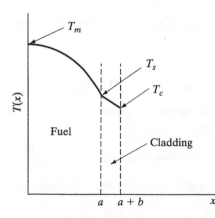

Figure 8.5 Temperature distribution across plate-type fuel element.

or

$$q = \frac{T_m - T_c}{\dfrac{a}{2k_f A} + \dfrac{b}{k_c A}}.$$ (8.42)

This formula gives the rate at which heat flows through one side of the fuel plate in terms of the difference in temperature between the center and surface of the plate. In the electrical analogue, $T_m - T_c$ is the difference in potential and

$$R = \frac{a}{2k_f A} + \frac{b}{k_c A}$$ (8.43)

is the total thermal resistance of the fuel and cladding.

Equation (8.43) shows that, when heat passes through a succession of two materials, the situation is analogous to an electrical potential across two resistances connected in series, where the total resistance is the sum of the two. This conclusion is readily generalized to a sequence of any number of materials—the total thermal resistance is the sum of the resistances of each material.

In the derivation of Eq. (8.42), it is assumed implicitly that the fuel and cladding are tightly bonded together at their point of contact. Some fuel elements, however, have a thin region of bonding material between the fuel and cladding, whereas others have a narrow region of gas between the two. The total thermal resistance is larger in this case and is obtained by adding an appropriate resistance term to Eq. (8.43). The temperature difference $T_m - T_c$ must then be greater for a given heat flow out of the surface of the fuel element.

Cylindrical Fuel Rod

Consider a long cylindrical fuel rod of radius a surrounded by cladding of thickness b. It is assumed again that heat is produced at the constant rate q''' within the rod and that there is no heat released in the cladding.

The temperature in the rod is only a function of the distance r from the axis of the rod, so that in cylindrical coordinates the heat conduction equation is

$$\frac{d^2T}{dr^2} + \frac{1}{r}\frac{dT}{dr} + \frac{q'''}{k_f} = 0.$$ (8.44)

The boundary conditions appropriate to this problem are

(i) T is nonsingular within the rod,
(ii) $T(0) = T_m$,

where T_m is the central temperature of the fuel.

It can be readily verified by substitution that the general solution to Eq. (8.44) is

$$T = -\frac{q''' r^2}{4k_f} + C_1 \ln r + C_2,$$

where C_1 and C_2 are constants to be determined. In view of the boundary conditions, it is evident that $C_1 = 0$ and $C_2 = T_m$. Thus, the temperature within the rod is

$$T = T_m - \frac{q''' r^2}{4k_f}. \tag{8.45}$$

The rate at which heat is produced within the rod, and therefore the rate at which it flows out of the rod, is equal to

$$q = \pi a^2 H q''', \tag{8.46}$$

where H is the length of the rod (approximately equal to the height of the core). Solving for q''' and introducing this into Eq. (8.45), evaluated at $r = a$ where the temperature is T_s, then gives

$$q = \frac{T_m - T_s}{1/(4\pi H k_f)}. \tag{8.47}$$

In view of Eq. (8.47), the thermal resistance of the fuel is clearly

$$R_f = \frac{1}{4\pi H k_f}. \tag{8.48}$$

The heat conduction equation for the cladding is

$$\frac{d^2 T}{dr^2} + \frac{1}{r}\frac{dT}{dr} = 0,$$

which has the solution

$$T = C_1 \ln r + C_2.$$

Using the boundary conditions

(i) $T(a) = T_s$,
(ii) $T(a + b) = T_c$

to determine the constants C_1 and C_2 gives

$$T = \frac{T_s \ln(a + b) - T_c \ln a - (T_s - T_c)\ln r}{\ln(1 + b/a)}. \tag{8.49}$$

From Fourier's law, the total heat flowing out of the cladding is

$$q = -2\pi(a+b)Hk_c\frac{dT}{dr},$$

evaluated at $r = a + b$. Carrying out the differentiation of Eq. (8.49) yields

$$q = \frac{2\pi Hk_c(T_s - T_c)}{\ln(1 + b/a)}. \tag{8.50}$$

The thermal resistance of the cladding is therefore

$$R_c = \frac{\ln(1 + b/a)}{2\pi Hk_c}. \tag{8.51}$$

Frequently, b is much less than a. In this case, since

$$\ln\left(1 + \frac{b}{a}\right) \simeq \frac{b}{a},$$

Eq. (8.51) can be written as

$$R_c \simeq \frac{b}{2\pi aHk_c}. \tag{8.52}$$

In terms of the overall temperature difference between the center of the fuel and the surface of the cladding, the heat flowing out of the rod is

$$q = \frac{T_m - T_c}{R_f + R_c}, \tag{8.53}$$

where R_f is given by Eq. (8.48) and R_c is given by Eq. (8.51) or (8.52). This result may be verified by solving Eq. (8.47) for $T_m - T_s$ and Eq. (8.50) for $T_s - T_c$ and adding the two expressions.

Space-Dependent Heat Sources

In both of the foregoing calculations of the temperature distribution and heat flow in plate-type and cylindrical fuel rods, it is assumed that the heat is produced at the constant rate q''' throughout the fueled portion of the rod. Accordingly, the temperature does not vary along the length of the rod, and it is possible to relate the total heat flow out of the *entire* rod to the difference in temperature between the center and surface of the rod.

However, it is shown in Section 8.2 that q''' does, in fact, vary approximately as (see Eqs. [8.12] and [8.13])

$$q''' = q'''_{\max} \cos \left(\frac{\pi z}{\widetilde{H}} \right),$$ (8.54)

where z is measured from the midpoint of the rod. There is negligible variation in q''' in the x and y directions of the plate-type element or in the radial direction of a cylindrical rod. A nonuniform heat distribution like that given in Eq. (8.54) would be expected to give rise to a nonuniform temperature distribution along the length of a fuel rod—and, indeed, to some extent this is the case. In a power reactor, however, the coolant passes along the fuel rods at a temperature considerably lower than the temperature of the fuel, and the temperature gradient is therefore much steeper across the diameter of the fuel than along the length of the rod. As a consequence, most of the heat generated in the fuel flows directly to the surface of the rod; there is little or no heat flow in the z direction.

It is possible to generalize the calculations made in this section to the case where q''' is a function of z. This can be done by dividing the calculated value of q, which is the the total heat flow out of the surface of a fuel rod, by the surface area of the rod. This gives the heat flux, q'', which is independent of position if q''' is constant, but varies with position if q''' is not. For instance, if Eq. (8.46) is divided by $2\pi(a + b)H$, the area of a clad fuel rod, the result is

$$q''(z) = \frac{a^2}{2(a + b)} q'''(z).$$ (8.55)

This is the heat flux at the surface of the rod as a function of z. Similarly, $q''(z)$ can be found as a function of the temperature difference $T_m - T_c$, where both temperatures may be functions of z, by dividing Eq. (8.53) by $2\pi(a + b)H$:

$$q''(z) = \frac{T_m(z) - T_c(z)}{2\pi(a + b)H(R_f + R_c)}.$$ (8.56)

Example 8.5

The fuel rods for the reactor described in Example 8.3 consist of a fueled portion 0.42 in. in diameter that is clad with Zircaloy-4, 0.024 in thick. Each rod is 12 ft long. Given that the center temperature of the fuel is 3,970 °F at the midpoint of the central rod, calculate at this point the (a) heat flux out of the rod; (b) outer temperature of the cladding.

Solution.

1. From Example 8.3, q''' at the midpoint of the central rod is 4.66×10^7 BTU/hr-ft^3. The radius of the fuel is $0.21/12 = 0.0175$ ft, whereas the cladding thickness is $0.024/12 = 0.002$ ft. Then using Eq. (8.55),

$$q'' = \frac{(0.0175)^2}{2(0.0175 + 0.002)} \times 4.66 \times 10^7$$

$$= 3.66 \times 10^5 \text{ Btu/hr-ft}^2$$

$$= 115.5 \text{W/cm}^2. \ [\textit{Ans.}]$$

2. From Table IV.6, $k_f = 1.1$ Btu/hr-ft-°F and $k_c \simeq 10$ Btu/hr-ft-°F. Equation (8.48) then gives for the thermal resistance of the fuel

$$R_f = \frac{1}{4\pi \times 12 \times 1.1} = 6.03 \times 10^{-3} \text{ °F-hr/Btu.}$$

For the cladding, using Eq. (8.51),

$$R_c = \frac{\ln(1 + 0.024/0.21)}{2\pi \times 12 \times 10} = 1.43 \times 10^{-4} \text{ °F-hr/Btu.}$$

The total resistance is then 6.17×10^{-3} °F-hr/Btu.
Introducing the prior parameters into Eq. (8.56) then gives

$$T_c = 3{,}970 - 2\pi(0.0175 + 0.002) \times 12 \times 6.17$$

$$\times 10^{-3} \times 3.66 \times 10^5 = 650\text{°F.} \ [\textit{Ans.}]$$

Exponential Heat Sources

It is often necessary to calculate the temperature distribution and heat transmission in reactor shields and pressure vessels in which radiation energy is deposited more or less exponentially. Consider a slab of thickness a whose surfaces are held at the temperatures T_1 and T_2, as shown in Fig. 8.6. If x is measured from the surface as

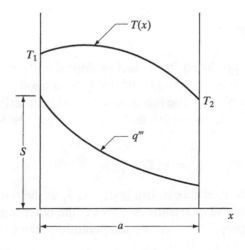

Figure 8.6   Slab containing exponentially distributed heat source.

indicated in the figure, the heat source distribution is given by

$$q''' = Se^{-\mu x},\tag{8.57}$$

where S and μ are constants. For γ-rays, μ may often be taken to be the linear absorption coefficient.

Introducing Eq. (8.57) into the one-dimensional heat conduction equation (Eq. 8.33) gives

$$\frac{d^2T}{dx^2} + \frac{S}{k}e^{-\mu x} = 0.\tag{8.58}$$

The general solution to this equation is easily found to be

$$T = C_1 x + C_2 - \frac{S}{k\mu^2}e^{-\mu x},$$

where the constants C_1 and C_2 are found from the boundary conditions

$$T(0) = T_1, \quad T(a) = T_2.$$

The final expression for the temperature distribution is

$$T = T_1 + (T_2 - T_1)\frac{x}{a} + \frac{S}{k\mu^2}\left[1 - e^{-\mu x} - \frac{x}{a}(1 - e^{-\mu a})\right].\tag{8.59}$$

This equation is the starting point for calculations of heat removal from media in which radiation is absorbed exponentially. It is also used to compute the thermal stresses that accompany such nonuniform temperature distributions. As is shown in the problems at the end of the chapter, under certain conditions the temperature in the slab may rise to a maximum value that is larger than either T_1 or T_2.

8.4 HEAT TRANSFER TO COOLANTS

As noted earlier, the heat produced in the fuel or deposited by radiation in other parts of a reactor is transferred to a coolant of one type or another, which, in turn, carries the heat outside the system. The fundamental relation describing the transfer of heat from a heated solid to a moving fluid (liquid or gas) is *Newton's law of cooling*:

$$q'' = h(T_c - T_b).\tag{8.60}$$

In this expression, q'' is the heat flux in Btu/hr-ft^2-°F, T_c is the temperature of the surface of the solid (in the case of reactor fuel, this is the outer temperature of the cladding), and T_b is an appropriate reference temperature of the fluid.

The numerical value of the heat transfer coefficient depends on many factors, including the nature of the coolant, the manner in which it flows by the heated surface, and the coolant temperature. Methods for calculating h from experimental data are presented later in this chapter. At this point, however, it is of interest to note the range of values of h that are applicable to the coolants in power reactors. For ordinary and heavy water, h usually lies between 5,000 and 8,000 Btu/hr-ft^2-°F (25 and 45 kW/m^2 K); for gases, h ranges between about 10 and 100 Btu/hr-ft^2-°F (55 and 550 W/m^2 K); and for liquid sodium, h is generally between 4,000 and 50,000 Btu/hr-ft ^{2}-°F (20 and 300 kW/m^2 K).

In most reactor heat transfer problems, the fluid flows along the fuel in well-defined coolant channels. Since the temperature of the fluid ordinarily varies with position across each channel, it is possible to define the temperature T_b in Eq. (8.60) in any number of ways. In most heat transfer calculations, T_b is taken to be the *mixed mean* or *bulk temperature* of the fluid. This is defined by the formula

$$T_b = \frac{\int \rho c_p v T \, dA_c}{\int \rho c_p v \, dA_c}, \tag{8.61}$$

where ρ is the fluid density, c_p is its specific heat, v is the fluid velocity, T is the temperature, dA_c is the differential of channel area, and the integrations are carried out over the cross-section of the channel. In general, each of the variables in the integrals may be a function of position across the channel. The temperature T_b defined by Eq. (8.61) is the temperature the fluid would achieve if it were allowed to mix adiabatically—for instance, if the fluid flowing out of the channel were collected and mixed in an insulated container. Measured values of h are normally given in terms of the bulk temperature of the fluid.

The total rate of heat flow across an area A between a solid and a fluid is

$$q = q''A = hA(T_c - T_b). \tag{8.62}$$

Written in the form of Ohm's law, Eq. (8.62) is

$$q = \frac{T_c - T_b}{1/ha}. \tag{8.63}$$

The denominator in this equation is the thermal resistance for convective heat transfer,

$$R_h = \frac{1}{hA}. \tag{8.64}$$

It is now possible to return to the fuel element problems considered in the last section and compute the rate at which heat is transferred to a coolant, for a given difference in temperature between the center of the fuel and the fluid. For the cladded plate-type fuel element, the total thermal resistance is the sum of Eqs. (8.43) and (8.64):

$$R = \frac{a}{2k_f A} + \frac{b}{k_c A} + \frac{1}{hA}, \tag{8.65}$$

and the total heat flow through one side of the fuel element to the fluid is

$$q = \frac{T_m - T_b}{\frac{a}{2k_f A} + \frac{b}{k_c A} + \frac{1}{hA}}. \tag{8.66}$$

Similarly, from Eqs. (8.48) and (8.51), the total thermal resistance for a clad cylindrical fuel rod is

$$R = \frac{1}{4\pi H k_f} + \frac{\ln(1 + b/a)}{2\pi H k_c} + \frac{1}{hA}, \tag{8.67}$$

where $A = 2\pi (a + b) H$. If $b \ll a$,

$$R = \frac{1}{4\pi H k_f} + \frac{b}{2\pi a H k_c} + \frac{1}{hA}. \tag{8.68}$$

In either case, the rate of heat flow into the coolant is

$$q = \frac{T_m - T_b}{R}. \tag{8.69}$$

These results, which apply to the case of uniform heat production along the length of the fuel, can be generalized to the more realistic situation where q''' depends on z. Thus, as explained in the preceding section, most of the heat produced in the fuel flows directly to the coolant in a direction normal to the axis of the fuel rod. The expression for the heat flux, obtained by dividing q in Eq. (8.69) by the surface area of the rod, is therefore valid at every point along the rod. Thus, for the cylindrical rod, it follows that

$$q''(z) = \frac{T_m(z) - T_b(z)}{2\pi (a + b) H R}. \tag{8.70}$$

In terms of T_c, the outer temperature of the cladding,

$$q''(z) = \frac{T_c(z) - T_b(z)}{2\pi (a + b) H R_h}$$
$$= h[T_c(z) - T_b(z)],$$

which is simply Newton's law, Eq. (8.60).

Example 8.6

The fuel rods in Examples 8.3 and 8.5 are cooled with pressurized water. Given that the heat transfer coefficient is 7,500 Btu/hr-ft^2-°F, calculate the bulk temperature of the water opposite the midpoint of the hottest rod.

Solution. According to Example 8.5, at the point in question, $q'' = 3.66 \times 10^5$ Btu/hr-ft^2 and $T_c = 650$°F. From Eq. (8.60),

$$T_b = 650 - 3.66 \times 10^5/7,500$$

$$= 601°F. \; [Ans.]$$

For reasons discussed later, this is not the highest temperature of the coolant, although q''' is greatest at this point.

Temperature along a Coolant Channel

As the coolant moves along the fuel, it absorbs heat; as a result, its temperature continually increases. However, the temperature does not increase at a constant rate since the heat is released from the fuel nonuniformly—according to

$$q''' = q'''_{max} \cos \left(\frac{\pi z}{\widetilde{H}} \right). \tag{8.71}$$

This equation applies to the central ($r = 0$) rod; for other rods, q''' must be multiplied by $J_0(2.405r/\widetilde{R})$. A simple method is now discussed for estimating the variation in the temperature of the coolant and the fuel as a function of z, when the heat is produced in the fuel according to Eq. (8.71).

Before proceeding, it should be noted that, except for the fuel rods at the edge of the core, there is only one coolant channel associated with each fuel rod. This may be taken to be either the volume between neighboring fuel rods, with a portion of each rod contributing heat to the channel, or the volume of coolant in a unit cell surrounding one rod. As shown in Fig. 8.7 for a square array of rods, the volumes of coolant associated with the surface area of one rod are the same in both cases.

Consider a coolant slab of thickness dz as it moves along a coolant channel. The volume of this slab is equal to $A_c dz$, where A_c is the cross-sectional area of the channel, and its mass is $\rho A_c dz$, where ρ is the coolant density. As the slab progresses up the channel the distance dz, it absorbs heat from the fuel, which raises the temperature of the slab the amount dT_b.[6] The heat required to produce the change dT_b is equal to

$$\rho A_c dz c_p dT_b,$$

---

[6]Throughout this discussion, the coolant is assumed not to boil. If it is boiling, then the absorption of heat goes into vaporization of the coolant at constant temperature.

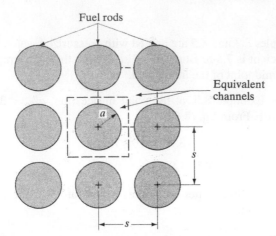

Fuel rods

Equivalent channels

Figure 8.7   Equivalent cooling channels in a square fuel lattice.

where c_p is the specific heat per unit mass. The *rate* at which heat is added to the flowing coolant is obtained by dividing this quantity by the time dt required for the coolant to move the distance dz. Thus,

$$dq = \rho A_c \frac{dz}{dt} c_p dT_b = \rho A_c v c_p dT_b,$$

where v is the velocity of the fluid. However, $\rho A - cv$ is equal to the rate of flow of the coolant through the channel, w, and so

$$dq = w c_p dT_b. \tag{8.72}$$

As noted earlier, most of the heat produced in the fuel flows directly into the coolant and not along the interior of the fuel. Therefore, the heat dq in Eq. (8.72) is generated in a fuel volume of $A_f dz$, where A_f is the cross-sectional area of the fuel portion of a fuel rod. Thus,

$$w c_p dT_b = q''' A_f dz.$$

Introducing the function q''' from Eq. (8.71) and integrating from the point of entry of the coolant at $z = -H/2$, where its temperature is T_{b0}, to an arbitrary point z along the channel gives

$$T_b = T_{b0} + \frac{q'''_{max} A_f H}{\pi w c_p} \left[1 + \sin\left(\frac{\pi z}{\widetilde{H}} \right) \right]$$

$$= T_{b0} + \frac{q'''_{max} V_f}{\pi w c_p} \left[1 + \sin\left(\frac{\pi z}{\widetilde{H}} \right) \right], \tag{8.73}$$

where V_f is the volume of the fueled portion of the rod.

Equation (8.73) gives the temperature of the coolant as a function of position along the central or *hottest* channel. The temperature along other channels can be obtained by multiplying the second term in Eq. (8.73) by the factor $J_0(2.405r/\widetilde{R})$. According to Eq. (8.73), the coolant temperature increases along the channel and reaches a maximum value

$$T_{b,\max} = T_{b0} + \frac{2q'''_{\max}V_f}{\pi w c_p} \tag{8.74}$$

at the exit of the channel. This behavior of T_b is illustrated in Fig. 8.8. The temperature of the fluid leaving channels other than the central, hottest channel is also highest at the exit, but necessarily lower than the temperature of the coolant leaving the hottest channel.

The temperature T_c of the surface of the cladding can now be found as a function of position along the channel by observing that the heat transferred from a length dz of the fuel rod to the coolant is $hC_c dz\,(T_c - T_b)$, where C_c is the circumference of the clad rod. This is equal to the heat generated in the length dz of the fuel—namely, $q'''_{\max}A_f dz \cos(\pi z/\widetilde{H})$. It follows that

$$hC_c(T_c - T_b) = q'''_{\max}A_f \cos\left(\frac{\pi z}{\widetilde{H}}\right).$$

Inserting T_b from Eq. (8.73) and solving for T_c gives

$$T_c = T_{b0} + \frac{q'''_{\max}V_f}{\pi w c_p}\left[1 + \sin\left(\frac{\pi z}{\widetilde{H}}\right)\right] + \frac{q'''_{\max}A_f}{hC_c}\cos\left(\frac{\pi z}{\widetilde{H}}\right).$$

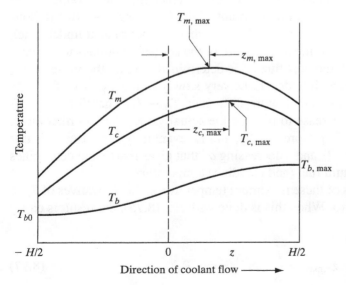

Figure 8.8 Axial temperature of fuel, T_m, temperature of cladding surface, T_c, and bulk temperature of coolant, T_b, as functions of distance along a coolant channel.

This may also be written as

$$T_c = T_{b0} + \frac{q'''_{max} V_f}{\pi w c_p} \left[1 + \sin\left(\frac{\pi z}{\widetilde{H}}\right) \right] + q'''_{max} V_f R_h \cos\left(\frac{\pi z}{\widetilde{H}}\right), \qquad (8.75)$$

where R_h is the resistance for convective heat transfer (see Eq. [8.64]).

The temperature T_m along the center of the fuel can also be calculated from Eq. (8.73). It is assumed that all of the heat produced in a slab of fuel of thickness dz flows directly into the coolant. This rate of heat flow is equal to the temperature difference $T_m - T_b$ divided by the total thermal resistance of a section of fuel and coolant dz thick. The resistance of such a section is simply the resistance calculated earlier for the full length of the fuel multiplied by the ratio dz/H. Equating heat production and heat flux then gives

$$\frac{T_m - T_b}{H} = q'''_{max} A_f \cos\left(\frac{\pi z}{\widetilde{H}}\right),$$

where R is given by Eq. (8.65) for plate-type fuel elements and by Eq. (8.67) or Eq. (8.68) for cylindrical fuel rods. Solving for T_m and introducing T_b from Eq. (8.73) yields finally

$$T_m = T_{b0} + \frac{q'''_{max} V_f}{\pi w c_p} \left[1 + \sin\left(\frac{\pi z}{\widetilde{H}}\right) \right] + q'''_{max} V_f R \cos\left(\frac{\pi z}{\widetilde{H}}\right). \qquad (8.76)$$

A plot of Eqs. (8.75) and (8.76) is given in Fig. 8.8. It is evident from the figure that both T_c and T_m rise along the channel and reach maximum values in the upper part of the channel beyond the midpoint of the fuel. There are two reasons that the maximum temperature of the fuel occurs there, rather than at midchannel, where q''' is the greatest. First, the temperature of the coolant continues to increase past the midpoint. Second, the heat flux q'' is determined only by the value of q''', and this, being a cosine function, decreases very slowly in the vicinity of $z = 0$. But q'', in turn, specifies the temperature difference $T_m - T_b$. Therefore, with T_b increasing, T_m must also increase to provide the actual value of q''. Further along the channel, q'' begins to drop more rapidly, and T_m eventually decreases. It is this combined effect of a rising T_b and a decreasing q'' that gives rise to and determines the position of the maximum fuel (and cladding) temperature.

To find the locations of these maximum temperatures, the derivatives of T_c or T_m are placed equal to zero. When this is done with Eq. (8.75), the result is easily found to be

$$z_{c,max} = \frac{\widetilde{H}}{\pi} \cot^{-1}(\pi w c_p R_h). \qquad (8.77)$$

Similarly, from Eq. (8.76),

$$z_{m,\max} = \frac{\tilde{H}}{\pi}\cot^{-1}(\pi w c_p R).$$ (8.78)

The actual maximum values of T_c and T_m can be found by introducing Eqs. (8.77) and (8.78) into Eqs. (8.75) and (8.76). With a little trigonometry, it is easy to obtain the following results:

$$T_{c,\max} = T_{b0} + q_{\max}''' V_f R_h \left[\frac{1+\sqrt{1+\alpha^2}}{\alpha}\right],$$ (8.79)

where

$$\alpha = \pi w c_p R_h$$ (8.80)

and

$$T_{m,\max} = T_{b0} + q_{\max}''' V_f R \left[\frac{1+\sqrt{1+\beta^2}}{\beta}\right],$$ (8.81)

where

$$\beta = \pi w c_p R.$$ (8.82)

It must be emphasized that the previous calculations of T_b, T_c, and T_m as functions of z are highly approximate and give only the most qualitative behavior of these temperatures. Furthermore, the results are not applicable to liquid metal coolants, where there may be appreciable heat conduction in the z direction. The derivations also do not apply to coolants that are undergoing bulk boiling, although they are valid for locally boiling coolants.

Example 8.7

Pressurized water enters the core of the reactor described in Examples 8.3 and 8.5 at a temperature of 543°F and passes along the fuel rods at the rate of 3,148 lb/hr per channel. (a) What is the exit temperature of the coolant from the hottest channel? (b) What is the maximum temperature of the cladding and the fuel in this channel?

Solution.

1. The volume of the fueled portion of the fuel rod is

$$V_f = 1.15 \times 10^{-2}\text{ft}^2.$$

From Example 8.3, $q_{\max}''' = 4.66 \times 10^7$ Btu/hr-ft³, and from Table IV.3, c_p has a value of approximately 1.3 Btu/lb-°F. Then using Eq. (8.74),

$$T_{b,\max} = 543 + \frac{2 \times 4.66 \times 10^7 \times 1.15 \times 10^{-2}}{\pi \times 3148 \times 1.13}$$

$$= 543 + 83 = 626°F. \ [Ans.]$$

2. From Eq. (8.64) and Example 8.6, the total convective resistance is

$$R_h = 1/7,500 \times 2\pi(0.0175 + 0.002) \times 12 = 9.07 \times 10^{-5}°F\text{-hr/Btu.}$$

The location of the maximum cladding temperature is then given by Eq. (8.77):

$$z_{c,\max} = \frac{12}{\pi}\cot^{-1}(\pi \times 3148 \times 1.3 \times 9.07 \times 10^{-5})$$

$$= \frac{12}{\pi}\cot^{-1}(1.17) = 2.71 \text{ ft.}$$

Thus, the maximum value of T_c occurs 2.71 ft beyond the midpoint of the channel. The value of α in Eq. (8.80) is 1.17 and from Eq. (8.79), the maximum cladding temperature is

$$z_{c,\max} = 543 + 4.66 \times 10^7 \times 1.15 \times 10^{-2}$$

$$\times 9.07 \times 10^{-5} \times \frac{1 + \sqrt{1 + (1.17)^2}}{1.17}$$

$$= 649°F. \ [Ans.]$$

For the fuel, from Example 8.5, $R_f + R_c = 6.17 \times 10^{-3}$, so the total resistance R is $6.17 \times 10^{-3} + 9.07 \times 10^{-5} = 6.26 \times 10^{-3} \,°F\text{-hr/Btu.}$ Then using Eqs. (8.78), (8.81), and (8.82), $\beta = 80.5$, $z_{m,\max} = 0.047$ ft (so that the temperature of the fuel is greatest at just about the midpoint of the fuel), and $T_{m,\max} = 3,940°F.$ [Ans.]

The Heat Transfer Coefficient—Nonmetallic Coolants

The extent to which heat is transferred to a moving fluid, and hence the value of the heat transfer coefficient, depends on the details of the internal motions of the fluid as it flows along a coolant channel. If every portion of the fluid moves parallel to the walls of the channel, then the heat travels radially into the fluid largely by conduction. The flow, in this case, is said to be *laminar.* In contrast, if there are significant radial components of velocity fluctuations within the fluid, the heat is picked up at the wall by portions of the fluid and carried directly into the interior of the channel. This is the description of *turbulent* flow. Clearly, other things being equal, heat is more readily transferred to a fluid undergoing turbulent flow than laminar flow. As a general rule, in those reactors in which the coolant is pumped through the system (as opposed to reactors cooled by natural convection), the coolant flows under turbulent conditions.

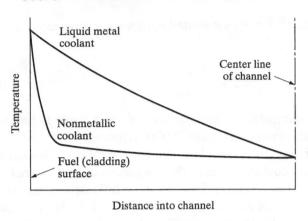

Figure 8.9   Temperature distributions in a nonmetallic coolant and in a liquid metal coolant undergoing turbulent flow.

One of the consequences of the internal motions of the coolant undergoing turbulent flow is that the temperature distribution tends to be more or less uniform over much of the interior region of a coolant channel. Thus, as shown in Fig. 8.9, the temperature drops rapidly with distance in the vicinity of the fuel and quickly reaches the bulk temperature of the fluid. This is in marked contrast to the situation in conduction, where the temperature changes more uniformly with position. Even under turbulent conditions, however, some heat is always transferred to the interior of a coolant by conduction, but for the nonmetallic coolants this contribution is negligibly small. With ordinary water, for example, less than 1% of the heat transfer occurs by conduction when the water is undergoing turbulent flow.

It is possible to characterize the flow of a fluid in terms of a dimensionless parameter known as the *Reynolds number,* which is defined as

$$\text{Re} = \frac{D_e v \rho}{\mu}, \tag{8.83}$$

where D_e is the *equivalent diameter* of the coolant channel discussed later, v is the average velocity of the fluid, ρ is its density, and μ is the fluid viscosity. The value of D_e is to be computed from the formula

$$D_e = 4 \times \frac{\text{cross-sectional area of coolant channel}}{\text{wetted perimeter of coolant channel}}. \tag{8.84}$$

The term *wetted perimeter* in this expression refers to that portion of the perimeter of the channel that is structural and that therefore creates drag to passing coolant. For a hollow pipe carrying a coolant, the wetted perimeter is simply the interior perimeter of a section of the pipe perpendicular to its axis. Thus, with a circular pipe of inside radius a, the numerator in Eq. (8.84) is πa^2 and the denominator is $2\pi a$, so that $D_e = 2a$, the actual pipe diameter. For a bundle of rods of radius a in a square array of pitch s, as shown in Fig. 8.7, the cross-sectional area of a single coolant channel is equal to $s^2 - \pi a^2$, the wetted perimeter is $2\pi a$, and Eq. (8.84) gives

$$D_e = 2 \times \frac{s^2 - \pi a^2}{\pi a}. \tag{8.85}$$

It has been found experimentally that the flow of most fluids is laminar up to a value of the Reynolds number of about 2,000. Between 2,000 and 10,000, the flow is partly laminar and partly turbulent, the fraction of each depending on the structural details of the coolant channel, the roughness of the channel walls, and other factors. Above Re = 10,000, a fluid moves in fully developed turbulent flow. In summary, a high value of the Reynolds number implies a large amount of turbulence, a high value of the heat transfer coefficient, and a high rate of heat flow into the coolant for a given difference in temperature between the cladding and the coolant.

Example 8.8

The fuel rods described in Example 8.5 are placed in a square array with a pitch of 0.600 in. The rods are cooled by pressurized water ($P = 2,000$ psi), which is flowing at a speed of 15.6 ft/sec. Calculate the Reynolds number for this coolant flow assuming the water temperature is 600°F.

Solution. The radius of the fuel rods is $0.210 + 0.024 = 0.234$ in. From Eq. (8.85), the equivalent diameter is

$$D_e = 2 \times \frac{(0.6)^2 - \pi(0.234)^2}{\pi \times 0.234} = 0.512 \text{ in} = 0.0427 \text{ ft.}$$

The flow velocity is $15.6 \times 3,600 = 56,200$ ft/hr. From Table IV.3, at 600°F and 2,000 psi, $\rho = 42.9$ lb/ft^3 and $\mu = 0.212$ lb/hr-ft. Then substituting into Eq. (8.83) gives

$$\text{Re} = \frac{0.0427 \times 56,200 \times 42.9}{0.212} = 486,000. \text{ [Ans.]}$$

The water is clearly flowing under turbulent conditions.

The numerical value of the heat transfer coefficient h, Eq. (8.60), is a function of the physical properties of the fluid, its rate of flow, and the diameter or effective diameter of the coolant channel. These parameters, including h, can conveniently be grouped together into three dimensionless quantities—the Reynolds number, already discussed; the *Nusselt number,* Nu, defined as

$$\text{Nu} = \frac{hD_e}{k}; \tag{8.86}$$

and the *Prandtl number,* Pr, defined as

$$\text{Pr} = \frac{c_p \mu}{k}. \tag{8.87}$$

In these formulas, D_e is the effective diameter of the channel, k and μ are the conductivity and viscosity of the fluid, and c_p is the specific heat. All of these parameters must be specified, of course, in consistent units.

It can be shown both from an analysis of experiments and from theory that convective heat transfer data can be correlated in terms of three dimensionless numbers—Re, Nu, and Pr. In particular, for ordinary water, heavy water, organic liquids, and most gases, all flowing through long straight channels under turbulent conditions, these data can be represented by an equation of the form

$$\text{Nu} = C\text{Re}^m\text{Pr}^n, \tag{8.88}$$

where C, m, and n are constants. The value of h can be obtained from Eq. (8.88) using the definition of the Nusselt number. Thus,

$$h = C\left(\frac{k}{D_e}\right)\text{Re}^m\text{Pr}^n. \tag{8.89}$$

In using correlations of data of the type expressed by Eqs. (8.88) and (8.89)—or any correlations, for that matter—care must be exercised to determine the reference temperature at which the fluid properties are to be evaluated. Frequently, this is the bulk temperature of the fluid T_b. It should also be noted that these equations are not valid for liquid metals, which must be considered separately.

With ordinary water, heavy water, organic liquids, and gases flowing through long, straight, and *circular* tubes, the following values have been recommended for the constants appearing in Eqs. (8.88) and (8.89): $C = 0.023$, $m = 0.8$, and $n = 0.4$. [With the constants appearing in Eq. (8.88), Eq. (8.88) is known as the *Dittus-Boelter equation.*] These constants are often used in computing h for non-circular coolant channels by introducing the appropriate value of D_e. However, significant errors in h may result if the channels deviate substantially from the circular configuration.

In the important case of ordinary water flowing through a lattice of rods, parallel to the axis of the rods, recommended constants are: $m = 0.8$, $n = 1/3$, and C is given by[7]

$$C = 0.042\frac{P}{D} - 0.024 \tag{8.90}$$

---

[7]Weisman, J., "Heat Transfer to Water Flowing Parallel to Tube Bundles," *Nucl. Sci. & Eng.* 6:78, 1979.

for square lattices with $1.1 \leq \frac{P}{D} \leq 1.3$ and

$$C = 0.026\frac{P}{D} - 0.024 \qquad (8.91)$$

for triangular lattices with $1.1 \leq \frac{P}{D} \leq 1.5$. The quantities P and D are, respectively, the lattice pitch and rod diameter.

Example 8.9

Calculate the heat transfer coefficient for the water flowing through the lattice described in Examples 8.5 and 8.8.

Solution. From Example 8.8, the lattice pitch is 0.6 in, the fuel rod radius is 0.234 in, and C is

$$C = 0.042\frac{P}{D} - 0.024 = 0.0299.$$

According to Table IV.3, $c_p = 1.45$ Btu/lb-°F, $\mu = 0.212$ lb/hr-ft, and $k = 0.296$ Btu/hr-ft°F. Introducing these values into Eq. (8.87) gives

$$Pr = \frac{1.45 \times 0.212}{0.296} = 1.039.$$

From Example 8.8, $D_e = 0.0427$ ft and Re = 486,000. Equation (8.89) then gives

$$h = 0.0299\left(\frac{0.296}{0.0427}\right)(486,000)^{0.8}(1.039)^{1/3}$$

$$= 7436 \text{ Btu/hr-ft}^2\text{-°F. } [Ans.]$$

The Heat Transfer Coefficient—Liquid Metals

Heat transfer to liquid metal coolants is strikingly different from the transfer of heat to ordinary fluids largely because the thermal conductivities of liquid metals are so much higher than those of other types of coolants. At 400°F, for instance, the thermal conductivity of liquid sodium is 46.4 Btu/hr-ft-°F, whereas it is only 0.381 Btu/hr-ft-°F for ordinary water and 0.115 Btu/hr-ft-°F for helium at 1 atm. Thus, at this temperature, the conductivity of sodium is 122 times that of water and 400 times that of helium.

One important effect of the high conductivity of liquid metals is that, even when they are flowing under turbulent conditions, these coolants absorb heat mostly by conduction. This is in marked contrast to the situation with nonmetallic coolants, where, it is recalled, heat transfer occurs largely as the result of the internal motions of the fluid.

Since heat flows into a liquid metal primarily by conduction, the temperature distribution within a coolant channel containing a liquid metal resembles the temperature distribution in a solid conductor whose axis and circumference are held at different temperatures. Thus, as shown in Fig. 8.9, the temperature varies more slowly across the channel with a liquid metal than with a nonmetallic coolant. In both cases, the bulk temperature of the coolant is given by Eq. (8.61).

The heat transfer coefficient for liquid metals has been studied extensively in recent years, and correlations have been given that cover most of the situations encountered in practice. For the case of a liquid metal flowing under turbulent conditions through a hexagonal lattice of rods, parallel to the rods, Dwyer (see References at the end of the chapter) has given the following correlation:

$$\text{Nu} = 6.66 + 3.126(s/d) + 1.184(s/d)^2 + 0.0155(\overline{\Psi}\text{Pe})^{0.86}, \qquad (8.92)$$

where s/d is the ratio of lattice pitch to rod diameter, $\overline{\Psi}$ is a function given graphically by Dwyer that has been fitted by Hubbard[8] by the expression

$$\overline{\Psi} = 1 - \frac{0.942(s/d)^{1.4}}{\text{Pr}(\text{Re}/10^3)^{1.281}}, \qquad (8.93)$$

and Pe is the so-called *Peclet number*, which is given by

$$\text{Pe} = \text{Re} \times \text{Pr} = \frac{D_e v \rho c_p}{k}. \qquad (8.94)$$

Equation (8.92) is valid only for lattices with $s/d > 1.35$. Values of the Nusselt number for more closely spaced hexagonal lattices are given in tabular form by Dwyer, Berry, and Hlavac (see References).

Equations (8.92) and (8.93) may be used for square lattices by replacing the ratio s/d in these expressions by $1.075(s/d)_s$, where $(s/d)_s$ is the value of s/d for the square array. For tightly packed square lattices, the following correlation may be used:

$$\text{Nu} = 0.48 + 0.0133(\text{Pe})^{0.70}. \qquad (8.95)$$

8.5 BOILING HEAT TRANSFER

Up to this point, it has been assumed that a liquid coolant does not undergo a change in phase as it moves along a coolant channel absorbing heat from the fuel. It was assumed that the coolant does not boil. However, there are some distinct advantages in permitting a reactor coolant to boil. For one thing, the coolant pressure is much lower when the coolant is allowed to boil than when boiling must be

---

[8]F. R. Hubbard III, private communication.

prevented. In addition, for a given flow rate and heat flux, lower cladding and fuel temperature are required for a boiling than for a nonboiling coolant. For these reasons, boiling of a restricted nature is now permitted in pressurized water reactors, although steam is not produced directly in these reactors. With boiling water reactors not only is advantage taken of higher heat transfer rates, but, as pointed out in Chapter 4, by producing steam within these reactors, the entire secondary coolant loop of the PWR can be eliminated.

Boiling coolants other than ordinary water have been considered in a number of reactor concepts, but none of these has reached a practical stage of development. Therefore, the following discussion pertains largely to water-cooled reactors. However, the principles to be considered also apply to other types of liquid-cooled reactors.

Boiling Regimes

To understand the phenomenon of boiling heat transfer, consider an experiment in which the heat flux (Btu/hr-ft^2 or W/cm^2) from heated fuel rods to a flowing liquid coolant is measured as a function of the temperature of the surface of the rods for a given system pressure and flow rate. The results of such an experiment are shown in Fig. 8.10. As indicated in the figure, the heat flux increases slowly as the rod

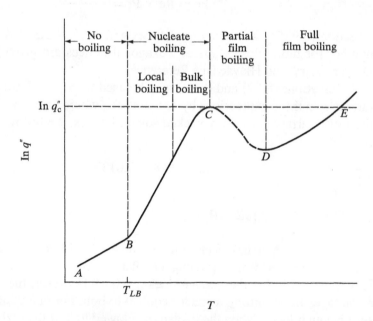

Figure 8.10   The logarithm of the heat flux into a flowing coolant as a function of the surface temperature of the coolant channel.

temperature is increased at low values. In this temperature range, between points A and B in Fig. 8.10, heat is transferred to the coolant by ordinary convection with no change in phase, and the heat transfer coefficient is determined by the correlations given in the preceding section.

As the surface temperature of the fuel is increased further, a point is eventually reached where bubbles of vapor begin to form at various imperfections on the surface of the fuel rods. This occurs at about the point B in Fig. 8.10, and is a form of boiling called *nucleate boiling*. As the bubbles are formed, they are carried away from the rods and into the body of the coolant as a result of the turbulent motions of the fluid. However, as long as the bulk temperature of the coolant is less than its saturation temperature, the vapor in the bubbles soon condenses to the liquid state and the bubbles disappear from the coolant. There is no net production of steam under these circumstances, and the boiling process is termed *subcooled nucleate boiling* or *local boiling*. If and when the bulk temperature of the coolant reaches its saturation temperature, the bubbles persist within the coolant stream, there begins to be a net output of steam, and the system is now said to be undergoing *saturated nucleate boiling* or *bulk boiling*.

The onset of local boiling and the transition to bulk boiling are shown in the lower two portions of Fig. 8.11 for a fluid flowing vertically through a heated pipe.

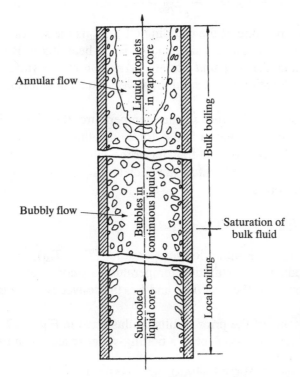

Figure 8.11   Flow patterns in a vertical heated channel. (From L. S. Tong, *Boiling Heat Transfer and Two-Phase Flow*. New York: Wiley, 1967.)

With local boiling, as indicated in the figure, bubbles exist only near the surface of the pipe, and most of the pipe is filled with liquid. In the region with bulk boiling, however, the bubbles are distributed throughout the fluid, which is said to move in *bubbly flow*. Under certain conditions—namely, at high-flow velocities and a large concentration of bubbles (large void fractions)—the bubbles combine to form a void space along the center of the channel. This is called *annular flow*.

In any event, with the onset of nucleate boiling, heat moves readily into the liquid. At every temperature in this region between B and C in Fig. 8.10, heat transfer is more efficient than by ordinary convection. There are two reasons for this. First, heat is removed from the rods both as heat of vaporization and as sensible heat. Second, the motions of the bubbles lead to rapid mixing of the fluid. The rapid increase in the heat flux with temperature, shown by the steep slope of the nucleate boiling region in Fig. 8.10, is explained by the fact that the density of bubbles forming at and departing from the rod surface increases rapidly with surface temperature.

For ordinary water at pressures between 500 and 2,000 psia undergoing nucleate boiling, either local or bulk, the heat flux can be computed from the following correlation:[9]

$$T_c - T_{\text{sat}} = \frac{60(q''/10^6)^{1/4}}{e^{P/900}}, \qquad (8.96)$$

where T_c is the surface temperature of the cladding (°F), T_{sat} is the saturation temperature (°F), P is the system pressure (psia), and q'' is the heat flux in Btu/hr-ft^2. Equation (8.96) is known as the *Jens and Lottes correlation* and is valid for any channel geometry. This correlation is not valid, however, if the coolant is undergoing annular flow.

It is important to estimate the fuel surface temperature T_{LB} at which local boiling begins to know which correlation to use for calculating heat flow into the coolant—a correlation for convective heat transfer or the Jens and Lottes equation. Although there is no single, fixed temperature at which boiling starts, T_{LB} is usually computed from the simple formula

$$T_{\text{LB}} = T_{\text{sat}} + (T_c - T_{\text{sat}})_{\text{JL}} - \frac{q''}{h}. \qquad (8.97)$$

In this equation, T_{sat} is again the saturation temperature, $(T_c - T_{\text{sat}})_{\text{JL}}$ is the difference between the cladding and saturation temperatures as computed from the Jens and Lottes correlation, q'' is the heat flux, and h is the convective heat transfer coefficient.

The application of some of the prior results is illustrated in Fig. 8.12, which shows the parameters of the water coolant in a boiling-water reactor as a function

[9]W. H. Jens and P. A. Lottes, USAEC Report ANL-4627, 1951.

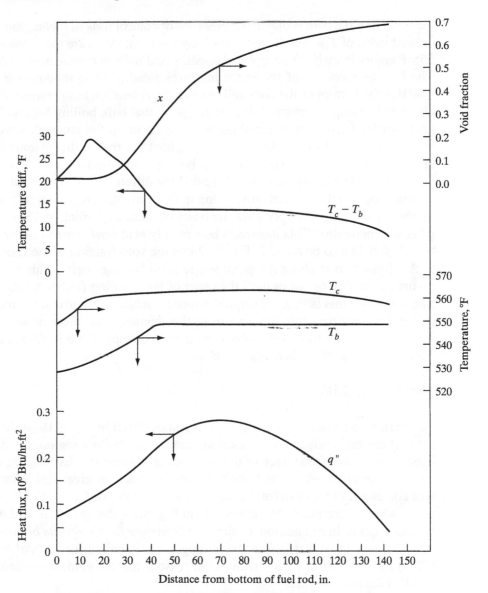

Figure 8.12   Thermal analysis of the hottest channel of a BWR. Here, x refers to void fraction not quality. (Courtesy of F. R. Hubbard III.)

of distance up the hottest channel. In this reactor, the fuel rods are about 12 ft long and the system pressure is 1,035 psia. At this pressure, the saturation temperature of water is 548.7°F, which can be estimated from the data in Table IV.2. The lower curve shows the actual heat flux into the channel. The skewing of q'' away from the bottom of the core is due to the presence of control rods in this region. The calculated value of T_{LB} and the outer cladding temperature T_c are the same—namely, 558°F approximately 13 in up the channel. Local boiling commences at this point. The bulk temperature of the water increases steadily from its entering value of 526°F at the bottom of the core and reaches the saturation temperature of 548.7°F about 45 in along the channel. It is at this point that bulk boiling begins. Thus, the heat transfer is convective for about the first 13 in up the channel, local boiling occurs between 13 in and 45 in, and throughout the rest of the channel the water undergoes bulk boiling. (The flow never becomes annular.)

It should be observed in Fig. 8.12 that the difference in temperature between the cladding and the coolant, after rising in the convective region in response to an increasing q'', drops sharply after the onset of nucleate boiling at 13 in, although q'' is still increasing. This illustrates how readily heat flows into a boiling coolant.

It should also be noted in Fig. 8.12 that the void fraction of the coolant starts to rise from zero at about the point where local boiling begins, although there is no net production of steam until the onset of bulk boiling further up the channel. The reason for this is that, although the steam bubbles that form at the rod surface collapse as they pass into the interior of the channel, their place at the surface is immediately taken by other bubbles. As a consequence, there is always a steady-state distribution of voids across the channel.

The Boiling Crisis

To return to the discussion of the experiment depicted in Fig. 8.10, if the temperature of the fuel surface is increased in the nucleate boiling region, the density of bubbles at or near the surface of the fuel rods also increases. Eventually, however, a point is reached where the bubble density becomes so great that adjacent bubbles coalesce and begin to form a vapor film across the surface of the rods. At this point, which corresponds to the point C in Fig. 8.10, the system is said to be in a *boiling crisis* or in a condition leading to a *departure from nucleate boiling* (DNB). The heat flux at or just before the boiling crisis is called the *critical heat flux*, (CHF) and denoted as q_c''. Sometimes, for reasons that are evident momentarily, q_c'' is called the *burnout flux*.

With the onset of the boiling crisis, the heat flux into the coolant begins to drop. This is due to the fact, that over the regions of the rods covered by vapor film, the heat is forced to pass through the vapor into the coolant by conduction and radiation, both of which are comparatively inefficient mechanisms for heat transfer. The heat flux continues to drop, more or less erratically (as indicated by the dotted lines in Fig. 8.10), with increasing fuel temperatures as the total area of the film covering the fuel increases. In this region of Fig. 8.10, the system is said to be experiencing *partial film boiling*.

Eventually, when the rod surface temperature is high enough, the vapor film covers the entire rod and the heat flux to the coolant falls to a minimum value (point D). Beyond this point, any increase in temperature leads to an increase in the heat flux simply because heat transfer through the film, although a poor and inefficient process, nevertheless increases with the temperature difference across the film. The system, in this case, is said to be undergoing *full film boiling*.

The existence of the various boiling regimes and the boiling crisis is an important consideration in the design of a liquid-cooled reactor, as may be seen from the following example. Suppose that a water-cooled reactor is designed and/or operated in such a manner that at some point along a coolant channel the water undergoes nucleate boiling near DNB conditions. If the reactor power is suddenly increased so that the heat flux into the water rises above the DNB value of q_c'', partial film boiling will immediately begin in this channel. However, as explained previously, the formation of the film impedes the transfer of heat to the coolant. As a consequence, the heat confined, so to speak, within the fuel raises the fuel temperature and the surface temperature of the rods—forming the channel. This, in turn, leads to an increase in the area of the film, which leads to a further decrease in the heat flux, a further increase in rod surface temperature, and so on. In this way, the wall temperature rapidly increases along the boiling curve from point C to point E. Long before E is reached, however, the temperature of the fuel will attain such high values (several thousands of degrees Fahrenheit) that the fuel will partially melt, the cladding will rupture, and fission products will be released into the coolant. As noted earlier, these are occurrences that must be prevented at all costs. For this reason, it is important to know the value of q_c'' and to keep a reactor from operating near the DNB point.

It must not be implied from Fig. 8.10 that it is necessary to have saturated nucleate boiling (bulk boiling) before the onset of the boiling crisis. The figure was intended merely to show that there are two types of nucleate boiling and that it is possible to make the transition from subcooled to saturated boiling if the bulk temperature of the coolant exceeds the saturation temperature. Bubbles at the surface of the fuel can combine to form areas of vapor film, the beginning of the boiling crisis, although the bubbles would condense if they traveled to the interior of the coolant stream.

A great many correlations have been developed from data on q_c'' (see Tong in the References for a tabulation of these correlations). As would be expected, these correlations differ depending on whether the boiling crisis is reached from subcooled or bulk boiling conditions. For subcooled boiling, the following correlation by Jens and Lottes[10] has been widely used:

----

[10]W. H. Jens and P. A. Lottes, *ibid.*

TABLE 8.1   PARAMETERS FOR JENS AND
LOTTES CORRELATION

P (psia)	C	m
500	0.817	0.160
1,000	0.626	0.275
2,000	0.445	0.500

$$q_c'' = C \times 10^6 \left(\frac{G}{10^6}\right)^m \Delta T_{\text{sub}}^{0.22}, \tag{8.98}$$

where C and m are pressure-dependent parameters given in Table 8.1, G is the coolant mass flux in lb/hr-ft^2, and ΔT_{sub} is the difference between the saturated and local temperatures in °F.

Another correlation for subcooled boiling was obtained by Bernath.[11] This is a combination of the three equations:

$$q_c'' = h_c(T_{wc} - T_b) \tag{8.99}$$

$$T_{wc} = 102.6 \ln P - \frac{97.2P}{P + 15} - 0.45v + 32 \tag{8.100}$$

$$h_c = 10,890 \left(\frac{D_e}{D_e + D_i}\right) + \frac{48v}{D_e^{0.6}}. \tag{8.101}$$

In these expressions, T_{wc} is the wall (cladding) temperature at the onset of the boiling crisis, T_b is the bulk temperature, P is the pressure in psia, v is the coolant velocity in ft/sec, D_e is the equivalent diameter in feet, and D_i is defined as the heated perimeter of a channel in feet divided by π. The Bernath correlation is valid for pressures between 23 and 3,000 psia, fluid velocities between 4.0 and 54 ft/sec, and for D_e between 0.143 and 0.66 in.

Example 8.10

For the PWR lattice described in Examples 8.3 through 8.8, calculate for the onset of the boiling crisis the: (a) cladding temperature, (b) heat transfer coefficient, and (c) critical heat flux.

Solution.

1. Using Eq. (8.99) of the Bernath correlation,

$$T_{wc} = 102.6 \ln 2,000 - \frac{97.2 \times 2,000}{2,000 + 15} - 0.45 \times 15.6 + 32 = 708°F. \text{ [Ans.]}$$

[11]Bernath, L., *Transactions A.I.Ch.E.*, 1955.

2. From Example 8.8, $D_e = 0.0427$ ft. The heated perimeter of the channel is $2\pi a$, where $a = 0.234$ in. Thus,

$$D_i = 2 \times 0.234/12 = 0.039 \text{ ft.}$$

Then from Eq. (8.101),

$$h_c = 10,890 \times \frac{0.0427}{0.0427 + 0.039} + \frac{48 \times 15.6}{(0.0427)^{0.6}}$$

$$= 10,659 \text{ Btu/hr-ft}^2\text{-}°\text{F. } [Ans.]$$

3. With a bulk water temperature of about 600°F from Example 8.6, the critical heat flux is

$$q_c'' = 10,659 \times (708 - 600) = 1.15 \times 10^6 \text{Btu/hr-ft}^2\text{-}°\text{F. } [Ans.]$$

Equations (8.89) through (8.101) provide estimates of the curve that best represents the mass of necessarily rather scattered heat transfer data. The curve passes more or less through the center of gravity of the data points. In the case of the approach to DNB from bulk boiling conditions, workers at the General Electric Company (GE) have developed correlations that form an envelope of the lowest measured values of q_c''. With bulk boiling, such correlations depend on the *flow quality* of the coolant denoted by the symbol χ, which is defined as

$$\chi = \frac{\text{mass flow rate of vapor (lb/hr-ft}^2)}{\text{mass flow rate of vapor-liquid mixture (lb/hr-ft}^2)}. \tag{8.102}$$

The GE correlation, due to Janssen and Levy,[12] is as follows:

$$\frac{q_c''}{10^6} = 0.705 + 0.237(G/10^6), \quad \chi < \chi_1 \tag{8.103}$$

$$\frac{q_c''}{10^6} = 1.634 - 0.270(G/10^6) - 4.710\chi, \quad \chi_1 < \chi < \chi_2 \tag{8.104}$$

$$\frac{q_c''}{10^6} = 0.605 - 0.164(G/10^6) - 0.653\chi, \quad \chi_2 < \chi, \tag{8.105}$$

where

$$\chi_1 = 0.197 - 0.108(G/10^6), \tag{8.106}$$

$$\chi_2 = 0.254 - 0.026(G/10^6), \tag{8.107}$$

---

[12]E. Janssen and S. Levy, General Electric Company Report APED-3892, 1962. The General Electric Company has also published a portion of a more recent (1966) correlation due to Hench and Levy.

and G is again the mass flux in lb/hr-ft^2. These equations refer to a system pressure of 1,000 psia. For other pressures, q_c'' can be found from

$$q_c''(P) = q_c''(1,000 \text{ psia}) + 400(1,000 - P). \qquad (8.108)$$

The GE correlation is valid for pressures between 600 and 1450 psia, G from 0.4×10^6 to 6.0×10^6 lb/hr-ft^2, values of χ up to 0.45, D_e from 0.245 to 1.25 in, and for coolant channels between 29 and 108 in. in length.

All of the foregoing correlations for the critical heat flux were developed largely from data for circular pipes. When used to determine q_c'' for noncircular channels or for coolant flowing through lattices of fuel rods, these correlations can be counted on to provide only the roughest approximations for q_c''. Various correction factors for the various geometries have been developed, however, but these are too detailed to be discussed here (see Tong in the References).

8.6 THERMAL DESIGN OF A REACTOR

As pointed out earlier in this chapter, reactors must be designed in such a way that the fission products remain confined within the fuel at all times—throughout the operational lifetime of the core, during shutdown, and under accident conditions, when the fuel may be denied normal cooling. This places upper limits on the temperature of the fuel and/or its cladding. The currently accepted design criterion is that the integrity of the cladding must be maintained through all operating conditions. Since the expansion of the fuel on melting can rupture the cladding, this design criterion is essentially equivalent to the requirement that the fuel must not melt.

The melting point of UO$_2$ which depends somewhat on burnup—that is, on the fraction of the fissile atoms that have undergone fission—is generally between 5,000°F and 5,100°F (2,760 °C and 2,815°C) for commercially available UO$_2$. In most reactors fueled with UO$_2$, the maximum fuel temperature is somewhat below 4,500°F (2,480°C).

In the HTGR, where the fuel consists of small particles of uranium and thorium dicarbide in a carbonaceous binder, the maximum permitted fuel temperature is about 6,500°F (3,600°C).

Natural or enriched uranium metal melts at 2,070°F (1,132°C), but it also undergoes two changes in solid phase (alteration in crystalline structure)—first at 1,234°F (668°C) and again at 1,425°F (774°C). However, above about 750°F (400°C), the strength of the metal decreases rapidly. This permits fission product gases to gather and diffuse into pockets in the fuel, which in turn leads to the expansion of the fuel and to cladding failure. Therefore, it is usual practice to design uranium metal-fueled reactors with maximum fuel temperatures below 750°F. This low temperature is not as serious a drawback in transferring heat to the coolant as it would at first appear since the thermal conductivity of uranium metal is much higher than, say, that of UO_2. At 600°F, for example, k_f is 6.5 times higher for uranium than for UO_2.

The DNB Ratio

To prevent the penetration of the cladding at any point in a water-cooled reactor due to the onset of film boiling, this type of reactor must be designed so that the heat flux q'' is always below the critical (burnout) heat flux q''_c. For this purpose, it is convenient to define the *DNB ratio* as

$$\text{DNBR} = \frac{q''_c}{q''_{\text{actual}}}. \tag{8.109}$$

In this ratio, q''_c is the critical heat flux computed as a function of distance along the hottest coolant channel from the appropriate correlation given in the preceding section, and q''_{actual} is the actual surface heat flux at the same position along this channel. If heat flow parallel to the fuel can be ignored, then

$$q''_{\text{actual}} = \frac{q''' A_f}{C_c}, \tag{8.110}$$

where q''' is the heat production per unit volume, A_f is the cross-sectional area of the fuel, and C_c is the circumference of a heated rod. These two functions and the DNB ratio are shown in Fig. 8.13 for a 1,500-MW boiling-water reactor. BWRs are currently designed with a minimum DNBR of 1.9.[13] By setting this lower limit on the DNB ratio, there can be reasonable certainty that burnout conditions will never be reached anywhere in the reactor, even during transient, overpower situations. The minimum design value for the DNBR for PWRs is 1.3. Needless to say, the establishment of a minimum DNB ratio provides a major limitation on the design of water-cooled reactors.

Hot Channel, Hot Spot Factors

With water-cooled reactors, the maximum heat flux anywhere in the core is limited by the DNB ratio. In gas-cooled reactors, the maximum q'' is determined solely by the requirement that the fuel temperature remain well below the melting point for this value of q''_{max}. In *any* reactor, the extent to which q''_{max} exceeds the average heat flux in the core is given by the *hot channel factor,* also called the *hot spot factor.*

---

[13]The minimum DNB ratio is also called the *minimum critical heat flux ratio* (MCHFR).

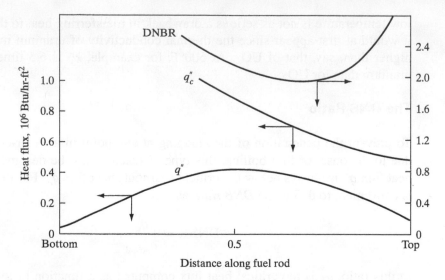

Figure 8.13 The actual heat flux, the computed critical heat flux, and the DNB ratio for a boiling-water reactor. (Courtesy of General Electric Company.)

This is defined by the relation

$$F = \frac{q''_{\max}}{q''_{\text{av}}}, \tag{8.111}$$

where q''_{av} is the average heat flux in the core.

There are several reasons that $q''_{\max}$ differs from q''_{av}, and therefore why F is different from unity. The most important of these is that, as seen earlier, the power distribution across the core is not flat. If this were the only reason, F could be computed directly for any given reactor design in more or less the same way as the maximum-to-average flux ratio Ω was calculated in Chapter 6. Thus, $q''_{\max}$ is a function of various nuclear parameters that determine the power distribution, and q''_{av} is given by the formula

$$q''_{\text{av}} = P/A, \tag{8.112}$$

where P is the thermal power output of the core and A is the total heat transfer area of the fuel (with cladding). The previous computation yields what is known as the *nuclear hot channel factor* F_N.[14]

---

[14]For a uniformly fueled reactor at the beginning of core life and in the absence of control rods, F_N is the same as Ω.

In addition to the effect of the nonflat power distribution, q_{max}'' may differ from the computed value of q_{max}'' as the result of various statistical factors over which the reactor designer has little or no control. For example, the amount of fissile material included in the fuel pellets of a BWR or PWR at the time of manufacture varies slightly from pellet to pellet because of the inherently statistical nature of the manufacturing process. Pellets containing more fissile material produce more power. If such a pellet was located at the point where q'' is highest, the value of q_{max}'' would be higher than calculated.

In a similar way, manufacturing tolerances in fuel assemblies may result in slight bowing of the fuel rods, leading to reduced coolant flow and excessive heating of a portion of the fuel. Similarly, fluctuations in the thickness of the cladding may give rise to hot spots where the cladding is thinnest.[15] Also, certain aspects of the coolant flow are inherently statistical in character and tend to give fluctuations in the heat flux.

These various mechanisms, and others by which q_{max}'' can differ from its computed value, taken together are described by the *engineering hot channel factor F_E*. The overall hot channel is then

$$F = F_N F_E. \qquad (8.113)$$

Each of the individual mechanisms is described, in turn, by an engineering hot channel *factor, $F_{E,x}$*.

The engineering subfactors are obtained from data on fabricated reactor components or from tests, such as coolant flow tests on assembled portions or mockups of the reactor. Consider, for instance, the amount of fissile material m_l per unit length of the fuel rods. For a given thermal flux, q_{max}'' is very nearly proportional to m_l. When measurements of m_l are carried out on manufactured rods, it is found that the actual values of m_l form a normal distribution about some mean or average value $\overline{m_l}$ like that shown in Fig. 8.14. With such a distribution, all values of m_l are presumably possible. However, specific values of m_l become less and less likely with increasing deviation from the mean. In particular, it can be shown that the probability that m_l exceeds $\overline{m_l}$ by more than 3σ, where σ is the standard deviation of the distribution, is only 1.35 per 1,000—an unlikely occurrence.

In those situations like the one just described, in which q_{max}'' is proportional to a normally distributed engineering variable x, the engineering subfactor is defined as

$$F_{E,x} = \frac{\overline{x} + 3\sigma(x)}{\overline{x}} = 1 + \frac{3\sigma(x)}{\overline{x}}. \qquad (8.114)$$

---

[15]This is because thin cladding leads to a larger gap (and hence a larger ΔT between the fuel pellet and the cladding).

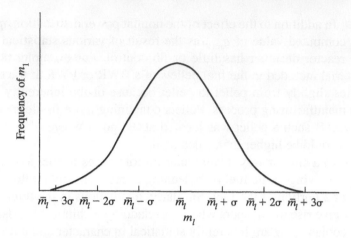

Figure 8.14   Normal distribution of measured values of m_l.

Here $\bar{x}$ is the mean value of x and $\sigma(x)$ is the standard deviation in the measurement of x. For example, in the case of the amount of contained fissile material, the subfactor is

$$F_{E,m_l} = 1 + \frac{3\sigma(m_l)}{\bar{m}_l}. \qquad (8.115)$$

Example 8.11

In a certain PWR, the average linear density $\bar{m}_l$ of UO_2 in a fuel rod is 0.457 lb/ft. The standard deviation in the measured values of m_l is 0.0122 lb/ft. What is F_{E,m_l}?

Solution. It is merely necessary to substitute these values into Eq. (8.115). This gives

$$F_{E,m_l} = 1 + \frac{3 \times 0.0122}{0.457}$$

$$= 1.08. \ [Ans.]$$

This result means that, due to statistical fluctuations in the amount of fissile material included in fuel rods during manufacture, there are 1.35 chances in 1,000 that q''_{max} will exceed its computed value by more than 8%.

Calculations of other engineering subfactors, especially those that do not involve a direct proportionality between q''_{max} and the statistical variable, are more complicated and require an analysis of the relationship between the variable in question and q''_{max}. As noted earlier, such subfactors are often obtained from engineering test data.

Once the various engineering subfactors have been determined, it is possible to calculate F_E. This can be done in a number of ways. The most obvious thing is to multiply all of the subfactors together, and this was done in the case of some early reactors. However, the resulting value of F_E leads to an unnecessarily conservative reactor design since, in using the product of subfactors, it is implicitly assumed that several statistically unlikely events occur simultaneously. The preferred method for computing F_E is based on an analysis of overlapping statistical distributions that is too complex to be reproduced here (see Tong and Weisman in the References). Needless to say, with $F_{E,x}$ known, F_E can be computed.

Reactor Design

In designing a reactor, there is considerable overlap between those aspects of the problem that are specifically nuclear—the neutron flux and power distributions, transient behavior, and so on—and those factors involving the removal of heat from the system. The final reactor necessarily represents a series of compromises between nuclear and thermal-hydraulic factors, all made in the interest of safety and economy. One noted reactor designer frequently refers to reactor design as "the art of compromise."[16]

Of course, most of the reactors manufactured today are not designed from scratch. They are either identical to or improved versions of existing plants. Only rarely is an engineer called on to design a reactor from the beginning. Nevertheless, the several steps in such a design study are now considered to illustrate some of the concepts developed earlier.

Consider the design of a specified type of power reactor having a given thermal power output. The first problem is the selection of the core materials. These are usually chosen on the basis of previous experience, availability, and cost. The amount and enrichment of the fuel are determined in a preliminary way by various nuclear considerations, including estimates of core lifetime and radiation damage to the fuel elements.

With this preliminary choice of materials, a fuel lattice arrangement is next proposed. The operating parameters of the reactor—the neutron flux, the power and temperature distributions, the coolant density and/or void distribution, the temperature and heat flux along the hot channel—are then computed for the proposed lattice. It should be noted that such computations are inherently iterative, at least for some types of reactors. With the PWR or BWR, for instance, it is necessary to know the water density as a function of position to compute the neutron flux and, from this, the power distribution. However, the power distribution must be known before the water density can be found since the density is determined by the heat flux distribution, which in turn depends on the power distribution. What

---

[16]B. Minkler, private communication.

must be done in this case is to assume some density distribution to start with—say, unity throughout the core. The flux and power distributions are then computed with this water density, and a new density distribution is determined from the computed heat flux. The neutron flux and power distributions are now computed using this new density distribution, and this in turn leads to a third density distribution. The calculations are continued in this way until convergence is obtained—that is, until the density distribution provides a power distribution that gives the same density distribution. Computer programs are available by which such iterative calculations can be carried out automatically. Incidentally, during this preliminary stage in the design of the core, the neutron flux is usually obtained from a multigroup diffusion calculation using only a few groups. To reduce computation costs, multigroup calculations with many groups are normally performed in the final stages of the design.

If the reactor is water cooled, the DNB ratio is next computed as a function of distance along the hottest channel. As pointed out earlier in this section, the minimum DNBR must be equal to some preset limiting value. If the computed DNBR is not equal to this value—it would only be luck if it were at this point—then the parameters of the lattice or the coolant flow are changed, the previous calculations of the core are repeated, and a new DNBR is obtained. These computations are repeated until the desired $(DNBR)_{min}$ is found. This also determines the maximum heat flux, although, as indicated in Fig. 8.13, q''_{max} does not necessarily occur at the point where the DNBR is smallest. If the reactor is not water cooled, q''_{max} is determined solely by the limiting value of the fuel temperature. For either type of reactor, water cooled or not, a preliminary calculated value of q''_{max} has now been found.

At this point, the overall hot channel factor is obtained using the methods discussed earlier. The average heat flux throughout the core can then be computed from Eq. (8.111):

$$q''_{av} = \frac{q''_{max}}{F}. \tag{8.116}$$

By the simple device of designing the reactor to operate at this average heat flux, the actual value of q''_{max} is reduced below its calculated value by the factor F_E. This provides the reactor with the desired margin of safety.

Once q''_{av} is known, the total heat transfer area can be found from Eq. (8.112). Finally, from the heat transfer and the dimensions of the fuel rods, it is possible to compute the total number of rods required in the reactor. This number will undoubtedly be different from the one used at the start of the design, and therefore many of the foregoing calculations must be repeated until a completely self-consistent reactor is obtained. At this point, detailed multigroup and two-dimensional neutron diffusion calculations can be undertaken, along with various transient analyses and lifetime-burnup computations, to refine the design.

Example 8.12

The core of a PWR consists of a lattice of fuel rods 12 ft long and 0.5 in. in diameter. The reactor operates at a thermal power of 3,000 MW; it has a maximum calculated heat flux of 539,000 Btu/hr-ft² and an overall hot channel factor of 2.80. Calculate (a) the total heat transfer area; (b) the number of fuel rods in the core.

Solution.

1. The average heat flux is given by Eq. (8.112):

$$q_{av}'' = \frac{539,000}{2.8} = 193,000 \text{ Btu/hr-ft}^2.$$

With a reactor power of 3,000 MW = 3,000 × 3.412 × 10^6 Btu/hr, the heat transfer area is

$$A = \frac{3,000 \times 3.412 \times 10^6}{193,000} = 53,000 \text{ ft}^2. \text{ [Ans.]}$$

2. The surface area of each fuel rod is 12 × (0.5π/12) = 0.5π ft². The total number of rods is then

$$n = \frac{53,000}{0.5\pi} = 33,740. \text{ [Ans.]}$$

REFERENCES

General

Eckert, E. R. G., and R. M. Drake, Jr., *Analysis of Heat and Mass Transfer*. Washington, D.C.: Hemisphere Pub., 1986.

Gebhart, B., *Heat Transfer*, 2nd ed. New York: McGraw-Hill, 1971.

Kreith, F., *Principles of Heat Transfer,* 5th ed. Boston, MA.: PWS Publishers, 1994.

McAdams, W. H., *Heat Transmission*, 3rd ed. New York: McGraw-Hill, 1954.

Conduction Heat Transfer

Carslaw, H. S., and J. C. Jaeger, *Conduction of Heat in Solids*, 2nd ed. London: Oxford, 1986.

Özisik, M. N., *Boundary Value Problems of Heat Conduction*. New York: Dover Publications, 1989.

Convective Heat Transfer

Kays, W. M., *Convective Heat and Mass Transfer*, 3rd ed. New York: McGraw-Hill, 1994.

Knudsen, J. G., and D. L. Katz, *Fluid Dynamics and Heat Transfer.* New York: McGraw-Hill, 1958.

Bejan, A., *Convection Heat Transfer.* New York: John Wiley & Sons, 1994

Liquid Metal Heat Transfer

Dwyer, O. E., *Liquid Metals Handbook, Sodium and NaK Supplement.* Washington, D.C.: U.S. Atomic Energy Commission, 1970, Chapter 5.

Dwyer, O. E., H. C. Berry, and P. J. Hlavac, "Heat Transfer to Liquid Metals Flowing Turbulently and Longitudinally through Closely Spaced Rod Bundles." *Nuclear Engineering and Design* **23**, 273 and 295 (1972).

Boiling Heat Transfer

Collier, J. G., *Convective Boiling and Condensation*, 3rd ed. Reprint, Oxford University Press, 1996.

Tong, L. S., *Boiling Crisis and the Critical Heat Flux.* U.S. Atomic Energy Commission Report TID-25887, 1972.

Tong, L. S., *Boiling Heat Transfer and Two-Phase Flow.* Bristol: Taylor & Francis, 1997.

Reactor Heat Transfer

Bonilla, C. F., *Nuclear Engineering.* New York: McGraw-Hill, 1957, Chapters 8 and 9. (Available from University Microfilms, Cat. No. OP33393, Ann Arbor, Michigan.)

Bonilla, C. F., *Nuclear Engineering Handbook.* H. Etherington, Editor. New York: McGraw-Hill, 1958, Section 9–3.

El-Wakil, M. M., *Nuclear Energy Conversion.* La Grange Park, Ill.: American Nuclear Society, 1982.

El-Wakil, M. M., *Nuclear Heat Transport.* La Grange Park, Ill.: American Nuclear Society, 1981.

Glasstone, S., and A. Sesonske, *Nuclear Reactor Engineering*, 4th ed. Chapman & Hall, 1994, Chapter 9.

Sesonske, A., *Nuclear Power Plant Design Analysis.* U.S. DOE, 1973, Chapter 4.

Tong, L. S., and J. Weisman, *Thermal Analysis of Pressurized Water Reactors*, 3rd ed. La Grange Park, Ill.: American Nuclear Society, 1996.

Lahey, R. T., and F. J. Moody, *The Thermal Hydraulics of a Boiling Water Nuclear Reactor*, 2nd ed. La Grange Park, Ill.: American Nuclear Society, 1993.

Todreas, N. E., and M. S. Kazimi, *Nuclear Systems.* Washington, D.C.: Hemisphere Publishing, 1989.

PROBLEMS

1. The nuclear ship *Savannah* was powered by a PWR that operated at a pressure of 1,750 psia. The coolant water entered the reactor vessel at a temperature of 497°F, exited at 519°F, and passed through the vessel at a rate of 9.4×10^6 lb/hr. What was the thermal power output of this reactor?

2. An experimental LMFBR operates at 750 MW. Sodium enters the core at 400 °C and leaves at 560°C. At what rate must the sodium be pumped through the core?

3. Show that if water passes through the core of a BWR at the rate of w lb/hr, then w_g lb/hr of steam will be produced, where

$$w_g = \frac{q - w(h_f - h_{in})}{h_{fg}};$$

q is the thermal power of the reactor; h_f and h_{in} are, respectively, the specific enthalpies of the saturated water leaving the core and the water entering the core; and h_{fg} is the heat of vaporization per pound.

4. A BWR operates at a thermal power of 1,593 MW. Water enters the bottom of the core at 526°F and passes through the core at a rate of 48×10^6 lb/hr. The reactor pressure is 1,035 psia. Using the result of the previous problem, compute the rate in lb/hr at which steam is produced in this reactor.

5. Starting with the value of the specific enthalpy of saturated water at 350°F given in Table IV.1, calculate the specific enthalpy of saturated water at 600°F. [*Hint:* Integrate the specific heat using Simpson's rule.]

6. The decommissioned Fort St. Vrain HTGR was designed to produce 330 MW of electricity at an efficiency of 39.23%. Helium at a pressure of 710 psia enters the core of the reactor at 760°F and exits at 1,430°F. What is the helium flow rate through the core?

7. A small PWR plant operates at a power of 485 MWt. The core, which is approximately 75.4 in. in diameter and 91.9 in high, consists of a square lattice of 23,142 fuel tubes of thickness 0.021 in and inner diameter of 0.298 in on a 0.422-in pitch. The tubes are filled with 3.40 w/o-enriched UO_2. The core is cooled by water, which enters at the bottom at 496°F and passes through the core at a rate of 34×10^6 lb/hr at 2,015 psia. Compute (a) the average temperature of the water leaving the core; (b) the average power density in kW/liter; (c) the maximum heat production rate, assuming the reactor core is bare.

8. The core of a BWR consists of 764 fuel assemblies, each containing a square array of 49 fuel rods on a 0.738-in pitch. The fuel rods are 175 in long, but contain fuel over only 144 in of their length. The outside diameter of the fuel rods is 0.563 in, the cladding (fuel tube) is 0.032 in thick, and the UO_2 fuel pellets are 0.487 in in diameter. This leaves a gap of $(0.563 - 2 \times 0.032 - 0.487)/2 = 0.006$ in between the pellets and the cladding. The UO_2 has an average density of approximately 10.3 g/cm^3. The radius of the core is 93.6 in, and the reactor is designed to operate at 3,293 MW. The peak-to-average power density is 2.62. Calculate for this reactor: (a) the total weight of

UO_2 and uranium in the core; (b) the specific power in kW/kg U; (c) the average power density in kW/liter; (d) the average linear rod power q''_{av} in kW/ft; (e) the maximum heat production rate.

9. The core of an LMFBR consists of a square lattice of 13,104 fuel rods 0.158 in in diameter, 30.5 in long, on a 0.210-in pitch. The fuel rods are 26 *w/o*-enriched uranium clad in 0.005-in stainless steel. Liquid sodium enters the core at approximately 300°C and passes through the core at an average speed of 31.2 ft/sec. The core produces 270 MW of thermal power, with a maximum-to-average power density of 1.79. Calculate: (a) the maximum heat production rate; (b) the maximum neutron flux.

10. The variation of the neutron flux and/or heat production rate along the z direction in the core of a reflected reactor is often approximated by the function

$$\phi = \text{constant} \times \cos\left(\frac{\pi z}{\widetilde{H}}\right),$$

where $\widetilde{H}$ is the distance between the extrapolated boundaries and is somewhat larger than H, the actual height of the core. Show that the maximum-to-average heat production ratio in the z direction Ω_z is given by

$$\Omega_z = \frac{\pi H/2\widetilde{H}}{\sin(\pi H/2\widetilde{H})}.$$

[*Note:* A similar approximation can be made for ϕ and/or q^m in the radial direction by writing $\phi = \text{constant} \times J_0(2.405r/\widetilde{R})$.]

11. Show that the maximum-to-average flux and/or power distribution ratios in the axial and radial direction are related to the overall ratio by

$$\Omega_z\Omega_r = \Omega,$$

where the notation is that of Problem 8.10.

12. The core of a fast reactor is a cylinder 38.8 cm in radius and 77.5 cm high. Two-dimensional (r, z) multigroup calculations show that the power density distribution in the core can be represented approximately by the expression

$$P(r, z) = P_0\left[1 - \left(\frac{r}{51}\right)^2\right]\cos\left(\frac{\pi z}{109}\right),$$

where P_0 is a constant and r and z are the distances in centimeters from the axis and the midplane of the core, respectively. (a) Evaluate P_0 in terms of the total core power P. (b) What is the maximum-to-average power ratio in the core? (c) Calculate the maximum-to-average power ratios in the radial and axial directions.

13. A BWR power plant operating at an efficiency of 34% has an electrical output of 1,011 MW. (a) What is the maximum fission product decay energy in the reactor at shutdown? (b) What is the decay energy 6 months after shutdown?

14. A 1,000-MW reactor is operated at full power for 1 year, then at 10% power for 1 month, and then shut down. Calculate the fission product decay heat at shutdown and for 1 month after shutdown.

15. A power reactor fueled with ^{235}U operates at the power P_0 for 1 year and is then scrammed by the insertion of \$9.63 in negative reactivity. (a) Calculate the level to which the fission power of the reactor immediately drops. (b) To what power level does the reactor actually drop? (c) Compute and plot the reactor power from 0.1 sec to 0.5 hr after shutdown.

16. A thermal reactor fueled with slightly enriched ^{235}U is operated at a power of 2,800 MW for a period of 2 years and then shut down for refueling. The reactor has a conversion factor of 0.82. (a) What is the fission product decay energy at shutdown and 1 month later? (b) How much decay heat is due to ^{239}U and ^{239}Np? (c) What is the activity of the fission products and the ^{239}U and ^{23p}Np at the prior times?

17. The reactor core in Problem 8.12 produces 400 MW, of which 8 MW is due to γ-ray heating of the coolant. The total heat transfer area of the fuel is 1,580 ft^2. Compute (a) the average power density in the core in kW/liter and kW/ft^3; (b) the average heat flux; (c) the maximum heat flux.

18. The plate-type fuel element described in Example 6.10 is placed in a test reactor in a region where the thermal flux is 5×10^{13} neutrons/cm^2-sec. Calculate: (a) the heat production rate in the fuel; (b) the heat flux at the cladding surface; (c) the difference in temperature between the center of the fuel and the cladding surface.

19. The temperature at the center of the fuel rod described in Problem 8.9, where q'' is the largest, is 1,220°F. Calculate the temperatures at the fuel-cladding interface at the outer surface of the cladding.

20. Derive the general solution to Eq. (8.44). [*Hint:* Write the first two terms of the equation as

$$\frac{1}{r}\frac{d}{dr} r \frac{dT}{dr}$$

and then perform the integration.]

21. Show that Eq. (8.47) can be written as

$$T_m - T_s = \frac{q'}{4\pi k_f},$$

where q' is the heat generated per unit length of a fuel rod. [*Note:* This result shows that the temperature difference across a fuel rod is directly proportional to the linear power density q'. The quantity q' is an important parameter in reactor design.]

22. The gap between UO$_2$ pellets and the inside of a fuel tube can be taken into account in calculations of heat flow out of the fuel by defining a *gap conductance* h_{gap} by the relation

$$q'' = h_{gap}\Delta T,$$

where q'' is the heat flux and ΔT is the temperature difference across the gap. Values of h_{gap} of about 1,000 Btu/hr-ft^2-°F are typical for PWR and BWR fuel. Show that the thermal resistance of a gap is

$$R_{\text{gap}} = \frac{1}{h_{\text{gap}} A},$$

where A is the area of the fuel.

23. At one point in the reactor described in Problem 8.8, the heat flux is 280,000 Btu/hr-ft^2, and the outer temperature of the cladding is 563°F. Compute and plot the temperature distribution in the fuel rod at this location (a) ignoring the presence of the gap between the pellets and the cladding; (b) using the gap conductance (see Problem 8.22) of 1,000 Btu/hr-ft^2-°F.

24. The fuel rod in the reactor described in Problem 8.8 with the highest heat flux contains uranium enriched to 2.28 w/o. (a) Calculate the maximum heat-generation rate in this rod when the reactor is operating at design power. (b) What is the maximum thermal neutron flux in this rod? [*Note:* In part (b), use the average fuel temperature from Problem 8.23.]

25. A slab of iron 5 cm thick, forming part of the thermal shield of an LWR, is exposed to γ-rays having an average energy of 3 MeV and an incident intensity of 1.5×10^{14} γ-rays/cm^2-sec. The side on which the radiation is incident is held at 585°F; the opposite side is at 505°F. (a) At what rate is energy deposited per ft^2 of the slab? (b) Compute and plot the temperature distribution within the slab. (c) Calculate the heat fluxes at both faces of the slab. [*Note:* For simplicity, assume that the γ-ray absorption is exponential with $\mu = 0.18$ cm^{-1}.]

26. Verify that Eq. (8.59) is a solution to Eq. (8.58) and satisfies the given boundary conditions.

27. γ-radiation is incident on a slab of thickness a. (a) Derive an expression giving the condition under which the temperature in the slab passes through a maximum. (b) In the special case where the two sides of the slab are held at the same temperature, show that the maximum temperature occurs at

$$x_m = \frac{a}{2}\left(1 - \frac{\mu a}{12}\right)$$

provided $\mu a \ll 1$.

28. The velocity profile of 1,000-psia water flowing through an insulated 8-in I.D. (inner diameter) coolant pipe of a research reactor can be fit by the expression

$$v = 60 - 540r^2,$$

where T is in °F. (a) Compute the bulk temperature of the water. (b) Compute the average velocity of the water in the pipe. (c) Compute the value of the Reynolds number. (d) How much water, in lb/hr, is flowing through the pipe? The water density is constant across the pipe at 61.8 lb/ft^3 and $\mu = 1.270$ lb/hr-ft.

29. Compute and plot the temperatures of the water, the outer surface of the cladding, and the fuel center as a function of distance along the hottest channel of the reactor described in Examples 8.3 and 8.5 under the conditions given in Example 8.7.

30. For the hottest channel in the reactor described in Problem 8.9, calculate the: (a) exit temperature of the sodium; (b) maximum temperatures of the fuel and the cladding surface. [*Note:* Take the heat transfer coefficient to be 35,000 Btu/hr-ft^2-°F.]

31. For the hottest channel in the reactor described in Problem 8.7, calculate, assuming the reactor is bare, the (a) exit temperature of the water; (b) maximum temperatures of the fuel and the cladding surface. [*Note:* Take $h = 7,600$ Btu/hr-ft^2-°F.]

32. The heat flux at the surface of fuel rods in an experimental reactor varies with position as indicated in Fig. 8.15. The fuel rods are 0.5 in. in diameter and 6 ft long. Water enters the reactor core at a temperature of 150 °F and flows through the core at a rate of 350 lb/hr per rod. The heat transfer coefficient is 1,000 Btu/hr-ft^2-°F. Calculate the temperatures of the coolant, fuel rod surface, and fuel rod center at the entrance of the channel, 3 ft up the channel, and at its exit.

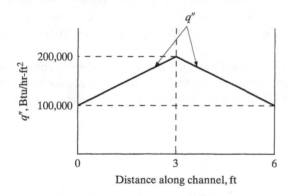

Figure 8.15 Heat flux v. distance for the fuel rod in Prob. 8.32.

33. Water at about 2,000 psia and 613°F leaves a PWR vessel via a 29-in I.D. primary coolant pipe with an average flow velocity of 49.5 ft/sec. (a) What is the mass flow rate out this pipe in lb/hr? (b) What is the Reynolds number for the water in the pipe? (c) If the water returns to the reactor vessel at the same pressure and velocity but at 555°F, how large should the return pipe be?

34. Convert each of the factors in the computation of the Reynolds number given in Example 8.8 to SI units and recompute Re.

35. Calculate the mass flow rate in lb/hr at which the coolant passes up a single channel of the lattice described in Example 8.8 under the conditions given in that example.

36. For the reactor in Problem 8.7, compute the: (a) average flow velocity in ft/sec; (b) equivalent channel diameter; (c) Reynolds number entering and leaving the hottest channel; (d) heat transfer coefficient at the locations in part (c).

37. For the LMFBR core described in Problem 8.9, calculate the: (a) equivalent diameter of a coolant channel; (b) total coolant flow area; (c) sodium flow rate in lb/hr; (d) average sodium exit temperature; (e) Reynolds number for the sodium entering and leaving (at the temperature in part [d]) the core; (f) average heat transfer coefficient for the sodium entering and leaving the core.

38. Water enters the core of the BWR described in Problem 8.8 at 525°F and flows through the core at a rate of 106.5×10^6 lb/hr. The heat production rate in the fuel along the hottest channel can be approximated as

$$q'''(z) = q'''_{max} \cos(\pi z/165),$$

where z is in inches. The reactor pressure is 1,020 psia. Compute the (a) maximum-to-average power density in the z direction; (b) total flow area in the core; (c) average water velocity near the bottom of the core; (d) water/fuel volume ratio; (e) equivalent diameter of a coolant channel; (f) Reynolds number near the entrance to the hottest channel; (g) heat transfer coefficient for the convective part of the hottest channel; (h) location of the onset of local boiling.

39. Referring to the PWR described in Examples 8.3 through 8.9: (a) Does boiling of any type occur in this reactor? (b) Does bulk boiling occur? (c) Does boiling occur in a channel 30 in from the axis of the reactor?

40. Starting with the curves of q'' and T_b shown in Fig. 8.12, calculate the curves T_c and $T_c - T_b$.

41. (a) Using the results of Problem 8.31 and the Bernath correlation, compute and plot q''_c as a function of position along the hottest channel of the reactor in Problem 8.7. (b) Compute and plot the DNBR for this channel. (c) What is the minimum DNBR?

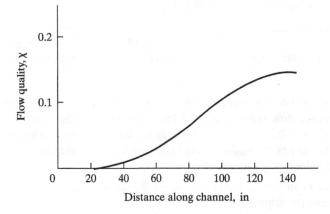

Figure 8.16 Flow quality v. position along the hot channel in Prob. 8.43.

42. Figure 8.16 shows the flow quality as a function of position along the hot channel of the reactor in Fig. 8.12. The mass flux for this channel is 2.3×10^6 lb/hr-ft². (a) Compute and plot q''_c and the DNBR along the channel. (b) What is the minimum DNBR?

43. The maximum heat flux in the HTGR described in Problem 8.6 is 120,000 Btu/hr-ft² and the hot channel factor is 2.67. (a) What is the average heat flux? (b) What is the total heat transfer area?

44. The hot channel factor for the LMFBR in Problem 8.9 is 1.85. Calculate the: (a) total heat transfer area in the core; (b) average heat flux; (c) maximum heat flux.

45. The power distribution in the axial direction of the 12-ft-long fuel rods of a PWR is well represented by

$$q''' = \text{constant} \times \cos(\pi z/144),$$

where z is in inches. The minimum DNBR is 2.1 and occurs 90 in up the hottest channel, where the critical heat flux is computed to be 1.08×10^6 Btu/hr-ft^2. The hot channel factor for the reactor is 2.78, and the total heat transfer area is 48,000 ft^2. Calculate (a) q''_{max}; (b) q''_{av}; (c) the operating power of the reactor.

9

Radiation Protection

Since radiation is present in reactors and virtually every other device that utilizes nuclear energy, the designers of these systems must include radiation shielding where necessary for the protection of operating personnel and of the public at large. The responsibility for continuing surveillance of the facility can then be turned over to *health physicists*,[1] whose task it is to assure that no individual receives an unnecessary or hazardous exposure to radiation. Criteria for the design of radiation shielding and for the establishment of a sound health physics program are based upon the effects of radiation on the human body. These effects, together with some of the calculations encountered in radiation problems, form the subject of this chapter. Radiation shielding is discussed in Chapter 10.

The literature of health physics contains many biological and medical terms that often confound the layman. In the present chapter, many of these terms are introduced in order to provide the reader with some of the background necessary to comprehend the more definitive literature in the field. A glossary of these terms is given at the end of the chapter.

---

[1]*Health physics* is the name given to the study of problems related to the protection of man from exposure to radiation. That this term is inappropriate and misleading could not be clearer. Only a small part of what is involved in health physics can reasonably be called physics; at the same time, there are many more physics problems associated with health than those arising from exposure to radiation.

466

9.1 HISTORY OF RADIATION EFFECTS

Throughout history, mankind has been exposed to various forms of natural background radiation. However, these radiation levels were too low to suggest the deleterious effects of radiation, which did not become evident until the late 1800s, when the discovery of x-rays by Roentgen and of radioactivity by Becquerel led to the availability of high-intensity radiation sources. Soon after, many incidents of obvious radiation injury were reported. More subtle, long-term effects of radiation were recognized many years later.

Radiation effects are usually divided into two categories, depending upon whether the damage arises from *external exposure* to radiation (i.e., to radiation incident on the body from the outside) or from *internal exposure*, due to radiation emitted from radionuclides absorbed into the body. It was noted as early as 1890 that external exposure to x-rays can produce *erythema* (abnormal redness of the skin), *edema* (swelling due to abnormal accumulation of fluids) and *epilation* (abnormal loss of hair). By 1897, one investigator reported 69 cases of x-ray burns, as these injuries are called, and both Becquerel and Mme. Curie suffered similar burns from small tubes containing radioactive materials. The *carcinogenic* (cancer-inducing) effects of radiation were observed soon thereafter. By 1911, 94 cases of x-ray induced tumors were reported, 50 of these in *radiologists* (physicians who specialize in the use of radiation for various purposes), and by 1922, it was estimated that 100 radiologists had died as a result of radiation-induced cancer. Somewhat later, it was shown that the incidence of *leukemia* (roughly speaking, cancer of the blood-forming organs, or the *bone marrow*) was significantly higher among radiologists than among other physicians. Evidence also accumulated suggesting that the average lifespan of radiologists was somewhat shorter than that of other physicians. Fortunately, recent studies of mortality in radiologists indicate that the increased incidence of leukemia and life shortening have disappeared. This is generally attributed to an increased awareness on the part of radiologists of the deleterious effects of radiation and to radiologists' adoption of improved radiation safety practices.

It has long been known that the miners who worked in the cobalt mines of Saxony in southeastern Germany and in the pitchblende mines in Czechoslovakia, both of which contain large concentrations of uranium, suffered from a high incidence of lung cancer (about 30 times the normal). Today, it is recognized that these workers were the victims of internal radiation exposure caused by the inhalation of the radioactive gas *radon* and its daughters, which are decay products of uranium released from the walls of the mine tunnels. The incidence of lung cancer is also substantially higher among uranium miners in the United States. Improved ventilation of the mines and recently adopted rigid standards for the protection of miners is expected to eliminate this source of radiation injury.

A number of other cases of injury due to internal exposure have been reported over the years. From about 1915 to 1930, it was thought in some medical circles that radium and thorium had therapeutic value, and many unfortunate persons received injections of tonics of these deadly substances. From about 1928 to 1945, a large number of persons were injected with suspensions of thorium dioxide, which was used as a contrast material for diagnostic x-rays. A significant increase in the incidence of cancer, especially cancer of the liver, was observed among this group. However, the most celebrated cases of internal exposure were those of radium-dial painters at the US Radium Corporation in Orange, New Jersey. These women painted the numerals on watch dials with paint containing radium, and in order to obtain sharp numerals, they were in the habit of tipping their paint-laden brushes with their lips. The associated ingestion of radium led to the death, by various forms of cancer, of some 50 women out of a total of approximately 2,000 workers exposed.

As a result of these and other tragic episodes in the use of x-rays and radioactive materials, a great deal is now known about the effects of radiation on the human species. The importance of safety in the utilization of radiation is now generally recognized, and, except for the use of radiation in medicine, strict standards of radiation exposure have been developed. This is fortunate, because so many of the devices of modern technology—accelerators, nuclear reactors, various radionuclide devices, television sets, high-flying airplanes, and space travel—all represent potential sources of radiation exposure. The nuclear industry, however, has contributed very little in the way of radiation injury either to its own personnel or to the general public. As will be shown later in this chapter, the principal source of human exposure to radiation is from medical x-rays.

9.2 RADIATION UNITS

To make meaningful calculations in radiation protection problems, it is necessary to have a set of units that allows radiation fields and radiation effects to be expressed quantitatively. Two systems of units are being used in the world today: the so-called *conventional system*, which in various forms and refinements has been used for several decades, and the new Système International (*SI*), which is gradually replacing the older system. This book, for the most part, will use conventional units, which are firmly imbedded in nuclear engineering practice and in governmental standards and regulations, at least in the United States.

Radiation units are promulgated by an international scientific body, the International Commission on Radiation Units and Measurements (ICRU).The discussion that follows is based on ICRU report 19 (1971, the conventional system) and ICRU report 33 (1980, SI).

Exposure

The term *exposure* is used in the conventional system to describe a γ-ray[2] field incident on a body at any point. As discussed in Chapter 3, γ-rays produce ions and electrons when they interact with air. The presence of γ-rays at the surface of a body can therefore be measured in terms of the numbers of ions or electrons produced in the adjacent atmosphere. This measure of radiation field is especially reasonable in the context of radiation protection. As will be shown later, the biological effect of γ-rays is a function of the ionization they produce in a body. Thus, the biological effects arising from a given exposure can be expected to increase more or less uniformly with exposure.

Accordingly, the ICRU has defined exposure, which is given the symbol X, by the relation

$$X = \frac{\Delta q}{\Delta m},\qquad(9.1)$$

where Δq is the sum of the electrical charges on all the ions of one sign produced in air when all the electrons ($+$ and $-$), liberated by photons in a volume of air whose mass is Δm, are completely stopped in air. It should be noted that the electrons released from Δm need not, and in most cases do not, stop in the volume containing Δm. These electrons may actually travel a considerable distance from Δm. From Eq. (9.1) it is seen that exposure has dimensions of coul/kg of air.

The Roentgen

The unit of exposure in the conventional system is the *roentgen*, abbreviated R, which is defined in the conventional system as

$$1\,R = 2.58 \times 10^{-4}\ \text{coul/kg},\qquad(9.2)$$

which can be shown to be equivalent to the production of 1 electrostatic unit (1 esu $= 3.33 \times 10^{-10}$ coul) of charge of one sign from the interaction of gamma radiation in 0.001293 g of air (1 cm^3 of air at 1 atmosphere pressure at 0°C). One one-thousandth of a roentgen, a *milliroentgen*, is denoted as mR.

In SI, the unit is the *exposure unit*, X. It is defined as the production of 1 coulomb of charge in 1 kg of air. One can show that

$$1\,X\ \text{unit} = 3{,}876\,R.$$

---

[2] As in Chapter 3, x-rays and γ-rays will both be referred to as γ-rays.

Exposure Rate

The rate at which Δq in Eq. (9.1) is liberated as the result of interactions in Δm is called the *exposure rate*. This is written as $\dot{X} = dX/dt$ and has units of roentgens per unit time—for example, R/sec, mR/hr, etc.

It must be emphasized that the concept of exposure and exposure rate applies only to γ-rays (or x-rays) in air at a point outside a body. These terms must not be used in a technical sense to describe neutrons or charged particles or interactions taking place within a body. It must also be remembered that these concepts and roentgen units of measurement are used only in the conventional system. They are not included in SI.

Example 9.1

In a 1-mR/hr γ-ray field, at what rate are ions being produced per cm^3/sec?

Solution. Only singly charged ions are usually produced when γ-rays interact in air. The charge on each ion is therefore the electronic charge 1.60×10^{-19} coul. Since 1 R releases 2.58×10^{-4} coul/kg, this is equivalent to

$$2.58 \times 10^{-4}/1.60 \times 10^{-19} = 1.61 \times 10^{15} \text{ ions/kg}$$

$$= 1.61 \times 10^{12} \text{ ions/g},$$

and so 1 mR produces 1.61×10^9 ions/g. There are 0.001293 g/cm^3 of air, so that in 1 cm^3, $1.61 \times 10^9 \times 0.001293 = 208 \times 10^6$ ions are produced. If the exposure rate is 1 mR/hr, these ions are formed at a rate of

$$2.08 \times 10^6/3{,}600 = 578 \text{ ions/cm}^3\text{-sec. } [\textit{Ans.}]$$

Imparted Energy

It will be shown later that the biological effect of radiation is a function of how much energy is deposited in a body. Consider, therefore, a volume ΔV containing the mass Δm of any substance. Let E_{in} and E_{out} be the sum of the kinetic energies of all particles—photons, neutrons, or charged particles—incident upon and emerging from ΔV, respectively. In the case of photons, E_{in} and E_{out} are their total energies as given by Eq. (2.13). The *imparted energy*, ΔE_D, is defined by the formula

$$\Delta E_D = E_{in} - E_{out} + Q, \tag{9.3}$$

where Q is the sum of the Q-values of whatever (if any) nuclear reactions that take place within ΔV. It is necessary to include Q in defining ΔE_D, because the concept of imparted energy is intended to reflect the deposition of energy that is available for producing biological effects, and nuclear reactions may increase the deposited energy, since one effect of the reaction is merely to the masses of atomic nuclei, but this is not a biologically significant process.

Absorbed Dose

The *absorbed dose* to the mass Δm, denoted by D, is defined as the imparted energy per unit mass. In symbols,

$$D = \frac{\Delta E_D}{\Delta m}.$$
(9.4)

The absorbed dose at a point is the limit of Eq. (9.4) as Δm approaches zero. The conventional unit of absorbed dose is the *rad*. One rad, which is the acronym for *r*adiation *a*bsorbed *d*ose, is equal to an absorbed dose of 0.01 joule/kg; that is,

$$1 \text{ rad} = 0.01 \text{ J/kg} = 100 \text{ ergs/g}.$$
(9.5)

A *millirad*, 0.001 rad, is abbreviated mrad.

The SI unit of absorbed dose is called the *gray*, abbreviated Gy. This is defined as an absorbed dose of 1 joule/kg; that is,

$$1 \text{ Gy} = 1 \text{ J/kg}.$$
(9.6)

In view of the definition of the rad, it follows that

$$1 \text{ Gy} = 100 \text{ rads}.$$
(9.7)

A *milligray* is abbreviated mGy.

Absorbed Dose Rate

The *absorbed dose rate*, the rate at which an absorbed dose is received, is denoted by D. In conventional units, it is measured in rad/sec, mrad/hr, etc., and in SI units, in Gy/sec, mGy/hr, etc.

Kerma

Another concept that is widely encountered, especially in connection with calculations involving fast neutrons, is that of *kerma*, denoted by K. Kerma is defined as the sum of the initial kinetic energies of all charged, ionizing particles released by indirectly ionizing radiation (mostly neutrons) per unit mass of substance. The energy absorbed in a material due to the attenuation of neutrons is often specified as the kerma or kerma dose. The units of K are the same as those of D, namely, rads, or ergs/gram, for the conventional system and grays, or joules/Kg, for the SI system. Kerma has more utility in and of itself; however, it is easier both to measure and compute kerma than it is dose, which is the real quantity of interest. Fortunately, the two can be shown to be virtually equal in a wide variety of situations.

Relative Biological Effectiveness

It would certainly simplify matters if the biological effect of radiation was directly proportional to the energy deposited by radiation in an organism. In this case, the absorbed dose would be a suitable measure of biological injury, regardless of the nature of the incident radiation or its energy. Unfortunately, the actual situation is considerably more complicated. Biological effects depend not only on the total energy deposited per gram or per cm^3, but also on the way in which this energy is distributed along the path of the radiation. In particular, the biological effect of any radiation increases with the LET (see Section 3.9) of the radiation. Thus, for the *same absorbed dose*, the biological damage from high LET radiation, α-particles, or neutrons is much greater than from low LET radiation, β-rays or γ-rays.

This result is not surprising, on the basis of a number of radiochemical observations. It has been found, for example, that when ^{210}Pb α-particles (5.3 MeV) are absorbed by a solution of ferrous sulfate, 1.5 molecules of free H_2 are formed per 100 eV absorbed in the solution. By contrast, ^{60}Co γ-rays (1.17 and 1.33 MeV) produce only 0.4 H_2 molecules per 100 eV absorbed in the same solution. Since radiation-induced changes in *chemical* systems are not solely a function of energy absorption, radiation-induced effects in *biological* systems could hardly be expected to be so. The term *quality* is used to describe the fact that energy may be deposited by radiation along its track in different ways. Radiations of different types or energies are said to differ in quality.

The fact that radiations of different types or energies, in general, give different biological effects for the same absorbed dose is described in terms of a factor known as the *relative biological effectiveness* (RBE). This quantity is determined in the following type of experiment. A tissue or organ is first irradiated with 200-KeV γ-rays, whose RBE is arbitrarily taken to be unity, and the resulting effect is observed for a given absorbed dose. The experiment is now repeated with another form of radiation. If, in this case, it is found that the same biological effect, whatever that may be, occurs with only one-tenth the previous absorbed dose, then the second type of radiation is evidently 10 times more effective than the γ-rays per absorbed dose, and its RBE is taken to be 10. It must be emphasized that there is no one RBE for a given type or energy of radiation. The RBE depends on the tissue, the biological effect under consideration, the dose, and, in some cases, also the dose rate. The RBE is now used almost exclusively in radiobiology, but it is considered to be too detailed and too specific a parameter to be used for ordinary radiation protection purposes.

Quality Factor and Radiation Weighting Factor

To take into account the differences in biological effect of different radiations and at the same time to simplify radiation protection calculations, the *quality factor*

TABLE 9.1   QUALITY FACTOR AS A
FUNCTION OF LET

LET, keV/micron	Q
3.5 or less	1
7	2
23	5
53	10
175 and above	20

of radiation was defined, denoted by the symbol Q, according to the US Nuclear Regulatory Commission. This parameter is a somewhat arbitrary approximation to RBE values as a function of the linear energy transfer. In contrast to the RBE, which is always determined by experiment, Q is simply assigned, but, of course, after consideration of RBE values. The quality factor as a function of LET is given in Table 9.1; values of Q for various radiations are given in Table 9.2. The *radiation weighting factor*, symbolized by W_R, is related to the type and energy of the incident radiation. It is approximately equal to the average quality factor $\overline{Q}$.

TABLE 9.2   QUALITY FACTORS FOR VARIOUS TYPES
OF RADIATION*

Type of radiation	Q	W_R
x-rays and γ-rays	1	1
β-rays, $E_{max} > 0.03$ MeV	1†	
β-rays, $E_{max} < 0.03$ MeV	1.7†	
Naturally occurring α-particles	10	
Heavy recoil nuclei	20	20
Neutrons:		
Thermal to 1 keV	2	5
10 keV	2.5	10
100 keV	7.5	10
500 keV	11	20
1 MeV	11	20
2.5 MeV	9	5
5 MeV	8	5
7 MeV	7	5
10 MeV	6.5	5
14 MeV	7.5	5
20 MeV	8	5
Energy not specified	10	

*Based on 10CFR20 (Q) and ICRP 60 (W_R).
†Recommended in ICRP Publication 9.

Equivalent Dose

The *equivalent dose*, denoted by H, is defined as the product of the absorbed dose and the radiation weighting factor W_R; that is

$$H(\text{equivalent dose}) = D(\text{absorbed dose}) \times W_R(\text{radiation weighting factor}). \tag{9.8}$$

The equivalent dose is also loosely called the *biological dose* (formerly it was called the RBE dose) and differs from the dose-equivalent H for tissues or organs. According to the NCRP,[3] the equivalent dose is an average absorbed dose in the specific tissue or organ weighted by the radiation weighting factor. The *dose-equivalent* is the absorbed dose at a point in tissue weighted by a quality factor determined from the LET of the radiation at that point.

Because weighting factors are, in effect, simplified versions of the RBE, it is clear that the equal equivalent doses from different sources of radiation, delivered to any tissue in the body, should produce more or less the same biological effect. This does not mean, of course, that a given equivalent dose produces the same effects in different parts of the body. It will be shown in the next section that a dose to the hand, for example, may have quite a different (and less serious) effect than the same dose to the blood-forming organs.

The US NRC uses the term dose-equivalent, which is also symbolized by the symbol H. The conventional unit of equivalent dose is the *rem* and is used by the NRC.[4] If the quality factor is unity, as it is for γ-rays, then an absorbed dose of 1 rad gives a dose-equivalent of 1 rem. With α-particles whose Q is 20, 1 rad of absorbed dose results in a dose-equivalent of 20 rems. A *millirem* is written mrem.

The SI unit of dose is the *sievert*, abbreviated Sv. This is the dose-equivalent arising from an absorbed dose of 1 gray. Since 1 Gy = 100 rads, it follows that

$$1 \text{ Sv} = 100 \text{ rem}. \tag{9.9}$$

A *millisievert* is abbreviated mSv. The dose-equivalent or equivalent dose is usually referred to simply as *dose*.

Equivalent Dose Rate

The rate at which equivalent dose is received is denoted by $\dot{H}$. This is determined from the absorbed dose rate by

---

[3]NCRP report No. 116, 1993.

[4]This is the acronym for *roentgen equivalent man*, although this identification is not encouraged, since the roentgen is now a unit of exposure, not dose.

$$\dot{H} = \dot{D} \times W \tag{9.10}$$

and is expressed as Sv/h, Sv/sec, etc.

Dose equivalent rate is also denoted by $\dot{H}$. Like equivalent dose rate, this is determined from the absorbed dose rate by

$$\dot{H} = \dot{D} \times Q$$

and is expressed in rem/sec, mrem/hr, etc.

Example 9.2

It can be shown that a beam of 1 MeV γ-rays having an intensity of 10^5 γ-rays/cm²-sec deposits in tissue approximately 5×10^{-3} ergs/g-sec. Calculate the absorbed dose rate and the dose-equivalent rate.

Solution. It takes a deposition of 100 ergs/g-sec, to give 1 rad/sec, so the absorbed dose rate is here

$$\dot{D} = 5 \times 10^{-3}/100 = 5 \times 10^{-5} \text{ rad/sec} = 5 \times 10^{-7} \text{ Gy/sec}.$$

This is a small number and is better written as

$$\dot{D} = 5 \times 10^{-5} \times 3{,}600 = 0.180 \text{ rad/hr} = 180 \text{ mrad/hr} = 1.80 \text{ mGy/hr.} \text{ [}Ans.\text{]}$$

From Table 9.2, $Q = 1$ for γ-rays, and it follows that

$$\dot{H} = 180 \text{ mrem/hr} = 1.80 \text{ mSv/hr.} \text{ [}Ans.\text{]}$$

Population Dose

Frequently, it is useful to specify the total dose-equivalent to a given group of people. This is called the *population dose*,[5] denoted as H_{pop}. The units of H_{pop} are *man-rems*, or *person-rems*, and in SI units are *person-sieverts*. Calculations of H_{pop} are carried out in an obvious way. Thus, if in some nuclear incident a town of 2,000 people is exposed to radiation in such a way that one-half receives 2 rems, while the other half receives 1 rem, the population dose is

$$H_{pop} = 1{,}000 \times 2 + 1{,}000 \times 1 = 3000 \text{ man-rem}$$

$$= 30 \text{ person-sievert.}$$

To generalize this procedure, let $N(H)dH$ be the number of persons in a total population N who receive doses between H and dH. Then,

---

[5]The term *collection dose* is frequently used instead of population dose, because of the possible misinterpretation of the latter to infer an average, or mean, dose to the population.

$$N = \int_0^\infty N(H)dH.$$

The population dose is the integral over the number of people who receive a given dose multiplied by that dose; that is,

$$H_{\text{pop}} = \int_0^\infty N(H)HdH. \qquad \qquad (9.11)$$

9.3 SOME ELEMENTARY BIOLOGY

In the last analysis, the effect of radiation on living things is due to the excitation or ionization of various molecules contained in the cells that make up a living system. Before pursuing these matters further, it is appropriate to review some elementary biology of cells, especially human cells.

There are approximately 4×10^{13} cells in the average adult person. These are not all identical, however, either in function or in size. Brain cells obviously perform a different function than liver cells. Most cells are quite small, on the order of 10^{-3} cm in diameter; nerve cells, by contrast, may be a meter long.

Cells are divided into two broad classes: *somatic cells* and *germ cells*. Almost all of the cells in the body are somatic cells. These are the cells that make up the organs, tissues, and other body structures. The germ cells, which are also called *gametes*, function only in reproduction. It is the union of gametes from different sexes that is the starting point of a new individual. The gametes also carry the hereditary material of the species that makes children look more like their parents than their neighbors, and ensures that the foibles of humanity pass with little change from generation to generation.

Modern research has shown that the living cell is an enormously complex system. A typical somatic cell of the kind found in animals is shown in Fig. 9.1. Cells contain a number of organlike structures called *organelles*, each of which performs specific functions for the cell as the organs do in the body as a whole. These organelles are suspended in the *cytoplasm*, a transparent, dilute mixture of water and various molecules and electrolytes, which comprises the bulk of the cell volume.

Other major parts of the cell shown in Fig. 9.1 are as follows:

Nucleus. This is the large, generally spherical body that functions as the control center of the cell; it contains the *chromatin*.

Chromatin. This is the genetic material of the cell.

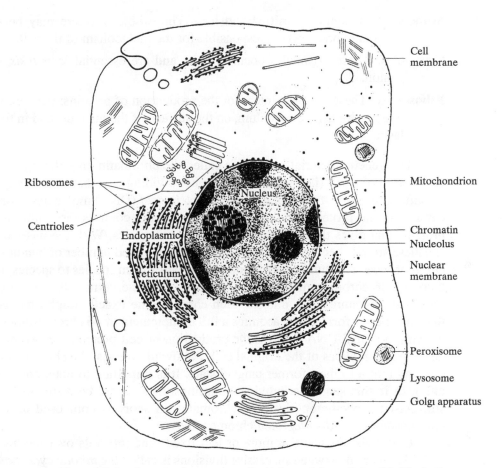

Figure 9.1   A typical somatic cell. (Adapted from John W. Kimball, *Biology*, 3d ed. Reading, Mass.: Addison-Wesley, 1974.)

Nucleoli (singular, **nucleolus**). These are spherical bodies (there may be as many as four located in the nucleus) that are important in the metabolism of certain chemicals.

Endoplasmic reticulum. This is a complex network of flattened tubules that serves to transport materials within the cell and is also a source of important metabolic enzymes.

Golgi apparatus. This organelle's functions are not entirely understood. The Golgi apparatus apparently concentrates and modifies certain chemical substances.

Lysosomes and **peroxisomes**. These organelles contain enzymes for producing various chemicals.

Mitochondria (singular, **mitochondrion**). These objects (there may be several thousand in one cell) are responsible for the metabolism of the cell.

Centrioles. These organelles occur in pairs and are essential in *mitosis*, or cell division.

Ribosomes. These are the centers for the production of proteins; they are located on the endoplasmic reticulum, on the surface of the nucleus, and in the cytoplasm.

The genetic material in a cell is called the chromatin only during the quiescent period of a cell (i.e., when the cell is not dividing). At that time, it appears as a confused mass of strands of *deoxyribonucleic acid* (DNA) molecules, along with certain nuclear proteins. In this stage, the chromatin controls the synthesis of the proteins that gives the cell its distinctive characteristics. With the onset of mitosis, these strands become untangled and coil into a fixed number of bundles called *chromosomes*. The number of chromosomes varies from species to species; in man, there are 46 chromosomes in every cell, except in the germ cells, which contain only half this number. In mitosis, each chromosome exactly duplicates itself, so that each newly formed cell acquires a full complement of 46 chromosomes. Since the chromosomal DNA controls the production of cell protein, the two new cells develop into replicas of the original cell. Following mitosis, the chromosomes uncoil and return to their former tangled state. It is important to note, however, that although in this stage the individual chromosomes cannot be distinguished, they still exist as specific entities. The chromatin, in short, is composed of *uncoiled* chromosomes, not *disintegrated* chromosomes.

Cell division occurs at more or less fixed time intervals over the life of the cell. The interval between successive divisions is called the *mitotic cycle time*. The cell cycle is further divided into three other intervals, as illustrated in Fig. 9.2. In

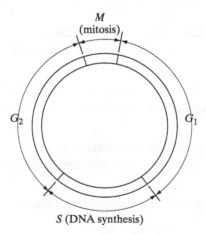

Figure 9.2   The mitotic cycle of a cell.

the figure, M is the time during which mitosis is taking place. During the interval S, at the middle of the cycle, DNA is synthesized in the nucleus. The intervals G_1 and G_2 are known as gaps in the cycle. The cell nucleus is relatively inactive at these times. In lung and ovarian cells of newborn hamsters, which are commonly used in radiobiological experiments, $M = 1$ hr, $S = 6$ hrs, $G_1 = 1$ hr and $G_2 = 3$ hrs. The total cycle time is therefore 11 hrs. Cells from other animals and tissues have different cycle times. In the basal stem cells of the human intestine, the mitotic cycle time is about 24 hrs. On the other hand, nerve and muscle cells, for example, are highly differentiated; they have long lifetimes and essentially never undergo mitosis.

The structure of germ cells is much different from that of somatic cells and will not be discussed here. In any case, there are two varieties of germ cells. These are called, respectively, *sperm cells* and *egg cells*, or *ova* (singular, *ovum*). The former are produced in the testes of the male, the latter in the ovaries of the female. Collectively, the testes and ovaries are called *gonads*. In reproduction, the union of a sperm and an ovum, each with 23 chromosomes, produces a *zygote*, the first bona fide cell of the new individual. The zygote contains 46 chromosomes. The mitosis of the zygote and its progeny eventually produces a full-size offspring.

For many years, it was thought that hereditary characteristics stem from the actions of individual *genes*, which were pictured as particle-like entities strung along the chromosomes. It is now known that the genes are actually segments of DNA molecules that provide a kind of code controlling the synthesis of proteins. Nevertheless, the term gene is still used to describe the point of origin of recognizable characteristics, and changes in the chromosomes that result in new, inheritable characteristics are called *gene mutations*.

9.4 THE BIOLOGICAL EFFECTS OF RADIATION

It is usual to divide the biological effects of radiation into two broad classes, depending upon whether the effects are inherently *stochastic*[6] or *nonstochastic* in nature. Stochastic effects are those whose *probability* of occurrence, as opposed to *severity*, are determined by dose. Cancer and genetic mutations are examples of stochastic effects. As discussed next, there is a definite probability that cancer will result from a given radiation dose, but it is by no means certain that cancer will result from that dose. Furthermore, the ultimate magnitude or severity of the cancer is unrelated to the original dose—it is determined by events and circumstances that occur long after the initiating event. With stochastic processes, there is no reason

---

[6]Stochastic is a general term referring to events or processes that are determined by a probability or a sequence of probabilities.

to suppose that there is some level of radiation below which no effect is possible. Presumably, any dose, however small, is capable of initiating a stochastic effect.

Nonstochastic, or deterministic, effects of radiation are entirely predictable, and their severity is an inevitable consequence of a given dose. Examples of nonstochastic effects are nonmalignant skin damage (*erythema*), the formation of *cataracts* (opaqueness of the lens of the eye), *hematological* effects (changes in the composition of the blood), and so on. Such nonstochastic processes may require some sort of a threshold dose before they become manifest, as in the case of cataract formation. But, whether there is a threshold or not, the severity of the injury ultimately increases with the level of the dose.

Mechanisms of Radiation Effects

Because the cell is such a complex system, there can be no single effect of radiation, even at the cellular level. Whatever effects do occur necessarily depend upon which organelles are involved and on their importance to the functioning of the cell. There are two fundamental mechanisms by which radiation can affect an organelle. First, radiation may lead to the breakage of molecules, rupturing their bonds by the ionizing effect of the radiation. This is known as the *direct effect* of radiation. Second, radiation, again because of its ionizing power, may result in the production of new chemicals, such as the highly reactive oxy (O) and hydroxyl (OH) radicals, which interact chemically within the cell. This is called an *indirect effect* of radiation. Recent experimental evidence suggests that the indirect effect is more important by far in producing biological effects. In any event, it is important to recognize that both processes are fundamentally chemical in nature—a broken molecule is necessarily two other molecules. The biological effects of radiation are therefore not essentially different from the biological effects of various chemicals. Radiation is merely a different and seemingly bizarre means for producing the chemicals in question.

The end result of these chemical transformations on a cell depends upon which cellular molecules are affected. The details of these processes are not entirely understood at the present time. Nevertheless, there is overwhelming evidence that, with respect to observable stochastic and nonstochastic effects, the most sensitive portion of the cell lies within the cell nucleus, probably with the chromosomes. Other organelles can be affected by radiation, of course. If one of the larger molecular structures in a mitochondrion is damaged, for instance, the functioning of this organelle can be disrupted. However, because there are many mitochondria in most cells, the malfunctioning of one of these structures would normally have no observable effect on the overall behavior of the cell. But events taking place within the nucleus are crucial to the well-being of the cell as a whole and may be transmitted to the cell progeny in mitosis.

Cell Killing

It is possible to remove portions of tissue from various animal organs, to extract individual cells, and to grow these into healthy, multiplying cell colonies. This is usually done in flat, shallow containers called petri dishes, which contain a suitable culture medium. When such colonies are exposed to radiation over a short period (i.e., when they are given *acute* doses of radiation), a fraction of the exposed cells cease to reproduce and eventually disintegrate. In this event, the cells are said to have been killed by the radiation; more accurately, they have suffered from what is called *reproductive death.*

When the fraction of surviving cells is plotted versus the radiation dose, *survival curves* like those shown in Fig. 9.3 are obtained. In all cases, the surviving fraction decreases with increases in the absorbed dose, as would be expected. However, as indicated in the figure, there is a marked difference for a given absorbed dose between the survival fractions with high and low LET radiation. With high LET radiation, the fraction of surviving cells drops sharply (exponentially), but with low LET radiation the survival curve exhibits some flatness or "shoulder" at low doses and only drops steeply at high doses.

A precise explanation of cell survival curves is not possible at the present time. It is clear, of course, that the effect of high LET radiation is greater than

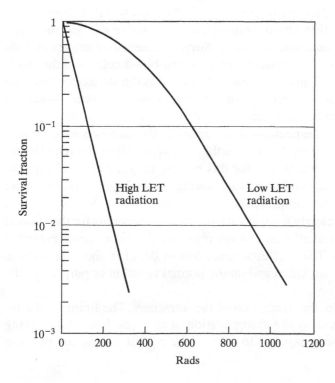

Figure 9.3   Survival curves for cells exposed to acute doses of radiation.

that of low LET radiation per absorbed dose, which is another way of saying that the RBE is larger for high LET radiation. The existence of the shoulder with low LET radiation at low doses suggests the existence of some sort of threshold for this type of radiation, possibly due to repair mechanisms within the cell that are overwhelmed at high doses. At least it is clear that dose-effect relationship is far less direct at low doses than at high doses. These general characteristics of high and low LET radiation are found in virtually all data involving the biological effects of radiation.

It will be recalled from the preceding section that cells have a natural mitotic cycle during which various activities occur at specified times. Using a number of sophisticated techniques, biologists can produce colonies of cells whose cycles are all in phase. That is, all of the cells undergo mitosis at the same time, DNA production at the same time, and so on. Such cells or colonies are said to be *synchronous*. When synchronous cells are irradiated at various points in their mitotic cycles, it is found that the effect of the radiation, both high LET and low LET, is somewhat more pronounced just before or during mitosis. Cells irradiated at other points in the cycle tend to survive the effect of radiation exposure until the onset of mitosis, when they fail to reproduce and die a reproductive death.

The experiments with synchronous cells account for the phenomenon of *radiosensitivity*, which is observed for both stochastic and nonstochastic effects of radiation on whole organisms, including man. Thus, certain tissues, namely, those whose cells reproduce most frequently, are normally the most sensitive to radiation. These cells often die when they attempt mitosis; if they survive mitosis, they may be sensitized for the later induction of cancer. Such radiosensitive tissues include the red blood marrow, which continually produces new blood cells, and the tissue beneath the lining of the gastrointestinal tract that contains the stem cells that continually replace the lining of the tract. By contrast, the more static tissues—muscle and bone, for example—are far more immune to the effects of radiation.

The foregoing remarks, especially with regard to the radiosensitive tissues, relate only to the body after birth. Prior to birth, or *in utero*, all of the fetal tissues are undergoing rapid development, and for this reason, they are particularly sensitive to radiation. This is especially the case during the embryonic period, from fertilization through about the eighth week.

Cell killing, which is clearly a nonstochastic process, accounts for the *clinical effects* (i.e., the gross medical effects) that are observed following acute exposure of individuals to radiation. These effects, described in detail in the next section, depend on how many cells are killed and on the normal function or purpose of the cells.

Consider, for example, the irradiation of the intestines. The lining of the intestine is continually renewed by the multiplication of cells just beneath the lining surface. When the intestine is exposed to γ-rays, the rate of production of these cells is slowed. Up to moderate radiation doses, enough new cells are formed to maintain the intestinal wall. However, at high doses, the surface cannot be maintained, and it disintegrates. As a consequence, various body fluids enter the intestines, while bacterial and toxic materials from the intestine pass into the blood stream. The overall effect on the individual is diarrhea, dehydration, infection and toxemia (blood poisoning). As this example shows, total-body involvement results from events at the cellular level.

Cancer

Cancer is the second-ranking cause of death in the world today. In the United States, approximately 38% of the population contracts the disease, and 17% succumbs to it. Of the 275 million people living in this country in 2000, about 47 million of them will die of cancer if cancer mortality rates do not change. Current mortality rates for various types of cancer in the United States are given in Table 9.3.

Cancer is not now viewed as a single disease, but rather as about one hundred closely related diseases, all characterized by rapid, uncontrolled cell division. There are a thousand or more chemicals, called *carcinogens*, that are known to produce cancer. More than 90 percent of all human cancers are believed to be the result of carcinogens in the environment, the bulk of which are naturally occurring, not man-made. Since various chemicals can produce cancer, it is not surprising that radiation is also capable of doing so, in view of the underlying mechanism of radiobiological effects described earlier.

TABLE 9.3   US CANCER MORTALITY RATES IN 1992–1996 (DEATHS PER HUNDRED THOUSAND PERSONS PER YEAR)*

Cancer type	Mortality
Breast	14.2
Leukemia	6.3
Lung, respiratory system	49.5
Pancreas	8.4
Stomach	4.2
Prostate	25.6
Thyroid	0.3
All sites	170.1

*Ries LAG, Kosary BF, et al., SEER Cancer Statistics Review, 1973–1996, National Cancer Institute, Bethesda, MD, 1999. The total US population in 1996 was 266 million.

Radiation-induced cancers are no different from cancers arising from other causes. The extent of the cancer-inducing effect of radiation must therefore be deduced from the (often small) increase in the observed incidence of cancer in groups of persons exposed to radiation. This is especially difficult at low doses where the effect, if it exists at all, may be hidden in statistical fluctuations.

The precise causes of cancer are not known as of this writing. There is, however, increasing evidence that cancer induction is a two-stage process. In the first, or *initiation*, stage, a lesion (i.e., some injury or other deleterious effect) is produced in DNA in one or a number of cells. The affected cells are thereby transformed into potentially cancerous cells, poised to undergo uncontrolled division. However, they are prevented from doing so by some as yet unidentified hormonal, immunologic, or other protective agent. Such agents are probably inherited characteristics of every individual.

During the second, or *promotion*, stage, the protective mechanism goes awry for some reason and permits the initiated cells to multiply without restraint. Presumably, the promotion stage can be triggered by such diverse factors as viral infections, chemical irritants, failure of the immunological system, or physiological changes in the body that occur naturally with aging.

The two-state theory of cancer explains the fact that there is a *latent period* of several years between the time of irradiation and the appearance of many types of cancers, during which the probability of a cancer occurring is essentially zero. It also accounts for the fact that cancer seems to run in families; that is, it is to some extent an inherited trait. This also implies, of course, that there may be segments of the population that are especially susceptible to radiation-induced cancer, a circumstance that has only recently been recognized. Similarly, there are probably segments of the population that are especially resistant to cancer induced by radiation.

It should be noted that the cell or cells triggered into malignancy during the promotion stage generally are not the same cells initiated during irradiation. Rather, they are progeny of the original cells, often many generations removed. It is clear, therefore, that whatever initiating character is conferred on the irradiated cell must necessarily be carried from generation to generation. Again this speaks of a chromosomal origin of cancer.

Genetic Effects

If radiation succeeds in disrupting a DNA molecule in a chromosome, the result may be a mutation. If this mutation occurs in a *somatic* cell of a fully developed individual, no macroscopic effect may be observed unless a great many cells are similarly involved. This is because the DNA determines the structure of the proteins manufactured by the cell, and a mutation thus interferes with the production of the proteins necessary for proper cellular functioning. As a result, mutations are usually harmful to cells, and the progeny of a mutant cell, if not the parent cell itself, simply die out. If the mutation occurs in a *germ* cell, the affected cell is usually incapable of being fertilized. However, if the mutant gamete *is* successfully fertilized and the zygote develops into a live offspring, then the mutation is carried into the progeny. For this reason, radiation exposure to the gonads is of special concern, at least through the breeding age.

Mutations occur spontaneously in the human population from unknown causes. Indeed, about 10% of all newborn infants either suffer directly from a disorder or malfunction of genetic origin or carry such defects, which are expressed later in life or in their progeny. Such defects can be as obvious as *albinism* (congenital deficiency in skin pigment) and *Down's syndrome* (mongoloid features and associated mental retardation caused by the presence of an extra chromosome), or so slight that they can only be detected by laboratory tests. Radiation-induced genetic defects are no different from those normal to the species. The effect of radiation is therefore to increase the naturally occurring mutation rate.

9.5 QUANTITATIVE EFFECTS OF RADIATION ON THE HUMAN SPECIES

A considerable effort has been made over the years to determine the effects of radiation on the human body. Since it is not possible to perform radiation experiments on people, current knowledge of radiation effects is based on data from radiation accidents and overexposures of the types discussed in Section 9.1; on *epidemiological*[7] studies of radiation-induced diseases such as leukemia and lung cancer; on studies of the casualties and survivors of World War II atomic bombings in Japan; and on numerous experiments with laboratory animals. The current status of these studies may be summarized as follows:

a) There is well-documented information on the effects of large, *acute* (short-term) radiation doses, in excess of 10 to 20 rems.

b) Because the effects are so rare, if they exist at all, there is only limited data showing positive effects of

   i) acute doses up to 10 or 20 rems and not repeated;

   ii) acute doses of a few rems and repeated occasionally; and

   iii) *chronic* (continuing for a long time) doses of the order of millirems per day.

Only categories (a) and (b-iii) will be considered here, as these are the most important to nuclear engineers. Large acute doses may be received accidentally at

---

[7]*Epidemiology* is the study of epidemics, diseases afflicting large numbers of persons.

a nuclear installation, while the low doses in the second category may be the rule at such facilities.

Large Acute Doses: Early Effects

In discussing the effects of acute doses, it is usual to distinguish between *early effects*, which are evident within 60 days of the exposure, and *late effects*, which become evident after 60 days. Early effects are generally nonstochastic in nature; late effects arise from both stochastic and nonstochastic processes. Table 9.4 shows the average clinical early effects observed following acute whole-body doses up to the order of 1,000 rems. It will be noted that no serious deleterious effects are normally seen for doses of less than about 75 rems. At doses larger than 75 rems, the exposed individual is said to suffer from *acute radiation syndrome*, or ARS, as it is known in medical circles. All of the symptoms listed in the table are the result of simultaneous damage to several body organs resulting, in turn, from injuries to individual cells as discussed in the preceding section.

The incidence of death given in Table 9.4 is for individuals who have not undergone treatment. In this case, fatalities begin to be observed with acute doses

TABLE 9.4   PROBABLE EARLY EFFECTS OF ACUTE WHOLE-BODY RADIATION DOSES*†

Acute dose (rems)	Probable observed effect
5 to 75	Chromosomal aberrations and temporary depression of white blood cell levels in some individuals. No other observable effects.
75 to 200	Vomiting in 5 to 50% of exposed individuals within a few hours, with fatigue and loss of appetite. Moderate blood changes. Recovery within a few weeks for most symptoms.
200 to 600	For doses of 300 rems or more, all exposed individuals will exhibit vomiting within 2 hours. Severe blood changes, with hemorrhage and increased susceptibility to infection, particularly at the higher doses. Loss of hair after 2 weeks for doses over 300 rems. Recovery from 1 month to a year for most individuals at the lower end of the dose range; only 20% survive at the upper end of the range.
600 to 1,000	Vomiting within 1 hour. Severe blood changes, hemorrhage, infection, and loss of hair. From 80% to 100% of exposed individuals will succumb within 2 months; those who survive will be convalescent over a long period.

*The whole-body doses given in this table are those measured in soft tissue near the body surface; because of energy absorption in the body, the interior (or vertical midline) doses, which are sometimes quoted, are about 70% of the values in the table.
†From S. Glasstone and A. Sesonske, *Nuclear Reactor Engineering*, 3d ed. Princeton, NJ.: Van Nostrand, 1981. (By permission US Department of Energy.)

TABLE 9.5 AVERAGE CONCENTRATIONS OF
FORMED ELEMENTS OF HUMAN BLOOD

Formed elements	Concentration (per cubic millimeter)
Erythrocytes	$(4.5\text{–}5.5) \times 10^6$
Leukocytes	6,000–10,000
Platelets	$(2\text{–}8) \times 10^5$

of approximately 200 rems. Persons who receive medical treatment have a somewhat better chance of survival; fatalities in this group begin to occur at about 500 rems. The whole-body acute dose that leads, without therapy, to the death of 50% of an exposed group within T days of exposure is called the LD_{50}/T dose.[8] For example, an $LD_{50}/60$ dose will kill one-half of an exposed population in 60 days. The $LD_{50}/60$ for man is not known accurately, but is thought to be approximately 340 rems. For most mammals it is about the same value. Bacteria and adult insects, by contrast, have an $LD_{50}/60$ on the order of 10,000 rads.

In the range of doses from about 100 to 1,000 rems, the most important effects are those associated with the blood, or more properly, the blood-forming organs, especially the red bone marrow. The patient in this case is said to exhibit *hematopoietic syndrome*.

Human blood consists of three *formed elements*: (1) *red blood cells*, or *erythrocytes*, (2) *white blood cells*, or *leukocytes*, and (3) *platelets*, all of which are suspended in a fluid called *plasma*. The red cells carry oxygen, which is necessary for life, through the body. The white cells consist of several different types of cells, the most populous of which are the *polymorphonuclear neutrophils* (50%–75%) and the *lymphocytes* (20%–40%). Among other functions they perform, the white cells act to repel or reduce infection in the body. The platelets play an essential role in the clotting of blood.

The normal range of concentrations of the formed elements of blood (i.e., the normal *blood count*) is given in Table 9.5. Following acute whole-body irradiation, the blood count changes; the extent depends on how large a dose was received. Figure 9.4 shows a typical blood-count history following an acute dose of 300 rems. Because of the drop in the number of leukocytes, the resistance of the exposed individual to infection is lowered. At the same time, the fall in platelet count prevents normal clotting of the blood, which, in severe cases, may lead to *hemorrhage*, or profuse bleeding.

It should be pointed out that with the exception of the lymphocytes, which are unusually sensitive to radiation, the other formed elements of blood are fairly

---

[8]LD stands for "lethal dose."

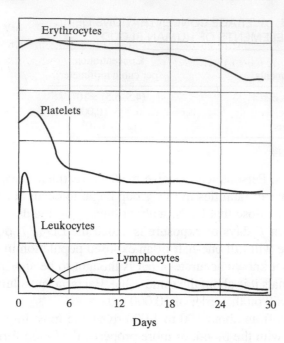

Figure 9.4 Blood count (on arbitrary scale) following acute dose of 300 rems.

radiation resistant.[9] Changes of the type shown in the figure are due, therefore, to the destruction of the blood-forming stem cells in the bone marrow, not to the destruction of the blood cells themselves. Treatment for a patient suffering from hematopoietic syndrome includes isolation in a sterile environment, the administration of antibiotics, and bone-marrow transplants.

An acute whole-body dose of between about 1,000 and 5,000 rem leads to what is known as the *gastrointestinal syndrome*. In this case, the dominant effect and ultimate cause of death is failure of the intestinal wall due to depletion of its stem cells as described in the preceding section.

The patient remains in satisfactory condition for a few days while the existing wall cells continue to function, but as these are sloughed off, the patient succumbs to infection. The exposed individual usually dies within two weeks.[10]

Persons who receive doses in excess of 5,000 rems die within a few hours of exposure. The cause of death is not entirely clear, but probably involves the rapid accumulation of fluid in the brain. These symptoms are known as *central nervous system syndrome*.

[9]This is due, at least in part, to the fact that erythrocytes and platelets have no nuclei. They are bodies contained in the blood, but are not cells in the ordinary sense of the term.

[10]Death from the gastrointestinal syndrome beginning in the neighborhood of 1,000 rems requires simultaneous failure of the blood-forming organs, which occurs with whole-body irradiation at this level and removes an important mechanism for fighting infection. With irradiation of the intestines alone, the onset of death begins around 2,000 rems with and $LD_{50/60}$ of about 3,500 rems.

The preceding discussion refers to whole-body irradiation. If only a portion of the body is exposed, the resulting early effects depend upon which portion is irradiated, although it may be said in general that the accompanying injury is always less severe, because fewer interacting organs and systems are involved. However, the exposure may result in serious late effects of the type to be described shortly. For example, a hand receiving an absorbed dose of 200 to 300 rads of x-rays will exhibit only erythema similar to mild sunburn. More serious burns, comparable to scalding or to chemical burns, occur with doses of thousands of rads. Although in both cases the affected region will heal, it becomes predisposed to the later development of skin cancer.

Large Acute Doses: Late Effects

There are considerable data showing late radiation effects in persons who receive a large acute whole-body dose at least once in their lifetime. Space permits only a brief summary of each effect.

Cancer   After an acute radiation dose, there is generally a period when there is no apparent increase in the probability of incurring cancer. The length of the *latent period* depends on the age at which irradiation occurs and the ultimate site of the malignancy. The latent period is followed by an interval, called the *plateau*, where the risk of a cancer is approximately constant. Here, the term risk is defined quantitatively as the number of cases of cancer to be expected per year following irradiation per million man-rems of dose. Beyond the plateau period, the risk of cancer drops essentially to zero. A simplified model of cancer risk is depicted in Fig. 9.5. Table 9.6 lists the lengths of the latent period and plateau for various tissues and gives the corresponding risk values (*risk coefficients*).

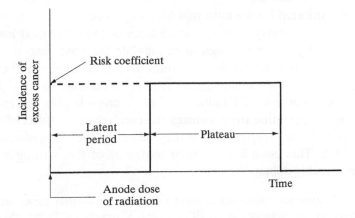

Figure 9.5   Simplified model of radiation-induced cancer.

TABLE 9.6   DATA ON RADIATION-INDUCED CANCER*

Type of cancer	Age at time of irradiation	Latent period years	Plateau period (years)	Risk coefficient (deaths/10^6/yr/rem)
Bone	0–19.9	10	30	0.4
	20+	10	30	0.2
Breast	10+	15	30	1.5
Leukemia	In utero	0	10	15
	0–9.9	2	25	2
	10+	2	25	1
Lung, respiratory system	10+	15	30	1.3
Pancreas	10+	15	30	0.2
Stomach	10+	15	30	0.6
Rest of alimentary canal	10+	15	30	0.2
Thyroid	0+	10	30	0.43
All other	In utero	0	10	15
	0–9.9	15	30	0.6†
	10+	15	30	1‡

*From Reactor Safety Study, WASH-1400, US Nuclear Regulatory Commission, October 1975, Appendix VI.
†"All other" includes all cancers except leukemia and bone.
‡"All other" includes all cancers except those specified in table.

Further explanation is necessary regarding the risk coefficients in Table 9.6. Data on human exposure to acute radiation doses above 100 rem indicate that the excess incidence of cancer, corrected for cell-killing early effects, increases approximately linearly with doses for both low LET and high LET radiation. A plot of the excess cancer incidence versus dose, called the *dose–response curve*, is therefore a straight line at high doses, as shown in Fig. 9.6. At lower doses, the data on human exposure are much less conclusive, and the relation of cancer incidence to dose is inferred for the most part from experiments with laboratory animals.

Until recently there was an absence of human data at low doses. Hence, the traditional practice has been to extrapolate the dose–response curve linearly to a zero dose, as indicated in the figure. This procedure is referred to as the *linear hypothesis*. It is generally agreed that such an extrapolation is proper for high LET radiation. For low LET radiation, there is considerable evidence to suggest that a linear extrapolation overestimates the carcinogenic effects of low doses, and that the actual dose–response curve may lie below the extrapolated curve, as shown in Fig. 9.6. This issue has not been settled as of this writing and is the subject of considerable debate.[11]

---

[11] Arguments supporting different varieties of dose-effect curves are discussed later in this chapter and are presented in the BEIR III and V reports and in the references at the end of the chapter.

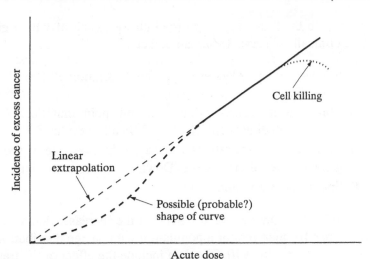

Figure 9.6   Dose–response curves for acute radiation doses.

The values of risk given in Table 9.6 are based on the linear hypothesis. Calculations using these values may very well exaggerate the effects of radiation exposure at low doses. In any case, Table 9.6 probably provides an upper boundary for estimating the consequences of radiation accidents. Calculations of this sort are described in the next section.

Mutations   Genetic effects of radiation exposure in man have not been demonstrated at the present time. In lieu of human data, the genetic risk of radiation for people is estimated from data on laboratory mice. Several significant facts emerge from these data. First, the male mouse is considerably more sensitive to the genetic effects of radiation than the female. Exposure of male mice at high dose rates produces many more mutations per rem than does the same dose at lower rates; in other words, a given dose does more damage if delivered over a shorter period. Thus, damage to genetic material is clearly repairable. But the data also suggest that the most radiogenetically sensitive portions of the male mouse are the spermatozoa and their precursor spermatids, which survive as individual entities over only a small fraction of the reproductive cycle. Because the same is true in man, this implies that the transmission of genetic damage from acute doses of radiation can be reduced by delaying conception until new sperm cells have matured from cells in a less sensitive stage at the time of exposure.

The basis for quantitative estimates of the genetic affects of radiation on the human species is complicated and beyond the scope of this book. (The reader should consult the references at the end of the chapter; see, in particular, the reports of the National Academy of Sciences.) In any event, it is usual to divide such effects into the following four broad classes, each of which gives rise to one or more clinically observable human defects:

a) Single-gene disorders arising from a mutation at a specific point on a chromosome.

b) Multifactorial disorders due to multiple point mutations. (The effect of radiation in producing such defects is difficult to evaluate.)

c) Chromosomal aberrations caused by the presence of too much or too little genetic material in the cells (Down's syndrome, for example).

d) Spontaneous abortions.

Table 9.7 summarizes the data on the increased incidence of these genetic effects per 10^6 man-rem of a population with an age distribution similar to that of the United States in 1970. The data include the effect of the transmittal of genetic defects to subsequent generations. Such propagation dies out after about three generations. The wide uncertainty in the effect on multifactorial disorders should be noted.

Cataracts   Vision-impairing opacity of the lenses of the eyes caused by radiation, or *radiation cataracts*, have a latent period of about 10 months and may appear as long as 35 years after exposure. They are a threshold phenomenon and do not occur with absorbed doses below the 200-to-500-rad range for low LET radiation. Where the population is exposed to whole-body doses of this order of magnitude, there would be few survivors with radiation cataracts. The corresponding threshold for fission neutrons is in the vicinity of 75 to 100 rads.

Fertility   The probable effects on human fertility, both male and female, of acute doses of γ-rays to the gonads are given in Table 9.8; comparable data for neutrons are not available.

TABLE 9.7   NUMBERS OF NATURALLY OCCURRING AND RADIATION-INDUCED GENETIC DISORDERS PER 10^6 MAN-REM*

Disorder	Normal incidence per 10^6 population per year†	Eventual number of genetic disorders
Single-gene disorders	166	42
Multifactorial disorders	560	8.4–84
Chromosomal disorders	75	6.4
Spontaneous abortions	780	42
Total	1581	99–175

*Based on WASH-1400; see references.
†Based on 14,000 live births per 10^6 population per year.

TABLE 9.8 PROBABLE EFFECT ON FERTILITY OF
SINGLE DOSES OF γ-RAYS TO THE HUMAN
GONADS

Dose, rads	Probable effect
150	Brief sterility
250	Sterility, 1 to 2 years
500–600	Permanent sterility, many persons
800	Permanent sterility, everybody

Degenerative Effects   Another long-term effect of radiation exposure is an increase in the incidence of degenerative conditions in various body organs due to the failure of the exposed tissues to regenerate properly. This leads to permanent, although not necessarily debilitating, impairment of organ function. The occurrence of such degenerative effects is to be expected in view of the damage that large doses can do to tissues and organs, an example of which was given in the preceding section.

Life Shortening   The overall effect of radiation exposure may be seen in its influence on the life span of exposed individuals. This is clearly to be expected, if only on the basis of the other long-term effects just discussed. Much of the evidence of life shortening has come from studies of the obituaries of physicians. It has been found, for example, that the average life span of radiologists who died between 1935 and 1944 was 4.8 years shorter than for other medical specialists who were not exposed to radiation, but who died in the same period. At the present time, it is generally believed that any observed life shortening arising from radiation exposure is due solely to the appearance of radiation-induced malignancies.

Chronic Low Doses

As already noted, doses of a few millirems per day, which accumulate up to a few rems per year, are of considerable concern in the development of nuclear power. This is the dose level permitted under current standards of radiation protection, which is discussed in Section 9.8. It is not surprising, therefore, that an enormous effort has been made over the years to determine what, if any, deleterious effects accompany chronic exposure to low levels of radiation. To date, the results are inconclusive. Thus, according to the Committee on the Biological Effects of Ionizing Radiations of the US National Academy of Sciences,[12]

[12]BEIR V report, p. 181. (See references.)

Derivation of risk estimates for low doses and low dose rates through the use of any type model involves assumptions that remain to be validated .... Moreover, epidemiologic data cannot rigorously exclude the existence of a threshold in the millisievert dose range. Thus, the possibility that there may be no risks from exposures comparable to the external natural background cannot be ruled out .... At such low dose rates, it must be acknowledged that the lower limit of the range of uncertainty in the risk estimates extends to zero.

In view of the evident lack of adequate human data on the effects of low levels of radiation, these effects are estimated, or *postulated*, by various dose–response models. For simplicity and conservatism, the linear hypothesis is often used. There exists a large body of evidence that strongly suggests the linear model overestimates the consequences of chronic exposure to low levels of radiation. In fact, the data suggest that doses in the range of 10–20 cGy (10–20 Rad) may be beneficial in lowering the chances of cancer in children and adults. Doses above 10 cGy are still potentially harmful to fetuses at 8–15 weeks gestation, based on analysis of Hiroshima and Nagasaki bomb survivors.

The linear hypothesis entirely ignores the capability of biological systems to repair themselves, even at the molecular level. For instance, it has been found that if only one strand of a DNA molecule (which consists of two strands twisted in the form of a double helix) is broken, the molecule may remain intact and the broken strand reconstituted in its original form. Such repair mechanisms may function normally at low rates of radiation exposure, but may be overwhelmed at high dose rates.

The theory of *radiation hormesis* suggests that organisms respond positively to low doses of radiation. The responses include inhibition of cancer and reduction in the aging process. While this theory is still under investigation, it is completely consistent with the observed data from the victims of the atomic bomb and with many of the other radiation case studies.

Notwithstanding the possibility that low doses result in a positive effect, the BEIR Committee and the regulatory agencies at this time are still utilizing the linear hypothesis and will likely continue to do so.[13] There is a growing consensus that the linear hypothesis is too conservative, and this may eventually lead to the acceptance of a threshold dose or even the application of radiation for preventive reasons.

---

[13]The BEIR V report acknowledges the possibility of at least a threshold of radiation health effects when it states, "For combined data (Hiroshima and Nagasaki), the rate of mortality is significantly elevated at 0.4 Gy (40 Rad) and above, but not at lower dose. At bone marrow doses of 3–4 Gy, the estimated dose–response curve peaks and turns downward." p. 242, op cit.

9.6 CALCULATIONS OF RADIATION EFFECTS

The data on radiation effects developed in the preceding sections will now be applied to specific cases of radiation exposure. As done earlier, acute and chronic exposures will be considered separately.

Acute Exposures

The probable short-term consequences of acute exposure to an individual can be determined directly from the information in Table 9.4. As indicated in the table, there are no serious, clinically observable effects below whole-body doses of 75 rems. Calculating the late effects of an acute exposure—in particular, the probability of the individual contracting radiation-induced cancer—is somewhat more complicated, because this probability depends on the location of the malignancy and the age of the individual at the time of the exposure. Nevertheless, the calculation is relatively straightforward; it is best described by a specific example.

Example 9.3

In a radiation accident, a 30-year-old male worker receives an acute whole-body dose of 25 rems. What is the probability that the worker will die from cancer as a result of this exposure? [*Note:* In the United States, a 30-year-old male, on the average, lives to an age of 77 years.]

Solution. The desired probability is the sum of the probabilities of death from cancer at any site in the worker's body. First, consider bone cancer. According to Table 9.6, the latent period is 10 years, so the probability of his getting this cancer from the radiation is zero until he is 40 years old. For 30 years thereafter, his chances of dying as a result of bone cancer are roughly constant at 0.2 per year per 10^6 rems. Because his dose was 25 rems, the probability of his dying over this period is

$$\frac{25}{10^6} \times 30 \times 0.2 = 1.5 \times 10^{-4},$$

or 1.5 chances in 10,000. He reaches the end of the risk plateau at the age of 70, and has, on the average, 7 remaining years with zero probability of contracting bone cancer due to his original exposure.

In a similar way, the probabilities of other cancers can easily be computed. The total probability is 4.7×10^{-3}. [*Ans.*]

The probability of dying from cancer at the age of 30 years or older is approximately 0.25, so the radiation increased this probability by $4.7 \times 10^{-3}/0.25 = 0.019$ or about 2%.

In this example, the entire risk plateau period lay within the worker's expected life span. Had the plateau extended beyond his life span, only that portion of the plateau within his life span would be used in the calculation.

It must also be noted that the foregoing solution is not entirely exact, since the total probability of death was taken simply to be the risk per year multiplied by the length of the plateau. This clearly overstates the risk, for in order for the worker to die of radiation-induced cancer near the end of the plateau, he must not have succumbed to this disease before that time. The exact solution is similar to the calculation of the probability that a particle can move a given distance in a medium with or without a collision, as discussed in Section 3.3. The present simplification is possible, because the probability of survival over the entire plateau is very close to unity.

In a radiation accident involving the public at large, people of all ages may be irradiated. Unless a priori knowledge of the age distribution within the irradiated group is available, it is the usual practice to assume that this distribution coincides with the average age distribution in the country as a whole. To determine the long-term effect of an acute exposure, it is then necessary to compute the probability that a cancer of any type will appear in each cohort in the population. A portion of this computation is given in the next example.

Example 9.4

In a hypothetical radiation accident, one million persons each receive a dose of exactly 1 rem. How many cases of leukemia can be expected to result in this population?

Solution. The solution to this problem is given in Table 9.9. The first three columns give the fractions of the US population in specified cohorts and the corresponding additional life expectancies.

Column 4 gives the years at risk (i.e., the number of years for which the risk is nonzero). This is either the full duration of the plateau or the life expectancy less the latent period, whichever is smaller.

TABLE 9.9   CALCULATION OF EXPECTED CASES (DEATHS) OF LEUKEMIA PER 10^6 MAN-REMS

Age group	Fraction	Expectancy	Years at risk	Risk coef.	Deaths
in utero	0.011	71.0	10.0	15	1.65
0–0.99	0.014	71.3	25.0	2	0.70
1–10	0.146	69.4	25.0	2	7.30
11–20	0.196	60.6	25.0	1	4.90
21–30	0.164	51.3	25.0	1	4.10
31–40	0.118	42.0	25.0	1	2.95
41–50	0.109	32.6	25.0	1	2.73
51–60	0.104	24.5	22.5	1	2.34
61–70	0.080	17.1	15.1	1	1.21
71–80	0.044	11.1	9.1	1	0.40
80+	0.020	6.5	04.5	1	0.09
Total					28.36

TABLE 9.10 EVENTUAL CANCER DEATHS IN
NOMINAL POPULATION PER 10^6 MAN-REMS

Cancer type	Deaths
Bone	6.9
Breast	25.6
Leukemia	28.4
Lung, respiratory system	22.2
Pancreas	3.4
Stomach	10.2
Rest of alimentary canal	3.4
Thyroid	13.4
All other	21.6
Total	135.0

Column 5 indicates the risk per year per 10^6 rems as taken from Table 9.6. The last column shows the number of leukemia cases expected over the lifetime of each cohort. For instance, the number of deaths in the in utero cohort is $0.011 \times 10 \times 15 = 1.65$; the number in the 51–60 cohort is $0.104 \times 22.5 \times 1 = 2.34$; and so on. The total number of leukemia cases (and deaths) is 28.4. [*Ans.*]

When calculations similar to those in Example 9.4 are carried out for other types of cancer, the resulting numbers of eventual deaths are as shown in Table 9.10. These calculations give the consequences of a dose of 1 rem to 10^6 people. However, since the risk coefficients used in the computations are based on the linear hypothesis, the number of deaths would be the same if 0.5×10^6 people received 2 rems, 0.25×10^6 people received 4 rems, and so on, as long as the total number of man-rems remained the same. This may be seen from the next argument.

Let $P(H)$ be the absolute probability that an individual will eventually die of cancer following a dose of H rems. The function $P(H)$ is the dose–response function plotted in Fig. 9.6. If $N(H)dH$ persons in the population N receive doses of between H and dH rems, the number of eventual cancer deaths is

$$n = \int_0^\infty N(H)\,P(H)dH. \tag{9.12}$$

According to the linear hypothesis, $P(H)$ is given by the simple function

$$P(H) = \alpha H, \tag{9.13}$$

where α is a constant. The number of deaths is then

$$n = \alpha \int_0^\infty N(H)HdH = \alpha H_{\text{pop}}, \tag{9.14}$$

where use has been made of Eq. (9.11). It follows from Eq. (9.14) that the total number of cancer deaths, on the basis of the linear hypothesis, is independent of the magnitude of the individual doses and depends only on the population dose, H_{pop}. This considerably simplifies calculations of the consequences of radiation accidents to the public. Thus, if the population in question has the same age distribution as the one used in calculating Table 9.10, the number of eventual cancer deaths from a population dose H_{pop} is simply

$$n = 135 H_{pop}. \tag{9.15}$$

As remarked in the preceding section, the linear hypothesis is widely believed to overestimate the late effects of radiation exposure, especially at low doses. A more accurate representation of the dose–response function is often taken to be of the form

$$P(H) = \beta_1 H + \beta_2 H^2, \tag{9.16}$$

where the values of β_1 and β_2 are the subject of continuing debate; but, in any case, β_1 is less than α in Eq. (9.13). Unfortunately, functions such as that given in Eq. (9.16) are somewhat more difficult to apply in practice, since the number of cancer deaths in an irradiated population depends on the distribution of doses to different individuals, and not on the population dose alone.

Chronic Exposure

Because it is necessary to take into account the manner in which a dose is received over time, the calculation of the late effects of chronic low doses is especially tedious if based on the linear-quadratic form of $P(H)$ given in Eq. (9.16). With such a function, it makes a difference whether an annual dose of 5 rems is received in one day or spread out over the year at a rate of 14 mrems per day. With the linear model, the fractionation of the dose is irrelevant, and only the total dose over a given interval need be considered.

Example 9.5

A radiation worker takes his first job at the age of 18. He continues to be so employed for 50 years and retires at the age of 68. If he receives an average annual occupational dose of 5 rems (which is the maximum acceptable dose according to current regulations; see Section 9.8) throughout his working career, what is the probability that he will incur and die of bone cancer as the result of this exposure? Use the linear hypothesis.

Solution. An accurate solution to this problem requires life-expectancy data on an annual basis from age 18 to 68. Such data are available but too lengthy to be reproduced here. An estimate of the probability can be made from Table 9.9 by assuming, where necessary, that all the radiation to a given cohort is received at the beginning of the cohort interval. This tends to overestimate the years at risk toward the end of the worker's career. The computation is similar to that of Example 9.3, except that, in this problem, the worker is assumed to receive a sequence of annual acute doses of 5 rems. The total probability of his dying from bone cancer is then readily calculated to be 1.2×10^{-3}, or 1.2 chances in 1,000.

9.7 NATURAL AND MAN-MADE RADIATION SOURCES

Throughout history, mankind has been exposed to radiation from the environment. This has come from two sources: *cosmic rays*, highly energetic radiation bombarding the earth from outer space; and *terrestrial radiation*, originating in radionuclides found in the earth and in our own bodies. In recent years, these natural radiation sources have been augmented by medical x-rays, nuclear weapons, nuclear reactors, television, and numerous other radiation-producing devices. It is important to know the magnitude of the doses that the public receives from these sources in order to place in perspective the standards of radiation exposure established by various regulatory bodies. Average annual doses to residents of the United States from the most significant sources are given in Table 9.11.

Cosmic Rays

The primary cosmic radiation incident on the earth consists of a mixture of protons ($\sim$87%), α-particles ($\sim$11%), and a trace of heavier nuclei ($\sim$1%) and electrons ($\sim$1%). The energies of these particles range between 10^8 and 10^{20} eV, with the bulk lying between 10^8 and 10^{11} eV. There is no known mechanism for the production of such highly energetic radiation—in short, the origin of cosmic rays is not understood.

The primary cosmic rays are almost entirely attenuated as they interact in the first few hundred g/cm^2 of the atmosphere. Large numbers of secondary particles, in particular, neutrons, additional protons, and charged pions (short-lived subnuclear particles), are produced as a result of these interactions. The subsequent decay of the pions results in the production of electrons, muons (other subnuclear particles), and a few photons. The resulting particle fluxes, which depend somewhat on the geomagnetic latitude, are given in Table 9.12 for sea level in the northern part of the United States.

The annual cosmic ray dose at sea level is between 26 and 27 mrems. The dose rate increases with altitude. Persons living in Denver, Colorado (the "mile-high city"), receive approximately twice the annual dose from cosmic rays as people living at sea level. When the fractions of the population living at different altitudes and latitudes are taken into account, the average annual dose due to cosmic rays in the United States is about 31 mrems. Since some of this radiation is shielded by buildings, this dose is reduced to the 28 mrems shown in Table 9.11.

TABLE 9.11 AVERAGE ANNUAL INDIVIDUAL DOSES IN MREMS FROM NATURAL
AND MAN-MADE RADIATION SOURCES*

Source	Exposed group	Number exposed	Body portion exposed	Individual dose
Natural Radiation				
External sources				
Cosmic rays	Total population	226×10^6	Whole Body	28
Terrestrial γ-rays	Total population	226×10^6	Whole Body	26
Internal sources				
^{40}K	Total population	226×10^6	Gonads	19
			Bone marrow	15
Heavy elements	Total population	226×10^6	Gonads	8
			Bone marrow	8.5
^{14}C	Total population	226×10^6	Gonads	0.7
			Bone marrow	0.7
^{87}Rb	Total population	226×10^6	Gonads	0.3
			Bone marrow	0.6
^{222}Rn	Total population	226×10^6	Lungs	200
			Bone marrow	0.6
Man-made radiation				
Medical x-rays	Adult patients	105×10^6	Bone marrow	103
	Medical personnel	195,000	Whole body	300–350
Dental x-rays	Adult patients	105×10^6	Bone marrow	3
	Dental personnel	171,000	Whole body	50–125
Radiopharmaceuticals	Patients	$10–12 \times 10^6$	Bone marrow	300
	Medical personnel	100,000	Whole body	260–350
Nuclear weapons fallout	Total population	226×10^6	Whole body	4–5
Nuclear power plants	Population < 10 miles away	$< 10 \times 10^6$	Whole body	$\ll 10$
	Workers	67,000	Whole body	400
Building materials	Persons living in brick or masonry buildings	110×10^6	Whole body	7
Air travel	Passengers	35×10^6	Whole body	3
	Crew	40,000	Whole body	160
Television	Viewers	100×10^6	Gonads	0.2–1.5
Tobacco	Smokers	50×10^6	Bronchial epithelium	8,000

*Based on "The Effects on Population of Exposure to Low Levels of Ionizing Radiation." BEIR III, National Academy of Sciences, 1980, "Health Risks of Radon and Other Internally Deposited Alpha Emitters," BEIR IV , 1988, and NCRP report 93, 1987.

TABLE 9.12 FLUXES OF COSMIC RAY PARTICLES AT SEA LEVEL

Particles	Muons	Neutrons	Electrons	Protons	Charged Pions
Flux ($cm^{-2}sec^{-1}$)	1.90×10^{-2}	6.46×10^{-3}	4.55×10^{-3}	1.71×10^{-4}	1.34×10^{-5}

*From "Natural Background Radiation in the United States," National Council on Radiation Protection and Measurements report 45, 1975.

Terrestrial Radiation

There are approximately 340 naturally occurring nuclides found on earth. Of these, about 70 are radioactive. Radioactivity is, therefore, everywhere; there is no escape from radiation exposure due to natural radioactivity in the environment or in the human body.

It is usual to divide the natural radionuclides into two groups, depending upon their origin: *primordial* radionuclides, those that have been here since the earth was formed, and *cosmogenic* radionuclides, those that are continually being produced by the action of cosmic rays. It goes without saying that primordial nuclides must be very long-lived. A nuclide with a half-life even as long as 10 million years would have passed through 450 half-lives over the approximately 4.5×10^9 years since the earth was formed, and its original activity would have diminished by a factor of 10^{135}. The nuclide, in short, would have vanished entirely from the earth. The most common primordial nuclides are $^{238}U(T_{1/2} = 4.5 \times 10^9$ yrs), $^{235}U(T_{1/2} = 7.1 \times 10^8$ yrs), $^{232}Th(T_{1/2} = 1.4 \times 10^{10}$ yrs), $^{87}Rb(T_{1/2} = 4.8 \times 10^{10}$ yrs), and $^{40}K(T_{1/2} = 1.3 \times 10^9$ yrs). The first three of these nuclides are the parents of long decay chains, such as the one shown in Fig. 9.7 for ^{238}U. As discussed later, some of the daughters in such chains may be of considerable biological significance. A dozen or so other primordial radioisotopes are known, but none of these has an important impact on the human radiation environment.

The presence on earth of naturally occurring short-lived radionuclides, such as $^{14}C(T_{1/2} = 5,730$ yrs), is due to their production by cosmic rays. Obviously, all of the ^{14}C would have disappeared billions of years ago if it were not continually replenished. About 25 other cosmogenic radionuclides have been identified, but only ^{14}C leads to significant radiation doses. This nuclide is formed primarily in the interaction of thermalized cosmic-ray neutrons with nitrogen in the atmosphere via the exothermic reaction $^{14}N(n, p)^{14}C$.

External exposure to terrestrial radioactivity originates with the γ-rays emitted following the decay of uranium, thorium, and their daughter products. These radionuclides are widely, but unevenly, distributed about the world. In the United States, for instance, there are three broad areas of differing terrestrial γ-ray levels, as shown in Fig. 9.8. The Colorado Plateau lies atop geological formations rich in uranium and radium; as a result, this region tends to have a much higher radiation

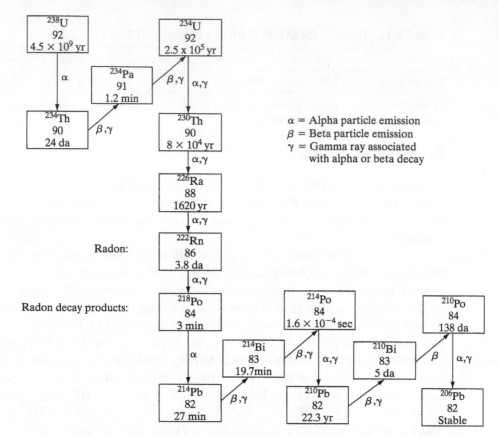

Figure 9.7   The ^{238}U decay series.

level than other parts of the country. There are also locations in the world, particularly in Brazil and India, where the presence of thorium-bearing monozite sands leads to radiation levels that are especially high (up to 3mR/hr). The population-averaged annual external terrestrial dose in the United States is 26 mrems.

The principal source of *internal* terrestrial exposure is from primordial ^{40}K. This peculiar nuclide decays both by negative β-decay to ^{40}Ca and by positive β-decay or electron capture to ^{40}Ar. Its isotopic abundance is 0.0118 a/o, so there is about 0.0157 g of ^{40}K from a total of 130 g of potassium in an average person weighing 70 kg. The total activity of the ^{40}K in the body is therefore approximately 0.11 μCi.

The heavy primordial nuclides and their daughters enter the body by ingestion of drinking water or foodstuffs in which they are distributed in various trace amounts. Heavy radionuclides also enter the body as the result of inhalation of ^{222}Rn($T_{1/2} = 3.8$ days) and its daughter products, especially ^{210}Pb($T_{1/2} = 21$ yrs).

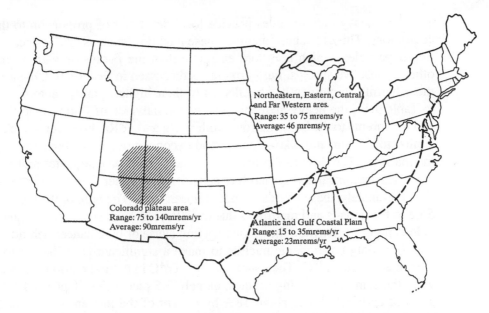

Figure 9.8   Terrestrial γ-ray doses in the United States. (From "The Effects on Populations of Exposure to Low Levels of Ionizing Radiation." National Academy of Sciences, 1980.)

^{222}Rn is the immediate daughter of the decay of ^{226}Ra, and it is therefore produced in radium-bearing rocks, soil, and construction materials. Since ^{222}Rn is a noble gas, it tends to diffuse into the atmosphere, where it may travel large distances before it decays via several short-lived species to ^{210}Pb. The half-life of ^{222}Rn is long compared with the residence time of air in the lungs, so that relatively little radon decays during respiration. What is more important, the chemical inertness of radon prevents its long-term retention within the body. As a result, ^{222}Rn itself contributes very little to the internal body dose. However, ^{210}Pb is not inert, and soon after its formation from ^{222}Rn, it becomes attached to moisture or dust particles in the atmosphere. When these particles are inhaled, some of the ^{210}Pb is retained by the body. ^{210}Pb may also enter the body through ingestion or by the decay of ingested parents.[14] In any event, if this were not already complicated enough, the ^{210}Pb does not itself lead to significant internal doses, since it is only a weak β-ray emitter. Rather, it is ^{210}Po($T_{1/2} = 138$ days), the decay product of ^{210}Pb, that emits a powerful 5.3-MeV α-particle that provides the ultimate dose. Thus, the ^{222}Rn and ^{210}Pb can be viewed as different sorts of carriers for ^{210}Po, the actual source of radiation damage.

---

[14] ^{210}Pb and ^{210}Po can also enter the body via cigarette smoke; see the next section.

The heavy radionuclides provide local doses out of proportion to their concentrations. This is because many of these nuclides (such as ^{210}Po) decay by emitting α-particles, which are more energetic than the β-rays or γ-rays emitted by other radioactive species. Furthermore, as discussed in Section 9.2, α-particles are more harmful biologically, as reflected in their higher quality factors. According to Table 9.11, the heavy elements give an annual dose of 8 mrems to the gonads and 8.5 mrems to the bone marrow. Both ^{226}Ra and ^{228}Ra, however, are chemically similar to calcium and, like calcium, tend to concentrate in the bone (as opposed to the marrow). The dose to the structural bone tissue (the osteocytes and the tissues in the Haversion canals) is therefore considerably higher than the dose to the marrow. Fortunately, these tissues are not radiosensitive. The bulk of the heavy element dose to the gonads and marrow is due to ^{210}Po, although some studies suggest that radon and its projeny are the second leading cause of lung cancer behind smoking.

The only cosmogenic nuclide to make a significant contribution to internal human exposure is ^{14}C. The concentration of ^{14}C in natural carbon has been found to be the same in all living species, namely 7.5 picocuries (1 pCi = 10^{-12}Ci) per gram of carbon.[15] Since about 18% by weight of the human body is carbon, the total body burden of an average 70-kg man is approximately 0.1 μCi. This gives an estimated dose of 0.7 mrem/year.

Man-Made Sources

Medical Exposures
By far, the largest nontrivial-localized dose received by the public at large from man-made sources is in connection with the healing arts. This dose includes contributions from medical and dental diagnostic radiology, clinical nuclear medicine (the use of radionuclides for various purposes), radiation therapy, and occupational exposure of medical and dental personnel. The dose rates, given in Table 9.11, are based on studies by the Bureau of Radiological Health of the US Department of Health, Education and Welfare. (See references at end of chapter.)

Fallout
Fallout from a nuclear weapon consists of fission fragments and neutron activation products in weapon debris that become attached to dust and water particles in the atmosphere. The larger of these particles soon come to earth near the site of the detonation, but the smaller ones may remain aloft in the upper atmosphere for five years or more. In time, they become distributed more or less uniformly around the world, and contribute to the general level of environmental radiation. The long-term exposure from fallout is mostly internal, from fission

---

[15]When a plant or animal dies, this equilibrium concentration is no longer maintained, due to the decay of the ^{14}C. By measuring the ^{14}C/C ratio, it is therefore possible to determine the age of nonliving organic materials, a technique known as *carbon dating*.

products that have been and continue to be ingested into the body. In the absence of further atmospheric testing of nuclear weapons,[16] the annual whole-body dose from fallout will be between 4 and 5 mrems through the year 2000.

Nuclear Power   The environmental effects of nuclear power will be discussed in full in Chapter 11. It may be said at this point, however, that the conclusion seems inescapable that the increasing use of nuclear power will lead to a small, but increasing, radiation dose to the general public. This dose is due not only to radiation released from power plants themselves, but also from uranium mines, mills and fabrication plants, and fuel-reprocessing facilities. In any case, the population-averaged dose in the United States was less than 1 mrem/year in 1980. The occupational dose to workers in the commercial nuclear industry is computed from industry data submitted to and published by the US Nuclear Regulatory Commission. (See Table 9.11.)

Building Materials   Many building materials, especially granite, cement, and concrete, contain a few parts per million of uranium and thorium, together with their radioactive daughters, and ^{40}K. Exposure to the radiation emanating directly from the walls of brick or masonry structures gives their occupants an annual dose of approximately 7 mrems.

Building materials are also a source of ^{222}Rn, and occupants of brick and masonry buildings, especially those with poor ventilation, may receive substantial lung doses from the subsequent decay of the ^{210}Po. Radon also enters buildings from other natural sources through basement openings; through windows and other openings; from the burning of natural gas, which frequently contains large amounts of radon; and from the spraying of radon-rich water from shower heads. Exceedingly high levels of radon, far in excess of established standards (discussed in the next section), have been measured in well-insulated houses that necessarily lack adequate ventilation. The average radon-related dose to an individual is estimated to be 200 mrem/yr.

Air Travel   The radiation dose from air travel stems from the fact that modern jet aircraft fly at high altitudes, from 9 to 15 km, where the cosmic ray dose rate is much greater than it is on the ground. For instance, at 43° north latitude (just north of Chicago) and at an altitude of 12 km, the dose rate is 0.5 mrem/hr. A 10-hour round trip flight across the United States can result, therefore, in a total dose of as much as 5 mrems. It is of interest to note in Table 9.11 that the average dose to aircraft flight crews is a significant fraction of the dose to occupational workers in the nuclear industry.

---

[16]Atmospheric testing of nuclear weapons was terminated by Great Britain, the United States, and the Soviet Union in 1962, by France in 1974, and later by all nuclear weapons states.

Television   Radiation is released in the form of x-rays from at least three sources within color television receivers. With increasing numbers of television viewers exposed to this radiation, the US Congress in 1968 passed standards requiring that the exposure rate averaged over 10 cm^2 at any readily accessible point 5 cm from the surface of a television receiver not exceed 5 mR/hr. With the adoption of this regulation, the individual dose rate has fallen over the past decade, and the average annual dose to the gonads of viewers is now between 0.2 and 1.5 mrems.

Tobacco   As noted earlier, ^{222}Rn diffuses from the earth to the atmosphere, where it decays into ^{210}Pb, which subsequently falls to the earth attached to dust or moisture particles. If these particles fall onto leafy vegetables or pasture grasses, the ^{210}Pb may enter directly into the food chain. Of perhaps greater significance is the fact that if the particles fall onto broadleaf tobacco plants, the ^{210}Pb and its daughter ^{210}Po may be incorporated into commercial smoking materials. Measurements show that there are on the order of 10 to 20 pCi of both ^{210}Pb and ^{210}Po in an average pack of cigarettes. The inhalation of cigarette smoke deposits these radionuclides on the tracheobronchial tree, where the ^{210}Po irradiates the radiosensitive basal cells of the bronchial tissue. The annual local dose to this tissue for an average cigarette smoker (1.5 packs per day) is estimated to be as high as 8 rems (8,000 mrems) and proportionately higher for heavy smokers. Many researchers believe that this radiation is the origin of the high incidence of lung cancer among smokers.

Other Man-Made Sources   Segments of the public are exposed to several other, often unsuspected, sources of radiation. Clocks and wristwatches with luminous dials, eyeglasses or porcelain dentures containing uranium or thorium, smoke detectors with α-emitting sources, fossil-fueled power plants that emit radioactive ash, and many other man-made devices result in generally small wholebody doses, but occasionally high local doses. For example, porcelain teeth and crowns in the United States contain approximately 0.02% uranium by weight. This is estimated to give an annual local tissue dose of about 30 rems, mostly from α-particles.

9.8 STANDARDS OF RADIATION PROTECTION

In 1928, in response to the growing recognition of the hazards of radiation, the Second International Congress of Radiology established the International Commission on Radiological Protection (ICRP) to set standards of permissible exposure to radiation. Shortly thereafter, in 1929, the National Council on Radiation Protection and Measurements (NCRP) was set up in the United States at the national level to perform the functions undertaken internationally by the ICRP.[17] In 1964, NCRP was given a federal charter by the US Congress to assure that the organization would remain an independent scientific agency and not be subject to governmental control.

Until 1970, official US Government policy regarding permissible radiation exposure was determined by the Federal Radiation Council (FRC), which was established to "advise the President with regard to radiation matters directly or indirectly affecting health ... ." The FRC was composed of the Secretaries of Agriculture; Commerce; Defense; Health, Education and Welfare; Labor; and the Chairman of the Atomic Energy Commission; together with a technical staff. In 1970, the functions and staff of the FRC were transferred to the newly formed Environmental Protection Agency (EPA). Standards for exposure to radiation recommended by the EPA, and earlier by the FRC, are called *radiation protection guides* (RPGs). When these guides have been approved by the President, they have the force and effect of law, as all agencies of the Government are required to follow the guides. The regulations of the US Nuclear Regulatory Commission, which are applicable to its licensees, must be consistent with the RPGs. In most countries other than the United States, official radiation policy is based directly on ICRP recommendations.

Not surprisingly, the recommended standards of the ICRP, the NRCP, and the EPA are not identical. Furthermore, these standards are continually revised, as the membership on these bodies changes and as new information on the biological effects of radiation becomes available. Nevertheless, the standards-setting bodies are now in agreement on the following ground rules for radiation exposure:

a) Since any exposure to radiation may be potentially harmful, no deliberate exposure is justified unless some compensatory benefit will be realized.

b) All exposure to radiation should be kept "as low as reasonably achievable" (ALARA), taking into account the state of technology, the economics of reducing exposures relative to the benefits to be achieved, and other relevant socioeconomic factors. (The ALARA concept is also used in connection with the emission of effluents from nuclear power plants; see Section 11.9.)

c) Radiation doses to individuals should not exceed certain recommended values. For many years, these limiting doses were known as *maximum permissible doses* (MPDs), and this term persists in general usage. More recently, the ICRP has preferred the term *dose limit* to MPD. Both the FRC and EPA have always used the RPG standards instead of the MPD.

---

[17]More precisely, from 1929 to 1946, the NCRP was known as the Advisory Committee on X-ray and Radium Protection; from 1946 to 1956, the National Committee on Radiation Protection; from 1956 to 1964, the National Committee on Radiation Protection and Measurements; and from 1964 until the present time, the National Council on Radiation Protection and Measurements.

Although there is no evidence of harm from low levels of exposure, scientists are reluctant to assume that there is no effect at all. To fill this gap, the standards-setting bodies arbitrarily have made the following assumptions:

a) There is a linear dose–effect relationship for all radiation effects from high dose levels in the ranges of several hundreds of rads down to zero dose.

b) There is no threshold radiation dose above which an effect may occur, but below which it does not.

c) Low doses delivered to an organ are additive, no matter at what rate or at what intervals they may be delivered.

d) There is no biological recovery from radiation effects at low doses.

As a practical matter, MPDs or RPGs have traditionally been established from the consensus judgments of members of the standards-setting bodies. Because such judgments are necessarily qualitative, the precise effect of adopting a particular standard can never be ascertained. The ICRP, using dose–effect computations such as those discussed in the preceding section, compared the number of fatalities among radiation workers from continued exposure to ICRP dose limits with the comparable number of fatalities in other occupations. It was found that the average risk of death, which in the case of radiation workers is due almost entirely to the eventual contraction of cancer, is actually smaller than the risk of job-related death in the safest of other major occupational categories, namely, the retail trades. Furthermore, there are essentially no injuries to radiation workers due to radiation exposure—the allowed dose limits are far below the threshold for observable nonstochastic effects. Thus, for the first time, a quantitative basis for radiation standards has been established.

The standards are regularly reviewed, both in the United States and elsewhere. The current standard in the United States is based largely on ICRP Publications 26 and 30. The criteria were adopted by the EPA and then, as required by law, by the US NRC in 1994. More recent recommendations were promulgated by the ICRP in 1990 in Publication 60. The later requirements have yet to be adopted in the United States. For completeness, both the recommendations of Publications 26 and 60 are included in this section.

An abridged summary of recommended dose limits for external and internal exposures is presented in Table 9.13. As noted in the table, dose limits are given for two categories: occupational exposure to radiation workers and exposure to the public at large.

With regard to nonuniform external exposure (i.e., exposure of one portion of the body more than another), the ICRP has adopted the principle that the risk of developing a fatal malignancy from a regional exposure should be the same as the risk when the whole body is irradiated uniformly. Because the ICRP whole-body

TABLE 9.13   DOSE-LIMITING RECOMMENDATIONS OF STANDARDS-SETTING BODIES; DOSES IN REMS*

Type of exposure-group	EPA (1987)	NCRP (1971)	ICRP (1977)	ICRP (1990)
Occupational Exposure				
Whole body				
prospective	5/yr	5/yr	5/yr	2/yr averaged over 5 yrs
retrospective	—	10–15 any year	—	Max. 2/yr
to N years of age	100 total career	$5(N-18)$	—	
Skin		15/yr	*(see text)*	50
Hands	50/yr	75/yr; 25/qtr	*(see text)*	50
Forearms	—	30/yr; 10/qtr	*(see text)*	
Gonads	5/yr	5/yr	*(see text)*	
Lens of eye	15/yr	5/yr	*(see text)*	15
Thyroid	—	15/yr	*(see text)*	
Any other organ	50/yr	15/yr; 5/qtr	*(see text)*	
Pregnant women	0.5 gestation	0.5 gestation	‡	
General population				
Individual	0.1/yr	0.5/yr	0.5/yr	0.1/yr
Average	—	5/30 yrs	—	

*Based on EPA notice in Federal Register, 46 FR 7836, 1981; NCRP report 39, 1971; ICRP report 26, 1977; ICRP report 60, 1990.

†Several alternative standards proposed.

‡Less than 0.3 times normal occupational dose from discovery of pregnancy through gestation.

annual dose limit is 5 rems, this principle can be expressed as

$$\sum_{T} w_T H_T \le 5 \text{ rems}, \tag{9.17}$$

where H_T is the annual dose to the Tth tissue or organ and where w_T is a weighting factor reflecting the relative radiosensitivity of the tissue or organ. Recommended values of w_T are given in Table 9.14; the value shown for "other organs" may be used for as many as five organs not listed (those receiving the highest dose should be used).

Equation (9.17) can be interpreted in the following way: using the ICRP 26 value of w_T for the lung, an annual dose of $5/0.12 = 42$ rems to this organ alone should carry the same risk as an annual dose of 5 rem to the whole body. Similarly, a dose of $5/0.03 = 167$ rems to the thyroid alone should be equivalent to 5 rems whole body (but see the next paragraph!). And so on.[18]

[18]The USNRC adopted the tissue-weighting factors of ICRP 26 in 1991 which became effective in 1994.

TABLE 9.14   VALUES OF THE WEIGHTING
FACTOR w_T

Tissue	ICRP*	ICRP 60†
Gonads	0.20	—
Breast	0.15	0.05
Red bone marrow	0.12	0.12
Lung	0.12	0.12
Thyroid	0.03	0.05
Bone surfaces	0.03	0.01
Skin	—	0.01
Other organs	0.06	0.05

*From ICRP publication 26, *op. cit.*
†From ICRP Publication 60, *op. cit.*

The standards given in Table 9.13 are predicated on limiting stochastic events (i.e., cancer). Nonstochastic effects, according to the ICRP, should be prevented by limiting the annual dose to 50 rems to any one organ, except the lens of the eye, which is permitted to receive only 30 rems. Thus, based on ICRP standards, the 42 rems to the lung computed would be acceptable; the 167 rems to the thyroid would not be acceptable, since it exceeds 50 rems.

Limits have also been established by the standards-setting bodies for the maximum allowable intake of radionuclides into the body, based on the doses that these materials give to various body organs. These limits, and the methods by which they are calculated, are discussed in Section 9.9.

Incidentally, it may be noted that an annual MPD of 5 rems is approximately equivalent to 100 mrems/week, and the MPD is often given in these terms. Also, 100 mrems distributed over a 40-hour week is equal to a dose-equivalent rate of 2.5 mrems/hr. Under US Nuclear Regulatory Commission regulations, any space accessible to personnel where a person can receive a dose in excess 5 mrems/hr is defined as a *radiation area* and is required to be posted with warning signs.

The MPD for individuals in the general population is much higher than the average dose allowed for the population as a whole. The MPD for individuals has been universally set (except by the EPA, which has not proposed guides for nonoccupational exposure) at 0.5 rem/year; the maximum average population dose is usually taken to be 5 rems/30 years, or about 170 mrems/year. The individual dose is based on reducing the risk of cancer, whereas the population dose rests on the risk of genetic damage, which was once thought to be of greater significance. According to the ICRP, recent data indicate that the opposite is true, and the ICRP no longer considers it necessary to set the average dose limit for the population lower than the limit for individuals.

It must be emphasized that the dose limits in Table 9.13 refer to radiation doses over and above those received from normal background radiation and from the so-called healing arts. Finally, the reader is cautioned to be alert for changes in the radiation standards that will certainly be made from time to time in the future. The ICRP has recommended considerable reduction in the dose limits and weighting factors. (See Table (9.13) and (9.14).) The reductions are based on a conservative analysis of the atomic bomb victims. The standards-setting bodies never sleep.

9.9 COMPUTATIONS OF EXPOSURE AND DOSE

It is often necessary to compute the exposure, absorbed dose, and dose equivalent, or their rates, resulting from various configurations of radiation sources. Such computations are somewhat different for γ-rays, neutrons, and charged particles, and they depend upon whether the exposure is external or internal.

External Exposure to γ-Rays

From Section 9.2, it will be recalled that exposure to γ-rays is measured in roentgens (R), where 1 R corresponds to the liberation of 2.58×10^{-4} coul of charge of either sign, when the γ-rays interact with 1 kg of air. For computing exposure, it is convenient to relate the roentgen to the energy that must be deposited in the air in order to liberate this charge. Repeated calculations involving electrical charge can then be avoided.

To produce 2.58×10^{-4} coul requires the formation of $2.58 \times 10^{-4}/1.60 \times 10^{-19} = 1.61 \times 10^{15}$ ion pairs (an ion and an ejected electron), where 1.60×10^{-19} is the charge in coulombs of either the ion or the electron. It has been found experimentally that approximately 34 eV (between 32 and 36 eV) must be deposited by γ-rays in air to produce an ion pair. The liberation of 2.58×10^{-4} coul thus requires the absorption of $1.61 \times 10^{15} \times 34 = 5.47 \times 10^{16}$ eV. An exposure of 1 R corresponds, therefore, to an energy deposition of

$$1\,R = 5.47 \times 10^{16} \text{ eV/kg} = 5.47 \times 10^{10} \text{ MeV/kg}$$

$$= 5.47 \times 10^{7} \text{ MeV/g.} \tag{9.18}$$

Since 1 MeV $= 1.60 \times 10^{-6}$ ergs, it follows that

$$1\,R = 5.47 \times 10^{7} \times 1.60 \times 10^{-6} = 87.5 \text{ ergs/g.} \tag{9.19}$$

To compute the γ-ray exposure from a specified radiation field, it is necessary only to determine the energy absorbed from the γ-rays in the air, and then to convert to roentgens using one of the above energy equivalents. Similarly, the exposure rate can be found from the rate at which γ-ray energy is deposited. It was shown in Section 3.8 that the energy deposition rate per unit mass is given by $IE(\mu_a/\rho)^{\text{air}}$, where I is the γ-ray intensity, E is the γ-ray energy, and $(\mu_a/\rho)^{\text{air}}$ is the mass absorption coefficient of air at the energy E. Values of $(\mu_a/\rho)^{\text{air}}$ are given in Table II.5. By making use of Eq. (9.19), it follows that the exposure rate $\dot{X}$ is given by

$$\dot{X} = IE(\mu_a/\rho)^{\text{air}}/5.47 \times 10^7$$
$$= 1.83 \times 10^{-8}IE(\mu_a/\rho)^{\text{air}} \text{ R/sec.} \tag{9.20}$$

In this equation, I must be expressed in photons/cm^2-sec, E is in MeV, and $(\mu_a/\rho)^{\text{air}}$ is in cm^2/g. For many practical problems, it is more appropriate to express $\dot{X}$ in mR/hr, rather than in R/sec. Then, since

$$1 \text{ R/sec} = 3.6 \times 10^6 \text{ mR/hr,}$$

Eq. (9.20) can be written as

$$\dot{X} = 0.0659IE(\mu_a/\rho)^{\text{air}} \text{ mR/hr.} \tag{9.21}$$

It is evident from Eqs. (9.20) and (9.21) that the exposure rate depends both on the intensity of the γ-rays and on their energy. Figure 9.9 shows the intensity (or flux; cf. later in this section) necessary to give an exposure rate of 1 mR/hr. The curve, in part, reflects the energy dependence of $(\mu_a/\rho)^{\text{air}}$, which is large at low energies, owing to photoelectric absorption, has a minimum at about 0.07 MeV, and then rises because of pair production. Less γ-ray intensity is obviously needed to provide a given exposure rate where the absorption coefficient is large.

Equation (9.21) applies only to a monoenergetic beam. If the γ-rays have a distribution of energies, it is necessary to integrate Eq. (9.21) over the spectrum if this is continuous, or to sum over the spectrum if it is a discrete (line) spectrum. For instance, in the latter case, the exposure rate is

$$\dot{X} = 0.0659 \sum_i I_i E_i (\mu_a/\rho)_i^{\text{air}} \text{mR/hr,} \tag{9.22}$$

where I_i is the intensity of γ-rays of energy E_i and $(\mu_a/\rho)_i^{\text{air}}$ is the mass absorption coefficient at E_i.

To obtain the total exposure over a time T, the preceding formulas must be integrated with respect to time. If time is measured in seconds, then from Eq. (9.20),

$$X = \int_0^T \dot{X}\, dt = 1.83 \times 10^{-8} E(\mu_a/\rho)^{\text{air}} \int_0^T I(t)\, dt$$

$$= 1.83 \times 10^{-8} \Phi E(\mu_a/\rho)^{\text{air}} \text{R.} \tag{9.23}$$

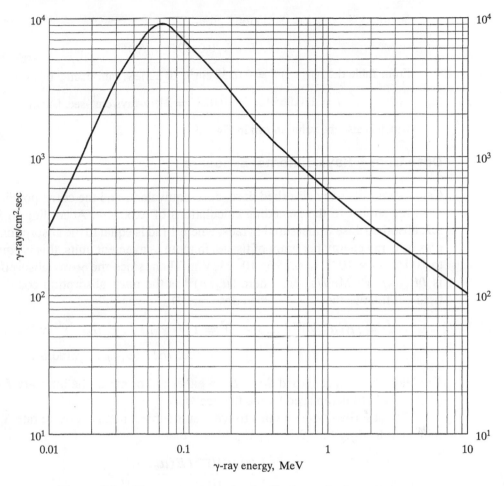

Figure 9.9   Gamma-ray intensity (or flux) required to give exposure rate of 1 mR/hr as a function of γ-ray energy.

Here,

$$\Phi = \int_0^T I(t)\, dt \tag{9.24}$$

is called the γ-ray *fluence* and has the dimensions of photons/cm^2. The term ΦE in Eq. (9.23) is called the *energy fluence*.

Example 9.6

What beam intensity of 2-MeV γ-rays is required to give an exposure rate of 1 mR/hr?

Solution. Solving Eq. (9.21) for I gives

$$I = \dot{X}/0.0659E(\mu_a/\rho)^{\text{air}}.$$

From Table II.5, $(\mu_a/\rho)^{\text{air}} = 0.0238$ cm^2/g at 2 MeV. Introducing this value gives

$$I = 1/0.0659 \times 2 \times 0.0238 = 319\gamma\text{-rays/cm}^2\text{-sec, } [\textit{Ans.}]$$

which is also the value shown in Fig. 9.9.

Dose from γ-Rays: External Exposure

Calculations of γ-ray absorbed dose, dose-equivalent, and the corresponding dose rates are similar to the preceding calculations of exposure. According to Section 9.2, absorbed dose is measured in rads, where 1 rad is equal to the absorption of 100 ergs of γ-ray energy per gram of tissue. In more convenient units, this is equivalent to $100/1.60 \times 10^{-6} = 6.25 \times 10^7$ MeV/g. Then, since the γ-ray absorption rate is $IE(\mu_a/\rho)^{\text{tis}}$ MeV/g-sec, where $(\mu_a/\rho)^{\text{tis}}$ is the mass absorption coefficient of tissue, it follows that the absorbed dose rate $\dot{D}$ is given by

$$\dot{D} = 1E(\mu_a/\rho)^{\text{tis}}/6.25 \times 10^7 = 1.60 \times 10^{-8}IE(\mu_a/\rho)^{\text{tis}} \text{ rad/sec} \quad (9.25)$$

$$= 0.0576IE(\mu_a/\rho)^{\text{tis}} \text{ mrad/hr.} \quad (9.26)$$

To obtain the total absorbed dose, D, in either rad or mrad, the intensity I in Eqs. (9.25) or (9.26) is replaced by the fluence Φ.

The absorbed dose rate in a tissue that is subject to an exposure rate $\dot{X}$ can be found by dividing Eq. (9.25) by Eq. (9.20); that is,

$$\dot{D} = \frac{1.60 \times 10^{-8}IE(\mu_a/\rho)^{\text{tis}}}{1.83 \times 10^{-8}EI(\mu_a/\rho)^{\text{air}}} \dot{X}$$

$$= 0.874\frac{(\mu_a/\rho)^{\text{tis}}}{(\mu_a/\rho)^{\text{air}}} \dot{X}. \quad (9.27)$$

Since this relationship between $\dot{D}$ and $\dot{X}$ depends on the absorption coefficient of the tissue, the same value $\dot{X}$ can give quite different $\dot{D}$s in different tissues. It is convenient therefore to write Eq. (9.27) as

$$\dot{D} = f\dot{X}, \quad (9.28)$$

where f is an energy-dependent function that depends on the composition of the tissue. Figure 9.10 shows this function for bone, muscle, and fat tissue. At low energy, the curve is generally higher for bone than for muscle or fat, because bone contains higher Z constituents that show stronger photoelectric absorption. Since mass absorption coefficients are independent of time, the time integrals of both

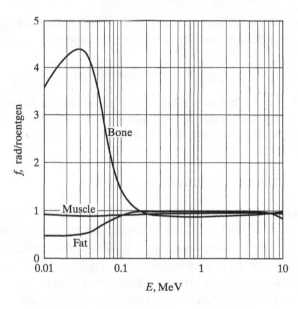

Figure 9.10   The parameter f as a function of γ-ray energy. (From K. Z. Morgan and J. E. Turner, *Principles of Radiation Protection*. New York: Wiley, 1967.)

sides of Eqs. (9.27) and (9.28) are also equal. These equations are therefore valid for total dose and total exposure when $\dot{D}$ and $\dot{X}$ are replaced by D and X, respectively.

To obtain the dose-equivalent or its rate, it is merely necessary to multiply D or $\dot{D}$ by the approximate quality factor. However, according to Table 9.2, $Q = 1$ for γ-rays, so that in the present case, D in rads is numerically equal to H in rems, and the foregoing expressions for D and $\dot{D}$ can be used directly for calculating H or $\dot{H}$.

The formulas for X, D, and H and their rates have been derived in this section in terms of the intensity I of a monodirectional beam of γ-rays. In many situations of interest, however, the γ-rays are not monodirectional; they move in several or all directions like the neutrons in a reactor. It is then more appropriate to express X, D, and H in terms of the γ-ray flux ϕ_γ. This is defined in the same way as the neutron flux ϕ (see Section 5.1) and has the same units as intensity, namely, γ-rays/cm²-sec. Equations (9.21) and (9.26), for example, can then be written as

$$\dot{X} = 0.0659\phi_\gamma \; E(\mu_a/\rho)^{\text{air}} \; \text{mR/hr} \qquad (9.29)$$

and

$$\dot{D} = 0.0576\phi_\gamma \; E(\mu_a/\rho)^{\text{tis}} \; \text{mrad/hr}. \qquad (9.30)$$

Example 9.7

The flux near a 50-kV x-ray machine is 2.4×10^5 x-rays/cm²-sec. What is the dose-equivalent rate in bone, muscle, and fat to an operator standing nearby?

Solution. From Fig. 9.9, a flux of approximately 8×10^3 x-rays/cm^2-sec gives an exposure rate of 1 mR/hr at 50 keV. The exposure rate to the operator is therefore

$$\frac{2.4 \times 10^5}{8 \times 10^3} \times 1 = 30 \text{ mR/hr.}$$

According to Fig. 9.10, $f = 3.3$ for bone, 0.93 for muscle, and about 0.90 for fat. The operator would thus receive absorbed dose rates and, since $Q = 1$, dose-equivalent rates of $3.3 \times 30 = 99$ mrem/hr to bone, $0.93 \times 30 = 28$ mrem/hr to muscle, and $0.90 \times 30 = 27$ mrem/hr to fat. [*Ans.*]

Dose from Neutrons—External Exposure

Dose computations for neutrons are somewhat more difficult than for γ-rays, because of the complex way in which neutrons interact with matter. It will be recalled from Chapter 3 that neutrons may undergo elastic or inelastic scattering, radiative capture, and various reactions. In these interactions, the struck nucleus may recoil with sufficient energy to be stripped of a portion of its electron cloud, whereupon it becomes a highly ionizing, charged particle. With inelastic scattering and radiative capture, γ-rays are also emitted, and in reactions, charged particles may be produced. The deposition of energy from these neutron-produced, secondary ionizing radiations must be determined in order to calculate neutron dose.

Because of the complexity of neutron interactions, the energy deposition must be computed numerically, and this is most conveniently done using the *Monte Carlo method*. In this method, the actual histories of the neutrons and the secondary radiation to which they give rise are duplicated on a high-speed computer as these radiations move about in a tissue medium. Calculations of this type have been carried out for beams of neutrons incident upon a slab or cylinder having a composition similar to that of the human body.[19] In determining the dose-equivalent, it is necessary to take into account the fact that the quality factor is a function of the energy of an ionizing particle. Thus, when a fast neutron undergoes repeated collisions and produces recoiling nuclei of various energies, the energy deposition from these nuclei must be multiplied by the appropriate quality factor as the calculation proceeds.

It is somewhat easier to discuss the results of such computations for a slab of tissue than for a cylinder, because of the simpler geometry. Figures 9.11(a) and (b) show the dose-equivalent as a function of penetration for thermal and for 5-MeV neutrons, respectively. Also shown in these figures are the contributions to the dose due to recoiling protons, recoiling heavier nuclei, and capture γ-rays. For the 5-MeV beam, most of the dose is from recoil protons produced by neutron collisions with hydrogen; very little of the dose is due to capture γ-rays. On the other

---

[19]Cylindrical or other shaped models used to simulate the human torso are called *phantoms*.

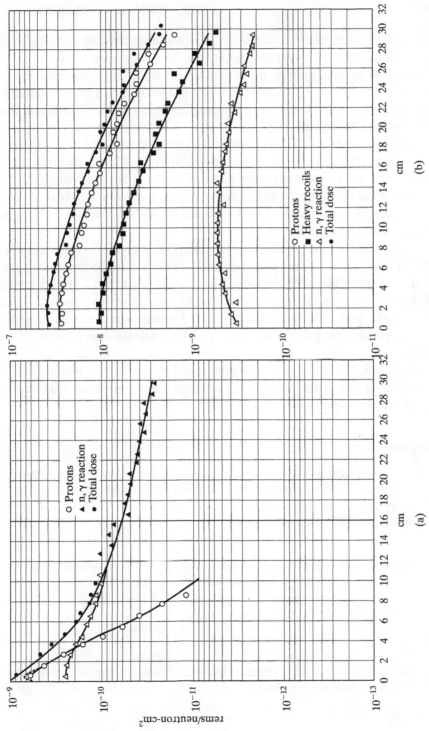

Figure 9.11 Dose equivalent as a function of penetration into tissue for (a) thermal neutrons, (b) 5-MeV neutrons. (From *Protection against Neutron Radiation up to 30 MeV*. National Bureau of Standards Handbook No. 63, 1957.)

hand, with the thermal neutron beam, the proton dose only exceeds the γ-ray dose over the first 3 cm. Beyond that point, the dose is primarily due to γ-rays. Incidentally, at thermal energy, the protons arise from the exothermic ($Q = 0.63$ MeV) reaction ^{14}N(n, p)^{14}C, which has a cross-section of 1.81 b at 0.0253 eV. At this energy, the neutrons are not sufficiently energetic to produce recoil protons in collisions with hydrogen.

When the maximum dose rates from curves like those in Fig. 9.11 are plotted versus energy, the result is the curve shown in Fig. 9.12, which gives the neutron flux as a function of neutron energy required to give a dose of 1 mrem/hr. It should be noted in the figure that at the higher energies, a smaller flux is required to produce a given dose than at the lower energies. In other words, neutron for neutron, the faster neutrons give the larger dose.

It is of some interest to note that, except at very low energy, the γ-ray flux required to produce a given dose-equivalent is always much larger than the corresponding neutron flux. For instance, from Figs. 9.9 and 9.12, it may be seen that at 1 MeV, a γ-ray flux of approximately 550 γ-rays/cm^2-sec gives an $\dot{H}$ in tissue of 1 mrem/hr, while a neutron flux of only 7.5 neutrons/cm^2-sec will give the same $\dot{H}$. Thus, particle for particle, neutrons give larger doses than photons of the same energy.

Figure 9.12 can be used to compute either the dose-equivalent rate from a given neutron flux or the total dose-equivalent from a given neutron fluence. This is illustrated by the next two examples.

Example 9.8

At a point near a neutron source, the fast (greater than 1 MeV) neutron flux is 20 neutrons/cm^2-sec, and the thermal flux is 300 neutrons/cm^2-sec. There are no γ-rays. How long may a radiation worker remain at this point if he is not to exceed his normal MPD dose?

Solution. According to Fig. 9.12, a fast neutron flux of about 7 neutrons/cm^2-sec gives 1 mrem/hr. The worker's fast neutron dose rate is then

$$\frac{20}{7} \times 1 = 2.8 \text{ mrem/hr.}$$

Similarly, from Fig 9.12, a thermal flux of 260 neutrons/cm^2-sec gives 1 mrem/hr, so that the thermal neutron dose is

$$\frac{300}{260} \times 1.1 \text{ mrem/hr.}$$

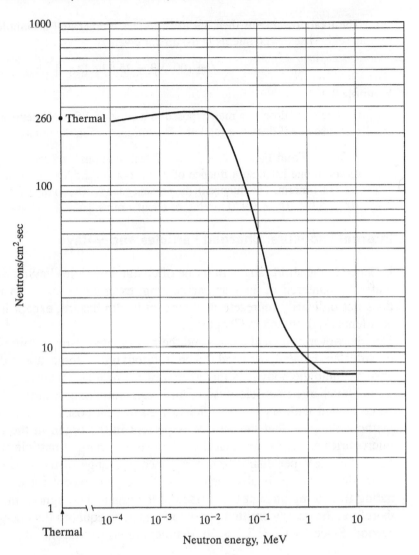

Figure 9.12   Neutron flux that gives dose-quivalent rate of 1 mrem/hr as a function of neutron energy. (Based on Appendix 6, ICRP Publication 21.)

The total dose rate is then 3.9 mrems/hr. A worker will accumulate the permitted weekly dose of 100 mrems in the time, t given by

$$t = 100/3.9 = 25.6 \text{ hr. } [Ans.]$$

Example 9.9

Compute the dose to a man exposed to a burst of 2.5-MeV neutrons from an accelerator source if there is a total of 10^8 neutrons/cm^2 in the burst.

Solution. From Fig. 9.12, a flux of 7 fast neutrons/cm^2-sec will give a dose of 1 mrem in one hour, so a fluence of $7 \times 3,600 = 2.52 \times 10^4$ neutrons/cm^2 gives a dose of 1 mrem. The fluence in the present case is 10^8, so the man receives a dose of $10^8/2.52 \times 10^4 = 3,970$ mrems, or about 4 rems. [*Ans.*]

Internal Exposure: Charged Particles and γ-Rays

Because of the short ranges in tissue of charged particles having the energies normally encountered in nuclear engineering, external exposure to these radiations does not ordinarily represent a significant health hazard, except in major nuclear accidents as discussed in Chapter 11.

However, if α- and β-rays and their associated γ-rays emitted by radioactive nuclides are inhaled, ingested (taken in through the stomach and digested), or are absorbed into the body in some other way, the results can be very serious.

The charged-particle dose to an organ containing a radioactive nuclide can be computed with accuracy sufficient for most purposes by the following simple method: Suppose that the nuclide is present in the organ in the amount of $C(t)$ microcuries at time t and that it emits a single charged particle with the average energy $\bar{E}$ MeV per disintegration.[20] Since the range of the radiation is so short, all of the energy $\bar{E}$ is absorbed within the organ, except for the radiation originating from a region near the surface of the organ. Throughout most of the organ, therefore, the energy emitted per second must be equal to the energy absorbed per second. Since 1 μCi $= 3.7 \times 10^4$ disintegrations per second, it follows that

$$\text{energy absorbed/sec} = 3.7 \times 10^4 C(t)\bar{E} \text{ MeV/sec}$$

$$= 5.92 \times 10^{-2} C(t)\bar{E} \text{ erg/sec}$$

dividing by the mass of the organ in grams

$$= \frac{5.92 \times 10^{-2} C(t)\bar{E}}{M} \text{ erg/g-sec,}$$

---

[20]In the case of a nuclide that decays by the emission of a single negative β-ray, $\bar{E}$ is approximately equal to 0.3 E_{max}, where E_{max} is the maximum β-ray energy. For positive β-decay, $\bar{E} \simeq 0.4 E_{max}$.

where M is the mass of the organ in grams. The absorption of 100 ergs/g gives 1 rad, so that the absorbed dose rate is

$$\dot{D} = \frac{5.92 \times 10^{-4} C(t) \bar{E}}{M} \text{ rad/sec.}$$

The dose-equivalent rate is obtained by multiplying $\dot{D}$ by the quality factor, so that

$$\dot{H} = \frac{5.92 \times 10^{-4} C(t) \bar{E} Q}{M} \text{ rem/sec,} \qquad (9.31)$$

and the equivalent dose is found by multiplying $\dot{D}$ by the radiation weighting factor w_t.

Most radionuclides decay by the emission of a number of α- or β-rays, and these are normally accompanied by several γ-rays, along with internal conversion or Auger electrons, as discussed in Section 2.7. Since these electrons are charged particles, essentially all their energy is absorbed within the organ. The γ-rays, however, do not have a finite range in tissue, and much of their energy is deposited in regions external to the organ.

Using Monte Carlo methods, computations of the fraction of the γ-ray energy retained within an organ as a function of energy and the fraction of energy deposited in other organs have been carried out for a variety of organs.[21] The Medical Internal Radiation Dose Committee (MIRD) of the Society for Nuclear Medicine has developed a methodology that accounts for both of these energy depositions. The method is general enough that it may be used for both γ-rays and for α and β radiation.

These computations can be summarized in the following way: first, let ϕ_i be the fraction of the ith radiation's energy absorbed in the target organ. For γ-rays, this will be less than one. For α- and β-rays, it will equal zero or unity, depending on whether the source and target organ are one in the same. If the energy emitted, including charged particles, by a radionuclide is E_i ($\bar{E}_i$ in the case of β-rays), then the energy absorbed is

$$E = n_i \phi_i E_i,$$

where n_i is the number of particles emitted with energy E_i. The dose rate for each type of radiation is calculated using the appropriate quality or weighting factor and the emission rate of the particles. For the case where the target organ is the organ in which the radionuclide resides, the dose rate may be calculated as follows. Define

---

[21]The computations in question are described in ICRP 30 and in several supplements to the Journal of Nuclear Medicine, which also has the energy absorbed in one organ from radiation emitted in another. (See references.)

ξ, *the effective energy equivalent*, as

$$\xi = \sum f_i F_i E_i Q_i, \tag{9.32}$$

where f_i is the fraction of the radiation emitted with an energy E_i, including charged particles, by a radionuclide. The quantity F_i is defined to be the average fraction of energy emitted in the ith mode that is retained in the organ and Q_i is the quality factor associated with the radiation in this ith mode. Values of ξ are given in Table 9.15 for several nuclides and organs.

If the disintegration rate $C(t)$ is in Ci, then $\dot{H}$ is

$$\dot{H} = \frac{592 C(t) \xi Q}{M} \text{ rem/sec.} \tag{9.33}$$

Finally, in some applications, it is appropriate to use $C(t)$ in μCi, but to express $\dot{H}$ in rem/day. In this case

$$\dot{H} = \frac{51.1 C(t) \xi Q}{M} \text{ rem/day.} \tag{9.34}$$

A radionuclide may enter into an organ if it is inhaled, ingested, injected, or otherwise taken into the body and is subsequently carried to the organ in question. It is well known in physiology that certain chemicals tend to accumulate in particular

TABLE 9.15   PHYSICAL AND BIOLOGICAL DATA FOR RADIONUCLIDES*

Radionuclide	Half-life, days Radioactive	Half-life, days Biological	Organ	ξ MeV	q Ingestion	q Inhalation
^{3}H (in water)	4.5×10^3	11.9	Total Body	0.01	1.0	1.0
^{14}C	2.1×10^6	10	Total body	0.054	1.0	0.75
		40	Bone	0.27	0.025	0.02
^{24}Na	0.63	11	Total body	2.7	1.0	0.75
^{41}A	0.076		Submersion†	1.8		
^{60}Co	1.9×10^3	9.5	Total body	1.5	0.3	0.4
^{87}Kr	0.053		Submersion†	2.8		
^{90}Sr	1.1×10^4	1.3×10^4	Total body	1.1	0.3	0.4
		1.8×10^4	Bone	5.5	0.09	0.12
^{131}I + ^{131}Xe	8.04	138	Total body	0.44	1.0	0.75
		138	Thyroid	0.23	0.3	0.23
^{137}Cs + ^{137}Ba	1.1×10^4	70	Total body	0.59	1.0	0.75
		140	Bone	1.4	0.04	0.03
^{226}Ra + daughters	5.9×10^5	1.64×10^4	Bone	110	0.04	0.03
^{235}U	2.6×10^{11}	300	Bone	230	1.1×10^{-5}	0.028
^{239}Pu	8.9×10^6	7.3×10^4	Bone	270	2.4×10^{-5}	0.2

*Based on ICRP publication 2.

†Value of ξ for calculation of dose to person surrounded by cloud of gas; see footnote to Table 9.16.

"target" organs. Thus, iodine accumulates in the thyroid gland, calcium in the bone, and so on. It follows that radioactive isotopes of iodine go to the thyroid, and the calcium-like radioisotopes of radium and plutonium tend to accumulate in bone. The fraction of a given substance that goes to a particular organ is denoted by the parameter q. The numerical value of q normally depends on whether the substance is inhaled or ingested. In the case of plutonium, for example, q is 0.2 for inhalation and 2.4×10^{-5} for ingestion. Thus, the inhalation of plutonium is a serious matter, although it is virtually harmless if swallowed. Other values of q are given in Table 9.15.

No substance, radioactive or not, remains in a body organ indefinitely. Normal biological processes tend to remove, or *clear*, the substance from the organ. In many cases, this clearance has been found to occur at a roughly exponential rate, so that the amount remaining varies with time as

$$e^{-\lambda_b t},$$

where λ_b is called the *biological decay constant*. The quantity

$$(\mathrm{T}_{1/2})_b = \frac{0.693}{\lambda_b} \tag{9.35}$$

is termed the *biological half-life* of the substance in the organ. Several values of $(\mathrm{T}_{1/2})_b$ are also given in the table.

If the amount of C_0 of a nonradioactive substance is taken into the body at time $t = 0$, it follows that the amount remaining in a particular organ at time t later is given by

$$R(t) = C_0 q e^{-\lambda_b t}. \tag{9.36}$$

The function $R(t)$ is known as the *retention function*.

Computations of internal doses to organs usually fall into two categories: an individual either inhales or ingests a specified amount of radionuclides in a single intake, or the intake occurs at a more or less constant rate over a long period. These two cases will be discussed in turn.

Case 1. Single intake: dose commitment Consider the intake of $C_0 \mu$Ci of a radionuclide at the time $t = 0$. The activity in the target organ rises quickly as the nuclide accumulates there and then dies away due both to its biological clearance and the radioactive decay of the nuclide. The activity at time t can be written as

$$C(t) = C_0 q e^{-\lambda_b t} \times e^{-\lambda t} = C_0 q e^{-\lambda_e t}, \tag{9.37}$$

where

$$\lambda_e = \lambda + \lambda_b \tag{9.38}$$

is the *effective decay constant* in the organ. Equation (9.38) can also be put in the form

$$\frac{1}{(T_{1/2})_{\text{eff}}} = \frac{1}{T_{1/2}} + \frac{1}{(T_{1/2})_b},\tag{9.39}$$

where $(T_{1/2})_{\text{eff}}$ is called the *effective half-life* of the nuclide in the organ in question. From Eq. (9.34), the dose-equivalent rate to the organ is

$$\dot{H} = \frac{51.1 C_0 \xi q}{M} e^{-\lambda_e t}\text{rem/day},\tag{9.40}$$

decreasing in time, as one would expect. The total dose to the organ from the intake of C_0 is called the *dose commitment*, because, aside from death or surgery, there is normally no way to prevent the organ from receiving that dose. This dose can be found by integrating Eq. (9.40) over the expected lifetime, t_l, of the individual following the intake:

$$H = \frac{51.1 \, C_0 \xi q}{M} \int_0^{t_l} e^{-\lambda_e t} \, dt$$

$$= \frac{51.1 \, C_0 \xi q}{M \lambda_e} (1 - e^{-\lambda_e t_l})\text{rem}.\tag{9.41}$$

If either the biological or radiological half-lives of the nuclide are short compared to the expected lifetime of the person, then Eq. (9.41) reduces to

$$H = \frac{51.1 \, C_0 \xi q}{M \lambda_e}\text{rem}.\tag{9.42}$$

It should be observed that the dose commitment is directly proportional to the magnitude of the intake and, in the case of Eq. (9.42), is independent of the point in the person's lifetime when the intake occurs. It follows that Eq. (9.42), where it applies, can be used for any time-varying dose where the total intake is C_0, provided that the duration of intake is short compared to the subject's lifetime. Thus, it clearly makes no difference in his total dose commitment whether an individual takes in a total of 1 μCi of ^{131}I, for which Eq. (9.42) applies, in one shot or spreads his intake of the 1 μCi over a year. The dose commitment will be the same.

The quantity H/C_0 computed from either Eq. (9.41) or (9.42) is called the *dose-commitment factor*, or DCF. Using Eq. (9.42), this factor is given by

$$\text{DCF} = \frac{51.1 \, \xi q}{M \lambda_e}\text{rem}/\mu\text{Ci}.\tag{9.43}$$

In view of the derivation of this formula, ξ has units of MeV, M is in g, and λ_e is in (day)$^{-1}$. This factor, which is widely tabulated, is handy for making dose calculations, as illustrated in the next example.

Example 9.10

(a) Compute the inhalation dose-commitment factor for ^{131}I for the thyroid. In an adult individual, this organ has a mass of 20 g. (b) Using the results from (a), compute the dose commitment to the thyroid when a person, breathing at a normal rate of 2.32×10^{-4} m^3/sec, stands for 2 hours in a radiation plume from a nuclear power plant containing ^{131}I at a concentration of 2×10^{-9} Ci/m^3.

Solution.

a) From Table 9.15, $T_{1/2} = 8.04$ days and $(T_{1/2})_b = 138$ days. Thus,

$$\lambda = \frac{0.693}{8.04} = 0.0862 \text{ day}^{-1},$$

$$\lambda_b = \frac{0.693}{138} = 0.00502 \text{ day}^{-1},$$

and from Eq. (9.38),

$$\lambda_e = 0.0862 + 0.00502 = 0.0912 \text{ day}^{-1}.$$

Also, from Table 9.15, $\xi = 0.23$ MeV and $q = 0.23$ for inhalation. Using Eq. (9.43) then gives

$$DCF = \frac{51.1 \times 0.23 \times 0.23}{20 \times 0.0912} = 1.48 \text{ rem}/\mu\text{Ci. } [Ans.]$$

b) At the given breathing rate, the person would inhale a total of

$$2.32 \times 10^{-4} \frac{m^3}{\text{sec}} \times 2 \times 10^{-9} \frac{\text{Ci}}{m^3} \times 2 \text{ hr} \times 3,600 \frac{\text{sec}}{\text{hr}} = 3.34 \times 10^{-9} \text{ Ci}$$

$$= 3.34 \times 10^{-3} \mu\text{Ci.}$$

This person's dose commitment is then

$$H = 1.48 \times 3.34 \times 10^{-3} = 4.94 \times 10^{-3} \text{ rem}$$

$$= 4.94 \text{ mrem. } [Ans.]$$

Case 2. Continuous intake   Consider next the situation in which $C_d\,\mu$ Ci/day of a radionuclide is inhaled or ingested on a continuous basis. During the interval $d\tau$ at τ, $C_d d\tau$ μCi is taken into the body. The amount of the radionuclide remaining in a given organ at the later time t is then

$$C_d q e^{-\lambda_e(t-\tau)} d\tau,$$

where use has been made of Eq. (9.37). The total quantity of the radionuclide in the organ at the time t due to the daily intake from $\tau = 0$ to $\tau = t$ is

$$C(t) = C_d q \int_0^t e^{-\lambda_e(t-\tau)} \tau$$

$$= \frac{C_d q}{\lambda_e}(1 - e^{\lambda_e t}). \qquad (9.44)$$

The dose rate to the organ at the time t is then found by introducing $C(t)$ into Eq. (9.34), which gives

$$\dot{H} = \frac{51.1 \, C_d \xi q}{M \lambda_e}(1 - e^{\lambda_e t}) \text{ rem/day}. \qquad (9.45)$$

According to Eq. (9.45), the dose rate rises steadily from the beginning of the radionuclide intake. If both $T_{1/2}$ and $(T_{1/2})_b$ are short compared to a person's expected lifetime, the exponential in Eq. (9.45) eventually dies out, and $\dot{H}$ reaches the steady-state value

$$\dot{H} = \frac{51.1 C_d \xi q}{M \lambda_e} \text{ rem/day}. \qquad (9.46)$$

This is the maximum dose rate that can be obtained from a steady intake of C_d μCi/day.

These results will now be applied to the determination of standards for the intake of radionuclides.

9.10 STANDARDS FOR INTAKE OF RADIONUCLIDES

As described in Section 9.8, standards-setting bodies have established maximum doses that may be accumulated over specified periods for each body organ or combination of organs. Based on these standards, it is possible to specify the maximum intake of radionuclides, which if not exceeded, will not violate the dose standards. These maximum intakes can be determined in two different ways.

Maximum Permissible Concentration

Intake standards are often stated in terms of *maximum permissible concentration* (MPC). For radiation workers this is defined as the maximum concentration of a radionuclide in air or water that at no time over a 50-year period (as long as any working lifetime starting at age 18) from the onset of constant daily intake provides a dose equivalent rate to any organ in excess of the maximum permissible dose rate for that organ. According to Eq. (9.45) and as illustrated in Fig. 9.13, the dose rate from radionuclides taken on a daily basis, whose radioactive *or* biological half-lives are much less than 50 years, will reach the maximum $\dot{H}$ = MPD long

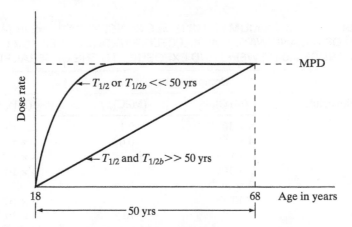

Figure 9.13   Dose rate to a body organ as a function of time after the onset of continuous intake of a radionuclide.

before the end of the 50-year intake period. On the other hand, radionuclides whose radioactive *and* biological half-lives are much longer than 50 years will deliver the maximum $\dot{H}$ on the final day of the 50 years. In either case, $\dot{H}$ decreases following the end of the permitted intake period.

To obtain an explicit expression for MPC, it is necessary to place $t = t_0 = 50$ years in Eq. (9.45), set $\dot{H}$ equal to the appropriate MPC and write the daily intake C_d in terms of the MPC. For the MPC of a radionuclide in water written $(\text{MPC})_w$, C_d is

$$C_d = (\text{MPC})_w V_w, \tag{9.47}$$

where V_w is the average daily intake of water via ingestion. Similarly,

$$C_d = (\text{MPC})_a V_a, \tag{9.48}$$

where V_a is the daily intake of air via respiration. Introducing these quantities into Eq. (9.45) and solving for MPC gives

$$(\text{MPC})_w = \frac{M\lambda_e(\text{MPD})}{51.1 V_w \xi q}(1 - e^{-\lambda_e t_0})^{-1} \tag{9.49}$$

and

$$(\text{MPC})_a = \frac{M\lambda_e(\text{MPD})}{51.1 V_a \xi q}(1 - e^{-\lambda_e t_0})^{-1} \tag{9.50}$$

The exponentials in these formulas disappear, of course, in cases where $\lambda_e t_0 \gg 1$.

In deriving Eqs. (9.49) and (9.50), it was implicitly assumed that the most sensitive target organ for the radionuclide is known a priori, and indeed this is

TABLE 9.16 MAXIMUM PERMISSIBLE CONCENTRATIONS IN MICROCURIES PER CM^3 OF AIR AND WATER OF SELECTED RADIONUCLIDES FOR OCCUPATIONAL EXPOSURE (40-HR WEEK) AND EXPOSURE TO THE GENERAL PUBLIC

Radionuclide	Occupational exposure		General public	
	$(MPC)_a$	$(MPC)_w$†	$(MPC)_a$	$(MPC)_w$†
^{3}H	5×10^{-6}	0.1	2×10^{-7}	3×10^{-3}
^{14}C	4×10^{-6}	0.02	1×10^{-7}	8×10^{-3}
^{24}Na	10^{-6}	6×10^{-3}	4×10^{-8}	2×10^{-4}
^{41}Ar	$2 \times 10^{-6}*$	*	$4 \times 10^{-8}*$	*
^{60}Co	3×10^{-7}	10^{-3}	10^{-8}	5×10^{-5}
^{85}Kr	$10^{-5}*$	*	$3 \times 10^{-7}*$	*
^{90}Sr	10^{-9}	10^{-5}	3×10^{-11}	3×10^{-7}
^{131}I	9×10^{-9}	6×10^{-5}	10^{-10}	3×10^{-7}
^{137}Cs	6×10^{-8}	4×10^{-4}	2×10^{-9}	2×10^{-5}
^{226}Ra	3×10^{-11}	4×10^{-7}	3×10^{-12}	3×10^{-8}
^{235}U	5×10^{-10}	8×10^{-4}	2×10^{-11}	3×10^{-5}
^{239}Pu	2×10^{-12}	10^{-4}	6×10^{-14}	5×10^{-6}

*Noble gases are not soluble in water; $(MPC)_a$ is based on the dose a person would receive if surrounded by an infinite hemispherical cloud of radioactive gas. The radiation from the cloud delivers a higher dose than that from gas held in the lungs or other internal organs. (See Section 11.5.)
†For water-soluble form of the radionuclide.

often the case. However, if the radionuclide goes to different organs with different MPDs, then the actual $(MPC)_w$ and $(MPC)_a$ are given by the smallest of the values computed from Eqs. (9.49) and (9.50) for the various organs.

It should also be noted that Eqs. (9.49) and (9.50) are based on the simplifying assumption that nuclides clear body organs exponentially (i.e., that the retention function is given by Eq. [9.36]). In fact, the movement of radionuclides within the body is generally much more complex, and accurate calculations of MPDs often require the use of complicated physiological transport models. An abridged tabulation of $(MPC)_{(w)}$ and $(MPC)_a$, originally computed by the ICRP[22] and later modified and adopted by the US Nuclear Regulatory Commission,[23] is given in Table 9.16. It will be observed that these MPCs are specified for persons whose occupational exposure is 40 hrs/wk and also for the general public. The use of Eqs. (9.49) and (9.50) in these situations is described in the next example.

[22]ICRP publication 2. (See References.)
[23]Code of Federal Regulations. (See References.)

Example 9.11

It has been found that tritium, whether inhaled or ingested, ultimately appears in equal concentrations in the blood, urine, sweat, and saliva. Therefore, Tritium, as HTO, can be assumed to mix freely with all of the body water and to metabolize in the same way as ordinary water. If the Standard Man (as defined by the ICRP) contains 43 liters of H_2O, drinks 2,200 cm^3 of H_2O per day, and produces internally another 300 cm^3 by oxidation every day, what is the $(MPC)_w$ for tritium for occupational exposure and for the general public?

Solution. Tritium undergoes β-decay with a half-life of 12.3 years. There are no γ-rays. The value of E_{max} is 0.018 MeV, so that $\bar{E} \simeq 0.3 \times 0.018 = 0.0054$ MeV. Because E_{max} is less than 0.03 MeV, $Q = 1.7$ (see Table 9.2), and $\xi = 0.0054 \times 1.7 = 0.0092$, or about 0.01, as given in Table 9.15.

Because the tritium is uniformly distributed throughout the body water, it gives, in effect, a whole-body dose. According to Table 9.13, the relevant MPD is then 0.1 rem/week, or 0.01428 rem/day, delivered to 43 liters of water having a mass $M = 4.3 \times 10^4$ g.

The retention function $R(t)$ can be found in the following way: The clearance of tritium from the body can be represented as in Fig. 9.14 by a tank containing $V = 43$ liters of water with a water throughput of $r = 2,500$ cm^3/day. Now, let m be the total mass of tritium in V at any time following a single intake of tritium at $t = 0$. Then, it is not difficult to see that m decreases according to the rate equation

$$\frac{dm}{dt} = -\frac{mr}{V}. \tag{9.51}$$

This equation has the same form as the equation determining the decay of a radioactive nuclide, and it follows that tritium is cleared exponentially. Thus, $R(t)$ is given by

$$R(t) = e^{\lambda_b t}, \tag{9.52}$$

where λ_b is the factor multiplying m on the right side of Eq. (9.51); that is,

$$\lambda_b = \frac{r}{V} = \frac{2,500}{4.3 \times 10^4} = 0.0581 \text{ day}^{-1},$$

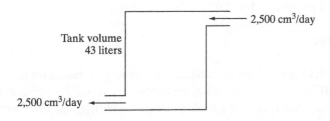

Figure 9.14   Diagram of throughput of water in the body.

which corresponds to a biological half-life of

$$T_b = \frac{0.693}{0.0581} = 11.9 \text{ days.}$$

Because $T_{1/2}$ and $(T_{1/2})_b$ are much less than 50 years, the exponential in Eq. (9.49) can be ignored. Furthermore, $\lambda_b \gg \lambda$, so that $\lambda_e \simeq \lambda_b$. Equation (9.49) can then be written as

$$(\text{MPC})_w = \frac{\lambda_b M (\text{MPD})}{51.1 \, V_w \xi q}. \tag{9.53}$$

Introducing numerical values ($q = 1$) gives

$$(\text{MPC})_w = \frac{0.0581 \times 4.3 \times 10^4 \times 0.01428}{51.1 \times 2{,}200 \times 0.01}$$

$$= 0.032 \; \mu\text{Ci/cm}^3.$$

This result is based on the assumption that the intake of tritium is continuous over a 7-day = 7×24 = 168-hour week, and indeed the official $(\text{MPC})_w$ for this type of occupational exposure is 0.03 $\mu\text{Ci/cm}^3$, as computed. If, however, a worker is only exposed during a 40-hour week, the permissible concentration may be higher. It has been found that a worker consumes about one-half of his daily intake of water during an 8-hour working day, so that his exposure while at work can be doubled over that just computed for the same MPD. Furthermore, a worker normally works only 5 out of 7 days in the week. The MPC for the 40-hour week is therefore $2 \times 7/5$ times the MPC for the 168-hour week; that is,

$$(\text{MPC})_w = 0.03 \; \mu\text{Ci/cm}^3 \text{ for a 168-hr week; [Ans.],}$$

and

$$(\text{MPC})_w = 0.03 \times 2 \times 7/5 = 0.084 \; \mu\text{Ci/cm}^3 \text{ for a 40-hr week. [Ans.]}$$

The official $(\text{MPC})_w$ for the 40-hour week is rounded off to 0.1 $\mu\text{Ci/cm}^3$.

According to Table 9.13, the whole-body MPD for any member of the general public is one-tenth the MPD for occupational exposure. The MPC for the public at large is therefore one-tenth of the MPC for radiation workers and must be calculated on the basis of the full 168-hour week. This is because the public is presumed to be exposed on a continuous basis, if exposed at all. Thus, in the case of tritium, the $(\text{MPC})_w$ for the general public is $0.1 \times 0.003 \; \mu\text{Ci/cm}^3$. [Ans.]

Annual Limit on Intake

The MPC values in Table 9.16 and others in general use are largely based on calculations performed by the ICRP during the 1950s and reported in ICRP publication 2. Similar calculations using more modern biological and radiological data were carried out in the 1970s and described in ICRP publication 30. So, in the more recent report, the ICRP has abandoned the term "maximum permissible concentration," because in the view of the ICRP:

> Although the Commission emphasized in ICRP Publication 2 that the rate of intake of a radionuclide could be varied provided that the intake in any quarter was no greater than that resulting from continuous exposure to the appropriate MPC for 13 weeks, the concept of MPC has been misused to imply a maximum concentration in air or water that should never be exceeded under any circumstances.

In short, governmental regulatory agencies have tended to take too literally the words "maximum permissible" in MPC.

The ICRP now computes and recommends what it calls the *annual limit on intake* (ALI) of radionuclides either by ingestion or inhalation. The EPA refers to the ALI as the *radioactivity intake factors* (RIF). The ICRP also specifies a new quantity, the *derived air concentration* (DAC), by the relation

$$\text{DAC} = \frac{\text{ALI}}{\text{air intake per year}}. \tag{9.54}$$

The DAC is essentially the $(\text{MPC})_a$ with another name. No comparable concentration is defined for water, since ICRP argues that water is only one of many sources of ingested materials.

The ICRP defines and computes the ALI in a unique and clever fashion. Suppose for the moment that a given radionuclide when ingested or inhaled goes only to one organ and that the entire dose from the radionuclide is confined to that organ. According to ICRP standards, the annual dose to the organ must not exceed some maximum value, $H_{\max}$, which is either the $5/w_T$ (cf. Eq. [9.17]) or 50 rems, whichever is smaller. If the radionuclide is taken into the body at the constant rate of I units per year for 50 years, the annual dose to the organ will increase in time until it reaches its maximum value. This will either occur in the 50th year or at some earlier time, depending on the magnitude of its radioactive and biological half-lives. In any event, the dose in no year may exceed $H_{\max}$. If $H_{1,50}$ is the dose in year 50 from unit intake in year 1 and $H_{2,50}$ is the dose in year 50 from unit intake in year 2, and so on, the total dose in the 50th year is then

$$H = I(H_{1,50} + H_{2,50} + H_{2,50} + \cdots + H_{50,50}). \tag{9.55}$$

The ALI is that value of I for which $H = H_{\max}$.

Because the annual intakes are all the same, Eq. (9.55) can be simplified. The dose in year 50 from the intake during year 2 must be identical to the dose in year 49 from the intake during year 1. That is,

$$H_{2,50} = H_{1,49}.$$

Similarly,

$$H_{3,50} = H_{1,48}, \ H_{4,50} = H_{1,47},$$

and so on. It follows that Eq. (9.55) can also be written as

$$H - I(H_{1,50} + H_{1,49} + H_{1,48} + \cdots + H_{1,1}). \tag{9.56}$$

The sum in the parentheses, however, is simply the 50-year dose commitment arising from the intake during the first year—or from the intake during *any* year, since each 50-year dose commitment is the same. If this 50-year dose commitment is denoted by H_{50}, then the ALI is determined from the equation

$$H_{\max} = (\text{ALI})H_{50}. \tag{9.57}$$

In fact, a radionuclide taken into the body is generally dispersed to a number of organs or tissues. Furthermore, any organ containing a radionuclide necessarily irradiates neighboring organs. In this case, according to the discussion in Section 9.8, the doses H_T to the complex of organs must satisfy the conditions

$$\sum_T w_T H_t \leq 5 \text{ rems}$$

and

$$H_T \leq 50 \text{ rems}$$

for all organs. Equation (9.57) can be enlarged to include the more realistic situation by first defining $H_{50,T}$ to be the 50-year dose commitment to tissue or organ T per unit intake of radionuclide. The ALI is then defined as the largest value of the intake I which satisfies both of the following relations:

$$I \sum_T W_T H_{T,50} \leq 5 \text{ rems} \tag{9.58}$$

and

$$I \ H_{T,50} \leq 50 \text{ rems}. \tag{9.59}$$

Values of ALI for a number of radionuclides based on Eqs. (9.58) and (9.59) are given in Table 9.17. It should be pointed out that the ICRP computations of these ALI are quite complex. The models of biological clearance used are far more complicated (and realistic) than the simple exponential discussed earlier. At the same time, the computation of doses from one organ to another is not a trivial undertaking.

What is more, in the computation of the ALI, the ICRP has taken into account the fact that different chemical forms of a given radionuclide may behave

TABLE 9.17 ANNUAL LIMIT ON INTAKE BY WORKERS OF SELECTED RADIONUCLIDES, IN MICROCURIES*

Radionuclide	Oral	Inhalation
^{3}H (tritiated water)	8.1×10^4	8.1×10^4
^{60}Co	190	27
^{90}Sr	27	2.7
^{131}I	27	54
^{137}Cs	110	160
^{226}Ra	1.9	0.54
^{235}U	14	0.054
^{239}Pu	5.4	0.0054

*From ICRP publication 30, 1979. ALI are rounded to two significant figures from ICRP values given in becquerels.

quite differently in the body. For example, the absorption of plutonium by the gastrointestinal tract is 10 times smaller for the oxide and hydroxide of plutonium than for other commonly occurring forms of this element. The ALI for the oral intake of plutonium oxides or hydroxides is therefore a factor of 10 larger than that of other chemical forms. Similarly, the uptake and retention of various inhaled chemical compounds may also differ, and the ICRP ALI for inhalation generally depends on the nature of the inhaled compound. The values of ALI given in the table are only the smallest of the ICRP values, so they should be used only when the chemical form of the ingested or inhaled radionuclide is known.

Radon Standards

Setting standards for the permissible concentration or allowable intake of radon is complicated, because, as discussed in Section 9.7, the bulk of the radiation dose originating in the presence of radon comes from its daughter products, not from the radon itself. What is more, a given concentration of radon does not guarantee any particular concentration of the daughters, for this depends on the extent to which the daughters have reached radioactive equilibrium with the parent radon. In an isolated environment furnished with a constant radon concentration, the daughters eventually come to equilibrium, so that the activities of all the radionuclides in the radon decay chain (see Fig. 9.7) are equal. However, in confined environments such as mines or buildings, only partial equilibrium is ever attained, because the longer-lived daughters become attached to moisture or dust particles and either diffuse to the walls or are swept out by ventilation in times that are short compared with their half-lives.

The concentration of radon and its daughters is frequently stated in terms of a *working level* (WL), a unit that was developed specifically in connection with measurements of radon/daughter levels in uranium mines. It is defined as any combination of the short-lived daughters of ^{222}Rn in 1 liter of air that will ultimately emit a total of 1.3×10^5 MeV in α-particle energy. It is easily shown that, in equilibrium, a radon concentration of 100 pCi/l will produce 1 WL. In practical situations, the radon daughters are usually not in equilibrium with the ^{222}Rn, so this radon concentration normally gives less than 1 WL. In a typical American home, for instance, 100 pCi/l of ^{222}Rn provides approximately 0.5 WL, corresponding to about 50% equilibrium.

Total exposure to radon daughters in uranium mines is measured in *working level months* (WLM). Inhalation of air containing a concentration of 1 WL of radon daughters for a total formation 170 hours (the average number of working hours in a month) gives an exposure of 1 WLM. According to EPA estimates based on the linear dose–effect hypothesis, a one-year exposure to 1 pCi/l of radon in equilibrium with its daughters (i.e., 0.01 WL) gives an added lifetime risk of lung cancer of 100 cases per million exposed individuals. This exposure corresponds to 0.01 WL $\times$ 8, 760 hrs/year $\div$ 170 hrs/month = 0.515 WLM. Current US standards limit the occupational exposure of uranium miners to no more than 4 WLM per year.

For determination of radon in homes, it is customary to use concentrations of pCi/l where 4pCi/l is the average outdoor concentration.

Example 9.12

In certain parts of the United States, the radon concentration has been measured at 1 pCi/l in older uninsulated homes and about 6 pCi/l in more modern, well-insulated homes. If a person spends an average of 3,500 hours in his home per year, roughly how many additional cases of lung cancer would result over a period of 50 years from insulating (and not ventilating) one million homes, each with four residents.

Solution. At 50% equilibrium, the increased exposure per insulated home would be

$$0.5 \times 5 \text{ pCi/l} \times \frac{3500 \text{ hrs exposure}}{8760 \text{ hrs/year}} = 1 \text{ pCi-year}$$

per person. Since 4×10^6 persons are exposed, the total number of cancers over 50 years would be

$$4 \times 10^6 \frac{\text{persons exposed}}{\text{year}} \times 50 \text{ yr}$$

$$\times \frac{100 \text{ cases}}{10^6 \text{ persons-pCi-yr}} \times 1 \text{ pCi-yr} = 20,000. \text{ [\textit{Ans.}]}$$

External and Internal Exposure

Occasionally, circumstances arise in which a radiation worker is exposed to an external dose of radiation and simultaneously exposed to an internal dose from the ingestion or inhalation of radiation. Both the ICRP and EPA have recommended that in these cases, the combined risk from these doses be no greater than the maximum risk permitted from either dose alone. This requirement is met, in the notation of the ICRP, if

$$\frac{H_{\text{ext}}}{H_{\text{wb,L}}} + \sum_{j} \frac{I_j}{\text{ALI}_j} \leq 1, \tag{9.60}$$

where H_{ext} is the external dose (assumed here to be uniform), $H_{\text{wb,L}}$ is the annual whole body dose limit of 5 rems, I_j is the annual intake of radionuclides of type j, and ALI_j is the annual limit of intake of that radionuclide. In the notation of the EPA, Eq. (9.60) is

$$\frac{H_{\text{ext}}}{\text{RPG}} + \sum_{j} \frac{I_j}{\text{RIF}_j} \leq 1, \tag{9.61}$$

where RPG = 5 rems, $\text{RIF}_j = \text{ALI}_j$ and the other symbols have the same meaning as in Eq. (9.60). If the external radiation is not uniform, then H_{ext} in Eq. (9.61) is replaced by the usual sum:

$$H_{\text{ext}} = \sum_{T} w_T H_T. \tag{9.62}$$

Example 9.13

During one year, a radiation worker anticipates receiving a uniform external dose of 3 rems and simultaneously inhaling some airborne ^{131}I. How much of the radionuclide may the worker inhale if he is not to exceed recommended limits?

Solution. From Table 9.17, the ALI for the inhalation of ^{131}I is 54 μCi. Substituting this value together with $H_{\text{ext}} = 3$ rems and $H_{\text{wb,L}} = 5$ rems into Eq. (9.60) gives

$$\frac{3}{5} = \frac{I}{54} = 1.$$

Solving for I yields $I = 22\mu\text{Ci}$. [*Ans.*]

9.11 EXPOSURE FROM γ-RAY SOURCES

It is often necessary to calculate exposures and exposure rates from specified distributions of γ-ray sources. Corresponding values of the absorbed dose and dose-equivalent can then be found using the formulas developed in the preceding section. In this section, a few simple γ-ray sources will be considered. Similar calculations for neutron sources, which are always heavily shielded, are more complicated; they will be discussed in the next chapter.

In the calculations that follow, it will be assumed that the sources are bare (un-shielded) and that the points of interest are so close to the sources that attentuation in the air can be neglected. This is the case for most γ-ray sources for distances up to about 30 meters. According to the formulas derived in Section 9.9, the quantities X, D, and H and their rates are proportional to the γ-ray flux ϕ_γ. It is necessary, therefore, to compute ϕ_γ for each source distribution.

Point Source

Consider first a point source emitting S γ-rays per second isotropically. At the outset, it should be noted that since all the γ-rays move in the radial direction, the γ-ray flux and intensity are equal. To compute the flux at the distance r from the source, it is observed that in the absence of attenuation, the total number of γ-rays passing through a sphere of radius r with the source as center, must be a constant, independent of r. This number is $4\pi r^2 I(r)$, where $I(r)$ is the intensity at r, and is equal to the source strength S. Thus,

$$4\pi r^2 I(r) = \text{constant} = S,$$

and so

$$I(r) = \phi_\gamma(r) = \frac{S}{4\pi r^2}. \tag{9.63}$$

Equation (9.63) may also be used for small spherical sources, provided that there is no attenuation within the source, but only at distances that are large compared with the radius of the source. At points near a spherical source, all the γ-rays do not move radially, which is the assumption upon which Eq. (9.63) is based.

Line Source

Consider next an infinite line source emitting S γ-rays/sec isotropically per centimeter of its length. The γ-rays emitted from the element of length dz (see Fig.

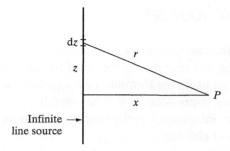

Figure 9.15 Diagram for computing flux from infinite line source.

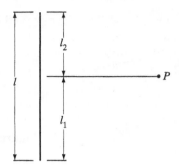

Figure 9.16   The finite line source.

9.15) appear at the point P as if emitted from a point source, and they make a contribution to the flux at P that is equal to

$$d\phi_\gamma(P) = \frac{S\,dz}{4\pi r^2}.$$

The total flux is then

$$\phi_\gamma(P) = \frac{S}{4\pi} \int_{-\infty}^{+\infty} \frac{dz}{r^2}. \tag{9.64}$$

Making the substitution $r^2 = x^2 + z^2$ and carrying out the integration gives

$$\phi_\gamma(P) = \frac{S}{4x}. \tag{9.65}$$

For a line source of length l as shown in Fig. 9.16, the γ-ray flux at P can be found by evaluating the integral in Eq. (9.64) between the limits of $-l_1$ and l_2. It is then easily found that

$$\phi_\gamma(P) = \frac{S}{4\pi x} \left[\tan^{-1}\left(\frac{l_2}{x}\right) + \tan^{-1}\left(\frac{l_1}{x}\right) \right]. \tag{9.66}$$

This formula is often used to estimate the γ-ray exposure from a bare fuel element when it is withdrawn from a reactor.

Ring Source

Suppose now that the source consists of a ring of radius R and length $l = 2\pi R$, which emits isotropically S γ-rays/sec per cm of length. The $S\,dl$ γ-rays emitted from the length dl of the ring (see Fig. 9.17) emerge as from a point source and give the flux at P

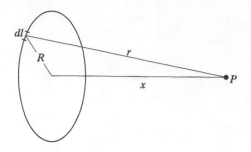

Figure 9.17   The ring source.

$$d\phi_\gamma(P) = \frac{S\,dl}{4\pi r^2}.$$

The total flux at P is then

$$\phi_\gamma(P) = \frac{Sl}{4\pi r^2}$$

$$= \frac{SR}{2r^2}. \qquad (9.67)$$

Disc Source

As a final example, imagine a planar disc of radius R, as shown in Fig. 9.18, which emits S γ-rays isotropically per cm²-sec. The flux along the axis of the disc can easily be computed; it is somewhat more difficult to calculate the flux at other points.

Consider the sources located in the annular region between z and $z+dz$. These emit $2\pi Sz\,dz$ γ-rays/sec, which is equivalent to the emission of Sl γ-rays/sec from the source just derived. Thus, from Eq. (9.67), these contribute the flux $d\phi_\gamma$ at P equal to

$$d\phi\gamma(P) = \frac{2\pi Sz\,dz}{4\pi r^2} = \frac{Sz\,dz}{2r^2}.$$

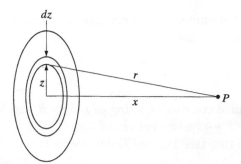

Figure 9.18   Diagram for computing flux from a disc source.

The total flux at P is therefore

$$\phi_\gamma(P) = \frac{S}{2} \int_0^R \frac{z \, dz}{r^2}. \tag{9.68}$$

The integral can be carried out by substituting $r^2 = x^2 + z^2$ with the result

$$\phi_\gamma(P) = \frac{S}{4} \ln\left(1 + \frac{R^2}{x^2}\right). \tag{9.69}$$

When $x \ll R$ (i.e., at points close to the disc), Eq. (9.69) reduces to

$$\phi_\gamma(P) = \frac{S}{2} \ln\left(\frac{R}{x}\right), \quad x \ll R. \tag{9.70}$$

It may be noted from either Eq. (9.69) or Eq. (9.70) that $\phi_\gamma(P)$ increases without limit with increasing R, and it might be concluded that the flux from an infinite sheet of sources is infinite. This is indeed the case in a vacuum (i.e., in the absence of attenuation). So, when attenuation in the air between the sources and the observation point is included, the flux remains finite, as will be shown in the next chapter.

GLOSSARY

Absorbed dose. The ratio $\Delta E_D / \Delta m$, where ΔE_D is the energy imparted by radiation in a volume element of matter of mass Δm; measured in rads.

Acute radiation syndrome (ARS). The symptoms that characterize a person suffering from the effects of radiation; "radiation sickness."

Biology. A branch of knowledge that deals with living organisms and vital processes.

Carcinoma. A form of cancer; cancer of the tissues that line body cavities.

Cataract. A condition in which the lens of the eye becomes increasingly opaque.

Centrioles Two small entities situated in the centrosome that play a role in controlling cell division.

Centrosome. A structure, contiguous to the nucleus of a cell, that contains the centrioles.

Chromatin. The genetic material of the cell; the mass of uncoiled chromosomes as seen, except during mitosis.

Chromosomes. The rod-shaped bodies in the nucleus, evident at mitosis, that carry the genetic material of a cell.

Chronic. Of long duration.

Cytoplasm. The fluid outside the nucleus that makes up most of an animal cell.

Dose equivalent. The absorbed dose multiplied by the quality factor, which accounts for the biological effects observed with a given absorbed dose.

Early effects (of radiation exposure). Effects that appear within 60 days of an acute exposure.

Edema. Swelling due to abnormal accumulation of fluid.

Epidemiology. The study of epidemics, diseases affecting large numbers of persons.

Epilation. Abnormal loss of hair.

Erythema. Abnormal redness of the skin.

Erythrocyte. Red blood corpuscle; red blood cell.

Exposure. The ratio $\Delta q / \Delta m$, where Δq is the sum of the electrical charges on all ions of one sign produced in air when all the electrons (+ and −), liberated from a volume of air whose mass is Δm, are completely stopped in air; measured in units of coulombs per kilogram.

Exposure rate. Exposure per unit time; measured in R/sec, mR/hr, etc.

Fluence. The time-integrated flux of particles per unit area; usually from a pulse or burst of radiation.

Formed elements (of the blood). The erythrocytes, leukocytes, and platelets in the blood.

Gametes. The germ cells—sperm of the male and ova of the female.

Gene. Originally, the entity or body, carried with the chromosome, responsible for a particular hereditary characteristic; now known to be a coded segment of a chromosome.

Gonad. The organ, of which there are usually two, in which germ cells are produced; in humans, the testes of the male and ovaries of the female.

Gray. The SI unit of absorbed dose; equal to 1 J/kg or 100 rads.

High dose. Acute whole-body dose-equivalent of more than 50 rems.

Imparted energy. Energy that is deposited by radiation in matter and that is available for producing biological effects.

In utero. In the uterus; prior to birth.

In vitro. In glass; in a laboratory vessel.

In vivo. In a living animal.

Kerma. Initial kinetic energy of charged particles released per unit mass of matter.

Late effects (of radiation exposure). Effects that appear after 60 days following an acute exposure.

LD$_{50}$/T. The acute whole-body dose (in rems) that leads on the average to the death of 50% of the exposed individuals within T days after exposure.

Leukemia. Cancer of the blood-forming organs, characterized by a great overproduction of white blood cells.

Leukocyte. A white blood cell of any type.

Linear energy transfer (LET). The energy deposited locally, in the form of excitation or ionization, per unit path of a charged particle.

Low dose. Acute whole-body dose-equivalent of less than 50 rems.

Lymphocyte. One type of white blood cell.

Meiosis. The process in which a germ cell divides into two new cells.

Mid-lethal dose. That acute whole-body dose that leads ultimately to the death of 50% of the exposed individuals.

Mitochondria (singular: mitochondrion). Small, oval structures distributed in the cytoplasm of a cell; they are important in controlling cell metabolism.

Mitosis. The process in which a somatic cell divides into two cells.

Mutation. A change in genetic material leading to unexpected changes in progeny.

Plasma (of the blood). The fluid component of the blood.

Polymorphonuclear neutrophil. The most abundant type of white blood cell.

Quality. A term describing the distribution of the energy deposited by a particle along its track; radiations that produce different densities of ionization per unit intensity are said to have different "qualities."

Quality factor. A factor by which the absorbed dose is multiplied to obtain dose-equivalent; used in radiation protection to take into account the fact that equal absorbed doses of radiation of different qualities have, in general, different biological effects.

Rad. The unit of absorbed dose; 1 rad is equal to the absorption of 100 ergs per gram.

Radiology. The branch of medicine in which patients are exposed to radiation for diagnosis and therapy.

Relative biological effectiveness (RBE). Ratio of the absorbed dose of γ-rays (whose RBE $= 1$) to the absorbed dose of another type of radiation that produces the same biological effect as γ-rays; now used in radiobiology only

Rem. The unit of equivalent dose.

Roentgen. The unit of exposure; equal to 2.58×10^{-4} coul/kg.

Sievert. The SI unit of dose-equivalent; equal to 100 rems.

Somatic cells. Body cells (as opposed to germ cells).

Zygote. The first bona fide cell formed by the union of an egg and sperm.

REFERENCES

General

Attix, F. H., and W. C. Roesch, Editors, *Radiation Dosimetry*. San Diego: Academic Press, to be published.

Cember, H., *Introduction to Health Physics*. 3rd ed. New York: McGraw-Hill, 1996.

Eisenbud, M., *Environmental Radioactivity*, from Natural, Industrial, and Military Sources, 4th ed. San Francisco: Morgan Kaufman, 1997.

Glasstone, S., and A. Sesonske, *Nuclear Reactor Engineering*, 4th ed. New York: Chapman & Hall, 1994, Chapter 9.

Hallenbeck, W. H., *Radiation Protection*. Lewis Publishers, 1994.

Hendee, W. R., *Medical Radiation Physics*, 2d ed. Chicago: Year Book Medical, 1979.

Morgan, K. Z., and J. E. Turner, Editors, *Principles of Radiation Protection*. New York: Wiley, 1967, Chapters 1–4, 10–14.

Taylor, L. S., *Radiation Protection Standards*. Elkins Park: Franklin Book Co., 1971. (This volume contains a thorough review of the history of radiation protection standards together with excerpts from the most important ICRP, ICRU, NCRP and FRC documents.)

Radiation Units

Radiation Quantities and Units, International Commission on Radiation Units and Measurements Report 19, 1971. This report defines important radiation quantities and conventional units. A supplement to this report clarifying the concept of dose-equivalent was issued by the ICRU in 1973.

Radiation Quantities and Units, Internal Commission on Radiation Units and Measurements Report 33, 1980. This is an updated version, in SI units, of ICRU report 19.

SI Units in Radiation Protection and Measurements, National Council on Radiation Protection and Measurements Report 82, 1985. This is the NCRP report recommending adoption of the SI units developed by the ICRU.

Biological Effects

The Effects on Population of Exposure to Low Levels of Ionizing Radiation. Washington, D.C.: National Academy of Sciences, 1972 (BEIR I) and 1980 (BEIR III).

Reactor Safety Study, US Nuclear Regulatory Commission Report WASH-1400, 1975, Appendix VI.

Health Effects of Exposure to Low Levels of Ionizing Radiation. Washington DC: National Research Council, National Academy of Sciences, BEIR V, 1990.

Radiation Exposure Sources

The Effects on Populations of Exposure to Low Levels of Ionizing Radiation, op. cit., Chapter III Hurwitz, H., Jr., "The Indoor Radiological Problem in Perspective," General Electric Report 81CR DO25, February 1981. This report discusses and compares radon exposure within dwellings with other sources of radiation.

Evaluation of Occupational and Environmental Exposure to Radon and Radon Daughters in the United States, National Council on Radiation Protection and Measurement Report 78, 1984.

Natural Background Radiation in the United States, National Council on Radiation Protection and Measurements Report 94, 1987.

Radiation Exposure from Consumer Products and Miscellaneous Sources, National Council on Radiation Protection and Measurements Report 56, 1977.

Radiological Quality of the Environment of the United States, US Environmental Protection Agency Report EPA 520/1-77-009, 1977.

Public Radiation Exposure from Nuclear Power Generation in the United States, National Council on Radiation Protection and Measurement Report 92, 1987.

Radiation Standards

Basic Radiation Protection Criteria, National Council on Radiation Protection and Measurements Report 39, 1971.

Federal Radiation Protection Guidance for Occupational Exposures, etc. Washington, D.C.: Federal Register, January 31, 1981. This notice describes the proposed regulations of the US Environmental Protection Agency.

Limitations of Exposure to Ionizing Radiation, National Council on Radiation Protection and Measurements Report 116, 1993. Supercedes NCRP Report 91.

Limits on Intakes of Radionuclides by Workers, International Commission on Radiation Units and Measurements Publication 30, 1979 and supplements. This report updates ICRP publication 2, described below.

General Concepts for the Dosimetry of Internally Deposited Radionuclides, National Council on Radiation Protection and Measurements 84, 1985.

Permissible Dose for Internal Radiation, International Commission on Radiation Units and Measurements Publication 2. Elmsford, N.W.: Pergamon Press, 1959. This document is the source of maximum concentrations of radionuclides allowed in air and water in the United States.

Problems Involved in Developing an Index of Harm, International Commission on Radiation Units and Measurements Publication 27. Elmsford, N.Y.: Pergamon Press, 1977. This report compares the risk of death or injury in radiation-related and nonradiation-related occupations.

Radiation Protection Standards, Federal Radiation Council Report 1. 1960.

Recommendations of the International Commission on Radiological Protection, ICRP Publication 26. Elmsford, N.Y.: Pergamon Press, 1977. These are the latest standards recommended by the ICRP.

Standards for Protection Against Radiation. Title 10, Code of Federal Regulations, Part 20. Washington, D.C.: continuously updated. These are the radiation regulations for the US Nuclear Regulatory Commission.

PROBLEMS

1. Demonstrate that the definition of the roentgen given in Eq. (9.2) is equivalent to the release of 1 esu of charge per cm^3 of air at 1 atmosphere pressure and at 0°C.

2. Fallout from a nuclear blast reached a man's home at the time t_0 after detonation. The man immediately enters a fallout shelter where there is zero exposure. He spends the time t_s in the shelter and then emerges. (a) Show that the factor by which his total dose is reduced by having gone into the shelter is given by

$$\frac{H(\text{shelter})}{H(\text{no shelter})} = (1 + t_s/t_0)^{-0.2}.$$

(b) Evaluate this ratio for $t_0 = 8$ hrs and $t_s = 2$ wks. [*Note:* The result in part (a) is based on the assumption that the man does not undertake decontamination procedures upon emerging from the shelter. Actually, the purpose of a fallout shelter is not to reduce the total, integrated dose, but to protect the individual from the high initial exposure rates. The period t_s in the shelter is supposed to be long enough that the exposure rate upon emerging from the shelter is low enough to permit decontamination.]

3. What is the probability that a 1 MeV γ-ray will interact in a cell? Take the cell diameter to be 10^{-3} cm, and assume that the cell has the density of water.

4. A beam of 1 MeV neutrons deposits energy in tissue at the rate of approximately 5×10^6 MeV per gram per sec. What dose-equivalent is this tissue receiving?

5. A desiccated sample of fossil contains 0.25 g of carbon. Its activity is measured to be 0.93 pCi. Approximately how old is the fossil?

6. At the age of 30, a man takes a job (his first) as a radiation worker. Five years later, his firm goes bankrupt, and his radiation exposure files are destroyed. If his next job is again as a radiation worker, what maximum doses may he receive in subsequent years at the new company? Show his radiation history graphically.

7. During emergency situations, it may be necessary for radiation workers to receive doses in excess of the MPD. The ICRP has recommended that emergency work "shall be planned on the basis that the individual will not receive a dose in excess of 12

rem. This shall be added to the occupational dose accumulated up to the time of the emergency exposure. If the sum then exceeds the maximum value permitted by the formula, $H = 5(N - 18)$, the excess will be made up by lowering the subsequent exposure rate, so that within a period not exceeding 5 years, the accumulated dose will conform with the limits set by the formula." Suppose that a man is first employed in the nuclear industry at the age of 18. During the next three years, he receives an average annual dose of 4 rems. At the end of this time, he engages in an emergency rescue operation and receives an acute dose of 12 rems. If ICRP recommendations are followed, what average dose may this man receive in subsequent years? Show the radiation history of this man graphically.

8. Calculate the fluence of 1 MeV γ-rays required to give an exposure of 1 mR.

9. (a) What intensity of 1 MeV γ-rays gives an exposure rate which provides the normal MPD for radiation workers? (b) What is the weekly γ-ray fluence corresponding to this exposure?

10. A 75 kV x-ray machine gives an exposure of 200 mR for a 0.2-sec chest x-ray. (a) What is the x-ray fluence during this chest x-ray? (b) What is the energy fluence? (c) What is the dose-equivalent delivered to bone, muscle, and fat?

11. What flux of 40 keV x-rays delivers a dose of 50 mrems/sec to bone?

12. Approximately how long may a person expose a hand to the x-ray beam of Problem 9.10 for the hand to develop (a) erythema? (b) blistering?

13. Consider the explosion of a 20 kiloton nuclear warhead. If there was no attenuation of the γ-rays in the atmosphere and all of the fission γ-rays escape from the warhead, approximately what would be the γ-ray dose received by a person standing (a) 1,000 meters and (b) 5,000 meters from the point of the blast? [*Note:* Take the energy release per fission to be 200 MeV, of which 7 MeV is in the form of fission γ-rays. Also, 1 kiloton $= 2.6 \times 10^{25}$ MeV.]

14. At a certain point near a nuclear reactor the flux of fast neutrons is 15 neutrons/cm^2-sec, the flux of thermal neutrons is 200 neutrons/cm^2-sec, and the γ-ray flux (assumed to be 2 MeV) is 400 γ-rays/cm^2-sec. How long may a worker remain at this point per week if he is not to exceed the normal weekly MPD dose?

15. During a criticality accident at a reactor facility, it is estimated that over a very short time a worker is exposed to the following radiation fluences: fast neutrons, 2×10^{10}/cm^2; thermal neutrons, 5×10^9/cm^2; γ-rays (2.5 MeV), 6×10^8/cm^2. (a) What total dose-equivalent did the worker receive? (b) What radiation sickness symptoms is the worker likely to develop? (c) What is the prognosis for his recovery?

16. Carbon-14 decays by the emission of a single negative β-ray having a maximum energy of 0.156 MeV. There are no γ-rays. The yearly gonad (internal) dose from ^{14}C is approximately 0.7 mrem. Compute the concentration of ^{14}C in pCi/g in these important organs.

17. Radium-226 and its daughters, taken together, emit a variety of α, β, and γ-rays. The effective energy equivalent ξ for this chain in bone is 110 MeV. (See Table 9.15.) If the total mass of the skeleton of a man is 7 kg, what amount of ^{226}Ra in bone gives a dose of 5 rems/year?

18. Whether inhaled or ingested, iodine, in most chemical forms, quickly enters the blood-stream, and much of this element flows to and is taken up by the thyroid gland. Shortly after an injection of iodine, the retention function for the thyroid is given approximately by

$$R(t) = 0.3e^{-0.00502t},$$

where t is in days. Suppose that a patient is given an injection of 1 mCi of ^{131}I for diagnosis of a thyroid condition. (a) What is the biological half-life of iodine in the thyroid? (b) What total dose will the patient's thyroid receive from the single injection? The mass of the thyroid is 20 g.

19. Compute the $(MPC)_w$ of ^{24}Na for the general public, based on the whole-body dose.

20. The maximum permissible concentrations of ^{90}Sr are based on the dose for bone. Using the data in Table 9.15, compute (a) the effective half-life in years of ^{90}Sr in bone and (b) the $(MPC)_w$ of ^{90}Sr for the general public.

21. Calculate the $(MPC)_w$ of ^{239}Pu for occupational exposure, based on the dose received in bone.

22. The drinking water at a certain laboratory is contaminated with tritium at a concentration five times the MPC. If workers drink this water for one week before its contamination is discovered, what total dose does each receive from this week's exposure?

23. A bare point source emits 5×10^7 fast neutrons and 5×10^8 thermal neutrons per sec. How long may a radiation worker stand at a distance of 3 meters from the source if he is not to exceed his normal weekly allowed dose?

24. A 1-Ci ^{226}Ra-Be source produces 10^7 neutrons/sec. The energy of the neutrons ranges from about 1 MeV to 13 MeV, with an average energy of approximately 5 MeV. The source also emits 3,500 γ-rays per neutron with an average energy of 1 MeV. In connection with an experiment, a worker must move the bare source from one location to another. He uses tongs so that the source is never closer than 6 ft to him. If the move takes one minute, how much of a dose does he receive?

25. Three isotropic point sources of γ-rays, each emitting S γ-rays/sec, are located in a vacuum at the corners of an equilateral triangle of side a. Derive an expression for the γ-ray flux at the center of the triangle.

26. Using the data given in Fig. 2.6, calculate the exposure rate and dose equivalent rate in tissue from an unshielded 1-Ci^{60}Co isotropic point source: (a) 1 cm from the source and (b) 1 m from the source.

27. A thin rod 1 m long, located in a vacuum, contains uniformly distributed neutron sources which emit 10^8 neutrons/sec per cm length of the rod. (a) Calculate the neutron flux at a point located 50 cm from the center of the rod. (b) What is the dose-equivalent rate in tissue at that point?

28. A fuel rod 1 m long and of small diameter contains 40 g of ^{235}U uniformly distributed throughout the rod. As a test, this rod is placed in a reactor having a uniform 2,200 m/sec flux of 1×10^{13} neutrons/cm^2-sec for two months and then removed. (a) Calculate the γ-ray exposure rate in R/sec at a point located 1 m from the center of the rod, two months after removal. (b) Estimate the γ-ray absorbed dose rate in rads/sec

in tissue at that point. (c) What is the γ-ray dose equivalent rate in rems/sec in tissue? [*Note:* For simplicity, take the average energy of the γ-rays to be 1 MeV.]

29. The sterilization of bacon requires an absorbed dose of approximately 5 million rads. What uniform concentration of ^{60}Co on a planar disc 5 ft in diameter is required to produce this dose 1 ft from the center of the disc after 1 hr exposure? [*Note:* For simplicity, assume that ^{60}Co emits two 1.25 MeV γ-rays per disintegration.]

30. Show that in the limit when $x \gg R$, the expression for the γ-ray flux along the axis of a disc source approached that for a point source.

31. Show that the γ-ray flux at the distance x from the center of a square planar source of side a emitting S photons/cm^2-sec is given approximately by

$$\phi_g = \frac{Sa^2}{4\pi x^2} \left(1 - \frac{a^2}{6x^2} \right),$$

provided that $x \gg a$. Compare this result with the similar formula for the photon flux from a circular disc.

32. A thin spherical shell of radius R has isotropic γ-ray sources distributed uniformly over its surface emitting S γ-rays/cm^2-sec. Calculate the γ-ray flux at the center of the shell.

33. A spherical shell of inner radius R_1 and outer radius R_2 contains a radioactive gas emitting S γ-rays/cm^3-sec. (a) Derive an expression for the γ-ray flux at the center of the shell. (b) What is the flux in the limit where either, $R_1 \rightarrow 0$ or $R_2 \rightarrow \infty$?

10

Radiation Shielding

Radiation shielding serves a number of functions. Foremost among these is reducing the radiation exposure to persons in the vicinity of radiation sources. Shielding used for this purpose is called *biological shielding*. This chapter is devoted largely to shielding of this type. Shields are also used in some reactors to reduce the intensity of γ-rays incident on the reactor vessel, which protects the vessel from excessive heating due to γ-ray absorption and reduces radiation damage due to neutrons. These shields are called *thermal shields*. Sometimes shields are used to protect delicate electronic apparatus that otherwise would not function properly in a radiation field. Such *apparatus shields* are used, for example, in some types of military equipment.

Ordinarily, it is necessary to shield only against γ-rays[1] and neutrons, not against α-particles or β-rays. This is because, as shown in Section 3.9, the ranges of charged particles in matter are so short. When α-particle or β-ray shielding is required, the formulas given in Section 3.9 can be used to calculate thickness.

This chapter is divided into two parts. The first is devoted to the biological shielding of γ-rays. The second part is concerned with the shielding of reactors—a problem that involves both γ-rays and neutrons. In all cases, the central problem

[1]Here, as in Chapter 9, the term γ-ray refers to both x-rays and γ-rays.

548

is to determine the thickness and/or composition of shielding material required to reduce biological dose rates to predetermined levels. Frequently these are the maximum permissible dose rates at certain points, as discussed in Section 9.8, for either occupational or general population exposure.

10.1 GAMMA-RAY SHIELDING: BUILDUP FACTORS

Consider a monodirectional beam of γ-rays of intensity (or flux)[2] Φ_0 and energy E_0, which is incident on a slab shield of thickness a as shown in Fig. 10.1. If there is no shield, then according to Eq. (9.20), the exposure rate at P in conventional units is

$$\dot{X}_0 = 0.0659\phi_0 E_0 (\mu_a/\rho)^{\text{air}} \text{mR/hr}, \tag{10.1}$$

where ϕ_0 is in units of γ-rays/cm^2-sec, E_0 is in MeV, and $(\mu_a/\rho)^{\text{air}}$ is in cm^2/g and is evaluated at the energy E_0. It is convenient to write Eq. (10.1) in the form

$$\dot{X}_0 = C\phi_0, \tag{10.2}$$

where

$$C = 0.0659 E_0 (\mu_a/\rho)^{\text{air}} \tag{10.3}$$

is a function of E_0. With the shield in place, the γ-ray flux ϕ emerging from the shield is different from ϕ_0, and the problem of determing the actual value of $\dot{X}$ at P reduces to computing ϕ.

It would be easy to compute ϕ if every time a photon interacted with matter it disappeared. Then ϕ would be equal to the flux of *uncollided* γ-rays

$$\phi_u = \phi_0 e^{-\mu a}, \tag{10.4}$$

where μ is the total attenuation coefficient at the energy E_0. Unfortunately, γ-rays do *not* disappear at each interaction. In the Compton effect, for instance, they are

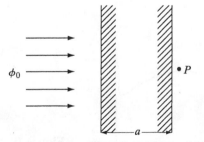

ϕ_0

$\bullet P$

Figure 10.1 Monodirectional beam of γ-rays incident on slab shield.

a

---

[2]Because the beam is monodirectional, the flux and intensity are equal in this example.

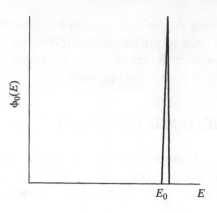

Figure 10.2 Energy spectrum of incident γ-ray beam.

merely scattered with a loss in energy. Even in the photoelectric effect and in pair production, although the incident photon is absorbed, x-rays are usually produced subsequent to the photoelectric effect, and annihilation radiation inevitably follows after pair production. As a result, a monoenergetic beam incident on a shield as in Fig. 10.1, with an energy spectrum like that depicted in Fig. 10.2, emerges from the shield with a continuous spectrum as shown in Fig. 10.3. In this figure, the sharp peak at E_0 corresponds to the unscattered photons and is reduced in size over the peak in Fig. 10.2 by the exponential factor in Eq. (10.4). For the most part, the continuous part of the spectrum is due to Compton-scattered photons, with some contribution from photoelectric x-rays and annihilation radiation.

The exact calculation of a spectrum like that shown in Fig. 10.3 is difficult and well beyond the scope of this text. It must suffice at this point merely to state that such calculations of $\phi(E)$ have been carried out for a variety of shielding materials as a function of the incident γ-ray energy and shield thickness. The computed values of $\phi(E)$ are then used to compute the exposure rate from the formula

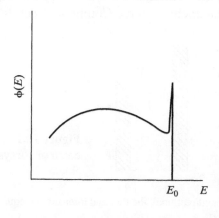

Figure 10.3 Energy spectrum of γ-rays emerging from shield.

$$\dot{X} = 0.0659 \int_0^{E_0} \phi(E)E(\mu_a/\rho)^{\text{air}} dE. \tag{10.5}$$

The results of these computations are written as

$$\dot{X} = \dot{X}_0 B_m(\mu a)e^{-\mu a}, \tag{10.6}$$

where $\dot{X}_0$ is the exposure rate in the absence of the shield as given by Eq. (10.1), $B_m(\mu a)$ is called the *exposure buildup factor* for a monodirectional beam, and μ is again the total attenuation coefficient at the energy E_0.

Values of B_m are given in Table 10.1 as a function of energy for several materials.[3] It is observed from this table that the numerical values of B_m can be very large, which shows the importance of the buildup of scattered radiation in shielding calculations. With a water shield, for example, which is $\mu a = 10$ mean free paths thick[4] at 2 MeV, B_m is approximately equal to 10. This means that if buildup is omitted from a calculation of the effectiveness of such a shield, the resulting exposure rate is in error by a factor of 10. Therefore, the buildup factor cannot be ignored. Since B_m is a continuous function of atomic number, values of B_m for materials not given in the table can be obtained by plotting B_m versus Z for the energy in question and interpolating from the curve. Equation (10.6) gives the actual exposure rate at the point P, behind the shield, in Fig. 10.1.

According to Eq. (10.1), $\dot{X}_0$ is proportional to ϕ_0, and it is reasonable, by analogy, to write $\dot{X}$ in the same form—namely,

$$\dot{X} = C\phi_b, \tag{10.7}$$

where C is again given by Eq. (10.3). The quantity ϕ_b is called the *buildup flux* and is clearly equal to the flux of monoenergetic γ-rays of energy E_0. E_0 gives the same exposure rate at P in Fig. 10.1 as the actual γ-ray flux at that point. Introducing Eqs. (10.2) and (10.7) into Eq. (10.6) shows that

$$\phi_b = \phi_0 B_m(\mu a)e^{-\mu a} \tag{10.8}$$

or equivalently

$$\phi_b = \phi_u B_m(\mu a),$$

where ϕ_u is the uncollided flux of Eq. (10.4).

---

[3]These buildup factors were originally called *dose buildup factors* and were actually computed for infinite media. However, the differences between the buildup factors for infinite media and slabs of finite thickness are ordinarily negligible except for light media, such as H_2O, and low energies— less than about 1 MeV.

[4]Since $\mu a = a/\lambda$, where λ is the photon mean free path, it follows that μa is equal to the thickness of the shield in mean free paths computed at the energy of the incident photons.

TABLE 10.1 EXPOSURE BUILDUP FACTOR FOR PLANE MONODIRECTIONAL
SOURCE*

| Material | E_0, MeV | \multicolumn{6}{c}{$\mu_0 x$} |
		1	2	4	7	10	15
Water	0.5	2.63	4.29	9.05	20.0	35.9	74.9
	1.0	2.26	3.39	6.27	11.5	18.0	30.8
	2.0	1.84	2.63	4.28	6.96	9.87	14.4
	3.0	1.69	2.31	3.57	5.51	7.48	10.8
	4.0	1.58	2.10	3.12	4.63	6.19	8.54
	6.0	1.45	1.86	2.63	3.76	4.86	6.78
	8.0	1.36	1.69	2.30	3.16	4.00	5.47
Iron	0.5	2.07	2.94	4.87	8.31	12.4	20.6
	1.0	1.92	2.74	4.57	7.81	11.6	18.9
	2.0	1.69	2.35	3.76	6.11	8.78	13.7
	3.0	1.58	2.13	3.32	5.26	7.41	11.4
	4.0	1.48	1.90	2.95	4.61	6.46	9.92
	6.0	1.35	1.71	2.48	3.81	5.35	8.39
	8.0	1.27	1.55	2.17	3.27	4.58	7.33
	10.0	1.22	1.44	1.95	2.89	4.07	6.70
Tin	1.0	1.65	2.24	3.40	5.18	7.19	10.5
	2.0	1.58	2.13	3.27	5.12	7.13	11.0
	4.0	1.39	1.80	2.69	4.31	6.30	
	6.0	1.27	1.57	2.27	3.72	5.77	11.0
	10.0	1.16	1.33	1.77	2.81	4.53	9.68
Lead	0.5	1.24	1.39	1.63	1.87	2.08	
	1.0	1.38	1.68	2.18	2.80	3.40	4.20
	2.0	1.40	1.76	2.41	3.36	4.35	5.94
	3.0	1.36	1.71	2.42	3.55	4.82	7.18
	4.0	1.28	1.56	2.18	3.29	4.69	7.70
	6.0	1.19	1.40	1.87	2.97	4.69	9.53
	8.0	1.14	1.30	1.69	2.61	4.18	0.08
	10.0	1.11	1.24	1.54	2.27	3.54	7.70
Uranium	0.5	1.17	1.28	1.45	1.60	1.73	
	1.0	1.30	1.53	1.90	2.32	2.70	3.60
	2.0	1.33	1.62	2.15	2.87	3.56	4.89
	3.0	1.29	1.57	2.13	3.02	3.99	5.94
	4.0	1.25	1.49	2.02	2.94	4.06	6.47
	6.0	1.18	1.37	1.82	2.74	4.12	7.79
	8.0	1.13	1.27	1.61	2.39	3.65	7.36
	10.0	1.10	1.21	1.48	2.12	3.21	6.58

*From H. Goldstein, *Fundamental Aspects of Reactor Shielding*. Reading, Mass.: Addison-Wesley, 1959; now available from Johnson Reprint Corp. New York.

Example 10.1

A monodirectional beam of 2-MeV γ-rays of intensity 10^6 γ-rays/cm^2-sec is incident on a lead shield 10 cm thick. At the rear side of the shield calculate the: (a) uncollided flux; (b) buildup flux; (c) exposure rate.

Solution.

1. From Table II.4, $\mu/\rho = 0.0457$ cm^2/ g for lead at 2 MeV. Then since $\rho = 11.34$ g/cm^3, $\mu = 0.0457 \times 11.34 = 0.518$ cm^{-1}, and $\mu a = 0.518 \times 10 = 5.18$. From Eq. (10.4),

$$\phi_u = 10^6 e^{-5.18} = 10^6 \times 5.63 \times 10^{-3} = 5.63 \times 10^3 \gamma\text{-rays/cm}^2\text{-sec. [Ans.]}$$

2. The buildup factor at 2 MeV for $\mu a = 5.18$ must be found by interpolating between the values given in Table 10.1 for $\mu a = 4$ and $\mu a = 7$. The result is $B_m = 2.78$. The buildup flux is therefore

$$\phi_b = \phi_u B_m(\mu a) = 5.63 \times 10^3 \times 2.78 = 1.56 \times 10^4 \gamma\text{-rays/cm}^2\text{-sec. [Ans.]}$$

3. The value of $\dot{X}$ is then

$$\dot{X} = C\phi_b = 0.0659 E_0 (\mu_a/\rho)^{\text{air}} \phi_b,$$

where $E_0 = 2$ MeV and $(\mu_a/\rho)^{\text{air}}$ is evaluated at this energy. From Table II.5, $(\mu_a/\rho)^{\text{air}} = 0.0238$ cm^2/ g, so that

$$\dot{X} = 0.0659 \times 2 \times 0.0238 \times 1.56 \times 10^4 = 48.8 \text{ mR/hr. [Ans.]}$$

An alternate and probably easier way to obtain the result in part (c) is to note from Fig. 9.9 that, at 2 MeV, a γ-ray flux of 3.2×10^2 gives an exposure rate of 1 mR/hr. In the present case, therefore,

$$\dot{X} = \frac{1.56 \times 10^4}{3.2 \times 10^2} = 48.8 \text{ mR/hr. [Ans.]}$$

The prior discussion pertains exclusively to a *monodirectional* beam normally incident on a slab shield, and the buildup factor $B_m(\mu a)$ can be used only in problems of this type. Buildup factors have also been computed for other types of shielding problems. For instance, consider a point isotropic source emitting S γ-rays/sec surrounded by a spherical shield of radius R. In this case, the exposure rate at a point on the surface of the shield is written as

$$\dot{X} = \dot{X}_0 B_p(\mu R) e^{-\mu R}. \tag{10.10}$$

Here $B_p(\mu R)$ is the *point isotropic* exposure buildup factor, and $\dot{X}_0$ is the exposure rate in the absence of the shield—specifically,

$$\dot{X}_0 = C\phi_0, \tag{10.11}$$

where according to Eq. (9.63)

$$\phi_0 = \frac{S}{4\pi R^2} \tag{10.12}$$

is the flux from a bare point source. The uncollided flux in this problem is

$$\phi_u = \frac{Se^{-\mu R}}{4\pi R^2}, \tag{10.13}$$

and the buildup flux is

$$\phi_b = \frac{SB_p(\mu R)e^{-\mu R}}{4\pi R^2}. \tag{10.14}$$

The computed values of B_p are shown in Table 10.2. It should be especially noted that B_p and B_m are entirely different functions, which must be used only in appropriate problems.

TABLE 10.2   EXPOSURE BUILDUP FACTOR FOR ISOTROPIC POINT SOURCE*

Material	E_0, MeV	$\mu_0 r$						
		1	2	4	7	10	15	20
Water	0.255	3.09	7.14	23.0	72.9	166	456	982
	0.5	2.52	5.14	14.3	38.8	77.6	178	334
	1.0	2.13	3.71	7.68	16.2	27.1	50.4	82.2
	2.0	1.83	2.77	4.88	8.46	12.4	19.5	27.7
	3.0	1.69	2.42	3.91	6.23	8.63	12.8	17.0
	4.0	1.58	2.17	3.34	5.13	6.94	9.97	12.9
	6.0	1.46	1.91	2.76	3.99	5.18	7.09	8.85
	8.0	1.38	1.74	2.40	3.34	4.25	5.66	6.95
	10.0	1.33	1.63	2.19	2.97	3.72	4.90	5.98
Aluminum	0.5	2.37	4.24	9.47	21.5	38.9	80.8	141
	1.0	2.02	3.31	6.57	13.1	21.2	37.9	58.5
	2.0	1.75	2.61	4.62	8.05	11.9	18.7	26.3
	3.0	1.64	2.32	3.78	6.14	8.65	13.0	17.7
	4.0	1.53	2.08	3.22	5.01	6.88	10.1	13.4
	6.0	1.42	1.85	2.70	4.06	5.49	7.97	10.4
	8.0	1.34	1.68	2.37	3.45	4.58	6.56	8.52
	10.0	1.28	1.55	2.12	3.01	3.96	5.63	7.32
Iron	0.5	1.98	3.09	5.98	11.7	19.2	35.4	55.6
	1.0	1.87	2.89	5.39	10.2	16.2	28.3	42.7
	2.0	1.76	2.43	4.13	7.25	10.9	17.6	25.1
	3.0	1.55	2.15	3.51	5.85	8.51	13.5	19.1
	4.0	1.45	1.94	3.03	4.91	7.11	11.2	16.0
	6.0	1.34	1.72	2.58	4.14	6.01	9.89	14.7
	8.0	1.27	1.56	2.23	3.49	5.07	8.50	13.0
	10.0	1.20	1.42	1.95	2.99	4.35	7.54	12.4

TABLE 10.2 *(CONTINUED)*

Material	E_0, MeV	$\mu_0 r$						
		1	2	4	7	10	15	20
Tin	0.5	1.56	2.08	3.09	4.57	6.04	8.64	
	1.0	1.64	2.30	3.74	6.17	8.85	13.7	18.8
	2.0	1.57	2.17	3.53	5.87	8.53	13.6	19.3
	3.0	1.46	1.96	3.13	5.28	7.91	13.3	20.1
	4.0	1.38	1.81	2.82	4.82	7.41	13.2	21.2
	6.0	1.26	1.57	2.37	4.17	6.94	14.8	29.1
	8.0	1.19	1.42	2.05	3.57	6.19	15.1	34.0
	10.0	1.14	1.31	1.79	2.99	5.21	12.5	33.4
Tungsten	0.5	1.28	1.50	1.84	2.24	2.61	3.12	
	1.0	1.44	1.83	2.57	3.62	4.64	6.25	(7.35)
	2.0	1.42	1.85	2.72	4.09	5.27	8.07	(10.6)
	3.0	1.36	1.74	2.59	4.00	5.92	9.66	14.1
	4.0	1.29	1.62	2.41	4.03	6.27	12.0	20.9
	6.0	1.20	1.43	2.07	3.60	6.29	15.7	36.3
	8.0	1.14	1.32	1.81	3.05	5.40	15.2	41.9
	10.0	1.11	1.25	1.64	2.62	4.65	14.0	39.3
Lead	0.5	1.24	1.42	1.69	2.00	2.27	2.65	(2.73)
	1.0	1.37	1.69	2.26	3.02	3.74	4.81	5.86
	2.0	1.39	1.76	2.51	3.66	4.84	6.87	9.00
	3.0	1.34	1.68	2.43	3.75	5.30	8.44	12.3
	4.0	1.27	1.56	2.25	3.61	5.44	9.80	16.3
	5.1	1.21	1.46	2.08	3.44	5.55	11.7	23.6
	6.0	1.18	1.40	1.97	3.34	5.69	13.8	32.7
	8.0	1.14	1.30	1.74	2.87	5.07	14.1	44.6
	10.0	1.11	1.23	1.58	2.52	4.34	12.5	39.2
Uranium	0.5	1.17	1.20	1.48	1.67	1.85	2.08	
	1.0	1.31	1.56	1.98	2.50	2.97	3.67	
	2.0	1.33	1.64	2.23	3.09	3.95	5.36	(6.48)
	3.0	1.29	1.58	2.21	3.27	4.51	6.97	9.88
	4.0	1.24	1.50	2.09	3.21	4.66	8.01	12.7
	6.0	1.16	1.36	1.85	2.96	4.80	10.8	23.0
	8.0	1.12	1.27	1.66	2.61	4.36	11.2	28.0
	10.0	1.09	1.20	1.51	2.26	3.78	10.5	28.5

*From H. Goldstein, *Fundamental Aspects of Reactor Shielding*. Reading, Mass.: Addison-Wesley, 1959; now available from Johnson Reprint Corp. New York.

Example 10.2

An isotropic point source emits $10^8 \gamma$-rays/sec with an energy of 1 MeV. The source is to be shielded with a spherical iron shield. What must the shield radius be if the exposure rate at its surface is to be 1 mR/hr?

Solution. In a problem of this type, the first step is to compute the buildup flux that will give the required exposure rate. Thus,

$$\phi_b = \dot{X}/0.0659 E (\mu_a/\rho)^{\text{air}}.$$

Here $\dot{X} = 1$ mR/hr, $E = 1$ MeV, and from Table II.5 $(\mu_a/\rho)^{\text{air}} = 0.0280$ cm^2/g. Introducing these values gives

$$\phi_b = 1/0.0659 \times 1 \times 0.0280 = 5.42 \times 10^2 \gamma\text{-rays/cm}^2\text{-sec}.$$

Then from Eq. (10.14),

$$5.42 \times 10^2 = \frac{10^8 B_p(\mu R)e^{-\mu R}}{4\pi R^2},$$

which may be written as

$$1 = 1.47 \times 10^4 \frac{B_p(\mu R)e^{-\mu R}}{R^2}.$$

This equation must be solved for R, which is easily done by plotting the right-hand side (RHS) of the equation as a function of μR. To express RHS in terms of μR, the numerator and denominator are multiplied by μ, obtained from Table II.4 ($\mu = 0.0595 \times 7.86 = 0.468$ cm^{-1}). This gives

$$1 = 3.22 \times 10^3 \frac{B_p(\mu R)e^{-\mu R}}{(\mu R)^2}. \tag{10.15}$$

As shown in Fig. 10.4, the RHS of Eq. (10.15) is almost a straight line when plotted on a semilog graph. From the figure, RHS = 1 for $\mu R = 6.55$, and so $R = 6.55/0.468 = 14.0$ cm. [*Ans.*]

For computational purposes, it is convenient to express the tabulated point buildup factor as a mathematical function, and several such functions have been developed. One of the most useful of these is the sum of exponential—namely,

$$B_p = A_1 e^{-\alpha_1 \mu r} + A_2 e^{-\alpha_2 \mu r} = \Sigma A_n e^{-\alpha_n \mu r}, \tag{10.16}$$

where A_1, A_2, α_1, and α_2 are functions of energy. Equation (10.16) is known as the *Taylor form* of the buildup factor, and it is sufficiently accurate to be used in many practical shielding problems.[5] As r goes to zero, B_p must approach unity since there can be no buildup of scattered radiation in a shield of zero thickness. It

[5]A more accurate representation of B_p is the *Berger* form:

$$B_p = 1 + C\mu r e^{-\beta \mu r},$$

where C and β depend on energy. Unfortunately, this expression is somewhat more difficult to use in analytical calculations than the Taylor form.

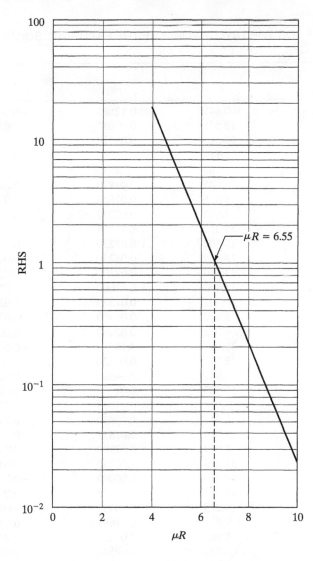

$\mu R = 6.55$

Figure 10.4   RHS of Eq. (10.15) as a function of μR.

follows that

$$A_1 + A_2 = 1. \tag{10.17}$$

Thus, it is sufficient to specify A_1, which hereinafter is denoted as A; then A_2 is equal to $1 - A$. Values of A, α_1, and α_2 are given in Table 10.3.

The use of this form of the buildup factor is illustrated in the next sections for a number of important shielding configurations. In each case, the uncollided and buildup fluxes are calculated. From the value of ϕ_b, the exposure rate is found using Eqs. (10.7) and (10.3) or Fig. 9.9.

TABLE 10.3   PARAMETERS FOR THE TAYLOR FORM OF THE EXPOSURE BUILDUP
FACTOR FOR POINT ISOTROPIC SOURCE*

Substance	Energy (MeV)	A	$-\alpha_1$	α_2
Water	0.5	100.845	0.12687	−0.10925
	1.0	19.601	0.09037	−0.02522
	2.0	12.612	0.05320	0.01932
	3.0	11.110	0.03550	0.03206
	4.0	11.163	0.02543	0.03025
	6.0	8.385	0.01820	0.04164
	8.0	4.635	0.02633	0.07097
	10.0	3.545	0.02991	0.08717
Concrete	0.5	38.225	0.14824	−0.10579
	1.0	25.507	0.07230	−0.01843
	2.0	18.089	0.04250	0.00849
	3.0	13.640	0.03200	0.02022
	4.0	11.460	0.02600	0.02450
	6.0	10.781	0.01520	0.02925
	8.0	8.972	0.01300	0.02979
	10.0	4.015	0.02880	0.06844
Aluminum	0.5	38.911	0.10015	−0.06312
	1.0	28.782	0.06820	−0.06312
	2.0	16.981	0.04588	0.00271
	3.0	10.583	0.04066	0.02514
	4.0	7.526	0.03973	0.03860
	6.0	5.713	0.03934	0.04347
	8.0	4.716	0.03837	0.04431
	10.0	3.999	0.03900	0.04130
Iron	0.5	31.379	0.06842	−0.03742
	1.0	24.957	0.06086	−0.02463
	2.0	17.622	0.04627	−0.00526
	3.0	13.218	0.04431	−0.00087
	4.0	9.624	0.04698	0.00175
	6.0	5.867	0.06150	−0.00186
	8.0	3.243	0.07500	0.02123
	10.0	1.747	0.09900	0.06627
Tin	0.5	11.440	0.01800	0.03187
	1.0	11.426	0.04266	0.01606
	2.0	8.783	0.05349	0.01505
	3.0	5.400	0.07440	0.02080
	4.0	3.496	0.09517	0.02598
	6.0	2.005	0.13733	−0.01501
	8.0	1.101	0.17288	−0.01787
	10.0	0.708	0.19200	0.01552

TABLE 10.3 (*CONTINUED*)

Substance	Energy (MeV)	A	$-\alpha_1$	α_2
Lead	0.5	1.677	0.03084	0.30941
	1.0	2.84	0.03503	0.13486
	2.0	5.421	0.03482	0.04379
	3.0	5.580	0.05422	0.00611
	4.0	3.897	0.08458	−0.02383
	6.0	0.926	0.17860	−0.04635
	8.0	0.368	0.23691	−0.05864
	10.0	0.311	0.24024	−0.02783

*From H. Goldstein, *Fundamental Aspects of Reactor Shielding*. Reading, Mass.: Addison-Wesley, 1959; now available from Johnson Reprint Corp. New York.

10.2 INFINITE PLANAR AND DISC SOURCES

Consider an infinite planar source emitting S γ-rays/cm^2-sec isotropically from its surface, which is located at the back of a shield of thickness a as shown in Fig. 10.5. The sources located on the differential ring of width dz at z emit $2\pi S z \, dz$ γ-ray/sec and make a contribution to the uncollided flux at P, which is equal to

$$d\phi_u = \frac{2\pi S z \, dz \, e^{-\mu r}}{4\pi r^2}$$

$$= \frac{S z e^{-\mu r} \, dz}{2r^2}. \qquad (10.18)$$

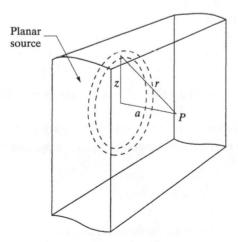

Figure 10.5   Isotropic planar source behind slab shield.

The total uncollided flux at P is then

$$\phi_u = \frac{S}{2} \int_0^\infty \frac{ze^{-\mu r}\,dz}{r^2}. \tag{10.19}$$

At this point, it is convenient to change the integration variable from z to r. Then, since $r^2 = a^2 + z^2$ and $r\,dr = z\,dz$, it follows that

$$\phi_u = \frac{S}{2} \int_2^\infty \frac{e^{-\mu r}}{r}\,dr. \tag{10.20}$$

Finally, letting $\mu r = t$, the integral may be put in the form

$$\phi_u = \frac{S}{2} \int_{\mu a}^\infty \frac{e^{-t}}{t}\,dt. \tag{10.21}$$

The integral in Eq. (10.21) cannot be evaluated analytically, but can be expressed in terms of one of the exponential integral functions, E_n, defined by the integral[6]

$$E_n(x) = x^{n-1} \int_x^\infty \frac{e^{-t}}{t^n}\,dt, \tag{10.22}$$

where n is an integer. Comparing Eqs. (10.21) and (10.22) shows that in the present case $n = 1$, so that

$$\phi_u = \frac{S}{2}E_1(\mu a). \tag{10.23}$$

The function $E_1(x)$ occurs in many shielding problems. Figure 10.6 shows this function, together with $E_2(x)$, for values of x up to $x = 14$. For larger values of x, $E_n(x)$ can be computed with sufficient accuracy for most purposes from the following approximate formula:

$$E_n(x) \simeq e^{-x}\left[\frac{1}{x+n} + \frac{n}{(x+n)^3}\right]. \tag{10.24}$$

When $x = 10$, this formula gives a value of $E_1(x)$, which is only .11% too large; when $x = 20$, the error is .04%.

To return to the problem of the shielded planar source, it is possible to calculate the buildup flux by noting that the contribution to ϕ_b, due to the differential ring at z, is

$$d\phi_b = B_p(\mu r)\,d\phi_u, $$

---

[6]For tabulations of $E_n(x)$, see references at the end of the chapter.

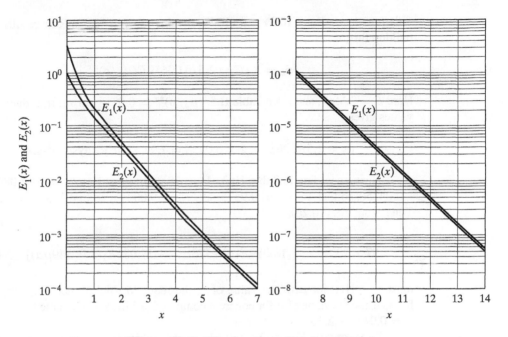

Figure 10.6   The functions $E_1(x)$ and $E_2(x)$.

so that from Eq. (10.18)

$$d\phi_b = \frac{SB_p(\mu r)ze^{-\mu r}\,dz}{2r^2}.$$

The total buildup flux is therefore

$$\phi_b = \frac{S}{2}\int_0^\infty \frac{B_p(\mu r)ze^{-\mu r}\,dz}{r^2}. \tag{10.25}$$

Introducing $B_p(\mu r)$ from Eq. (10.16) gives

$$\phi_b = \frac{S}{2}\sum A_n \int_0^\infty \frac{ze^{-(1+\alpha n)\mu r}\,dz}{r^2}.$$

The integral can be transformed, as in the uncollided case, and written in terms of the E_1 function. The final result is

$$\phi_b = \frac{S}{2}\sum A_n E_1[(1+\alpha_n)\mu a]. \tag{10.26}$$

Example 10.3

An infinite isotropic planar source emits 10^9 γ-rays/cm^2-sec with an energy of 2 MeV. What thickness of concrete is required to reduce the exposure rate to 2.5 mR/hr? The concrete density is 2.35 g/cm^3.

Solution. From Eqs. (10.3) and (10.7), a value of $\dot{X} = 2.5$ mR/hr results from a buildup flux of

$$\phi_b = \dot{X}/C = 2.5/0.00659 E (\mu_a/\rho)^{\text{air}}.$$

Here $E = 2$ MeV, and from Table II.5, $(\mu_a/\rho)^{\text{air}} = 0.0238$ cm^2/g. Introducing these values gives

$$\phi_b = 2.5/0.0659 \times 2 \times 0.0238 = 7.97 \times 10^2 \gamma\text{-rays/cm}^2\text{-sec}.$$

Next, according to Table 10.3, $A_1 = 18.089$; $A_2 = 1 - A_1 = -17.089$; $\alpha_1 = -0.04250$, so that $1 + \alpha_1 = 0.9575$; and $\alpha_2 = 0.00849$, so that $1 + \alpha_2 = 1.00849$. Thus, from Eq. (10.26),

$$7.97 \times 10^2 = \frac{10^9}{2}[18.089 E_1(0.9575\mu a) - 17.089 E_1(1.00849\mu a)] \quad (10.27)$$

must be solved for μa. The value of a, the required shield thickness, can then be found using the value of μ for concrete computed at 2 MeV from Table II.4—namely, $\mu = 0.0445 \times 2.35 = 0.1046$ cm^{-1}.

Equation (10.27) can be solved either graphically or numerically. If the former method is adopted, the computations can be speeded up by first simplifying the equation—namely,

$$1 = 1.13 \times 10^7[E_1(0.9575\mu a) - 0.94 E_1(1.00849\mu a)] \quad (10.28)$$

and then plotting the RHS of the equation as a function of μa on a semilog graph as in Example 10.2. As indicated in Fig. 10.7, the resulting curve is almost linear, at least over short distances. According to the figure, RHS = 1 for $\mu a = 13.6$, and so

$$a = 13.6/0.1046 = 130 \text{ cm or about } 4\tfrac{1}{4} \text{ ft. } [Ans.]$$

In the foregoing calculations, the flux was evaluated at a point on the shield surface. Suppose, however, that the point P is actually located at the distance s_1 from the shield and that the shield, in turn, is some distance x_2 from the source as shown in Fig. 10.8. Then the uncollided flux from a ring of width dz at z is

$$d\phi_u = \frac{2\pi S z \, dz e^{-\mu r_s}}{4\pi r^2}, \quad (10.29)$$

where r_s is that portion of r located within the shield, and so ϕ_u is

$$\phi_u = \frac{S}{2} \int_0^\infty \frac{z e^{-\mu r_s} \, dz}{r^2}.$$

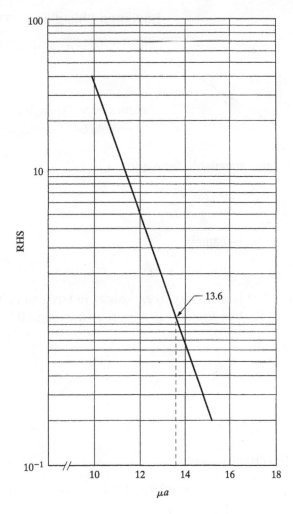

Figure 10.7 RHS of Eq. (10.28) as a function of μa.

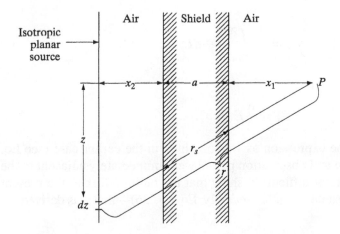

Figure 10.8 Isotropic planar source at some distance from slab shield.

Figure 10.9 Triangles for evaluating r_s.

Transforming the integration variable from z to r gives

$$\phi_u = \frac{S}{2} \int_y^\infty \frac{e^{-\mu r_s}}{r} \, dr, \tag{10.30}$$

where y is the smallest value of r—namely,

$$y = x_1 + a + x_2.$$

To evaluate this integral, it is necessary to express r_s in terms of r. This can easily be done by comparing the two similar triangles shown in Fig. 10.9. Thus, it is seen that

$$\frac{r_s}{r} = \frac{a}{y},$$

and so

$$r_s = \frac{a}{y} r.$$

Introducing this result in Eq. (10.30) gives

$$\phi_u = \frac{S}{2} \int_y^\infty \frac{e^{-\mu a r/y}}{r} \, dr,$$

which is equivalent to

$$\phi_u = \frac{S}{2} \int_{\mu a}^\infty \frac{e^{-t}}{t} \, dt.$$

Thus, it follows that

$$\phi_u = \frac{S}{2} E_1(\mu a),$$

which is exactly the same expression as was obtained in the earlier case (see Eq. [10.23]), when the source and observation points were immediately adjacent to the shield. Furthermore, it is not difficult to show that the buildup flux in the present case is also given by the same formula—namely, Eq. (10.26)—that was derived for the previous problem.

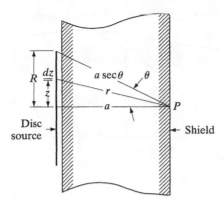

Figure 10.10   Isotropic disc source behind slab shield.

Consider next a disc source of radius R like that discussed in Section 9.11, emitting S photons/cm^2-sec, but now shielded by a slab of thickness a as shown in Fig. 10.10. The uncollided and buildup fluxes at the point P on the axis of the disc can be computed as in the infinite planar case by integrating the contributions from differential rings. Thus, Φ_u is given by Eq. (10.19), but with an upper integration limit of $z = R$ rather than infinity—that is,

$$\Phi_u = \frac{S}{2}\int_0^R \frac{ze^{-\mu r}\,dz}{r^2}.$$

Changing the integration variable from z to r gives

$$\Phi_u = \frac{S}{2}\int_a^{a\sec\theta} \frac{e^{-\mu r}}{r}\,dr, \tag{10.31}$$

where $a\sec\theta$[7] is the largest value of r as indicated in the figure. The integral can again be expressed in terms of the E_1 function:

$$\Phi_u = \frac{S}{2}[E_1(\mu a) - E_1(\mu a \sec\theta)]. \tag{10.32}$$

In a similar way, the buildup flux is easily found to be

$$\Phi_b = \frac{S}{2}\sum A_n\{E_1[(1+\alpha_n)\mu a] - E_1[(1+\alpha_n)\mu a\sec\theta]\}. \tag{10.33}$$

Again, as in the case of the infinite source, these expressions for Φ_u and Φ_b are also valid when the source, shield, and observation point are separated by regions of air.

---

[7]$\sec\theta = 1/\cos\theta$.

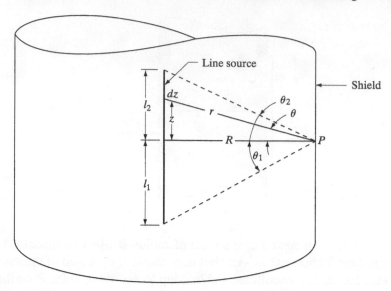

Figure 10.11   Isotropic line source imbedded in a cylindrical shield.

10.3 THE LINE SOURCE

Figure 10.11 shows a line source of length l surrounded by a cylindrical shield of radius R. If there are S γ-rays emitted isotropically per unit length of the source, then from sources in dz the uncollided flux at P, is

$$d\Phi_u = \frac{Se^{-\mu r}\,dz}{4\pi r^2},$$

and Φ_u is

$$\Phi_u = \frac{S}{4\pi}\int_{-l_1}^{l_2}\frac{e^{-\mu r}}{r^2}\,dz, \tag{10.34}$$

where l_1 and l_2 are the lower and upper portions of l, respectively.

It is convenient to change the integration variable in Eq. (10.34) from z to the angle θ. Then since

$$r = R\sec\theta,$$
$$z = R\tan\theta,$$
$$dz = R\sec^2\theta\,d\theta,$$

the uncollided flux becomes

$$\Phi_u = \frac{S}{4\pi R}\int_{-\theta_1}^{\theta_2}e^{-\mu R\sec\theta}\,d\theta, \tag{10.35}$$

The integral in Eq. (10.35) cannot be evaluated analytically, but it can be expressed in terms of the *Sievert integral function*. This is a function of two variables defined as

$$F(\theta, x) = \int_0^\theta e^{-x \sec \theta} \, d\theta, \tag{10.36}$$

where θ is restricted to values less than $\pi/2$. Figure 10.12 shows $F(\theta, x)$ as a function of x for several values of θ.[8] It is observed that $F(\theta, x)$ decreases more or less exponentially with x and increases with θ. For large values of θ (near $\pi/2$) and x, $F(\theta, x)$ can be computed from the formula

$$F(\theta, x) \simeq \sqrt{\frac{\pi}{2x}} e^{-x} \left(1 - \frac{5}{8x} \right). \tag{10.37}$$

Finally, it should be noted that, since $\sec(-\theta) = \sec \theta$, the integrand in Eq. (10.36) is an even function of θ; since the integral of an even function is always an odd function, it follows that

$$F(-\theta, x) = -F(\theta, x). \tag{10.38}$$

In view of these properties of $F(\theta, x)$, Eq. (10.35) for the uncollided flux at P can be written as

$$\phi_u = \frac{S}{4\pi R} [F(\theta_1, \mu R) + F(\theta_2, \mu R)]. \tag{10.39}$$

The calculation of the buildup flux is essentially the same in this problem as in the slab and disc problems discussed earlier. Thus, ϕ_b is given by

$$\phi_b = \frac{S}{4\pi R} \sum A_n \int_{-\theta_1}^{\theta_2} e^{-(1+\alpha_n)\mu R \sec \theta} \, d\theta$$

$$= \frac{S}{4\pi R} \sum A_n \{ F[\theta_1, (1+\alpha_n)\mu R] + F[\theta_2, (1+\alpha_n)\mu R] \}. \tag{10.40}$$

Numerical calculations utilizing Eq. (10.40) can be carried out in much the same way as illustrated in Example 10.3 for the slab shield. In particular, if the shield thickness must be determined, then it is usually helpful to plot the log of the RHS of Eq. (10.40). The resulting curve is then virtually a straight line.

If the observation point P is located beyond the surface of the shield at the distance R from the source, and the shield radius is a where $a < R$, then it is not difficult to show that

$$\Phi_u = \frac{S}{4\pi R} [F(\theta_1, \mu a) + F(\theta_2, \mu a)]. \tag{10.41}$$

---

[8]Consult the references for tabulations of this function.

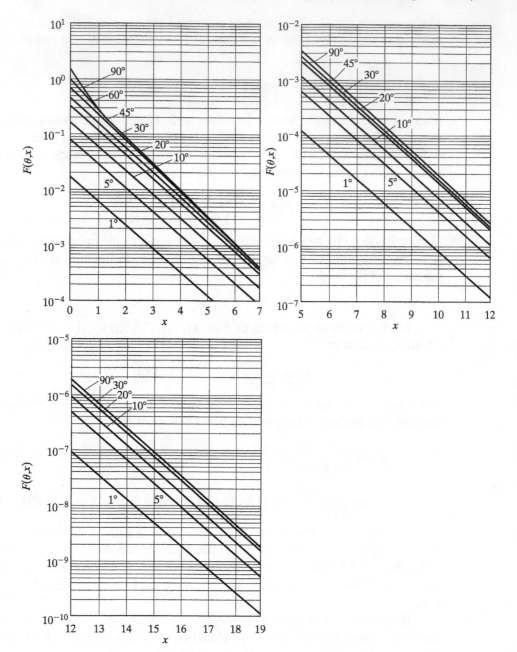

Figure 10.12 The function of $F(\theta, x)$.

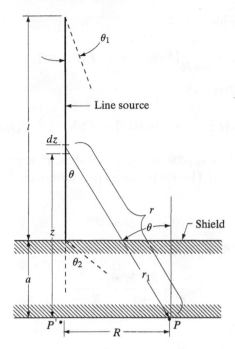

Figure 10.13   Shield at end-of-line source.

The buildup flux is then

$$\Phi_b = \frac{S}{4\pi R} \sum A_n \{ F[\theta_1, (1+\alpha_n)\mu a] + F[\theta_2, (1+\alpha_n)\mu a] \}. \qquad (10.42)$$

Suppose the shield is located at the bottom of the line source as in Fig. 10.13. Then the uncollided flux at P is

$$\Phi_u = \frac{S}{4\pi} \int_a^{l+a} \frac{e^{-\mu r_s}}{r^2} dz. \qquad (10.43)$$

To change the integration variable from z again to θ, it is observed that

$$r = R \csc \theta,$$

$$r_2 = a \sec \theta,$$

$$z = R \cot \theta,$$

$$dz = -R \csc^2 \theta \, d\theta.$$

Introducing these identities into Eq. (10.43) gives

$$\Phi_u = \frac{S}{4\pi R} \int_{\theta_1}^{\theta_2} e^{-\mu a \sec \theta} d\theta,$$

which when evaluated yields

$$\Phi_u = \frac{S}{4\pi R}[F(\theta_2, \mu a) - F(\theta_1, \mu a)]. \tag{10.44}$$

The buildup flux is then given by

$$\Phi_b = \frac{S}{4\pi R}\sum A_n\{F[\theta_2, (1+\alpha_n)\mu a] - F[\theta_1, (1+\alpha_n)\mu a]\}. \tag{10.45}$$

In the limit in which R approaches zero—that is, P approaches P'—θ_1 and θ_2 also approach zero, and Eq. (10.45) becomes indeterminate. However, according to Problem 10.9,

$$F(\theta, \mu a) \simeq \theta e^{-\mu a}.$$

Thus, for small values of θ, Φ_u behaves as

$$\Phi_u \xrightarrow[R\to 0]{} \frac{Se^{-\mu a}}{4\pi}\frac{\theta_2 - \theta_1}{R}.$$

From Fig. 10.13,

$$\frac{R}{a} = \tan\theta_2 \simeq \theta_2$$

and

$$\frac{R}{l+a} = \tan\theta_1 \simeq \theta_1;$$

thus it follows that

$$\Phi_u \xrightarrow[R\to 0]{} \frac{Sle^{-\mu a}}{4\pi a(l+a)}. \tag{10.46}$$

In a similar way, Φ_b goes to

$$\Phi_b \xrightarrow[R\to 0]{} \frac{SlB_p(\mu a)e^{-\mu a}}{4\pi a(l+a)}, \tag{10.47}$$

where $B_p(\mu a)$ is the point buildup factor.

The various formulas derived in this section can be used to perform preliminary designs for cylindrical fuel casks. These are heavily shielded containers that are used for carrying or shipping radioactive fuel assemblies. Equation (10.40) determines the thickness of the cylindrical sides. The thickness of the ends can be estimated from Eq. (10.47), since this gives the largest value of Φ_b along the top or bottom of the cask. Today, such calculations are typically performed using the Monte Carlo methods discussed in Section 10.10.

10.4 INTERNAL SOURCES

Circumstances often arise in which γ-ray sources are distributed in the interior of attenuating media. This occurs, for example, in shields for reactors. Energetic fission neutrons inelastically scattered within the shield give rise to sources of inelastic γ-rays, while thermalized neutrons undergo radioactive capture and produce capture γ-rays. Reactor shields must include provisions for the shielding against such secondary γ-rays.

Consider a slab of thickness a containing sources of γ-rays emitting $S(x)$ photons/cm^3-sec at the distance x as shown in Fig. 10.14. It is required to find the exposure rate at P. The uncollided flux at P from the planar source of thickness dx at x is given by Eq. (10.23)—specifically

$$d\phi_u = \frac{S(x)}{2} E_1[\mu(a - x)]\, dx,$$

and the total uncollided flux is then

$$\phi_u = \frac{1}{2} \int_0^a S(x) E_1[\mu(a - x)]\, dx. \tag{10.48}$$

As a special case, suppose that $S(x) = S$ is a constant. Then Eq. (10.48) becomes

$$\phi_u = \frac{S}{2} \int_0^a E_1[\mu(a - x)]\, dx. \tag{10.49}$$

To evaluate this integral, let $t = \mu(a - x)$ so that $dx = -dt/\mu$. Equation (10.49) then becomes

$$\phi_u = \frac{S}{2\mu} \int_0^{\mu a} E_1(t)\, dt. \tag{10.50}$$

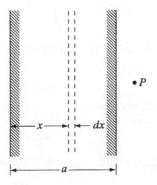

Figure 10.14   Slab containing γ-ray sources.

In view of the identity[9]

$$\int_0^x E_1(t)\, dt = 1 - E_2(x),$$

Eq. (10.50) reduces to

$$\phi_u = \frac{S}{2\mu}[1 - E_2(\mu a)]. \tag{10.51}$$

The buildup flux, computed in the usual way, is

$$\phi_b = \frac{S}{2\mu} \sum A_n\{1 - E_2[(1+\alpha_n)\mu a]/(1+\alpha_n)\}. \tag{10.52}$$

A source distribution that more realistically reproduces the γ-ray sources in a shield, where the neutron flux decreases approximately exponentially, is the function

$$S(x) = Se^{-\kappa x}, \tag{10.53}$$

where κ is a constant. The evaluation of Eq. (10.48) with this source distribution is rather complicated and is not given here. The resulting formula for ϕ_u is

$$\phi_u = \frac{Se^{-\kappa a}}{2\kappa}\left\{ e^{\kappa a} E_1(\mu a) - E_1[(\mu-\kappa)a] + \ln\left(\frac{\mu}{\mu-\kappa}\right)\right\}, \tag{10.54}$$

provided μ is greater than κ. If μ is less than κ, then ϕ_u is given by

$$\phi_u = \frac{Se^{-\kappa a}}{2\kappa}\left\{ e^{\kappa a} E_1(\mu a) + E_i[(\kappa-\mu)a] + \ln\left(\frac{\mu}{\kappa-\mu}\right)\right\}, \tag{10.55}$$

where $Ei(x)$ is the exponential integral function[10]

$$Ei(x) = -\int_{-x}^{\infty} \frac{e^{-t}}{t}\, dt. \tag{10.56}$$

The buildup flux outside the slab can be found in the usual way by replacing μ by $(1+\alpha_n)\mu$ and summing over n.

[9]This identity follows from the relation

$$\frac{dE_n}{dx} = -E_{n-1},$$

which is obtained from Eq. (10.22) by differentiation followed by integration by parts.
[10]See the references.

For an arbitrary source distribution $S(\mathbf{r}')$ in a volume of arbitrary shape, the γ-ray flux at the point $\mathbf{r}$ on the surface can be found by evaluating the integral

$$\phi_u(\mathbf{r}) = \frac{1}{4\pi} \int \frac{S(\mathbf{r}')e^{-\mu|\mathbf{r}-\mathbf{r}'|}}{|\mathbf{r}-\mathbf{r}'|^2} \, dV' \tag{10.57}$$

over the source volume. This can be done either on a computer or by a hand calculation using the following straightforward method. First, the source volume is divided into convenient subvolumes V_i, which need not be the same size. The number of γ-rays S_i emitted from V_i is then computed from the source function. The contributions to the flux from each volume are then added, taking S_i to be a point source. The result for the buildup flux is

$$\phi_b = \frac{1}{4\pi} \sum \frac{S_i B_p(\mu r_i)e^{-\mu r_i}}{r_i^2}, \tag{10.58}$$

where r_i is the distance from some point in V_i to the observation point.

10.5 MULTILAYERED SHIELDS

The γ-ray shields discussed so far have all consisted of single regions of shielding materials. It is now necessary to consider shields composed of more than one layer of material. For example, suppose that a monodirectional beam of γ-rays of energy E_0 is incident on a shield having two layers of different materials as shown in Fig. 10.15. It is not difficult to see that the uncollided flux at the point P is

$$\phi_u = \phi_0 e^{-(\mu_1 a_1 + \mu_2 a_2)}, \tag{10.59}$$

where μ_1 and μ_2 are the total attenuation coefficients of the first and second layers and a_1 and a_2 are their respective thicknesses.

The calculation of the buildup flux at P poses a more difficult problem. This stems from the fact that buildup factors have been computed only for monoenergetic γ-rays incident on various materials. Although these buildup factors can be

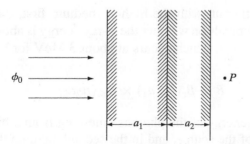

Figure 10.15 γ-rays incident on layered shield.

used to describe the radiation leaving the first layer in Fig. 10.15, they are not appropriate for the radiation leaving the second layer. The radiation incident on this layer has a continuous spectrum resulting from interactions in the first layer. Furthermore, the buildup flux at P can clearly be expected to depend on which of the two layers comes first. A monoenergetic flux incident on the first region, and producing a continuous spectrum incident on the second, gives one value of ϕ_b at P. There is no reason to believe that if the media are interchanged, then ϕ_b will be the same. Indeed, the two values of ϕ_b are, in general, quite different.

Consider, for example, 0.5-MeV γ-rays incident on layers composed of lead and water. From Table 10.1, it is seen that the buildup factor of water is much larger than that of lead for the same thickness. It follows that there is a greater buildup of low-energy scattered radiation in water than in lead. This is because, in water, the low-energy radiation propagates with little photoelectric absorption. Owing to its high value of Z, photoelectric absorption is important in lead at low energies and a buildup of low-energy radiation is not possible. If the water is placed before the lead, the buildup radiation from the water is absorbed as it passes through the lead, and the overall buildup through the shield is small. However, if the lead is placed first, the subsequent buildup of radiation in the water will reach the observer and the overall buildup is larger.

There is no simple way to obtain the exact value of the buildup factor for a layered shield in terms of the buildup factors of each layer. Approximate values can be obtained using the following recipes, which are based on qualitative arguments. In every instance, the method can be expected to give reasonable answers.

1. If the media are alike—that is, their Zs do not differ by more than 5 to 10—then use the buildup factor of the medium for which this factor is the larger and compute the overall buildup factor as $B(\mu_1, a_1 + \mu_2 a_2)$. This rule is based on the observation that, except at low energy, the buildup factors do not vary rapidly with the Z of the medium.

2. If the media are different, with the low-Z medium first, then use the buildup factor of the second medium as if the first medium is not there—that is, use $B_{z_2}(\mu_2 a_2)$. This is because the low-energy buildup radiation from region 1 is absorbed in region 2.

3. If the media are different, with the high-Z medium first, the procedure to follow in this case depends on whether the γ-ray energy is above or below the minimum in the μ-curve, which occurs at about 3 MeV for heavy elements. If $E < 3$ MeV, then

$$B = B_{z_1}(\mu_1 a_1) \times B_{z_2}(\mu_2 a_2). \qquad (10.60)$$

This is because the energy of the photons emerging from a high-Z shield is little different from that of the source, and in the second medium the photons can be treated as if they were source γ-rays. However, if $E > 3$ MeV, then

$$B = B_{z1}(\mu_1 a_1) \times B_{z2}(\mu_2 a_2)_{\min}, \qquad (10.61)$$

where $B_{z2}(\mu_2 a_2)$ is the value of B_{z2} at 3 MeV. Equation (10.61) is based on the fact that the γ-rays penetrating the first layer have energies clustered about the minimum in μ, so that their penetration through the second layer is determined by this energy rather than by the actual source energy.

It must be emphasized again that the previous recipes give only approximate results for the attenuation of γ-rays in layered shields. They are to be used in the absence of exact calculations.

Example 10.4

A monodirectional beam of 6-MeV γ-rays of intensity 10^2 γ-rays/cm^2-sec is to be shielded with 100 cm of water and 8 cm of lead. Compute the exposure rate if (a) the water is placed before the lead, (b) the water follows the lead.

Solution. In both cases, the buildup flux is given by

$$\phi_b = 10^6 B_m e^{-(\mu_w x_w + \mu_{Pb} x_{Pb})},$$

where B_m is the appropriate monodirectional buildup factor—$x_w = 100$ cm and $x_{Pb} = 8$ cm. From Table I.4, $\mu_w = 0.0275$ and $\mu_{Pb} = 0.4944$, so that $\mu_w x_w = 2.75$ and $\mu_{Pb} x_{Pb} = 3.96$. For case (a), the value of B_m is that of lead alone

$$B_m = B_{Pb}(3.96) = 1.86,$$

where use has been made of Table 10.1. Then Φ_b is

$$\Phi_b = 10^6 \times 1.86 e^{-(2.75 + 3.96)}$$

$$= 2.27 \times 10^3 \gamma\text{-rays/cm}^2\text{-sec}.$$

According to Fig. 9.9, a flux of 150 6-MeV γ-rays/cm^2-sec gives an exposure rate of 1 mR/hr. Thus, in case (a),

$$\dot{X} = \frac{2.27 \times 10^3}{150} \times 1 = 15.1 \text{ mR/hr. } [Ans.].$$

In case (b), B_m is given by Eq. (10.61):

$$B_m = B_{Pb}(3.96) \times B_W(2.75),$$

where $B_W(2.75)$ is evaluated at about 3.2 MeV—the minimum in the μ_{Pb} curve. Thus, from Table 10.1,

$$B_W(2.75) = 2.72$$

and $B_{Pb}(3.96)$ is again 1.86, so that

$$B_m = 1.86 \times 2.72 = 5.06.$$

The buildup flux is then

$$\phi_b = 10^6 \times 5.06e^{-(2.75+3.96)}$$

$$= 6.17 \times 10^3 \gamma\text{-rays/cm}^2\text{-sec.}$$

The exposure rate in this case is then

$$\dot{X} = \frac{6.17 \times 10^3}{150} \times 1 = 41.1 \text{ mR/hr. } [Ans.]$$

The exposure rate is thus smaller when the γ-rays traverse the water before the lead.

10.6 NUCLEAR REACTOR SHIELDING: PRINCIPLES OF REACTOR SHIELDING

This part of the chapter concerns the design of the biological shields that must be placed around a nuclear reactor and at various points in a nuclear power plant to protect operating personnel and the public at large from the radiations emanating from the installation. The following sources of radiation are to be considered.

Prompt Fission Neutrons   Emitted from the core, these constitute the most important radiation to be considered in designing a biological shield.

Delayed Fission Neutrons   Ordinarily these are not an important consideration because, on the one hand, there are so few of them and, on the other, their energies are so low (on the order of 400 keV). Delayed neutrons can lead to special shielding problems in circulating fuel reactors like the molten salt breeder reactor (see Section 4.5). In that system, the delayed neutron precursors are carried outside of the core with the moving fuel, and some of the delayed neutrons are emitted there.

Prompt Fission γ-Rays   Emitted in the core, these γ-rays are largely attenuated by the materials in the core.

Fission Product Decay γ-Rays   Emitted from the fuel, these γ-rays are a continuing source of radiation after the shutdown of a reactor.

Inelastic γ-Rays   These are γ-rays emitted by nuclei that have been left in excited states as the result of inelastic neutron scattering. Since inelastic scattering occurs only with energetic neutrons, the inelastic γ-rays tend to be emitted from the reactor core and from inner portions of the shield where the neutrons are most energetic.

Capture γ-Rays   These are emitted each time a neutron is absorbed by a nucleus in a radiative capture reaction. This reaction is most probable at resonance and thermal energies. Since thermal neutrons are usually present in a reactor shield, the shield may be a source of capture γ-rays.

Activation ***γ*-Rays** These γ-rays are emitted by radioactive nuclides formed as the result of neutron absorption. Much of the internal structure of a reactor becomes radioactive in this way, and so does the coolant and any extraneous atoms contained in the coolant. Such radionuclides, contained in and/or picked up and carried along by the coolant, may be deposited along the piping and in equipment located in the primary coolant loop. As a result, these parts become sources of activation γ-rays that require shielding.

In practice, it is found that, among all of these radiation sources, the prompt fission neutrons are the most difficult to shield against. Unfortunately, it is not possible to absorb fast neutrons. Absorption cross-sections are simply too small at high energies. What is done, therefore, is first to slow the fast neutrons down to thermal energies and then to absorb the thermal neutrons. In this connection, it is recalled from Section 3.5 that, on average, neutrons lose 50% of their energy in an elastic collision with hydrogen—more than with any other nucleus. For this reason, hydrogen is one of the principal components of every reactor shield. Water, with its high hydrogen content, and other materials containing water are often employed to shield reactor installations. Most of the shielding in stationary nuclear power plants is ordinary concrete, which contains approximately 10 w/o water and has a hydrogen atom density equal to about one quarter that of water. Concrete is inexpensive, structurally sound, and readily formed into any desired shape.

Shields designed to attenuate very fast neutrons, such as those emitted from accelerator sources or those emitted from fusion reactions (see Example 2.8 and Section 2.11), often contain moderately heavy or heavy materials to slow down the neutrons by inelastic scattering. This process is surprisingly effective. Thus, it can be shown that the average energy $\bar{E}'$ of a neutron emerging from an inelastic collision with a nucleus of mass number A is given approximately by the formula

$$\bar{E}' = 6.4\sqrt{\frac{E}{A}}, \tag{10.62}$$

where E is the energy of the incident neutron and both $\bar{E}'$ and E are in MeV. Suppose that a 14-MeV neutron is inelastically scattered by iron ($A = 56$). Then from Eq. (10.62), $\bar{E}' = 3$ MeV. This represents an average energy loss of 11 MeV, which is substantially more than the neutron would have lost in an elastic collision even with hydrogen.

Iron and other heavy materials have also been used extensively in reactor shields—either in solid sheets or distributed uniformly in concrete, but more for the purpose of attenuating γ-rays than for slowing down neutrons. The average energy of fission neutrons is evidently too low to have inelastic scattering play a significant role in slowing down neutrons in reactor shields. It can be shown that when a shield is divided into alternating regions of, say, iron and a hydrogenous material such as concrete or polyethylene, so that the attenuation of the γ-rays and neutrons proceeds at the same rate through successive layers, the overall size and weight of the shield is less than that of a concrete or polyethylene shield alone. This can be an important consideration for mobile nuclear power systems where space and weight are at a premium. With stationary power plants, space is not a crucial matter, and simple concrete shielding is cheaper to build.

Once the fast neutrons have been thermalized in a shield, they must then be absorbed. However, when thermal neutrons are captured in water, a 2.2-MeV γ-ray is emitted as the result of the $^1\text{H}(n,\gamma)^2\text{H}$ reaction. If they are captured by iron, however, a 7.6-MeV and a 9.3-MeV γ-ray are emitted. The exposure rates from these γ-rays outside the shield can be quite significant especially if the shield is not thick. To reduce the intensity of such γ-rays, boron—either natural or enriched— is sometimes added to reactor shields. Boron has a high thermal absorption cross-section (759 b), which is due mostly to the $^{10}\text{B}(n,\alpha)^7\text{Li}$ reaction. Although γ-rays do accompany this reaction, they have a relatively low energy of about 0.5 MeV.

10.7 REMOVAL CROSS-SECTIONS

Important experiments bearing on the attenuation of fission neutrons in water were carried out some years ago at the Oak Ridge National Laboratory and have since been repeated elsewhere. The experimental setup is shown in Fig. 10.16. Thermal neutrons from a reactor fall onto a disc of enriched uranium, about 28 in. in diameter, which is located at the end of a large tank of water. The thermal neutrons induce fissions in the uranium, and in this way the disc becomes a source of fission neutrons in water. Uranium sheets or discs used in this manner are called *fission*

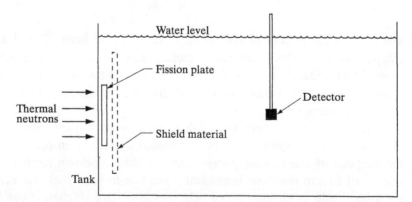

Figure 10.16 Experimental arrangement for measuring the attenuation of fission neutrons in water to compute removal cross-sections.

plates. The flux of fast neutrons—that is, neutrons with energies greater than about 1 MeV—is then measured in the water as a function of distance from the source.

From these measurements, it is possible to infer the fast flux in water from a point isotropic fission source. Thus, let $G(r)$ be the flux at the distance r from a point fission source emitting one fission neutron per sec isotropically. This function is known as the *point water kernel.* Using the methods given earlier in this chapter (see Section 10.2), it is easy to see that the flux at the distance x from the center of a fission plate of radius R emitting neutrons per cm²/sec is given by

$$\phi(x) = 2\pi S \int_0^R G(r)z \, dz, \tag{10.63}$$

where z is the distance from the center of the disc to a ring of width dz emitting $2\pi S z \, dz$ neutrons/sec. By inverting this integral, $G(r)$ can be found from measured values of $\phi(x)$.

The results of these measurements and calculations are shown in Fig. 10.17, where $4\pi r^2 G(r)$ is given as a function of distance from the source. It is observed that, beyond about 40 cm, the curve becomes almost a straight line so that $G(r)$ can be written as

$$G(r) = \frac{Ae^{-\Sigma_{RW}r}}{4\pi r^2}. \tag{10.64}$$

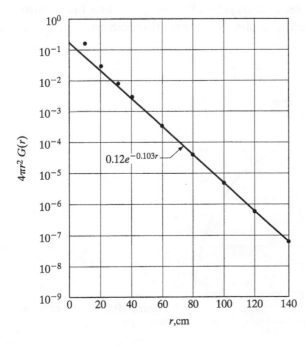

Figure 10.17 Measured values of $4\pi r^2 G(r)$ for point source emitting one fission neutron per second and the function $0.12 \exp(-0.103r)$.

The constants have the values $A = 0.12$ and $\Sigma_{RW} = 0.103$ cm^{-1}. This form of $G(r)$ suggests that, beyond about 40 cm, the neutrons are absorbed with an effective macroscopic absorption cross-section of Σ_{RW}. What is happening, however, is that the neutrons are being *removed* from the group of neutrons with energies greater than 1 MeV as the result of scattering in the water, and Σ_{RW} is called the *macroscopic removal cross-section* of water.

To understand these results physically, consider those fission neutrons emitted from a point source and headed in the direction of the detector. If one of these neutrons suffers a collision with hydrogen, its energy, on average, is reduced by one-half. Owing to the rapid increase in the cross-section of hydrogen with decreasing energy, as indicated in Fig. 10.18, the mean free path for its next collision is greatly shortened. As a consequence, the neutron has its second collision near the site of its first, whereon its energy and mean free path are again reduced, and so on. In this way, the energy of the neutron eventually falls below 1 MeV. However, because of the continually decreasing mean free path, this occurs not far from the location of the first collision. Thus, it follows that a single collision with hydrogen effectively removes a neutron from among the fast neutrons entering the detector.

Scattering from the oxygen in water is complicated by the fact that this nucleus tends to scatter neutrons preferentially through small angles. Neutrons scattered at such angles may continue to the detector. However, neutrons scattered at large angles by oxygen are lost because they subsequently suffer collisions with hydrogen and are thereby removed. Thus, a single collision with hydrogen and all but forward scattering collisions with oxygen effectively remove fast neutrons from water. The same phenomenon is observed with all media containing substantial

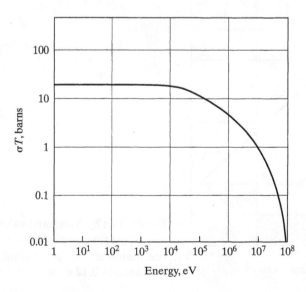

Figure 10.18   The cross-section of hydrogen.

amounts of hydrogen. In particular, Eq. (10.64) is also valid for concrete, although the value of the removal cross-section is sensitive to the amount of water present in this material.

To return to the Oak Ridge experiments, these were repeated with slabs of various thicknesses placed in front of the fission plate as shown in Fig. 10.16. When these data are related to an equivalent experiment with a point fission source surrounded by varying thicknesses of materials, it is found that the fast neutron flux can be expressed by the formula

$$\phi(r) = SG(r)e^{-\Sigma_R t}. \tag{10.65}$$

Here S is the source strength, $G(r)$ is the point water kernel, Σ_R is the *macroscopic removal cross-section* for the material in question, and t is its thickness. Measured values of Σ_R for several materials are given in Table 10.4.

The explanation of Eq. (10.65) is similar to the interpretation of the water kernel given earlier. Thus, it is only those fission neutrons emitted in the direction of the detector that have the best chance of reaching it. Any of the neutrons that interact in the material surrounding the source, except those that are scattered through small angles, are removed from the beam heading for the detector. Their second and succeeding collisions are with water, and they slow out of the fast flux in these collisions by the mechanism discussed earlier. The removal cross-section is thus equivalent to an absorption cross-section, at least as far as the fast neutrons are concerned. In this connection, it is important to recognize that the removal cross-

TABLE 10.4   REMOVAL CROSS-SECTIONS

Material	Macroscopic cross-section, cm^{-1}	Microscopic cross-section, b
Hydrogen		1.00
Deuterium		0.92
Beryllium	0.132	1.07
Boron		0.97
Carbon	0.065	0.81
Oxygen		0.92
Sodium	0.032	1.26
Iron	0.168	1.98
Zirconium	0.101	2.36
Lead	0.118	3.53
Uranium	0.174	3.6
Water	0.103	
Heavy Water	0.092	
Concrete*	0.089	

*Containing 6% water by weight.

section is defined for a substance only when it is followed by a medium containing sufficient hydrogen. It is of such a thickness to ensure that a neutron interacting in the material is, in fact, lost from the fast flux.

If the atom density of the shielding material is N, then it is possible to write

$$\Sigma_R = N\sigma_R, \tag{10.66}$$

where σ_R is the *microscopic removal cross-section*. Values of σ_R for several nuclei are given in Table 10.4. The macroscopic removal cross-section for a mixture of elements is obtained in the usual way from the formula

$$\Sigma_R = \sum_i N_i \sigma_{Ri}, \tag{10.67}$$

where N_i and σ_{Ri} are the atom density and microscopic removal cross-section for the ith species.

Example 10.5

A fission plate 28 in. in diameter in a tank of water operates at a fission density of 4×10^7 fissions/cm²-sec. (a) Calculate the fast flux at a point P, 75 cm from the center of the plate. (b) Repeat the calculation with a 3-in slab of iron placed in front of the plate.

Solution.

1. It is first necessary to derive an expression for the fast flux as a function of distance from the plate. This is given by Eq. (10.63):

$$\phi(x) = 2\pi S \int_0^R G(r) z\, dz.$$

From Fig. 10.19, it is seen that $x^2 + z^2 = r^2$, so that $z\, dz = r\, dr$ and $\phi(x)$ becomes

$$\phi(x) = 2\pi S \int_x^{x \sec \theta} G(r) r\, dr.$$

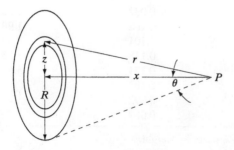

Figure 10.19 Geometry of fission plate calculation.

Next, introducing G(r) from Eq. (10.64) gives

$$\phi(x) = \frac{SA}{2} \int_x^{x\sec\theta} \frac{e^{-\Sigma_{RW}r}}{r}\, dr$$

$$= \frac{SA}{2}[E_1(\Sigma_{RW}x) - E_1(\Sigma_{RW}x \sec\theta)].$$

To evaluate $\phi(x)$, the values of S and $\sec\theta$ are required. Since $\nu = 2.42$ fission neutrons are emitted per fission, it follows that $S = 2.42 \times 4 \times 10^7 = 9.68 \times 10^7$ neutrons/cm^2-sec. Also, $\tan\theta = 14 \times 2.54/75 = 0.474$, so that $\theta = 0.443$ radians and $\sec\theta = 1.107$. Then $\Sigma_{RW}x = 0.103 \times 75 = 7.73$ and $\Sigma_{RW}x \sec\theta = 8.55$.

Thus, the fast flux is equal to

$$\phi(P) = \frac{9.68 \times 10^7 \times 0.12}{2}[E_1(7.73) - E_1(8.55)]$$

$$= 177 \text{ neutrons/cm}^2\text{-sec. } [Ans.]$$

2. With the iron slab inserted, $\phi(x)$ becomes

$$\phi(x) = 2\pi S \int_x^{x\sec\theta} e^{-\Sigma_R r_s} G(r) r\, dr,$$

where r_s is the slant distance along r in the shield. By similar triangles,

$$\frac{r_s}{t} = \frac{r}{x},$$

where t is the thickness of the slab. Inserting this expression for r_s, together with $G(r)$, gives

$$\phi(x) = \frac{SA}{2} \int_x^{x\sec\theta} \frac{e^{-(\Sigma_{RW}+\Sigma_R t/x)r}}{r}\, dr$$

$$= \frac{SA}{2}[E_1(\Sigma_{RW}x + \Sigma_R t) - E_1(\Sigma_{RW}x \sec\theta + \Sigma_R t \sec\theta)].$$

From Table 10.4, $\Sigma_R = 0.168$ cm^{-1} for iron, and so $\Sigma_R t = 0.168 \times 3 \times 2.54 = 1.28$. Finally,

$$\phi(P) = \frac{9.68 \times 10^7 \times 0.12}{2}[E_1(9.01) - E_1(9.97)]$$

$$= 46.6 \text{ neutrons/cm}^2\text{-sec. } [Ans.]$$

The foregoing problem demonstrates the efficacy of moderately heavy materials in reducing the flux of fast neutrons, assuming that the materials are followed by a sufficiently thick layer of hydrogenous material that the scattered neutrons are indeed lost from the fast flux. Shielding material placed near the outer surface of a water shield is not nearly as effective in attenuating fast neutrons.

10.8 REACTOR SHIELD DESIGN: REMOVAL–ATTENUATION CALCULATIONS

The preliminary design of a reactor shield can be carried out using the point kernel and the removal cross-sections discussed in the preceding section. Such point kernel *removal–attenuation* calculations are best performed on a computer, and methods for doing so are discucssed at the end of this section. However, it is shown first how the gross properties of a reactor shield can be calculated analytically using the simple exponential kernel given in Eq. (10.64).

Consider a spherical reactor consisting of a core of radius R surrounded by a hydrogenous shield of thickness a as shown in Fig. 10.20. The core consists of a mixture of metal and water, with the volume fraction of metal equal to f. It is assumed for simplicity that the fission neutrons are being produced uniformly throughout the coefficient at the rate of S neutrons/cm^3-sec.

The fast neutron flux at the point P at the surface of the shield can be computed by adding up the contributions to the flux from volume elements dV within the core. In view of the symmetry of the problem, the appropriate volume element is that of a ring (see Fig. 10.20) whose volume is

$$dV = 2\pi r^2 \sin\theta \, d\theta \, dr.$$

Since the removal cross-section is equivalent to an absorption cross-section, the flux at P from the element dV is given by

$$d\phi(P) = \frac{SA \, dV}{4\pi r^2} e^{-n(r)}, \tag{10.68}$$

where $n(r)$ is the total number of removal mean free paths from dV to P. In the core water, there are

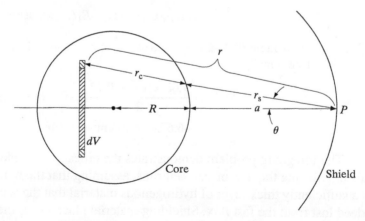

Figure 10.20   Spherical reactor core surrounded by a spherical shield.

$$(1 - f)\Sigma_{RW}r_c$$

removal mean free paths, where $(1 - f)\Sigma_{RW}$ is the macroscopic removal cross-section of the water at density $1 - f$ and r_c is the distance shown in Fig. 10.20. The metal in the core accounts for another

$$f\Sigma_{RM}r_c$$

mean free paths, where Σ_{RM} is the macroscopic removal cross-section of the metal at its normal density. Finally, the number of mean free paths in the shield is

$$\Sigma_{Rs}r_s,$$

where Σ_{Rs} is the removal cross-section of the shield. The total number of mean free paths from dV to P is therefore

$$(1 - f)\Sigma_{RW}r_c + f\Sigma_{Rm}r_c + \Sigma_{Rs}r_s.$$

Since

$$r_c = r - r_s,$$

this number can also be written as

$$[(1 - f)\Sigma_{RW} + f\Sigma_{Rm}]r + [\Sigma_{Rs} - (1 - f)\Sigma_{RW} - f\Sigma_{Rm}]r_s$$

or

$$\alpha r + \beta r_s,$$

where

$$\alpha = (1 - f)\Sigma_{Rw} + f\Sigma_{Rm} \qquad (10.69)$$

and

$$\beta = \Sigma_{Rs} - (1 - f)\Sigma_{RW} - f\Sigma_{Rm}. \qquad (10.70)$$

The flux at P from dV is then

$$d\phi(P) = \frac{SA\,dV}{4\pi r^2}e^{-(\alpha r + \beta r_s)}. \qquad (10.71)$$

Introducing dV and integrating over the core gives

$$\phi(P) = \frac{SA}{2}\int_0^{\theta_{max}}\sin\theta\,d\theta\int_{r_s}^{r_{max}}e^{-(\alpha r + \beta r_s)}\,dr, \qquad (10.72)$$

where θ_{max} and r_{max} are the maximum values of θ_r, respectively. From Fig. 10.21,

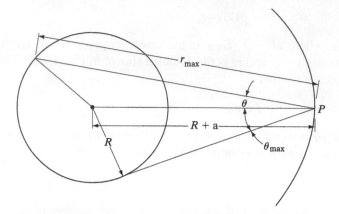

Figure 10.21   Diagram for computing r_{max}.

$$\theta_{max} = \sin^{-1}\left(\frac{R}{R+a}\right)$$

and

$$r_{max} = (R+a)\cos\theta + \sqrt{R^2 - (R+a)^2 \sin^2\theta}.$$

Although r_s is a function of θ, it is independent of r and is therefore held constant during the r-integration in Eq. (10.72). Carrying out this integration yields

$$\phi(P) = \frac{SA}{2\alpha} \int_0^{\theta_{max}} (e^{-\alpha r_s} - e^{-\alpha r_{max}}) e^{-\beta r_s} \sin\theta \, d\theta. \qquad (10.73)$$

The remaining integral cannot be performed analytically and must be calculated numerically. However, in the case of a small core and a thick shield, the integral can be approximated by taking

$$\sin\theta \simeq \theta$$

and

$$\theta_{max} \simeq \frac{R}{R+a}.$$

Furthermore,

$$r_s \simeq a$$

so that

$$r_{max} \simeq 2R + a.$$

Equation (10.73) then reduces to

$$\phi(P) = \frac{SA}{2\alpha} e^{-(\alpha+\beta)a} (1 - e^{-2\alpha R}) \int_0^{\theta_{max}} \theta \, d\theta.$$

Performing the integration and noting from Eqs. (10.69) and (10.70) that

$$\alpha + \beta = \Sigma_{R_s}$$

yields the final result:

$$\phi(P) = \frac{SA}{4\alpha} \left(\frac{R}{R+a}\right)^2 e^{-\Sigma_{Rs}a} (1 - e^{-2\alpha R}). \qquad (10.74)$$

If there is a layer of metal—say iron—of thickness t between the reflector and shield, where $t \ll a$, then $\phi(P)$ is reduced by the factor $e^{-\Sigma_{RFe}t}$. That is,

$$\phi(P)_{\text{withiron}} = e^{-\Sigma_{RFe}t} \phi(P)_{\text{noiron}}. \qquad (10.75)$$

Example 10.6

The core of a certain research reactor, operating at a power of 250 kilowatts, consists of an assembly of uranium rods and water, with a metal volume fraction of 0.75. The core has a volume of 32 liters and is surrounded by a water reflector 15 cm thick. It is proposed to shield this reactor by 150 cm of unit-density water. Estimate the fast neutron dose rate at the surface of this shield.

Solution. With a recoverable energy per fission of 200 meV $= 3.2 \times 10^{-11}$ joule, there are

$$250{,}000 \text{ joules/sec} \div 3.2 \times 10^{-11} \text{ joules/fission} = 7.81 \times 10^{15} \text{ fissions/sec}$$

in the core. This gives a fission density of $7.81 \times 10^{15}/32 \times 10^3 = 2.44 \times 10^{11}$ fissions/cm^3-sec. Assuming that the fissions occur in ^{235}U, then $\nu = 2.42$ and

$$S = 2.42 \times 2.44 \times 10^{11} = 5.90 \times 10^{11} \text{ neutrons/cm}^3\text{-sec}.$$

The radius of the core, which is taken to be spherical, is

$$R = \sqrt[3]{\frac{3 \times 32{,}000}{4\pi}} = 19.7 \simeq 20 \text{ cm}.$$

Since the reflector and shield are both water, the effective thickness of the shield is $a = 165$ cm.

From Table 10.4, the macroscopic removal cross-section of uranium is 0.174 cm^{-1}. Then from Eq. (10.69),

$$\alpha = 0.25 \times 0.103 + 0.75 \times 0.174 = 0.156 \text{ cm}^{-1},$$

$$2\alpha R = 2 \times 0.156 \times 20 = 6.24,$$

and

$$\Sigma_{Rs} a = 0.103 \times 165 = 17.0.$$

Introducing these values into Eq. (10.74) gives

$$\theta = \frac{5.90 \times 10^{11} \times 0.12}{4 \times 0.156} \left(\frac{20}{20 + 165} \right)^2 e^{-17.0} (1 - 3^{-6.24})$$

$$= 54.8 \text{ neutrons/cm}^2\text{-sec.}$$

From Fig. 9.12, a flux of 6.8 fast neutrons/cm²-sec gives a dose equivalent rate of 1 mrem/hr. The dose rate in the present example is therefore

$$\dot{H} = \frac{54.8}{6.8} \times 1 = 8.1 \text{mrem/hr. [}Ans.\text{]}$$

The reactor would need additional shielding, at least as far as the neutrons are concerned, if access to the region near the surface of the shield was permitted.

10.9 THE REMOVAL–DIFFUSION METHOD

The removal–attenuation method discussed in the last section takes into account the most energetic neutrons. The complete design of a shield usually also requires a knowledge of the spatial distributions of neutrons of all energies, including thermal neutrons, which detemine the sources of inelastic and capture γ-rays. These distributions are provided by the *removal–diffusion method*,[11] which is a combination of a removal–attenuation calculation and a multigroup calculation of the type given in Section 6.7.

Several versions of the removal–diffusion method have been developed over the years. In most of these, the neutrons in a shield are divided into three broad classes. The first class consists of the fast neutrons—those with energies exceeding 6 MeV. It is these neutrons that penetrate deepest into the shield. Lumped together, these are called the *removal group neutrons*. The neutrons in the second class are those that have been knocked out of the removal group by collisions in the shield, together with the neutrons that have leaked from the core. Neutrons in this class undergo diffusion and slowing down as the result of successive collisions, just like the neutrons in the reactor. These are called *intermediate neutrons*. The third class is composed of those neutrons that have thermalized and diffuse about as thermal neutrons.

In a removal–diffusion calculation, the removal group neutrons are divided into N groups and the intermediate neutrons are divided into M groups; there is

---

[11] This is also called the *Spinney method*.

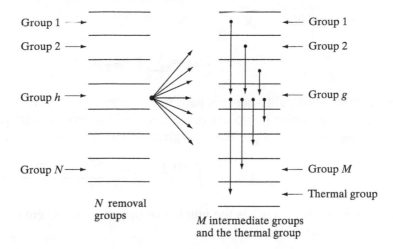

Figure 10.22   Energy groups for removal–diffusion calculations.

also the thermal group, as illustrated in Fig. 10.22. Consider now an arbitrary gth intermediate group. As in Section 6.7, neutrons are lost from this group as the result of absorption or scattering to a lower group while neutrons enter the group from collisions occurring in the higher intermediate groups and in the removal groups. Let $S_{r \to g}$ be the number of neutrons entering the gth group from the hth removal group. Then using the multigroup notation of Section 6.7, the diffusion equation for the gth group is

$$D_g \nabla^2 \phi_g - \Sigma_{ag} \phi_g - \sum_{g' > g} \Sigma_{g \to g'} \phi_g + \sum_{g' < g} \Sigma_{g' \to g} \phi_g + \sum_{h=1} N S_{h \to g} = 0. \quad (10.76)$$

The second term in this equation gives the true absorption of neutrons in the gth group; the third term gives the total number of neutrons scattered out of this group; the fourth gives the neutrons scattered into the gth group as the result of collisions in intermediate groups at higher energy; and the last term is the source of neutrons scattered into the gth group from collisions of the primary neutrons in all of the removal groups. The equation for the thermal neutrons is

$$\bar{D} \nabla^2 \phi_T - \bar{\Sigma}_a \phi_T + \sum \Sigma_{g \to T} \phi_g = 0. \quad (10.77)$$

The last term in this equation is the number of neutrons entering the thermal group per cm³/sec.

The quantities $S_{h \to g}$ are computed in the following way. First, removal group transfer cross-sections, $\Sigma_{R, h \to g}$, are defined. These are equal to the probability-per-unit-path-length that a neutron in the hth removal group will have a collision in which its energy drops into the gth intermediate group. The total removal cross-section for the hth group,

$$\Sigma_{Rh} = \sum_{g=1}^{N} \Sigma_{R,h \to g},$$

is then the probability-per-unit-path that a neutron in the hth group will have a collision that carries it into any of the intermediate groups. With these definitions, $S_{h \to g}$, as a function of position $\mathbf{r}$, is given by the integral

$$S_{h \to g}(\mathbf{r}) = \frac{1}{4\pi} \int \frac{s(\mathbf{r}') \Sigma_{R,h \to g} e^{-(\Sigma_{Rh}|\mathbf{r}-\mathbf{r}'|)_{\text{total}}} dV'}{|\mathbf{r} - \mathbf{r}'|^2}. \tag{10.78}$$

In this expression, $s(\mathbf{r}')$ is the number of fission neutrons emitted per cm^3/sec at the point $\mathbf{r}'$ in the core, and

$$(\Sigma_{Rh}|\mathbf{r} - \mathbf{r}'|)_{\text{total}}$$

is equal to the total number of mean free paths from the point $\mathbf{r}'$ to the point $\mathbf{r}$ for neutrons in the hth group. Equation (10.78) merely states that $S_{h \to g}$ is equal to the uncollided flux of neutrons in the hth group at the point $\mathbf{r}$ multiplied by the cross-section $\Sigma_{R,h \to g}$ for a collision that drops the neutron into the gth group.

As in any multigroup calculation, the actual computations must be carried out on a computer. These proceed roughly as follows. First, the quantities $S_{h \to g}$ are computed by evaluating the integral in Eq. (10.78). Next, the first of the intermediate group equations, which involves only the single group flux ϕ_1, is solved numerically for ϕ_1. Then this value of ϕ_1 is introduced into the second group flux equation and is solved for ϕ_2. This procedure is repeated until all of the group fluxes, including the thermal flux, have been determined. In this computation, it is the usual practice to start the calculations at the core–reflector interface, matching boundary conditions of continuity of flux and current with the results of multigroup calculations used to determine the criticality and internal fluxes of the reactor. It is in this way that the leakage of intermediate and thermal neutrons from the core into the shield are taken into account.

The removal–diffusion method has been used successfully for the design of many reactor shields. Variants of the method described here are widely used by shielding designers.

10.10 EXACT METHODS

A number of methods have been developed to calculate the penetration of neutrons and γ-rays through shields. These methods may be divided into two general categories. In the first of these, the *transport equation*, which describes the distribution of radiation within the shield, is solved by one of several numerical techniques. A discussion of such techniques is beyond the scope of this book, and the reader should consult the references at the end of the chapter. In any case, all of the methods based on solving the transport equation require the use of a computer, and many computer problems have been written to calculate the penetration of radiation as a function of shield configuration, composition, and thickness.

The second type of attenuation calculation uses the *Monte Carlo method*,[12] which is alluded to in Chapter 9. The idea behind this method can be understood by the following analogy. Suppose it is desired to determine the probability of winning the card game solitaire ("patience"). There are two approaches one can take. First, all of the probabilities of the various hands can be computed. Then, step by step, as the game is played, these probabilities are combined to give a final answer. This is a fairly involved computation. A far more simple and pleasant procedure is merely to play a few hands of the game and record the number of times the game is won. This number divided by the total number of games played is an estimate of the probability of winning. Naturally, the more hands that are played, the better this estimate is.

In much the same way, a Monte Carlo calculation of the penetration of radiation proceeds by compiling the life histories of individual neutrons (or γ-rays) as they move about—from the point where they enter the shield to the point where they are either absorbed in the shield or pass through it. The number of neutrons that successfully penetrate the shield divided by the total number of histories is then an estimate of the probability that a neutron will not be stopped by the shield.

To be specific, consider a Monte Carlo calculation of the attenuation of a parallel beam of monenergetic neutrons with energy E_0 by a slab shield of thickness a. To start the history of the first neutron, it is necessary to determine where, if at all, the neutron has its first collision. This is done by sampling the first collision probability distribution function (see Section 3.3)

$$p(x) = \Sigma_t e^{-\Sigma_t x},$$

where Σ_t is evaluated at the energy E_0. Methods are available for sampling such a distribution function in a random manner. Suppose that the value of x selected is x_1. It is next necessary to test whether x_1 is greater than or less than a. If $x_1 > a$, then the neutron has penetrated the shield without a collision; this fact is registered, and the history is terminated. However, if $x_1 < a$, a collision has occurred within the shield at the point x_1.

The type of collision must now be determined. If it is assumed that only absorption and elastic scattering are possible, their probabilities of occurrences are $p_a = \Sigma_a/\Sigma_t$ for absorption and $p_s = \Sigma_s/\Sigma_t$ for scattering, where all cross-sections are computed at the energy E_o and, of course, $p_a + p_s = 1$. The nature of

[12]*Monte Carlo* was the code name given to this method during World War II.

the collision is obtained by comparing p_a with a number R selected from a group of random numbers having values between 0 and 1. Then if

$$R < p_a,$$

the collision is assumed to be an absorption. If

$$R > p_a,$$

it is taken to be elastic scattering. The rationale behind this procedure is that the random numbers R lie uniformly between 0 and 1 so that the fraction of the interactions chosen to be absorption reactions will necessarily be equal to p_a, as required.

If the collision results in absorption, then the history is terminated and another started. If it is an elastic collision, then the angle at which the neutron is scattered must be determined by randomly sampling the experimental angular distribution of scattering. Suppose that this yields the angle ϑ_1. Then the energy of the scattered neutron can be computed from Eq. (3.28):

$$E_1 = \frac{E_0}{(A+1)^2} \left[\cos \vartheta_1 + \sqrt{A^2 - \sin^2 \vartheta_1} \right],$$

where A is the mass number of the struck nucleus. The history of this neutron is now continued by again sampling the collision distribution function $p(x)$ to determine where the neutron has its second collision, taking note of the fact that the neutron started from the point x_1 and moved in the direction ϑ_1. Again a test is made as to whether this collision occurs inside the shield. If it does, the nature of the collision is again determined, and so on. When this computation is programmed for a computer, each of the previous steps is a subroutine, which is used repeatedly as the calculation proceeds.

A set of Monte Carlo computations of particular interest is that of Clark, et al.[13] These authors calculated the attenuation of beams of neutrons normally incident on slabs of ordinary concrete for several neutron energies. Sample results of these calculations are shown in Fig. 10.23(a) and (b), where the dose equivalent rate[14] per unit incident beam intensity is given as a function of the slab thickness for two-beam energies. Curves like these are useful in designing shields for accelerator-produced neutrons and for other sources of neutrons with known energy spectra.

[13]Clark, F. H., N. A. Betz, and J. Brown, *Monte Carlo Calculations of the Penetration of Normally Incident Neutron Beams through Concrete*. Oak Ridge National Laboratory Report ORNL-3926, 1966.

[14]Clark, et al. actually determined the energy disposition per mass of tissue rather than dose equivalent. The curves in Fig. 10.23 are based on a quality factor of 10.

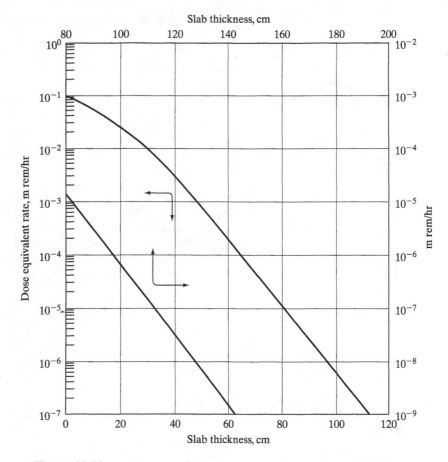

Figure 10.23   (a) Dose equivalent rate per unit intensity of 2-MeV neutrons incident on concrete slabs.

Example 10.7

A certain neutron generator produces 14-MeV neutrons isotropically at the rate of 10^9 neutrons/sec. How thick must a concrete wall be to reduce the dose equivalent rate to 1 mrem/hr if it is located 10 ft from point of emission of the neutrons?

Solution. The intensity of the beam incident on the wall is

$$I = \frac{10^9}{4\pi(10 \times 12 \times 2.54)^2} = 857 \text{ neutrons/cm}^2\text{-sec,}$$

where attenuation in the air has been neglectcd. If the dose equivalent rate shown in Fig. 10.23(b), which is normalized to unit intensity, is denoted as $\dot{H}_0(x)$, then $\dot{H}$

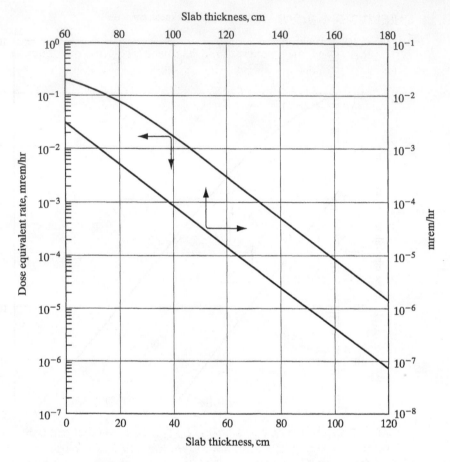

Figure 10.23 (b) Dose equivalent rate per unit intensity of 14-MeV neutrons incident on concrete slabs.

behind the wall is

$$\dot{H} = 857\dot{H}_0(x) = 1 \text{ mrem/hr.}$$

Thus, $\dot{H}_0(x)$ is

$$\dot{H}_0(x) = 1/857 = 1.17 \times 10^{-3}.$$

From Fig. 10.23(b), this value of $\dot{H}_0(x)$ occurs with $x = 70$ cm. [*Ans.*]

It must be emphasized that the dose equivalent rates given in Fig. 10.23 are for the neutrons that penetrate the slab; they do not include the dose due to γ-rays produced in the slab as the result of inelastic scattering and radiative capture. This γ-ray dose, which may actually exceed the neutron dose, is discussed in the next section.

10.11 SHIELDING γ-RAYS

Up to this point, the reactor shield has been designed to attenuate only the neutrons, and it is necessary next to consider the γ-rays. A method for calculating the penetration of γ-rays from a given distribution of γ-ray sources and determining the shielding necessary to reduce the γ-ray flux to desired levels was discussed in the first part of this chapter. It is recalled that for a general source distribution this involves the evaluation of the integral in Eq. (10.57), which is best done with a high-speed computer. The problem of shielding reactor γ-rays thus reduces to that of finding the space and energy distribution of these γ-rays.

Most of the γ-rays emanating from a reactor—the prompt fission γ-rays, the fission product decay γ-rays, and those from radiative capture and inelastic scattering—appear with more or less continuous energy spectra. For attenuation calculations these spectra are divided into energy groups, and the computations are carried out, group by group, by lumping together all of the γ-rays emitted in each group and by assuming that these γ-rays all have a single, average energy of the group.

Consider first the prompt fission γ-rays. Their spectrum rises sharply from very low values to a peak at 0.3 MeV and then diminishes slowly according to the distribution function

$$\chi_p(E) = 26.0e^{-2.3E}, \quad 0.3 < E < 1,$$
$$= 8.0e^{-1.1E}, \quad\quad 1 < E < 7,$$

where E is in MeV. The function $\chi_p(E)$ is defined so that $\chi_p(E)\,dE$ is the number of prompt γ-rays emitted per fission with energies between E and $E + dE$. To determine the number of γ-rays, χ_{pn}, emitted in the nth group, lying between the energies E_n and E_{n+1}, the total energy emitted by these γ-rays is divided by the average energy of the group $\bar{E}_n = (E_n + E_{n+1})/2$. Thus,

$$\chi_{pn} = \frac{1}{\bar{E}_n} \int_{E_n}^{E_{n+1}} E\chi_p(E)\,dE. \tag{10.79}$$

Table 10.5 gives values of χ_{pn} obtained in this way.[15] The source density at the point $\mathbf{r}$ of prompt γ-rays in the nth group is then given by

$$s_n(\mathbf{r}) = \chi_{pn}S(\mathbf{r}),$$

where $S(\mathbf{r})$ is the number of fissions occurring per cm^3/sec at the point $\mathbf{r}$.

---

[15]Equation (10.79) is an energy-weighted average. It is appropriate to use such an average because the later computation of the exposure rate from these γ-rays involves an integration of the γ-ray flux *times* the γ-ray energy, as shown by Eq. (10.5).

TABLE 10.5 NUMBERS OF PROMPT FISSION γ-RAYS
AND FISSION PRODUCT DECAY γ-RAYS EMITTED PER
FISSION

Group number	Energy interval, MeV	Prompt (χ_{pn})	Decay (χ_{dn})	Total
1	0–1	5.2	3.2	8.4
2	1–3	1.8	1.5	3.3
3	3–5	0.22	0.18	0.40
4	5–7	0.025	0.021	0.046

The spectrum of the fission product decay γ-rays changes with time from the instant the fission products are formed. In general, the more energetic γ-rays are emitted more rapidly than those with lower energy so that the spectrum tends to soften as time passes. In an operating reactor, however, where new fission products are continually being formed, the fission product γ-ray spectrum comes more or less to equilibrium. This is approximately the same as the prompt fission spectrum—namely,

$$\chi_d(E) = 6.65e^{-1.1E} \qquad (10.80)$$

γ-rays per fission per MeV. The average numbers of γ-rays χ_{dn} emitted in the various energy groups are also given in Table 10.5. The total source of prompt fission γ-rays and fission product decay γ-rays is then

$$s_n(\mathbf{r}) = (\chi_{pn} + \chi_{dn})S(\mathbf{r}). \qquad (10.81)$$

The spectra of capture γ-rays, which are produced primarily by the radiative capture of thermal neutrons, depend on the capturing nucleus. Table 10.6 gives these spectra for a few of the nuclei found in many reactors. More complete tabulations are given in the references at the end of the chapter. If the number of photons emitted in the nth group by neutron capture in the mth nucleus is denoted by χ_{Tmn}, then the source density of capture γ-rays at the point $\mathbf{r}$ is

$$s_n(\mathbf{r}) = \sum_m \chi_{Tmn} \bar{\Sigma}_{\gamma m}(\mathbf{r})\phi_T(\mathbf{r}), \qquad (10.82)$$

where $\bar{\Sigma}_{\gamma m}(\mathbf{r})$ is the macroscopic thermal capture cross-section for the mth element at $\mathbf{r}$ and $\phi_T(\mathbf{r})$ is the thermal flux.

The situation with inelastic γ-rays is somewhat more complicated than with capture γ-rays because the spectrum depends not only on the atom from which the neutron is scattered, but also on the energy of the incident neutron. Let χ_{imnh} be the number of γ-rays emitted with the energy of the nth group as the result of

TABLE 10.6 γ-RAYS FROM THERMAL NEUTRON CAPTURE IN SEVERAL ELEMENTS

Target nucleus	Photons per 100 captures					
	0–1 MeV	1–3 MeV	3–5 MeV	5–7 MeV	7–9 MeV	>9 MeV
H	0	100	0	0	0	0
D	0	0	0	100	0	0
C	0	0	100	0	0	0
Na	> 96	314	70	31	0	0
Al	> 236	264	62	19	19	0
Si	> 100	93	89	11	4.1	0.1
Fe	> 75	87	23	25	38	2.1
Pb	0	0	0	7	93	0
U	254	269	34	0	0	4

inelastic scattering by the mth atomic species of neutrons in the hth group of a removal–diffusion calculation. Extensive tables of χ_{imnh} are available and are not reproduced here. The source density of inelastic γ-rays in the nth group is then

$$s_n(\mathbf{r}) = \sum_{mh} \chi_{imnh} \Sigma_{imh}(\mathbf{r}) \phi_h(\mathbf{r}). \qquad (10.83)$$

Here $\Sigma_{imh}(\mathbf{r})$ is the macroscopic inelastic cross-section for the mth atom for the hth group of a removal–diffusion calculation, and $\phi_h(\mathbf{r})$ is the hth group flux.

With the γ-ray source distributions known, the exposure rate outside the shield can be determined from Eq. (10.57) as described earlier. If it is found that the exposure level is unacceptable, then more shielding must be added beyond that required for the neutrons alone, or a change in the composition of the shield may be indicated.

For instance, if it turns out that the exposure rate from capture γ-rays emanating from the shield is too high, it may be necessary to add boron to the shield or add a thickness of γ-ray shielding to the outside of the shield. The importance of such γ-rays is illustrated in Fig. 10.24. This figure shows the ratio of the dose rate from γ-rays produced in slabs of concrete to the dose rate from neutrons that have penetrated the slabs—for various slab thicknesses and as a function of the energy of the incident neutrons. It is observed that, with an incident beam of 4-MeV neutrons, the γ-ray and neutron dose rates are about equal for a 100-cm slab. With boron added to the concrete, the ratios shown in the figure would be substantially smaller.

In any event, once changes in the shield design have been made, new neutron calculations must be carried out to determine new multigroup neutron fluxes. The γ-ray calculations are then repeated, new adjustments to the shield are made, and

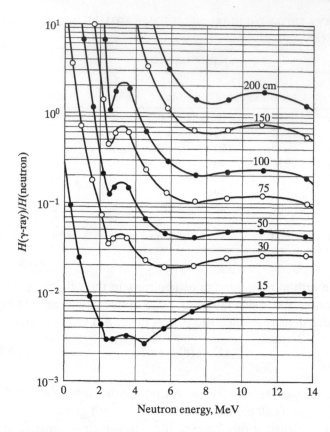

Figure 10.24   Ratio of γ-ray dose to neutron dose for neutrons incident on slabs of concrete based on $Q = 10$ for neutrons. (From F. H. Clark, N. A. Betz, and J. Brown, *Monte Carlo Calculations of the Penetration of Normally Incident Neutron Beams through Concrete*. Oak Ridge National Laboratory Report ORNL-3926, 1966.)

so on. This procedure is continued until the most appropriate shield is obtained, providing the desired attenuation for both neutrons and γ-rays.

Example 10.8

A teaching reactor consists of a hydrogenous core surrounded by a water reflector and water shield. The distance from the center of the core to the surface of the shield is 7 ft. By treating the core as a point source, estimate the exposure rate from the prompt fission γ-rays at the surface of the shield when the reactor is operating at a power of 10 watts.

Solution. It is first necessary to determine the rate at which fissions are occurring in the core. Assuming a recoverable energy per fission of 200 MeV $= 3.2 \times 10^{-11}$ joules, this is

$$\text{fission rate} = 10 \text{ joules/sec} \div 3.2 \times 10^{-11} \text{joules/fissions}$$

$$= 3.13 \times 10^{11} \text{ fissions/sec.}$$

The γ-rays in the first group in Table 10.5 make a negligible contribution to the exposure rate, so consider those in the second group. There are

$$S = 1.8 \times 3.13 \times 10^{11} = 5.63 \times 10^{11}\gamma\text{-rays/sec}$$

emitted in this group, all of which are assumed to have an energy of 2 MeV. The buildup flux at the shield surface from these γ-rays is then

$$\phi_b = \frac{S}{4\pi R^2} B_p(\mu R)e^{-\mu R}.$$

Here $R = 7\,\text{ft} = 7 \times 12 \times 2.54 = 213$ cm. From Table II.4, $\mu = 0.0493$ cm^{-1} so that $\mu R = 10.5$. Then from Table 10.2, $B_p = 13.1$ and

$$\phi_b = \frac{5.63 \times 10^{11}}{4\pi (213)^2} \times 13.1 \times e^{-10.5} = 356\gamma\text{-rays/cm}^2\text{-sec}.$$

From Fig. 9.9, this gives an exposure rate of 1.1 mR/hr. [*Ans.*]

The γ-rays in other groups give the following exposures: first group, 0.0013 mR/hr; third group, 2.3mR/hr; and fourth group, 0.93 mR/hr. The total exposure rate is 4.3 mR/hr. [*Ans.*]

10.12 COOLANT ACTIVATION

Radioactivity in the coolant is an important source of radiation external to a reactor. This radioactivity originates in several ways: (a) by neutron activation of the coolant; (b) by activation of impurity atoms contained in the coolant; and (c) by picking up the coolant of radioactive atoms from the walls of the coolant channels in the reactor.

The most important reactions by which coolants are activated are listed in Table 10.7. All of the nuclides formed in these reactions undergo β^- decay, and the radiation listed in the table is emitted immediately owing the emission of the

TABLE 10.7 ACTIVATION REACTIONS IN COOLANTS

Reaction	Cross-section, b	Half-life	Energy of radiation, MeV
$^{16}O(n,p)^{16}N$	$1.9 \times 10^{-5}*$	7.1 s	6.13, 7.12 (γ-rays)
$^{17}O(n,p)^{17}N$	$5.2 \times 10^{-6}*$	4.14 s	1.2, 0.43 (neutrons)
$^{18}O(n,\gamma)^{18}O$	$2.1 \times 10^{-4\dagger}$	29 s	0.20, 1.36 (γ-rays)
$^{23}Na(n,\gamma)^{24}Na$	$0.53^\dagger$	15.0 h	2.75, 1.37 (γ-rays)
$^{40}Ar(n,\gamma)^{41}Ar$	$0.53^\dagger$	1.83 h	1.29 (γ-rays)

*Average cross-section over fission spectrum; see text.
†Thermal (0.0253 eV) cross-section.

β-rays. The energies of the β-rays are not included in the table since these ordinarily do not present a radiation hazard except in the case of an accident.

The first three reactions in the table are responsible for the radioactivity of the water in water-cooled reactors. Of these, the $^{16}O(n,p)^{16}N$ reaction is the most important despite that it is an endothermic reaction with a threshold energy of 10.2 MeV. The $^{17}O(n,p)^{17}N$ reaction is also endothermic; it has a threshold of 8.5 MeV. It should be noted that this peculiar reaction leads to the emission of neutrons. This occurs because 95% of the β-decays of the ^{17}N go to the excited states of ^{17}O having energies greater than the binding energy of the last neutron in this nucleus. Since ^{17}O consists of a tightly bound core of ^{16}O and a loosely attached neutron, this neutron is immediately emitted when it has sufficient energy to do so. The $^{18}O(n,\gamma)^{19}O$ reaction is exothermic and occurs at all energies, including thermal energies. However, it is a less important source of γ-rays than the $^{16}O(n,p)^{16}N$ reaction because the isotopic abundance of ^{18}O is so low (0.204 a/o) compared to that of ^{16}O (99.8 a/o), and because the γ-rays are less energetic.

The activation of sodium coolants is due entirely to the $^{23}Na(n,\gamma)^{24}Na$ reaction as sodium consists only of the single isotope ^{23}Na. This reaction occurs with neutrons of all energies. The 2.75 MeV and 1.37 MeV γ-rays are emitted in cascade—that is, both appear with each disintegration of ^{24}Na.

When air is used as a reactor coolant, it becomes radioactive as the result of the activation of argon, which is present to the extent of 1.3 w/o in air. Reactions with oxygen and nitrogen do not contribute significantly to the activation of air. Argon is also present in small amounts in CO_2, which is used to cool many of the British reactors, and its activation accounts for the radioactivity of the CO_2 coolant. Argon activation is also a problem in teaching and research reactors of the swimming pool type, in which the reactor is located at the bottom of an open tank of water. A certain amount of air containing argon becomes dissolved in the pool water; on passing through the core, the argon becomes activated. Later this radioactive argon emerges from the pool surface and passes into the atmosphere above the reactor.

The activity acquired by a coolant via the prior reactions depends on the flow pattern of the coolant and its residence times within and without a neutron flux. Figure 10.25 shows the simplest type of coolant flow, the coolant spending, the time t_i in the reactor flux, and t_o in the outer circuit with no flux. Consider a unit volume of coolant leaving the reactor. If its activity at this point is α, then at the time t_o later, when it reenters the reactor, its activity will have fallen to

$$\alpha e^{-\lambda t_0},$$

where λ is the decay constant. In passing through the reactor, the already activated atoms continue to decay by the amount

$$e^{-\lambda t_i}.$$

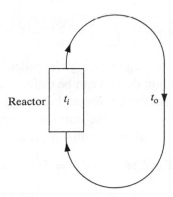

Figure 10.25   A simple coolant circuit.

In the reactor, however, a unit volume, exposed to an average flux ϕ_{av}, acquires in the time t_i the additional activity

$$\Sigma_{act}\phi_{av}(1 - e^{-\lambda t_i}),$$

where Σ_{act} is the average macroscopic activation cross-section in the flux ϕ_{av}. The total activity leaving the reactor is thus

$$\alpha e^{-\lambda(t_i + t_o)} + \Sigma_{act}\phi_{av}(1 - e^{-\lambda t_i}).$$

In equilibrium, after the coolant has made many passes around the circuit, this activity must be equal to α—that is,

$$\alpha = \alpha e^{-\lambda(t_i + t_o)} + \Sigma_{act}\phi_{av}(1 - e^{-\lambda t_i}).$$

Solving for α gives

$$\alpha = \frac{\Sigma_{act}\phi_{av}(1 - e^{-\lambda t_i})}{1 - e^{-\lambda(t_i + t_o)}}. \tag{10.84}$$

This is the activity of the coolant leaving the reactor—the point where the activity is the highest. If the residence times t_i and t_o are small compared with the half-life of the induced activity, as is the case with sodium-cooled reactors, then the exponentials in Eq. (10.84) can be expanded with the result

$$\alpha = \Sigma_{act}\phi_{av}\frac{t_i}{t_i + t_o}. \tag{10.85}$$

The quantity $\Sigma_{act}\phi_{av}$, the activation rate per unit volume in the reactor, must be chosen appropriately for the activation reaction in question. Thus, for the ^{18}O $(n,\gamma)^{19}$O and ^{40}Ar$(n,\gamma)^{41}$Ar reactions, Σ_{act} is the macroscopic thermal activation cross-section and ϕ_{av} is the space-averaged thermal flux. For the threshold reactions in oxygen, the activation rate is given by the integral

$$\Sigma_{act}\phi_{av} = \int_{E_t}^{\infty} \Sigma_{act}(E)\phi_{av}(E)\,dE, \tag{10.86}$$

where $\phi_{av}(E)$ is the space-averaged energy-dependent flux and E_t is the reaction threshold energy. The function $\phi_{av}(E)$ can be estimated by assuming that any collision, except small-angle scattering, drops the energy of a fission neutron below threshold. Then in the steady state, the number of neutrons scattered out of the energy interval dE at E is equal to the number produced in dE—that is,

$$\Sigma_R(E)\phi_{av}(E)\,dE = S(E)\,dE,$$

where $\Sigma_R(E)$ is the removal cross-section at the energy E, and $S(E)$ is the number of fission neutrons produced per unit energy at E, which is assumed to be independent of position in the reactor. Solving for $\phi_{av}(E)$ gives

$$\phi_{av}(E) = \frac{S(E)}{\Sigma_R(E)}. \tag{10.87}$$

Introducing this function into Eq. (10.86) gives

$$\Sigma_{act}\phi_{av} = \int_{E_t}^{\infty} \frac{\Sigma_{act}(E)S(E)\,dE}{\Sigma_R(E)}$$

$$\simeq \frac{1}{\Sigma_R} \int_{E_t}^{\infty} \Sigma_{act}(E)S(E)\,d(E), \tag{10.88}$$

where Σ_R is the macroscopic removal cross-section for fission neutrons computed from Table 10.4.

The activation cross-sections for the threshold reactions given in Table 10.7 are averages over the fission neutron spectrum,—that is,

$$\langle \sigma_{act} \rangle = \frac{\int_{E_t}^{\infty} \sigma_{act}(E)S(E)\,dE}{\int_0^{\infty} S(E)\,dE}.$$

Writing

$$\langle \Sigma_{act} \rangle = N \langle \sigma_{act} \rangle,$$

where N is the atom density of the isotope in question, and combining these definitions with Eq. (10.88), yields

$$\Sigma_{act}\phi_{av} = \frac{\langle \Sigma_{act} \rangle}{\Sigma_R} \int_0^{\infty} S(E)\,dE$$

$$= \frac{\langle \Sigma_{act} \rangle S}{\Sigma_R}, \tag{10.89}$$

where S is the total number of fission neutrons emitted per cm^3-sec in the reactor. The application of these formulas is illustrated by the following example.

Example 10.9

A hypothetical water-moderated and water-cooled reactor operates at a power density of 55 watts/cm^3. The core consists of uranium rods and water; the metal volume fraction of the rods is 0.33. The water spends an average of 3 sec in the reactor and 2 sec in the external coolant circuit. Estimate the equilibrium activity of the water leaving the reactor due to the $^{16}O(n,p)^{16}N$ reaction.

Solution. The atom density of oxygen in water at normal density is 0.033×10^{24} per cm^3, and virtually all of this is ^{16}O. In the present case, the effective density of the water is 0.67 g/cm^3 so that $\langle\Sigma_{act}\rangle$ is

$$\langle\Sigma_{act}\rangle = 0.67 \times 0.033 \times 1.9 \times 10^{-5} = 4.2 \times 10^{-7} cm^{-1}.$$

Assuming a recoverable energy of 200 MeV $= 3.2 \times 10^{-11}$ joules per fission, the fission density is $55/3.2 \times 1^{-11} = 1.72 \times 10^{12}$ fissions/cm^3-sec, so that with 2.42 neutrons emitted per fission,

$$S = 2.42 \times 1.72 \times 10^{12} = 4.2 \times 10^{12} \text{ neutrons/}cm^3\text{-sec.}$$

From Table 10.4, Σ_R for water is 0.103 cm^{-1} and for uranium is 0.174 cm^{-1}. For the uranium–water mixture, therefore,

$$\Sigma_R = 0.67 \times 0.103 + 0.33 \times 0.174 = 0.126 \ cm^{-1}.$$

By use of Eq. (10.89), the activation rate is then found to be

$$\Sigma_{act}\phi_{av} = \frac{4.2 \times 10^{-7} \times 4.2 \times 10^{12}}{0.126} = 1.40 \times 10^7 \text{ atoms/}cm^3\text{-sec.}$$

Introducing this result into Eq. (10.84), with $t_i = 3$ sec, $t_o = 2$ sec, and $\lambda = 0.693/71 = 0.0976$ sec^{-1}, gives

$$\alpha = \frac{1.40 \times 10^7(1 - e^{-3\times0.0976})}{1 - e^{-5\times0.0976}}$$

$$= 9.20 \times 10^6 \text{ disintegrations/}cm^3\text{-sec}$$

$$= 249\mu Ci/cm^3. \ [Ans.]$$

Sodium is used mostly to cool fast reactors, and its activation is due to the absorption of fast and intermediate energy neutrons. Thus, the activation rate per cm^3 is given by

$$\Sigma_{act}\phi_{av} = \int \Sigma_{act}(E)\phi_{av}(E) \, dE, \tag{10.90}$$

where the integration is carried out over the entire neutron spectrum. In terms of the multigroup formula given in Section 6.7, this can be written as

$$\Sigma_{act}\phi_{av} = \sum_g \Sigma_{act}\phi_{av\,g}, \tag{10.91}$$

where $\Sigma_{act\,g}$ is the macroscopic activation (radiative capture) cross-section of sodium for the gth group and $\phi_{av\,g}$ is the space-averaged flux in this group.

The activation of coolant atoms discussed earlier is ordinarily the major source of radioactivity in reactor coolant loops while the reactor is in operation. It is far more important than the activation of impurities in the coolant or the acquisition of radioactive atoms from the reactor by the coolant. However, when the reactor is shut down, the coolant activity quickly dies away except in the case of sodium (see Table 10.7), and the residual coolant activity is due to these extraneous atoms. Furthermore, these atoms tend to deposit along the walls of the primary coolant piping and in valves, pumps, and heat exchangers. This gives rise to the radioactivity of these components even if the coolant is removed, which hampers maintenance and repairs. In any event, it is difficult to predict the level of this activity since it depends on countless numbers of materials present in the reactor and in the primary coolant system.

10.13 DUCTS IN SHIELDS

Passages or ducts through a reactor shield are necessary to accommodate instrumentation cables and coolant channels, as well as to provide access for refueling and maintenance. Penetrations are normally also present in the shielded walls of the reactor room and in other compartments in a nuclear power plant, and in the walls of rooms containing accelerators or other sources of radiation. Any penetration of this type can be a serious source of radiation leakage through the shield.

The worst possible situation is that of a straight duct containing air or some other gas. In this case, the radiation passes directly through the duct without attenuation—a process known as *radiation streaming*. To avoid streaming, ducts are designed with sharp bends where possible, as indicated in Fig. 10.26. In such

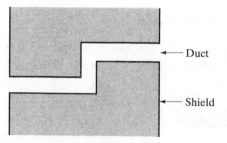

Duct

Shield

Figure 10.26   A duct through a shield.

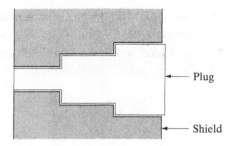

Figure 10.27 A shield plug inserted in a shield.

ducts, only the radiation that has been appropriately scattered from the walls of the duct can successfully penetrate the shield.

Nevertheless, it is sometimes necessary to provide passages directly through a shield, which is plugged while the reactor is in operation. This is the case, for instance, with refueling ports. The duct and plug are then made in a series of steps as shown in Fig. 10.27. This effectively reduces the streaming of radiation through the space between the plug and wall.

The computation of the penetration of radiation through ducts with bends and/or steps is quite complicated and is not discussed here. Nevertheless, the presence of ducts is an important consideration in the design of any shield.

REFERENCES

General

Chilton, A. B., *Principles of Radiation Shielding*. Paramus: Prentice-Hall, 1983.

Glasstone, S., and A. Sesonske, *Nuclear Reactor Engineering*, 4th ed. New York: Chapman & Hall, 1994, Chapter 6.

Goldstein, H., *Fundamental Aspects of Reactor Shielding*. Reading, Mass.: Addison-Wesley, 1959 (out of print); now available from Johnson Reprint Corp., New York. This volume contains an authoritative discussion of the calculation of buildup factors.

Jaeger, R. G., Editor, *Engineering Compendium on Radiation Shielding*, Vol. I. New York: Springer-Verlag, 1968.

Rockwell, T., Editor, *Reactor Shielding Design Manual*. Oak Ridge, Tenn.: U.S. Department of Energy, 1956. A dated but still useful publication.

Schaeffer, N. M., Editor, *Reactor Shielding for Nuclear Engineers*. Oak Ridge, Tenn.: U.S. Department of Energy, 1973.

Shielding Information

The Radiation Shielding Information Center at the Oak Ridge National Laboratory, Oak Ridge, Tennessee, publishes state-of-the-art reviews, data compilations, and a monthly newsletter, all aimed at providing the latest information on the shielding of all types of radiation sources. In addition, the Center, which is sponsored by several government agencies, collects and makes available computer programs that have been written for shielding calculations. Information from the center is accessible over the Internet.

Tables of Functions

An excellent discussion of the various functions mentioned in this chapter is available on the web at `mathworld.wolfram.com`.

The $E_n(x)$ Function

Etherington, H., Editor, *Nuclear Engineering Handbook*. New York: McGraw-Hill, 1958, pp. 1–120.

Goldstein, H., *op. cit.*, p. 355.

Abramowitz, M., and I. A. Stegun, Editors, *Handbook of Mathematical Functions with Formulas, Graphs, and Mathematical Tables*, 9th printing. New York: Dover, 1972, pp. 227–233.

The Exponential Integral $Ei(x)$

Etherington, H., *op. cit.*, pp. 1–121.

Abramowitz, M., and I. A. Stegun, Editors, *op. cit.*

The Sievert Integral Function $F(\theta, x)$

This function is not widely tabulated. See Abramowitz, M., and I. A. Stegun, Editors, *op. cit.*, pp. 1000–1001. Extensive graphs of $F(\theta, x)$ are given in:

Jaeger, R. G., *op. cit.*

Rockwell, T., *op. cit.*

PROBLEMS

1. A monodirectional beam of 500-kilovolt x-rays having an intensity of 10^7 x-rays/cm²-sec is incident on an iron slab 20 cm thick. Calculate: (a) the exposure rate in the absence of the iron; (b) the uncollided flux; (c) the buildup flux; (d) the exposure rate behind the slab.

2. How thick must a lead shield be to reduce the exposure rate from a monodirectional beam of 1-MeV γ-rays of intensity 2×10^5 γ-rays/cm²-sec to 2.5 mR/hr?

3. An isotropic point source emits 10^{10} γ-rays/sec with an energy of 1 MeV. The source is surrounded by a lead shield 10 cm thick. Calculate at the surface of the shield:

(a) the flux in the absence of the shield; (b) the uncollided flux; (c) the buildup flux; (d) $\dot{X}$ in the absence of the shield; (e) $\dot{X}$ without the buildup of scattered radiation; (f) $\dot{X}$ with buildup.

4. A 1-Ci radium-beryllium source (see Problem 9.24) is to be placed in a spherical lead shield. What is the minimum thickness of lead that will reduce the γ-ray exposure rate at its surface to no more than 1 mR/hr?

5. At the time t_0, 5 hours after the detonation of a nuclear warhead, the fission products are distributed as fallout uniformly over a certain area at a density of 6.2×10^{-5} Ci/cm^2. A man enters a fallout shelter (shown in Fig. 10.28) at this time and remains there for t_s hours. (a) Show that if his total exposure is not to exceed XR, then his initial exposure rate in the shelter must not exceed

$$\dot{X}_0 = \frac{X}{5t_0}\left[1 - \left(\frac{t_0}{t_0 + t_s}\right)^{0.2}\right]^{-1}.$$

(b) How thick must the concrete roof of the shelter be for the man to receive an exposure of 2R over 2 weeks? [*Note:* For simplicity, take the roof to be infinite in extent and assume the γ-rays have an energy of 0.7 MeV.]

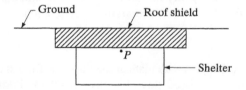

Figure 10.28   Fallout shelter diagram for Prob. 10.5

6. It is desired to place the disc source described in Problem 9.29 in a tank of water as shown in Fig. 10.29. What thickness of water is required to reduce the exposure rate at the midpoint (P) of the tank to 1 mR/hr?

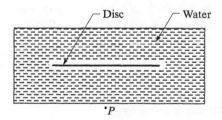

Figure 10.29   Disk source in water for Prob. 10.6.

7. By repeated integration by parts, show that

$$E_1(x) = \frac{e^{-x}}{x}\left[1 - \frac{1}{x} + \frac{2!}{x^2} + \frac{3!}{x^3} + \cdots\right].$$

Discuss the convergence of this series.

8. Sodium coolant with a specific activity of 0.05 Ci/cm^3 passes through a long pipe 8 in. in diameter. By treating the pipe as a line source and ignoring attenuation in the sodium, estimate the thickness of a cylindrical concrete shield, wrapped around the pipe, that reduces the dose rate to 2.5 mrem/hr. [*Note:* The activity of sodium is due to ^{24}Na (see Table 10.7).]

9. Show that, for small θ, $F(\theta, x)$ is given by

$$F(\theta, x) \simeq \theta e^{-x}.$$

10. Suppose that, instead of letting it cool for 2 months, the fuel rod in Problem 9.28 is immediately placed in a cylindrical lead cask for transport to a hot lab. Calculate the thickness of the cask wall and ends that will reduce the exposure rate at its surface to no more than 1 mR/hr. [*Note:* Make all calculations for 1 minute after removal from the reactor.]

11. (a) Had the cask in the preceding problem been constructed of iron instead of lead, how much would it weigh? (b) If lead costs $.20 a pound and iron costs $20 a ton, which of these two materials yields the cheaper cask in materials?

12. Thermal neutrons are incident on a slab of water 30 cm thick. As the neutrons diffuse about in the slab, they disappear by radiative capture in hydrogen according to the reaction ^{1}H(n,γ)^{2}H, where the γ-ray has an energy of approximately 2 MeV. Given that the thermal flux in the slab dies off according to the relation

$$\phi(x) = \phi_0 e^{-x/L},$$

where $\phi_0 = 10^8$ and $L = 2.85$ cm, calculate at the far side of the slab (a) the thermal neutron dose rate; (b) the uncollided flux of γ-rays; (c) the buildup flux of γ-rays; (d) the γ-ray dose rate.

13. A spherical cloud of fallout of radius R contains uniformly distributed particles emitting S γ-rays per cm^3/sec. Derive an expression for (a) the unscattered γ-ray flux at the center of the cloud; (b) the buildup flux at the center of the cloud.

14. A radioactive substance is dissolved in a large body of water so that S γ-rays are emitted per cm^3/sec throughout the water. (a) Show that the uncollided flux at any point in the water is given by

$$\Phi_u = \frac{S}{\mu}.$$

(b) Show that the buildup flux is given by

$$\Phi_b = \frac{S}{\mu} \sum \frac{A_n}{1 + \alpha_n},$$

where A_n and α_n are parameters for the Taylor form of the buildup factor.

15. An isotropic point source emitting 10^{10} 1-MeV γ-rays/sec is located at the center of a spherical container of water 50 cm in radius. What thickness of lead outside of the container is required to reduce the exposure rate at the surface to 1 mR/hr?

16. If the source in the preceding problem had been surrounded first with a sphere of lead followed by 50 cm of water, what thickness of lead would be required? Compute the weights of the shields in the two cases.

17. An 8-MeV neutron is inelastically scattered from a barium nucleus. What is the average energy of the emergent inelastic neutrons?

18. Calculate the macroscopic removal cross-section of uranium dioxide (UO_2) having a density of 10 g/cm^3.

19. An infinite planar source emits S fission neutrons/cm^2-sec. Adjacent to the source, there is a slab of iron a cm thick, followed by a slab of water b cm thick as shown in Fig 10.30. Derive an expression for the fast neutron flux at the point P located c cm from the surface of the water.

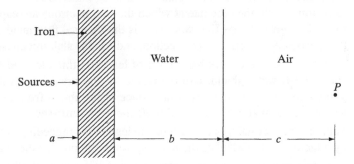

Figure 10.30   Energy spectrum of incident γ-ray beam.

20. A fission plate 30 in in diameter and 0.06 in thick is composed of uranium enriched to 20.8% in ^{235}U. The plate is located at one end of a large tank of water and is exposed to a thermal flux of 1×10^8 neutrons/cm^2-sec. (a) Calculate the fast flux and fast neutron dose equivalent rate in the water at a distance of 1 meter from the center of the fission plate. (b) Repeat the calculations when slabs of iron 2, 4, and 6 in thick are placed in front of the plate.

21. The core of a shipboard reactor that operates at a power of 90 megawatts has a volume of approximately 1,380 liters. The composition of the core in volume percent is as follows: uranium, 1%, zirconium, 40%, water (density 0.8 g/cm^3), 59%. Estimate the thickness of a unit-density water shield that is required to reduce the fast neutron flux at the surface of the shield to 10 neutrons/cm^2-sec.

22. A small prototype fast breeder reactor operates at a total power of 200 megawatts at an average power density of 600 watts/cm^3. The core composition in volume percent is as follows: ^{239}Pu, 10%; ^{238}U, 20%; iron, 25%; sodium, 45%. The core is surrounded by a blanket 50 in thick of the following composition: ^{238}U, 45%; iron, 20%; sodium, 35%. Estimate the thickness of a concrete shield surrounding the reactor that reduces the fast neutron dose equivalent rate to 1 mrem/hr. [*Note:* Take $v = 3$ and assume σ_R of plutonium is the same as that of uranium.]

23. Devise a scheme using random numbers for selecting the type of interaction occurring at a collision site in a Monte Carlo calculation when three interactions are energetically possible.

24. A point source emitting 10^9 neutrons/sec at an energy of 2 MeV is located in the center of a cubical experimental chamber 25 ft on a side. (a) How thick must the concrete walls of the chamber be if the dose equivalent rate from the neutrons may exceed 2.5 mrems/hr at no point on the outer surface of the chamber? (b) What is the additional dose due to the γ-rays produced in the concrete by the neutrons? (c) What concrete thickness is required if the total dose rate is not to exceed 2.5 mrems/hr?

25. Verify the values of χ_{pn} and χ_{dn} given in Table 10.5 for the group $n = 2$.

26. Verify the exposure rates given in Example 10.8.

27. Show that if the thickness of a slab of diffusing material is much less than the thermal diffusion length, then the rate at which thermal neutrons are captured per cm^2/sec of the slab is given by $\phi_T \bar{\Sigma}_a t$, where ϕ_T is the thermal flux incident on the slab, $\bar{\Sigma}_a$ is the macroscopic thermal cross-section, and t is the slab thickness.

28. A certain research reactor has a layer of lead 3 in thick located between the reflector and a 2.5-ft water shield. Use the results of the preceding problem to estimate the γ-ray exposure rate at the exterior surface of the shield from radiative capture in the lead if the thermal flux at the lead is 10^5 neutrons/cm^2-sec.

29. Suppose the thermal flux incident on the lead in the preceding problem were increased by a factor of 10. Calculate the new exposure rate outside the shield.

30. A certain pressurized-water reactor operates at a power of 100 megawatts (thermal). The coolant water, which has an average density of 0.75 g/cm^3, spends approximately 0.27 sec traversing the reactor core (of volume 1,700 li) and 6.88 sec traversing the rest of the coolant circuit. The reactor is fueled with highly enriched uranium, which is contained in zirconium fuel elements at very low concentration. The metal-to-water volume ratio is 0.25. Calculate the equilibrium γ-ray and neutron specific activities of the water at the exit and entrance to the reactor as the result of the $^{16}O(n,p)^{16}N$ and $^{17}O(n,p)^{17}N$ reactions.

31. Given that the average thermal flux in the reactor described in the preceding problem is 2×10^{13} neutrons/cm^2-sec, calculate the activity of the water due to the $^{18}O(n,\gamma)^{19}O$ reaction at the entrance and exit to the reactor.

32. In some early gas-cooled reactors, the gas made only one pass through the reactor before it was exhausted to the atmosphere. Show for such a reactor that if the gas spends the time t_i on the average in the flux ϕ_{av}, the rate at which activated atoms having half-life $T_{1/2}$ are exhausted is given by

$$\text{Rate exhausted} = \frac{1.87 \times 10^{-11} W \Sigma_{act} \phi_{av} t_i}{T_{1/2}} \text{ Ci/ sec,}$$

where W is the flow rate of the gas in cm^3/sec and provided $t_i \ll T_{1/2}$.

33. The first research reactor at Brookhaven National Laboratory was an air cooled system. The air made a single pass through the reactor and was then exhausted from a

stack. If the air flow rate was 270,000 cu ft per minute, the air volume in the reactor was 960 cu ft and the average thermal flux was 2×10^{12} neutrons/cm^2-sec, at what rate was radioactivity exhausted from the stack?

34. In many water-cooled reactors, the water acts as moderator, reflector, and coolant. The coolant circuit is as shown in Fig. 10.31, where the fraction f_e of the coolant entering the reactor passes through the core, with the remaining fraction f_r going to the reflector. The average activation rates per unit volume in the core and reflector are $(\Sigma_{act}\phi_{av})_c$ and $(\Sigma_{act}\phi_{av})_r$, respectively, and the water spends the times t_c, t_r, and t_o in the core, reflector, and outer circuit. Show that the specific activity leaving the reactor is given by

$$\alpha = \frac{(\Sigma_{act}\phi_{av})_c(1 - e^{-\lambda t_e}) + (\Sigma_{act}\phi_{av})_r(1 - e^{-\lambda t_r})}{1 - f_c e^{-\lambda(t_c + t_o)} - f_r e^{-\lambda(t_r + t_o)}}.$$

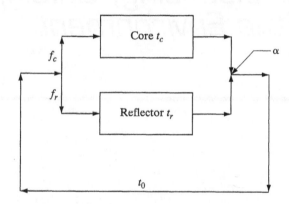

Figure 10.31 Schematic representation of coolant circuit showing flow split between core and reflector for Prob. 10.34.

35. A long straight air duct of radius R passes through a slab shield of thickness x. Given that the radiation incident on the shield is isotropic, show that the dose rate H at the end of the duct due to streaming is related to the dose rate H_0 at the beginning of the duct by

$$\dot{H} = \frac{\dot{H}_0 R^2}{4x^2}.$$

11

Reactor Licensing, Safety, and the Environment

All engineering structures and devices—buildings, bridges, dams, vehicles, manufacturing plants, electrical machinery, and so on—inherently present some element of risk to their owners or operators and to the public at large. Nuclear reactors are no exception in this regard. However, at times, the risk of nuclear reactors has been unduly exaggerated, in some measure because of an innate fear of nuclear radiation on the part of the public, and also because of the memory of the awesome effects of the nuclear weapons employed in World War II.

In fact, the risks from nuclear power are extremely low when compared with risks commonly accepted in other forms of human endeavor. To date, the safety record of the nuclear power industry has been excellent. This is the result of the recognition by both industry and government of the central role that safety must play in the design and operation of every nuclear installation.

In terms of current technology, nuclear power is one of the least, environmentally harmful ways of producing large amounts of electrical power. However, nuclear power plants do discharge small amounts of radioactivity to their environments. This aspect of nuclear power is discussed in the closing section of the chapter.

11.1 GOVERNMENTAL AUTHORITY AND RESPONSIBILITY

The ultimate responsibility for protection of the public in any country rests with the local or national government. The manner in which this responsibility is exercised in the case of nuclear energy varies somewhat from country to country. The present discussion is confined to procedures currently in effect in the United States. It is reasonable to outline the governmental function first, before considering the technology of reactor safety, for much of the philosophy and most of the technical requirements related to safety stem from government regulations.

The Nuclear Regulatory Commission

In the United States, the authority for federal government involvement in and regulation of any activity originates in acts of Congress. The first legislation related to the development and regulation of nuclear energy was the *Atomic Energy Act of 1946*, also known as the *McMahon Act*. This act was greatly expanded by the *Atomic Energy Act of 1954*. Both acts established a five-man *Atomic Energy Commission* and spelled out in some detail the mandate of the Commission in nuclear matters. In general terms, the Acts also provided for the establishment of an organization or agency—loosely called the "AEC" to distinguish the agency from the five-man Commission—to carry out the various functions of the Commission, including "the regulation of the production and utilization of atomic energy and the facilities related thereto...."

The Atomic Energy Act of 1954 also stated that "it is the purpose of this Act to effectuate ... a program to encourage the widespread participation in the development and utilization of atomic energy...." This provision of the Act was increasingly criticized over the years for placing the Commission in the position of being simultaneously the promoter and regulator of nuclear energy—roles that conceivably could conflict with one another. To correct this situation, Congress enacted the *Energy Reorganization Act of 1974*. This act (1) abolished the Atomic Energy Commission; (2) consolidated into a new agency, the *Energy Research and Development Administration (ERDA)*, the research and development activities of the AEC and several other government agencies; and (3) transferred to a new *Nuclear Regulatory Commission* (and its agency, the NRC) all of the licensing and regulatory functions previously performed by the AEC.

In 1977, in an effort to further consolidate the energy programs of the federal government, Congress merged ERDA and several other energy-related agencies into a new, Cabinet-level *Department of Energy (DOE)*. The structure and functions of the NRC were not affected by this reorganization. The NRC remains an independent agency, not included within DOE. Thus, federally sponsored research and development programs in nuclear energy are carried out under the auspices of the DOE. The NRC continues to license and regulate nuclear activities in the

United States under the provisions of the 1954 and 1974 Acts, as amended. It also sponsors confirmatory research in support of these activities.

Like most acts of Congress, the 1954 and 1974 Acts are written in general terms. In order to specify in detail how it conducts its business, the NRC promulgates its own rules and regulations. These become part of the *Code of Federal Regulations (CFR)*, a master manual for the operations of the executive departments and agencies of the federal government. These regulations have the force and effect of law.

The CFR is divided into 50 sections called *titles*, each of which represents a broad subject area of federal regulation. Each title, in turn, is divided into *chapters*, which usually bear the name of the issuing agency. Title 10, Chapter 1, is reserved for the regulations of the NRC. Each chapter of the CFR is further subdivided into *parts* covering specific regulatory areas. For example, Part 20 of Title 10 establishes standards for radiation protection; Part 50 gives the regulations whereby the Commission licenses reactors; and so on. At the time of this writing, there are 37 parts (not numbered consecutively) to Title 10. Those parts most frequently encountered in connection with nuclear power plants are described in Table 11.1.

Each part of Title 10 is denoted in abbreviated form, as in 10 CFR 20, 10 CFR 50, etc. Specific paragraphs or topics within each part are also numbered. For example, 10 CFR 20.4 defines units of radiation dose; 10 CFR 50.34 describes the technical information that must be furnished to the NRC by an applicant for a reactor license; and so on.

In addition to the regulations published in 10 CFR, the NRC occasionally issues *Regulatory Guides*. The purpose of these guides is to (1) describe the methods the NRC regulatory staff finds acceptable in implementing the regulations contained in 10 CFR; (2) discuss the techniques used by the staff in evaluating specific problems or postulated nuclear accidents; and (3) provide general guidance to applicants for NRC licenses. The guides are not substitutes for the regulations, and compliance with them is not required. However, from a practical standpoint, they are becoming an increasingly important part of the overall NRC regulatory process.

11.2 REACTOR LICENSING

The responsibility of the NRC for the regulation of nuclear energy extends over the whole fuel cycle, from the mining and processing of uranium ore through the utilization of the fuel in reactors, the transportation and reprocessing of spent fuel, and, finally, the disposal of radioactive wastes. The present discussion is largely restricted to the regulation of nuclear power plants.

NRC regulatory authority is implemented through the issuance of various licenses. Considerable effort is normally required, both by the applicant for a reactor

TABLE 11.1   MOST FREQUENTLY USED PARTS OF CFR TITLE 10, CHAPTER 1

Part	Name	Applicability
1	Statement of Organization and General Information	Describes the organization of the NRC.
2	Rules of Practice	Governs the conduct of proceedings for granting, revoking, amending, or taking other actions regarding an NRC license; the imposition of civil fines; rulemaking procedures; patent matters.
20	Standards for Protection Against Radiation	Establishes standards and regulations for protection against radiation.
50	Licensing of Production and Utilization Facilities	Gives the requirements of obtaining a construction permit or operating license for a reactor, including a detailed account of the information that must be supplied by an applicant for such a permit or license.
51	Licensing and Regulatory Policy and Procedures for Environmental Protection	Give the procedures to be followed by an applicant for an NRC license to satisfy the National Environmental Policy Act.
55	Operator's Licenses	Gives procedures and criteria for the issuance of licenses to operators of NRC-licensed reactor facilities.
70	Domestic Licensing of Special Nuclear Material	Establishes regulations and criteria for the issuance of licenses to own and use special nuclear material (roughly speaking, any fissile material or uranium enriched in ^{235}U).
100	Reactors' Site Criteria	Defines the criteria the NRC uses to evaluate proposed sites for nuclear power reactors.

license and by the NRC, before any license is issued. Although the granting of an NRC license is usually considered to be the major licensing action in the case of a nuclear power plant, all the various permits and licenses issued by other federal, state, and local authorities for construction of ordinary fossil fuel plants must also be secured. Often, this entails 40 or more separate licensing actions. As a result, resolution of licensing problems associated with the permits or licenses needed to construct or to operate nuclear power plants usually requires several years.

Several groups within the NRC are involved in reactor licensing, including the following:

Regulatory Staff.   As in any other large enterprise, the NRC staff is divided into a number of separate offices that perform specially assigned functions within the organization. In practice, an applicant proposing to build a nuclear plant interacts primarily with staff members in the *Office of Nuclear Reactor Regulation (NRR)* and the *Office of Inspection and Enforcement (IE)*. NRR personnel perform all the licensing functions associated with the construction and operation of nuclear reactors and with the receipt, possession, ownership, and use of nuclear fuel at these facilities. NRR staff reviews applications and issues licenses for reactor facilities after evaluating the health, safety, and environmental aspects of the facilities. NRR is also responsible for the licensing of reactor operators.

When and if a plant reaches the construction stage, IE personnel inspect the project on a continuing basis to assure that the plant is constructed according to the NRC construction permit and related specifications. Later, when the plant is operational, other IE personnel monitor its operations to see that these are carried out in keeping with the provisions of the operating license.

Advisory Committee on Reactor Safeguards (ACRS).   This committee, which consists of a maximum of 15 members, was established by Congress in 1957, in an amendment to the Atomic Energy Act of 1954, to review all power reactor license applications with regard to potential hazards and to report thereon to the Commission. The members of ACRS are not regular employees of the NRC.

Atomic Safety and Licensing Boards (ASLB).   These boards conduct hearings and make decisions with respect to the granting, suspending, revoking, or amending of any license authorized by the Atomic Energy Act. Each board consists of three members. One member (usually an attorney) must be qualified to conduct a hearing and serves as the chairman of the board. The other two members are drawn from the ranks of technology. (Because of the increasing importance of environmental questions in power plant hearings, one of the two technical members of the board is frequently a life scientist of one type or another.)

Steps in Licensing

The NRC has adopted two different approaches to the licensing of nuclear power plants. The original method was derived from the Atomic Energy Act of 1954. This process involves a construction phase and an operating phase. Under this approach, before a reactor can be built at a particular site, a *construction permit* (CP) must be obtained from the NRC. Before the reactor can be operated, an *operating license* (OL) must be obtained. The licensing process is thus divided into two stages, normally referred to as the *CP stage* and the *OL stage*. A long sequence of steps, which have become formalized in NRC procedures, are involved in each stage. In condensed form, these steps, which are illustrated in Fig. 11.1, are as follows:

The CP Stage

Informal Site Review.   The selection of the site for a nuclear power plant is the responsibility of the applicant proposing to build the plant. However, the NRC must ultimately approve the site, on the basis of criteria to be discussed in Section 11.6. Once a site has been selected, the applicant is encouraged, but not required, to ask the NRC for an informal review of the site before proceeding with a formal application for a license. In the Environmental Report (see Step 2), the applicant must compare the merits of at least two alternate sites.

Application for License.   The applicant files an application with the NRC for an operating license for the facility, although at this stage, only a construction permit can be received. The application is not a formidable document. It contains such information as the names and addresses of the major officers of the applicant's organization, the state of incorporation, evidence of financial qualifications to undertake the project, and so on. The information required for an application is given in 10 CFR 50.33.

In support of the application, the applicant must submit technical information, appropriate data for antitrust evaluation, and an environmental report. The largest and most difficult items to prepare are the following two documents: the *Preliminary Safety Analysis Report* (PSAR) and the *Applicant's Environmental Report* (AER).[1]

The PSAR is a major technical document, usually on the order of five volumes in length. The required content in a PSAR is specified in 10 CFR 50.34 and Regulatory Guide 1.70. Roughly speaking, the PSAR contains comprehensive data on the proposed site; a description of the proposed facility, with special consideration of safety-related features; a discussion of hypothetical accident situations and their consequences; a preliminary plan for the organization of personnel at the facility and for the conduct of operations; and plans for coping with emergencies.

The requirement for the submission of the AER stems from the National Environmental Policy Act of 1969 (NEPA, Public Law 91–190, 1970).

---

[1]The NRC publishes two documents that are valuable aids for those preparing or reviewing these reports. The Standard Review Plan outlines the Commission's safety criteria and the Environmental Standard Review Plan lists environmental guidelines. The NRC staff now must identify and document any deviations from these criteria; in the future, this responsibility will be shifted to the applicant.

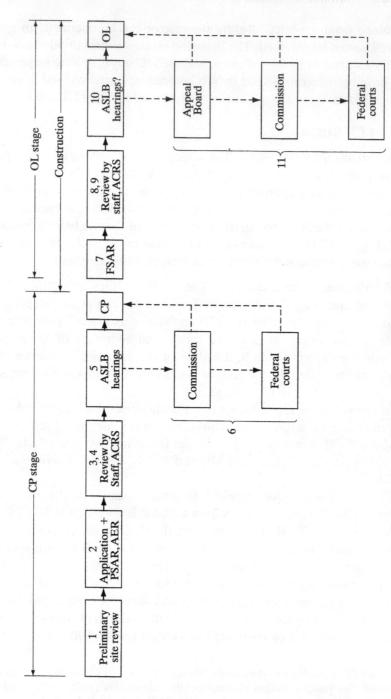

Figure 11.1 Schematic of steps in the NRC reactor licensing program.

According to this Act ... all agencies of the Federal Government shall ... include in every recommendation ... for major Federal actions significantly affecting the quality of the human environment, a detailed statement by the responsible official on—

(i) the environmental impact of the proposed action,

(ii) any adverse environmental effects which cannot be avoided should the proposal be implemented,

(iii) alternatives to the proposed action,

(iv) the relationship between local short-term uses of man's environment and the maintenance and enhancement of long-term productivity, and

(v) any irreversible and irretrievable commitments of resources which would be involved in the proposed action should it be implemented.

The issuance of a construction permit or operating license by the NRC is a "major Federal action," and, accordingly, the NRC requires each applicant to submit the AER, from which the NRC prepares its Environmental Statement, discussed next. The contents of the AER are based largely on items (i) through (v), together with evidence that the proposed facility complies with various environmental standards and regulations. The detailed content of an AER is given in 10 CFR 51 and Regulatory Guide 4.2.

Once the application and its supporting documentation have been received, checked for completeness, and accepted ("docketed"), the NRC publishes a notice of its receipt in the *Federal Register*[2] and places copies of all the documentation and correspondence in Public Document Rooms in Washington, D.C., and in the locale of the proposed plant. Simultaneously, copies of the documentation are distributed to members of the ACRS and to cognizant members of the NRC regulatory staff. The applicant is required to serve a copy of the application and supporting documents on the chief executive of the municipality or county in which the facility is to be built and to notify the chief executives of municipalities or counties that are identified as locations of alternative sites.

Review by Regulatory Staff.   The NRC regulatory staff reviews the applicant's documentation to determine whether the facility can be constructed without undue risk to the health and safety of the public and without unacceptable environmental impact. If, in the opinion of the staff, some aspect of the facility is unacceptable, the applicant is requested to make the necessary modifications. If the applicant cannot or will not do so, the conflict in views must be resolved

[2]The Federal Register, a journal of the regulatory activities of the departments and agencies of the executive branch of the government, is published by the National Archives and Records Service of the General Services Administration. New regulations are published first in the Federal Register. When extracted and codified, they become part of the CFR.

prior to issuance of the construction permit or operating license. In the extreme case, the application could be denied by the Commission. Following the review of the applicant's documentation, the staff then issues a *Safety Evaluation Report* (SER), which summarizes its findings and its conclusion that the facility can, in its judgment, be constructed without undue risk to the health and safety of the public.

The staff also conducts a review of the AER (this is called the NEPA review), and issues a *Draft Environmental Statement* (DES), summarizing the results of its evaluation of the environmental impact of the plant. The DES is then circulated to other federal, state, and local government agencies concerned with environmental matters for their comment or corroboration. Subsequently, the staff issues its Final Environmental Statement (FES), incorporating these comments and issues its conclusions.

Review by ACRS.

The documentation pertaining to the application is first assigned to a subcommittee of ACRS members. When the subcommittee has completed its review, and usually after the NRC staff has issued its SER, the full committee considers the project. Under its present charter, the ACRS is not responsible for reviewing the environmental impact of the proposed facility. The findings and recommendations of the committee are reported by letter to the chairman of the commission.

Following the meetings of the ACRS, the regulatory staff issues a *Supplement to the Safety Evaluation Report* (SSER). This supplemental report addresses any safety issues raised by the ACRS and includes any new information that has become available since the issuance of the SER.

Public Hearings.

After the staff's FES and SSER and the ACRS letter have been issued, a public hearing is held before an Atomic Safety and Licensing Board to consider whether the Commission should issue a CP. Public notice of these hearings, which are held in the vicinity of the proposed site, is issued and sent to local news media.

According to the provisions of the 1954 Act, any person whose interests may be affected by issuance of the CP may intervene in and become a party to the proceedings. To do so, the person must file a petition identifying what aspect of the project to be intervened in and explaining the interest and the basis of the contentions. The ASLB receives comments on the intervenor's petition from the regulatory staff and the applicant. If it then rules in favor of the intervention, the hearings are said to be contested.

The public hearings may be carried out in two stages. First, hearings are held to consider environmental and site-related issues. If, at the conclusion of these hearings, the Board determines (1) that the plant can be constructed without undue harm to the environment and (2) that there is reasonable assurance that the site is a suitable location for the plant, the Board may provide the applicant with a Lim-

ited Work Authorization (LWA). The LWA permits the applicant to begin preparation of the site for construction of the facility, including erection of construction
support facilities, excavation for plant structures, construction of roadways and
railroad spurs, installation of lighting and sewage systems, and other site-related
matters. Following the environmental hearings, the Board holds hearings on the
safety-related issues concerning the plant.

In uncontested hearings, the applicant's testimony consists of the application
and supporting documentation, including the PSAR, AER, and amendments, to
each that inevitably are added during the course of the staff review. The staff's
testimony includes its SER, SSER, and FES. In contested hearings, the applicant, the staff, and the intervenors provide additional testimony on points at issue.
Courtroom-style adjudicatory procedures are followed, including the right of all
parties to present and cross-examine witnesses.

At the close of the hearings, the Board considers the record that has been
compiled, together with findings of fact and conclusions of law filed independently
by the parties. The Board then issues what is known as its *initial decision*. If the
decision of the Board is favorable and the hearings are uncontested, the NRC Director of Nuclear Regulation, acting for the Commission, issues a CP, and work can
begin in earnest on the plant. Frequently, the CP is issued with conditions, based
on issues that are not entirely resolved to the Board's satisfaction, but for which
there is every reason to believe can be resolved at a later date.

Appeals.   Any party not satisfied with the initial decision may independently "file exceptions" (i.e., appeal the decision). The Commission handles
the appeal. The Commission reviews the record of the Licensing Board with
regard to fact and law. This is the last step available to a dissatisfied party
within the administrative (executive) branch of the federal government. If, at
this point, the Commission does not overrule the Licensing Board, the aggrieved
party may take his case to court—a federal court, as the issues involve federal
law.

The OL Stage   Much of the procedure that must be followed at this stage
of the licensing process is a repeat of what was done earlier at the CP stage.

Submittal for Operating License.   If a CP has been successfully obtained, the applicant has presumably begun to build the plant. When the construction has reached the point where all of the design details have been worked out and
the plans for operation of the facility have been finalized, the applicant submits to
the NRC an amendment to the original application in the form of a Final Safety
Analysis Report (FSAR), an updated Environmental Report, and a request for the
issuance of an operating license.

Review by Regulatory Staff.   The thrust of the staff review at this point is: (1) to determine whether any new information has come to light since the issuance of the CP, either regarding the specific facility itself or in nuclear technology, in general, that could have a bearing on the safe or environmentally sound operation of the facility; (2) to establish, by further review of design, operating plans, organization, and site inspections, that the facility has, in fact, been constructed according to the plans on which the issuance of the permit was based; (3) to assure the plant will operate in conformity with the application (as amended) and the rules and regulations of the Commission; (4) to gain reasonable assurance that operations are conducted without danger to the public; and (5) to verify the technical and financial qualifications of the applicant. Following its evaluation, the Staff issues a new Safety Evaluation Report.

Review by ACRS.   The ACRS makes another independent safety review of the facility and again sends its recommendations to the chairman of the Commission.

Hearings.   A public hearing before an ASLB is not required by law at the OL state. However, public notice is required on the pending issuance of an OL, to give another opportunity for an expression of public interest by way of a petition to intervene. If there is substantial public interest in the facility or if a hearing is requested by someone whose interest may be affected, then the Commission will order that a hearing be held. The decision-making process at this stage, including the right of intervention, is essentially the same as at the CP stage. If the board's decision is favorable, the OL is issued.

Appeals.   If the initial decision of the ASLB is not satisfactory in the view of an intervenor, several levels of appeal are open. These are the same at the OL stage as they are at the CP stage.

Licensing under 10 CFR 52

Recently, the Commission adopted a new approach for licensing reactors which is defined in 10 CFR Part 52. Under this approach, the NRC seeks early resolution of safety issues to facilitate standardization and simplification of the licensing process. The new approach utilizes the issuance of early site permits, design certification rulemaking, and combined construction and operating permits.

A critical element in this new process is the concept of design certification. Under 10 CFR 52, a rulemaking may be requested that accepts a specific nuclear power plant design. Anyone who then decides to purchase and build this design may reference the certification in the application for a construction permit and for an operating license or for a combined construction permit and operating license.

This means that the purchaser need not be subjected to a complete review of the design, as was the case under 10 CFR 50.

Another new aspect of the licensing process under 10 CFR 52 is the combined construction permit and operating license. The combined license enables an applicant to avoid issuance of a construction permit and then a separate operating license. The application is still subject to safety, environmental, and antitrust review, as in the case of the separate permitting and licensing process of 10 CFR 50.

11.3 PRINCIPLES OF NUCLEAR POWER PLANT SAFETY

The unique feature of nuclear power plants, as distinct from other power-generating facilities, is the presence of large amounts of radioactive materials, primarily the fission products. As has been noted previously in this book, the central safety problem in the design of a nuclear plant is to assure that, insofar as is possible or practical, these fission products remain safely confined at all times—during the operation of the plant, refueling of the reactor, and preparation and shipping of spent fuel.

Multiple Barriers

To prevent the escape of radioactivity, nuclear plants are designed using the concept of *multiple barriers*. These barriers represent a sequence of obstacles (not all of them physical) to block the passage of radioactive atoms from the fuel, or wherever they may originate, to the surrounding population. The barriers normally present are the following:

The Fuel   Except in the MSBR (see Section 4.5), the fissile and fissionable material is located within solid fuel elements—as natural or enriched uranium, in an oxide or carbide form, or in a dilute alloy of a structural material such as zirconium, aluminum, or stainless steel. Because fission fragments are emitted in fission as highly ionized particles, they are strongly attenuated, and, except for those fragments originating near the surface of the fuel, they all come to rest within the fuel. Subsequently, however, those fission products that are gases, namely, the isotopes of iodine, xenon, and krypton (the latter two are noble gases), undergo diffusion and may escape from the fuel. In PWR and BWR fuel rods, the gases that escape from the surface of the UO_2 pellets are held in the pellet-cladding gap and are collected in a small plenum provided at the end of each fuel rod.

The release of fission-product gases from UO_2 is highly temperature dependent and is also a function of the burnup of the fuel. Up to a burnup of approximately 20,000 megawatt-days/ton of UO_2, less than 1% of the gases escape from a pellet, provided that the temperature is less than 3,043°F. At temperatures above 3,043°F, recrystallization of UO_2 occurs and between 50 and 100% of the gases may be released from the fuel before it melts. Above the melting point, about 5,000°F, all of the fission product gases are released.

Cladding   To prevent the fission product gases from escaping, and to confine fission fragments emitted near the surface of the fuel, the fuel is surrounded by a layer of cladding. In some reactors, this cladding is bonded directly onto the fuel, as in the HTGR, where each fuel particle is coated with layers of pyrolytic carbon and silicon carbide (both ceramics). In other reactors, namely, the PWR and BWR, the cladding consists of hollow metal tubes into which the fuel pellets are inserted. Although Zircaloy is widely used for cladding, stainless steel cladding has also been used in some commercial reactors.

During the normal operation of a reactor, leaks can be expected to develop in the cladding of a few fuel elements (recall that a PWR or BWR of moderate size has several tens of thousands of fuel rods), despite the care with which the fuel and cladding are fabricated. This ultimately leads to the escape of small amounts of radioactivity into the environment, as will be discussed in Section 11.9. Should any part of the cladding reach its melting temperature, all of the fission product gases that have accumulated behind the cladding can escape into the coolant.

Closed Coolant System   In all modern power reactors, the primary coolant (i.e., the coolant that comes in contact with the fuel) moves in one or more closed loops. Fission products that have escaped from the fuel, activated atoms picked up by the coolant, and activated atoms of the coolant itself, are thus confined within the coolant system. Furthermore, in most reactors, a portion of the coolant is diverted on a continuing basis into a coolant purification (cleanup) system, where most of the fission products and other radioactive atoms are removed.

Reactor Vessel   Because they represent an obvious barrier to the release of radioactivity, reactor vessels are required to be designed, manufactured, and tested to meet the highest standards of quality and reliability. These requirements are spelled out in Section III of the American Society of Mechanical Engineers Boiler and Pressure Vessel Code.

LWR reactor vessels are fabricated from low carbon steels to prevent embrittlement that could lead to a sudden catastrophic rupture of a vessel. Great care must also be exercised in the rate of heating and cooling of a reactor vessel; changes are usually limited to about 25°C per hour. The time required to start up a power reactor from cold shutdown conditions or return it to cold shutdown from operational power level is determined by the heating or cooling of the reactor vessel.

Containment   All reactors are required to be entirely enclosed by a structure of one type or another to contain radioactivity, should this be released from either the coolant system or from within the reactor vessel itself.

Figure 11.2 shows a typical containment structure for a PWR plant. This structure also serves as the building to house the entire PWR nuclear steam supply system. Most present-day PWR containment structures are made of reinforced concrete with a steel liner. Their size and thickness are dictated by the maximum temperature and pressure that would result if all of the pressurized water in the primary system were released as steam into the structure as a result of a loss-of-coolant accident (LOCA), which is described in detail in Section 11.7. In some PWR plants, the pressure in the containment space is kept at slightly below

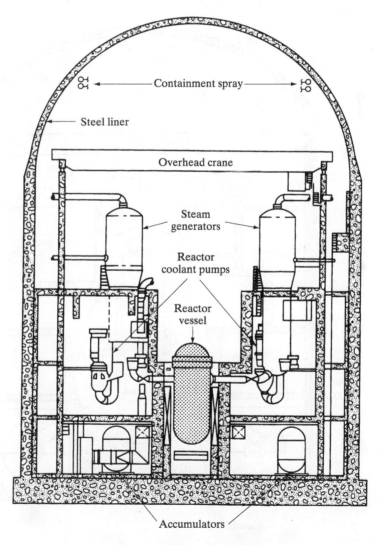

Figure 11.2   Typical PWR containment.

atmospheric pressure, so that leakage, at most times, through the containment wall is from the outside inward. Some plants utilize cold-water sprays near the top of the containment structure to condense the steam released during a LOCA. This reduces the pressure within the structure, thereby reducing the leakage of radioactivity from the building. Stored ice is also used for this purpose in some PWR facilities. Another measure often used is to circulate the containment atmosphere through various types of filters and absorption beds to remove airborne radioactivity.

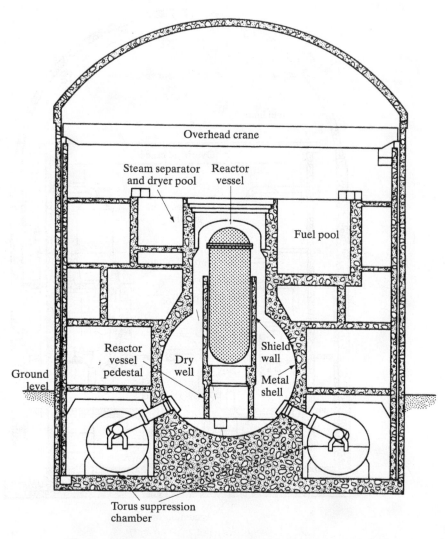

Figure 11.3 Light bulb and torus containment for a BWR.

BWR plants provide two levels of containment: the *primary containment*, which encloses the reactor, and the *secondary containment*, which, as does the PWR containment, more or less coincides with the reactor building. Several versions of primary containment are presently in use. Figure 11.3 shows the "light bulb and torus" system. The "light bulb," which is also called a *dry well*, since it is free of water, is a hollow metal shell, surrounded by concrete. Large ducts lead from the bottom of the dry well to the torus, which is normally about half-filled with water. In the event of a LOCA, the steam would be released into the dry well, pass through the ducts, and be condensed in the torus, thus relieving the pressure in the containment and reducing the likelihood of the escape of radioactivity. Figure 11.4 shows a more recent version of BWR containment in which the light bulb and torus are replaced by a large inner dry well and a suppression pool. Two levels of containment are possible with the BWR because of the compact nature of this reac-

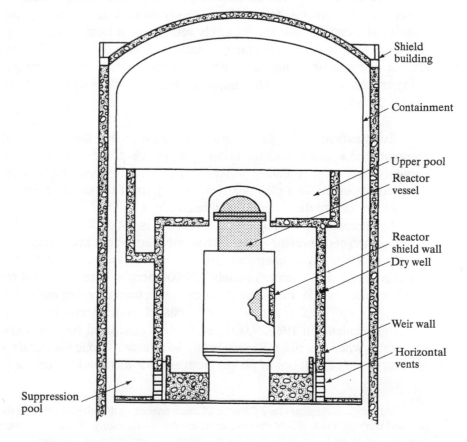

Figure 11.4 A recent form of a BWR containment. (Courtesy of General Electric Company.)

tor compared with the PWR, whose primary system includes the steam generators and pressurizer.

The HTGR also has two levels of containment. The prestressed concrete reactor vessel, holding the reactor core, heat exchangers, pumps (circulators), and so on, constitutes the primary containment. Secondary containment is the same as with the PWR and BWR, namely, a heavily built reactor building.

Together, the closed cooling system and the containment structure have proved very effective in preventing release of radioactivity to the environment in nuclear plant accidents. For example, at the Three Mile Island accident in March, 1979, only between 13 and 17 Ci of radioactive iodine were released. About 10×10^6 Ci of radioactive iodine was released from the fuel when it overheated, but only about 15 Ci of this radioactivity was released to the environment.

Site Location　　The technical aspects of reactor site selection are discussed in Section 11.6. At this point, it may merely be said that nuclear plants are constructed at locations that are relatively remote from large masses of people and where the plant and, in particular, the reactor vessel and the containment structures are not likely to be damaged by natural phenomena, such as earthquakes. Such siting criteria represent another important barrier to the exposure of the public to radioactivity.

Evacuation　　The final barrier, in the event that the physical barriers are compromised or prove inadequate during an accident, is the evacuation of the local populace from areas receiving or likely to receive excessively high radiation doses. An adequate evacuation plan is an important part of the documentation that must be approved before the issuance of either a CP or an OL.[3]

The evacuation, even of large populations, is not as difficult as sometimes perceived. Emergency evacuations are frequently carried out everywhere in the world, usually as the result of chemical spills, natural disasters, or wartime conditions. For example, in 1979, approximately 250,000 people were evacuated from a town near Toronto, Canada, following derailment of a train carrying chemical tank cars. Some years ago, 25,000 people were evacuated from Cicero, Illinois, because of a similar accident. In 1980, 9,000 people were evacuated from Somerville, Massachusetts, when a minor rail crash led to the release of toxic chemicals. Of course, the evacuation of millions of people from a major metropolitan center would pose

---

[3] Although an applicant for a CP or an OL must provide such a plan, developing the evacuation plan and carrying it out, should the need arise, are the responsibility of the local governmental authorities, not the applicant. If the need for evacuation appears imminent, personnel at a nuclear plant simply notify appropriate local authorities, in essentially the same way as they would do in the case of a fire. Such services are the *quid pro quo* for local taxes.

more serious difficulties. For this reason, nuclear plants are not located near such centers.

Three Levels of Safety   The use of multiple, successive barriers to the escape of radioactivity is basic to the design of nuclear power plants. In order to assure that none of these barriers are compromised as the result of such abnormal occurrences as equipment failure, human error, or natural phenomena, the NRC has adopted the concept of the *three levels of safety* as its safety philosophy. These are intended to provide a kind of in-depth defense to each of the radioactivity barriers. Naturally, there is some overlap between these levels, and the division between levels in some instances tends to be arbitrary.

The NRC defines each safety level by specific *precepts* or rules.

The First Level of Safety.   Precept: "Design for maximum safety in normal operation and maximum tolerance for system malfunction. Use design features inherently favorable to safe operation; emphasize quality, redundancy, inspectability, and testability prior to acceptance for sustained commercial operation and over the plant lifetime."

In brief, this first level of safety addresses the prevention of accidents by virtue of the design, construction, and surveillance of the plant. Some of the considerations involved at this level are the following:

1. The reactor should have a prompt negative temperature coefficient and a negative void coefficient.
2. Only materials whose properties are known to be stable under the operating conditions of the plant, including radiation exposure, should be used for the fuel, coolant, and safety-related structures.
3. Instrumentation and controls should be provided so that the plant operators know and have control over the status of the plant at all times. Sufficient redundancy must be included so that loss of key instruments or controls does not deprive operators of needed information or does not prevent shutdown of the plant.
4. The plant must be built and equipment installed in a manner that satisfies the highest standards of engineering practice.
5. Components should be designed and installed to permit continual or periodic monitoring and inspection for signs of wear and incipient failure and to permit periodic testing of the components.

The Second Level of Safety.   Precept: "Assume that incidents will occur in spite of care in design, construction and operation. Provide safety systems to protect operators and the public and to prevent or minimize damage when such incidents occur."

It is prudent to anticipate, despite the care taken at the first level of safety, that some failure may occur somewhere in the plant that could affect the safety of the facility. The object of the second level of safety is to protect plant personnel and the public from the consequences of such failures through the use of various safety devices and systems. Examples of considerations at this level include the following:

1. The reactor must be provided with an emergency core cooling system (ECCS) to prevent meltdown of the fuel and release of fission products due to fission product heating following a LOCA.

2. The reactor must have redundant capability for fast shutdown in the event that some of the control rods cannot be inserted, either because they are physically stuck or because of a malfunction of electrical circuitry.

3. The plant must be furnished with sources of power that are independent of the operation of the reactor to operate the ECCS, if necessary; to provide power for the continued operation of instrumentation; and to be used for other emergency situations in the plant. Such emergency power includes *off-site power*, supplied by two physically separated access circuits, and *on-site power* from generators driven by fast-starting, physically separated, and redundant-in-number diesel engines. On-site DC power for instrumentation is normally also supplemented with storage batteries.

The Third Level of Safety.   Precept: "Provide additional safety systems as appropriate, based on the evaluation of effects of hypothetical accidents, where some protective systems are assumed to fail simultaneously with the accident they are intended to control."

This third level of safety supplements the first two by adding a margin of safety in the event of extremely unlikely or unforeseen events. The need for additional engineered safety features is determined by analytically evaluating the effect on the plant, its associated personnel, and the public, of severe incidents arising from the simultaneous failure of various components of the facility and some of the redundant safety systems. Such events, used in this way to evaluate the overall safety of a plant and to point up the need for supplementary safety systems, are called *design basis accidents* (DBA) . The analysis of DBAs, discussed in Section 11.7, plays an important role in the design and licensing of a nuclear power plant.

Design Criteria   The NRC has translated the preceding safety philosophy into a set of *General Design Criteria* which establish the design, fabrication, construction, testing, and performance requirements for all the structures, systems, and components that are important to the safety of the plant. These criteria, which are too lengthy to be reproduced here, are given in Appendix A of 10 CFR 50.

11.4 DISPERSION OF EFFLUENTS FROM NUCLEAR FACILITIES

All nuclear plants emit small amounts of radioactivity, mostly fission product gases, during their normal operation. They may release considerably more radioactivity during the course of an accident. It is necessary to be able to calculate the doses to the public from such releases in order to evaluate the environmental impact of the normally operating plant, to ensure that this is within acceptable standards, and to ascertain the radiological consequences of reactor accidents. Such computations also play an important role in determining the acceptability of a proposed reactor site.

Before dose calculations can be carried out, however, it is necessary to determine how the concentration of the radioactive effluent varies from point to point following its emission into the atmosphere. This question is considered in this section. Dose calculations are discussed in Section 11.5.

Meteorology of Dispersion

Consider a volume of air of thickness dz and cross-sectional area A that is in equilibrium (motionless) at the altitude z as shown in Fig. 11.5. The volume is supported in place by the pressure difference between the top and bottom. Balancing forces on the volume means that

$$[P(z) - P(z + dz)]A = \rho g A \, dz, \tag{11.1}$$

where ρ is the air density within the volume and g is the acceleration of gravity. Simplifying Eq. (11.1) by taking the limit as dz tends to zero gives

$$-\frac{dP}{dz} = \rho g. \tag{11.2}$$

To a good approximation, air may be taken to be an ideal gas, in which case the ideal gas law

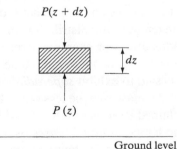

$P(z + dz)$

dz

$P(z)$

Ground level

Figure 11.5 Differential volume of air in equilibrium in the atmosphere.

$$PV = n_M RT, \qquad (11.3)$$

where n_M is the number of moles of gas in V and R is the gas constant, is valid. Dividing Eq. (11.3) by V and noting the n_M/V is proportional to the density of the gas then gives

$$P = \text{constant} \times \rho T. \qquad (11.4)$$

Air is not a good conductor. Furthermore, the normal motions of volumes of air within the atmosphere are so rapid that there is little exchange of heat from one volume to another. As a consequence, it is possible to assume that atmospheric motions are *adiabatic*. This, in turn, means that as the pressure changes, the temperature varies according to

$$T = \text{constant} \times P^{(\gamma-1)/\gamma}, \qquad (11.5)$$

where γ is the usual ratio of the specific heats at constant pressure and constant volume.

When ρ and P are eliminated from Eq. (11.2) using Eqs. (11.4) and (11.5), the following simple equation results:

$$-\frac{dT}{dz} = C.$$

Here C is a constant. The solution to this equation is

$$T = T_0 - Cz, \qquad (11.6)$$

where T_0 is the temperature at $z = 0$, ground level. It will be observed from Eq. (11.6) that T decreases linearly with increasing altitude. For this reason, T, which in Eqs. (11.3) and (11.5) has to be given on an absolute scale, can be expressed in either °C or °F in Eq. (11.6). The constant C is equal to the rate of change in temperature per unit altitude and is called the *adiabatic lapse rate*. This lapse rate is about 5.4°F/1,000 ft, or 1°C/100 m.

In the absence of vertical heat transfer, an atmosphere that is well mixed exhibits an adiabatic lapse rate. However, atmospheric conditions are continually changing, and, as a result, the temperature distribution in the atmosphere frequently differs substantially from adiabatic. For example, over certain ranges in altitude, the rate of temperature decrease may be more rapid than the adiabatic rate, and, in this case, the stratum of air is said to exhibit *superadiabatic* behavior. On the other hand, the rate of temperature decrease may be less rapid than the adiabatic rate, and the stratum, for reasons explained later, is said to be *stable*. An important situation of this kind occurs when the temperature *increases* over some range in altitude, a condition called an *inversion*, or when the temperature remains constant across a

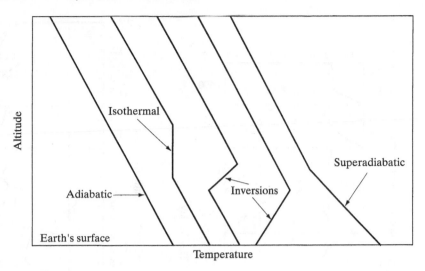

Figure 11.6 Examples of low-level temperature distribution in the atmosphere.

stratum, which is then known as *isothermal*. Examples of these temperature distributions are illustrated in Fig. 11.6.

The actual temperature profile of the atmosphere at any time is determined by a number of factors, including the heating and cooling of the earth's surface, the movement of large air masses, the existence of cloud cover, and the presence of local topographic obstacles. The earth is heated by the sun during the daytime, but at night, in the absence of sunshine, it cools down as it radiates energy. On clear days and with light winds, superadiabatic conditions may occur in the first few hundreds of meters of the atmosphere, due to the heat transferred to the air from the hot surface of the earth. On the other hand, on a cloudless night, when the earth radiates energy most easily, the earth's surface may cool down faster than the air immediately above it, and the result is a *radiation inversion*. This diurnal variation of the temperature profile of the atmosphere up to 450 m is shown in Fig. 11.7. The shifting from superadiabatic to inversion conditions and back again is plainly evident in the figure.

As is commonly shown on a weather map, regions of the atmosphere with identifiable characteristics occur throughout the world. Regions that are at higher pressure than the surrounding atmosphere are called *highs*; those at lower pressure are called *lows*. Most of the variations in weather and the local temperature profiles of the atmosphere are associated with the existence or movements of these highs and lows.

Consider a column of air as it sinks within the atmosphere as shown in Fig. 11.8. When the column falls to lower altitudes, it is heated adiabatically as its pres-

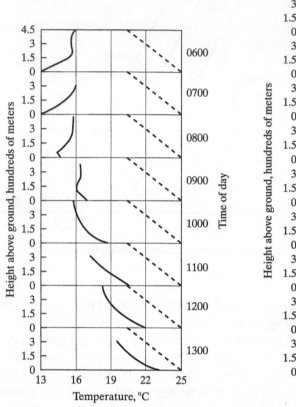 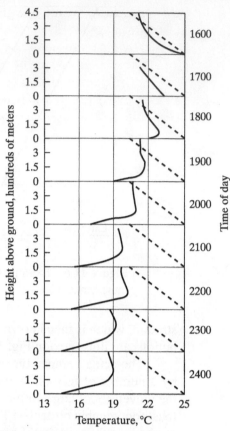

Figure 11.7 The average diurnal variation of the vertical temperature profile at the Oak Ridge National Laboratory during the period of September–October, 1950. The dashed line in each panel represents the adiabatic lapse rate. (From J. Z. Holland. USAEC report ORO-99, 1953.)

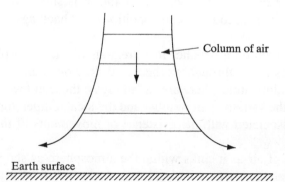

Figure 11.8 Sinking of a column of air to form a subsidence inversion.

sure increases. However, in a given interval of time, the top of the column moves a greater distance downward than the bottom of the column, since the latter is constrained by the presence of the earth's surface to move laterally. As a result, if the air falls far enough, the top of the column may reach a higher temperature than the bottom. This condition is known as a *subsidence inversion*. These types of inversions are found within the large high-pressure areas, characteristic of fair weather, that continually move across the North American and Asian continents. They also occur along the edges of oceanic stationary highs found at lower latitudes.

The degree to which pollutants are dispersed in the atmosphere depends to a large extent on the atmospheric temperature profile. Consider first the case of dispersion in a superadiabatic atmosphere. If a small volume of polluted air (called a *parcel* of air by meteorologists, since its volume generally changes after release) is released at some altitude h and at the same temperature T_a as the atmosphere, as indicated in Fig. 11.9(a), then according to the derivation of Eq. (11.6), the parcel will remain in static equilibrium at that point if not disturbed. Suppose, however, that a fluctuation in the atmosphere moves the parcel upward. The parcel will cool as it rises, but adiabatically; that is, the temperature of the parcel will follow the adiabatic curve, shown by dashed lines in the figure. It is clear from the figure that although the temperature of the parcel decreases in absolute terms, the parcel, as it rises, becomes increasingly hotter than the surrounding superadiabatic atmosphere. In turn, this means that the parcel becomes increasingly buoyant, causing it to move more rapidly upward, away from its point of release. On the other hand, if the parcel is initially pushed downward, its temperature will fall more rapidly than that of the surrounding atmosphere, and the air will become increasingly dense, which will accelerate the motion of the parcel downward. It is clear that superadiabatic temperature conditions are inherently *unstable* and are highly favorable for the dispersing of pollutants.

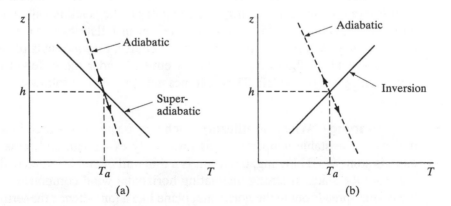

Figure 11.9 Movement of a parcel of air in (a) a superadiabatic profile and (b) an inversion profile.

By contrast, suppose the parcel is released into an inversion profile as shown in Fig. 11.9(b). If a fluctuation moves the parcel upward, its temperature falls more rapidly than that of the atmosphere, the parcel becomes denser, its upward motion stops, and it returns to its release point. If the parcel is moved downward, its temperature increases above that of the atmosphere, the parcel becomes more buoyant, and it again returns to the release point. The atmosphere in this case is said to be *stable*, obviously not a desirable state of affairs for pollutant dispersion. This same argument also applies for releases into an isothermal profile.

Finally, if a parcel is released in an adiabatic profile, its movements are not accompanied by changes in its relative buoyancy. The parcel simply moves a short distance up or down the adiabatic, subject to whatever local fluctuations occur in the atmosphere. For this reason, an adiabatic profile is referred to as *neutral*.

In the preceding discussion, it was assumed that the parcel of air was released at the same temperature as the atmosphere. If, as is frequently the case, the parcel is hotter than its surroundings, the parcel will immediately rise, due to its greater buoyancy. With unstable or neutral conditions, the parcel rises a distance Δh until it has mixed with and achieved temperature equilibrium with the atmosphere. From this point on, it disperses in a manner to be discussed later in this section. On the other hand, when a hot parcel is released in stable air, it rises until it reaches the inversion and then fans out as described next.

The distance Δh is called *plume rise*. A number of formulas, too complex to be reproduced here, have been given in the literature to estimate Δh. In any case, if plume rise is ignored in calculations of dispersion, the concentration of an effluent at ground level is overestimated, which results in conservative (higher than actual) estimates of radiation doses.

Plume Formation   The column or cloud of smoke emanating from the mouth of a continuously emitting chimney or smoke stack is called a *plume*. Radioactive effluents emitted from nuclear power installations behave in essentially the same way as ordinary smoke, except, of course, they are invisible to the naked eye. Figure 11.10 shows typical plumes emitted under the different temperature conditions given on the left. These plumes are classified as follows:

Fanning   When the effluent, which may be hot, is released in an inversion, it rises to the stable temperature, as previously explained, and remains there. If there is some wind blowing, the result is a thin trail of smoke at fixed altitude leading from the stack. If strong fluctuating horizontal wind components are present, the plume spreads out in the horizontal plane like a fan—hence the term *fanning*. In the absence of strong fluctuations, the plume becomes a long, meandering ribbon of smoke.

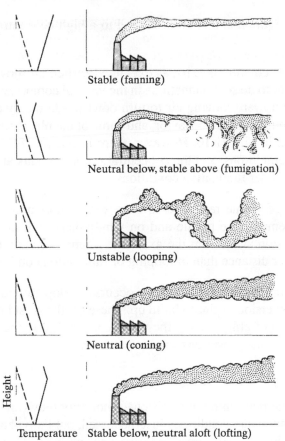

Stable (fanning)

Neutral below, stable above (fumigation)

Unstable (looping)

Neutral (coning)

Temperature   Stable below, neutral aloft (lofting)

Height

Figure 11.10 Various types of smoke plume patterns observed in the atmosphere. The dashed curves in the left-hand column of diagrams show the adiabatic lapse rate, while the solid lines are the actual profiles. (From D. H. Salde, Editor, *Meteorology and Atomic Energy—1968*. Washington, D.C.: US Atomic Energy Commission, 1968.)

A fanning plume is not necessarily an unfavorable condition for the dispersion of effluents. For one thing, such a plume does not touch the ground. However, if the plume is blocked in any way, the effluents necessarily remain in the area. This is the case in the Los Angeles region, which is surrounded by mountains to the north and east and which lies under the subsidence inversion from the stationary Pacific High. This combination of circumstances accounts for the persistence of smog in that region.

Fumigation   Shortly after the sun rises on a clear morning, the inversion due to the nighttime radiating of the earth begins to dissipate as the surface of the earth heats up. Starting at ground level, the inversion is replaced by an adiabatic profile that moves slowly upward. (See Fig. 11.7.) Effluents emitted after the new profile has reached the top of the stack are confined by the inversion overhead, but can be dispersed toward the ground as the result of turbulence developed in the newly heated air. Such *fumigation* conditions can lead to a high concentration of effluents at ground level.

Looping   If unstable conditions exist, as explained earlier, air moving upward or downward continues to do so. Fluctuations in the vertical component of the wind, however, may cause upward moving air to turn downward, and vice versa. As a result, the plume exhibits a random behavior, and some of the plume may even touch the ground, as shown in Fig. 11.10. However, there is usually considerable turbulence under unstable conditions, and the concentration of effluents at ground level is generally less than under fumigation conditions.

Coning   This type of plume resembles a cone with a horizontal axis. It occurs with an adiabatic temperature profile and on windy days with turbulence providing the radial dispersion away from the axis. The plume normally touches the ground at a much greater distance than in looping or fumigation conditions.

Lofting   The conditions under which lofting occurs develop around sunset, as the nighttime radiation inversion begins to build up. These are the most favorable conditions for the dispersion of effluents, for they are kept away from the ground and are dispersed at great distances over large volumes of air.

Diffusion of Effluents

An effluent released at some point into the atmosphere not only moves in a gross way, due to the various temperature conditions just discussed, but individual particles in the effluent become increasingly separated from one another as the result of local atmospheric turbulence. This process is called *turbulent diffusion*. This type of diffusion is fundamentally different from the diffusion of a solute in a solvent or ordinary neutron diffusion, despite the fact that they are all described by essentially the same equations. Thus, the latter types of diffusion result from the successive collisions of individual particles, while the diffusion of pollutants is due to the cumulative effects of turbulent eddies in the atmosphere.

Let χ be the concentration of some effluent as a function of space and time. If the atmosphere is isotropic and at rest, then χ is determined by the usual diffusion equation (see Eq. 5.16)

$$K \nabla^2 \chi = \frac{d\chi}{dt}, \tag{11.7}$$

where K is the diffusion coefficient. For the more usual case of a nonisotropic atmosphere, the diffusion equation is

$$K_x \frac{d^2\chi}{dx^2} + K_y \frac{d^2\chi}{dy^2} + K_z \frac{d^2\chi}{dz^2} = \frac{d\chi}{dt}. \tag{11.8}$$

With a wind blowing at an average speed $\bar{v}$ in the x-direction, the diffusion equation must be altered to account for the fact that the entire medium in which the diffusion is taking place is in motion. The equation then becomes[4]

$$K_x \frac{d^2\chi}{dx^2} + K_y \frac{d^2\chi}{dy^2} + K_z \frac{d^2\chi}{dz^2} = \frac{d\chi}{dt} + \bar{v}\frac{d\chi}{dx}. \tag{11.9}$$

Consider now a point source located at the origin of coordinates emitting effluent at the constant rate of Q' units per unit of time. The concentration χ is then not a function of time. Further, it has been found experimentally that most of the movement of an effluent in the direction of the wind is due to the wind itself and not to diffusion. Thus, diffusion in the x-direction can be ignored, which can be accomplished by placing K_x equal to zero. Equation (11.9) then reduces to

$$K_y \frac{d^2\chi}{dy^2} + K_z \frac{d^2\chi}{dz^2} = \bar{v}\frac{d\chi}{dx}. \tag{11.10}$$

The solution to Eq. (11.10) that satisfies all the usual boundary conditions can be shown to be[5]

$$\chi = \frac{Q'}{4\pi x\sqrt{K_y K_z}} \exp\left[-\frac{\bar{v}}{4x}\left(\frac{y^2}{K_y} + \frac{z^2}{K_z}\right)\right]. \tag{11.11}$$

According to Eq. (11.11), the effluent moving along the x-direction spreads out in Gaussian distributions in the y- and z- directions. The standard deviations of these distributions are given by

$$\sigma_y = \left(\frac{2xK_y}{\bar{v}}\right)^{1/2}; \ \sigma_z \left(\frac{2xK_z}{\bar{v}}\right)^{1/2}. \tag{11.12}$$

For purposes of matching Eq. (11.11) with experimental data, it is convenient to write the equation in terms of σ_y and σ_z, which, it will be noted, are functions of x. Thus, Eq. (11.11) becomes

$$\chi = \frac{Q'}{2\pi \bar{v}\sigma_y\sigma_z} \exp\left[-\left(\frac{y^2}{2\sigma_y^2} + \frac{z^2}{2\sigma_z^2}\right)\right]. \tag{11.13}$$

In the present context, σ_y and σ_z are called, respectively, the horizontal and vertical *dispersion coefficients*.

Up to this point, it has been assumed that the effluents are emitted at the origin of coordinates into an infinite atmosphere. In fact, they are generally emitted

---

[4]See, for example, P. M. Morse and H. Feshbach, *Methods of Theoretical Physics*, Burr Ridge: McGraw-Hill, 1953, p.153.

[5]See, for instance, LaMarsh's *Nuclear Reactor Theory, op. cit.*, p. 192. Equation (11.10) is identical in form to the Fermi age equation describing the slowing down of fast neutrons.

at some altitude h into an atmosphere that exists only above the ground. The solution to the diffusion equation in this case can easily be found using the method of images, familiar from electrostatics. With z at the vertical coordinate, this solution is

$$\chi = \frac{Q'}{2\pi \bar{v}\sigma_y\sigma_z} \left\{ \exp\left[-\left(\frac{y^2}{2\sigma_y^2} + \frac{(z+h)^2}{2\sigma_z^2} \right) \right] \right.$$

$$\left. + \exp\left[-\left(\frac{y^2}{2\sigma_y^2} + \frac{(z-h)^2}{2\sigma_z^2} \right) \right] \right\}. \tag{11.14}$$

From this result, the concentration at ground level, $z = 0$, is

$$\chi = \frac{Q'}{\pi \bar{v}\sigma_y\sigma_z} \exp\left[-\left(\frac{y^2}{2\sigma_y^2} + \frac{h^2}{2\sigma_z^2} \right) \right]. \tag{11.15}$$

The value of χ is largest along the centerline of the plume (i.e., where $y = 0$). The concentration there is

$$\chi = \frac{Q'}{\pi \bar{v}\sigma_y\sigma_z} \exp\left(-\frac{h^2}{2\sigma_z^2} \right). \tag{11.16}$$

Furthermore, because the exponential factor in Eq. (11.16) is never greater than unity, it follows that the effluent concentration at all points is always greater along the plume with a ground-level release ($h = 0$) than when the effluents are released at some altitude. In this case, Eq. (11.16) reduces to

$$\chi = \frac{Q'}{\pi \bar{v}\sigma_y\sigma_z}. \tag{11.17}$$

To assure conservative estimates (i.e., overestimates) of effluent concentration, it is normal practice to use Eq. (11.16) when the altitude of emissions is known and Eq. (11.17) when it is not.

In many contexts, it is convenient to divide these equations by Q'. The LHS of the equation, namely, χ/Q', is then called the *dilution factor*.

Occasionally, radioactivity is emitted from nuclear power plants in puffs, rather than at a constant rate. The effluent concentration at any point at ground level then rises in time to some maximum value and subsequently falls to zero as the puff passes. It will be shown in the next section that the total radiation dose from such a puff is proportional to the time integral of χ over the passage of the puff that is, to χ_T, where

$$\chi_T = \int \chi(t)\, dt. \tag{11.18}$$

Because the emission and dispersion of a pollutant are separate and distinct processes, χ_T can be written as

$$\chi_T = \frac{Q}{\pi \bar{\upsilon} \sigma_y \sigma_z} \exp\left(-\frac{h^2}{2\sigma_z^2}\right). \tag{11.19}$$

Here, Q is the total amount of effluent released in the puff (i.e., the time interval of Q').

Pasquill conditions. According to Eq. (11.12), both σ_y and σ_z should increase as $\sqrt{x}$ from the point of emission. In fact, experimental data show that σ_y and σ_z increase much more rapidly with distance. This means that the diffusion model for atmospheric dispersion is not an exact description of the phenomenon. The time-averaged distribution of effluents in the y- and z-directions, however, has been found to be approximately Gaussian. It has become standard practice, therefore, to use experimental values of σ_y and σ_z in Eq. (11.13) or (11.16) to calculate effluent concentration. Naturally, these functions depend on atmospheric conditions. Thus, σ_z, which relates to the dispersion in the vertical direction, can be expected to increase more rapidly with distance under unstable conditions than under stable conditions.

Working from experimental data, Pasquill[6] obtained a set of curves for σ_y and σ_z for six different atmospheric conditions. These are given in Figs. 11.11 and 11.12. A seventh stability condition, type G, extremely stable, may be approximated by the following relations:

$$\sigma_z(G) = \frac{3}{5}\sigma_z(F); \quad \sigma_y(G) = \frac{2}{3}\sigma_y(F). \tag{11.20}$$

As shown in the figures, the less stable conditions have higher values of both σ_y and σ_z than stable conditions, at all distances from the source.

It is possible to estimate the stability conditions in the lower atmosphere by simply measuring the temperature at two or more heights on a meteorological tower. The slope of the temperature profile can then be computed by dividing the temperature difference ΔT by the difference in height Δz of the measurements. The relationship between the Pasquill stability categories and the observed $\Delta T/\Delta z$ is given in Table 11.2. In the United States, the NRC requires the temperature (as well as wind speed and direction) to be continually monitored at two points, usually at 10 m and 60 m, on a tower or mast near every operating nuclear plant.

The Pasquill conditions can also be determined by monitoring the fluctuations in the angle of a wind vane. On days when the atmosphere is unstable, a wind vane tends to fluctuate more widely than on days when the atmosphere is stable. The correlation between the standard deviation of the angle of the vane, σ_θ, to

---

[6]See the references at the end of this chapter.

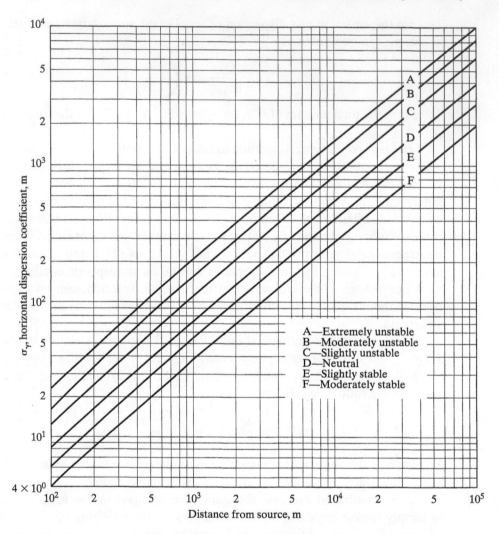

Figure 11.11   Horizontal dispersion coefficient σ_y as a function of distance from source for the various Pasquill conditions. (From D. H. Slade, Editor, *Meteorology and Atomic Energy—1968*. Washington, D.C.: US Atomic Energy Commission, 1968.)

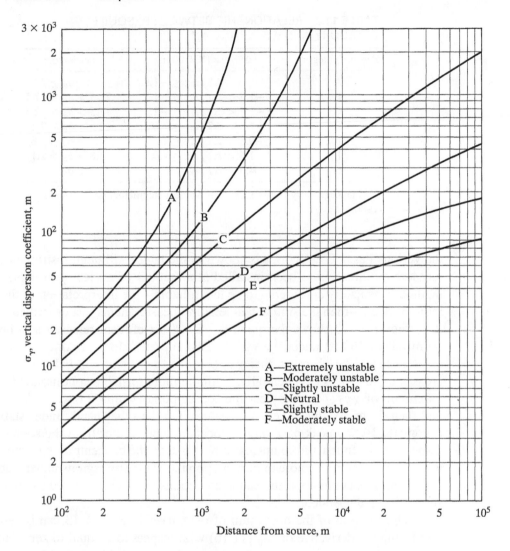

Figure 11.12   Vertical dispersion coefficient σ_z as a function of distance from source for the various Pasquill conditions. (From D. H. Slade, Editor, *Meteorology and Atomic Energy—1968*. Washington, D.C.: US Atomic Energy Commission, 1968.)

TABLE 11.2 RELATIONSHIP BETWEEN PASQUILL
CATEGORY AND $\Delta T/\Delta z$ AND σ_θ*

Pasquill category	$\Delta T/\Delta z$ (°C/100 m)	σ_θ (degrees)
A	$\Delta T/\Delta z \leq -1.9$	$\sigma_\theta \geq 22.5$
B	$-1.9 < \Delta T/\Delta z \leq -1.7$	$22.5 > \sigma_\theta \geq 17.5$
C	$-1.7 < \Delta T/\Delta z \leq -1.5$	$17.5 > \sigma_\theta \geq 12.5$
D	$-1.5 < \Delta T/\Delta z \leq -0.5$	$12.5 > \sigma_\theta \geq 7.5$
E	$-0.5 < \Delta T/\Delta z \leq 1.5$	$7.5 > \sigma_\theta \geq 3.8$
F	$1.5 < \Delta T/\Delta z \leq 4.0$	$3.8 > \sigma_\theta \geq 2.1$
G	$4.0 < \Delta T/\Delta z$	$2.1 > \sigma_\theta$

*From Regulatory Guide 1.23, U.S. Nuclear Regulatory Commission,
1980.

the various Pasquill categories is shown in the table. Although instrumentation to measure σ_θ directly has been installed at a number of nuclear power plants, this method is generally considered to be less reliable—the interpretation of the data is more difficult—than the simple temperature measurements just described.

Figure 11.13 shows the quantity $\chi \bar{v}/Q'$ for effluent released at a height of 30 m (about 100 ft) under the various Pasquill conditions as computed from Eq. (11.16). It will be observed that $\chi \bar{v}/Q'$ rises to a maximum value and then decreases more or less exponentially. With the more unstable conditions (A, B), the maximum of $\chi \bar{v}/Q'$ occurs near the source point (within a few hundred meters) and then drops rapidly to very low values. On the other hand, under stable conditions (E, F), the peak of $\chi \bar{v}/Q'$ is located much further from the source. In the dispersion of effluents from nuclear power plants, the concentration of the effluent is usually higher in the more important, populated off-site regions under stable than under unstable conditions, and stable conditions are often assumed calculations of such effluent dispersion.

The location of the maximum of the curves in Fig. 11.13 can be estimated by placing the derivative of Eq. (11.16) with respect to x equal to zero. However, because all of the dependence of χ upon x is contained in σ_y and σ_z, it is necessary, before differentiating, to assume some functional relationship between these parameters. For simplicity, let

$$\sigma_y = a\sigma_z,$$

where a is a constant. Taking the logarithms of both sides of Eq. (11.16) prior to differentiating gives

$$\ln \chi = -2 \ln \sigma_z - \frac{h^2}{2\sigma_z^2} + \ln C,$$

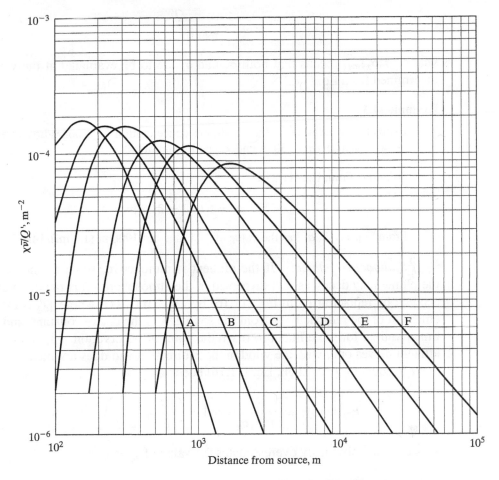

Figure 11.13   The quantity $\chi \bar{v}/Q'$ at ground level, for effluents emitted at a height of 30 m, as a function of distance from the source. (From D. H. Slade, Editor, *Meteorology and Atomic Energy—1968*. Washington, D.C.: US Atomic Energy Commission, 1968.)

where C is a composite of constants. Then,

$$\frac{1}{\chi}\frac{d\chi}{dx} = \left(\frac{2}{\sigma_z} + \frac{h^2}{\sigma_z^3}\right)\frac{d\sigma_z}{dx} = 0,$$

and so

$$h^2 = 2\sigma_z^2. \tag{11.21}$$

This condition determines the location of the maximum. Finally, substituting Eq. (11.21) into Eq. (11.16) gives

$$\chi_{\text{max}} = \frac{Q'}{\pi e \bar{v} (\sigma_y \sigma_z)_{\text{max}}}, \qquad (11.22)$$

where $(\sigma_y \sigma_z)_{\text{max}}$ means that both σ_y and σ_z are to be evaluated at the value of x determined from Eq. (11.21).

Example 11.1

Estimate the location of the maximum concentration of a nonradioactive effluent released at a height of 30 m under type-F conditions.

Solution. From Eq. (11.21),

$$\sigma_z = h/\sqrt{2} = 21.2 \text{ m}.$$

From Fig. 11.12, σ_z has the value of 21.1 at about 1900 m (1.2 mi). [*Ans.*]

Radioactive effluent . If the effluent is radioactive, some of the activity in the plume may decay as it is dispersed in the atmosphere. This may be taken into account by replacing Q' in Eq. (11.16) by $Q_0' \exp(-\lambda t)$, where Q_0' is the rate of emission of the radioactivity from the source, λ is the decay constant, and t is the time required for the effluent to reach the point of observation. Assuming that the effluent moves only with the wind in the x-direction and does not meander, we see that t is equal to $\chi/\bar{v}$. Then, Eq. (11.16) becomes

$$\chi = \frac{Q_0'}{\pi v \sigma_y \sigma_z} \exp\left[-\left(\frac{\lambda x}{\bar{v}} + \frac{h^2}{2\sigma_z^2}\right)\right]. \qquad (11.23)$$

This equation obviously overestimates the value of χ at x.

Deposition and Fallout

The amount of radioactivity in an effluent plume may also decrease with distance from the source as some of the radioactivity falls out or diffuses out of the plume and deposits on the ground. This effect is most pronounced for some types of radionuclides, in particular, the isotopes of iodine, during periods of rain. Water droplets tend to pick up the radioactivity and carry it directly to the ground, a process known as *washout*, or *wet deposition*. Once radioactivity is deposited on the ground, it may provide a significant source of radiation exposure. Furthermore, radioactivity falling on food stuffs, pasture land, or bodies of water may enter into the human food chain.

The mathematical description of wet deposition is qualitatively similar to that for ordinary, or dry, deposition, although the underlying phenomena are fundamentally different. Thus, for both, the deposition rate per unit area of the ground is taken to be proportional to the effluent concentration and is written as

$$R_d = \chi v_d,$$ (11.24)

where v_d is a proportionality constant. If R_d has the units of Ci/m^2-sec and χ is in Ci/m^3, then v_d has the units of m/sec and is called the *deposition velocity*. In Eq. (11.24), χ is evaluated at or near ground level, and the value of v_d is obtained from experiment.

The dry-deposition velocity for iodine ranges from about 0.002 to 0.01 m/sec; it is somewhat smaller for particulate fallout. The value of v_d for wet deposition depends on the height of the plume ($\bar{z}$ in the calculations presented next). For a nominal height of 10^3 m, v_d is approximately 0.2 m/sec for iodine and 0.1 m/sec for particulates. The noble gases are not subject to either dry or wet deposition.

Consider now the concentration of an effluent subject to deposition as a function of distance from the source. Along the centerline of the plume, χ will be a function of only x and z. The total amount of effluent (radioactivity) in a volume element of the plume of unit width in the y-direction and thickness dx is given by the integral

$$dx \int_{z=0}^{\infty} \chi(x, z)\, dz.$$

The rate at which this effluent decreases due to deposition is then

$$\frac{d}{dt} dx \int_{z=0}^{\infty} \chi(x, z)\, dz = -\chi(x, 0) v_d dx.$$ (11.25)

Noting that $dx/dt = \bar{v}$, we find that the average wind speed, Eq. (11.25), can be written as

$$d \int_{0}^{\infty} \chi(x, z)\, dz = -\frac{\chi(x, 0) v_d dx}{\bar{v}}.$$ (11.26)

Next, define $\bar{z}$ by the relation

$$\bar{z} = \frac{1}{\chi(x, 0)} \int_{0}^{\infty} \chi(x, z)\, dz.$$ (11.27)

This parameter is called the *effective height* of the plume. It is easy to show that if $\chi(x, z)$ is given by Eq. (11.14), then

$$\bar{z} = \sqrt{\frac{\pi}{2}} \sigma_z \, \exp(h^2/2\sigma_z^2).$$ (11.28)

Introducing $\bar{z}$ into Eq. (11.26) gives

$$d\chi(x, 0) = -\Lambda \chi(x, 0)\, dx,$$ (11.29)

where

$$\Lambda = \frac{v_d}{vz}.$$ (11.30)

Equation (11.29) shows that at ground level, χ decreases exponentially with distance, due to deposition. Combining this result with Eq. (11.23) gives finally

$$\chi = \frac{Q_0'}{\pi \bar{v} \sigma_y \sigma_z} \exp\left[-\left(\Lambda + \frac{l}{\bar{v}} x - \frac{h^2}{2\sigma_z^2}\right)\right].$$ (11.31)

The Wedge Model

The diffusion model with empirical dispersion coefficients gives reasonably accurate values of effluent concentration up to the order of 10^4 m $= 10$ km from the source. At much larger distances, however, diffusion theory cannot be expected to be valid. Fluctuations in atmospheric conditions tend to disperse an effluent at long distances in an unpredictable way. While the anticipated radiation dose to an individual located far from a nuclear plant is usually negligible, even after a major accident, the total population dose (man-rems or people-sieverts) may be significant. For calculations of population doses a simple method, the *wedge model* of atmospheric dispersion, is often used.

Consider a source emitting effluent at the rate of Q' units per time. The assumption underlying the wedge model is that the effluent concentration becomes uniform within a wedge of angle θ and height a as indicated in Fig. 11.14. In the absence of radioactive decay or deposition on the ground, all of the effluent emitted in the time dt, namely, $Q'dt$, eventually appears in the volume element dV between r and $r + dr$, where

$$dV = ar\theta dr.$$

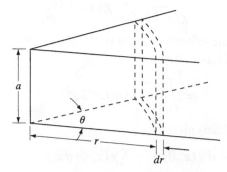

Figure 11.14 Diagram for calculating the wedge model.

If $\chi(r)$ is the concentration at r, then it follows that

$$\chi(r)dV = Q'dt$$

and

$$\chi(r) = \frac{Q'dt}{ar\theta dr}.$$

However, dr/dt is equal to the wind speed $\bar{v}$, so that

$$\chi(r) = \frac{Q'}{ar\theta\bar{v}}. \tag{11.32}$$

When radioactive decay and deposition are included, Eq. (11.32) becomes

$$\chi(r) = \frac{Q'}{ar\theta\bar{v}} \exp\left[-\left(\Lambda + \frac{\lambda}{\bar{v}}\right)r\right], \tag{11.33}$$

where Λ is given by Eq. (11.30).

Releases from Buildings

In a reactor accident, radionuclides may be released into the containment structure, which may then leak at one or more points. If a wind is blowing, the presence of the structure creates a wake of turbulent eddies on the down-wind side of the building. This, in turn, tends to disperse the effluents immediately upon their release from the building. The effect of this initial dispersion can be included in the preceding formula in the following way.

First, the *building dilution factor* is defined as

$$D_B = cA\bar{v}, \tag{11.34}$$

where c is an empirical constant, sometimes called the *shape factor*; A is the cross-sectional area of the building; and $\bar{v}$ is the average wind speed. Note that D_B has units of m^3-sec^{-1}. Experiments show that c has a value of between 0.50 and 0.67. The dispersion coefficients σ_y and σ_z are then replaced in all of the formulas derived earlier by the parameter Σ_y and Σ_z, which are defined by the relations:

$$\Sigma_y^2 = \sigma_y^2 + \frac{D_B}{\pi\bar{v}}, \quad \Sigma_z^2 = \sigma_z^2 + \frac{D^B}{\pi\bar{v}}. \tag{11.35}$$

Thus, for example, Eq. (11.16) becomes

$$\chi = \frac{Q'}{\pi\bar{v}\Sigma_y\Sigma_z} \exp\left(-\frac{h^2}{2\Sigma_z^2}\right). \tag{11.36}$$

This equation is applicable to releases near or at ground level, and to be conservative, h is often placed equal to zero. Another essentially equivalent formula that is often used to compute χ from ground-level releases from buildings is

$$\chi = \frac{Q'}{\pi \bar{v} \sigma_y \sigma_z + D_B}. \qquad (11.37)$$

In the immediate vicinity of a leaking structure, both σ_y and σ_z are equal to zero, for the effluent at this point has not begun to form a plume. In this case, both Eq. (11.36), with $h = 0$, and Eq. (11.37) reduce to

$$\chi = \frac{Q'}{D_B}. \qquad (11.38)$$

This formula can also be used to compute the concentration of radionuclide in the neighborhood of an exhaust vent. If the concentration of the radionuclide in the vent is χ_0 Ci/m^3 and the vent exhausts at the rate of V m^3/sec, then $Q' = V\chi_0$, and Eq. (11.38) gives

$$\chi = \left(\frac{V}{D_B}\right)\chi_0. \qquad (11.39)$$

11.5 RADIATION DOSES FROM NUCLEAR PLANTS

Nuclear plants may give rise to a number of sources of radiation exposure to persons residing in their vicinity. From their gaseous radioactive effluent, there is: (1) an external dose from radiation emitted from the plume; (2) an internal dose from the inhalation of radionuclides; (3) an external dose from radiation emitted by radionuclides deposited on the ground; and (4) an external dose from radionuclides deposited on the body and clothing. The last of these doses is highly variable and will not be considered here.

Radiation doses may also be received from the ingestion of foodstuffs contaminated by gaseous or liquid effluents from a nuclear plant. The calculation of these doses is covered in Section 11.9.

Finally, there is a *direct dose* from γ-rays emitted from within the plant. In a normally operating plant, these may emanate from the reactor itself, from coolant piping, or from other components of the plant containing radioactivity. Following a reactor accident in which fission products are released into the containment building, the building becomes a source of direct radiation dose.

External Dose from Plume: γ-Rays

In making calculations of external doses, it is usual to assume that the plume is infinitely large. This assumption simplifies the computations and gives conservative answers; that is, it gives doses that are larger than they actually are.

Consider an infinite uniform cloud located above ground level, containing a single radionuclide at a concentration of χ Ci/cm^3, emitting a single γ-ray of energy E MeV. According to Eq. (9.20) and the discussion in Section 10.1, the exposure rate is

$$\dot{X} = 1.83 \times 10^{-8} \phi_{\gamma b} E (\mu_a/\rho)^{\text{air}} \text{ R/sec.} \tag{11.40}$$

where $\phi_{\gamma b}$ is the γ-ray buildup flux. To compute $\phi_{\gamma b}$, let dV be a small volume element in a hemispherical shell of radius r and thickness dr centered at the point P as shown in Fig. 11.15. If S γ-rays are emitted per cm^3/sec, the buildup flux at P from dV is

$$d\phi_{\gamma b} = \frac{S\,dV}{4\pi r^2} B_p(\mu r)e^{-\mu r},$$

where $B_p(\mu r)$ is the point buildup factor for the air. The total flux from all elements dV in the shell is then

$$d\phi_{\gamma b} = \frac{S}{4\pi r^2} \times 2\pi r^2\,dr \times B_p(\mu r)e^{-\mu r}$$

$$= \frac{S}{2} B_p(\mu r)e^{-\mu r}\,dr, \tag{11.41}$$

and $\phi_{\gamma b}$ is given by

$$\phi_{\gamma b} = \frac{S}{2} \int_0^\infty B_p(\mu r)e^{-\mu r}\,dr. \tag{11.42}$$

The value of the integral in Eq. (11.42) is independent of the form of $B_p(\mu r)$, as can be seen by the next argument. Suppose that a point source emits S γ-rays/sec

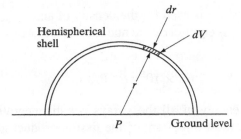

Figure 11.15   Hemispherical shell in a cloud.

of energy E into an infinite atmosphere. The buildup flux at the distance r is then

$$\phi_{\gamma b} = \frac{S}{4\pi r^2} B_p(\mu r) e^{-\mu r}. \tag{11.43}$$

The energy absorbed out of this flux in a spherical shell of thickness dr at r is

$$\phi_{\gamma b} \mu_a E \, dV = \frac{S \mu_a E}{4\pi r^2} \times B_p(\mu r) e^{-\mu r} \times 4\pi r^2 dr$$

$$= S \mu_a E B_p(\mu r) e^{-\mu r} \, dr,$$

where μ_a is the absorption coefficient of air. Since the atmosphere is infinite, all of the energy emitted by the source per second, namely, SE, must be absorbed somewhere. It follows that

$$SE = \int_0^\infty \phi_{\gamma b} \mu_a E \, dV.$$

Introducing $\phi_{\gamma b}$ from Eq. (11.43) gives

$$1 = \mu_a \int_0^\infty B_p(\mu r) e^{-\mu r} \, dr. \tag{11.44}$$

In view of Eq. (11.44), Eq. (11.42), giving the buildup flux at the center of the cloud becomes

$$\phi_{\gamma b} = \frac{S}{2\mu_a}. \tag{11.45}$$

With a radionuclide concentration in the cloud of χ Ci/cm³, S is equal to $3.7 \times 10^{10} \chi$, provided that there is only one γ-ray emitted per disintegration. Introducing these results into Eq. (11.40) yields

$$\dot{X} = 1.83 \times 10^{-8} \times \frac{3.7 \times 10^{10} \chi}{2\mu_a} \times E \left(\frac{\mu_a}{\rho}\right)^{\text{air}}$$

$$= 2.62 \times 10^5 \chi E \text{ R/sec}, \tag{11.46}$$

where $\rho = 1.293 \times 10^{-3}$ g/cm³ is used for the density of air.

If more than one γ-ray is emitted by the nuclide, Eq. (11.46) should be written as

$$\dot{X} = 2.62 \times 10^5 \chi \bar{E}_\gamma \text{ R/sec}, \tag{11.47}$$

where $\bar{E}_\gamma$ is the average energy of all the γ-rays per disintegration. Table 11.3 gives values of $\bar{E}_\gamma$ for the most important of the fission product gases. These are

TABLE 11.3 AVERAGE DECAY ENERGIES PER DISINTEGRATION OF FISSION PRODUCT GASES*

Nuclide†	Half-life‡	$\overline{E}_\gamma$, MeV	$\overline{E}_\beta$, MeV
^{85m}Kr	4.4 h	0.151	0.223
^{85}Kr	10.76 y	0.00211	0.223
^{87}Kr	76 m	1.37	1.05
^{88}Kr	2.79 h	1.74	0.341
^{133m}Xe	2.26 d	0.326	0.155
^{133}Xe	5.27 d	0.030	0.146
^{135m}Xe	15.7 m	0.422	0.0974
^{135}Xe	9.2 h	0.246	0.322
^{131}I	8.04 d	0.371	0.197
^{132}I	2.28 h	2.40	0.448
^{133}I	20.8 h	0.477	0.423
^{134}I	52.3 m	1.94	0.455
^{135}I	6.7 h	1.78	0.308

*From J. S. Moore and R. Salvatori, *Lectures on Nuclear Safety*, University of Pittsburg, 1974.

†Superscript *m* refers to a nuclide in an isomeric state (see Section 2.8).

‡m = minutes, h = hours, d = days, y = years.

the gases that are released to the environment in small quantities during the normal operation of a reactor and in somewhat larger amounts in an accident.

Equation (11.47) can also be derived in a less complicated way. At any point in an infinite uniform cloud, as many γ-rays must be absorbed in the steady state as are emitted per cm^3/sec. For a radionuclide at a concentration of χ Ci/cm^3 emitting an average of $\bar{E}_\gamma$ MeV per disintegration, the total amount of energy emitted is $3.7 \times 10^{10} \chi \bar{E}_g \times 1.6 \times 10^{-6}$ erg/cm^3-sec. This energy is absorbed in air having a mass of 1.293×10^{-3} g/cm^3, so that the energy absorption rate is

$$\frac{3.7 \times 10^{10} \chi \bar{E}_\gamma \times 1.6 \times 10^{-6}}{1.293 \times 10^{-3}} = 4.58 \times 10^7 \chi \bar{E}_\gamma \text{ erg/g-sec.}$$

From Eq. (9.19), the absorption of 87.5 erg/g gives an exposure of 1 R. Thus, the exposure rate in conventional units is

$$\dot{X} = \frac{4.58 \times 10^7 \chi \bar{E}\gamma}{87.5} = 5.23 \times 10^5 \chi \bar{E}_\gamma \text{ R/sec.} \tag{11.48}$$

Because a person on the ground is exposed to only one-half of the cloud, it is necessary to multiply Eq. (11.48) by 0.5, which yields Eq. (11.47).

In the preceding equations, χ has units of Ci/cm^3. If χ is given in Ci/m^3, as is frequently the case in view of the units of σ_y and σ_z in Eq. (11.16), then Eq. (11.47) becomes

$$\dot{X} = 0.262\chi\,\bar{E}_\gamma \text{ R/sec.} \tag{11.49}$$

To obtain the dose equivalent (biological dose) rate in conventional units from the preceding equations, $\dot{X}$ must be multiplied by the f-factor from Fig. 9.10 and by the quality factor. Both of the factors are approximately unity, however, so that

$$\dot{H} \simeq 0.262\chi\,\bar{E}_\gamma \text{ rem/sec,} \tag{11.50}$$

where χ is in Ci/m^3. In SI units, this becomes

$$\dot{H} \simeq 2.62\chi\,\bar{E}_\gamma \text{ Sv/sec.}$$

Example 11.2

A PWR power plant releases an average of 2×10^3 Ci of ^{133}Xe per year from a vent 100 m above the ground. Calculate the external γ-ray dose rate at a point 10^4 m from the plant under type F conditions and with a wind speed of 1 m/sec

Solution. The average rate of emission of the ^{133}Xe is $2 \times 10^3/3.16 \times 10^7 = 6.33 \times 10^{-5}$ Ci/sec. From Figures 11.11 and 11.12, $\sigma_y = 275$ m and $\sigma_z = 46$ m. Then using Eq. (11.16) with $h = 100$ m gives

$$\chi = \frac{6.33 \times 10^{-5}}{\pi \times 1 \times 275 \times 46} \exp\left[-\frac{10^4}{2 \times (46)^2}\right] = 1.50 \times 10^{-10} \text{ Ci/m}^3.$$

Substituting this value and $\bar{E}_g = 0.030$ MeV from Table 11.3 into Eq. (11.50) yields

$$\dot{H} = 0.262 \times 1.50 \times 10^{-10} \times 0.030 = 1.18 \times 10^{-12} \text{ rem/sec}$$

$$= 0.0373 \text{ mrem/yr. } [Ans.]$$

In SI units, this is

$$\dot{H} = 2.62 \times 1.50 \times 10^{-10} \times 0.030 = 1.18 \times 10^{-12} \text{ Sv/sec}$$

$$= 0.373 \text{ mSv/yr. } [Ans.]$$

According to Eq. (11.50), the dose rate from a given radionuclide is directly proportional to the concentration of the radionuclide in the plume. Thus, $\dot{H}$ can be written as

$$\dot{H} = C_\gamma \chi,$$

where C_γ is a constant that depends only on the characteristics of the nuclide, not on its concentration. The quantity C_γ is called a *dose-rate factor*. Clearly, C_γ is simply the dose rate from external exposure per unit density of radionuclide in the plume. This factor has been computed and tabulated for the most significant radionuclides. (See the references at the end of the chapter.)

Example 11.3

Compute the dose-rate factor for external γ-ray exposure to ^{85}Kr.

Solution. From Table 11.3, $\bar{E}_\gamma = 0.00211$ MeV, so that

$$C_\gamma = 0.262 \times 0.00211 = 5.53 \times 10^{-4} \text{ rems/sec per Ci/m}^3.[Ans.]$$

External Dose from Plume: β-Rays

The external dose rate due to β-rays is usually computed in two steps. First, the absorbed dose rate is determined in the air immediately outside the body. The dose within the surface layer of the body is then estimated on the basis of an empirical relation between the air and epidermal doses.

The dose rate in air, $\dot{D}_{air}$, can be found using the second method for computing $\dot{X}$ in the infinite cloud previously described. The only difference is that $\dot{D}_{air}$ is given in rads/sec and $\dot{X}$ is in R/sec. Because one rad corresponds to the absorption of 100 ergs/g, while one roentgen corresponds to only 87.5 ergs/g, it follows that

$$\dot{D}_{air} = \frac{87.5}{100} \times 0.262\chi \bar{E}_\beta$$

$$= 0.229\chi \bar{E}_\beta \text{ rad/sec}, \tag{11.51}$$

where χ is in Ci/m^3 and $\bar{E}_\beta$ is the average energy of the emitted β-rays in MeV.

The dose equivalent rate in tissue is then given by

$$\dot{H} = 0.229\chi \bar{E}_\beta \times f(d, E_{max}) \text{ rem/sec}, \tag{11.52}$$

where f is an experimentally determined function of d, distance into the tissue, and E_{max}, the maximum energy of the emitted β-rays. The dose rate is largest at the surface of the skin, where $f = 1$, and decreases rapidly with distance into the tissue. To be conservative, therefore, the skin dose is usually computed with $f = 1$.[7]

The dose-rate factor for β-rays is then

$$C_\beta = 0.229\bar{E}_\beta f(d, E_{max}). \tag{11.53}$$

---

[7]For external exposure, the NRC frequently calculates whole-body γ-ray doses at a depth of 5 cm into the body and skin doses at a depth of 7 mg/cm^2. The 5 cm corresponds to the average depth of the blood-forming organs; 7 mg/cm^2 is the depth of the epidermal basal cells, the living cells in the skin.

Example 11.4

What is the yearly external β-ray dose rate from ^{133}Xe under the conditions of Example 11.2?

Solution. From Table 11.3, $\bar{E}_\beta = 0.146$ MeV. Then, using $\chi = 1.50 \times 10^{-10}$Ci/m^3 in Eq. (11.52) with $f = 1$ gives

$$\dot{H} = 0.229 \times 1.50 \times 10^{-10} \times 0.146 = 5.02 \times 10^{-12} \text{ rem/sec}$$

$$= 0.158 \text{ mrem/yr. } [Ans.]$$

The total γ-ray or β-ray dose received over a period of t_0 sec is obtained by integrating Eq. (11.50) or Eq. (11.52) from zero to t_0. In the case where χ is constant, the total dose is simply $\dot{H}$ multiplied by t_o. On the other hand, if χ decreases with the half-life of the radionuclide,

$$\chi = \chi_0 e^{-\lambda t},$$

then the total dose is

$$H = \int_0^{t_0} \dot{H}(t)\, dt$$

$$= \frac{\dot{H}_0}{\lambda}(1 - e^{-\lambda t_0}), \tag{11.54}$$

where $\dot{H}_0$ is the initial dose rate in the concentration χ_0. When t_0 is long compared with the half-life of the nuclide, Eq. (11.54) reduces to

$$H = \frac{\dot{H}_0}{\lambda} = 1.44 \dot{H}_0 T_{1/2}. \tag{11.55}$$

For a puff of effluent, the total γ-ray and β-ray doses are given by

$$H = 0.262 \chi_T \bar{E}_\gamma \text{ rem} \tag{11.56}$$

and

$$H = 0.229 \chi_T \bar{E}_\beta \text{rem}, \tag{11.57}$$

where χ_T is defined in Eq. (11.19).

Internal Dose from Inhalation

It is common knowledge that a person's breathing rate (i.e., intake of air) is a function of the activity in which he is engaged. Thus, the average breathing rate during an 8-hr working day for a standard man (as defined by the ICRP) is taken to be

$$B = 10 \text{ m}^3/8 \text{ hrs} = 3.47 \times 10^{-4} \text{ m}^3/\text{sec}, \qquad (11.58)$$

while the average for a full 24-hr day is

$$B = 20 \text{ m}^3/24 \text{ hr} = 2.32 \times 10^{-4} \text{ m}^3/\text{sec}. \qquad (11.59)$$

For short periods, a person may breathe more rapidly (especially if he has reason to think that he is in a radiation field), and breathing rates as high as $5 \times 10^{-4} \text{ m}^3/\text{sec}$ are used in some reactor safety calculations.

In computing internal doses, it will be recalled from Chapter 9 that a radionuclide introduced into the body does not contribute to the dose until some time later, when, provided that it has been retained by the body, the radionuclide ultimately decays. Consider first the problem of determining the *dose rate* received by a particular body organ at the end of a period of inhalation of duration t_0. During the time $d\tau$ at τ, the amount $B\chi(\tau)\,d\tau$ Ci of radionuclide is inhaled. The activity remaining in the organ at the time t_0 is

$$B\chi(\tau)R(t_0 - \tau)e^{-\lambda(t_0-\tau)}d\tau,$$

where $R(t_0 - \tau)$ is the retention function, discussed in Sec. 9.9, for the organ in question. The total activity in the organ as the result of inhalation from $\tau = 0$ to $\tau = t_0$ is then

$$B\int_0^{t_0} \chi(\tau)R(t_0 - \tau)e^{-\lambda(t_0-\tau)}d\tau.$$

This activity gives a dose rate at t_0 of

$$\dot{H} = \frac{592B\xi}{M}\int_0^{t_0} \chi(\tau)R(t_0 - \tau)e^{-\lambda(t_0-\tau)}\,d\tau \text{ rem/sec.} \qquad (11.60)$$

The usual retention function is given by Eq. (9.36), namely,

$$R(t_0 - \tau) = C_0 q e^{-\lambda_b(t_0-\tau)},$$

where q is the fraction of the radionuclide that goes to the particular organ and λ_b is the decay constant for biological clearance from that organ. Introducing $R(t_0 - \tau)$ into Eq. (11.60) gives

$$\dot{H} = \frac{592B\xi q}{M}\int_0^{t_0} \chi(\tau)e^{-\lambda_e(t_0-\tau)}\,d\tau \text{ rem/sec}, \qquad (11.61)$$

where

$$\lambda_e = \lambda + \lambda_b \qquad (11.62)$$

is the effective decay constant of the radionuclide in the organ.

In the special case where χ is not time dependent, Eq. (11.61) can be integrated directly, giving

$$\dot{H} = \frac{592\xi q\chi}{M\lambda_e}(1 - e^{-\lambda_e t_0}) \text{ rem/sec,} \qquad (11.63)$$

or, in SI units of Sv/sec,

$$\dot{H} = \frac{5920\xi q\chi}{M\lambda_e}(1 - e^{-\lambda_e t_0}) \text{ Sv/sec.}$$

When the duration of the inhalation is greater than either $T_{1/2}$ or $T_{1/2b}$, the exponential in Eq. (11.63) becomes very small. The dose rate then reduces to

$$\dot{H} = \frac{592B\xi q\chi}{M\lambda_e} \text{ rem/sec.} \qquad (11.64)$$

This formula gives the steady-state, equilibrium dose rate for persons continually inhaling a radionuclide at the concentration χ. The *dose-rate factor for inhalation* is given by the combination of parameters multiplying χ in either Eq. (11.63) or Eq. (11.64).

Example 11.5

A BWR plant emits an average of 1.23 Ci of ^{131}I per year from a vent 30 m above ground level. Assuming type-E stability conditions and an average wind speed of 1.2 m/sec, calculate (a) the equilibrium dose to an adult thyroid at a point on the ground 2,000 m from the plant; (b) the annual dose to the thyroid at this point. [*Note:* For ^{131}I, $T_{1/2} = 8.04$ days, $T_{1/2b} = 138$ days, $\xi = 0.23$ MeV[8], and $q = 0.23$. The mass of an adult thyroid is 20 g.[9]]

Solution.

1. A release rate of 1.23 Ci/year is equal to $1.23/3.16 \times 10^7 = 3.89 \times 10^{-8}$ Ci/sec. From Fig. 11.13, $\chi \bar{v}/Q' = 6.0 \times 10^{-5}/\text{m}^2$ at 2,000 m for type-E conditions. Thus,

$$\chi = \frac{6.0 \times 10^{-5} \times 3.89 \times 10^{-8}}{1.2} = 1.95 \times 10^{-12} \text{ Ci/m}^3.$$

Using the normal breathing rate from Eq. (11.59) gives

$$C_d = 1.95 \times 10^{-12} \times 2.32 \times 10^{-4} = 4.52 \times 10^{-16} \text{ Ci/sec.}$$

---

[8]The values of ξ for the other isotopes of iodine are given in Table 11.17 and Problem 11.17.
[9]The mass of an infant (one-year old) thyroid is generally taken to be 2 g. Thus, the dose rate and dose commitment to the infant thyroid is 10 times that to the adult thyroid.

The decay constants are

$$\lambda = \frac{0.693}{8.04 \times 86400} = 9.98 \times 10^{-7} \text{ sec}^{-1},$$

$$\lambda_b = \frac{0.693}{138 \times 86400} = 5.81 \times 10^{-8} \text{ sec}^{-1},$$

and so

$$\lambda_e = 9.98 \times 10^{-7} + 5.81 \times 10^{-8} = 1.06 \times 10^{-6} \text{ sec}^{-1}.$$

Substituting these values into Eq. (11.64) gives

$$\dot{H} = \frac{592 \times 4.52 \times 10^{-16} \times 0.23 \times 0.23}{20 \times 1.06 \times 10^{-6}} = 6.68 \times 10^{-10} \text{ rem/sec. } [Ans.]$$

2. The dose over one year would be

$$H = 6.68 \times 10^{-10} \times 3.16 \times 10^7 = 0.0211 \text{ rem}$$

$$= 21.1 \text{ mrem. } [Ans.]$$

This problem can also be solved by comparing the concentration of ^{131}I at the point inhaled with the (MPC)$_\alpha$ for ^{131}I and its corresponding annual dose. Thus, in ICRP Publication 2 (see the references in Chapter 9), the (MPC)$_\alpha$ of ^{131}I for occupational exposure is given as $3 \times 10^{-9} \mu$Ci/cm^3 $= 3 \times 10^{-9}$ Ci/m^3 for continuous inhalation at a rate of 20 m^3/24 hr. This was computed from an annual thyroid dose of 30 rems. It follows that the continuous inhalation of air with ^{131}I at the concentration χ would give a yearly dose of

$$H = \frac{\chi}{3 \times 10^{-9}} \times 30 \text{ rem.}$$

Using the value $\chi = 1.95 \times 10^{-12}$ Ci/m^3 of this example yields

$$H = \frac{1.95 \times 10^{-12}}{3 \times 10^{-9}} \times 30 \times 19.5 \text{ mrem,}$$

which is essentially the same answer as obtained earlier.

Consider next the total future dose, or *dose commitment*, that is received by a body organ subsequent to the inhalation of $B\chi(\tau)\,d\tau$ Ci in the time $d\tau$. The activity in the organ at time t is

$$Bq\chi(\tau)d\tau e^{-\lambda_e(t-\tau)},$$

and the dose received in the time dt at t is

$$d^2H = \frac{592B\xi q}{M}\chi(\tau)d\tau e^{-\lambda_e(t-\tau)}dt \text{ rem.}$$

The total dose over the lifetime of the individual (gratuitously taken to be infinite) is then

$$dH = \frac{592 B \xi q}{M} \chi(\tau) \, d\tau \int_{\tau}^{\infty} e^{-\lambda_e(t-\tau)} dt$$

$$= \frac{592 B \xi q}{M \lambda_e} \chi(\tau) \, d\tau \text{ rem.} \tag{11.65}$$

The dose commitment from the inhalation of a total of

$$B \int \chi(\tau) \, d\tau$$

curies of radionuclide is simply

$$H = \frac{592 B \xi q}{M \lambda_e} \int_{\tau}^{\infty} \chi(\tau) \, d\tau, \tag{11.66}$$

where the integral is carried over the period of intake.

The *dose-commitment factor*, C_{dc}, is the dose commitment per unit of radioactivity inhaled (or ingested). From either Eq. (11.65) or (11.66), this is

$$C_{dc} = \frac{592 \xi q}{M \lambda_e}. \tag{11.67}$$

These factors have also been tabulated for various body organs and most radionuclides. (See references.)

It should be noted from Eq. (11.66) that the dose commitment from the inhalation (the same is true for ingestion) of radionuclides depends only on the total amount of radionuclides taken into the body and is independent of the time interval over which the inhalation occurs. One curie inhaled in one second ultimately gives the same dose as one curie inhaled over a period of a year.[10]

Dose from Ground-Deposited Radionuclides

Significant doses may be received from radionuclides deposited on the ground from the plume following a major, although improbable, nuclear plant accident. Thus, it has been estimated that, in the absence of evacuation, approximately 80% of the total dose received by the public in a core meltdown accident with breech of containment would be due to ^{137}Cs ($T_{1/2} = 30.17$ years) that escapes from the

---

[10]The NRC computes dose-commitment factors over a period of 50 years, rather than for an infinite time, beginning with the onset of a one-year intake of the radionuclide. This leads to slightly smaller dose-commitment factors for nuclides with long radiological and biological lives than factors computed from Eq. (11.65). Also, in calculations of dose-commitment factors for an infant or child, it is necessary to account for changing organ size and breathing rate as the individual grows.

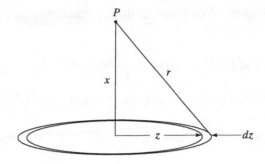

Figure 11.16   Diagram for computing dose from ground source.

plant and falls to the ground over a large area. In calculations of doses from ground-deposited radionuclides, it is usual to consider only the γ-ray dose. The β-rays are frequently ignored, since these provide only a skin dose, which, in any case, is normally much smaller than the corresponding γ-ray dose.

Consider then an infinite planar surface containing radionuclides at a density of S Ci/m^2 emitting one γ-ray of energy E MeV per disintegration. The buildup flux at point P in Fig. 11.16 from sources located on a ring of radius z and thickness dz is

$$d\phi_{\gamma b} = \frac{3.7 \times 10^{10} S \times 10^{-4} \times 2\pi z dz}{4\pi r^2} B_p(\mu r) e^{-\mu r}. \qquad (11.68)$$

The factor of 10^{-4} is necessary, because fluxes used in this book have units of cm^{-2} and S is in m^{-2}. Because $r^2 = x^2 + z^2$, where x is the height of the observation point[11], $z dz = r dr$, and Eq. (11.68) becomes

$$d\phi_{\gamma b} = 1.85 \times 10^6 S \frac{B_p(\mu r) e^{-\mu r}}{r} dr.$$

The total buildup flux is then

$$\phi_{\gamma b} = 1.85 \times 10^6 S \int_x^\infty \frac{B_p(\mu r) e^{-\mu r}}{r} dr. \qquad (11.69)$$

Introducing the Berger form of the buildup factor from Problem 11.9

$$B_p(\mu r) = 1 + C \mu r e^{-\beta \mu r},$$

where C and β are given in Table 11.16, gives

$$\phi_{\gamma b} = 1.85 \times 10^6 S \left[\int_x^\infty \frac{e^{-\mu r}}{r} dr + C \mu \int_x^\infty e^{-(1-\beta)\mu r} dr \right].$$

---

[11]The NRC calculates ground doses for $x = 1$ meter, a little high for gonadal doses, but reasonable for computing whole body doses.

The first integral can be expressed in terms of the E_1 function (see Section 10.2) and the second is directly integrable. The result is

$$\phi_{\gamma b} = 1.85 \times 10^6 S \left[E_1(\mu x) + \frac{C}{1-\beta} e^{-(1-\beta)\mu x} \right]. \tag{11.70}$$

When Eq. (11.70) is substituted in Eq. (11.40), the exposure rate is

$$\dot{X} = 3.39 \times 10^{-2} SE(\mu_a/\rho)^{\text{air}} \left[E_1(\mu x) + \frac{C}{1-\beta} e^{-(1-\beta)\mu x} \right] \text{ R/sec} \tag{11.71}$$

as usual, $\dot{X} \simeq \dot{H}$.

In most problems, μx is much less than unity. Calculations of $\dot{X}$ or $\dot{H}$ can then be simplified by using the following expansion

$$E_1(x) = -\gamma + \ln\left(\frac{1}{x}\right) + x - \frac{x^2}{4} + \frac{x^3}{18} - \cdots. \tag{11.72}$$

Here $\gamma = 0.57722$ is *Eulers constant*. Dose-rate conversion factors (i.e., the quantity multiplying S in Eq. (11.71)), are tabulated in the references at the end of the chapter.

Example 11.6

Calculate the γ-ray dose-rate conversion factor for ground exposure for ^{137}Cs at a height of 1 m. [*Note:* The half-life of ^{137}Cs is 30.2 years, and ^{137}Cs emits only one γ-ray with an energy of 0.66 MeV in 93.5% of its disintegrations.]

Solution. Interpolating Table II.5 gives $(\mu_a/\rho)^{\text{air}} = 0.0294$ cm^2/g. Table II.4 gives $(\mu/\rho)^{\text{air}} = 0.0775$, and $\mu = 0.775 \times 1.293 \times 10^{-3} = 1.00 \times 10^{-4}$ cm^{-1}. Thus, $\mu x = 0.0100$, which is much less than unity. From Eq. (11.72),

$$E_1(0.91) = -0.577 + \ln 100 + 0.01 - \cdots = 4.04.$$

Using $C = 1.41$ and $\beta = 0.0857$ from Table 11.16 gives directly

$$\frac{\dot{H}}{S} = 3.39 \times 10^{-2} \times 0.66 \times 0.0294 \left[4.04 + \frac{1.41}{0.914} e^{-0.00914} \right]$$

$$= 3.66 \times 10^{-3} \text{ rems/sec per Ci/m}^2$$

$$= 13.2 \text{rems/hr per Ci/m}^2. \text{ [}Ans.\text{]}$$

(*Note:* $S = 0.035$ of actual number of curies, because of the branching ratio.)

Leakage from Buildings

In reactor accidents, radionuclides, especially fission product gases, are first released into the containment building and subsequently may leak to the atmosphere.

Among other things, the magnitude of the resulting plume depends on the rate at which this leakage occurs.

The leakage rate from a building is specified as a percentage of the gas in the building that leaks out per day. A reasonably tight reactor building, for example, will leak less than 0.1% per day of its contained gases. Now suppose that in an accident C_0 units of some *stable* fission product are released into the building, and let C be the amount of the nuclide remaining in the building at time t. Then, C is determined by the equation

$$\frac{dC}{dt} = -0.01pC, \tag{11.73}$$

where p is the leakage rate in percent. Equation (11.73) is the same as the equation for radioactive decay with the decay constant

$$\lambda_l = 0.01p. \tag{11.74}$$

The solution to Eq. (11.73) can then be written as

$$C = C_0 e^{-\lambda_l t}. \tag{11.75}$$

If, as is the usual case, the fission product is radioactive, then

$$C = C_0 e^{-(\lambda + \lambda_l)t}, \tag{11.76}$$

where λ is the radioactive decay constant and C and C_0 are in appropriate units of radioactivity.

Because λ_l is equivalent to a decay constant, the rate at which the fission product is released from the building is

$$Q' = \lambda_l C$$
$$= \lambda_l C_0 e^{-\lambda_c t}, \tag{11.77}$$

where

$$\lambda_c = \lambda + \lambda_l \tag{11.78}$$

is the total decay constant of the fission product in the building.

Equation (11.77) is the source term in the dispersion relations derived in Sec. 11.4. For example, with ground level release, Eq. (11.16) gives

$$\chi = \frac{\lambda_l C_0 e^{-\lambda_c t}}{\pi \bar{\upsilon} \sigma_y \sigma_z}, \tag{11.79}$$

where deposition from the plume and building wake effects have been ignored.

The total external dose to a person who stands in the plume given by Eq. (11.79) for the time t_0 can be computed as in the derivation of Eq. (11.54). For γ-rays,

$$H = 0.262 \bar{E}_\gamma \int_0^{t_0} \chi \, dt.$$

Introducing χ then gives

$$H = \frac{0.262 \bar{E}_\gamma \lambda_l C_0}{\pi \bar{\upsilon} \sigma_y \sigma_z \lambda_c} (1 - e^{-\lambda_c t_0}) \text{rem.} \tag{11.80}$$

When $\lambda_c t_0 \ll 1$, the exponential in Eq. (11.80) can be expanded to give

$$H = \frac{0.262 \bar{E}_\gamma \lambda_l C_0 t_0}{\pi \bar{\upsilon} \sigma_y \sigma_z} \text{rem.} \tag{11.81}$$

If the individual stands in the plume indefinitely, then placing $t_0 = \infty$ in Eq. (11.80) gives

$$H = \frac{0.262 \bar{E}_\gamma \lambda_l C_0}{\pi \bar{\upsilon} \sigma_y \sigma_z \lambda_c} \text{rem.} \tag{11.82}$$

The dose rate in an internal body organ after standing in the plume for the time t_0 is obtained by substituting Eq. (11.79) into Eq. (11.61). Thus,

$$\dot{H} = \frac{592 B \xi q \lambda_l C_0}{\pi \bar{\upsilon} \sigma_y \sigma_z M} \int_0^{t_0} e^{-\lambda_c \tau} \times e^{-\lambda_e (t_0 - \tau)} \, d\tau$$

$$= \frac{592 B \xi q \lambda_l C_0}{\pi \bar{\upsilon} \sigma_y \sigma_z M (\lambda_c - \lambda_e)} (e^{-\lambda_e t_0} - e^{-\lambda_c t_0}) \text{ rem/sec.} \tag{11.83}$$

If both $\lambda_e t_0 \ll 1$ and $\lambda_c t_0 \ll 1$, then Eq. (11.83) reduces to

$$\dot{H} = \frac{592 B \xi q \lambda_l C_0 t_0}{\pi \bar{\upsilon} \sigma_y \sigma_z M} \text{ rem/sec.} \tag{11.84}$$

Finally, the dose commitment for inhalation up to t_0 is found by introducing Eq. (11.79) into Eq. (11.54) and carrying out the integration. The result is

$$H = \frac{592 B \xi q \lambda_l C_0}{\pi \bar{\upsilon} \sigma_y \sigma_z M \lambda_e \lambda_c} (1 - e^{-\lambda_c t_0}) \text{rem.} \tag{11.85}$$

When $\lambda t_0 \ll 1$, this becomes

$$H = \frac{592 B \xi q \lambda_l C_0 t_0}{\pi \bar{\upsilon} \sigma_y \sigma_z M \lambda_e} \text{rem,} \tag{11.86}$$

and when either $\lambda_l t_0$ or $\lambda t_0 \gg 1$,

$$H = \frac{592 B \xi q \lambda_l C_0}{\pi \bar{\upsilon} \sigma_y \sigma_z M \lambda_e \lambda_c} \text{rem.} \qquad (11.87)$$

It should be noted that in the preceding discussion, the calculation of the dose received by an organ over a given period of inhalation, t_0, was omitted. This is simply because such a calculation would be of little value, as the organ is destined to receive its full dose commitment in any case, beyond t_0, barring death or surgery.

Example 11.7

In an accident in a 1,000-MW reactor, 6.25×10^6 Ci of ^{135}I (25% of its equilibrium activity) is released into the containment building, which leaks at the rate of 0.1% per day. Calculate the following doses for a person during the first 2 hrs after the accident who is standing at a distance of 2,000 m from the reactor building. ^{131}I is assumed to be released at ground level, under type F dispersion conditions and with a wind speed of 1 m/sec: (a) the total external γ-ray dose; (b) the dose rate to the thyroid at the end of the 2 hrs; (c) the dose commitment to the thyroid.

Solution.

1. From Eq. (11.74),

$$\lambda_l = 0.01 \times 0.1 = 10^{-3} \text{ day}^{-1}$$
$$= 10^{-3}/86,400 = 1.16 \times 10^{-8} \text{ sec}^{-1}.$$

From Example 11.5, the radioactive decay constant is $9.98 \times 10^{-7} \text{ sec}^{-1}$. Using Eq. (11.78), we see that

$$\lambda_c = 9.98 \times 10^{-7} + 1.16 \times 10^{-8} = 1.00 \times 10^{-6} \text{ sec}^{-1}.$$

Then,

$$\lambda_c t_0 = 1.00 \times 10^{-6} \times 2 \times 3,600 = 7.2 \times 10^{-3},$$

which is $\ll 1$, so that Eq. (11.81) applies. Introducing the appropriate values gives

$$H = \frac{0.262 \times 0.371 \times 1.16 \times 10^{-8} \times 3.25 \times 10^6 \times 2 \times 3600}{\pi \times 1 \times 70 \times 21}$$
$$= 1.10 \times 10^{-2} \text{ rems. } [Ans.]$$

2. Equation (11.84) applies in this case. Using the foregoing values and data from Example 11.5 yields

$$\dot{H} = \frac{592 \times 2.32 \times 10^{-4} \times 0.23 \times 1.16 \times 10^{-8} \times 6.25 \times 10^6 \times 2 \times 3600}{\pi \times 1 \times 70 \times 21 \times 20}$$

$$= 4.11 \times 10^{-5} \text{ rem/sec} = 148 \text{ mrem/hr. } [Ans.]$$

3. The dose commitment is given by Eq. (11.87), and it will be observed that this is simply the dose rate at 2 hrs divided by λ_e. From Example 11.5, $\lambda_e = 1.06 \times 10^{-6} \text{ sec}^{-1}$, so that

$$H = \frac{4.11 \times 10^{-5}}{1.06 \times 10^{-6}} = 38.8 \text{ rem. } [Ans.]$$

This example shows that ^{131}I gives a negligible whole-body γ-ray dose. The dose commitment to the thyroid, however, is already substantial at only 2 hrs.

Direct γ-Ray Dose

The exposure rate from the γ-rays emitted from within the reactor building can be approximated by taking the building to be a point source. Then, if shielding by the building is ignored, the exposure rate is given by Eq. (11.40) with the buildup flux

$$\phi_{\gamma b} = \frac{SB_p(\mu r)e^{-\mu r}}{4\pi r^2}, \tag{11.88}$$

where S is the number of γ-rays emitted per second and $B_p(\mu r)$ is the point buildup factor for air. (See problem 11.9.) Introducing Eq. (11.88) into Eq. (11.40) gives

$$\dot{X} = 1.46 \times 10^{-9} SE \left(\frac{\mu_a}{\rho}\right)^{\text{air}} \frac{B_p(\mu r)e^{-\mu r}}{r^2} \text{R/sec.} \tag{11.89}$$

As earlier, $\dot{H} \simeq \dot{X}$.

If the direct dose originates with γ-rays emitted by radionuclides released into the building in an accident, then the factor S in Eq. (11.89) is given by

$$S = 3.7 \times 10^{10} C_0 e^{-\lambda_c t}, \tag{11.90}$$

where C_0 is the total number of curies released, λ_c is the decay constant including leakage (see Eq. (11.78)), and t is the time after the accident. The total direct dose received up to the time t_0 is obtained by substituting Eq. (11.90) into Eq. (11.89) and integrating from $t = 0$ to $t = t_0$. The result is

$$H \simeq X = \frac{54.0 C_0}{\lambda_c} (1 - e^{-\lambda_c t_0}) E \left(\frac{\mu_a}{\rho}\right)^{\text{air}} \frac{B_p(\mu r)e^{-\mu r}}{r^2} \text{rem.} \tag{11.91}$$

In this formula, C_0 is in Ci, λ_c is in sec^{-1}, t_0 is in sec, μ is in cm^{-1}, r is in cm, E is in MeV, and $(\mu_a/\rho)^{\text{air}}$ is in cm^2/g.

Shielding by the reactor building may be included by using the appropriate prescription from Chapter 10. For example, if the building is concrete and a cm thick, then Eqs. (11.89) and (11.91) must be multiplied by the factor

$$B_{pc}(\mu_c a)e^{-\mu_c a},$$

where $B_{pc}(\mu_c a)$ is the buildup factor for concrete calculated using parameters given in Table 10.3.

Example 11.8

The fission product gas ^{88}Kr ($T_{1/2} = 2.79$ hr) emits several γ-rays, the most prominent of which has an energy of 2.40 MeV and is emitted in about 40% of the disintegrations. In an accident, 3×10^7 Ci of ^{88}Kr are released into a reactor building that leaks at a rate of 0.1%/day. Calculate the direct 2-hr dose from the 2.40-MeV γ-ray at a distance of 1,000 m from the building.

Solution. The effective number of curies in which the γ-ray is emitted is

$$C_0 = 3 \times 10^7 \times 0.40 = 1.2 \times 10^7 \text{Ci}.$$

The decay constant for ^{88}Kr is

$$\lambda = 0.693/2.79 \times 3600 = 6.90 \times 10^{-5} \text{ sec}^{-1}.$$

From Example 11.7, $\lambda_l = 1.16 \times 10^{-8}$ sec^{-1}, so that from Eq. (11.78),

$$\lambda_c = 6.90 \times 10^{-5} + 1.16 \times 10^{-8} \simeq 6.90 \times 10^{-5} \text{ sec}^{-1}.$$

This shows that with such a short-lived nuclide, the decrease in the concentration in the building due to leakage can be ignored.

From Table II.4 at 2.40 MeV, $\mu = 5.30 \times 10^{-5}$cm^{-1} and $\mu r = 5.3$. Then, using problem 11.9, $B_p(\mu r) = 4.7$. Also from Table II.5 $(\mu_a/\rho)^{\text{air}} = 0.0227$ cm^2/g. Substituting these values into Eq. (11.91) with $t_0 = 2$ hr $= 7,200$ sec then yields

$$H = \frac{54.0 \times 1.2 \times 10^7}{6.90 \times 10^{-5}}(1 - e^{-6.90 \times 10^{-5} \times 7,200}) \times 2.40 \times 0.0227 \times \frac{4.7 \times e^{-5.3}}{(10^5)^2}$$

$$= 0.468 \text{ rems. } [Ans.]$$

Population Doses

Up to this point, only doses to individuals have been considered. However, to evaluate the overall effect on the surrounding population of routine or accidental releases of radiation from a nuclear plant, it is also usual to compute the total *population dose*. This is the sum of the doses to individuals multiplied by the number of persons receiving those doses. The units of population dose are obviously person-rems. For example, if out of 1,000 people exposed to radiation, 500 receive 1 rem and the remainder receive 2 rems, the population dose is simply

$$H_{\text{pop}} = 500 \times 1 + 500 \times 2 = 1{,}500 \text{ person-rems.}$$

To obtain a general formula for H_{pop}, it will first be recalled that the total dose to an individual at the distance r from a plant from external exposure to a plume, external exposure to radionuclides deposited on the ground, or radionuclides that are inhaled, can be written as

$$H(r) = C\chi_T(r), \tag{11.92}$$

where C is an appropriate dose conversion factor and $\chi_T(r)$ is the time integral of the radionuclide concentration at r over the period of exposure. It is not ordinarily necessary to calculate the effect of direct exposure, as this is significant only near the plant itself. Equation (11.92) then includes all relevant exposures.

Population doses, especially at long distances, are best computed using the wedge model described in the preceding section. Then, from Fig. 11.14, if $P(r)$ is the population density at r, the number of people living between r and $r + dr$ is $P(r)r\theta dr$. If these people receive the dose $H(r)$, then the population dose between R_1 and R_2 is

$$H_{\text{pop}} = \int_{R_1}^{R_2} H(r)P(r)r\theta dr$$

$$= C\int_{R_1}^{R_2} \chi_T(r)P(r)r\theta dr. \tag{11.93}$$

With a total release of Q units of radioactivity and with deposition and inflight decay, both of which are very important in this type of calculation, $\chi_T(r)$ is given by

$$\chi_T(r) = \frac{Q}{ar\theta\bar{v}}e^{-(\Lambda+\lambda/\bar{v})r}. \tag{11.94}$$

Introducing this expression into Eq. (11.93) gives

$$H_{\text{pop}} = \frac{CQ}{a\bar{v}}\int_{R_1}^{R_2} P(r)e^{-(\Lambda+\lambda/\bar{v})r}dr. \tag{11.95}$$

In the special case where P is approximately constant,

$$H_{\text{pop}} = \frac{CQ}{a\bar{v}\left(\Lambda+\dfrac{l}{\bar{v}}\right)}[e^{-(\Lambda+\lambda/\bar{v})R_1} - e^{-(\Lambda+\lambda/\bar{v})R_2}]. \tag{11.96}$$

It is of interest to observe that the angle of the wedge, θ, does not appear in Eqs. (11.95) and (11.96). This is to be expected, because an increase in θ increases the number of people exposed, but decreases the dose per person, such that the result remains the same.

11.6 REACTOR SITING

The location of a nuclear reactor has an obvious bearing on the consequences of a reactor accident to the public. Thus, an accident in a reactor located in a heavily populated area could involve more of the public than the same accident in a remote location. Also, a nuclear power plant, like any other industrial installation, cannot help having some affect on the local environment. For these reasons, reactor siting is a major consideration in the design and licensing of a nuclear plant.

In an application for a construction permit, an applicant must show to the satisfaction of the NRC that the plant can be built and operated at a proposed site without undue risk to the health and safety of the public and with minimal effect on the environment. Furthermore, one must show that the proposed site is the most acceptable among a number of alternate sites. The present section is related to the *safety* aspects of reactor siting. Some of the *environmental* questions involved in siting are discussed in Section 11.9.

The NRC considers four factors in evaluating the suitability of a proposed reactor site.[12] First, it considers the reactor itself, its design characteristics, and its proposed mode of operation. Special attention is paid to the expected inventory of radioactivity and to the extent to which the reactor is designed to prevent and mitigate the consequences of a release of radioactivity. Needless to say, if the reactor involves a substantial departure from earlier technology or is a design where that experience is lacking, special siting requirements may be considered to ensure an ample margin of safety.

Second, the NRC evaluation considers the population density in and use characteristics of the environs of the site. This is an involved question and is discussed later.

Third, the NRC considers the physical characteristics of the site. These include the seismology, meteorology, geology, and hydrology of the area. Each of these topics is also discussed later in this section.

Finally, in the event that the population density or the physical characteristics of the site are unfavorable for the location of the reactor and the applicant proposes

---

[12]NRC regulations regarding reactor site criteria are contained in 10 CFR 100; the regulations in this part apply to power reactors and testing reactors (generally speaking, research reactors operating at a power in excess of 10MW). For the siting of research reactors, the NRC uses as criteria the general regulations of Title 10, pertinent sections of the Atomic Energy Act of 1954, as amended, and accumulated court decisions. See also Regulatory Guide 4.7, "General Site Suitability Criteria for Nuclear Power Stations."

to circumvent this problem by the use of appropriate engineering safeguards, the NRC will evaluate the adquacy of these safeguards. For example, suppose that a utility wishes to place a nuclear power plant near the end of an active airport runway. Such a site would be unacceptable if the probability of an airplane crash were above some minimum value. However, if the utility indicates that it will *harden* the plant against such crashes by reinforcing the structure in some way, the site might well be approved. The Three Mile Island Reactor is such a plant, located near an active airport. The reactor's design and construction incorporate unique safety features to reinforce the plant's structure and control systems.

Population Considerations

For the purpose of evaluating a proposed reactor site and for making comparisons with possible alternate sites, the NRC has defined two areas in the vicinity of the reactor as follows (10 CFR 100.3):

1. An *exclusion area*, or *exclusion zone*, is "that area surrounding the reactor in which the reactor licensee has the authority to determine all activities including exclusion or removal of personnel and property from the area .... Residents within the exclusion area shall normally be prohibited. In any event, residents shall be subject to ready removal in case of necessity ...."

2. A *low-population zone* (LPZ) is "the area immediately surrounding the exclusion area which contains residents, the total number and density of which are such that there is a reasonable probability that appropriate protective measures could be taken in their behalf in the event of a serious accident ...."

In short, the exclusion area is a piece of property usually belonging to the reactor licensee, which he can surround with a fence and from which he can exclude everyone not connected with the operation of the facility. The LPZ is the surrounding area that contains so few people that it can be easily evacuated should the need arise.

The boundaries of these two areas are defined in terms of the consequences of a serious, postulated reactor accident involving a major release of fission products. Again, quoting from 10 CFR 100, "an exclusion area [is] of such a size that an individual located at any point on its boundary for two hours immediately following onset of the postulated fission product release would not receive a total radiation dose to the whole body in excess of 25 rem or a total radiation dose in excess of 300 rem to the thyroid from iodine exposure."

On the other hand, the "low population zone [is] of such size that an individual located at any point on its outer boundary who is exposed to the radioactive cloud resulting from the postulated fission product release (during the entire period of its passage) would not receive a total radiation dose to the whole body in excess of 25 rem or a total radiation dose in excess of 300 rem to the thyroid from iodine exposure."

It should be noted that while the 25-rem and 300-rem doses generally fall within the guidelines of the NCRP for emergency exposure, the NRC does not imply that these are in any way acceptable doses to the public under accident conditions. They are used merely as guides for the evaluation of reactor sites. As will be shown in the next section, considerably smaller doses can actually be expected from realistic appraisals of reactor accidents.

As another aid in site evaluation, the NRC also defines the *population center distance*. This is defined as "the distance from the reactor to the nearest boundary of a densely populated center containing more than 25,000 residents."

It is impossible, of course, to predict precisely what the decision of the NRC will be with regard to the acceptability of any particular site. Such decisions naturally involve a certain amount of judgment on the part of the persons involved. Generally speaking, however, the NRC will not accept a site for which an applicant cannot provide an exclusion area of the necessary size. Ordinarily, the NRC also requires that the population center distance be no less than 1.33 times the radius of the LPZ. Furthermore, in evaluating two otherwise comparable sites, the NRC normally will accept the site having the smaller number of residents in the LPZ. Thus, the NRC has acted emphatically to encourage the siting of nuclear reactors where the fewest individuals could become involved.[13] This is also in accord with the policy of keeping the total man-rems received by the public as low as possible. However, the nature of the population in the LPZ must also be taken into account, as it relates to the ease with which the zone can be evacuated.

The size of the exclusion area and LPZ are determined by a postulated serious accident in the plant. This accident assumes that a significant fraction of the gaseous and volatile fission products are released into the containment and that they then leak to the environment at a specified rate determined by the characteristics of the containment structure. Until now, the fraction of radioactivity assumed to be released to the containment has been defined arbitrarily, rather than being based on analysis of a specific accident scenario. However, the technique for calculating the actual release for various accident scenarios has improved greatly in recent years, so the NRC has recently released Regulatory Guide 1.183 that allows the use of alternative source terms based on more realistic criteria.

The assumptions that the NRC makes in calculating the radii of the exclusion area and the LPZ are contained in Regulatory Guide 1.3 (for BWRs) and 1.4 (for

---

[13]The history of power reactor siting shows some cases where applicants ultimately abandoned or moved site locations because of population density considerations.

PWRs).[14] In simplified form, these are as follows:

1. Prior to the accident, the reactor has been operating at full power long enough for the fission product activity to come to equilibrium, except for the longest-lived nuclides.

2. The accident immediately releases into the reactor building (containment) 25% of the iodine inventory (91% are in elemental form, 5% are particulates, and 4% are organic iodides[15]) and 100% of the noble gas inventory.

3. The effect of containment sprays and recirculation filter systems in reducing the airborne activity in containment are taken into account if they are present in the plant. (These are discussed further in Section 11.7.)

4. The gas within the containment leaks at a rate characteristic of the structure.

5. The gas is released from containment at ground level and is dispersed under Pasquill F conditions for the first 8 hrs with a wind speed of 1 m/sec; after 8 hrs, the gas disperses under a specified mixture of conditions and wind speeds.

6. There is no depletion of the cloud as the result of deposition on the ground.

7. The fission products do not decay en route in the atmosphere.

These assumptions are used to compute the external and internal dose from the effluent cloud and the direct dose from nuclides within the building in the manner discussed in the preceding section. Ordinarily, the internal thyroid dose from iodine is the dominant mode of exposure.

To begin the computation, it is first necessary to determine the inventories of the various fission products. Suppose that the reactor has been operating at a power of P MW. If the recoverable energy per fission is taken to be 200 MeV, the total number of fissions occurring per second in the reactor is

$$\text{Fission rate} = P\,\text{MW} \times \frac{10^6\,\text{joule}}{\text{MW-sec}} \times \frac{\text{fissions}}{200\,\text{MeV}} \times \frac{\text{MeV}}{1.60 \times 10^{-13}\text{joule}}$$

$$= 3.13 \times 10^{16} P \text{ fissions/sec.}$$

---

[14] As of this writing, 10 CFR 100 still refers to the 1962 report TID-14844 (see the references at the end of this chaper) for guidance in calculating exclusion areas and LPZs. However, this report has been largely superceded by the Regulatory Guides, which contain more detailed assumptions regarding a LOCA and a new section for reactors built after 1997.

[15] Organic iodides, mostly methyl iodide, are formed by the reaction of iodine with oils, greases, and other organic materials located in the containment building. They are not subject to washout or plate-out within the building.

If the *cumulative yield*[16] of the ith fission product is γ_i atoms per fission, then the rate of production of this nuclide is

$$\text{rate of production} = 3.13 \times 10^{16} P \gamma_i \text{ atoms/sec.}$$

From Eq. (2.30), its activity at time t is

$$\alpha_i = 3.13 \times 10^{16} P \gamma_i (1 - e^{-\lambda_i t}) \text{ disintegrations/sec.}$$

Expressed in curies, this is[17]

$$\alpha_i = \frac{3.13 \times 10^{16} P \gamma_i}{3.7 \times 10^{10}} (1 - e^{-\lambda_i t})$$

$$= 8.46 \times 10^5 P \gamma_i (1 - e^{-\lambda_i t}) \text{ Ci.} \qquad (11.97)$$

If the activity saturates in the time t (i.e., if $\lambda_i t \gg 1$), Eq. (11.97) reduces to

$$\alpha_i = 8.46 \times 10^5 P \gamma_i \text{Ci.} \qquad (11.98)$$

Table 11.4 gives the inventories of the most important noble gases and iodine fission products computed for a typical 1,000 MWe plant at the end of a fuel cycle.

The amount of a fission product available for release to the atmosphere can be estimated by

$$C_0 = 8.46 \times 10^5 F_p F_b P \gamma_i (1 - e^{-\lambda_i t}) \text{ Ci,} \qquad (11.99)$$

where F_p is the fraction of the radionuclide released from the fuel into the reactor containment and F_b is the fraction of this that remains airborne and capable of escaping from the building.

Example 11.9

The fission yield of ^{131}I is 2.77%. Calculate the saturation activity of this nuclide in a reactor operating 1,000 MWe (3,200 MW) and the activity released in elemental form into the reactor containment in a meltdown under the assumptions given, ignoring assumption 3.

Solution. The value of γ is 0.0277. From Eq. (11.98),

$$\alpha_i = 8.46 \times 10^5 \times 3.2 \times 10^3 \times 0.0277 = 7.50 \times 10^7 \text{ Ci. [Ans.]}$$

[*Note:* This differs slightly from the value calculated in Table 11.4 by a more accurate method.]

---

[16]The cumulative yield of a fission product is the yield of the product itself plus the yields of all its short-lived precursors.

[17]This equation is only approximate, as it does not account for removal by neutron absorption or for changes in yield as the neutron energy spectrum changes. In most cases, such corrections are small.

TABLE 11.4   TYPICAL CORE INVENTORY OF
SELECTED VOLATILE FISSION PRODUCTS IN A 1,000
MWE PWR AT THE END OF A FUEL CYCLE ¶

Nuclide*	Half-life†	Fission yield‡	Curies($\times 10^8$)
^{85m}Kr	4.4 h	0.0133	0.24
^{85}Kr	10.76 y	0.00285	0.0056
^{87}Kr	76 m	0.0237	0.47
^{88}Kr	2.79 h	0.0364	0.68
^{133}Xe	5.27 d	0.0677	1.7
^{135}Xe	9.2 h	0.0672	0.34
^{131}I	8.04 d	0.0277	0.85
^{132}I	2.28 h	0.0413	1.2
^{133}I	20.8 h	0.0676	1.7
^{134}I	52.3 m	0.0718	1.9
^{135}I	6.7 h	0.0639	1.5

¶From "Reactor Safety Study" WASH 1400, 1975.

*Superscript m refers to a nuclide in an isomeric state (see Section 2.8).

†m = minutes, h = hours, d = days, y = years.

‡Cumulative yields in atoms per fission; equal to yield of nuclide plus cumulative yield of precursor. From M. E. Meek and B. F. Rider, "Compilation of Fission Product Yields," General Electric Company Report NEDO-12154, 1972.

From assumption 2, $F_p = 0.25 \times 0.91 = 0.2275$. Then, taking $F_b = 1$ gives

$$C_0 = 7.50 \times 10^7 \times 0.2275 = 1.71 \times 10^7 \text{ Ci. } [Ans.]$$

With the source of radiation known, the doses to the public can be calculated and the prospective exclusion and low-population radii evaluated in the following way: first, the 2-hr dose to the thyroid is computed as a function of distance from a reactor operating at 1 MW for all of the iodines, ^{131}I through ^{135}I, taken together, exactly as was done in Example 11.7 for ^{131}I. The defining dose, namely, 300 rems, is next divided by the computed dose. This quotient gives the reactor power at which the 300-rem dose is received in 2 hrs. For example, if 1 MW gives a dose of 0.5 rem at 1,000 m, then 300 rems would be received at the same point if the reactor operated at $300/0.5 = 600$ MW. When this procedure is carried out for a number of points, the result is a curve similar to that shown in Fig. 11.17, of the reactor power that gives 300 rems to the thyroid in 2 hrs as a function of distance.

It should be mentioned that in siting considerations at the CP licensing stage, the thyroid dose for defining the exclusion area and LPZ is taken to be 150 rems, rather than 300 rems, to allow for uncertainties in the plant design and in the local meteorology. Thus, the thyroid curve in the figure is also computed for 150 rems—

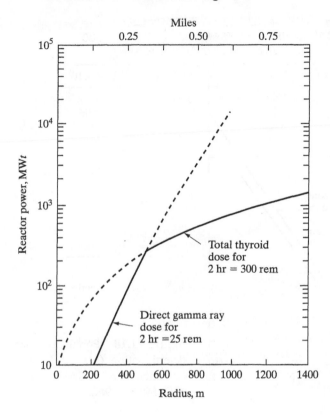

Figure 11.17 Exclusion area radius determination. (From J. J. DiNunno, *et al.*, USAEC Report TID-14844, 1962.)

and, indeed, it is common practice to compute a family of such curves for different defining doses as an aid in comparing alternate sites.

The 2-hr external and direct dose versus distance is also computed for 1 MW and then divided into 25 rems. This computation is somewhat more lengthy, as it must be made for all of the dozen or so important γ-emitting fission products released into and from the reactor building. The results of these computations are also shown in Fig. 11.17.[18]

For a given reactor power, both the internal and direct doses decrease with distance from the reactor. Therefore, the exclusion distance is determined by the defining dose, 300 rems internal or 25 rems direct, that is received *farthest* from the plant. As shown in the figure, the direct dose is controlling up to a power of about 300 MW, the thyroid dose controls beyond that point.

---

[18]In the calculations on which Fig. 11.17 is based, the external dose from the noble gases is ignored. On the other hand, the calculation does not include the effects of containment shielding, filters, or sprays that would further reduce the exclusion area.

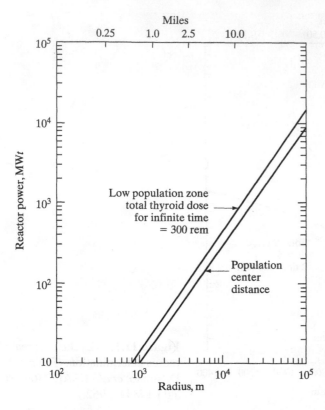

Figure 11.18 Population zone determination. (From J. J. DiNunno, *et al.*, USAEC Report TID-14844, 1962.)

Similar calculations are used to obtain the radius of the LPZ. The results are shown in Fig. 11.18. The direct dose is negligible at the distances involved and is not included in the figure.

Physical Characteristics of Site

Nuclear power plants must be designed and constructed in such a manner that all structures and systems important to safety can withstand the effects of earthquakes, tornadoes, hurricanes, floods, and other natural phenomena, without a loss of safety function. The physical characteristics of a proposed site must therefore be considered with some care in evaluating its acceptability.

Seismology   Geologists now believe that the surface of the earth is composed of large structures called *tectonic plates*. Figure 11.19 shows the major tectonic plates that have been identified. These plates are under pressure to move relative to one another as the result of the heat stored within the earth. They do not move smoothly, however, because the edges of the plates, also called *faults*,

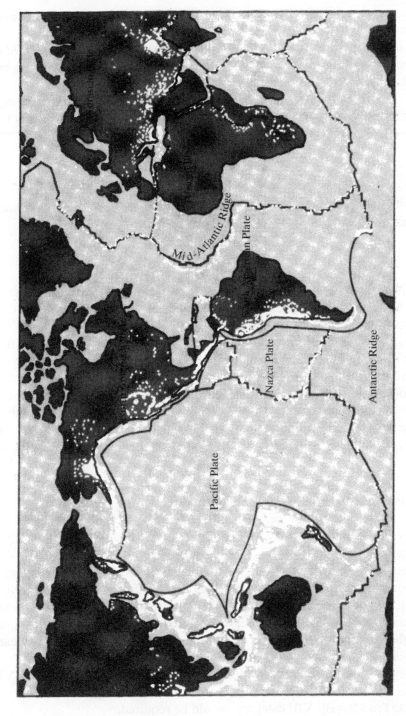

Figure 11.19 The earth's tectonic plates and earthquake belts (From C. Kissinger, "Earthquake Prediction," *Physics Today*, March, 1974.)

are rough and jagged. In time, when the stress energy along a fault becomes large enough, the plates undergo relatively sudden movement, and an earthquake is the result. The white dots in Fig. 11.19 show the centers of 42,000 earthquakes recorded between 1961 and 1969. It will be observed that the vast bulk of these occurred along the boundaries of the tectonic plates.

The *magnitude* of an earthquake (i.e., the total amount of energy released) is measured in terms of the *Richter scale*. The energy E in ergs is related to the magnitude M on this scale by the following empirical formula:

$$\log_{10} E = 11.4 + 1.5M. \tag{11.100}$$

It should be noted that this is a logarithmic scale, and the energy increases rapidly with M. For instance, if E_M and E_{M+1} are energies associated with earthquakes of magnitude M and $M + 1$, respectively, then from Eq. (11.100),

$$\frac{E_{M+1}}{E_M} = 10^{1.5} = 31.6.$$

Thus, an increase of 1 on the Richter scale corresponds to an increase in energy by a factor of almost 32. Comparing earthquakes of magnitudes 4 and 8, we find that the latter releases nearly one million (31.6^4) times more energy.

For purposes of structural design, it is necessary to know the maximum acceleration of the ground in an earthquake. This is called the *intensity* of the quake, and is measured by the *modified Mercalli scale*. This scale is divided into 12 categories, each category giving a somewhat subjective evaluation of the consequences of an earthquake. Table 11.5 shows these categories, together with the approximate associated acceleration. As indicated in the table, the Mercalli scale varies roughly logarithmically with acceleration. Thus, I on this scale, is given approximately by

$$I = 1 + 3\log_{10} a, \tag{11.101}$$

where a is the acceleration in cm/sec^2.

The technology of predicting the occurrence of earthquakes has not progressed to the point where it is possible to say precisely when or where an earthquake will occur. However, based on an analysis of the seismological history of the area in the vicinity of a site, its geological formation, and, in particular, the distance from the site to known faults, the likelihood and magnitude of an earthquake at the site can be reasonably assessed. In terms of the Mercalli scale, the maximum earthquake deduced from this analysis is called the *safe shutdown earthquake*. Safety-related portions of the plant must be functional following such an earthquake. In this regard, nuclear plants are usually designed conservatively. For example, if the largest earthquake ever recorded in a particular region is a Mercalli VII, and there are no known faults in the area, it is likely that a 0.2-g acceleration, which is well into the Mervalli VIII category, would be required.

TABLE 11.5 THE MODIFIED MERCALLI INTENSITY
SCALE AND THE CORRESPONDING GROUND
ACCELERATION

Modified Mercalli Intensity Scale, I	Ground Acceleration, a		
	$\dfrac{cm}{sec^2}$	$\dfrac{a}{g}$
I — Detected only by sensitive instruments.	1	
II — Felt by a few persons at rest, especially on upper floors; delicate suspended objects may swing.	2	
III — Felt noticeably indoors, but not always recognized as a quake; standing autos rock slightly, vibration like passing truck.	4 6	.005g
IV — Felt indoors by many, outdoors by a few; at night some awaken; dishes, windows, doors disturbed; motor cars rock noticably.	8 10	.01g
V — Felt by most people; some breakage of dishes, windows, and plaster; disturbance of tall objects.	20	
VI — Felt by all; many frightened and run outdoors; falling plaster and chimneys; damage small.	40 60	.05g
VII — Everybody runs outdoors; damage to buildings varies, depending on quality of construction; noticed by drivers of autos.	80 100	.1g
VIII — Panel walls thrown out of frames; fall of walls, monuments, chimneys; sand and mud ejected; drivers of autos disturbed.	200	
IX — Buildings shifted off foundations, cracked, thrown out of plumb; ground cracked; underground pipes broken.	400 600	.5g
X — Most masonry and frame structures destroyed; ground cracked; rails bent; landslides.	800 1000	1g
XI — New structures remain standing; bridges destroyed; fissures in ground; pipes broken; landslides; rails bent.	2000	
XII — Damage total; waves seen on ground surface; lines of sight and level distorted; objects thrown up into air.	4000 6000	5g

Meteorology   The importance of meteorological conditions on the dispersion of effluents from a nuclear plant was discussed in Section 11.4. Unfavorable conditions—for instance, the presence of frequent inversions—can militate against the location of a plant at a particular site.

Meteorological factors must also be considered to guarantee the integrity of safety-related structures in hurricanes and tornadoes. Hurricanes are defined as large storms, up to 600 miles in diameter, with winds from 75 to 200 mi/hr. They occur most frequently in the late summer and early autumn in the Caribbean and along the Gulf of Mexico and the eastern coast of the United States and in the late spring and early summer in the southern hemisphere. (They are called typhoons in the Near and Far East.)

Tornadoes, by contrast, are defined as smaller and more intense storms than hurricanes. Their diameters range from several feet to a mile. Horizontal winds rotate about the axis of a tornado with speeds of 100 to 300 mi/hr. There is also an updraft at the center with vertical winds as high as 200 mi/hr. Tornadoes most frequently occur in the central and eastern United States. They travel generally in a north-easterly direction at speeds of from 10 to 70 mi/hr and where they touch the ground they cause considerable destruction.

Reactor structures must be designed not only to withstand the direct force of tornado winds and the associated drop in pressure, but also to withstand the impact of objects that have been picked up by the winds and hurled against the structure. Typical tornado missiles used in safety studies are 1,500-lb utility poles moving at 180 ft/sec and 4,000-lb automobiles at 75 ft/sec.

Geology   Studies must be made of the geological structure of a proposed site in order to determine whether the area can firmly support the reactor building with all its internal components. The details of these studies are beyond the scope of this text, however, and will not be discussed further.

Hydrology   It is necessary to prevent large quantities of water from entering the site of a nuclear power plant, since water could compromise some of the safety-related systems of the plant. For example, offsite power could be jeopardized, especially if the water enters the plant from underground; the emergency diesel generators could be rendered inoperative; and so on. Rushing water conceivably could crush building walls, burst steam pipes, etc.

The particular types of hydrological phenomena that must be considered obviously depend upon the nature and location of the site. If, for instance, the site is on the seashore, it is necessary to estimate the largest tidal wave or tsunami or hurricane-driven wave that may be incident on the site. The maximum flood level must be ascertained for a plant sited along a lake, river, or estuary. The possibility that the bursting of a dam may lead to the sudden inundation of the plant must also be examined. Dikes of appropriate size or watertight structures must then be designed to withstand the maximum expected water invasion.

11.7 REACTOR ACCIDENTS

In the preliminary and final safety analysis reports (PSAR and FSAR), an applicant for a reactor license must analyze a set of postulated, severe accidents to show that the facility can be operated without undue risk to the health and safety of the public. These are called design basis accidents (DBA). For the purpose of evaluating the safety of the plant, these accidents are analyzed within the framework of highly conservative assumptions that tend to exaggerate the consequences of each accident. They are used as a basis for the design of plant safety features.

In the Environmental Report, the applicant must discuss the environmental effects of a broader range of possible accidents. Because the objective of this analysis is an accurate assessment of the environmental impact of the plant, these accidents may be considered in the light of more realistic and less conservative assumptions than those utilized in the safety analysis of the facility. For the purpose of standardization, the NRC has divided the spectrum of possible accidents into nine classes, in increasing order of severity. These are listed in an annex to Appendix D of 10 CFR 50; an outline of the classes is given in Table 11.6.

Class 1 accidents are trivial and involve releases of radioactivity that are not substantially different from those accompanying normal operation of the plant. Accidents of Class 2 through 8 entail the release of increasingly significant amounts of activity. Class 8 accidents include the DBAs analyzed in the PSAR and FSAR.

Class 9 accidents involve a broad range of events that, while more serious than those of Class 8, are so improbable that they can be ignored in analyzing both the safety and environmental aspects of the plant. Such accidents are not specifically defined and therefore do not appear in the table. They involve either an incredible sequence of unlikely events or a single improbable event of major proportions. An example of a Class 9 accident of the first type is the simultaneous failure of all off-site and on-site power at the moment of a LOCA. An accident of the second type could result from the sudden rupture of the pressure vessel. Another accident of this kind might occur if the containment building were struck by a heavily loaded civilian aircraft or by a military aircraft carrying bombs. The likelihood of these accidents is extraordinarily remote, and accordingly they are categorized as Class 9 events.

Loss-of-Coolant Accident

Any unexpected decrease in coolant flow through a reactor core can lead to serious consequences for the plant as a whole. Such a drop in flow can be caused by

TABLE 11.6   NRC ACCIDENT CLASSIFICATION

Class 1	Trivial incidents.
Class 2	Small releases outside containment.
Class 3	Radwaste system failures.
3.1	Equipment leakage or malfunction.
3.2	Release of waste gas storage tank contents.
3.3	Release of liquid waste storage tank contents.
Class 4	Fission products to primary system (BWR).
4.1	Fuel cladding defects.
4.2	Off-design transients that induce fuel failures above those expected.
Class 5	Fission products to primary and secondary systems (PWR).
5.1	Fuel cladding defects and steam generator leaks.
5.2	Off-design transients that induce fuel failure above those expected and steam generator leak.
5.3	Steam generator tube rupture.
Class 6	Refueling accidents.
6.1	Fuel bundle drop.
6.2	Heavy object drop onto fuel in core.
Class 7	Spent fuel handling accident.
7.1	Fuel assembly drop in fuel storage pool.
7.2	Heavy object drop onto fuel rack.
7.3	Fuel cask drop.
Class 8	Accident initiation events considered in design basis evaluation in safety analysis report.
8.1	Loss-of-coolant accidents.
8.1(a)	Break in instrument line from primary system that penetrates the containment.
8.2(a)	Rod ejection accident (PWR).
8.2(b)	Rod drop accident (BWR).
8.3(a)	Steam line breaks (PWR's outside containment).
8.3(b)	Steam line breaks (BWR).
Class 9	See text.

anything from a leak in a small coolant pipe to the complete severance (sometimes called a "guillotine" break) of a major coolant pipe. It is the latter that is the starting point of a design basis LOCA. In a PWR plant, the break in question occurs in one of the primary coolant loops, while in a BWR, the break is in a recirculation loop. With the HTGR, all of the primary loops (there are four or six loops in currently designed plants, depending on the reactor power) are located within the prestressed concrete reactor vessel (PCRV) as indicated in Fig. 4.25. The event corresponding to a pipe break in a PWR or BWR is the gross failure of one of the PCRV penetration closures. Although this would lead to the rapid escape of the helium coolant, such a LOCA for the HTGR is called a design basis depressurization accident (DBDA).

If the containment structure is not present or is improperly designed or if the emergency core cooling system (ECCS) is not present or is inoperable, the consequences of a LOCA is very serious, especially for a PWR or BWR plant. To begin with, the water is under great pressure, and on the breaking of the pipe, the water flows out and flashes to steam. More important, in the absence of an ECCS the uncovered fuel rods melt because of fission product heating. This, in turn, leads to the initiation of various exothermic chemical reactions between the molten material and the water-steam mixture, some of which produce hydrogen. Furthermore, the pool of molten fuel and structural material at the bottom of the reactor vessel might, in time, melt its way through the vessel, then through the concrete underlying the reactor building, and then sink into the ground. In the United States this phenomenon is often referred to as the *China syndrome* (and presumably in China as the *US syndrome*), because of the ancient myth that if one digs directly downward from a point in the United States one eventually reaches China.[19] Extensive measures are taken to prevent a core meltdown, because of the potential for releasing large amounts of fission products.

To avoid core meltdown and to lessen the consequences of a LOCA, as well as the assumptions underlying the analysis of a LOCA, the ECCS and associated systems are somewhat different for the PWR, BWR, and HTGR. Each will now be considered, in turn.

The PWR

LOCA Safety Systems   The ECCS is divided into an active and a passive system. The passive system consists of large accumulator tanks (one per primary loop) containing borated water under a blanket of nitrogen at a pressure of about 650 psi. When the pressure in the reactor vessel falls below this value, the reversal of pressure across a check valve causes the tanks to discharge into the cold legs of the primary coolant loops and then through the core. This passive system provides automatic rapid cooling of the core in the case of large breaks that might result in the uncovering and overheating of the core, before the electrically driven, active part of the ECCS can be put into operation.

The active system, in turn, is divided into high-pressure and low-pressure coolant injection systems. In the low-pressure system, pumps take water either from a large water storage tank[20] or from the containment sump and pass it through heat exchangers and into the hot legs of the primary coolant loops. Following a normal shutdown of the reactor, the low-pressure system is also used for the removal

---

[19]In fact, however, the molten material could only melt a rather short distance before solidifying.

[20]The water in these tanks is normally used for refueling the reactor, a procedure that must be done under water.

of residual fission product decay heat. The high-pressure system often utilizes the charging pumps which are in continuous operation for the chemical shim system when the reactor is critical. Upon an actuation signal, the suction from these pumps is quickly diverted to pumping water from the storage tank through the boron shim tank and into the cold legs of the primary coolant system.

Consider now the sequence of events following a postulated guillotine break of a PWR primary coolant pipe. Immediately, high-pressure, subcooled water rushes out of the break and flashes to steam in the containment structure. As a result, the pressure in the reactor vessel quickly drops to the saturation pressure, in the neighborhood of 1,500 psi, corresponding to an average water temperature of about 600°F. Thereafter, a mixture of steam and water flows out the break until the pressure in the reactor and in the containment building become equal (at a pressure somewhat under 45 psi, a typical design pressure for PWR containment). This is known as the *blowdown* phase of a LOCA.

For the first few seconds following the break, the fuel and cladding temperatures remain roughly constant, as indicated in Fig. 11.20, but with the onset of boiling, about 5 secs after the break, the temperature rises rapidly. For complex reasons, having to do with the flow of the water-steam mixture out of the vessel, the cladding temperature, after reaching a peak value of nearly 1,400°F, drops again as shown in the figure, because of the action of the ECCS. Incidentally, as soon as boiling occurs in the core, the reactor falls subcritical, because of the negative void coefficient.

The first action of the ECCS is the flow of water from the accumulator tanks, which commences when the pressure has dropped below about 650 psi. The cladding temperature soon levels off. However, the capacity of the accumulator tanks is limited, so to provide continued core cooling, the low-pressure pumps come into action, first carrying additional water from the refueling tank to the reactor vessel. Because of the break in the primary coolant pipe, most of the primary coolant water, the borated water from the accumulators, and the low-pressure water from the refueling tank ultimately passes out through the break. All of this water is collected in a large sump at the bottom of the containment building and is pumped through the low-pressure heat removal system after the level in the refueling tank falls to a predetermined level. In this way, the reactor is guaranteed an endless supply of recycled core cooling water. The heat absorbed from the core is carried outside the plant via the heat exchangers which have cooling water from an external source on their cold side in the low-pressure heat removal system.

In the case of a small break or a crack in the primary coolant loop, the blowdown and depressurization occur much more slowly, and the outflowing of primary coolant could lead to the uncovering of a portion of the core before the flow of water from accumulators and the low-pressure coolant system is initiated. This situation

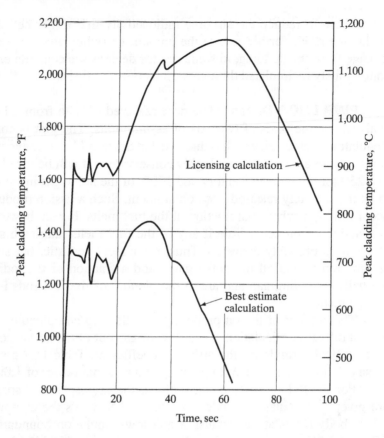

Figure 11.20 Typical calculations of peak cladding temperatures following a large break LOCA.

is handled by the high-pressure coolant injection system, which rapidly introduces water at a pressure high enough to overcome the pressure remaining in the primary coolant loop.

With the onset of a LOCA, several other components of a PWR plant in addition to the ECCS operate to lessen the effect of the incident. The containment structure is particularly important in this regard. This must be strong enough to hold the coolant that has flashed to steam and tight enough to contain whatever radioactivity is released in the accident. Several spray nozzles are located in the upper part of the building and are connected, with appropriate redundancy, to the refueling water storage tank and the containment sump. When these sprays are actuated, they perform two functions. First, they condense the steam and thereby limit the pressure in the containment atmosphere to below the design pressure for the building. Second, they remove (i.e., wash out) enough of the iodine and other nongaseous fusion products released from the fuel (see later) to the containment atmosphere to reduce the airborne radioactivity that might be released from the containment. To further reduce the amount of radioactive material available for leakage from the building, in some reactor designs the containment atmosphere is continuously recirculated through various types of filters.

PWR LOCA Doses   The dose received off-site from a LOCA is usually analyzed on the basis of two sets of assumptions. The first is contained in NRC Regulatory Guide 1.4, which is discussed in Section 11.6 in connection with reactor siting. These assumptions are highly conservative. It will be recalled, in particular, that 25% of the iodine inventory and 100% of the noble gas inventory are assumed to be immediately released into containment. Such a fission product release could occur only if a substantial fraction of the fuel melts. In fact, however, if the ECCS is functioning properly, little if any of the fuel melts, although some of the fuel cladding is probably damaged. Thus, a far more realistic, but still conservative, assumption is that all the noble gases and a fraction of the iodine contained in the pellet–cladding gap and are in the plenum of the fuel rods is released to the containment.

The amount of fission product gas in the gap and plenum can be estimated from a diffusion calculation of the flow of gas out of the surface of the UO_2, using an empirical formula for the diffusion coefficient. Table 11.7 gives the results of these calculations for a PWR operating at a nominal power of 1,000 MWe.

For PWR LOCA dose calculations, the atmospheric dispersion conditions are given in Regulatory Guide 1.4. Table 11.8 shows the computed thyroid and whole-body doses at the exclusion and low-population boundaries (indicated as EZB and LPZ, respectively) for a PWR operating at 3,216 MW. In the calculations, leakage from the building is 0.1% per day, and the spray system provides an iodine concentration decay constant of $32\,\mathrm{hr}^{-1}$, to be added to the radioactive and leakage decay constants. As seen in the table, the conservative assumption of total gap–plenum release gives small doses for this event.[21]

Three Mile Island Accident.   The accident at the Three Mile Island nuclear power station (TMI) near Harrisburg, Pennsylvania, in March 1979 is one of the worst that has occurred in a commercial nuclear power plant. The plant involved, TMI Unit 2, a 900 MWe PWR, was operating at nearly full power when the accident happened. During maintenance operations, the feedwater flow to the steam generator was lost, an event that can be expected to happen two or three

---

[21] It should be mentioned that the PWR installation in question has valves that effectively isolate the containment building from the rest of the plant. These valves are closed within approximately one minute after the onset of a LOCA. Beyond that time, there conceivably could be no leakage whatsoever from the building. With a building release lasting only one minute, the overall dose at the EZB is less than one-third of a rem and at the LPZ about 0.1 rem, both for the gap release.

TABLE 11.7 TYPICAL VOLATILE FISSION PRODUCTS IN GAP AND PLENUM OF A 1,000 MWE PWR AT THE END OF A FUEL CYCLE.*

Isotope	Fraction of core inventory	Curies in gap and plenum ($\times 10^5$)
^{85m}Kr	0.0020	0.328
^{85}Kr	0.2157	0.841
^{87}Kr	0.0020	0.400
^{88}Kr	0.0029	0.893
^{133}Xe	0.0185	10.60
^{135}Xe	0.0054	0.842
^{134}Cs	0.050	3.75
^{137}Cs	0.050	2.35
^{131}I	0.023	5.38
^{132}I	0.0026	0.907
^{133}I	0.0079	4.52
^{134}I	0.0016	0.971
^{135}I	0.0043	2.33

*Inventory fractions from the PSAR for Indian Point Nuclear Generating Unit No. 3, Consolidated Edison Company of New York.

times a year in a plant such as TMI 2. Because of the sudden loss of heat removal, pressure began to increase in the primary system. The pressure rise was terminated by inserting the control rods, but before this could take full effect, the pressure excursion caused the relief valve on the pressurizer (which acts as an expansion tank for the reactor coolant) to open. This was the expected sequence of events follow-

TABLE 11.8 DOSES FROM A LOCA IN A 3,216-MW PWR*

Assumed release	EZB (350 m)† 2 hr, rem	LPZ (0.7 mi) 30 days, rem
Regulatory Guide 1.4		
Thyroid	189	130
Whole body	9.74	8.54
Gap–plenum		
Thyroid	3.5	2.9
Whole body	0.04	0.07

*From the PSAR for Indian Point Nuclear Generating Unit No. 3, Consolidated Edison Company of New York.

†Actually, the site boundary, which lies somewhat within the EZB.

ing a loss of feedwater. However, once the pressure was reduced the relief valve should have closed. It failed to do so and thus created a small break LOCA in the primary system.

At this point, the system pressure began to drop, and this automatically turned on the high-pressure ECCS, which worked properly. After a short time, the operator thought that the primary system was once again full of water; to prevent over filling the system, he turned off the ECCS. Unfortunately, the reading on the water level indicator was not indicative of the water level in the reactor itself, and, therefore, the ECCS water should not have been turned off. In a short time, the water level in the reactor dropped below the top of the fuel, which became severely overheated. As a result, a significant fraction of the gaseous and volatile fission products were released from the fuel. In addition, the Zircaloy cladding began to react with the water to produce hydrogen.

Subsequent analysis has shown that about 50% of the noble gases, iodine, and cesium were released from the fuel. Almost all the cesium and iodine were trapped in the water that was present, but most of the noble gases were released to the containment air. The fraction of the iodine released to the atmosphere was about 2×10^{-7}, while that of the noble gases krypton and xenon was about 0.1 (10%). No cesium was measured outside of the plant. The maximum off-site radiation dose was less than 0.1 rad, and the total population dose was about 10 person-rems. This is below the level that would be expected to produce any health effects.

Despite the very serious nature of the accident, the safety systems worked very well to protect the public. There are, however, many important lessons to be learned from this event. (See Pigford article cited in the references at the end of this chapter.) Industry and the regulators have carefully reviewed the accident and have put into practice a number of changes that should reduce the likelihood of such events in the future.

The Chernobyl Accident. The Chernobyl reactor was a graphite-moderated boiling water pressure tube reactor of the RBMK-1,000 type discussed in Chapter 4. In this type reactor, the pressure vessel is replaced by a large number of pressure tubes. The fuel is located in these tubes. The Chernobyl Nuclear Power Plant, which consists of four RBMK 1,000 reactors, is situated near the Ukranian City of Kiev in a forest area (Polesje) directly on the banks of the river Pripjat.

At the time of the accident, four reactor units were operating at the site. The accident took place in Unit number four on April 26, 1986, when the plant was being shut down as planned. During the shutdown process, a test to demonstrate certain safety features of the plant was planned. Deficiencies in the test program, unexpected conditions during the performance of the test, as well as several unforeseeable events and unplanned actions by the operating staff, led to an extremely unstable operating condition. This eventually caused the reactor to undergo a prompt, supercritical power excursion that led to the disastrous failure of the reactor.

Several features of the RBMK-type reactor and several unique conditions at Chernobyl contributed to the accident. These features included a high positive void reactivity during operational conditions with high fuel burnups; a positive shut-down effect of fully withdrawn control rods; insufficient effectiveness of shutdown facilities; and a lower than expected shutdown reactivity margin.

The RBMK has a different reactivity behavior because of the utilization of graphite as a moderator and water as a coolant. The coolant water works as neutron moderator and neutron absorber at the same time. While the moderator effect of water is of minor importance compared with the one of graphite, the reactivity behavior of the reactor is considerably influenced by the neutron absorption of the water. A reduction of the coolant density (e.g., by voiding) or a loss of coolant leads to a significant reduction of neutron absorption in the reactor core. An increase of the steam content in the pressure tubes results in an increase in reactivity. The RBMK thus has a positive steam void reactivity effect or positive void coefficient. Furthermore, the void effect grows strongly positive in the course of operation as higher fuel burnups are achieved.

One special feature of RBMK reactors prior to the accident was a positive reactivity effect of the control rod motion. Of the control rods, 163 are inserted from the top of the reactor and consist of an absorber part of approximately 6 m in length and a displacement part made of graphite 4.5 m in length. When the control rod is completely withdrawn, the center graphite containing part of the control rod is approximately in the center of the reactor core. Above and below this part, the control rod channel is filled by columns of water. If a control rod is inserted into the core from this position, the neutron moderating part of the control rod that contains the graphite replaces the lower, more neutron absorbing, water column. Thus, the opposite of the intended effect is attained at first: The reactor power is not reduced by the control rod insertion; instead, it initially increases, due to decreased absorption and increased moderation.

As mentioned before, on the day of the accident a series of tests was planned. The accident occurred when the plant was being shut down as planned in prepa-ration for these tests. During the shutdown process, a test to demonstrate certain safety properties of the plant had been intended. Immediately prior to the initia-tion of the test, the plant was in an extremely unstable condition, as there was an unfavorable loading condition, a low power level with an unfavorable power dis-tribution, a high coolant flow rate in the core, a reduced feedwater flow rate to the reactor with increasing coolant temperature at the core inlet, and an unstable xenon spatial distribution.

At 1:23 am, the test was initiated. As planned, four main coolant pumps were slowed down, reducing coolant flow. The reduction of the coolant flow rate in the reactor core connected with the test, and the shutdown of the reactor actuated a short time later, coupled with the unstable state of the plant at first led to a power increase of more than 15% and then, within a few seconds, to an uncontrolled reactor power excursion. The reactor excursion led to a rapid increase of the energy release in the fuel elements and to the destruction of the reactor core. The heat stored in the fuel was very quickly transferred to the surrounding coolant, the coolant practically evaporated, which further increased power, due to the positive void reactivity effect. The resulting large pressure increase led to the explosion of the reactor.

The explosion destroyed large parts of the reactor building of Unit 4, the turbine building, and the connecting building. The walls of the reactor building were partially destroyed, and the roof was destroyed completely. Figure 11.21 shows the reactor building shortly after the accident. The energy release was sufficient to ignite the graphite blocks, resulting in a fire. The fire spread fuel and fission products throughout the reactor building and into the surrounding countryside. People residing in the nearby town were exposed to large doses of radiation and inhaled significant amounts of radioactive iodine and other radioisotopes. Because of the extent of the damage and the extreme radiation levels that exist even today at the damaged reactor, a Sarcophagus, Fig. 11.22, was built around the destroyed reactor unit. The Sarcophagus was built to limit the release of radioactive material into the environment and to safely enclose the radioactivity in the destroyed unit for 30 years.

It is estimated that about 3.5% of the fuel was ejected from the reactor, all of the radioactive noble gases, and approximately 33% of the core inventory of cesium, and about 50% of the iodine 131 inventory was released as well. Table 11.9 gives an estimate of the released fraction of many of the isotopes in the core. Prevailing winds carried the radioactive releases to northern Europe and to Scandinavia.

The operating staff of the reactor, firemen, and members of the army (e.g., helicopter pilots) were employed to immediately fight the fire and to cover the exposed reactor core. About 300 persons were taken to the hospital. Part of this group, 134 people, received very high dose rates and showed signs of acute radiation sickness with weakness, vomiting, and dizziness, as well as skin burns. Despite intensive medical efforts, including, with the help of American surgeons, bone-marrow transplants in specialized hospitals in Moscow and Kiev, 28 persons died of radiation sickness and fire injuries. The total-body doses were up to 13 Gy. At a body dose of 4 Gy, the chance of survival is about 50%.

The radiation exposure of the population caused by the reactor accident can essentially be attributed to the short-lived Iodine 131 and the long-lived Cesium 137 released in the explosion and subsequent fire. Immediately after the accident, Soviet authorities ordered emergency protective measures. The most severely af-

Figure 11.21   The Chernobyl Sarcophagous as it appears years after the accident. Courtesy GRS, IPSN and RRC KI.

TABLE 11.9   RADIONUCLIDE RELEASE FRACTION FROM THE CHERNOBYL ACCIDENT*

Radionuclide	Half-life value [days]	Core inventory [Bq]	Estimated released fraction [%]
Krypton 85	3930	$3.3 \times 10E16$	100
Xenon 133	5.27	$7.3 \times 10E18$	100
Iodine 131	8.05	$3.1 \times 10E18$	50
Tellurium132	3.25	$3.2 \times 10E18$	15
Cesium 134	750	$1.9 \times 10E17$	33
Cesium 137	11000	$2.9 \times 10E17$	33
Ruthenium 106	368	$2.0 \times 10E18$	3
Strontium 89	53	$2.3 \times 10E18$	4
Strontium 90	10200	$2.0 \times 10E17$	4
Plutonium 238	31500	$1.0 \times 10E15$	3
Plutonium 239	8900000	$8.5 \times 10E14$	3
Plutonium 240	2400000	$1.2 \times 10E15$	3
Plutonium 241	4800	$1.7 \times 10E17$	3
Curium 242	164	$2.6 \times 10E16$	3

*Chernobyl ten years after, CD by GRS, IPSN, RRC KI, 1996.

Figure 11.22   The Chernobyl reactor building shortly after the accident
of April 25–26, 1986. Courtesy GRS, IPSN and RRC KI.

fected areas, like the town of Pripjat, for example, were evacuated within a short
period. Further evacuations followed when additional exposed areas were found by
monitoring programs.

The areas in the Ukraine, Belorussia, and Russia affected by the reactor accident, report a general increase of diseases among the population, especially stress-
related illnesses such as depressions, anxieties, and other psychological problems.
Also, chronic bronchitis, high blood pressure, coronary diseases and diabetes are
being diagnosed more and more. According to present findings, these diseases

cannot be directly attributed to the radiation exposure. But they are, nevertheless, considered as indirect consequences of the accident and the expulsion of many thousands of people from their familiar circumstances. One reason for a part of these diseases may also be the general impairment of the social and economic situation after the collapse of the Soviet Union and the uncertainties connected therewith. The increase of contagious diseases (e.g., diptheria and tuberculosis) might also be connected with this impairment of the general living conditions and the insufficient medical treatment.

Nontheless, there are clear indications that the radiological consequences of the accident had a negative impact on the general health of the exposed population. Most evident is the increase in thyroid cancer among children in Belorussia and in the areas affected in the Ukraine and the Russian Federation. Epidemiologists have seen a highly statistically significant increase among this cohort during the period between 1986 and 1994. From 1981 to 1985, a total of 39 cases of thyroid cancer occurred in this region. Between 1986 and 1995, 565 children in these areas contracted thyroid cancer, an increase of 1,000%. Thyroid cancer of children can generally be treated by surgical removal with good success, but it does require long-term therapy. Despite the operation, some children have died in Belorussia.

The BWR

LOCA Safety Systems   BWRs, like PWRs, have multiple provisions for reducing the consequences of a LOCA. These provisions also vary somewhat from plant to plant. A typical BWR ECCS consists of a low-pressure coolant injection system and independent high- and low-pressure core spray systems. These systems, together with the containment and suppression pool, are designed to function in the following way in the event of a LOCA.

Suppose first that there is a small break in the reactor coolant system that does not result in a rapid depressurization of the reactor vessel above some preselected height and that additional water is provided by a high-pressure, turbine-driven pump, operated by steam diverted from the main steam line. Water is first supplied from a condensate water storage tank and later from the suppression pool. The water enters the reactor either by way of a special ring of nozzles located over the core or via the normal feedwater nozzles, depending on the model of BWR.

Should the high-pressure coolant injection system fail to maintain the water level, the reactor vessel is depressurized. This depressurization is accomplished by the automatic opening of pressure relief valves located on the main steam line piping. Steam from the vessel then enters the primary containment dry well, from which it is conducted to the suppression pool as explained in Section 11.3. If the original pipe break is large, the depressurization occurs without relief valve assistance by the rapid coolant blowdown through the break.

With the depressurization of the vessel, the low-pressure core spray system and the low-pressure coolant injection system are actuated. In the former, electric pumps carry water from the suppression pool to a ring of nozzles above the core similar to those in the high-pressure system. The low-pressure coolant injection system pumps water from the suppression pool directly into the reactor recirculation loops. In normal operation, this system is also used for the removal of fission product heat after shutdown of the reactor. As in the PWR plant described earlier, all of the water flowing out of the break in the pipe is collected—in the BWR case in the suppression pool—thus assuring a continuing source of cooling water.

Recent versions of BWR plants are also equipped with a filtration, recirculation, and ventilation system (FRVS) for reducing the release of airborne radioactivity to the environment. In the event of a LOCA (or a fuel-handling accident within the reactor building), this system recirculates the reactor building atmosphere through high-efficiency filters and exhausts through an elevated vent. The reactor building pressure is thus maintained slightly negative with respect to ambient outdoor pressure. The FRVS removes at least 90% of particles 0.30 micron or larger in the building atmosphere and at least 90% of the elemental and methyl iodine with each pass through the filters. This system greatly reduces off-site doses, as discussed next.

BWR LOCA Doses   Like those for a PWR LOCA, BWR LOCA doses are usually calculated on the basis of two sets of assumptions. The assumptions that the NRC uses in Regulatory Guide 1.3 to analyze the radiological consequences of a LOCA in a BWR are not substantially different from those for the PWR. Thus, 25% of the iodine and 100% of the noble gas inventory are assumed to be released into

TABLE 11.10   DOSES FROM LOCA IN 3,440-MW BWR

Assumed release	EZB ($\sim$ 0.5 mi) 2 hr, rems	LPZ 30 Days, rems
Regulatory Guide 1.3*		
Thyroid	≤ 1	< 1
Whole body	1.25	< 1
Gap–plenum†		
Thyroid	0.00059	0.00088
Whole body	0.00007	0.0011

*Supplement No. 3, Safety Evaluation Report for Hope Creek Generating Station, AEC Directorate of Licensing. Uses LPZ radius of 5 miles.

†PSAR for the Hope Creek Generating Station, Public Service Electric Gas Company. Uses LPZ radius of 1 mile.

containment, and no credit is taken for the retention of iodine in the suppression pool. The resulting off-site doses are essentially the same as those calculated earlier for the PWR in the absence of the filtration, recirculation, and ventilation system. With the FRVS in operation, the off-site doses at the EZB and LPZ drop to the values shown in Table 11.10.

Analysis by the General Electric Company indicates that only about 9% of the fuel rods would actually be perforated in a LOCA, and none of the fuel would melt. To be conservative, yet more realistic than Regulatory Guide 1.3, a BWR LOCA may be assumed to release into primary containment 25% of the fission product gases in the gap and plenum. The resulting doses for a 3,440-MW reactor are shown in the table. These doses are calculated on the assumption that the FRVS is 90% efficient.

The HTGR

Design Basis Design Accident (DBDA) Safety Systems.   Generally speaking, a depressurization accident (the DBDA) in a HTGR is much less serious than a LOCA in a light-water reactor. For one thing, the high heat capacity of the large mass of graphite in the HTGR (about three million pounds in a 3,000-MW reactor) guarantees that the core temperature transients are slow and controllable. Furthermore, with the use of a single-phase gas coolant, a sudden reduction in system pressure cannot lead to a loss of all of the coolant or prevent its circulation because of a change in phase.

The main cooling system of an HTGR consists of four (or six) separate loops, each with its own steam generator and steam-turbine-driven helium circulator. There is no single failure of an active component in the HTGR that will prevent the operation of enough of the main coolant loops to assure safe shutdown of the plant. Nevertheless, the HTGR is provided with a core auxiliary cooling system (CACS), which performs the same functions as the ECCS in a PWR or BWR. This CACS consists of two (or three) separate cooling loops, each containing a heat exchanger and an electrically driven circulator.

A DBDA in an HTGR presumably would occur in the following way: upon the failure of a prestressed concrete reactor vessel penetration, the helium coolant would immediately begin to leave the vessel. To prevent too sudden an outrush of the gas, which might damage some of the reactor components, the flow area is restricted to 100 in^2 or less by the design of the penetrations and by the addition of flow restrictors within the coolant loops. The reactor scrams on a low-pressure signal at about 10 sec, and the reactor power drops to decay heat levels about 35 sec later. By 100 sec from the time of the break, the helium flow has dropped to 4% of normal.

Sufficient steam continues to be available from the main coolant system to operate the main circulators for approximately 12 min. After this time, the circulators are driven by steam from low pressure (200 psig, 600°F) auxiliary boilers, which normally supply the steam required for reactor startup, postshutdown heat removal, and heating of the reactor building. The CACS is actuated if, and only if, this secondary steam is unavailable.

DBDA Doses   An analysis of this incident shows that even with only two CACS loops operating, nowhere in the core does the temperature rise high enough to release fission products from the fuel. Thus, the only activity released into containment in a DBDA is the activity normally circulating with the coolant and activity plated out of the coolant onto the walls of the coolant loops that is picked up by the escaping gas. The latter is known as *liftoff activity* and consists mostly of iodines and particulates containing isotopes of strontium and cesium. The activity in the circulating gas is mainly due to noble gases that have diffused out of the fuel particles. The total activity of these noble gases is comparable to that found in the gap and plenum of an LWR of the same power level; the iodine activity is 10 to 100 times smaller.

HTGR plants are also equipped with a filter and a recirculation, and ventilation system. However, because of the small release of iodine in a DBDA and in order to protect the filters from the initially high containment temperature, this system is not started until about 2 hrs after an accident.

The doses at the exclusion zone and LPZ boundaries are shown in Table 11.11. These are clearly well below 10 CFR 100 limits.

Other LWR and HTGR Accidents   Many other types of accidents in addition to the LOCA must be considered in the design and licensing of a nuclear power plant. Space permits only a brief description of a few of these. Some additional accidents are discussed in the problems at the end of the chapter.

TABLE 11.11   DOSES FROM A DBDA IN A 3,164-MW HTGR*

Exposure	EZB (762 m) 2 hr, rem	LPZ (1.5 mi) 30 days, rem
Thyroid	0.200	0.117
Whole body	0.0155	0.0108

*From the PSAR for the Fulton Generating Station, Philadelphia Electric Company.

BWR: Steam Pipe Break   The steam in a BWR plant is somewhat radioactive, since it is produced directly in the reactor. If a steam pipe should break anywhere outside of containment, the activity in the pipe escapes and quickly enters the environment. To assure that steam flow into the broken pipe does not continue, each steam line in a BWR plant has at least two isolation valves that close automatically with the decrease in pressure accompanying a break in the line.

In analyzing this accident, it is assumed (Regulatory Guide 1.5) that (1) the isolation valves close in the maximum time characteristic of the valves (about 10.5 sec); (2) all of the coolant in the broken steam line and its connecting lines at the time of the break, plus the steam passing through the valves prior to closure, is released; (3) the activity (including all the iodine and noble gases that may be present in the steam from leaking fuel rods) is released to the atmosphere within 2 hrs, at a height of 30 feet, under fumigation conditions. With these assumptions, the calculated thyroid doses for a 3,440-MW reactor at the EZB are on the order of 1 rem and somewhat lower at the LPZ. The whole-body doses are fractions of 1 mrem.

BWR: Rod Drop   It will be recalled that the control rods in a BWR enter from the bottom of the core and are inserted upwards. A number of failures in the control rod drive system of a BWR can be envisioned that could lead to the release of some activity into the containment. However, none is more serious than the uncoupling of a control rod from the drive mechanism when the rod is fully inserted in the reactor and the subsequent dropping of the rod out of the reactor.

Although the mechanical design of the control rod prevents the rod from falling more rapidly than 5 ft/sec, the resulting increase in reactivity leads to a power transient, releasing approximately 4×10^9 joules of energy, for a reactor originally operating at 3,440 MW, before the transient is terminated by the negative Doppler coefficient of the fuel and the insertion of other rods. This energy release is estimated to perforate the cladding of about 1% of the fuel rods ($\sim$ 330 rods), and to release all of the fission product gases contained in the gap and plenum of these rods.

All the noble gases are assumed to be transported to the steam dome at the top of the reactor vessel. However, most of the iodine isotopes remain behind in the water. The fission product gases in the dome are then carried by the steam down the steam lines, through the turbine, and into the condenser. The steam lines in a BWR plant are equipped with radiation monitors, and with the passing of the gases down the lines, these monitors trip the isolation valves, thus retaining the bulk of the fission product gases within containment. The gases reaching the condenser are assumed to be exhausted directly to the environment, although in normal operation the off-gas exhaust from the condenser is routed through time-delay holdup tanks and a filter system before exiting to the atmosphere.

The overall doses from this incident are small. For a 3,440-MW plant, the EZB and LPZ thyroid doses are 0.019 rem and 0.037 rem, respectively. The corresponding whole-body doses are 0.24 rem and 0.18 rem.

PWR: Rod Ejection It is conceivable, although highly unlikely, that a failure of the control rod housing could occur in such a way that high-pressure reactor coolant water might forcibly eject a cluster control rod assembly. This would result in a power transient similar to that in a BWR rod drop accident. Detailed calculations show that this would lead to the failure of about 15% of the fuel rods with an attendant release into containment of the fission product gases from the gap and plenum of these rods. This is a far smaller release than in a LOCA, and the off-site doses are consequently also smaller.

Fast-Reactor Accidents To date, only one experimental commercial fast reactor, the Fermi Reactor,[22] has been built in the United States. Thus, there has been little experience with the licensing of fast reactors in this country, and the NRC has not yet developed formal guides for the evaluation of accidents. Studies of the LMFBR systems both here and abroad indicate, however, that the worst postulated accident for this type of reactor involves the meltdown and subsequent disruption of a portion of the core. This type of accident is often referred to as a *core disruptive accident* (CDA) or *severe accident*.

A CDA can be initiated in several ways. Suppose, for example, that the flow of sodium into the core is partially blocked by some extraneous piece of metal that has become lodged at the bottom of the core. The sequence of events that follows such a blockage depends on many factors. The next scenario doubtless exaggerates the incident. First, those fuel rods in the core that are denied their normal cooling start to melt. This occurs about halfway up the core, at the point where the fuel temperature is initially the highest. The molten fuel then flows downward (and may solidify) in adjacent coolant channels. This, in turn, blocks the passage of sodium in these channels, causing the meltdown of additional fuel, which blocks other channels, and so on. This process continues until a portion of the core has been more or less eaten away as shown in Fig. 11.23(b).

The situation at this point is that upwards of one-third of the core is left momentarily hanging above the remainder of the core. The reactor, of course, is subcritical. However, the unsupported part of the core now drops, and as the two parts come together, the multiplication factor rapidly increases. As soon as k goes above unity, energy is produced in the fuel. This release of energy tends to counter the momentum of the falling fuel and eventually the system blows apart.

[22]The Fermi Reactor was an LMFBR designed to operate at 300 MW. It went critical in 1963, suffered a partial meltdown in 1966, was repaired and returned to critical in 1970, only to be decommissioned soon thereafter for lack of operating funds.

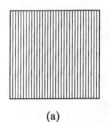
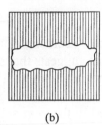
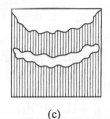

(a)   (b)   (c)

Figure 11.23   Stages in a core disruptive accident (CDA) in an LMFBR.

This chain of events is similar to the operation of a so-called gun-type nuclear bomb, in which a supercritical mass of fissile material is rapidly assembled by the shooting of one subcritical mass of material into another. Of course, the conditions within a fast reactor are not designed to optimize the explosive release of nuclear energy. Accordingly, the TNT equivalent of the CDA is much less than the yield of even a small nuclear bomb. With the Fermi Reactor, for instance, calculations showed that the worst postulated CDA would release the equivalent of about 700 lb (*not* tons) of TNT. Although this is not an insignificant amount of energy, the mass of that reactor, its pressure vessel, and its concrete shielding would have been more than adequate to contain such an energy release.

In actual fact, the foregoing model of a CDA probably grossly overestimates the severity of the accident. For one thing, it appears unlikely that a small blockage, affecting only a few fuel channels and leading to the melting of a few fuel rods, will propagate over much of the core. Such a blockage did occur in the Fermi Reactor when pieces of zirconium broke off from a conical flow guide and became lodged at the base of the core. Coolant flow to two fuel assemblies, each containing 144 rods, was reduced, and portions of these assemblies were melted. However, very few neighboring rods were involved.[23]

It is also unrealistic to assume that the whole of the upper portion of the core will remain in place while the center of the reactor melts out, especially if the reactor is large in diameter. Regions of the upper core can reasonably be expected to collapse as soon as the lower, supporting fuel, has melted. Such a gradual collapse considerably reduces the overall energy yield of a CDA.

It may be mentioned that a flow blockage is not the only way in which a CDA might be initiated. This accident can result, for example, from the sudden ejection of a control rod when the reactor is operating at power, from the formation and passage of large sodium bubbles through the core, or from the breaking of a major coolant pipe—a *bona fide* LOCA.

---

[23]This may not have been a fair test, however, as the blockage occured at a low power-density region of the core and while the reactor was being brought up to power.

Nonpower Accidents

The previous discussions have centered on reactor accidents that involve operation of the reactor at power. Recent analysis of events suggests that scenarios involving accidents with the reactor shut down or accidents involving fuel handling are much more likely to occur.

Accident risk analysis, which is discussed in the next section, suggests that low-power and shutdown accidents contribute significantly to the likelihood of core damage. In fact, analysis suggests that the contribution by such events to reactor core damage is equal to that of all other causes.

When the reactor power level decreases to 10% or less of full power, the automatic actuation of safety systems is often disabled either by the hardware or procedures. Thus, many normal safety systems that help to protect the reactor in the event of an accident are not easily available. The types of accidents that can occur at low power are similar to those at power. The typical accidents that are considered at shutdown or low power are as follows: loss of cooling, loss of coolant accidents, and reactivity events. The loss-of-cooling events occur when there is a loss of heat-removal capability. At shutdown, the reactor is cooled using an auxiliary system, the residual heat-removal system, to remove the decay heat from the core. Loss of cooling can occur from a malfunction in this system or from a loss of electric power to drive the pumps in the residual heat removal system.

LOCAs at low power are of concern for the same reason that they are at full power. At shutdown, the typical loss-of-coolant event of concern is the drain down of the system. Since the system is usually depressurized at shutdown, pipe breaks are unlikely. A drain down can occur from the inadvertent opening of a valve or the interconnection of a system to the primary system.

Reactivity events at shutdown can lead to a local or core-wide criticality. Examples of such events include the inadvertent removal of boron from the reactor coolant, an inadvertent control rod withdrawal, or refueling errors. Analysis of such events suggest that these accidents could lead to large pressure peaks that would threaten the integrity of the piping system leading to a loss-of-coolant event.

For PWR's, the most likely risk comes from the situation where the reactor vessel is partially drained. The lowered water level could lead to uncovering of the piping connection to the residual heat removal system. If this were to occur, then the accident would become a loss-of-cooling event. For a BWR, the most likely event is a cold-overpressurization event during filling of the system after refueling.

Other possible shutdown accidents are fire and accidents involving fuel handling. Fires can occur at anytime, but are particularly likely when maintenance of the system is being done, as during a shutdown. Fuel-handling accidents are of concern, since the fuel in a reactor is off-loaded to the spent fuel storage pool during refueling. The fuel has a large inventory of fission products and could be over-heated and melt if not properly cooled. Another possibility is an accident involving a criticality due to a mix-up in fuel types. While such events have not occurred at reactors, they have occurred in fuel-manufacturing facilities. The most recent event was the nuclear accident at the Tokai-Mura fuel-fabrication facility.

Tokai-Mura Nuclear Accident   The Tokai-Mura facility is operated by the Japan Nuclear Fuel Conversion Co. (JCO), a subsidiary of Sumitomo Metal Mining Co. The facility is a fuel-manufacturing plant that converts enriched uranium hexafluoride to uranium oxide in preparation for fuel fabrication.

On September 30, 1999, three workers were preparing a small batch of fuel for the JOYO experimental fast breeder reactor, using uranium enriched to 18.8% ^{235}U. It was JCO's first batch of fuel for that reactor in three years, and no proper qualification and training requirements appear to have been established to prepare the workers for the job. At around 10:35, when the volume of solution in the precipitation tank reached about 40 liters, containing about 16 kg U, a critical mass was reached.

The resulting criticality exposed the workers and nearby residents to significant doses of radiation. The three workers involved were exposed to doses of from 1,000 to 20,000 mSv, where 8,000 mSv is a lethal dose. Workers and residents received doses of 10 to 50 mSv. No significant contamination escaped the facility. The dose was due to the radiation emitted during the criticality accident.

The event occurred because the workers failed to follow proper procedures and lacked adequate safety training. The system was designed to preclude such an event, but various safety features were defeated by the workers.

11.8 ACCIDENT RISK ANALYSIS

Nuclear power reactors, with their enormous inventories of fission products, obviously carry the potential for major accidents involving many people. Fortunately, there has been only one such accident to date—Chernobyl[24]—and, as a result, it is not possible to fully evaluate the risk of this kind of accident as directly as, say, that of a fatal automobile accident, regarding which there is a superabundance of data. However, sophisticated methods of accident risk analysis, some of which were developed for the aerospace industry, have been applied to nuclear power plants in recent years.

---

[24]The accident at Three Mile Island did seriously damage the core, but did not result in a large release of radioactivity to the atmosphere.

The Meaning of Risk

"Risk" is a common word that may convey different meanings to different people. For the purpose of making comparisons between different types of risks, the risk associated with a specified event is given the technical definition as the *consequence of the event per unit time*. For example, there are on the order of 50,000 deaths from auto accidents per year in the United States. The total societal risk from such accidents is defined as 41,000 deaths/year. Because the US society contains approximately 270 million persons, the average individual risk is defined as

$$\frac{4.1 \times 10^4 \text{ deaths/year}}{270 \times 10^6 \text{ persons}} = 1.5 \times 10^{-4} \text{ deaths/person-year}.$$

The risk of an event can be computed in an obvious way from the frequency of the event and the magnitude of the consequences of the event:

$$\text{Risk} \left\{ \frac{\text{consequences}}{\text{unit time}} \right\} = \text{frequency} \left\{ \frac{\text{events}}{\text{unit time}} \right\}$$

$$\times \text{ magnitude} \left\{ \frac{\text{consequence}}{\text{event}} \right\}. \qquad (11.102)$$

In 1998, the number of automobile accidents in the US was about 6.3 million accidents per year, and one accident in 150 results in a fatality. The societal risk is therefore

$$\text{Risk} = 6.3 \times 10^6 \frac{\text{accidents}}{\text{year}} \times \frac{1 \text{ death}}{150 \text{ accidents}} = 41,000 \frac{\text{deaths}}{\text{year}},$$

as given earlier.

Table 11.12 presents some US accidental death statistics for the year 1969. It will be observed that the risk per person-year of accidental death ranges from about 10^{-4} (for automobiles and falls) to 10^{-7} (for lightning and storms). Except for certain hazardous occupations and sports, which involve a limited segment of the population, there are few if any activities engaged in by the general public that have associated risks as high as 10^{-3}/person-year.

Indeed, it appears that the public has developed a more or less consistent attitude with regard to the more familiar risks. Evidently, risks as large as 10^{-3}/person-year are not acceptable. With a risk of 10^{-4}/person-year, the public is willing to go to considerable trouble and expense to hold down the accident rate. Thus, various traffic programs are instituted: automobiles are required to be inspected; seat belts must be installed; fences and railings are placed at dangerous locations; money is spent on fire prevention and active fire control. Risks on the order of 10^{-5}/person-year are still acknowledged. Lifeguards are employed at swimming areas; special bottles with child-resistant caps are used for drugs; and so on.

TABLE 11.12 INDIVIDUAL RISK OF ACUTE FATALITY
BY VARIOUS CAUSES IN THE UNITED STATES IN 1969*

Accident	Total number for 1969	Approximate individual risk, acute fatality per person-year
Motor vehicles	55,791	3×10^{-4}
Falls	17,827	9×10^{-5}
Fires and hot substances	7,451	4×10^{-5}
Drowning	6,181	3×10^{-5}
Poisons	5,516	2×10^{-5}
Firearms	2,309	1×10^{-5}
Machinery (1968)	2,054	1×10^{-5}
Water transport	1,743	9×10^{-6}
Air travel	1,778	9×10^{-6}
Falling objects	1,271	6×10^{-6}
Electrocution	1,148	6×10^{-6}
Railway	884	4×10^{-6}
Lightning	160	5×10^{-7}
Hurricanes	93†	4×10^{-7}
Tornadoes	91‡	4×10^{-7}
All others	8,695	4×10^{-5}
All accidents	112,016	6×10^{-4}

*Based on "Reactor Safety Study," U.S. Nuclear Regulatory
Commission report WASH-1400, 1975.
†(1901–1972 avg.)
‡(1953–1971 avg.)

For risks of 10^{-6}/person-year or less, the public attitude seems to change. Accidents in this category tend to be viewed either as acts of God or as the individual's own fault—as in "everybody knows you shouldn't stand under a tree in a thunder storm." It appears that the public finds a risk of 10^{-6}/person-year or less in some sense acceptable—at least no cause for great concern.

However, the public acceptability of a given risk depends not only on the size of the risk, but also on the magnitude of the consequences of the event. Consider, for example, two accidents A and B. A occurs on the average of once a year and there is one fatality. The societal risk of A is therefore one death per year. B, on the other hand, occurs only once in 10,000 years, but results in 10,000 deaths. The risk associated with B is also one death per year. Nevertheless, it appears that the public generally views accident B less favorably than accident A. Evidently, faced with the possibility of a disaster involving the loss of 10,000 lives, the public tends to ignore the low probability of the event and assumes that if it *can* happen it *will* happen. This attitude on the part of the public is known as *risk aversion*.

Risk Determination

The calculation of the risk associated with accidents in a nuclear power plant is a three-step process. First, it is necessary to determine the probabilities of the various releases of radioactivity resulting from accidents at the plant. Second, the consequences to the public of these releases must be evaluated. Finally, the release probabilities and their consequences are combined to obtain the overall risk.

Because nuclear power plants are designed with carefully engineered safety features, some of which were discussed in Sections 11.3 and 11.7, the failure of a single plant component, in itself, cannot lead to a major release of radioactivity. Such a component failure sets in operation a sequence of safety systems intended to mitigate the effect of the failure. The ultimate release, if there is one, depends on the extent to which these safety systems perform their functions. Thus, any given initiating event can lead to a number of possible releases, each with its own probability.

The identification of the accident sequences leading to various releases is facilitated by the use of *event trees.* This technique can best be understood by a simple illustrative example. Consider a large pipe break that leads to a LOCA. To develop the appropriate event tree, it is first necessary to determine which plant systems are likely to have an effect on the subsequent events. In this case, these are the station electric power, the emergency core cooling system, the radioactivity removal system (sprays, filters, etc.), and the containment system (the containment and its heat removal system). These systems and the initiating pipe break are then written in a sequence according to the time they can be expected to affect the course of the accident, as indicated at the top of Fig. 11.24.

The event tree shown in the center of the figure consists of a series of branches for every possible sequence of events in which the indicated safety systems do or do not operate.[25] According to the *basic* tree, the single initiating event can lead by way of 16 different paths to 16 different outcomes. The probability of any particular outcome is obtained by multiplying together the succession of probabilities along the appropriate branch. The computation is simplified by the fact that the probability of failure of any safety system is small, and so the probability that it succeeds is essentially unity. The probability of each outcome is indicated to the right of the corresponding branch.

---

[25]Partial operation of a safety system is not included. To be conservative, a partially operating system is assumed not to be operating at all.

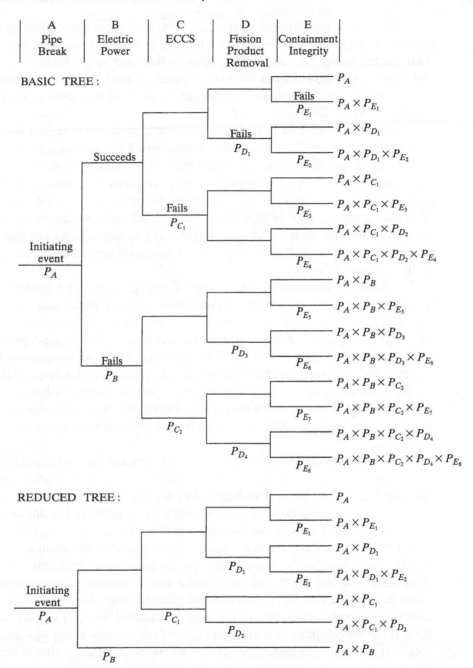

Figure 11.24 Basic and reduced event trees for a loss-of-coolant accident. (From "Reactor Safety Study," U.S. Nuclear Regulatory Commission report WASH-1400, 1975.)

Because of the physical relationship between some reactor safety systems, the basic tree shown in the figure has a number of illogical branches. For instance, if all electric power fails at the plant, none of the other safety systems can operate and none of the growth from the lower branch of the basic tree is possible. The *reduced event tree*, pruned, so to speak, of its illogical branches, is shown at the bottom of Fig. 11.24.

Once the event tree has been constructed, each outcome can be analyzed to determine how much radioactivity is released. This depends, in the case of a LOCA, on the extent of the meltdown of the fuel and on a number of other factors. The computation of the associated probability requires, as indicated in the figure, a knowledge of the failure probabilities of all the safety systems. The probabilities of initiating events, such as pipe breaks and vessel ruptures, can be otained from failure rate data on nuclear and other engineering systems. However, there is little data available on the failure rates of reactor safety systems. These are determined by the method of *fault tree analysis*.

A fault tree is essentially the reverse of an event tree. For a particular failure, the fault tree is used to identify the combinations and sequences of other failures that lead to the given failure. An example of this method is shown in Fig. 11.25, which gives a portion of the fault tree for the loss of all electrical power to the engineered safety features (ESFs). The operation of the ESFs requires both AC power, which provides the energy to run the pumps, fans, and so on, and DC power for the electrical systems that control the use of the AC power. Failure of *either* AC *or* DC power renders the ESFs inoperable. Hence, the fault tree in the figure starts at the first level with an "OR" gate, which is given the special symbol indicated in the figure.

If the probabilities of AC or DC power failures are P_{AC} and P_{DC}, respectively, then the probability P_{EP} of the loss of all electrical power to the ESFs is essentially the sum $P_{AC} + P_{DC}$. Thus, if sufficient data are available on the failure of AC and DC power, the probability P_{EP} can be computed directly. In the absence of such data, it is necessary to continue the tree to another level.

For instance, there are two sources of AC power in the event of a LOCA— off-site power supplied by power lines from the local power grid, and on-site power from standby diesel generators (and from a second reactor, if one is present). Because these power sources are independent of one another, *both* must fail to produce a loss of AC power, and accordingly they are coupled in Fig. 11.25 by an "AND" gate. If the probabilities of loss of on-site and off-site power are P_{NS} and P_{FS}, respectively, then the probability of loss of all AC power P_{AC} is clearly the product $P_{NS} P_{AC}$.

If experimental failure rates for P_{NS} or P_{FS} are not known, additional levels would have to be added to the fault tree. In this way, the fault tree is continued

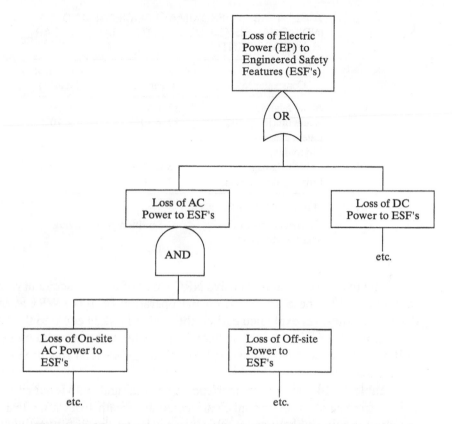

Figure 11.25   The development of a fault tree for determining the probability of a loss of power to engineered safety features. (From "Reactor Safety Study," U.S. Nuclear Regulatory Commission Report WASH-1400, 1975.)

until failures are identified for which statistical data are available. The result of this analysis is the probability P_{EP}, which is the same as P_B in the event tree in Fig. 11.24.

With the calculation of possible accidental releases of radioactivity and their probabilities, the first step in the evaluation of nuclear risk is completed. The doses that would be received by the public are computed next, based on the population distribution in the vicinity of the plant and on prevailing weather conditions. These doses are then translated into acute fatalities and various nonfatal health effects, using dose–effect relationships similar to those discussed in Chapter 9. The probabilities and the consequences are then combined using Eq. (11.102) to obtain the risks of nuclear plant accidents.

TABLE 11.13 APPROXIMATE SOCIETAL AND
INDIVIDUAL RISKS DUE TO POTENTIAL NUCLEAR
PLANT ACCIDENTS*

Consequence	Societal (year^{-1})	Individual (year^{-1})
Acute fatalities	†3×10^{-3}	2×10^{-10}
Acute illness	†2×10^{-1}	1×10^{-8}
Latent cancers	2	1×10^{-8}
Thyroid injury	20	1×10^{-7}
Genetic damage	3×10^{-1}	1×10^{-9}
Property damage ($)	2×10^{6}	—

*From WASH-1400, *op. cit.*

†Individual risk value based on 15 million people living
near the 100 plants.

The results of a comprehensive NRC study of nuclear accident risks are given
in Table 11.13. These are based on the operation of 100 LWRs at 68 currently
designated sites. As explained earlier, the societal risk is equal to the total number
of people in the United States that can be expected to be affected by the operation
of the reactors per year. The individual risk is the societal risk divided by the US
population.

Table 11.14 shows a comparison of societal and individual risk of acute fatalities from nuclear and nonnuclear accidents. Death from a nuclear accident is
clearly a highly unlikely event. To place nuclear risks in further perspective, the
NRC has also computed the frequency of nuclear accidents that would result in
more than a given number of fatalities and has compared this with similar data on
various man-caused and natural disasters. The results are given in Figs. 11.26 and
11.27, where the ordinate is the average number of events per year leading to more
than N fatalities. It will be observed from the figures that the expected number of
fatalities from nuclear accidents is orders of magnitude smaller than the number
from common man-caused accidents and is on the order of the number of fatalities
from meteors. It may also be noted that the possibility of large nuclear disasters—

TABLE 11.14 RISK OF ACUTE FATALITIES FROM
NUCLEAR AND NONNUCLEAR ACCIDENTS

Societal risk, acute fatalities per year		Individual risk, acute fatality per person-year	
Nonnuclear	Nuclear	Nonnuclear	Nuclear
115,000	4×10^{-2}	6×10^{-4}	2×10^{-10}

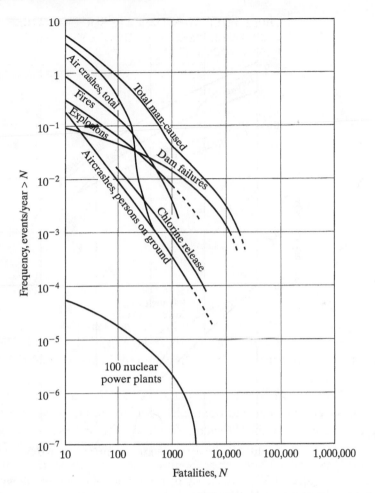

Figure 11.26   Frequency of human-caused and nuclear events with fatalities greater than N. (From "Reactor Safety Study," U.S. Nuclear Regulatory Commission Report WASH-1400, 1975.)

often alluded to by critics of nuclear power and the source of much of the public risk aversion with regard to nuclear power—is virtually nonexistent.

More recently, the NRC undertook to evaluate, on a plant by plant basis, the accident risk posed by each nuclear power plant. The NRC evaluated the risk posed by five operating nuclear power plants. The results reported in the study[26] generally support the previous findings of the Reactor Safety Study. Licensees have now

[26]Severe Accident Risks: An Assessment for Five U.S. Nuclear Power Plants, NUREG-1150, US NRC, 1990.

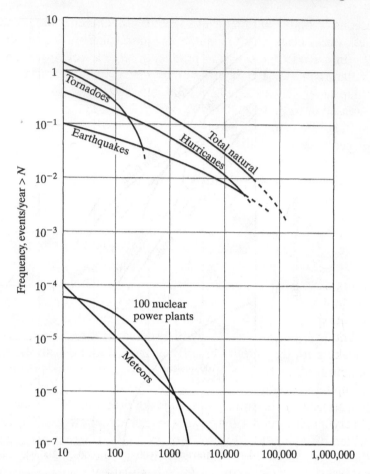

Figure 11.27   Frequency of natural and nuclear events with fatalities greater than N. (From "Reactor Safety Study," U.S. Nuclear Regulatory Commission Report WASH-1400, 1975.)

completed similar studies for all commercial nuclear power plants. These Individual Plant Examinations allow differences between plant designs and other plant specific factors to be accounted for.

11.9 ENVIRONMENTAL RADIATION DOSES

It is evidently not economically practical to reduce the emission of radioactivity from a nuclear power plant to zero, and it is necessary from time to time to release to the environment, small quantities of radionuclides. This effluent is in either gaseous or liquid form. However, the origin, amount, and composition of this effluent varies from plant to plant, even among plants using the same type of reactor.

In a PWR plant, fission product gases released from leaking fuel rods and circulating in the primary loops are purged from the coolant in the chemical shim holdup tanks. These gases are then stored in decay tanks until their activity has decreased to a point where they can be released to the environment. In some PWR plants, however, the main source of gaseous radioactive effluent arises simply from the need to purge the atmosphere of the containment building prior to entering. Liquid wastes originate in leaking valves, pumps, and other equipment. These wastes are collected in sumps within the reactor building and are also stored in decay tanks. The liquid is later diluted and discharged to a suitable body of water or solidified and then drummed for off-site disposal.

In a BWR plant, the most important source of gaseous radioactive effluent has traditionally been the main condenser. It will be recalled that the steam from the reactor, which in a BWR is radioactive, is condensed to water in the condenser after passing through the turbines. It can be shown that the efficiency of a steam cycle is highest when the condenser operates at as close to zero pressure as possible. With the condenser at zero pressure, some air inevitably leaks into it, and this must be continually exhausted. When the radioactive gases in the steam mix with the air leaking in, they are necessarily exhausted along with the air. To prevent this contaminated air from escaping directly into the environment, it is pumped into hold-up tanks where the gases are permitted to decay for a period of time before being released to the atmosphere. Some of the newer BWR plants are also equipped with cryogenic systems that liquefy the condenser off-gas. The noble gases are then distilled off and stored for upwards of 60 days before being released. All of the noble gases, except ^{85}Kr, have decayed by this time. Liquid wastes in the BWR originate in (and are handled in) essentially the same manner as those in the PWR.

Regulation of Effluents

In the United States, the authority for regulating effluent emissions from nuclear power plants rests with the NRC. However, such effluents, once they pass the site boundary, must meet rather complex standards established by the EPA and published in 40 CFR 190.10. In an effort to assure that these EPA standards are not exceeded by any of the plants it regulates, the NRC has adopted the policy (10 CFR 50.34a) of "keeping levels of radioactive material in effluents to unrestricted areas *as low as reasonably achievable*." According to the NRC, the term "as low as reasonably achievable" means "as low as is reasonably achievable, taking into account the state of technology and the economics of improvements in relation to benefit to the public health and safety . . . ."

The NRC has translated its "as low as reasonably achievable" doctrine into the following numerical guides, applicable only to water cooled reactors, for the limitation of effluents from a *single* reactor:[27]

1. The annual release of radioactive material above background in liquid effluents shall not result in an annual dose or dose commitment to any individual
   in an unrestricted area in excess of 3 mrems to the total body or 10 mrems to
   any organ.
2. The annual release of radioactive material above background in gaseous effluents shall not result in an annual air dose (i.e., D to air) in an unrestricted
   area in excess of 10 mrads for γ-rays and 20 mrads for β-rays. Larger releases are permitted, provided that the annual dose does not exceed 5 mrems
   to the total body or 15 mrems to the skin.
3. The annual release of radioactive iodine and radioactive material in particulate form shall not result in an annual dose or dose commitment to any organ
   in excess of 15 mrems.

All releases of radioactivity from a nuclear power plant are carefully monitored, and the NRC requires semiannual reporting of such releases from each licensee. The NRC, in turn, issues yearly reports summarizing release data. Figure
11.28 shows the average of the releases from 20 BWR and 133 PWR plants in the
United States during the period from 1972 to 1978. Only one HTGR was in operation at this time, and it is not included in the figure. As shown in Fig. 11.28,
BWRs over the years have consistently emitted more radioactive gaseous effluents
than PWRs. This has originated, as noted earlier, in the condenser off-gas. Presumably, newer methods of waste treatment, in particular the cryogenic separation
previously referred to, will reduce this effluent in future BWRs.

It will also be observed in Fig. 11.28 that both the BWR and PWR, especially
the latter, release relatively copious amounts of ^{3}H. This nuclide is produced directly in fission at the rate of approximately one atom per 10,000 fissions, and it is
also formed by the interaction of fast neutrons with ^{10}B via the reaction

$$^{10}B + n \rightarrow {}^3H + 2\,^4He. \tag{11.103}$$

Tritium is a pure β-emitter, it is not concentrated by biological species, and it
passes fairly quickly through the human body. Its MPC values for air and water
are therefore among the highest of any radionuclide. (See Table 9.16.)

It is of some interest to mention that fossil fuel power plants, especially those
burning coal, also emit radioactive effluents. The source of this activity is the fly
ash discharged from such plants, which contains trace amounts of uranium and
thorium and their decay products, primarily ^{226}Ra and ^{228}Th. The concentration of
these nuclides in fly ash varies from one specimen of coal to another, but a typical

[27]Extracted from 10 CFR 50, Appendix I.

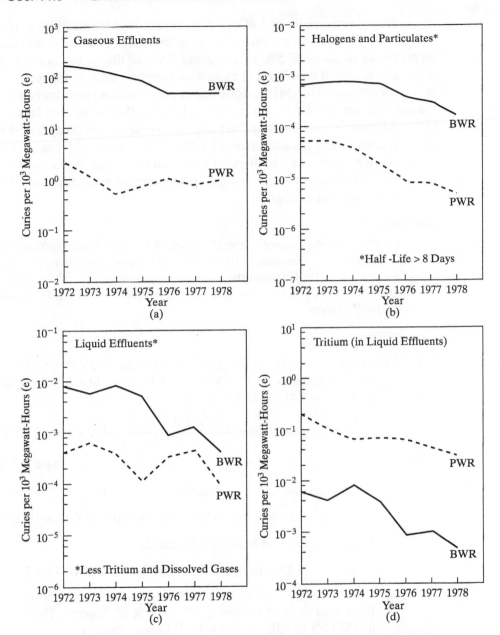

Figure 11.28 Average rate of emission of effluents from PWR and BWR power plants, 1972–1978. (From J. Tilcher and C. Benkovite, *Radioactive Materials Released from Nuclear Power Plants*, report NUREG/CR2227, 1981.)

ash contains on the order of 3 pCi/g of each nuclide. All coal-burning plants have
equipment for removing fly ash as it goes up the stack. This equipment usually has
an efficiency of about 97.5%, so that about 2.5% of the fly ash is exhausted from
the stack. A typical coal-burning plant may therefore emit as much as 5,000 tons
of fly ash per year. The (MPC)$_a$ for the isotopes of radium and thorium are among
the smallest for any radionuclide. This is because these elements have very long
biological half-lives in the body (they are bone seekers) and decay by the emission
of α-particles, whose quality factor is 10. The concentrations of these radionuclides
in the atmosphere near a coal-burning plant can thus reach levels much higher, in
terms of (MPC)$_a$, than those in the gaseous effluent of nuclear power plants. This
is shown by the next example.

Example 11.10

A 1,000-MWe plant operates at an efficiency of 38% on 13,000 Btu/lb coal with a 9%
ash content. The plant stack is 100 m high. (a) Estimate the activity of ^{226}Ra emitted
from the plant. (b) Compare the maximum concentration of ^{226}Ra at ground level
under Pasquill F, 1 m/sec dispersion conditions with the (MPC)$_a$ for this nuclide of
$3 \times 10^{-12} \mu$Ci/cm^3.

Solution.

1. The thermal power of the plant is $1,000/0.38 = 2,632$ MW. In 1 year, the
energy released is $2,632 \times 8,760 = 2.30 \times 10^7$ MW-hr. From Appendix I,
1 MW-hr $= 3.412 \times 10^6$Btu, and so

$$\text{Coal consumption} = \frac{2.30 \times 10^7 \times 3.412 \times 10^6}{13,000} = 6.04 \times 10^9 \text{ lb/yr}$$

$$= 2.74 \times 10^{12} \text{ g/yr.}$$

Because the coal is 97.5% ash and 2.5% of it escapes from the stack,

$$\text{Emitted ash} = 0.09 \times 0.025 \times 2.74 \times 10^{12} = 6.17 \times 10^9 \text{g/hr.}$$

In one year, the plant emits the total activity

$$6.17 \times 10^9 \text{ g} \times 3 \times 10^{-12} \text{ Ci/g} = 0.0185 \text{ Ci. } [Ans.]$$

2. The maximum in $\chi\bar{v}/Q'$ for a stack height of 100 m occurs at about 10^4 m.
For a wind speed of 1 m/sec, $\chi/Q' = 2.5 \times 10^{-6}$ sec/m^3. The value of Q' is
0.0185 Ci/$3.16 \times 10^7 = 5.85 \times 10^{-10}$ Ci/sec. Thus,

$$\chi = 5.85 \times 10^{-10} \times 2.5 \times 10^{-6} = 1.46 \times 10^{-15} \text{ Ci/m}^3$$

$$= 1.46 \times 10^{-15} \mu\text{Ci/cm}^3.$$

While this is a factor of $3 \times 10^{-12}/1.46 \times 10^{-15} = 2,050$ times smaller than
(MPC)$_a$ for ^{226}Ra, the concentrations of radioactive gases emitted from a nom-

inal PWR plant are typically 10^4 to 10^5 times smaller than $(MPC)_a$ for these gases under the same conditions.

Doses from Effluents

To evaluate the environmental impact of radioactive effluents from nuclear power plants, it is necessary to compute the doses received by the surrounding human population from these effluents.[28] The computations are somewhat different for gaseous effluents emitted to the atmosphere and liquid wastes discharged to bodies of water, and these two cases will be considered separately.

Gaseous Effluents   As noted earlier, the gaseous effluent from nuclear power plants consists mostly of the noble gases and the isotopes of iodine. There are several routes by which these gases can enter into or otherwise interact with the human population. The principal exposure pathways are shown schematically in Fig. 11.29.

As indicated in the figure, these gases, while in the atmosphere, give both a direct dose to the body and an internal dose from inhalation. The direct dose is largely due to the noble gases; the inhalation dose is mostly an iodine dose to the thyroid. Both of these doses can be computed in a straghtforward way from the formulas derived in Section 11.5. (See also Examples 11.2, 11.4, and 11.5.)

The noble gases remain airborne at all times, and they do not enter into other exposure pathways. The iodine isotopes, however, gradually deposit out of the plume onto soil and vegetation. This activity provides a direct ground dose to persons nearby. Methods for computing this dose are discussed in the problems. In any event, this dose is usually smaller than that from other pathways.

The deposition of the iodines carries these isotopes into the food chain. However, because of their short half-lives (the longest is that of ^{131}I and is only 8.04 days[29]) compared with the growing time of plants, very little activity enters the vegetation from the soil by way of the root system. Most of the iodine enters the food chain as the result of deposition on leafy vegetables and on forage grasses. A radiation dose is received directly when the vegetables are ingested and indirectly from the meat and dairy products—especially milk—from grazing animals.

To compute the radiation dose from ingested food, it is first necessary to determine the equilibrium Iodine activity on the foliage. The rate R_d at which Iodine deposits from a cloud is proportional to its concentration in the cloud, so that

$$R_d = \chi \upsilon_d, \tag{11.104}$$

[28] Doses to other animals in the vicinity of a nuclear plant must also be calculated, but only for the purpose of protecting the animal population as a whole, not individual animals.
[29] A small amount of ^{129}I ($T_{1/2} = 1.6 \times 10^7$ yr) is formed in and released from reactors. However, because of its long half-life, its activity is insignificant.

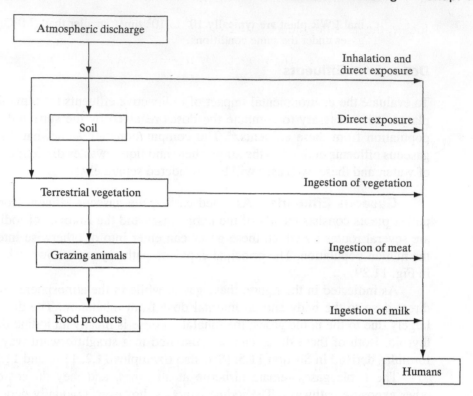

Figure 11.29 Principal exposure pathways to humans for radioactive gaseous effluent.

where υ_d is a proportionality constant. In this equation, R_d has units of Ci/m^2-sec and χ is in Ci/m^3. Thus, υ_d has units of m/sec and is called the *deposition velocity*. Experiments show that υ_d is approximately 0.01 m/sec. Because the emission of radioactive effluent is often stated in Ci/yr (or μCi/yr), it is convenient to rewrite Eq. (11.104) as

$$R_d = Q'_y(\chi/Q')\upsilon_d. \tag{11.105}$$

Here, R_d is in Ci/m^2-yr, Q'_y is in Ci/yr, (χ/Q') is the dilution factor in sec/m^3, and υ_d is in m/sec.

Once the Iodine has fallen on the foliage, it may disappear, because of weathering, or become diluted by new plant growth. It has been found that these are first-order decay processes, with a decay constant λ_f corresponding to a half-life of about 14 days. Therefore, if C_f is the activity on the foliage at any time in Ci/m^2, its total rate of disappearance is then $(\lambda + \lambda_f)C_f$, where λ is the usual radioactive decay constant. In equilibrium, with the plume providing a constant deposition of

Iodine, the rates of production and decay are equal, so that

$$R_d = Q'_y(\chi/Q')\upsilon_d = (\lambda + \lambda_f)C_f$$

and

$$C_f = \frac{Q'_y(\chi/Q')\upsilon_d}{\lambda + \lambda_f}. \qquad (11.106)$$

In this equation, λ and λ_f must be in yr^{-1} if Q'_y is in Ci/yr.

The concentration of Iodine in the milk of cows is proportional to the Iodine concentration on their pasture grasses. The proportionality factor has the value[30] 9.0×10^{-5} Ci/cm^3 of milk per Ci/m^2 of grass. Thus, the Iodine concentration in milk is

$$C_m = 9.0 \times 10^{-5}C_f = 9.0 \times 10^{-5}\frac{Q'_y(\chi/Q')\upsilon_d}{\lambda + \lambda_f}\text{Ci/cm}^3 \qquad (11.107)$$

$$= 90\frac{Q'_y(\chi/Q')\upsilon_d}{\lambda + \lambda_f}\mu\text{Ci/cm}^3. \qquad (11.108)$$

The dose from this Iodine can be obtained most simply by comparing the preceding concentration with the general population MPC for water, as milk is mostly water and is ingested in the same way. It will be recalled from Chapter 9 that a continuous intake of 2,200 cm^3/day of water containing a radionuclide at $(MPC)_w$ gives a dose rate, in equilibrium or after 50 yr, of 500 mrems/yr. An average person, however, consumes less milk than water, namely, about 1,000 cm^3/day. The $(MPC)_w$ is given in units of μCi/cm^3, so the dose rate from the continuous uptake of 1,000 cm^3/day of milk containing iodine at the concentration $C_m\mu$Ci/cm^3 is

$$\dot{H} = \frac{C_m}{(MPC)_w} \times \frac{1,000}{2200} \times 500 = \frac{227C_m}{(MPC)_w\text{mrem/yr}}. \qquad (11.109)$$

It was noted earlier that the mass of an infant thyroid is one-tenth the mass of the adult thyroid. Because an infant consumes about as much milk as an adult, a given concentration of radioiodine will result in a dose to the infant thyroid about 10 times higher than to an adult thyroid. It follows, therefore, that

$$\dot{H}_{\text{infant}} = \frac{2,270C_m}{(MPC)_w}\text{mrem/yr}. \qquad (11.110)$$

---

[30]This proportionality factor is highly variable, depending strongly on local feeding habits and on the time of year. The value given is a nominal, average value.

Example 11.11

A nuclear power plant emits 1.04×10^{-2} Ci/yr of ^{131}I. A dairy farm is located near the plant at a point where the annual average dilution factor is 4.0×10^{-8} sec/m^3. Calculate (a) the activity of ^{131}I on the vegetation at the farm; (b) the concentration of ^{131}I in the milk; and (c) the annual dose to an infant thyroid from the consumption of milk from this farm. [*Note:* (MPC) $= 3.0 \times 10^{-7}\mu$Ci/cm^3 for ^{131}I.]

Solution.

1. The decay constants λ and λ_f are given by

$$\lambda = \frac{0.693}{8.04} \times 365 = 31.5 \text{ yr}^{-1},$$

$$\lambda_f = \frac{0.693}{14} \times 365 = 18.1 \text{ yr}^{-1}.$$

From Eq. (11.106) the activity on the foliage is

$$C_f = \frac{1.04 \times 10^{-2} \times 4.0 \times 10^{-8} \times 0.01}{31.5 + 18.1} = 8.39 \times 10^{-14} \text{ Ci/m}^2$$

$$= 8.39 \times 10^{-8}\mu\text{Ci/m}^2. \text{ [Ans.]}$$

2. Using Eq. (11.107) then gives

$$C_m = 9.0 \times 10^{-5} \times 8.39 \times 10^{-8} = 7.55 \times 10^{-12}\mu\text{Ci/m}^3. \text{ [Ans.]}$$

3. Then, from Eq. (11.110), the annual dose rate is

$$\dot{H} = \frac{2{,}270 \times 7.55 \times 10^{-12}}{3.0 \times 10^{-7}} = 5.71 \times 10^{-2} \text{ mrem/yr. [Ans.]}$$

Liquid Effluents   There are several pathways by which man may become exposed to the radioactive waste discharged into bodies of water. For example, radioactive atoms may be absorbed directly if the body of water in question is used as a source of drinking water. The dose from this source can easily be computed by comparing the concentrations of the various radionuclides in the drinking water with the applicable (MPC)$_w$. Exposure may also be obtained by persons swimming in or boating on the body of water. These doses are usually very small.

A more important source of exposure is from the consumption of seafood.[31] This exposure occurs by way of the complex food chain shown in simplified form in Fig. 11.30. Starting on the left in the figure, the radionuclides are absorbed or ingested by *plankton*. These are small plants (*phytoplankton*) and animals (*zooplank-*

[31]The term "seafood" is used here to connote food originating in any body of water—oceans, rivers, lakes, etc.

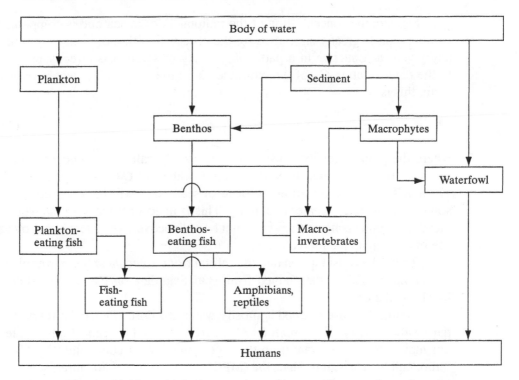

Figure 11.30   Principal exposure pathways to humans for radioactive liquid effluent.

ton) that drift about near the surface of the sea and freshwater bodies; they are the most abundant of all living organisms. These are eaten by fish, some of which, in turn, are eaten by other fish, by reptiles and amphibians, and by various *macroinvertebrates*, such as crabs, lobsters, and mollusks, all of which are consumed by man. Radionuclides are also absorbed from the water and the sediment by *benthos*, which are plants and animals that inhabit the bottoms of oceans, lakes, and rivers. These are eaten by fish and macroinvertebrates, and some activity eventually finds its way to man. Activity absorbed by sediment also enters into *macrophytes*, larger plants, and passes to man via macroinvertebrates. Finally, various waterfowl pick up activity directly from the water or from macrophytes.

The calculation of the radiation dose from contaminated seafood is a four-step process. First, the concentration of the radionuclides discharged from the plant is estimated from the discharge rate and dispersion characteristics of the receiving body of water. This part of the calculation is the most difficult and usually entails experimental and theoretical studies of the dilution of effluents in the particular body of water concerned.

Second, the concentration of the radionuclides in seafood is computed. This computation is greatly simplified by the observation that the concentration C_s of any given radionuclide in a particular variety of seafood is directly proportional to the concentration C_w of the radionuclide in the surrounding water. In equation form, this is

$$C_s = CF \times C_w, \qquad (11.111)$$

where the proportionality constant CF is usually called the *concentration factor* and sometimes the *bioaccumulation factor*. Values of CF for several nuclides are given in Table 11.15 for different forms of seafood. It will be observed that CF in some cases is much greater than unity. This is in accordance with the well-known fact that certain elements (they need not be radioactive) are concentrated by aquatic life in passing up the food chain.

Third, the consumption rate of seafood from waters near the power plant must be estimated. This is determined from statistical data on the eating habits of the local population.

Fourth, because seafood is mostly water, the dose rate can be found by comparing the activity of the seafood C_s in μCi/cm^3 and its consumption rate R_s in cm^3/day with the dose rate from the ingestion of water containing (MPC)$_w$ of the radionuclide in question. Since the (MPC)$_w$ gives a dose of 500 mrems/yr for a continuous intake of 2,200 cm^3/day of water, the dose rate received from the seafood

TABLE 11.15 NOMINAL CONCENTRATION FACTORS FOR AQUATIC ORGANISMS*

Element	Fresh water				Salt water			
	Fish	Crustacea	Mollusks	Plants	Fish	Crustacea	Mollusks	Plants
H	1	1	1	1	1	1	1	1
K	4,400	—	—	—	16	12	8	13
Ca	70	—	—	350	1.9	40	16.5	10
Mn	81	125,000	300,000	150,000	363	2,270	22,080	5,230
Fe	191	930	25,170	6,675	1,800	2,000	7,600	2,260
Co	1,615	—	32,408	6,760	650	1,700	166	554
Zn	1,744	1,800	33,544	3,155	3,400	5,300	47,000	900
Sr	14	—	—	200	0.43	0.6	1.7	21
Zr-Nb	—	—	—	—	86	51	81	1,119
Ru	—	—	—	—	6.6	382.2	448	
	9	—	320	69	10	31	5,010	1,065
Cs	3,680	—	—	907	48	18	15	51
Ce	81	600	1,100	3,180	99	88	240	1,610

*Based on data from M. Eisenbud, *Environmental Radioactivity,* 2d ed. New York: Academic Press, 1973.

is

$$\dot{H} = \frac{C_s}{(MPC)_w} \times \frac{R_s}{2200} \times 500$$

$$= 0.227 \frac{C_s R_s}{(MPC)_w} \text{mrems/yr.} \qquad (11.112)$$

An example of this type of calculation follows.

Example 11.12

The maximum probable consumption rates of seafood from the (brackish) waters near a BWR plant are determined to be: fish, 110 g/day; mollusks, 10 g/day; and crustacea, 10 g/day. The plant discharges 0.197 μCi of ^{59}Fe per year, and the dilution factor at a point 30 ft from the exit pipe is 6.29×10^{-16} yr/cm^3. The $(MPC)_w$ for ^{59}Fe is $6 \times 10^{-5} \mu$Ci/cm^3, based on exposure to the gastrointestinal (GI) tract. Estimate the yearly dose to the GI tract from seafood.

Solution. The concentration of ^{59}Fe 30 ft from the exit is

$$C_w = 0.197 \times 6.29 \times 10^{-16} = 1.24 \times 10^{-16} \mu\text{Ci/cm}^3.$$

From Table 11.15, the saltwater concentration factor of fish for iron is 1,800. Then, from Eq. (11.111), the activity in the fish is

$$C_s = 1800 \times 1.24 \times 10^{-16} = 2.23 \times 10^{-13} \mu\text{Ci/cm}^3.$$

Finally, from Eq. (11.112),

$$\dot{H} = \frac{2.23 \times 10^{-13}}{6 \times 10^{-5}} \times \frac{100}{2200} \times 500 = 8.45 \times 10^{-8} \text{mrems/yr. [\textit{Ans.}]}$$

For mollusks, the dose rate is 3.57×10^{-8} mrem/yr, and for crustaceans, it is 9.39×10^{-9} mrems/yr.

The total dose from seafood is obtained by repeating this computation for all radionuclides in the liquid waste effluent.

REFERENCES

Licensing

Joint Committee on Atomic Energy. *Atomic Energy Legislation through the 92nd Congress*, 2nd Session. 1973. This valuable document contains all legislation from 1946 through 1973.

Code of Federal Regulations, Title 10-Energy. US Government Printing Office. This document is updated and reissued every year. Supplements are issued through the year by the NRC and are available at the NRC website.

Ebbin, S., and R. Kasper, *Citizen Groups and the Nuclear Power Controversy*. Cambridge, Mass: M.I.T. Press, 1974.

Study of the Reactor Licensing Process, US Department of Energy Report, 1973.

U.S. Nuclear Regulatory Commission Regulatory Guides. Available from the U.S. Nuclear Regulatory Commission, Washington D.C. and on the web at the U.S. NRC website.

Dispersion of Effluents

Briggs., G. A. *Plume Rise*. US Department of Energy Report TID-25075, 1969.

Finleyson-Pitts, Atmospheric Dispersion in Nuclear Power Plant Siting. Lauham Berman Associates, 1980.

Lutgens, F. K., and E. Tarbuck. *Atmosphere*. Paramus: Prentice Hall, 1997.

Pasquill, F., and F. B. Smith. Study of Dispersion of Windbourne Material from Industrial and Other Sources. Paramus: Prentice Hall, 1983.

Slade, D. H., Editor, *Meteorology and Atomic Energy—1968*. U. S. Atomic Energy Commision Report TID-24190, 1968.

Williamson, S. J., *Fundamentals of Air Pollution*. Reading, Mass.: Addison-Wesley, 1973.

Nuclear Reactor Safety

"Chernobyl: The Soviet Report," Nuclear News **29**, no. 3 (Oct 1986) 59–66.

DiNunno, J. J., et al., *Calculation of Distance Factors for Power and Test Reactor Sites*, US Atomic Energy Commission Report TID-14844, 1962.

IAEA, The Safety of Nuclear Power Plants: Strategy for the Future. Vienna: IAEA, 1992.

OECD, Achieving Nuclear Safety Improvements in Reactor Safety Design and Operation. Paris: OECD, 1993.

Pershagen, B. *Light Water Reactor Safety*. New York: Elsevier Science, 1989.

Pigford, T. H., "The Management of Nuclear Safety: A Review of TMI After Two Years," *Nuclear News*, **24** 3, March 1981.

Principles and Standards of Reactor Safety. Vienna: International Atomic Energy Agency, 1973. Proceedings of a symposium.

Reactor Safety Study, U.S. Nuclear Regulatory Commission Report WASH-1400, 1975. This report summarizes the results of a two-year study of nuclear accident risks by a task force headed by Dr. N. C. Rasmussen.

Severe Accident Risks: An assessment for Five US Nuclear Power Plants, U.S. Nuclear Regulatory Commission Report NUREG-1150, 1990. This report summarizes the results of a study of nuclear accident risks posed by five US nuclear power plants and updates the results of WASH-1400.

The Safety of Nuclear Power Reactors (Light Water Cooled) and Related Facilities, US Department of Energy Report WASH-1250, 1973.

Sesonske, A., *Nuclear Power Plant Design Analysis*. US Department of Energy Report
TID-26241, 1973, Chapter 6.

Radioactive Effluents

Eisenbud, M., *Environmental Radioactivity*, from Natural, Industrial, and Military Sources.
San Francisco: Morgan Kaufman, 1997.

Electricity and the Environment. Vienna: International Atomic Energy Agency, 1991.

Environmental Aspects of Nuclear Power Stations. Proceedings of a symposium. Vienna:
International Atomic Energy Agency, 1971.

Russel, R. S., Editor, *Radioactivity and the Human Diet*. London: Pergamon, 1966.

PROBLEMS

1. (a) Show that the constant C in Eq. (11.6) is given by

$$C = \frac{Mg(\gamma - 1)}{R\gamma},$$

where M is the gram molecular weight of air and the other symbols are defined in the
text. (b) Evaluate C using $M = 29$ g, $R = 8.31 \times 10^7$ ergs/gram-mole K, $g = 980$
cm/sec^2, and $\gamma = 1.4$.

2. Compute and plot the dilution factor for ground release under Pasquill A and F conditions with 1 m/sec wind speed.

3. Verify the location of the maximum χ in Fig. 11.13 for Pasquill C conditions.

4. Had Example 11.1 been carried out for type G conditions, where would the maximum
concentration have occurred?

5. Argon-41 ($T_{1/2} = 1.83$ hr) is produced in reactors cooled with air and CO_2 and in
some reactors cooled with water (see Section 10.12). It decays by the emission of
approximately one β-ray per disintegration, which has a maximum energy of 1.20
MeV, and one γ-ray, with an energy of 1.29 MeV. Suppose that a research reactor
releases ^{41}Ar from a 100-ft vent at a rate of 1 mCi/sec. Calculate the external dose
rate at the point of highest concentration on the ground in a wind of 1.5 m/sec under
unfavorable (type F) dispersion conditions.

6. Equations (11.50) and (11.52) give the dose rate from external exposure to radioactive
gases at constant concentration. Show that if the concentration varies as

$$\chi = \chi_0 e^{-\lambda t},$$

the dose received in the time t_0 sec from γ-rays is

$$H = \frac{0.262\chi_0 \bar{E}_l}{\lambda}(1 - e^{-\lambda t_0}) \text{ rems}$$

and from β-rays is

$$H = \frac{0.229\chi_0\bar{E}_\beta}{\lambda}(1 - e^{-\lambda t_0}) \text{ rem.}$$

7. The air-cooled Brookhaven Research Reactor discharged as much as 750 Ci of ^{41}Ar per hour into the atmosphere through a 400-ft stack. Estimate the doses to persons in the town of Riverhead, located approximately 11 miles downwind from the reactor, under Pasquill F, 1 m/sec conditions. [*Note:* Under unfavorable meteorological conditions, this reactor, which is now decommissioned, was reduced in power because of the ^{41}Ar.]

8. Gas leaks from a containment structure at the constant rate of 0.2% per day. How long is it before 90% of the gas has escaped?

9. Parameters for the Berger form of the point source exposure buildup factor for air,

$$B_p = 1 + C\mu r e^{-\beta\mu r},$$

are given in Table 11.16. Another often-used approximation attributed to Goldstein is $B_p(\mu r) = 1 + k\mu r$ where $k = \frac{\mu}{\mu_a} - 1$. Compute and plot the Berger B_p and the Goldstein B_p as a function of distance up to 20 mean free paths for 1-MeV and 10-MeV γ-rays.

TABLE 11.16 PARAMETERS FOR BERGER FORM OF POINT SOURCE EXPOSURE BUILDUP FACTOR FOR AIR

E(MeV)	C	β	E(MeV)	C	β
0.5	1.5411	0.09920	4	0.6020	0.00323
1	1.1305	0.05687	6	0.5080	-0.00289
2	0.8257	0.02407	8	0.4567	-0.00349
3	0.6872	0.01002	10	0.4261	-0.00333

10. Show that the buildup flux at ground level from an infinite cloud of γ-emitting radionuclides computed using the Berger form of the point buildup factor is given by

$$\phi_b = \frac{S}{2\mu}\left[1 + \frac{C}{(1+\beta)^2}\right],$$

where C and β are defined in Problem 11.9.

11. Recompute the dose in Example 11.8, taking into account the reactor building, which is concrete and 30 in thick.

12. Radionuclides are distributed uniformly over the ground and emit S γ-rays/cm^2-sec at the energy E. (a) Show that the buildup flux at a point x cm above the ground is

given by

$$\phi_b(x) = \frac{S}{2} \int_x^\infty \frac{B_p(\mu r)e^{-\mu r}}{r} \, dr,$$

where $B_p(\mu r)$ is the point buildup factor. (b) Using the Berger form for $B_p(\mu r)$, show that

$$\phi_b(x) = \frac{S}{2}\left[E_1(\mu x) + \frac{C}{1+\beta}e^{-(1+\beta)\mu x} \right],$$

and using the Goldstein form given in Problem 11.9, show that

$$\phi_b(x) = \frac{S}{2}[E_1(\mu x) + ke^{-\mu x}].$$

[*Note:* In using these results, it is frequently necessary to evaluate the E_1 function for small argument. For this purpose, the following expansion is helpful:

$$E_1(x) = -\gamma + \ln\left(\frac{1}{x}\right) + x - \frac{x^2}{4} + \frac{x^3}{18} - \cdots,$$

where $\gamma = 0.57722$ is Euler's constant.]

13. A bad accident leaves ^{131}I deposited on the ground outside a nuclear plant at a density of approximately 0.01 Ci/m^2. Compute the gonadal dose from this radionuclide in mrem/hr that a person would receive if standing near the plant. [*Note:* For simplicity, assume that ^{131}I decays by emitting one γ-ray per disintegration with an energy of 0.37 MeV.]

14. A utility submits an application and PSAR for a nuclear plant to be located at site A, with site B as an alternative. During the CP hearings, an intervenor brings out the fact that a major housing development is being planned for the low population zone of site A. The NRC subsequently advises the utility to abandon site A in favor of site B. Hearing this, the mayor of the town encompassing site B writes to the NRC asking, in effect, "If the plant is not safe enough for the people at site A, why is it safe enough for us?" Draft a letter responding to the mayor.

15. Show that the equilibrium activity of ^{135}Xe in the thermal flux ϕ_T is

$$\alpha = 5.60 \times 10^4 P \left(1 + \frac{\phi_T}{\phi_X}\right)^{-1} \text{Ci,}$$

where P is in megawatts and ϕ_X is the parameter defined in Eq. (7.97).

16. An accident releases 25% of the inventory of the fission product gases from a 3,500-MW reactor into a containment building that leaks at 0.15% per day. Calculate for a point 2,000 m from the plant and under Pasquill F, 1 m/sec conditions (a) the 2-hr external dose; (b) the 2-hr thyroid dose. [*Note:* The effective energy equivalents of the iodine fission products are given in Table 11.17. The value of q is the same for all isotopes.]

TABLE 11.17   EFFECTIVE ENERGY EQUIVALENTS FOR IODINE ISOTOPES IN THE THYROID

Isotope	ξ
^{131}I	0.23
^{132}I	0.65
^{133}I	0.54
^{134}I	0.82
^{135}I	0.52

17. A reactor core consists of n identical fuel rods. Show that the largest fission product inventory in any of the rods is given by

$$\alpha_{i\max} = \frac{\alpha_i \Omega_r}{n},$$

where α_i is the total fission product inventory as given by Eq. (11.97) and Ω_r is the maximum-to-average power ratio in the r direction. (See Problem 8.11.)

18. Assuming that the off-site thyroid dose is entirely due to ^{131}I, calculate the radii of the exclusion area and the LPZ for a 3,000-MW reactor. [*Note:* Assume a 0.2%-per-day leakage rate and type F, 1 m/sec dispersion conditions.] What is the population center distance?

19. In a steam pipe accident in a BWR, 1.6 Ci of ^{131}I is released outside of containment before the isolation valve closes. (a) If it is assumed that this nuclide immediately escapes at ground level to the atmosphere under type F, 1m/sec dispersion conditions, what are the 2-hr thyroid doses at the 700-m exclusion zone boundary and 1-mi LPZ? (b) If credit is taken for the building wake effect, what are these doses? [*Note:* The cross-sectional area of the building is 2,240 m^2, and the shape factor is $C = 0.5$.]

20. The core of a PWR contains 39,372 fuel rods. The value of Ω_r (see Problem 11.17) is 1.9. The reactor is operated at a thermal power of 1,893 MW for 1,000 days and then shut down for repairs and refueling. In the course of these procedures all the fission product gases from one fuel rod escape into the 0.15%-per-day containment. What is the maximum dose from ^{131}I at the 800-m exclusion zone boundary, assuming unfavorable dispersion conditions and ground level release?

21. The containment sprays in a PWR plant lead to the exponential removal of Iodine from the containment atmosphere with a decay constant of 32 hr^{-1}. Recompute the 2-hr thyroid dose from the accident described in Example 11.7 with the sprays operating.

22. A guerrilla shoots a missile through the secondary containment of a 3,400-MW BWR power plant and severs a steam line. Discuss the consequences quantitatively.

23. Had the guerrilla in the preceding problem shot his missile through the containment of a 3,400-MW PWR plant and broken a primary coolant pipe, what would have been the consequences?

24. Recompute the exclusion zone boundary and LPZ for the reactor described in Problem 11.18 if the plant is equipped with the containment sprays described in Problem 11.21.

25. Experiments with the fuel of a research reactor indicate that up to a fuel temperature of 400°C, the fractional release of fission product gases into the gap between the fuel and cladding is approximately 1.5×10^{-5}. If the reactor has 62 fuel elements and operates at a power of 250 kW, what are the inventories of these gases in the gap of one fuel element? [*Note:* Assume uniform power distribution across the core.]

26. Suppose that in an accident all the fission product gases in the gap of one fuel element in the preceding problem escape into a cubical reactor room, 40 ft on a side. Evacuation of the room requires, at the most, 10 minutes. (a) What external exposure would be received at this time? [*Hint:* Use the result of Problem 11.6 for the short-lived isotopes.] (b) What thyroid dose would be received?

27. In the United States, 30 times as many persons are seriously injured as are killed in automobile accidents. Compute the societal and individual risks of serious injury from such accidents.

28. The probability of a major LOCA in a PWR plant, followed by failure of the ECCS, melting of the core, failure of the containment spray and heat removal systems, and consequent failure of containment due to overpressure is estimated to be 5×10^{-6} per year. The release of radioactivity accompanying this accident would lead to approximately 62 acute fatalities. Compute the risk from this accident.

29. If n independent and unrelated systems affect the outcome of an accident-initiating event, how many different consequences of the event are possible?

30. The Italian earthquake of 1980 registered 6.8 on the Richter scale. The San Francisco earthquake of 1906 is estimated to have had a magnitude of 8.3. How many times more powerful was the San Francisco earthquake?

31. The BWR is equipped with a total of 13 relief and safety valves that are designed to open when the reactor vessel pressure exceeds preselected levels. In the event of a

severe transient, 8 of these valves must open in order to prevent an overpressure in the vessel. If the probability that a single valve does not open is estimated to be 10^{-4}, what is the probability that fewer than 8 valves open? *Hint:* The probability $P(k, n)$ of a total of k successes out of n tries, with an individual success probability p and failure probability $q = 1 - p$, is given by the *binomial distribution*

$$P(k, n) = \frac{n!}{k! \, (n - k)!} p^k q^{n-k}.$$

32. Statistics show that off-site power to a nuclear plant can be expected to be interupted about once in 5 years. The probability of the loss of on-site (diesel) AC power is 10^{-2} per year. If the probability of the loss of on-site DC power is 10^{-5} per year, what is the probability of loss of electrical power to the engineered safety features of the plant?

33. Engine failures occur in a military aircraft at the rate of one per 900,000 miles. The ejection mechanism, which carries the pilot clear of the plane, fails once in 800 operations. Parachutes fail to open approximately once in 1,300 jumps, but pilots land safely, even with an open parachute, only 95% of the time. (a) Draw basic and reduced event trees for engine failure leading to death or survival of the pilot. (b) Compute the risk in deaths per mile from engine failure with this airplane.

34. The average annual releases of gaseous fission products from a PWR plant are given in the accompanying list. Compute the average annual external and internal doses to the public at a point 5,000 m from the plant under Pasquill type E, 1.5 m/sec conditions. [*Note:* The fission products are released from a vent 120 m high.]

Nuclide	Curies
^{85m}Kr	4.46×10^1
^{85}Kr	1.21×10^3
^{133}Xe	1.01×10^4
^{131}I	3.35×10^{-2}
^{133}I	1.5×10^{-3}

35. Compare the maximum annual average concentration of ^{131}I released from the PWR plant described in the preceding problem under type-E, 1.5 m/sec dispersion conditions, with the requirements of 10 CFR 50, Appendix I.

36. A dairy farm is located approximately 3,000 m from a nuclear plant that emits an average of 4.9×10^{-2} Ci/yr of ^{131}I from a vent 300 ft above ground level. The average observed meteorological conditions in the direction of the farm are as follows:

Condition	Wind speed (m/sec)	Percent
A	2	17
B	2	15
C	4	31
D	3.7	17
E	2.5	12
F	2	8

Calculate (a) the average deposition rate of ^{131}I onto vegetation on the farm; (b) the annual dose to an infant thyroid from drinking milk from the farm.

37. Argon-41 (see Problem 11.5) is released from the surface of a pool-type research-training reactor at the rate of 3.3×10^7 atoms/sec. The reactor is housed in a cubical reactor building 40 ft on a side, which is exhausted by fans at the rate of 1,000 cfm (cubic feet per minute). (a) What is the equilibrium activity of ^{41}Ar in the reactor building in Ci/cm^3? (b) What is the dose rate in mrems/hr from ^{41}Ar received by persons in the reactor building? (c) At what rate is ^{41}Ar exhausted from the building? (d) Compare the concentration of ^{41}Ar immediately outside the building with the nonoccupational (MPC)$_a$. [*Note:* Use a building shape factor of 0.5 and a wind speed of 1 m/sec.]

38. It is proposed to raise mussels in the warm freshwater discharged from a nuclear power plant. If the average concentration of ^{56}Mn in the diluted discharge is $6.3 \times 10^{-12} \mu$Ci/cm^3 and certain persons can be assumed to eat an average of 800 cm^3 of mussels per day, what annual dose would these people receive from this radionuclide alone? [*Note:* The general population (MPC)$_w$ for ^{56}Mn is $1 \times 10^{-4} \mu$Ci/cm^3.]

39. Calculate the inventory of fission-produced Tritium in a reactor that has been operating at a power of 3,400 MW for two years.

40. The average activity of ^{3}H in the discharge canal of a nuclear plant is $2 \times 10^{-8} \mu$Ci/cm^3. Calculate the annual dose to persons who eat fish from the canal at an average rate of 90 g/day. [*Note:* The general population (MPC)$_w$ of ^{3}H is $3 \times 10^{-3} \mu$Ci/cm^3.]

I

Units and Conversion Factors

UNITS

The final authority for the designation of the units to be used by any nation in its industry and commerce rests with the government of the region concerned. Following the development of the metric system of units by the French Academy of Sciences at the end of the 18th century, this system was gradually adopted by many nations around the world. In 1872, in response to the increasing need for international standardization of the metric system, an international conference was held in France. The result of this conference was an international treaty called the *Metric Convention*, which was signed by 17 countries including the United States. Among other things, the treaty established the General Conference of Weights and Measures (GCWM), an international body that meets every 6 years to consider improvements to the metric units system.

At its 1990 meeting, the GCWM undertook an extensive revision and simplification of the metric system. The resulting modernized metric system is called the *International System of Units (SI units)*, *SI* being the acronym for the French *Systéme International d'Unités*. SI units are now being used throughout most of the civilized world.

731

TABLE I.1 CORRESPONDING UNITS IN THREE-UNIT SYSTEMS

Quantity	SI Unit	Symbol	cgs Unit	Symbol	English Unit	Symbol
Length	meter	m	centimeter	cm	inch, foot	in, ft
Mass	kilogram	kg	gram	g	pound mass	lb or lb_m
Time	second	s	second	sec	second	sec
Temperature	kelvin	K	degree Kelvin	°K	degree Fahrenheit	°F
			degree Celsius	°C		
Force	newton	N	dyne	dyne	pound	lb or lb_f
Pressure	pascal	Pa	$dyne/cm^2$	$dyne/cm^2$	$pound/inch^2$	psi
Energy	joule	J	erg	erg	foot pound	ft lb
Power	watt	W	erg/sec	erg/sec	foot pound/sec	ft lb/sec
Heat	joule	J	calorie	cal	British thermal unit	Btu

This does not include the United States. As of this writing, English units remain in use in this country. However, it is generally assumed that the United States will eventually change to the new metric system. Indeed, most corporations and government agencies employ a dual system.

The US nuclear industry, stemming as it does from nuclear physics, on the one hand, and the traditional branches of engineering, on the other, currently employ a variety of units from different unit systems—metric or SI units, cgs units from nuclear physics, and the English units of conventional engineering. The units used to describe fundamental nuclear engineering quantities are given in Table I.1. Other quantities, such as power density, heat flux, and so on, are given by obvious combinations of these units. Multiples and submultiples of the SI units (and some of the cgs units) are formed by appending prefixes to the base units. These prefixes and their approved symbols are given in Table I.2.

TABLE I.2 SI PREFIXES

Factor	Prefix	Symbol	Factor	Prefix	Symbol
10^{12}	tera	T	10^{-1}	deci	d
10^9	giga	G	10^{-2}	centi	c
10^6	mega	M	10^{-3}	milli	m
10^3	kilo	k	10^{-6}	micro	μ
10^2	hecto	h	10^{-9}	nano	n
10^1	deka	da	10^{-12}	pico	p

CONVERSION FACTORS

Factors for converting one unit into another are given in Tables I.3 through I.14 for most of the units found in nuclear engineering. The use of these tables is fairly obvious. Thus, in Table I.3, for example, 1 centimeter is equal to 0.01 meters, 10^{-5} kilometers, 0.3937 inches; 1 meter is equal to 100 centimeters, 10^{-3} kilometers, 39.37 inches; and so on. In each table, the SI unit is indicated by an asterisk.

TABLE I.3   LENGTH

Centimeters	Meters*	Kilometers	Inches	Feet	Miles
1	0.01	10^{-5}	0.3937	0.03281	6.214×10^{-6}
100	1	10^{-3}	39.37	3.281	6.214×10^{-4}
10^5	10^3	1	3.937×10^4	3281	0.6214
2.540	0.0254	2.540×10^{-5}	1	0.08333	1.578×10^{-5}
30.48	0.3048	3.048×10^{-4}	12	1	1.894×10^{-4}
1.609×10^5	1609	1.609	6.336×10^4	5280	1

TABLE I.4   AREA

cm^2	m^2*	km^2	in^2	ft^2	mi^2
1	10^{-4}	10^{-10}	0.1550	1.076×10^{-3}	3.861×10^{-11}
10^4	1	10^{-6}	1550	10.76	3.861×10^{-7}
10^{10}	10^6	1	1.550×10^9	1.076×10^7	0.3861
6.452	6.452×10^{-4}	6.452×10^{-10}	1	6.944×10^{-3}	2.491×10^{-10}
929.0	0.09290	9.29×10^{-8}	144	1	3.587×10^{-8}
2.590×10^{10}	2.590×10^6	2.590	4.014×10^9	2.788×10^7	1

TABLE I.5   VOLUME

cm^3	Liters	m^3*	in^3	ft^3
1	10^{-3}	10^{-6}	0.06102	3.531×10^{-5}
10^3	1	10^{-3}	61.02	0.03531
10^6	10^3	1	6.102×10^4	35.31
16.39	0.01639	1.639×10^{-5}	1	1
2.832×10^4	28.32	0.02832	1728	1

TABLE I.6   MASS

Grams	Kilograms*	Pounds	Tons (short)	Tons (metric)
1	0.001	2.205×10^{-3}	1.102×10^{-6}	10^{-6}
1,000	1	2.205	1.102×10^{-3}	10^{-3}
453.6	0.4536	1	5.0×10^{-4}	4.536×10^{-4}
9.072×10^5	907.2	2,000	1	0.9072
10^6	1,000	2,205	1.102	1

TABLE I.7 TIME

Seconds*	Minutes	Hours	Days	Years
1	1.667×10^{-2}	2.778×10^{-4}	1.157×10^{-5}	3.169×10^{-8}
60	1	1.667×10^{-2}	6.994×10^{-4}	1.901×10^{-6}
3,600	60	1	0.04167	1.141×10^{-24}
8.64×10^4	1,440	24	1	2.738×10^{-3}
3.156×10^7	5.260×10^5	8,766	365.24	1

TABLE I.8 ENERGY

erg	Joule*	Kilowatt-hour	Gram-calorie	Btu	MeV
1	10^{-7}	2.778×10^{-14}	2.388×10^{-8}	9.478×10^{-11}	6.242×10^{-5}
10^7	1	2.778×10^{-7}	0.2388	9.478×10^{-4}	6.242×10^{12}
3.6×10^{13}	3.6×10^6	1	8.598×10^5	3412	2.247×10^{19}
4.187×10^7	4.187	1.163×10^{-6}	1	3.968×10^{-3}	2.613×10^{13}
1.055×10^{10}	1055	2.931×10^{-4}	252.0	1	6.586×10^{15}
1.602×10^{-6}	1.602×10^{-13}	4.450×10^{-20}	3.826×10^{-14}	1.518×10^{-16}	1

TABLE I.9 POWER

Watt*	Kilowatt	Megawatt	Btu/hr	MeV/sec
1	10^{-3}	10^{-6}	3.412	6.242×10^{12}
10^3	1	10^{-3}	3412	6.242×10^{15}
10^6	10^3	1	3.412×10^6	6.242×10^{18}
0.2931	2.931×10^{-4}	2.931×10^{-7}	1	1.829×10^{12}
1.602×10^{-13}	1.602×10^{-16}	1.602×10^{-19}	5.466×10^{-13}	1

TABLE I.10 POWER DENSITY—HEAT SOURCE DENSITY

Watt*/m³*	Watt/cm³	Cal/sec-cm³	Btu/hr-ft³	MeV/sec-cm³
1	10^{-6}	2.388×10^{-7}	0.09662	6.242×10^6
10^6	1	0.2388	9.662×10^4	6.242×10^{12}
4.187×10^6	4.187	1	4.045×10^5	2.613×10^{13}
10.35	1.035×10^{-5}	2.472×10^{-6}	1	6.461×10^7
1.602×10^{-7}	1.602×10^{-13}	3.826×10^{-14}	1.548×10^{-8}	1

TABLE I.11   HEAT FLUX—ENERGY FLUX

Watt*/m²*	Watt/cm²	Cal/sec-cm²	Btu/hr-ft²	MeV/sec-cm²
1	10^{-4}	2.388×10^{-5}	0.3170	6.242×10^{8}
10^4	1	0.2388	3170	6.242×10^{12}
4.187×10^4	4.187	1	1.327×10^4	2.613×10^{13}
3.155	3.155×10^{-4}	7.535×10^{-5}	1	1.969×10^9
1.602×10^{-9}	1.602×10^{-13}	3.826×10^{-14}	5.078×10^{-10}	1

TABLE I.12   THERMAL CONDUCTIVITY

Watt/m-K*	Watt/cm-°C	Btu/hr-ft-°F	
1	10^{-2}	2.388×10^{-3}	0.5778
10^2	1	0.2388	57.78
418.7	4.187	1	241.9
1.731	0.01731	4.134×10^{-3}	1

TABLE I.13   VISCOSITY

kg/m-sec*	Poise=g/cm-sec	Centipoise	lb (mass)/hr-ft
1	10	1000	2419
0.1	1	100	241.9
0.001	0.01	1	2.419
4.134×10^{-4}	4.134×10^{-3}	0.4134	1

TABLE I.14   PRESSURE

Pa*	Dyne/cm²	Bar	psi	atm
1	10	10^{-5}	1.450×10^{-4}	9.869×10^{-6}
0.1	1	10^{-6}	1.450×10^{-5}	9.869×10^{-7}
10^5	10^6	1	14.50	0.9872
6,895	6.895×10^4	.06895	1	0.06805
1.013×10^5	1.013×10^6	1.013	14.70	1

REFERENCES

The International System of Units. National Bureau of Standards Special Publication 330, 1972.

SI Units and Recommendations for the Use of Their Multiples. New York: American National Standards Institute, 1973.

Fundamental Constants
and Data

TABLE II.1   FUNDAMENTAL CONSTANTS

Quantity	Symbol or definition	Value
Atomic mass unit	amu	1.66054×10^{-24} g 931.494 MeV
Avagadro's number	N_A	0.6022137×10^{24} (g-mole)$^{-1}$
Boltzmann's constant	k	1.38066×10^{-23} J/°K 8.61707×10^{-5} eV/°K
Compton's wavelength of the electron	λ_C	2.42631×10^{-10} cm
Electron rest mass	m_e	9.10939×10^{-31} Kg 5.485799×10^{-4} amu 0.510999 MeV
Elementary charge	e	1.602192×10^{-19} coul
Neutron rest mass	M_n	1.674929×10^{-27} Kg 1.008665 amu 939.56563 MeV
Planck's constant	h	6.626075×10^{-34} J-sec 4.13572×10^{-15} eV-sec
Proton rest mass	M_p	1.67262×10^{-27} Kg 1.007276 amu 938.27231 MeV
Speed of light	c	2.997925×10^8 m/sec

*"Reviews of Particle Properties," *Phys. Rev. D* 80, No. 3 (1994).

TABLE II.2   CROSS-SECTIONS OF SOME IMPORTANT NUCLIDES IN NUCLEAR ENGINEERING

Atomic number	Nuclide	Abundance a/o	Half-life*	σ_a, † barns	σ_f, ‡ barns
0	n		12 m		
1	^{1}H	99.985		333 mb	
	^{2}H	0.015		0.53 mb	
	^{3}H		12.33 y		
3	^{6}Li	92.5		941	
	^{7}Li	7.42		45.7 mb	
5	^{10}B	19.6		3840	
	^{11}B	80.4		5.5 mb	
6	^{12}C	98.89		3.4 mb	
	^{12}C	1.11		1.37 mb	
	^{14}C		5736 y		
7	^{14}N	99.64		1.9	
	^{15}N	0.36		24 μb	
8	^{16}O	99.756		0.190 mb	
	^{17}O	0.039		0.239	
	^{18}O	0.204		0.16 mb	
53	^{135}I		6.7 h		
54	^{135}Xe		9.17 h	2.65×10^6‡	
61	^{143}Pm		53.1 h		
62	^{149}Sm	13.83		41,000‡	
90	^{232}Th	100	1.41×10^{10} y	5.13	
	^{233}Th		23.3 m	1465	15
92	^{233}U		1.592×10^5 y	575‡	529‡
	^{234}U	0.0055	2.46×10^5 y	103.47	0.465
	^{235}U	0.72	7.038×10^8 y	687.0‡	587‡
	^{236}U		2.34×10^7 y	5.2	
	^{238}U	99.27	4.68×10^9 y	2.73‡	
	^{239}U		23.5 m	36	14
94	^{239}Pu		24110 y	1020‡	749‡
	^{240}Pu		6564 y	289.5	0.064
	^{241}Pu		14.35 y	1378	1015
	^{242}Pu		3.733×10^5 y	10.3	<0.002

*m = minute, h = hour, y = year.

†Cross sections at 0.0253 eV or 2200 m/sec.

‡Non-1/v absorber, see table 3.2 for non-1/v factor.

TABLE II.3 PROPERTIES OF THE ELEMENTS AND CERTAIN MOLECULES

Element or Molecule	Symbol	Atomic Number	Atomic or Molecular Weight*	Nominal Density, g/cm³	Atoms or Molecules per cm³† (×10²⁴)	σ_a,‡barns	σ_s,‡ barns	Σ_a,†cm⁻¹	Σ_s,‡ cm⁻¹
Actinium	Ac	89	227			515			
Aluminum	Al	13	26.9815	2.699	0.06024	0.230	1.49	0.01386	0.08976
Antimony	Sb	51	121.75	6.62	0.03275	5.4	4.2	0.1769	0.1376
Argon	Ar	18	39.948	Gas		0.678	0.644		
Arsenic	As	33	74.9216	5.73	0.04606	4.3	7	0.1981	0.3224
Barium	Ba	56	137.34	3.5	0.01535	1.2		0.01842	
Beryllium	Be	4	9.0122	1.85	0.1236	0.0092	6.14	0.001137	0.7589
Bismuth	Bi	83	208.980	9.80	0.02824	0.033		0.0009319	
Boron	B	5	10.811	2.3	0.1281	759	3.6	97.23	0.4612
Bromine	Br	35	79.909	3.12	0.02351	6.8	6.1	0.1599	0.1434
Cadmium	Cd	48	112.40	8.65	0.04635	2450	5.6	113.56	0.2596
Calcium	Ca	20	40.08	1.55	0.02329	0.43		0.01001	
Carbon (graphite)	C	6	12.01115	1.60	0.08023	0.0034	4.75	0.0002728	0.3811
Cerium	Ce	58	140.12	6.78	0.02914	0.63	4.7	0.01836	0.1370
Cesium	Cs	55	132.905	1.9	0.008610	29.0		0.2497	
Chlorine	Cl	17	35.453	Gas		33.2			
Chromium	Cr	24	51.996	7.19	0.08328	3.1	3.8	0.2582	0.3165
Cobalt	Co	27	58.9332	8.8	0.08993	37.2	6.7	3.345	0.6025
Copper	Cu	29	63.54	8.96	0.08493	3.79	7.9	0.3219	0.6709
Deuterium	D	1	2.01410	Gas		0.00053			
Dysprosium	Dy	66	162.50	8.56	0.03172	930	100	29.50	3.172
Erbium	Er	68	167.26	9.16	0.03203	162	11.0	5.189	0.3523
Europium	Eu	63	151.96	5.22	0.02069	4600	8.0	95.17	0.1655
Fluorine	F	9	18.9984	Gas		0.0095	4.0		

(continued)

TABLE II.3 *(CONTINUED)*

Element or Molecule	Symbol	Atomic Number	Atomic or Molecular Weight*	Nominal Density, g/cm³	Atoms or Molecules per cm³† (×10²⁴)	σ_a, ‡barns	σ_s,‡ barns	Σ_a,†cm⁻¹	Σ_s,‡ cm⁻¹
Gadolinium	Gd	64	157.25	7.95	0.03045	49000		1492	
Gallium	Ga	31	69.72	5.91	0.05105	2.9	6.5	0.1480	0.3318
Germanium	Ge	32	72.59	5.36	0.04447	2.3	7.5	0.1023	0.3335
Gold	Au	79	196.967	19.32	0.05907	98.8		5.836	
Hafnium	Hf	72	178.49	13.36	0.04508	102	8	4.598	0.3606
Heavy water	D$_2$O		20.0276	1.105	0.03323	0.00133	13.6	4.420×10^{-5}	0.4519
Helium	He	2	4.0026	Gas		< 0.05			
Holmium	Ho	67	164.930	8.76	0.03199	66.5	9.4	2.127	0.3007
Hydrogen	H	1	1.00797	Gas		0.332			
Indium	In	49	114.82	7.31	0.03834	193.5		7.419	
Iodine	I	53	126.9044	4.93	0.02340	6.2		0.1451	
Iridium	Ir	77	192.2	22.5	0.07050	426	14	30.03	0.9870
Iron	Fe	26	55.847	7.87	0.08487	2.55	10.9	0.2164	0.9251
Krypton	Kr	36	83.80	Gas		25.0	7.50		
Lanthanum	La	57	138.91	6.19	0.02684	9.0	9.3	0.2416	0.2496
Lead	Pb	82	207.19	11.34	0.03296	0.170	11.4	0.005603	0.3757
Lithium	Li	3	6.942	0.53	0.04600	70.7		3.252	
Lutetium	Lu	71	174.97	9.74	0.03353	77	8	2.581	0.2682
Magnesium	Mg	12	24.3050	1.74	0.04310	0.063	3.42	0.002715	0.1474
Manganese	Mn	25	54.9380	7.43	0.08145	13.3	2.1	1.083	0.1710
Mercury	Hg	80	200.59	13.55	0.04068	375		15.26	
Molybdenum	Mo	42	95.94	10.2	0.06403	2.65	5.8	0.1697	0.3714
Neodymium	Nd	60	144.24	6.98	0.02914	50.5	16	1.472	0.4662
Neon	Ne	10	20.1797	Gas		0.038	2.42		
Nickel	Ni	28	58.71	8.90	0.09130	4.43	17.3	0.4045	1.579
Niobium	Nb	41	92.906	8.57	0.05555	1.15		0.06388	

Element	Symbol	Z	Atomic mass	Density					
Nitrogen	N	7	14.0067	Gas		1.85	10.6		
Osmium	Os	76	190.2	22.5	0.07124	15.3		1.090	
Oxygen	O	8	15.9994	Gas		0.00027	3.76		
Palladium	Pd	46	106.4	12.0	0.06792	6.9	5.0	0.4686	0.3396
Phosphorus (yellow)	P	15	30.9738	1.82	0.03539	0.180		0.006370	
Platinum	Pt	78	195.09	21.45	0.06622	10.0	11.2	0.622	0.7167
Plutonium	Pu	94	239.0522	19.6	0.04938	$\sigma_a = 1011.3$ $\sigma_f = 742.5$	7.7	49.93	0.3902
Polonium	Po	84	209	9.51	0.02727			36.66	
Potassium	K	19	39.095	0.86	0.01325	2.10	1.5	0.02783	0.01988
Praseodymium	Pr	59	140.907	6.78	0.02898	11.5	3.3	0.3333	0.09563
Promethium	Pm	61	145						
Protactinium	Pa	91	231.0359			210			
Radium	Ra	88	226.0254	5.0	0.01332	11.5		0.1532	
Rhenium	Re	75	186.2	20	0.06596	88	11.3	5.804	0.7453
Rhodium	Rh	45	102.905	12.41	0.07263	150		10.89	
Rubidium	Rb	37	85.47	1.53	0.01078	0.37	6.2	0.003989	0.06684
Ruthenium	Ru	44	101.07	12.2	0.07270	2.56		0.1861	
Samarium	Sm	62	150.35	6.93	0.02776	5800		161.0	
Scandium	Sc	21	44.956	2.5	0.03349	26.5	24	0.8875	0.8038
Selenium	Se	34	78.96	4.81	0.03669	11.7	9.7	0.4293	0.3559
Silicon	Si	14	28.086	2.33	0.04996	0.16	2.2	0.007994	0.1099
Silver	Ag	47	107.870	10.49	0.05857	63.6		3.725	
Sodium	Na	11	22.9898	0.97	0.02541	0.530	3.2	0.01347	0.08131
Strontium	Sr	38	87.62	2.6	0.01787	0.530	10	0.02162	0.1787
Sulfur (yellow)	S	16	32.064	2.07	0.03888	0.520	0.975	0.02022	0.03791
Tantalum	Ta	73	180.948	16.6	0.05525	21.0	6.2	1.160	0.3426
Technetium	Tc	43	99			19			
Tellurium	Te	52	127.60	6.24	0.02945	4.7		0.1384	
Terbium	Tb	65	158.925	8.33	0.03157	25.5	20	0.8050	0.6314

(continued)

TABLE II.3 (CONTINUED)

Element or Molecule	Symbol	Atomic Number	Atomic or Molecular Weight*	Nominal Density, g/cm³	Atomic or Molecules per cm³† (×10²⁴)	σ_a‡ barns	σ_s‡ barns	Σ_a,†cm⁻¹	Σ_s,‡ cm⁻¹
Thallium	Tl	81	204.37	11.85	0.03492	3.4	9.7	0.1187	0.3387
Thorium	Th	90	232.038	11.71	0.03039	7.40	12.67	0.2249	0.3850
Thulium	Tm	69	168.934	9.35	0.03314	103	12	3.413	0.3977
Tin	Sn	50	118.69	7.298	0.03703	0.63		0.02333	0.2268
Titanium	Ti	22	47.90	4.51	0.05670	6.1	4.0	0.3459	
Tungsten	W	74	183.85	19.2	0.06289	18.5		1.163	
Uranium	U	92	238.03	19.1	0.04833	$\sigma_a = 7.59$ $\sigma_f = 4.19$	8.90	0.3668 0.2025	0.4301
Vanadium	V	23	50.942	6.1	0.07212	5.04	4.93	0.3635	0.3556
Water	H₂O		18.0153	1.0	0.03343	0.664	103	0.02220	3.443
Xenon	Xe	54	131.30	Gas		24.5	4.30		
Ytterbium	Yb	70	173.04	7.01	0.02440	36.6	25.0	0.8930	0.6100
Yttrium	Y	39	88.906	5.51	0.03733	1.28	7.60	0.04778	0.2837
Zinc	Zn	30	65.37	7.133	0.06572	1.10	4.2	0.07230	0.2760
Zirconium	Zr	40	91.22	6.5	0.04291	0.185	6.40	0.007938	0.2746

*Based on ¹²C = 12.00000.

†Four-digit accuracy for computational purposes only; last digit(s) usually is not meaningful.

‡Cross-sections at 0.0253 eV or 2200 m/sec. The scattering cross sections, except for those of H₂O and D₂O, are measured values in a thermal neutron spectrum and are assumed to be 0.0253 eV values because σ_s is usually constant at thermal energies. The errors in σ_s tend to be large, and the tabulated values of σ_s should be used with caution. (From BNL-325, 3rd ed., 1973).

The value of σ_a given in the table is for pure graphite. Commercial, reactor-grade graphite contains verying amounts of contaminants and σ_a is somewhat larger, say, about 0.0048 barns, so that $\Sigma_a \cong 0.0003851$ cm⁻¹.

The value of σ_a given in the table is for pure D₂O. Commercially available heavy water contains small amounts of ordinary water and σ_a in this case is somewhat larger.

TABLE II.4 THE MASS ATTENUATION COEFFICIENT (μ/ρ) FOR SEVERAL MATERIALS, IN CM2/G*†

Material	Gamma-ray energy, MeV																	
	0.1	0.15	0.2	0.3	0.4	0.5	0.6	0.8	1.0	1.25	1.50	2	3	4	5	6	8	10
H	.295	.265	.243	.212	.189	.173	.160	.140	.126	.113	.103	.0876	.0691	.0579	.0502	.0446	.0371	.0321
Be	.132	.119	.109	.0945	.0847	.0773	.0715	.0628	.0565	.0504	.0459	.0394	.0313	.0266	.0234	.0211	.0181	.0161
C	.149	.134	.122	.106	.0953	.0870	.0805	.0707	.0636	.0568	.0518	.0444	.0356	.0304	.0270	.0245	.0213	.0194
N	.150	.134	.123	.106	.0955	.0869	.0805	.0707	.0636	.0568	.0517	.0445	.0357	.0306	.0273	.0249	.0218	.0200
O	.151	.134	.123	.107	.0953	.0870	.0806	.0708	.0636	.0568	.0518	.0445	.0359	.0309	.0276	.0254	.0224	.0206
Na	.151	.130	.118	.102	.0912	.0833	.0770	.1676	.0608	.0546	.0496	.0427	.0348	.0303	.0274	.0254	.0229	.0215
Mg	.160	.135	.122	.106	.0944	.0860	.0795	.0699	.0627	.0560	.0512	.0442	.0360	.0315	.0286	.0266	.0242	.0228
Al	.161	.134	.120	.103	.0922	.0840	.0777	.0683	.0614	.0548	.0500	.0432	.0353	.0310	.0282	.0264	.0241	.0229
Si	.172	.139	.125	.107	.0954	.0869	.0802	.0706	.0635	.0567	.0517	.0447	.0367	.0323	.0296	.0277	.0254	.0243
P	.174	.137	.122	.104	.0928	.0846	.0780	.1685	.0617	.0551	.0502	.0436	.0358	.0316	.0290	.0273	.0252	.0242
S	.188	.144	.127	.108	.0958	.0874	.0806	.0707	.0635	.0568	.0519	.0448	.0371	.0328	.0302	.0284	.0266	.0255
Ar	.188	.135	.117	.0977	.0867	.0790	.0730	.0638	.0573	.0512	.0468	.0407	.0338	.0301	.0279	.0266	.0248	.0241
K	.215	.149	.127	.106	.0938	.0852	.0786	.0689	.0618	.0552	.0505	.0438	.0365	.0327	.0305	.0289	.0274	.0267
Ca	.238	.158	.132	.109	.0965	.0876	.0809	.0708	.0634	.0566	.0518	.0451	.0376	.0338	.0316	.0302	.0285	.0280
Fe	.344	.183	.138	.106	.0919	.0828	.0762	.0664	.0595	.0531	.0485	.0424	.0361	.0330	.0313	.0304	.0295	.0294
Cu	.427	.206	.147	.108	.0916	.0820	.0751	.0654	.0585	.0521	.0476	.0418	.0357	.0330	.0316	.0309	.0303	.0305
Mo	1.03	.389	.225	.130	.0998	.0851	.0761	.0648	.0575	.0510	.0467	.0414	.0365	.0349	.0344	.0344	.0349	.0359
Sn	1.58	.563	.303	.153	.109	.0886	.0776	.0647	.0568	.0501	.0459	.0408	.0367	.0355	.0355	.0358	.0368	.0383
I	1.83	.648	.339	.165	.114	.0913	.0792	.0653	.0571	.0502	.0460	.0409	.0370	.0360	.0361	.0365	.0377	.0394
W	4.21	1.44	.708	.293	.174	.125	.101	.0763	.0640	.0544	.0492	.0437	.0405	.0402	.0409	.0418	.0438	.0465
Pt	4.75	1.64	.795	.324	.191	.135	.107	.0800	.0659	.0554	.0501	.0445	.0414	.0411	.0418	.0427	.0448	.0477
Tl	5.16	1.80	.866	.346	.204	.143	.112	.0824	.0675	.0563	.0508	.0452	.0420	.0416	.0423	.0433	.0454	.0484
Pb	5.29	1.84	.896	.356	.208	.145	.114	.0836	.0684	.0569	.0512	.0457	.0421	.0420	.0426	.0436	.0459	.0489
U	10.60	2.42	1.17	.452	.259	.176	.136	.0952	.0757	.0615	.0548	.0484	.0445	.0440	.0446	.0455	.0479	.0511
Air	.151	.134	.123	.106	.0953	.0868	.0804	.0706	.0636	.0567	.0517	.0445	.0357	.0307	.0274	.0250	.0220	.0202
NaI	1.57	.568	.305	.155	.111	.0901	.0789	.0657	.0577	.0508	.0465	.0412	.0367	.0351	.0347	.0347	.0354	.0366
H$_2$O	.167	.149	.136	.118	.106	.0966	.0896	.0786	.0706	.0630	.0575	.0493	.0396	.0339	.0301	.0275	.0240	.0219
Concrete	.169	.139	.124	.107	.0954	.0870	.0804	.0706	.0635	.0567	.0517	.0445	.0363	.0317	.0287	.0268	.0243	.0229
Tissue	.163	.144	.132	.115	.100	.0936	.0867	.0761	.0683	.0600	.0556	.0478	.0384	.0329	.0292	.0267	.0233	.0212

*From L. T. Templin, editor, *Reactor Physics Constants*, ANL–5800, 2nd ed., 1963; based on G. W. Grodstein National Bureau of Standards circular 583, 1957.

†Nominal densities of the elements are given in Table II.3. For air at 1 atm and 0°C, $\rho = 1.293 \times 10^{-3}$ g/cm³; ρ (NaI) = 3.67 g/cm³; ρ (tissue) $\simeq 1$ (H$_2$O) = 1 g/cm³; ρ (concrete) = 2.25 − 2.40 g/cm³.

TABLE II.5 THE MASS ABSORPTION COEFFICIENT (μ_A/ρ) FOR SEVERAL MATERIALS, IN CM^2/G*

Gamma-ray energy, MeV

Material	0.1	0.15	0.2	0.3	0.4	0.5	0.6	0.8	1.0	1.25	1.50	2	3	4	5	6	8	10
H	.0411	.0487	.0531	.0575	.0589	.0591	.0590	.0575	.0557	.0533	.0509	.0467	.0401	.0354	.0318	.0291	.0252	.0255
Be	.0183	.0217	.0237	.0256	.0263	.0264	.0263	.0256	.0248	.0237	.0227	.0210	.0203	.0164	.0151	.0141	.0127	.0118
C	.0215	.0246	.0267	.0288	.0296	.0297	.0296	.0289	.0280	.0268	.0256	.0237	.0209	.0190	.0177	.0166	.0153	.0145
N	.0224	.0249	.0267	.0288	.0296	.0297	.0296	.0289	.0280	.0268	.0256	.0236	.0211	.0193	.0180	.0171	.0158	.0151
O	.0233	.0252	.0271	.0289	.0296	.0297	.0296	.0289	.0280	.0268	.0257	.0238	.0212	.0195	.0183	.0175	.0163	.0157
Na	.0289	.0258	.0266	.0279	.0283	.0284	.0284	.0276	.0268	.0257	.0246	.0229	.0207	.0194	.0185	.0179	.0171	.0168
Mg	.0335	.0276	.0278	.0290	.0294	.0293	.0292	.0285	.0276	.0265	.0254	.0237	.0215	.0203	.0194	.0188	.0182	.0180
Al	.0373	.0283	.0275	.0283	.0287	.0286	.0286	.0278	.0270	.0259	.0248	.0232	.0212	.0200	.0192	.0188	.0183	.0182
Si	.0435	.0300	.0286	.0291	.0293	.0290	.0290	.0282	.0274	.0263	.0252	.0236	.0217	.0206	.0198	.0194	.0190	.0189
P	.0501	.0315	.0292	.0289	.0290	.0290	.0287	.0280	.0271	.0260	.0250	.0234	.0216	.0206	.0200	.0197	.0194	.0195
S	.0601	.0351	.0310	.0301	.0301	.0300	.0298	.0288	.0279	.0268	.0258	.0242	.0224	.0215	.0209	.0206	.0206	.0206
Ar	.0729	.0368	.0302	.0278	.0274	.0272	.0270	.0260	.0252	.0242	.0233	.0220	.0206	.0199	.0195	.0195	.0194	.0197
K	.0909	.0433	.0340	.0304	.0298	.0295	.0291	.0282	.0272	.0261	.0251	.0237	.0222	.0217	.0214	.0212	.0215	.0219
Ca	.111	.0489	.0367	.0318	.0309	.0304	.0300	.0290	.0279	.0268	.0258	.0244	.0230	.0225	.0222	.0223	.0225	.0231
Fe	.225	.0810	.0489	.0340	.0307	.0294	.0287	.0274	.0261	.0250	.0242	.0231	.0224	.0224	.0227	.0231	.0239	.0250
Cu	.310	.107	.0594	.0368	.0316	.0296	.0286	.0271	.0261	.0247	.0237	.0229	.0223	.0227	.0231	.0237	.0248	.0261
Mo	.922	.294	.141	.0617	.0422	.0348	.0315	.0281	.0263	.0248	.0239	.0233	.0237	.0250	.0262	.0274	.0296	.0316
Sn	1.469	.471	.222	.0873	.0534	.0403	.0346	.0294	.0268	.0248	.0239	.0233	.0243	.0259	.0276	.0291	.0316	.0339
I	1.726	.557	.260	.100	.0589	.0433	.0366	.0303	.0274	.0252	.0241	.0236	.0247	.0265	.0283	.0299	.0327	.0353
W	4.112	1.356	.631	.230	.1219	.0786	.0599	.0426	.0353	.0302	.0281	.0271	.0287	.0311	.0335	.0355	.0390	.0426
Pt	4.645	1.556	.719	.262	.138	.0892	.0666	.0465	.0375	.0315	.0293	.0280	.0296	.0320	.0343	.0365	.0400	.0438
Tl	5.057	1.717	.791	.285	.152	.0972	.0718	.0491	.0393	.0326	.0301	.0288	.0304	.0326	.0349	.0354	.0406	.0446
Pb	5.193	1.753	.821	.294	.156	.0994	.0738	.0505	.0402	.0332	.0306	.0293	.0305	.0330	.0352	.0373	.0412	.0450
U	9.63	2.337	1.096	.392	.208	.132	.0968	.0628	.0482	.0383	.0346	.0324	.0332	.0352	.0374	.0394	.0443	.0474
Air	.0233	.0251	.0268	.0288	.0296	.0297	.0296	.0289	.0280	.0268	.0256	.0238	.0211	.0194	.0181	.0172	.0160	.0153
NaI	1.466	.476	.224	.0889	.0542	.0410	.0354	.0299	.0273	.0253	.0242	.0235	.0241	.0254	.0268	.0281	.0303	.0325
H_2O	.0253	.0278	.0300	.0321	.0328	.0330	.0329	.0321	.0311	.0298	.0285	.0264	.0233	.0213	.0198	.0188	.0173	.0165
Concrete	.0416	.0300	.0289	.0284	.0297	.0296	.0295	.0287	.0278	.0272	.0256	.0239	.0216	.0203	.0194	.0188	.0180	.0177
Tissue	.0271	.0282	.0293	.0312	.0317	.0320	.0319	.0311	.0300	.0288	.0276	.0256	.0220	.0206	.0192	.0182	.0168	.0160

*From L. T. Templin, editor, *Reactor Physics Constants*, ANL–5800, 2nd ed., 1963; based on G. W. Grodstein, National Bureau of Standards c Circular 583, 1957.

III

Vector Operations in Orthogonal Curvilinear Coordinates

Three coordinate systems are ordinarily encountered in nuclear engineering problems—rectangular, cylindrical, and spherical coordinates. The last two are depicted in Figs. III.1 and III.2. The location of a point P in these coordinates is indicated by

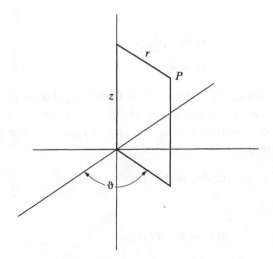

Figure III.1 Cylindrical coordinates.

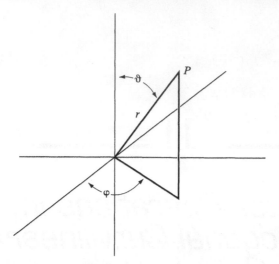

Figure III.2   Spherical coordinates.

(q_1, q_2, q_3), where $q_1 = x$, $q_2 = y$, and $q_3 = z$ in rectangular coordinates; $q_1 = r$, $q_2 = \vartheta$, and $q_3 = z$ in cylindrical coordinates; and $q_1 = r$, $q_2 = \vartheta$, and $q_3 = \varphi$ in spherical coordinates. These coordinate systems are called *orthogonal curvilinear coordinates*—orthogonal because the surfaces $q_1 = $ constant, $q_2 = $ constant, and $q_3 = $ constant intersect at right angles; curvilinear because the surfaces in question are generally curved.

Suppose that a point on the surface $q_1 = $ constant moves the distance ds_1 in a direction perpendicular to the surface, to a point on the surface $q_1 + dq_1 = $ constant (see Fig. III.3). In general, ds_1 is proportional to dq_1 so it is possible to write

$$ds_1 = h_1 dq_1.$$

In a similar manner,

$$ds_2 = h_2 dq_2,$$

$$ds_3 = h_3 dq_3.$$

The quantities, h_1, h_2, h_3 are called *scale factors*. In rectangular coordinates, $h_x = h_y = h_z = 1$; in cylindrical coordinates, it is easily shown that $h_r = 1$, $h_\vartheta = r$, and $h_z = 1$; and in spherical coordinates, $h_r = 1$, $h_\vartheta = r$, and $h_\varphi = r \sin \vartheta$.

The general expression for a volume element is

$$dV = ds_1 ds_2 ds_3 = h_1 h_2 h_3 dq_1 dq_2 dq_3.$$

Thus, in rectangular coordinates

$$dV = dx \, dy \, dz;$$

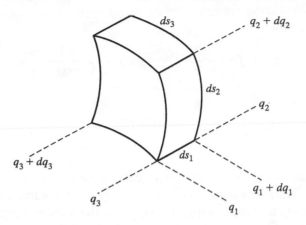

Figure III.3   Volume element in rectangular curvilinear coordinates.

in cylindrical coordinates

$$dV = r \, dr \, d\vartheta \, dz;$$

and in spherical coordinates

$$dV = r^2 \sin \vartheta \, dr \, d\vartheta \, d\varphi.$$

GRADIENT

The gradient of a scalar function f is defined as the vector whose components are equal to the rates of change of f along the direction of the component. Thus,

$$\operatorname{grad} f = \frac{\partial f}{\partial s_1} \mathbf{a}_1 + \frac{\partial f}{\partial s_2} \mathbf{a}_2 + \frac{\partial f}{\partial s_3} \mathbf{a}_3,$$

where $\mathbf{a}_1, \mathbf{a}_2, \mathbf{a}_3$ are unit vectors normal to the coordinate surfaces $q_1 = $ constant, $q_2 = $ constant, and $q_3 = $ constant. Because

$$\frac{\partial f}{\partial s_1} = \frac{1}{h_1} \frac{\partial f}{\partial q_1},$$

it follows that the components of the gradient are

$$\operatorname{grad}_x f = \frac{\partial f}{\partial x}, \qquad \operatorname{grad}_y f = \frac{\partial f}{\partial y}, \qquad \operatorname{grad}_z f = \frac{\partial f}{\partial z}$$

in rectangular coordinates;

$$\operatorname{grad}_r f = \frac{\partial f}{\partial r}, \qquad \operatorname{grad}_\vartheta f = \frac{1}{r} \frac{\partial f}{\partial \vartheta}, \qquad \operatorname{grad}_z f = \frac{\partial f}{\partial z}$$

in cylindrical coordinates; and

$$\text{grad}_r\, f = \frac{\partial f}{\partial r}, \qquad \text{grad}_\vartheta\, f = \frac{1}{r}\frac{\partial f}{\partial \vartheta}, \qquad \text{grad}_\varphi\, f = \frac{1}{r \sin \vartheta}\frac{\partial f}{\partial \varphi}$$

in spherical coordinates.

DIVERGENCE

The general formula for the divergence of a vector in curvilinear coordinates can be obtained by applying the divergence theorem to the infinitely small volume element $dV = ds_1\, ds_2\, ds_3$. According to this theorem, the integral of the normal component of a vector over a closed surface is equal to the integral of the divergence of the vector throughout the enclosed volume. In symbols,

$$\int_A \mathbf{F} \cdot \mathbf{n}\, dA = \int_V \text{div}\mathbf{F}\, dV,$$

where $\mathbf{n}$ is a unit normal vector pointing outward from the surface A bounding the volume V. If V is infinitely small, the volume integral is simply div $\mathbf{F}\, dV$ and so

$$\text{div}\,\mathbf{F}\, dV = \int_A \mathbf{F} \cdot \mathbf{n}\, dA.$$

Here, div $\mathbf{F}$ is the divergence of F at the location of dV, and the integral is over the surface of dV.

Carrying out the integral over the faces of dV shown in Fig. III.3 gives

$$\int_A \mathbf{F} \cdot \mathbf{n}\, dA = [(F_1\, ds_2\, ds_3)_{q_1+dq_1} - (F_1\, ds_2\, ds_3)_{q_1}]$$

$$+ [(F_2\, ds_1\, ds_3)_{q_2+dq_2} - (F_2\, ds_1\, ds_3)_{q_2}]$$

$$+ [(F_3\, ds_1\, ds_2)_{q_3+dq_3} - (F_3\, ds_1\, ds_2)_{q_3}].$$

The minus signs appear because if a component—say, F_1—points outward on the face at $q_1 + dq_1$, then it must point inward on the face at q_1. The first bracket can be written

$$[(F_1\, ds_2\, ds_3)_{q_1+dq_1} - (F_1\, ds_2\, ds_3)_{q_1}] = \frac{\partial}{\partial q_1}(F_1\, ds_2\, ds_3)\, dq_1$$

$$= \frac{\partial}{\partial q_1}(F_1\, h_2\, h_3)\, dq_1\, dq_2\, dq_3.$$

Similar expressions hold for the other brackets in the equation. Combining terms then gives

$$\text{div}\,\mathbf{F}\,dV = \left[\frac{\partial}{\partial q_1}(F_1 h_2 h_3) + \frac{\partial}{\partial q_2}(F_2 h_1 h_3) + \frac{\partial}{\partial q_3}(F_3 h_1 h_2)\right] dq_1\,dq_2\,dq_3.$$

Dividing by dV yields finally

$$\text{div}\,\mathbf{F} = \frac{1}{h_1 h_2 h_3}\left[\frac{\partial}{\partial q_1}(F_1 h_2 h_3) + \frac{\partial}{\partial q_2}(F_2 h_1 h_3) + \frac{\partial}{\partial q_3}(F_3 h_1 h_2)\right].$$

Introducing the values of the h's gives

$$\text{div}\,\mathbf{F} = \frac{\partial F_x}{\partial x} + \frac{\partial F_y}{\partial y} + \frac{\partial F_z}{\partial z}$$

in rectangular coordinates;

$$\text{div}\,\mathbf{F} = \frac{1}{r}\frac{\partial}{\partial r}(r F_r) + \frac{1}{r}\frac{\partial F_\vartheta}{\partial \vartheta} + \frac{\partial F_z}{\partial z}$$

in cylindrical coordinates; and

$$\text{div}\,\mathbf{F} = \frac{1}{r^2}\frac{\partial}{\partial r}(r^2 F_r) + \frac{1}{r\sin\vartheta}\frac{\partial}{\partial \vartheta}(\sin\vartheta\,F_\vartheta) + \frac{1}{r\sin\vartheta}\frac{\partial F_\varphi}{\partial \varphi}$$

in spherical coordinates.

LAPLACIAN

This operator is the divergence of the gradient of a scalar function. The component F_1 of the gradient is

$$F_1 = \frac{1}{h_1}\frac{\partial f}{\partial q_1},$$

with similar expressions for the other components. Inserting these components into the previous formula for the divergence gives for the Laplacian:

$$\nabla^2 f = \frac{1}{h_1 h_2 h_3}\left[\frac{\partial}{\partial q_1}\left(\frac{h_2 h_3}{h_1}\frac{\partial f}{\partial q_1}\right) + \frac{\partial}{\partial q_2}\left(\frac{h_1 h_3}{h_2}\frac{\partial f}{\partial q_2}\right) + \frac{\partial}{\partial q_3}\left(\frac{h_1 h_2}{h_3}\frac{\partial f}{\partial q_3}\right)\right].$$

The symmetry of this formula should be noted. In view of this result, the Laplacian is

$$\nabla^2 f = \frac{\partial^2 f}{\partial x^2} + \frac{\partial^2 f}{\partial y^2} + \frac{\partial^2 f}{\partial z^2}$$

in rectangular coordinates;

$$\nabla^2 f = \frac{1}{r}\frac{\partial}{\partial r}\left(r\frac{\partial f}{\partial r}\right) + \frac{1}{r^2}\frac{\partial^2 f}{\partial \vartheta^2} + \frac{\partial^2 f}{\partial z^2}$$

in cylindrical coordinates; and

$$\nabla^2 f = \frac{1}{r^2}\frac{\partial}{\partial r}\left(r^2\frac{\partial f}{\partial r}\right) + \frac{1}{r^2 \sin\vartheta}\frac{\partial}{\partial \vartheta}\left(\sin\vartheta\frac{\partial f}{\partial \vartheta}\right) + \frac{1}{r^2 \sin^2\vartheta}\frac{\partial^2 f}{\delta\varphi^2}$$

in spherical coordinates.

IV

Thermodynamic and Physical Properties

TABLE IV.1   PROPERTIES OF WATER AND DRY SATURATED STEAM AS A
FUNCTION OF SATURATION TEMPERATURE*

Temp., °F	Pressure, psia	Specific Volume, ft³/lb		Enthalpy, Btu/lb		
		Sat. Liquid	Sat. Vapor	Sat. Liquid	Evap.	Sat. Vapor
T	P	v_f	v_g	h_f	h_{fg}	h_g
32.0	0.08866	0.01602	3302	0	1075.4	1075.4
350	134.53	.01799	3.346	321.8	871.3	1193.1
360	152.92	.01811	2.961	332.4	862.9	1195.2
370	173.23	.01823	2.628	342.9	854.2	1197.2
380	195.6	.01836	2.339	353.6	845.4	1199.0
390	220.2	.01850	2.087	364.3	836.2	1200.6
400	247.1	.01864	1.866	375.1	826.8	1202.0
410	276.5	.01878	1.673	386.0	817.2	1203.1
420	308.5	.01894	1.502	396.9	807.2	1204.1
430	343.3	.01909	1.352	407.9	796.9	1204.8
440	381.2	.01926	1.219	419.0	786.3	1205.3
450	422.1	.01943	1.1011	430.2	775.4	1205.6
460	466.3	.01961	.9961	441.4	764.1	1205.5
470	514.1	.01980	.9025	452.8	752.4	1205.2
480	565.5	.02000	.8187	646.3	740.3	1204.6
490	620.7	.02021	.7436	475.9	727.8	1203.7
500	680.0	.02043	.6761	487.7	714.8	1202.5
510	743.5	.02067	.6153	499.6	701.3	1200.9
520	811.4	.02091	.5605	511.7	687.3	1198.9
530	884.0	.02117	.5108	523.9	672.7	1196.6
540	961.5	.02145	.4658	536.4	657.5	1193.8
550	1044.0	.02175	.4249	549.1	641.6	1190.6
560	1131.8	.02207	.3877	562.0	625.0	1187.0
570	1225.1	.02241	.3537	575.2	607.6	1182.8
580	1324.3	.02278	.3225	588.6	589.3	1178.0
590	1429.5	.02319	.2940	602.5	570.1	1172.5
600	1541.0	.02363	.2677	616.7	549.7	1166.4
610	1659.2	.02411	.2434	631.3	528.1	1159.4
620	1784.4	.02465	.2209	646.4	505.0	1151.4
630	1916.9	.02525	.2000	662.1	480.2	1142.4
640	2057.1	.02593	.1805	678.6	453.4	1131.9
650	2205	.02673	.1621	695.9	423.9	1119.8
660	2362	.02767	.1446	714.4	391.1	1105.5
670	2529	.02882	.1278	734.4	353.9	1088.3
680	2705	.03032	.1113	756.9	309.8	1066.7
690	2892	.03248	.09428	783.8	253.9	1037.7
700	3090	.03666	.07438	822.7	167.5	990.2
705.44	3204	.05053	.05053	902.5	0	902.5

*From J. H. Keenan, F. G. Keyes, P. G. Hill, and J. G. Moore, *Steam Tables.* New York: Wiley,
1969. (These tables are available in both English and metric units.)

TABLE IV.2 PROPERTIES OF WATER AND DRY SATURATED STEAM AS FUNCTION
OF SATURATION PRESSURE*

| Pressure, psia | Temp., °F | Specific volume, ft³/lb | | Enthalpy, Btu/lb | | |
		Sat. Liquid	Sat. Vapor	Sat. Liquid	Evap.	Sat. Vapor
T	P	v_f	v_g	h_f	h_{fg}	h_g
14.696	211.99	0.01672	26.80	180.15	970.4	1150.5
100	327.86	.01774	4.434	298.61	889.2	1187.8
200	381.86	.01839	2.289	355.6	873.7	1199.3
300	417.43	.01890	1.544	394.1	809.8	1203.9
400	444.70	.01934	1.162	424.2	781.2	1205.5
500	467.13	.01975	.9.83	449.5	755.8	1205.3
600	486.33	.02013	.7702	471.7	732.4	1204.1
700	503.23	.02051	.6558	491.5	710.5	1202.0
800	518.36	.02087	.5691	509.7	689.6	1199.3
900	532.12	.02123	.5009	526.6	699.5	1196.0
1000	544.75	.02159	.4459	542.4	650.0	1192.4
1100	556.45	.02195	.4005	557.4	631.0	1188.3
1200	567.37	.02232	.3623	571.7	612.3	1183.9
1300	577.60	.02269	.3297	585.4	593.8	1179.2
1400	587.25	.02307	.3016	598.6	575.5	1174.1
1500	596.39	.02346	.2769	611.5	557.2	1168.7
1600	605.06	.02386	.2552	624.0	538.9	1162.9
1700	613.32	.02428	.2358	636.2	520.6	1156.9
1800	621.21	.02472	.2183	648.3	502.1	1150.4
1900	628.76	.02517	.2025	660.1	483.4	1143.5
2000	636.00	.02565	.1881	671.9	464.4	1136.3
2100	642.95	.02616	.1749	683.6	445.0	1128.5
2200	649.64	.02670	.1627	695.3	425.0	1120.3
2300	656.09	.02728	.1513	707.0	404.4	1111.4
2400	662.31	.02791	.1407	718.8	383.0	1101.8
2500	668.31	.02860	.1306	730.9	360.5	1091.4
3000	695.52	.03431	.08404	802.5	213.0	1015.5
3203.06	705.44	.05053	.05053	902.5	0	902.5

*From J. H. Keenan, F. G. Keyes, P. G. Hill, and J. G. Moore, *Steam Tables.* New York: Wiley, 1969. (These tables are available in both English and metric units.)

TABLE IV.3   PROPERTIES OF ORDINARY WATER*

Temp., °F	Specific Heat, c_p, Btu lb-°F			Thermal Conductivity, k, Btu/hr/ft-°F			Viscosity, μ, lb/hr-ft			Density, lb/ft^3		
	Sat. Liquid	1,000 psia	2,000 psia	Sat. Liquid	1,000 psia	2,000 psia	Sat. Liquid	1,000 psia	2,000 psia	Sat.	1,000	2,000
32	1.008	1.003	1.000	0.3185	0.3198	0.3211	4.340	4.309	4.279	62.41	62.63	62.85
300	1.029	1.023	1.017	.3952	.3981	.4013	0.452	0.460	0.468	57.31	57.54	57.77
320	1.035	1.031	1.024	.3944	.3969	.3998	.420	.426	.433	56.65	56.86	57.11
340	1.046	1.040	1.032	.3921	.3947	.3977	.391	.396	.404	55.95	56.18	56.44
360	1.056	1.050	1.041	.3891	.3919	.3951	.366	.372	.278	55.22	55.45	55.74
380	1.067	1.061	1.051	.3857	.3885	.3919	.346	.351	.356	54.46	54.68	54.99
400	1.079	1.074	1.062	.3809	.3840	.3880	.327	.330	.335	53.65	53.91	54.23
420	1.094	1.087	1.075	.3753	.3787	.3833	.310	.312	.317	52.81	53.03	53.38
440	1.111	1.105	1.091	.3693	.3728	.3776	.294	.296	.301	51.92	52.15	52.53
460	1.132	1.124	1.109	.3640	.3664	.3713	.280	.282	.286	50.98	51.19	51.62
480	1.154	1.149	1.131	.3575	.3595	.3642	.267	.270	.273	50.00	50.16	50.63
500	1.186	1.176	1.154	.3494	.3510	.3562	.256	.257	.260	48.95	49.12	49.65
520	1.23	1.21	1.188	.3397	.3410	.3475	.246	.246	.249	47.82	47.94	48.54
540	1.28		1.225	.3298		.3371	.235	.235	.239	46.62	46.64	47.35
560	1.34		1.278	.3189		.3256	.225		.231	45.31		46.04
580	1.41		1.341	.3064		.3118	.217		.222	43.90		44.58
600	1.51		1.448	.2919		.2962	.210		.212	42.32		42.92
620	1.65		1.62	.2753		.2778	.200		.202	40.57		40.93
640	1.88			.2565			.190			38.57		
660	2.34			.2335			.177			36.14		
670	2.84			.2198			.169			34.70		
680	3.5			.2056			.161			32.98		
690	5.54			.1854			.148			30.79		

*From J. F. Hogerton and R. C. Grass, Editors, *The Reactor Handbook*, Vol. 3. Washington, D.C.: U.S. Dept. of Energy, 1955. Density calculated from values of specific volume given in J. H. Keenan, et. al., *op. cit.*

TABLE IV.4   PROPERTIES OF HELIUM*†

Temp., °K	Density, g/cm^3 × 10^{-2}	Specific Heat, c_p, J/g-°K	Enthalpy, J/g	Thermal Conductivity, k, W/cm-°K × 10^{-3}	Viscosity, μ g/cm-sec × 10^{-4}	Prandtl Number
300	0.7939	5.194	1589	1.57	2.01	0.664
350	0.6830	5.193	1849	1.74	2.23	0.665
400	0.5993	5.192	2109	1.91	2.44	0.665
450	0.5339	5.191	2368	2.07	2.65	0.665
500	0.4813	5.191	2628	2.22	2.85	0.665
600	0.4022	5.191	3147	2.52	3.23	0.665
700	0.3453	5.191	3666	2.81	3.60	0.665
800	0.3026	5.191	4185	3.08	3.95	0.665
900	0.2692	5.191	4704	3.35	4.29	0.665
1000	0.2425	5.191	5223	3.61	4.62	0.665
1100	0.2206	5.191	5742	3.86	4.94	0.665
1200	0.2023	5.191	6261	4.10	5.26	0.665
1300	0.1869	5.191	6781	4.34	5.56	0.665
1400	0.1736	5.192	7300	4.58	5.86	0.665
1500	0.1521	5.192	7819	4.81	6.16	0.665

*From R. D. McCarty, "Thermophysical Properties of Helium." National Bureau of Standards report NBS-TN-631, 1972.
†The tabulated values are for a pressure of 50 atm. However, all of the listed properties except density are relatively independent of pressure. The density, to a good approximation, is proportional to the pressure.

TABLE IV.5   PROPERTIES OF SODIUM*

Temp., °F	Density, lb/ft^3	Specific Heat, Btu/lb-°F	Enthalpy,† Btu/lb	Thermal Conductivity, k, Btu/hr-ft-°F	Viscosity, μ lb/ft-hr	Prandtl Number
212	57.87	0.3305			1.706	
302			239.9		1.309	
392	56.44	0.3200	268.9	47.11	1.089	0.0074
482			301.8		0.949	
572	55.06	0.3116	332.6	43.75	0.835	0.0059
752	53.63	0.3055	381.4	42.15	0.687	0.0051
932	52.07	0.3015	436.0	38.61	0.588	0.0046
1112	50.51	0.2998	490.1	36.24	0.508	0.0042
1292	48.88	0.3003	544.2	34.10	0.450	0.0040
1472	47.26	0.3030	598.5	31.62	0.399	0.0038
1652		0.3079	653.4		0.363	

*From R. N. Lyon, Editor, *Liquid Metals Handbook*. Washington: U.S. Dept. of Energy, 1952, 1955.
†Interpolated from values given by C. J. Meisl and A. Shapiro, "Thermodynamic Properties of Alkali Metal Vapors and Mercury." General Electric report R 60 FPD358-A, 1960.

TABLE IV.6   THERMAL CONDUCTIVITY (BTU/HR-FT-°F)

°F	Uranium	UO*$_2$	PuO$_2$	Aluminum	Steel	Zirconium	Zircaloy
200	15.80	4.2	3.60	119	10	11.8	< 10
400	17.00	3.5		124	10	11.2	
600	18.10	3.0		134	11	10.8	
800	19.20	2.6		148	11	10.6	
1000	20.25	2.4			13		
1200	21.20	2.1			13.5		
1400	22.00	1.9	1.57		15		
1600		1.7			16		
1800		1.5			18		
2000		1.4					
2500		1.3					
3000		1.2					
3500		1.1					
4000		1.1					

*Data on UO$_2$ are highly variable, especially above 3000°F.

V

Bessel Functions

Bessel's equation is

$$\frac{d^2\phi}{dx^2} + \frac{1}{x}\frac{d\phi}{dx} + \left(\alpha^2 - \frac{n^2}{x^2}\right)\phi = 0,$$

where α and n are constants. If n is an integer or zero, as is usually the case in practical problems, the two independent solutions to the equation are written as $J_n(\alpha x)$ and $Y_n(\alpha x)$. The general solution to Bessel's equation is then

$$\phi = AJ_n(\alpha x) + CY_n(\alpha x),$$

where A and C are constants. The functions $J_n(\alpha x)$ and $Y_n(\alpha x)$ are called *ordinary Bessel functions of the first and second kind,* respectively.

When α^2 is negative, Bessel's equation becomes

$$\frac{d^2\phi}{dx^2} + \frac{1}{x}\frac{d\phi}{dx} - \left(\alpha^2 + \frac{n^2}{x^2}\right)\phi = 0.$$

If n is an integer or zero, the independent solutions are written as $I_n(\alpha x)$ and $K_n(\alpha x)$, so that the general solution is

$$\phi \ AI_n(\alpha x) + CK_n(\alpha x),$$

where A and C are again constants. The functions $I_n(\alpha x)$ and $K_n(\alpha x)$ are known as *modified Bessel functions of the first and second kind.*

Table V.1 gives values of the Bessel functions over the usual range of interest in nuclear engineering problems. More complete tabulations are noted in the references.

REFERENCES

Abramowitz, M., and I. A. Stegun, Editors, *Handbook of Mathematical Functions.* Dover Publications, 1965.

TABLE V.1   BESSEL FUNCTIONS

x	J_0x	J_1x	Y_0x	Y_1x	I_0x	I_1x	K_0x	K_1x
0	1.0000	0.0000	$-\infty$	$-\infty$	1.000	0.0000	∞	$-\infty$
0.05	0.9994	0.0250	−1.979	−12.79	1.001	0.0250	3.114	19.91
0.10	0.9975	0.0499	−1.534	−6.459	1.003	0.0501	2.427	9.854
0.15	0.9944	0.0748	−1.271	−4.364	1.006	0.0752	2.030	6.477
0.20	0.9900	0.0995	−1.081	−3.324	1.010	0.1005	1.753	4.776
0.25	0.9844	0.1240	−0.9316	−2.704	1.016	0.1260	1.542	3.474
0.30	0.9776	0.1483	−0.8073	−2.293	1.023	0.1517	1.372	3.056
0.35	0.9696	0.1723	−0.7003	−2.000	1.031	0.1777	1.233	2.559
0.40	0.9604	0.1960	−0.6060	−1.781	1.040	0.2040	1.115	2.184
0.45	0.9500	0.2194	−0.5214	−1.610	1.051	0.2307	1.013	1.892
0.50	0.9385	0.2423	−0.4445	−1.471	1.063	0.2579	0.9244	1.656
0.55	0.9258	0.2647	−0.3739	−1.357	1.077	0.2855	0.8466	1.464
0.60	0.9120	0.2867	−0.3085	−1.260	1.092	0.3137	0.7775	1.303
0.65	0.8971	0.3081	−1.2076	−1.177	1.108	0.3425	0.7159	1.167
0.70	0.8812	0.3290	−0.1907	−1.103	1.126	0.3719	0.6605	1.050
0.75	0.8642	0.3492	−0.1372	−1.038	1.146	0.4020	0.6106	0.9496
0.80	0.9463	0.3688	−0.0868	−0.9781	1.167	0.4329	0.5653	0.8618
0.85	0.8274	0.3878	−0.0393	−0.9236	1.189	0.4646	0.5242	0.7847
0.90	0.8075	0.4059	−0.0056	−0.8731	1.213	0.4971	0.4867	0.7165
0.95	0.7868	0.4234	0.0481	−0.8258	1.239	0.5306	0.4524	0.6560
1.0	0.7625	0.4401	0.0883	−0.7812	1.266	0.5652	0.4210	0.6019
1.1	0.7196	0.4709	0.1622	−0.6981	1.326	0.6375	0.3656	0.5098
1.2	0.6711	0.4983	0.2281	−0.6211	1.394	0.7147	0.3185	0.4346
1.3	0.6201	0.5220	0.2865	−0.5485	1.469	0.7973	0.2782	0.3725
1.4	0.5669	0.5419	0.3379	−1.4791	1.553	0.8861	0.2437	0.3208
1.5	0.5118	0.5579	0.3824	−0.4123	1.647	0.9817	0.2138	0.2774
1.6	0.4554	0.5699	0.4204	−0.3476	1.750	1.085	0.1880	1.2406
1.7	0.3980	0.5778	0.4520	−0.2847	1.864	1.196	0.1655	1.2094
1.8	0.3400	0.5815	0.4774	−0.2237	1.990	1.317	0.1459	0.1826
1.9	0.2818	0.5812	0.4968	−0.1644	2.128	1.448	0.1288	0.1597
2.0	0.2239	0.5767	0.5104	−0.1070	2.280	1.591	0.1139	0.1399
2.1	0.1666	0.5683	0.5183	−0.0517	2.446	1.745	0.1008	0.1227
2.2	0.1104	0.5560	0.5208	−0.0015	2.629	1.914	0.0893	0.1079
2.3	0.0555	0.5399	0.5181	0.0523	2.830	2.098	0.0791	0.0950
2.4	0.0025	0.5202	0.5104	0.1005	3.049	2.298	0.0702	0.0837
2.5	−0.0484	0.4971	0.4981	0.1459	3.290	2.517	0.0623	0.0739
2.6	−0.0968	0.4708	0.4813	0.1884	3.553	2.755	0.0554	0.0653
2.7	−0.1424	0.4416	0.4605	0.2276	3.842	3.016	0.0493	0.0577
2.8	−0.1850	0.4097	0.4359	0.2635	4.157	3.301	0.0438	0.0511
2.9	−0.2243	0.3754	0.4079	0.2959	4.503	3.613	0.0390	0.0453
3.0	−0.2601	0.3391	0.3769	0.3247	4.881	3.953	0.0347	0.0402
3.2	−0.3202	0.2613	0.3071	0.3707	5.747	4.734	0.0276	0.0316
3.4	−0.6343	0.1792	0.2296	0.4010	6.785	5.670	0.0220	0.0250
3.6	−0.3918	0.0955	0.1477	0.4154	8.028	6.793	0.6175	0.0198
3.8	−0.4026	0.0128	0.0645	0.4141	9.517	8.140	0.0140	0.0157
4.0	−0.3971	−0.0660	−0.0169	0.3979	11.302	93759	0.0112	0.0125

Index